MW00856086

SAP PRESS e-books

Print or e-book, Kindle or iPad, workplace or airplane: Choose where and how to read your SAP PRESS books! You can now get all our titles as e-books, too:

- By download and online access
- For all popular devices
- And, of course, DRM-free

Convinced? Then go to www.sap-press.com and get your e-book today.

Production Planning and Control with SAP® ERP

SAP PRESS is a joint initiative of SAP and Rheinwerk Publishing. The know-how offered by SAP specialists combined with the expertise of Rheinwerk Publishing offers the reader expert books in the field. SAP PRESS features first-hand information and expert advice, and provides useful skills for professional decision-making.

SAP PRESS offers a variety of books on technical and business-related topics for the SAP user. For further information, please visit our website: *www.sap-press.com*.

Martin Murray, Jawad Akhtar
Materials Management with SAP ERP: Functionality and Technical Configuration (4th edition)
2016, 739 pages, hardcover and e-book
www.sap-press.com/4062

Jawad Akhtar
Quality Management with SAP
2015, 883 pages, hardcover and e-book
www.sap-press.com/3755

Uwe Goehring
Materials Planning with SAP
2016, 519 pages, hardcover and e-book
www.sap-press.com/3745

Chandan Jash, Dipankar Saha
Implementing SAP Manufacturing Execution
2016, 480 pages, hardcover and e-book
www.sap-press.com/3868

Jawad Akhtar

Production Planning and Control with SAP® ERP

Editor Meagan White
Acquisitions Editor Emily Nicholls
Copyeditor Julie McNamee
Cover Design Graham Geary
Photo Credit iStockphoto.com/67091973/© svedoliver
Layout Design Vera Brauner
Production Marissa Fritz
Typesetting SatzPro, Krefeld (Germany)
Printed and bound in the United States of America, on paper from sustainable sources

ISBN 978-1-4932-1430-3
© 2020 by Rheinwerk Publishing, Inc., Boston (MA)
2nd edition 2016, 2nd reprint 2020

Library of Congress Cataloging-in-Publication Data
Names: Akhtar, Jawad, author.
Title: Production planning and control with SAP ERP / Jawad Akhtar.
Description: Bonn ; Boston : Rheinwerk Publishing, [2016] | Includes index.
Identifiers: LCCN 2016016860| ISBN 9781493214303 (print : alk. paper) | ISBN 9781493214327 (print and ebook : alk. paper) | ISBN 9781493214310 (ebook)
Subjects: LCSH: Production planning--Data processing. | Production control--Data processing. | SAP ERP.
Classification: LCC TS155 .A2997 2016 | DDC 634.9/2830285--dc23 LC record available at https://lccn.loc.gov/2016016860

All rights reserved. Neither this publication nor any part of it may be copied or reproduced in any form or by any means or translated into another language, without the prior consent of Rheinwerk Publishing, 2 Heritage Drive, Suite 305, Quincy, MA 02171.

Rheinwerk Publishing makes no warranties or representations with respect to the content hereof and specifically disclaims any implied warranties of merchantability or fitness for any particular purpose. Rheinwerk Publishing assumes no responsibility for any errors that may appear in this publication.

"Rheinwerk Publishing" and the Rheinwerk Publishing logo are registered trademarks of Rheinwerk Verlag GmbH, Bonn, Germany. SAP PRESS is an imprint of Rheinwerk Verlag GmbH and Rheinwerk Publishing, Inc.

All of the screenshots and graphics reproduced in this book are subject to copyright © SAP SE, Dietmar-Hopp-Allee 16, 69190 Walldorf, Germany.

SAP, the SAP logo, ABAP, Ariba, ASAP, Concur, Concur ExpenseIt, Concur TripIt, Duet, SAP Adaptive Server Enterprise, SAP Advantage Database Server, SAP Afaria, SAP ArchiveLink, SAP Ariba, SAP Business ByDesign, SAP Business Explorer, SAP BusinessObjects, SAP BusinessObjects Explorer, SAP BusinessObjects Lumira, SAP BusinessObjects Roambi, SAP BusinessObjects Web Intelligence, SAP Business One, SAP Business Workflow, SAP Crystal Reports, SAP EarlyWatch, SAP Exchange Media (SAP XM), SAP Fieldglass, SAP Fiori, SAP Global Trade Services (SAP GTS), SAP GoingLive, SAP HANA, SAP HANA Vora, SAP Hybris, SAP Jam, SAP MaxAttention, SAP MaxDB, SAP NetWeaver, SAP PartnerEdge, SAPPHIRE NOW, SAP PowerBuilder, SAP PowerDesigner, SAP R/2, SAP R/3, SAP Replication Server, SAP S/4HANA, SAP SQL Anywhere, SAP Strategic Enterprise Management (SAP SEM), SAP SuccessFactors, The Best-Run Businesses Run SAP, TwoGo are registered or unregistered trademarks of SAP SE, Walldorf, Germany.

All other products mentioned in this book are registered or unregistered trademarks of their respective companies.

Contents at a Glance

Dear Reader,

In manufacturing, there's a maxim: "You can have it cheap, fast, or good. Pick two." In the publishing industry, where micro-level editing is seemingly endless, we usually say, "It's either published or perfect. Pick one."

For an editor, then, there is a unique pleasure in working on a second edition. It offers a chance to right past editorial wrongs (the missing comma on page 438 or the typo on page 822 that sneaked into the first edition). We can take the opportunity to re-examine a best-seller that has already reached bookshelves around the world—and find new ways to both reinvigorate its content and aim for that elusive perfection.

In my mission to create the perfect book, I could have asked for no better partner than expert and author Jawad Akhtar. His tireless effort, his attention to detail, and his commitment to a new edition of his first-ever book were second to none. While perfection may never truly be possible, I do believe we have come quite close!

As always, your comments and suggestions are the most useful tools to help us make our books the best they can be. Let us know what you thought about this second edition of *Production Planning and Control with SAP ERP*! Please feel free to contact me and share any praise or criticism you may have.

Thank you for purchasing a book from SAP PRESS!

Meagan White
Editor, SAP PRESS

Rheinwerk Publishing
Boston, MA

meaganw@rheinwerk-publishing.com
www.sap-press.com

Contents

PART III Production Planning Workflow by Production Type

PART IV Production Planning Workflow Tools

PART V Optimizing Production Planning

Acknowledgments

This book is dedicated to the loving memory of my parents, who have left for their eternal abodes. If you are blessed with one or both living parents, please take great care of them and spend as much time with them as possible, for they are the true paths to all professional and personal successes in life.

It all seems like yesterday when I embarked on an amazingly wonderful journey of authoring my first SAP PRESS book. Little did I know or imagine that the first edition of the *Production Planning and Control with SAP ERP* book would be so well received by all of you. I am truly humbled and wish to thank you so much for all the appreciation.

At Rheinwerk Publishing, I have always been greatly impressed by its managing director, Florian Zimniak, who treats his authors as the greatest asset, which ensures that authors keep coming back to do many more projects. My sincere and repeated appreciation goes out to Emily Nicholls, the acquisitions editor, who is an amazing human being, a top-notch professional, and a wonderful friend to have. Finally, to my editor, Meagan White: It's a sheer pleasure to collaborate on every book or e-book project. Under Meagan's able editorship, I have enormous satisfaction that my book (or E-Bite) is in great hands—hands that know how to present information in a structured and logical way that makes perfect sense and is easy to read.

I will remain eternally indebted to Frank Layer, who in his role as technical reviewer to the first edition of this book was a massive source of strength and guidance. He was also the most trusted advisor to offer an unmatchable blend of experience and expertise. He ensured a winning combination of putting together actual project experiences with information in the book. Frank will always remain both my mentor and highly regarded in my professional life.

PART I
Production Planning Core Concepts

1 Introduction

A company that is in the business of manufacturing a product and selling it to customers goes through the rigor of production planning and then production execution. SAP ERP Production Planning (referred to as PP throughout the book) plays a critical role in the logistics functions of the company to accomplish just this. This component enables the company to benefit from historical data to prepare a forecast that can then be used in sales and production planning. From an initial sales plan or sales orders from customers, to the highly integrated and complex chain of interdependent activities in logistics in the SAP ERP system, the PP component reflects its strength, both in planning and execution. It seamlessly integrates with sales, procurement, quality, maintenance, projects, human capital, finance, and controlling functions of the company. It also integrates with SAP Manufacturing Execution (SAP ME), as well as with SAP Manufacturing Integration and Intelligence (SAP MII).

1.1 Goals of This Book

The first goal of this book is to provide you with the step-by-step approach to configure and implement three different production types in PP: discrete, process, and repetitive manufacturing.

The book lays the initial foundation in the form of configuration and then explains how the configuration impacts actual business processes. The configuration to business process approach is maintained throughout the book.

The next goal is to provide comprehensive coverage to the PP workflow tools available. Further, there are significant "hidden" or lesser-used functionalities in PP that you can integrate even when (and long after) your SAP ERP system implementation is complete. These tools are covered to bring greater optimization to your business processes and greater return on your investment in the SAP ERP system.

The book offers several real-life examples and other modeling hints and tips to help you decide which option best meets the business needs of the company. Screenshots are used extensively and are duly supported by in-depth coverage of concepts and terminologies. SAP ERP 6.0 EHP 8 is used in the screenshots. The menu paths or transaction codes are given to perform each step. Where possible, a deliberate attempt is made to use the SAP's Internet Demonstration and Evaluation System (IDES), so you can configure and implement a solution in a training client. Where specific or unique data is used, all necessary prerequisites and hints are given to enable you to set up the data or meet the prerequisites before attempting to run a business process. While this book can only cover so much of a topic, we highly encourage you to explore and try out a large number of options, icons, menu paths, and other pointers to continue the process of self-learning and eventually become an expert in PP.

In this book, we also cover several cross-component functionalities that enable you to leverage their strengths not only in PP but also in other logistics components that are implemented in your company. For example, you can use the classification system, digital signature, Early Warning System (EWS), flexible planning standard analysis, information systems, and reporting in many other logistics components. In other words, this book goes beyond the PP component to help in optimizing business processes in other logistics components.

1.2 Target Audience

This book is intended for all readers who use PP in the SAP ERP system, such as the component's team leader, project team members in an SAP ERP system implementation, integration managers, production planners, or production controllers working in operational positions in the company. Because this book covers three different production types—discrete, process, and repetitive manufacturing—it tends to benefit those readers who are either transitioning or intending to transition from companies using different production types. Additionally, if the company is embarking on production and capacity expansion, then this book can help by facilitating the creation of the new enterprise structure needed in the SAP ERP system to support the expansion. Finally, this book can be an invaluable reference to SAP ERP system consultants and even business process owners who are considering the transition to a consulting career and need a comprehensive understanding of the required concepts and fundamentals.

1.3 Structure and Content

This book takes a deep-dive approach to deliver in-depth and comprehensive coverage of discrete, process, and repetitive manufacturing in SAP ERP. It begins by covering the enterprise structure that you need to set up in the PP component, which also reflects the interdependencies of the enterprise structures of other components. The configuration basics that you need to know for each production type are covered next. Similarities and differences in various production types are highlighted to enable you to comprehensively differentiate one from the other. The configuration of each production type is then put to actual use, in which we show the impact of the configuration on the business processes. You must understand a business process in a comprehensive way before modeling and configuring it in the SAP ERP system.

The book then transitions to cover the PP workflow tools available. You'll also learn how to optimize your production processes by using several latent features that are often not as frequently used to bring about business processes improvements. This book moves toward conclusion by covering the reporting capabilities, including the flexibility to create self-defined queries. Finally, the book concludes by broadly covering the integration of PP with some of the other SAP ERP components.

In summary, the following structure is used:

In Part I of this book, starting in **Chapter 2**, we cover the broad outline of the entire book and why you should implement a specific functionality or how it will benefit your business processes. We'll discuss the enterprise structure that you'll need to set up in PP, which at the same time also depends on the enterprise structures of other components. The enterprise structure forms the backbone of the SAP ERP system, in which all the important business processes of the company are mapped. Eventually, reporting also takes important elements from the enterprise structure.

In Part II of this book, we move forward with covering the configuration basics that you need to set up for each production type. However, the primary focus of the three chapters in this part is on the configuration basics only, whereas the actual and practical use of configuration basics are covered with the business processes in Part III. **Chapter 3** covers the configuration basics of discrete manufacturing, **Chapter 4** attends to the configuration basics of process manufacturing, and **Chapter 5** covers the configuration details for repetitive manufacturing.

Part III of this book discusses the PP workflow by each production type, and we make logical connections to the business processes of each production type for which we undertook the configuration in the relevant chapters of Part II. **Chapter 6** provides an in-depth coverage of the business processes of PP in discrete manufacturing. **Chapter 7** brings out the similarities and differences between discrete and process manufacturing, but remains primarily focused on the process industry-specific functionality known as Process Management. Process Management then matures to a user-friendly functionality known as Execution Steps (XSteps). XSteps can also be used in discrete manufacturing. In the same chapter, we also cover how to use the process manufacturing cockpit. The focus of **Chapter 8** is on the important business processes of repetitive manufacturing, in which, once again, we make consistent and logical links to the configuration chapter.

Part IV of this book covers the PP workflow tools. **Chapter 9** focuses on sales and operations planning (S&OP), in which we cover product group, flexible planning, and standard analysis in flexible planning. Forecasting as an invaluable planning tool is also covered in this chapter. **Chapter 10** is on SAP Demand Management, in which we cover planning strategies and production methods such as make-to-order (MTO) and make-to-stock (MTS). Material requirements planning (MRP) is covered in **Chapter 11**, in which we discuss the planning calendar and also MRP areas. In **Chapter 12**, you'll see how you can use MRP to successfully execute Long-Term Planning (LTP) to simulate what-if planning scenarios.

Part V is all about optimizing PP. **Chapter 13** covers special procurement types, such as subcontracting, phantom assembly, procurement or production at another plant, withdrawal from another plant, consignment, and pipeline materials. In **Chapter 14**, we show you how to manage the capacity requirements planning (CRP) in your SAP ERP system, including its evaluation and leveling. **Chapter 15** covers the versatile and dynamic functionality of the classification system, which is cross-modular and finds several applications not just in PP, but also in other logistics components. The co-products and by-products that the actual production process generates find comprehensive coverage in **Chapter 16**. Next, in **Chapter 17**, we show you the benefits of implementing the digital signature functionality in your business processes to eliminate or reduce the manual signature and approval process. Digital signature is also cross-modular.

The last part, Part VI, is all about monitoring and evaluating PP. In **Chapter 18**, you'll learn how to quickly set up alerts in your SAP ERP system with the Early Warning System (EWS) to closely monitor important deviations to your business processes and make quick decisions and actions. You can also set up EWS in other logistics functions, if needed. In **Chapter 19**, you'll learn the features, functionalities, menu paths, navigation tools, and many options available to run a large number of standard reports available in SAP ERP. The concepts you'll develop here will enable you to expand your knowledge horizon to explore standard reports available in other logistics components. In this chapter, we also cover how you can quickly create your own reports by using the SAP Query tools. Finally, in **Chapter 20**, we give you some "flavors" to the complex and highly interconnected world of PP integration with other logistics functions. Here, we provide five examples in which PP integrates with SAP ERP Materials Management (MM), SAP ERP Quality Management (QM), SAP ERP Project Systems (PS), and SAP ERP Plant Maintenance (PM). We also provide a roadmap you can use to ensure effective planning and comprehensive monitoring of cross-components integration during your SAP ERP system implementation project.

In the appendix, you'll find a comparison table of the production types (discrete, process, and repetitive), and a glossary of some of the more important terms used in PP.

While this book is certainly a significant expansion to the areas and functionalities that the PP offers, note that we don't cover the following in detail:

- Variant configuration
- Distribution resource planning
- Kanban

[«]

Note

Kanban is now covered in the E-Bite titled *Configuring Kanban in SAP ERP MM and PP*, which is available at *www.sap-press.com/4013*.

Let's now move on to Chapter 2, where we'll discuss the internal organizational structure of SAP ERP from a PP perspective.

SAP ERP Production Planning is a direct and in-depth reflection and mapping of the business processes that a company either currently follows as a part of industrial operations or will transition to when the implementation of the SAP ERP system is complete. We'll start your journey with a discussion of the organizational structure of all the core components.

2 Organizational Structures in SAP ERP

In this chapter, we'll help you get an overall understanding of how business functions and the SAP ERP system interact and work together. After you understand the basics, we'll slowly move into some specific details on how Production Planning (PP) works in the SAP ERP system. We'll then overview the three main types of manufacturing, which are a large focus of this book.

From a PP perspective, the important organizational units are company code, plant, and storage location. In the following sections, we'll review the structure as it applies to PP. We'll discuss the importance of the organizational units and explain how they work together to accomplish the organizational, legal, and reporting requirements of the company. We'll also explain the SAP calendar, which is an essential part of maintaining your entire system schedule.

2.1 Breaking Down the Structure into Units

During an SAP ERP system implementation, one of the first and highly intensive activities undertaken is the finalization of the organizational structure. This involves having inter-modular and intra-modular discussions and deliberations to ensure that SAP ERP can cover the legal aspect of the company's organizational structure, as well as attend to component-specific reporting needs. In other words, the business process owners, business analysts, and SAP ERP system consultants review the existing organizational structure of the company and then simultaneously begin mapping it in the SAP ERP system.

A practical approach to adopt while finalizing the organizational structure in the SAP ERP system is to ensure that the organizational structure isn't so generic that it loses its significance and prevents the business process owner from extracting the required information from the system, nor is it so minute or detailed that it becomes cumbersome to collate and consolidate the information. You should also keep a forward-thinking view of your organizational structure. If you foresee that you'll need certain organizational elements in your SAP ERP system in the future, for example, it's better to have them available in the system than to add them at a later point.

The organizational structure in the SAP ERP system is equally applicable to all manufacturing types—discrete, process, or repetitive. Take a look at Figure 2.1, which shows the client as the highest level of the organizational structure in the SAP ERP system. The profitability analysis of the company is performed at the *controlling area* level, and the cost center and profit center accountings are performed at that level as well.

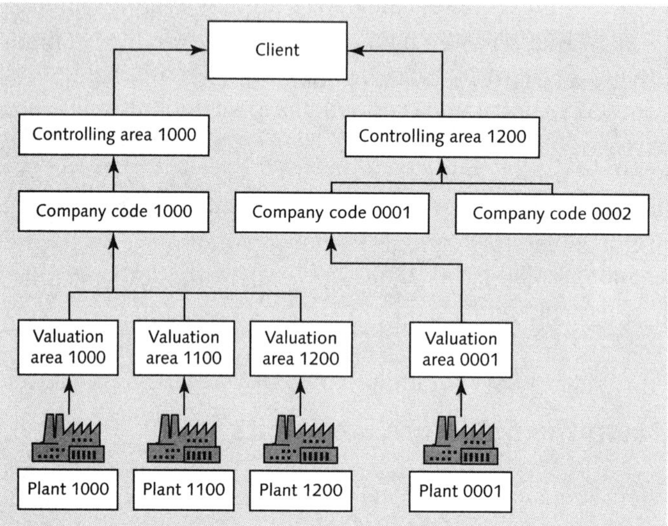

Figure 2.1 Organizational Structure in SAP ERP

A company can consist of several legal entities, each including separate, individual financial statements that must be prepared at the end of the financial year. This is reflected as a separate *company code* for each legal entity.

The *valuation area* represents the level at which the company values its material stock consistently. It's part of the logistics area of the SAP ERP system. A one-to-one relationship exists between the valuation area and the plant. For example, a material at one plant may have a different standard price than at another plant.

The diagram shown in Figure 2.2 represents the *organizational unit* of PP, wherein the company code attains the highest level. Within each company code, there can be one or multiple *plants*. Within each plant, there can be one or multiple physical and virtual *storage locations*.

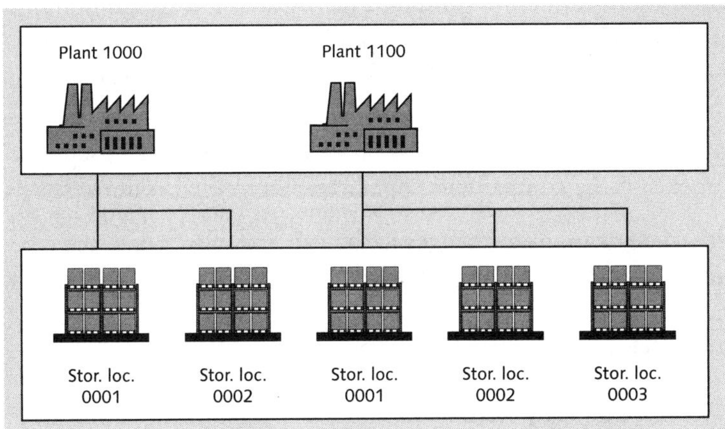

Figure 2.2 Production Planning and Control Organizational Unit

[«]

Note

See Chapter 11, where we cover further organizational units such as the MRP area and how it works.

In the next sections, we'll explain each unit in greater detail.

2.1.1 Client

A *client* represents the highest element of the SAP ERP system's organizational structure. Often, the client represents a company or a group of companies, within which there are several independent company units. An SAP ERP system can contain several clients in logical units. The additional organizational elements and the master and transaction data are created and managed within a client.

From the SAP ERP system's landscape perspective, you normally have three clients (systems): development (DEV), quality assurance (QAS), and production (PRD). The actual configuration of the SAP system takes place in the DEV system, which is then transported to the QAS system for testing and training. The final configuration eventually moves to PRD, which is the final and live system on which the business process owners of the company make real-time live entries.

2.1.2 Company Code

The company code is the level below the client in the SAP ERP system, and it reflects the level at which the company legally reports income statements and balance sheets. It's an organizational element (unit) of SAP ERP Financials (FI). You can have a separate company code for each line of business—for example, textile and chemicals—as long as the two are legally separate entities. Similarly, separate company codes can exist if the company has operations in foreign countries.

To create a company code or to make changes to the existing one, follow the configuration (Transaction SPRO) menu path, ENTERPRISE STRUCTURE • DEFINITION • FINANCIAL ACCOUNTING • EDIT, COPY, DELETE, CHECK COMPANY CODE • EDIT COMPANY CODE DATA (see Figure 2.3).

Figure 2.3 Company Code

[»] **Note**

In this book, whenever we refer to Transaction SPRO, it implies that the next step you need to take is to click on REFERENCE IMG or press F5, followed by the menu path given. Wherever possible or available, we've also given the relevant configuration transaction code to facilitate your configuration efforts.

[«]

Note

Your FI team decides and works on the creation of company codes in the system.

2.1.3 Plant

A *plant* is an organizational unit within the logistics component. You can classify a plant from the point of view of production, procurement, maintenance, warehouse, and planning. For example, the plant can be a manufacturing site, a head office, or a distribution center within a company. It organizes the tasks for the production logistics, and it can be a physical production site or the logical grouping of several sites in which materials are produced or goods and services are provided. Different production locations are mapped with the plant in the SAP ERP system. At the plant level, you can perform the following tasks:

- Managing inventory
- Evaluating and performing physical inventory of stocks
- Managing demand
- Planning production
- Executing and controlling production
- Performing material requirements planning (MRP)

In the organizational structure of the SAP ERP system, you can assign only one company code to a plant. However, you can assign multiple plants to the same company code.

To create a new plant or to make changes to the existing plant, follow the configuration (Transaction SPRO) menu path, Enterprise Structure • Definition • Logistics – General • Define, Copy, Delete, Check Plant (no transaction code available).

[«]

Note

Your SAP ERP Materials Management (MM) team decides and works on the creation of plants in the system.

Figure 2.4 shows the change transaction screen of plant 3000, with the provision to enter the complete address and other details. It's important to use the Factory

CALENDAR field to assign a factory calendar to a plant so that the system can plan out all the working and nonworking days of the plant.

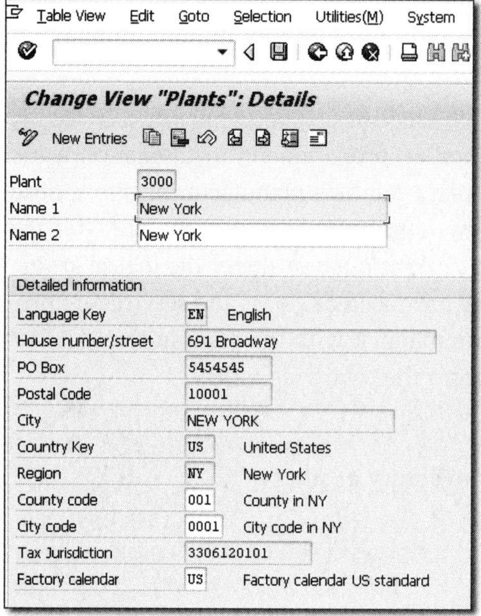

Figure 2.4 Plant

[»] **Note**

We cover the creation of the factory calendar in Section 2.4.3.

After the creation of the plant, the next step is to assign the plant to the company code. A plant can only be assigned to one company code, and you can assign multiple plants to the same company code. It's mandatory to assign a plant to a company code.

To make a plant–company code assignment, follow the configuration menu path, ENTERPRISE STRUCTURE • ASSIGNMENT • LOGISTICS – GENERAL • ASSIGN PLANT TO COMPANY CODE, or use Transaction OX18 (see Figure 2.5). Choose the NEW ENTRIES icon to create a new plant–company code assignment.

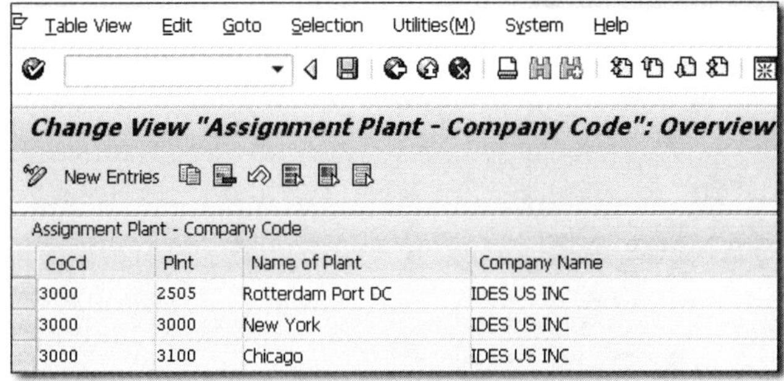

Figure 2.5 Assignment of Plant to Company Code

Note [«]

Your MM team creates the plant–company code assignment in the system.

2.1.4 Storage Location

A *storage location* is the physical or virtual storage site for the materials. Examples of physical storage locations include raw materials store, components store, returned goods store, finished goods store, and so on, whereas the virtual storage location can be self-defined and may be a scrap yard or a production shop floor in which semifinished goods are temporarily stored.

You can even treat storage tanks or silos for storing bulk chemicals, oils, or grains as storage locations in the system. However, the limitation is that a storage location in the SAP ERP system doesn't have the provision to define the maximum storage capacity of an individual tank or silo. This provision is available in SAP ERP Warehouse Management (WM).

You can create as many storage locations as needed within a plant, but you can assign a storage location to one plant only.

To create a new storage location or to make changes to the existing storage location, follow the configuration (Transaction SPRO) menu path, ENTERPRISE STRUCTURE • DEFINITION • MATERIALS MANAGEMENT • MAINTAIN STORAGE LOCATION, or use Transaction OX09. Figure 2.6 shows the PLANT popup screen ❶, in which you enter the plant value as "3000". You add both the storage location code with a DESCRIPTION ❷ and the complete address.

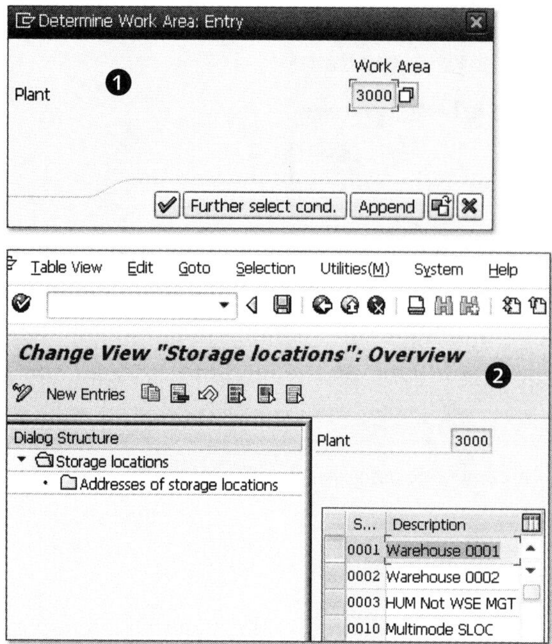

Figure 2.6 Storage Location

[»]

Note

Your MM team creates storage locations and assigns them to the plant(s).

2.1.5 Material Requirements Planning Controllers

An *MRP controller* can be an individual role or group of roles, performing the same task. For example, if three people in a company manage the packaging materials procurement, then it makes sense to define one MRP controller for this. The MRP controller is primarily responsible for attending to the requirements of materials. When defining the MRP controller, focus must remain on making sure it's defined based on responsibility, role, or area of working, instead of individuals. For example, you may have one MRP controller who is responsible for raw materials only, while another one may be for packaging material. You may have an MRP controller who is only responsible for managing consumables.

The MRP controller is assigned in the MRP 1 view of the material master. When you select the relevant MRP type in the material master by indicating that planning will be done on the material, the system prompts you to enter the MRP controller. Again, the MRP controller should be based on position or responsibility rather than on the person. Later, when you run several reports, you'll be able to use the MRP controller as a selection criterion, among others.

To configure the MRP controller in SAP ERP, follow the configuration (Transaction SPRO) menu path, PRODUCTION • MATERIAL REQUIREMENTS PLANNING • MASTER DATA • DEFINE MRP CONTROLLERS.

Note [«]

MRP controllers are extensively used by both production and procurement departments and the corresponding PP and MM components, respectively. MRP controllers for procurement may be raw materials, packaging materials, consumables, or spare parts. MRP controllers for production may be finished goods, semifinished goods, or assemblies.

The two teams (PP and MM) must coordinate in finalizing production and procurement MRP controllers because the bifurcation of PP and MM MRP controllers (including its transaction code) isn't obvious. It's best to mutually agree and make them available in the system.

2.1.6 Capacity Planners

A *capacity planner* or *capacity planner group* is responsible for evaluating the current work center's or resource's capacity and, if needed, also performing the capacity leveling. When you create a new work center, you also have to assign the person responsible in a specific field. The capacity planner can also handle the role of person responsible for the work center. For example, it may make sense to combine all of the packing units of similar products as one capacity planner if the same person is responsible for it. If a company produces 10 different sizes of tomato ketchup—from a packet to a gallon size—and the same capacity planner is responsible for ensuring that various machines' capacities for each packing size are available, then you can simply agree to have one capacity planner in the SAP ERP system, together with its code.

The capacity planners are assigned in the capacity header data of the work center (resource). Then, in all the capacity evaluation and leveling reports, the capacity planner is available as the selection criterion for the planner to choose from and enables the system to display only relevant information.

To create a capacity planner, follow the configuration (Transaction SPRO) menu path, PRODUCTION • CAPACITY PLANNING • MASTER DATA • CAPACITY DATA • SET UP CAPACITY PLANNER.

2.1.7 Production Schedulers

A production scheduler is responsible for ensuring that production execution and operation takes place per the production plan. The production scheduler immediately attends or takes immediate remedial action, where necessary. To define a production scheduler, follow the configuration (Transaction SPRO) menu path, PRODUCTION • SHOP FLOOR CONTROL • DEFINE PRODUCTION SCHEDULER, or use Transaction OPJ9 (see Figure 2.7). You assign production schedulers in the WORK SCHEDULING view of the material master.

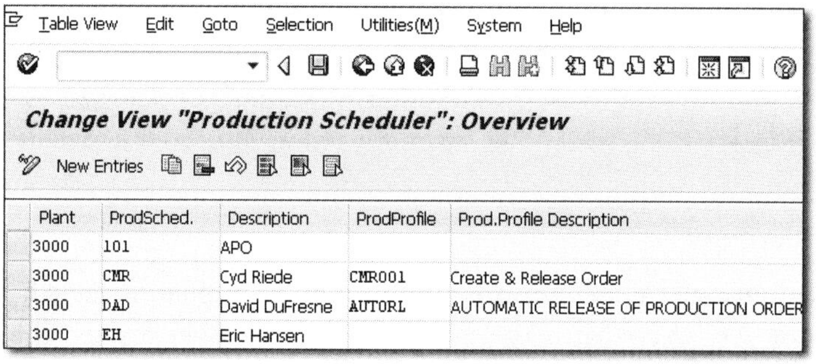

Table View Edit Goto Selection Utilities(M) System Help

Change View "Production Scheduler": Overview

New Entries

Plant	ProdSched.	Description	ProdProfile	Prod.Profile Description
3000	101	APO		
3000	CMR	Cyd Riede	CMR001	Create & Release Order
3000	DAD	David DuFresne	AUTORL	AUTOMATIC RELEASE OF PRODUCTION ORDER
3000	EH	Eric Hansen		

Figure 2.7 Production Scheduler with Production Profile Assignment

Now that you have an understanding of how the SAP ERP system works, we'll add another ingredient into the mix: PP.

2.2 Production Planning in SAP ERP

Production planning is the core of any manufacturing process. SAP ERP helps you set up and streamline your specific process to maximize efficiency in the workplace when working with different types of manufacturing.

Actually, you'll find that the SAP ERP system is made up of several different components, in addition to PP (see Figure 2.8). We'll go into the different integration of PP with the different components you see here in Chapter 20.

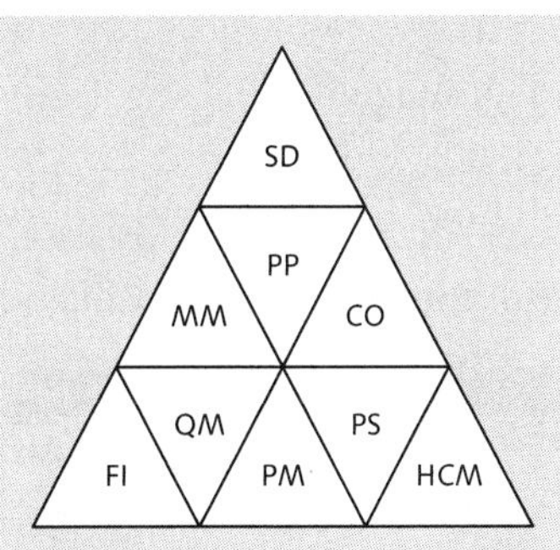

Figure 2.8 SAP ERP Components

In general, the entire process of production planning and control starts when you forecast the demand of a product and prepare a sales plan. The sales plan is synchronized with a production plan to take the project realities into account, such as capacity constraints. Various simulated models are considered, and the finalized production plan becomes the basis of MRP. Materials planning helps the production and the procurement planners know when to procure and produce a material for its eventual availability and dispatch to the customer. The production execution accounts for and records each production detail, including generation of scrap, co-products, or by-products, if any. Quality checks in the production processes ensure minimal customer returns or other rejections. The produced product is sold to a customer, and the production plan continues to be a monitoring barometer against the sales plan.

Of course, this information is great for providing a bird's-eye view of the production planning process. We'll help you understand how the individual objects you have to work with in the SAP system help streamline and manage your business processes in the following sections.

PP includes the following types and tools:

- **Master data**
  This includes the material master, work centers, resources, production lines, routings, master recipe, rate routing, bill of materials (BOM), and production version.
- **Sales and operations planning (S&OP)**
  You can use standard S&OP or flexible planning to forecast sales and production plans to meet customers' requirements for products.

- **Production Planning**
  This includes material forecasting, SAP Demand Management, Long-Term Planning (LTP), and Master Production Scheduling (MPS).
- **Material requirements planning (MRP)**
  This attends to standard and unique customers' requirements via various planning and production methods.
- **Discrete manufacturing or shop floor control (SFC)**
  Production orders processing, goods issuances and receipts, and confirmations are used for complex manufacturing processes in which there may be a need for intermediate or interim storage.
- **Process manufacturing or Production Planning for Process Industries (PP-PI)**
  Process orders processing, Process Management, material quantity calculation, goods issuances and receipts, and confirmations are used for production processes of liquid-based or flow-based materials.
- **Repetitive manufacturing (REM)**
  This adopts the lean manufacturing principle in which generally the production process is not only simple but also consistent over a considerable period of time.
- **Capacity requirements planning (CRP)**
  This consists of capacity evaluation and capacity leveling. Capacity evaluation reflects the load and overload at work centers/resources, whereas capacity leveling helps the planner optimize the production processes.
- **Product Costing (CO-PC)**
  This completely integrates with PP and is responsible for ensuring all production-related costs are accounted for, including overheads, variances, and work in process (WIP).
- **Kanban**
  This production type replenishes stocks based on a pull system by using Kanban cards. Kanban works well for both in-house produced materials and outside procured materials.
- **Distribution resource planning (DRP)**
  This enables planning the demand of products at distribution centers.
- **Reporting**
  A large number of information systems and standard and flexible analysis reporting options are available in PP.

In the following sections, we'll cover the features and characteristics of various production types as well as important business processes in production planning and control.

2.2.1 Characteristics of Production Types

A *production type* characterizes the frequency, complexity, or stability with which a product is produced in the production process. When implementing an SAP ERP system, one of the very first decisions a company makes is which production type to implement to reflect the complexity (or simplicity) of the production process. For example, if the production process is relatively simple with a linear production line involving one operation and one work center, then it makes sense to implement the REM production type to enable the company to benefit from lean manufacturing. Similarly, the process manufacturing production type is more suited to scenarios in which the product is generally in liquid form and flows or where the manufacturing process is generally continuous. The discrete manufacturing production type is used where the production process is order based, involves special procurement types, or when products are stored in interim storage locations between the production processes. Kanban is a demand-driven production type in which the demand triggers the replenishment and initiates the supply process. This production type enables minimal involvement of the Inventory Management function.

We discuss each of the main production types in the following sections. While this book will primarily cover discrete, process, and repetitive production types, this sections also briefly covers engineer-to-order (ETO) and Kanban to provide a comprehensive look at production types.

Discrete Manufacturing

The *discrete manufacturing* production type, which is also known as shop floor production, describes the production of a product on the basis of production orders. Discrete manufacturing is implemented where the products change frequently, the demand pattern is irregular, and production is workshop oriented in character. A range of master data is required for discrete manufacturing; the most important are the material, BOM, work center, and routing.

[Ex]

Example

In steel rerolling mills, the entire production process passes through five different production steps. However, customers can place orders based on a different level of the processed good. Hence, the company has to produce and also store a semifinished good at each production step to meet its customer's demand.

The production process in discrete manufacturing starts when a production order is created and processed. A production order can either be created manually or by converting a planned order that the system generated after running MRP. A production order is a request to the production department to produce the product at a specific time and in a specific quantity. It specifies the work centers and material components that are required for production. The creation of a production order automatically creates reservations for the required material components. Purchase requisitions are created for externally procured material components and services, and capacity requirements are created for the work centers at which each operation of the order will be executed. The discrete process is shown in Figure 2.9.

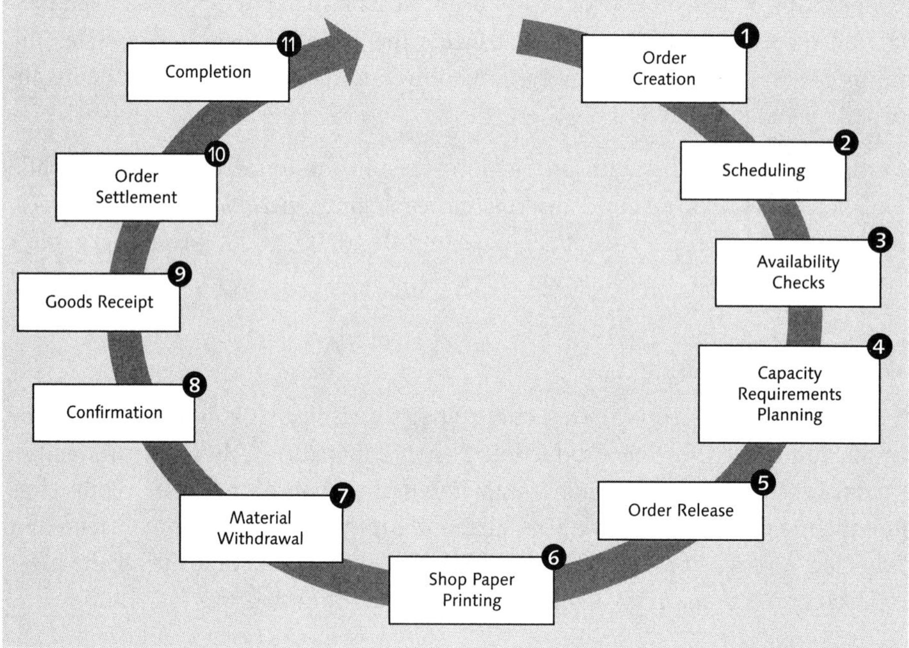

Figure 2.9 Discrete Manufacturing Process Flow

Production orders are released on the release date as long as the required materials and capacity are available. The production order-related documents (known as shop floor papers) are printed to prepare for production execution. The capacity situation is evaluated, and any required capacity leveling is carried out in any phase of production order processing, although this is usually ensured before the actual production starts.

The components required to produce the products are issued with reference to the production order, the product is produced on the basis of the production order, and the finished quantity is confirmed with reference to the production order. The product is put into a storage location, and the goods receipt is posted. Finally, the settlement of the production order is ensured.

Note [«]

Chapter 3 covers the configuration basics of discrete manufacturing, and Chapter 6 details the business processes of discrete manufacturing.

Process Manufacturing

Process manufacturing is the batch-oriented and recipe-oriented production of products or co-products in the process industry. Manufacturing can be in the form of continuous production, discontinuous production, or regulated production. In *continuous production*, the product is continuously produced, raw material is continuously supplied to the production line, and the plant and machinery are in continuous operation. An example of this is fertilizer manufacturing, where the production process is continuous, starting with production of ammonia from natural gas (methane) and continuing until the final urea/fertilizer is produced. The process may find an interim storage in the form of bulk urea being stored in the warehouse before the bagging process starts.

In *discontinuous production*, the products aren't produced in a continuous process. Instead, the material components are provided and weighed out as required for each step of the production process. Its greater application is found in industries such as food processing.

Regulated production is used if the product quality requirements are very stringent and specific industry standards must be met. This type of production is generally followed in pharmaceutical or cosmetics manufacturing. In regulated production, orders can be created only with approved recipes. If changes need to be made to master recipes, these are subject to master data change administration procedures.

The central master data elements in process manufacturing are the material, the BOM, the resource, and the master recipe.

The business process in process manufacturing starts when a *process order* is created and processed in accordance with a master recipe. A process order is a request to the production department to produce a product at a specific time and in a specific quantity. It specifies the resource and material components that are required for production.

A process order can be created either manually or when a planned order that was created in the PP process is converted. The creation of a process order automatically creates reservations for the required material components. The system automatically creates purchase requisitions for externally procured material components and services, and capacity requirements are created for the resources at which the order will be executed. Process orders are released on the release date, provided the required materials and capacity are available. At the time of release, you can run an automatic batch-determination process for components that are subject to a Batch Management requirement. The relevant documents in the process order can be printed to prepare for the execution of the process order. The process manufacturing flow is shown in Figure 2.10.

The capacity situation is evaluated, and any required capacity leveling can be carried out in any phase of the process order processing, although this is usually ensured before the actual production commences.

[»] **Note**

Refer to Chapter 14 in which we show you how to use capacity requirement planning (CRP) for evaluation and leveling.

The actual production can now begin, with or without the use of Process Management. If you implement Process Management to execute a process order, this serves as the interface between the SAP ERP system and process control. The flexible structure of this interface makes it possible to connect automated, semiautomated, and manually controlled plant and equipment to the production process.

Process Management makes extensive use of the classification system, which is cross-component.

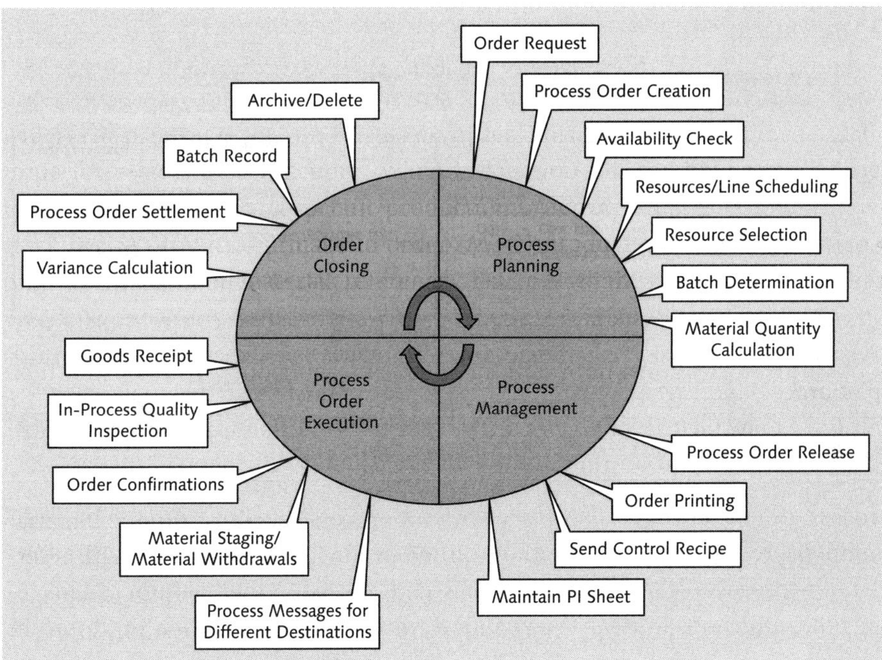

Figure 2.10 Process Manufacturing Process Flow

Note [«]

Chapter 15 shows you how to implement the classification system and then integrate it in Process Management. You can also integrate classification in other logistics components to bring better organization to your master data management.

After the process order or the relevant phases of the process order is released for production, control recipes are generated from the process instructions in the process order. The *control recipes* contain all the information required for the process control function to execute a process order. Next, the control recipes for the process control system are either automatically or manually sent to the relevant process operator in the form of process instruction sheets. In the process instruction sheet, the process operator can refer to operation's instructions, refer to the online instruction manual using the Document Management System (DMS), input process parameters, or write shift highlights.

When the process operator has entered all process parameters and is ready to mark the process instruction sheet as complete, the system can prompt the process operator to digitally sign the process instruction sheet to set it to completion status.

[»] **Note**

In Chapter 17, we show you how to implement digital signature and then integrate it in Process Management.

The process data that results from the execution of the process order are sent back to the SAP ERP system, are transferred to external function modules for further processing, or both. This data is transferred from the process control function to the various recipients by means of the process-coordination interface with the help of process messages. A material consumption message, for example, causes a goods issue to be posted for a component. Similarly a material-produced message triggers a goods receipt posting in the system.

If process order execution takes place without process coordination, the material components required to produce the finished product are withdrawn with reference to the process order, and the goods issue is posted in the Inventory Management subcomponent of MM. The required finished product is then produced in accordance with the process order. The quantities created and the products produced are confirmed to the process order, the finished product is put into storage, and the goods receipt is posted. In the final step, the product costing team ensures the order settlement.

[»] **Note**

Chapter 4 and Chapter 7 cover the configuration basics and business processes of process manufacturing, respectively.

Repetitive Manufacturing

Repetitive manufacturing (REM) is the interval-based and quantity-based creation and processing of production plans. With REM, a certain quantity of a stable product is produced over a certain period of time. The product moves through a work center, which may be a group of machines, in a continual flow, and intermediate products aren't put into intermediate storage (e.g., motherboard assembly in computer manufacturing).

The data entry efforts involved in production control with REM is significantly reduced when compared with single-lot and order-based production control. REM can be used for the make-to-stock (MTS) production method. In this case, production has no direct connection to a sales order. The requirements are created in the SAP Demand Management process, and the sales orders are supplied from stocks. Sales order-based production (i.e., the make-to-order [MTO] production method) is also possible in REM. In this case, production is directly related to a sales order or can even be directly trigger from a sales order. The most important master data in REM are material, BOM, production line, rate routing, and production version. The REM process flow is shown in Figure 2.11.

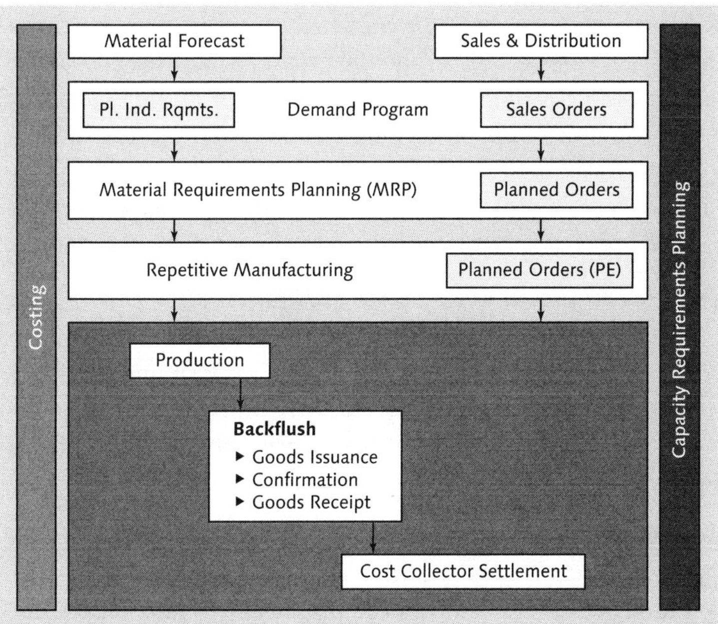

Figure 2.11 Repetitive Manufacturing Process Flow

There are significant similarities between the master data of REM when compared with discrete manufacturing or process manufacturing.

If a material is produced using the REM production type, it has to be flagged accordingly in the material master by setting the REM checkbox in the MRP 4 view of SAP ERP. Further, it's mandatory to assign a REM profile to the material. This profile determines the type of planning and confirmation by specifying, among other things, whether reporting points will be used, whether production activities will be posted to the cost collector for material confirmations, whether a decoupled confirmation will be used, whether a backflush will be carried out for the entry of actual data, and which transaction types will be used.

The BOM for the material specifies the quantities of components required for production. In REM, not every goods issue is recorded at the same time as the physical withdrawal of the material from stock. Usually, component usage/goods issuance (*backflush*) is automatically posted only when the finished product is received in the warehouse. To backflush a component, a storage location is specified in every BOM item, and the backflush is carried out from this storage location.

Work centers in REM are known as *production lines* because the product moves through the machines in a continuous flow, and the machines are usually spatially arranged in a line. These can be simple production lines, which often consist of just one work center, or complex production lines, which consist of several work centers. The individual processing stations are set up as individual production lines and are grouped into a line hierarchy. A production line determines the available capacity of the processing station and is assigned to a single cost center.

In REM, standard routing is known as *rate routing*. A rate routing contains the operations or the processes required to produce the material. Because the same product is produced over a long period of time in REM, very simple routing is used, often consisting of just one operation/process. This kind of process specifies the production rate, which, in turn, specifies the quantity per time unit that is produced on the line (e.g., 50 units per hour).

Because there are different BOMs and routings for a material depending on the production process, a *production version* is used to specify which BOM and which routing will be used to produce the material. The production version also specifies the lot size for which the production version is valid. It's important to set the REM ALLOWED checkbox in the production version. There has to be at least one production version of a material in REM. The costs incurred in REM are posted to a *product cost collector (PCC)*. In the process of entering actual data, the material costs and production costs are added to the PCC. The PCC is created for a material within a plant in a specific production version.

In REM, the planned orders for a material that result from the production and procurement planning process are managed in a *planning table*. In these tables, the planner can schedule the production quantities on the assembly lines. In REM, the term *run schedule quantity* is used instead of planned orders (as used in discrete or process manufacturing) to denote the quantity that you plan to produce. The components are supplied anonymously to the production line by using the pull list. The components required on a production line for a specific period are calculated in the pull list. The missing quantities that the system detects are replaced by means of direct stock transfers, for example, from the main storage location to the production storage location. This is known as *replenishment*.

The production of the product usually takes place in a continuous flow along the production line. Entry of actual data is carried out at regular intervals for each finished production quantity. Component use (backflush) and production activities are automatically posted when the finished product is received in the warehouse. For longer production lead times, the actual data is recorded with a *reporting point* within the production line to enable the system to post consumption data more promptly.

Note [«]

Chapter 5 covers the configuration basics of REM, and Chapter 8 details the business processes.

Engineer-to-Order

The *engineer-to-order* (ETO) production type attends to the complexities and challenges when a sales order-based MTO production method is unable to fulfill the requirements. In the MTO production method, the system is unable to make a distinction between the predecessor-successor relationships in the production process; for example, a material's production can't initiate (successor) until the production of the previous product (predecessor) is ensured. In ETO, the system uses work breakdown structure (WBS) and networks for scheduling and coordinating the production processes and also managing Cost Accounting. All produced goods are specific to the project, and the system maintains project-based inventory.

Note [«]

Chapter 20 illustrates the integration of PP with SAP ERP Project Systems (PS).

Kanban

Kanban involves a requirements-oriented production control procedure and uses material flow control that avoids time-intensive requirements planning. With Kanban, a material is produced or procured only when it's actually required. A specific quantity of the components required to produce a material are stored on-site and in containers. When a container is empty, this component is replenished according to a predefined replenishment strategy (in-house production, external procurement, or stock transfer). In the interval between the request for replenishment and the delivery of the refilled container, the other available containers simply do the work of the empty one.

The replenishment process is largely automatic in the Kanban procedure, thereby greatly reducing the amount of manual posting work required. The material isn't pushed through the production process as specified by an overall plan; rather, it's requested by one production level (consumer) from the previous production level (source) as and when needed. It adopts the "pull" strategy in the production process.

In Kanban processing, *production supply areas* (PSAs) divide the plant. The components required for production are stored in these PSAs, and various work centers take what they need from them. A Kanban control cycle is defined to specify how a material should be obtained within a PSA. The control cycle defines a replenishment strategy for the material that specifies, for example, whether the required material is to be produced in-house or procured externally. The control cycle also specifies the number of containers in circulation between the consumer and source, as well as the quantity per container.

Replenishment strategies specify how a material component should be replenished and which of the following replenishment elements should be created for this purpose:

- In-house production
- Manual Kanban
- Replenishment with run schedule quantity
- Replenishment with production order
- Replenishment by purchase order
- External procurement

- Replenishment with schedule agreement
- Replenishment with reservation
- Replenishment with direct transfer posting
- Replenishment with summarized just-in-time (JIT) call
- Stock transfer
- Replenishment by transport requirements of a WM administered storage location

The replenishment process with Kanban entails that a material is produced at a machine. The components required to produce it are available onsite in containers and are ready for withdrawal. If one of these containers is empty, the source that is responsible for its replenishment has to be informed. If Kanban processing without the SAP ERP system support is being used, the consumer sends a card to the work center (source). The card contains the information about which material is required, in what quantity, and where it should be delivered to.

Note [«]

The replenishment process gets its name from the Japanese word for these cards (*Kanban*).

The source can now produce or procure the material and then refill the container. If Kanban processing with SAP ERP support is in place, the containers are managed in the system and have a specific status. After the last component is withdrawn from a container, the status of that container is simply changed from "full" to "empty." This status change is the Kanban signal, and it can be set by passing a barcode reader over the card attached to the container. It's also possible to have the system display the containers in a production area in the form of a Kanban table and to make the status change there. The Kanban signal now triggers the replenishment process and creates, for example, a run schedule quantity in accordance with the replenishment strategy. The source then processes the run schedule quantity, and the finished material is sent to the container. The status of the container is set to "full" again (through barcode or Kanban table), and the goods receipt for the material is posted with reference to the procurement element.

The SAP ERP system also supports other kinds of Kanban procedures besides the more-prevalent procedure just mentioned. The Kanban process also works well with stock transfer replenishment (plant-to-plant and store-to-store stock transfer).

2.2.2 Processes in Production Planning and Control

We discuss the main processes in PP in the following sections.

Sales and Operations Planning

The S&OP process is used to determine the quantities for production. Sales planning, which is also known as demand planning, covers future requirements without considering stocks and available capacities. The historical sales figures serve as a basis for sales planning. Operations planning uses the results of the sales planning process to plan the production quantities and takes initial stocks and capacities into account.

[»] **Note**

Chapter 9 covers S&OP.

SAP Demand Management aligns sales planning with the customer requirements in accordance with the planning strategy and thus calculates the independent requirements for production. The planning methods that SAP Demand Management looks for are MTS, MTO, planning with final assembly, and several others.

[»] **Note**

Chapter 10 covers SAP Demand Management.

Material Requirements Planning

MRP is one of the most important functions of PP. The system performs net quantity calculation for component requirements while taking scrap and lot sizes into account. MRP calculates requirement coverage elements for all MRP levels such as plant, material, product group, and MRP areas, and it takes into account the lead times, lot sizes, and scrap quantities. MRP also enables capacity planning.

[»] **Note**

Chapter 11 covers MRP.

Long-Term Planning (LTP) is a simulation tool for MRP that examines how a change in planned independent requirements (PIRs) will affect capacity utilization, stocks, and external procurement. LTP is also suitable for short-term simulations.

Note [«]

Chapter 12 covers LTP.

Capacity Requirements Planning

For detailed production planning while taking available capacities into account, capacity requirements planning (CRP) schedules the worklist in detail, which usually consists of the processes for created or released production orders. CRP delivers a production sequence that is feasible from the capacity viewpoint. CRP consists of capacity evaluation and capacity leveling.

Note [«]

Chapter 14 covers CRP.

Production Control

The central controlling and recording element—the production process—is the production order in discrete manufacturing, the process order in process manufacturing, and the run schedule quantity in REM. While the previous processes dealt with production planning, production execution is concerned with how the actual production as specified in the production order is recorded and controlled, from material withdrawal to order confirmation to storage and invoicing.

2.3 Product Costing

Product Costing (CO-PC) is a subcomponent of SAP ERP Controlling (CO) and comprehensively integrates with PP. In fact, PP is unable to function completely until the Product Costing subcomponent is in place. Product Costing helps to ensure that the total cost of goods manufactured (COGM) and cost of goods sold (COGS) are completely accounted for. To calculate the COGM, you need to have first-hand information of the cost of all the raw materials and components used. Further, you also need to know the activity rates for each work center (resource). The material and activities costs are also known as *direct costs* and are individually assigned to the order without any allocation. *Overhead cost* is determined by overhead charges. Examples of overhead costs are the electricity consumed in the production process and the salaries of employees involved in the production of goods.

There is also a method of assigning a *costing sheet* to an order type, which for example, may contains details such as 2% of raw material cost will equal the electricity cost of producing a material. Before the actual business processes, such as order creation, in PP begins, the product costing team runs a *material cost estimate* in the SAP ERP system. When running the material's standard cost estimate, the system refers to the complete master data information of PP such as BOM, routing, work center, and production version. It draws information from CO, such as activity types and activity rates. The material cost estimate is first saved and then released. Then, when you create an order, the system performs planned cost calculations within the order. When you perform production execution activities such as goods issuance, confirmation, goods receipt, and recording of co-products or by-products, the system continuously updates the actual cost and presents a comparison of planned costs with actual costs.

[»] **Note**

Chapter 16 shows you how to manage co-products and by-products in the production processes.

The WIP, the overhead, the variances, and finally the settlement are some of the functions managed by the product costing team. When an individual order is settled, the system updates the material price based on the price control. If the price control in the material master (finished or semifinished good) is standard price, the system reflects all differences and variances to the price difference account. If the price control in the material master is a moving average, the system updates the material price. Order-based settlement in discrete and process manufacturing is a mandatory requirement.

The process differs slightly in REM, in which either the material's standard cost estimate is used or a PCC is created with infinite validity. A PCC is preliminary costing, and all the product costs are summed up in the PCC for a material before the actual settlement takes place. In REM, the settlement process isn't order-based but period-based.

2.4 SAP Calendar

For all of the planning and scheduling to effectively take place, it's imperative that a calendar exists in the system. This calendar is then assigned to the plant. You have to first define all of the national holidays, followed by combining all of the individual holidays in the holiday calendar. This holiday calendar then gets assigned to the factory calendar.

The SAP calendar creation function includes three individual steps:

- Defining holidays
- Creating a holiday calendar
- Defining a factory calendar and assigning a holiday calendar to it

To create a new calendar, follow the configuration (Transaction SPRO) menu path, SAP NetWeaver • General Settings • Maintain Calendar, or use Transaction SCAL (see Figure 2.12). Here you have the options you need to maintain the requisite details, such as Public holidays, Holiday calendar, and Factory calendar, in the same sequence.

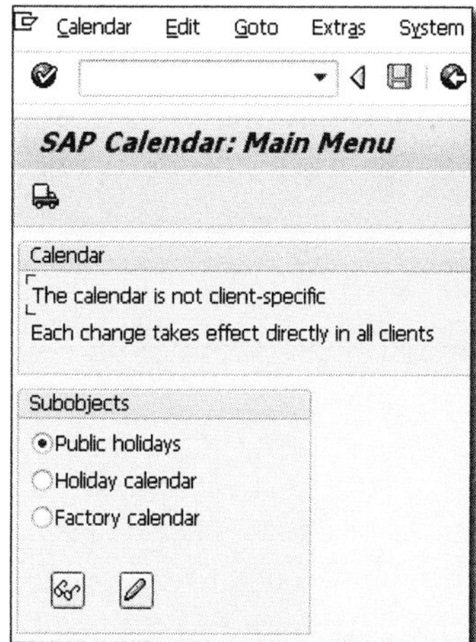

Figure 2.12 SAP Calendar

In the following sections, we'll go into more detail about the different calendar steps.

2.4.1 Public Holidays

Select the PUBLIC HOLIDAYS radio button shown previously in Figure 2.12, and then choose the CHANGE icon. Select the NEW HOLIDAY icon, so you can then select whether it's a fixed date or a floating public holiday. A floating public holiday depends on factors such as moon sighting to decide the holiday. In Figure 2.13, selecting FLOATING PUBLIC HOLIDAY ❶ leads to the FLOATING PUBLIC HOLIDAYS dialog box ❷, in which you can choose the holiday to be any specific date, day, or even with religious denominations, such as Buddhist, Christian, Islamic, or Jewish calendars.

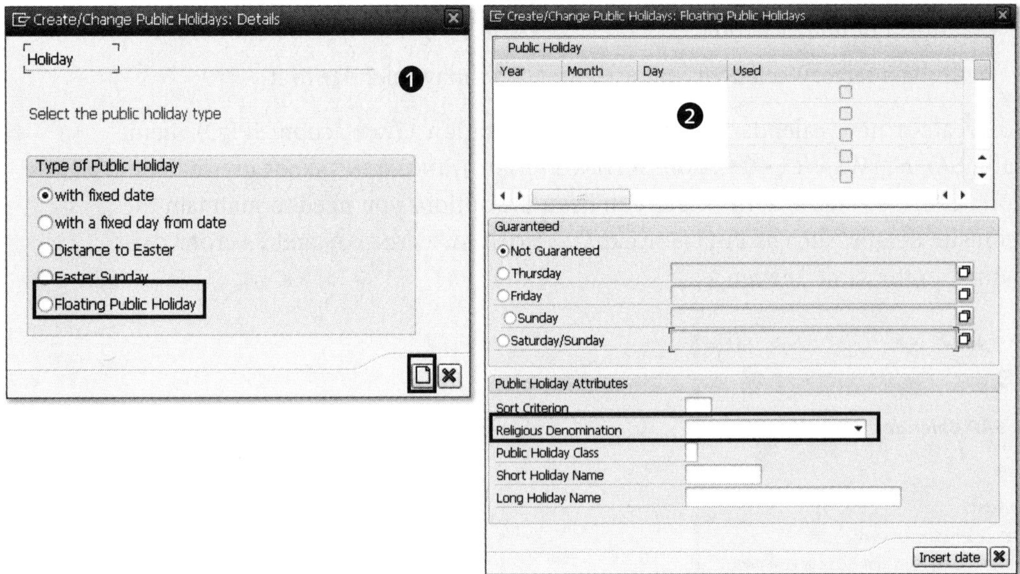

Figure 2.13 Public Holidays

2.4.2 Holiday Calendar

After defining and saving a public holiday, you'll again come back to the screen shown in Figure 2.12, where you select the HOLIDAY CALENDAR radio button, which consists of a list of all of the holidays defined so far. Choose the NEW ENTRY icon, which leads to the screen shown in Figure 2.14. After you provide the identification code and a short text for the holiday calendar in this screen, you define the validity of the holiday calendar. Next, select the ASSIGN HOLIDAY button, which leads to the pop-up in which you can select all of the relevant public holidays by choosing the relevant checkboxes, pressing Enter to confirm, and finally saving the holiday calendar. This takes you back to the original SAP CALENDAR screen shown earlier in Figure 2.12.

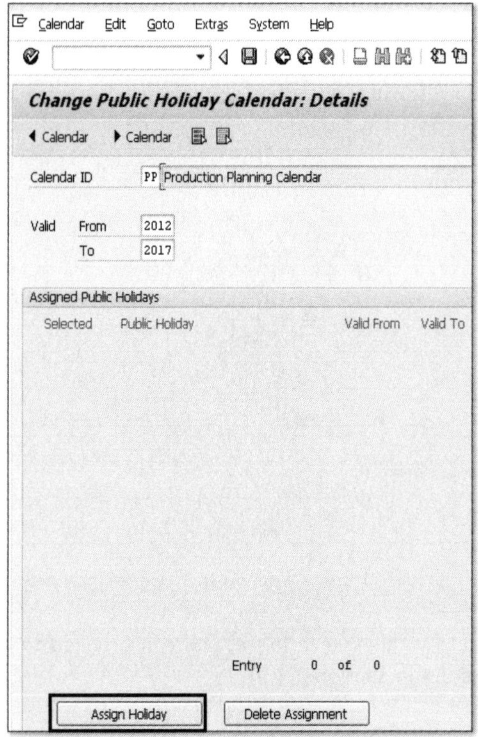

Figure 2.14 Public Holiday Calendar

2.4.3 Factory Calendar

Finally, in the screen shown earlier in Figure 2.12, select the FACTORY CALENDAR radio button, which leads to the screen shown in Figure 2.15, where you can enter the validity date of the factory calendar, assign a HOLIDAY CALENDAR ID, and define WORKDAYS. You can also define SPECIAL RULES to denote any holiday (off day) as a workday.

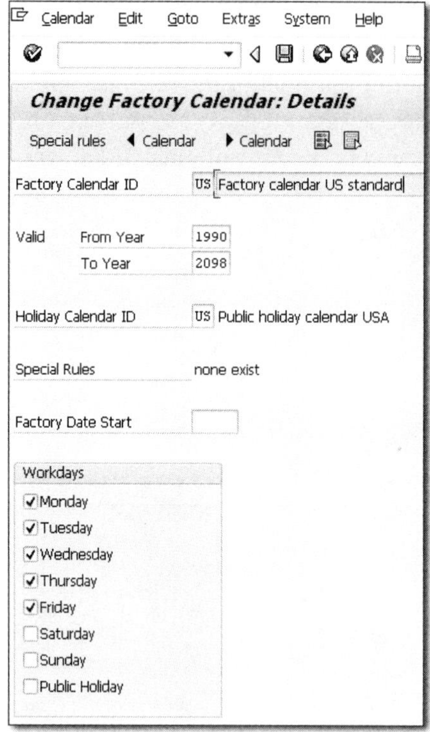

Figure 2.15 Factory Calendar

[»]

Note

With the necessary configuration of the factory calendar in place, you can proceed to assign the same in the plant (refer to Figure 2.4 in Section 2.1.3).

Table 2.1 provides a summarized view of configurations specific to PP, which will eventually form the basis of selection criteria in reporting (or the evaluation tools). Based on the various roles configured in this chapter, for example, capacity planner, MRP controller, or production scheduler, this table provides a broader description of which role makes use of which evaluation tools. For example, the MRP controller generally uses Transaction MD06 (MRP List/Collective Display Evaluation Tool); hence, the option to enter the MRP controller on the initial selection screen is available. Similarly, capacity evaluation tools (Transactions

CM01, CM02, etc.) are of more interest to capacity planners than, for example, to MRP controllers. Further, all these evaluation tools make use of the factory calendar defined earlier in this chapter.

Keep this table in mind as you read through the rest of the book.

Function	Transaction	MRP Controller	Capacity Planner	Production Scheduler
SAP Demand Management	MD73	✓		
MRP List/Collective Display	MD06	✓		
Stock/Requirements List/ Collective Display	MD07	✓		
Long-Term Planning (Requirements)	MS65	✓		
Mass Processing of Production/Process Orders	COHV COMAC	✓		✓
Production Order Information System	COOIS	✓		✓
Process Order Information System	COOISPI	✓		✓
Missing Parts Information System	CO24	✓		✓
Capacity Evaluation	CM01 CM02 CM04 CM05 CM07		✓	
Capacity Leveling	CM21 CM22		✓	
Production Resources and Tools	CF10 CF13			✓

Table 2.1 MRP Controllers, Capacity Planners, and Production Schedulers Available as Selection Criteria

2.5 Summary

This chapter explained the importance of mapping most, if not all, of the important actual business processes of the company in SAP ERP during its implementation. We also highlighted the importance of the enterprise structure of not just PP, but of the entire organization, along with their interdependencies. The reference table acts as an invaluable guide in helping understand the importance of several PP-specific configuration elements, such as MRP controllers, production schedulers, and capacity planners.

The next chapter begins Part II and covers the configuration basics of PP for discrete manufacturing.

PART II
Configuration Specifics for Manufacturing Types

Implementing discrete manufacturing, also known as shop floor control, involves a series of logical and sequential configuration steps to ensure complete mapping of configuration with the business processes of the company.

3 Configuration Basics of Discrete Manufacturing

During an SAP ERP implementation project, when it's established that discrete manufacturing will most closely serve the business needs of the company, the next logical step is to have intensive discussions and several workshops to agree on the configuration objects of discrete manufacturing. Configuration of the discrete manufacturing production type forms the basis on which the business processes of the company will run. For example, how should the system behave when it comes across a material or capacity shortage during production order creation or release? How should it behave when the actual production exceeds the defined underdelivery or overdelivery of the material? What should the system do if it's unable to schedule production within the defined basic dates? For each of these (and many more) questions, you can set the controls on the degree of freedom or flexibility (or strictness) that you want the system to allow when performing business functions. For example, you can configure the system to allow you to create a production order despite a component shortage, but to stop you from releasing it until the requisite components for production are available in stock.

In this chapter, we begin by covering the configuration basics needed to set up the master data used in discrete manufacturing. Next, we follow a step-by-step process to create a new production order type PP10, including assigning it a new number range. All of the subsequent configuration steps covered for this order type and in this chapter are sufficient to enable you to run end-to-end business processes in SAP ERP. In Chapter 6, we cover the business processes side of the configuration undertaken in this chapter.

If, as an SAP ERP system consultant or as a business process owner, this is the first time you're configuring and implementing SAP ERP Production Planning (PP), we suggest that you follow the step-by-step approach in this chapter. Because PP integrates with several other components such as SAP ERP Materials Management (MM), SAP ERP Quality Management (QM), and most importantly with SAP ERP Controlling's Product Cost Controlling (CO-PC), we suggest that you maintain close coordination all along by consulting the resources of these components.

[»] **Note**

The business processes of discrete and process manufacturing are also similar in a lot of ways. Any differences are specifically covered in the relevant chapters (process manufacturing is covered in Chapter 4).

[»] **Note**

Refer to the appendix for features comparison among discrete manufacturing, process manufacturing, and repetitive manufacturing, as well as how to decide which manufacturing type is most relevant for a given industry.

3.1 Material Master

The configuration of the material master is primarily managed within MM. During an SAP ERP system implementation, the MM team coordinates with the client to discuss and agree on a large number of MM-specific configuration objects, which also includes material types. A *material type* is a unique identification to distinguish materials used in various business processes. Some examples of material types are raw materials, semifinished goods, trading materials, packing materials, nonvaluated materials, spare parts, and consumables. However, the importance and involvement of PP can't be overemphasized here because the material requirements planning (MRP) and work scheduling views of the material master are very important to PP, both from a planning and execution perspective. The PP information is maintained in the MRP views of the material master to enable the system to perform reliable material planning. The work scheduling view controls how production execution should take place.

Apart from the option for quantity and value updates, you can also control the views that the system makes available to the end user during material master creation.

For example, the purchasing view isn't normally available for finished goods because the company doesn't purchase finished goods. Similarly, for raw materials, the sales views aren't available because the company normally doesn't sell its raw materials.

To set up the material type attributes, follow the configuration (Transaction SPRO) menu path, LOGISTICS – GENERAL • MATERIAL MASTER • BASIC SETTINGS • MATERIAL TYPES • DEFINE ATTRIBUTES OF MATERIAL TYPES.

Figure 3.1 shows the configuration view of MATERIAL TYPE FERT (FINISHED PRODUCT). On the lower-right side of the screen, you can control the views that you want the system to make available during material master creation. At the bottom of the screen is the PRICE CONTROL field, which enables you to select whether the material will have a moving average price or standard price.

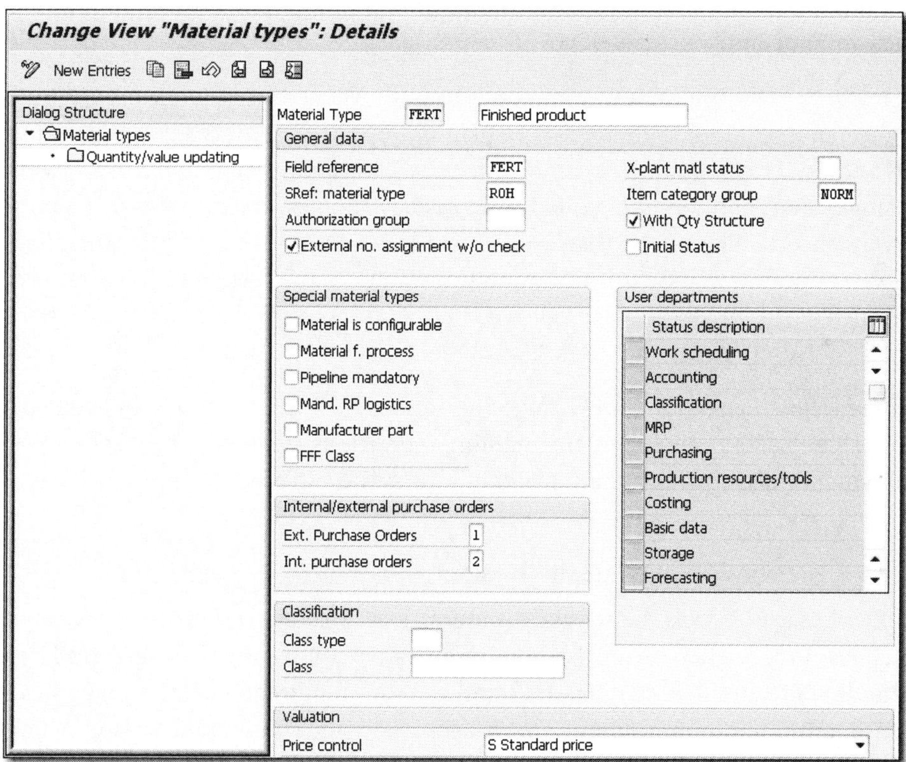

Figure 3.1 Material Types

3.2 Bill of Materials

Similar to the material master, a material's bill of materials (BOM) is used extensively in various areas of the supply chain, including planning, production, sales, and costing. A BOM is a formally structured list of components that you need to use to produce a material. These components may be raw materials or packing materials procured directly from vendors or subassemblies produced in-house.

The BOM has a large number of functions. You can have a BOM that is specific to engineering/design only, whereas you can have another BOM of the same material that you can use for costing purposes. You can have a production BOM and also a sales BOM. In a sales BOM, the system explodes the components and makes them an integral part of sales processing. For example, when a company sells a new car, it also includes the accessories such as a spare tire, the tire changing toolkit, and the owner's manual, among other things. These accessories are, in fact, components in a sales BOM.

A material BOM is a central component in MRP. When the system runs the MRP on a material, it looks for its BOM to plan not just at the finished goods level but also at the components level and raw materials level. The material BOM is always a single level, and you can explode and display the cascade of BOMs as a multi-level structure. The system displays a single-level BOM by showing its immediate next component or assembly. In a multi-level BOM, the system reflects comprehensive details of all the assemblies, components, associated quantities of assemblies and components, and their logical relationship to each other.

In the following sections, we'll cover the various business functions and controls available for BOM, including maintaining the various statuses of BOM as well as allowing or disallowing certain material type's usage in a BOM.

3.2.1 Define Bill of Material Usages

A BOM usage controls the activities and functions that the system can perform in business processes. You can also have a separate BOM usage for Long-Term Planning (LTP) to simulate the materials and capacity requirements for a given business scenario. To create a new BOM usage, follow the configuration (Transaction SPRO) menu path, Production • Basic Data • General Data • BOM Usage • Define BOM Usages, or use Transaction OS20.

Here you'll find several standard BOM usages. You can create a new BOM usage by choosing NEW ENTRIES and selecting the control functions to allow or disallow the business processes in which the BOM usage is applicable (see Figure 3.2).

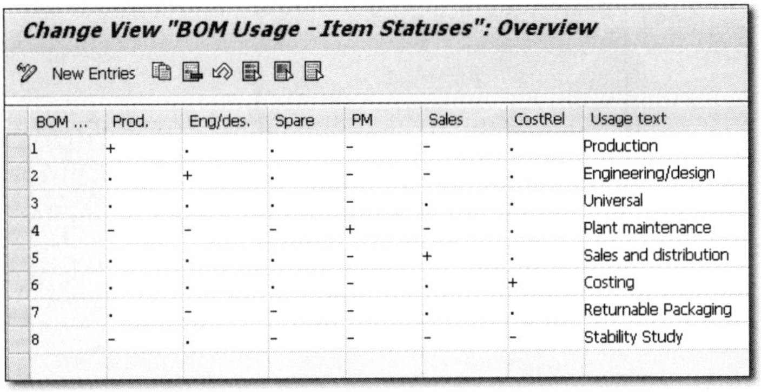

Change View "BOM Usage - Item Statuses": Overview

New Entries

BOM ...	Prod.	Eng/des.	Spare	PM	Sales	CostRel	Usage text
1	+	.	.	-	-	.	Production
2	.	+	.	-	-	.	Engineering/design
3	.	.	.	-	.	.	Universal
4	-	-	-	+	-	.	Plant maintenance
5	.	.	.	-	+	.	Sales and distribution
6	.	.	.	-	.	+	Costing
7	.	-	-	-	.	.	Returnable Packaging
8	-	.	-	-	-	-	Stability Study

Figure 3.2 BOM Usages

3.2.2 Allowed Material Types in the Bill of Materials Header

You can control the material types that the system allows for creation of a material BOM. For example, you normally don't create a material BOM for spare parts or consumable material types. This control on material types for BOM creation also helps prevent the creation of unnecessary or unwanted material BOMs. If a company has several company-specific material types, then you need to specifically identify and perform the necessary configuration for all of the material types that will have any BOM usage.

To specify the material types for a material BOM creation, follow the configuration (Transaction SPRO) menu path, PRODUCTION • BASIC DATA • BILL OF MATERIAL • GENERAL DATA • DEFINE MATERIAL TYPES ALLOWED FOR BOM HEADER, or use Transaction OS24.

Figure 3.3 shows that you can also specify the BOM usage for the material type at the header level. The * symbol denotes that a BOM can have all usage types and can also be used in all material types at the header level.

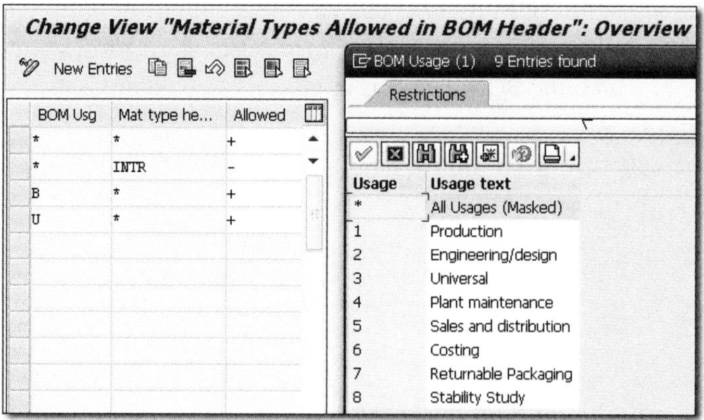

Figure 3.3 Allowed Material Types in the BOM Header

[»]

Note

In addition to maintaining the control function of the material type at the BOM header level, you can also do the same for a material at the BOM item level. To do so, follow the configuration (Transaction SPRO) menu path, Production • Basic Data • Bill of Material • Item Data • Define Material Types Allowed for BOM Items, or use Transaction OS14.

3.2.3 Bill of Material Status

You can control the different applications of a material BOM from its status. For example, during new product development, a material has a BOM status of Engineering/design. When the engineering/design departments approve it, the next status can be Costing to enable the product costing team to calculate the cost of the material. Finally, when the costing department also approves the material BOM, it can attain the status of Production. This status enables the production team to begin producing the material. When the BOM has a Production status, it becomes available during the production order creation, whereas when its status is either Engineering/design or Costing, it isn't available in production order creation. You can also set the status in which all functions are possible.

To create or set the BOM status, follow the configuration (Transaction SPRO) menu path, Production • Basic Data • General Data • Define BOM Status, or use Transaction OS23. As shown in Figure 3.4, you can control whether the BOM status should allow business functions such as being available during MRP explosion, for costing, or for work scheduling (production).

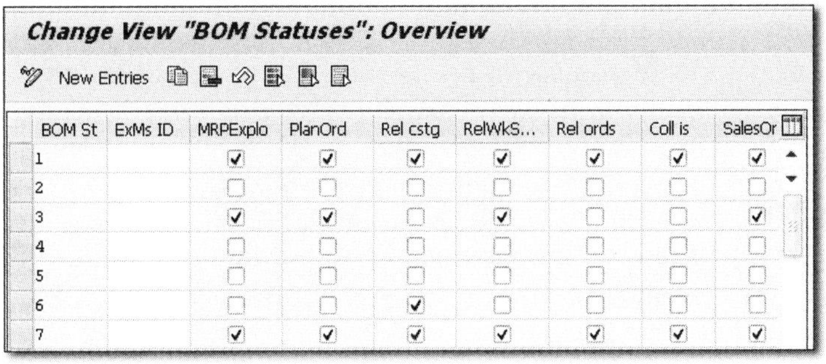

Change View "BOM Statuses": Overview

New Entries

BOM St	ExMs ID	MRPExplo	PlanOrd	Rel cstg	RelWkS...	Rel ords	Coll is	SalesOr
1		☑	☑	☑	☑	☑	☑	☑
2		☐	☐	☐	☐	☐	☐	☐
3		☑	☑	☐	☑	☐	☐	☑
4		☐	☐	☐	☐	☐	☐	☐
5		☐	☐	☐	☐	☐	☐	☐
6		☐	☐	☑	☐	☐	☐	☐
7		☑	☑	☑	☑	☑	☑	☑

Figure 3.4 BOM Statuses

3.2.4 Bill of Material with History Requirement

You can control whether changes made to the material BOM are with reference to a change number or Engineering Change Management (ECM). With a history requirement or change number, the system requires you to enter the change number before it allows you to make the desired changes, which adds a level of security.

To select the BOM usage and status combination for which you want to set the history requirements, follow the configuration (Transaction SPRO) menu path, PRODUCTION • BASIC DATA • GENERAL DATA • CONFIGURE HISTORY REQUIREMENT FOR BOMS, or use Transaction OS25. You can mark the required BOMs with a history requirement by selecting the checkbox.

3.2.5 Item Category in Bill of Material

The item category provides further divisions to the different BOM classes. While some item categories are relevant for production or for planning, others are merely to provide information.

Following are some of the most important predefined item categories:

- **L: Stock item**
  Stock items contain components that you store in your warehouse and include as a part of Inventory Management.

- **N: Nonstock item**
  A nonstock item is a material that isn't available in stock but is procured directly for the given production order. A nonstock item has direct relation to the procurement process. There is also no need to have a material master (item code) for nonstock material. If you use nonstock material, you also have to fill in the procurement details, such as cost element, purchasing group, material group, and price.
- **R: Variable-size item**
  In this item category, you can use the formula and also define the sizes of the variables to enable the system to perform calculations and suggest the component's quantity.
- **T: Text item**
  The text item has a descriptive character.
- **M: Intra material**
  This item category is commonly used in master recipes (process industry). Materials that are temporarily used in process engineering are recorded as components with this item category.

The material input parameter (MATINPT) indicates whether a material reference to the item exists. This isn't the case with document items or nonstock items. The Inventory Management parameter (INVMG) allows you to set that you can only use those materials whose quantities are managed in Inventory Management.

To maintain a new item category or make changes to the existing ones, follow the configuration (Transaction SPRO) menu path, PRODUCTION • BASIC DATA • ITEM DATA • DEFINE MATERIAL TYPES ALLOWED FOR BOM HEADER, or use Transaction OS24. Select or deselect the checkboxes to meet your business needs.

3.2.6 Variable Size Item Formulas

In the fabrication industry, components issued to produce an assembly are often based on a formula. For example, to produce the fuel tank of a motorcycle, the warehouse issues the steel sheet based on the formula, which calculates the requirement. When you assign the variable-size item in the BOM of the material, and with item category R, the system enables you to enter the variable-size details in the relevant area of the BOM's item details area.

[«]

Note

Before you create a formula for a variable size item, you can also self-define a unique unit of measure to denote the formula via Transaction CUNI.

To create a variable size item formula, follow the SAP ERP system configuration (Transaction SPRO) menu path, PRODUCTION • BASIC DATA • ITEM DATA • DEFINE VARIABLE-SIZE ITEM FORMULA, or use Transaction OS15. Figure 3.5 shows the list of available formulas that you can use, or you can create a new one.

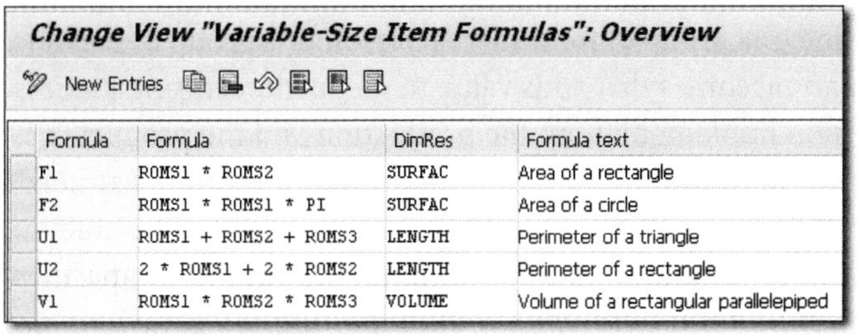

Change View "Variable-Size Item Formulas": Overview

New Entries

Formula	Formula	DimRes	Formula text
F1	ROMS1 * ROMS2	SURFAC	Area of a rectangle
F2	ROMS1 * ROMS1 * PI	SURFAC	Area of a circle
U1	ROMS1 + ROMS2 + ROMS3	LENGTH	Perimeter of a triangle
U2	2 * ROMS1 + 2 * ROMS2	LENGTH	Perimeter of a rectangle
V1	ROMS1 * ROMS2 * ROMS3	VOLUME	Volume of a rectangular parallelepiped

Figure 3.5 Variable-Size Item Formulas

3.2.7 Bill of Material Explosion Types

You can control how the system takes a specific component's explosion into account in the BASIC DATA view of the BOM creation screen. You can control whether direct production, a phantom assembly, or even LTP is deactivated. For example, if you don't want the system to plan a particular component in LTP, you can set its explosion type status in the BASIC DATA view of the material's component. If you don't find the desired configuration settings, then you can configure using the configuration (Transaction SPRO) menu path, PRODUCTION • BASIC DATA • ITEM DATA • DEFINE EXPLOSION TYPES.

3.2.8 Bill of Material Selection (Order of Priority)

You can control how the system makes an automatic selection of a BOM to incorporate it, for example, in a planned order during an MRP run. For example, during the MRP run, if the system is unable to find a material's BOM for production (BOM usage 1), then you can define the next BOM selection priority as universal (BOM usage 3).

To configure the BOM selection and its order of selection priority, follow the configuration (Transaction SPRO) menu path, PRODUCTION • BASIC DATA • BILL OF MATERIAL • ALTERNATIVE DETERMINATION • DEFINE ORDER OF PRIORITY FOR BOM USAGES, or use Transaction OS31. Here you define the selection ID to combine all BOMs with one unique ID. Then you define the selection priority of each BOM and finally assign the BOM usage, such as production or universal.

3.3 Work Center

A *work center* is a machine or a group of machines, a person or a group of persons, or a group of persons and machines that adds value to the manufacturing process. During an SAP ERP system implementation, the production and the product costing teams discuss and mutually agree on the number of work centers that needs to be available. The decision is primarily focused on ensuring that the production department is able to schedule and plan work centers and machine capacities, whereas the product costing team ensures that the reporting for both activities and cost centers is available. For example, if the packaging process incurs significant costs that the product costing team needs to monitor, then it makes sense to create a packaging work center and assign a separate cost center and associated activities to it. If these costs don't require separate monitoring, then the production line cost center is sufficient.

In the following sections, we explain how to make field selections in the work center so that during creation of the work center, the system either makes a field entry mandatory or optional. We also discuss how you can use a standard value key (SVK) to define which activities are important for an operation from a business perspective. You can define formulas for the work center that you can use in capacity requirements planning (CRP), scheduling, and costing. You can use the location groups to account for the time it takes to move a product from one work center to another, and the corresponding system considers this during scheduling. Finally, you can use a control key for operations as a control function to decide if, for example, scheduling or printing for an operation is allowed.

3.3.1 Work Center Category

A *work center category* is a control function that ensures which master data applications and business processes of discrete manufacturing you can use the work center in. For example, work center category 0007 is available for rate routing in REM, and in this category, you'll find the available application option for REM. In the same way, work center category 0008 is available and used for process manufacturing, and in this category, you'll find the master recipe application.

To create a work center category, follow the configuration (Transaction SPRO) menu path, PRODUCTION • BASIC DATA • WORK CENTER • GENERAL DATA • DEFINE WORK CENTER CATEGORY, or use Transaction OP40. Select the work center category 0001 used in discrete manufacturing, and double-click on the APPLICATION folder. You can see the available applications in the resulting screen in Figure 3.6 for CAT. (category) 0001.

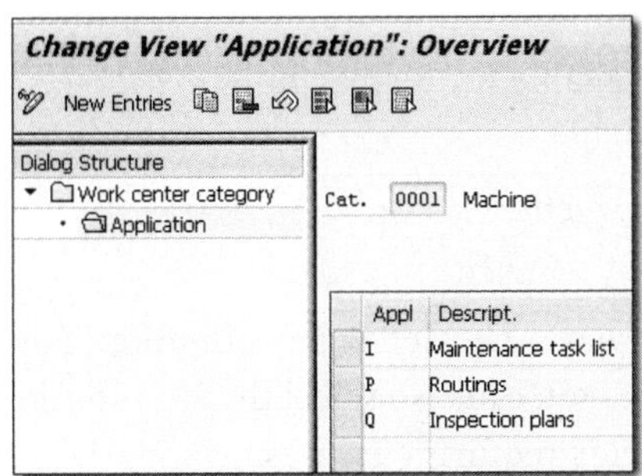

Figure 3.6 Application of Work Center Category

3.3.2 Field Selection in the Work Center

You can control the fields in the SAP ERP system for which entry is mandatory, optional, an input option, or is hidden from display. For example, during the work center creation, if you want the user to enter information in a specific field, you can select the REQ. radio button. When a user enters information in one field, you can also control how the system prompts the user to perform any dependent function. This option works when one modifiable field relates to the influencing fields. For this example, you select the work center category as 0001 as an influencing field, and make the BACKFLUSH field indicator (a modifiable field) as a mandatory entry. So, whenever a user is going to create a work center with category 0001, selecting the BACKFLUSH field is also mandatory.

[«]

Note

The field selection option isn't just restricted to work centers; you can also use it in BOM, routing, and confirmation.

To define field selection in a work center, follow the configuration (Transaction SPRO) menu path, PRODUCTION • BASIC DATA • WORK CENTER • GENERAL DATA • DEFINE FIELD SELECTION, or use Transaction OPFA. Figure 3.7 shows that the SCREEN GROUP BASIC DATA has several modifiable fields, such as BACKFLUSH or PERSON RESPONSIBLE. Notice that you have five options available in the modifiable fields:

- INPUT
  The entry in this field is optional.
- REQ. (required)
  The entry in this field is mandatory.
- DISP. (display)
  There's no entry in this field because it's available for display only.
- HIDE
  The system hides this field, and it isn't displayed.
- HILI (highlight)
  Any specific field can be highlighted if you want the user to pay attention. For example, when marking a field entry as REQ., you can also select the HILI checkbox to enable the user to quickly see the fields requiring entries.

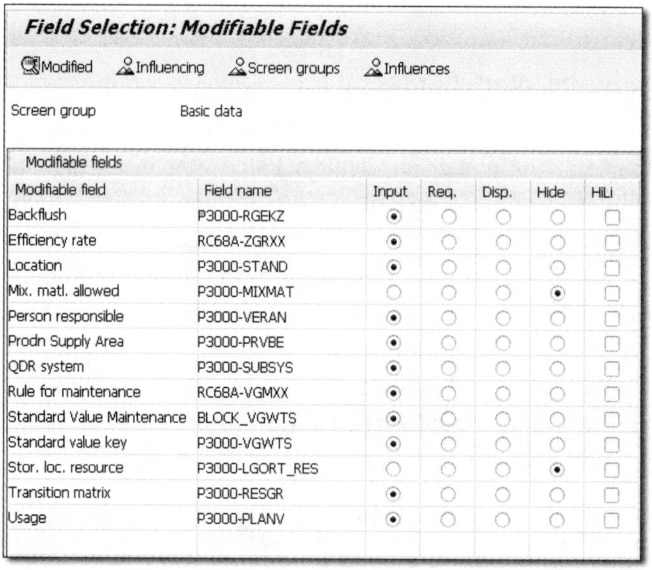

Field Selection: Modifiable Fields

Modified | Influencing | Screen groups | Influences

Screen group: Basic data

Modifiable fields

Modifiable field	Field name	Input	Req.	Disp.	Hide	HiLi
Backflush	P3000-RGEKZ	●	○	○	○	☐
Efficiency rate	RC68A-ZGRXX	●	○	○	○	☐
Location	P3000-STAND	●	○	○	○	☐
Mix. matl. allowed	P3000-MIXMAT	○	○	○	●	☐
Person responsible	P3000-VERAN	●	○	○	○	☐
Prodn Supply Area	P3000-PRVBE	●	○	○	○	☐
QDR system	P3000-SUBSYS	●	○	○	○	☐
Rule for maintenance	RC68A-VGMXX	●	○	○	○	☐
Standard Value Maintenance	BLOCK_VGWTS	●	○	○	○	☐
Standard value key	P3000-VGWTS	●	○	○	○	☐
Stor. loc. resource	P3000-LGORT_RES	○	○	○	●	☐
Transition matrix	P3000-RESGR	●	○	○	○	☐
Usage	P3000-PLANV	●	○	○	○	☐

Figure 3.7 Modifiable Fields of the Basic Data Screen Group

Double-click on the BACKFLUSH field or click on the MODIFIED button. In the screen that appears as shown in Figure 3.8, click on the NEW VALUES button. In the pop-up that appears, enter the work center category as "0001", and choose CONTINUE. Select the REQ. radio button to ensure that whenever a user creates a work center of category 0001, selecting the BACKFLUSH indicator is mandatory.

Field Selection: Modified Field

New values ... | Delete value | Influencing | Modifiable | Screen groups | Influences | View

Screen group: Basic data
Modifiable field: Backflush

Influencing fields

Influencing field	Contents	Input	Req.	Disp.	Hide	HiLi
Field selection	0001	○	◉	○	○	☐
Field selection	0008	○	○	○	◉	☐
Field selection	0015	○	○	○	◉	☐
Field selection		◉	○	○	○	☐

Figure 3.8 Modifiable Field with Influences

Repeat the same with the work center categories 0008 and 0015, but this time, select the HIDE radio button. The system won't show the BACKFLUSH field when the user proceeds to create a work center with work center categories 0008 and 0015. Save your entries.

3.3.3 Standard Value Key

During the course of an SAP ERP system implementation, one of the main areas where the production and the product costing teams collaborate is in defining the standard value key (SVK), which consists of individual parameters that are then grouped together as one SVK. You assign the SVK in the BASIC DATA view of the work center and also enter the formula that the system will use for each of the parameters. The following sequence of steps is used to define an SVK:

1. Define the parameters.
2. Assign the parameters to the SVK.
3. Create a formula for the work center.
4. Assign a formula against each parameter.

Let's explain this with an example. Suppose that in addition to monitoring and recording standard durations such as setup, machine, or labor, your product costing department also wants you to record the electricity and steam consumed in producing a product. The reason to record these two unique parameter values is that a significantly high cost is associated with these values. For example, in the caustic soda industry, electricity consumption is excessive and is closely monitored, so it's a critical cost component that the company wants to monitor and control. Similarly, in the steel rerolling industry, the electricity and natural gas form the significant cost of the finished product and therefore must be recorded and monitored.

When the user uses a specific work center (or resource) consisting of the SVK in the routing (or master recipe), the system requires the user to enter the standard consumptions. For this example, in the master recipe, the system prompts the user to define the standard electricity consumption in producing 1 metric ton of caustic soda. The product costing team will also have an associated cost (in the form of an activity type) assigned to this parameter (electricity). When the user performs the confirmation against the process order and enters the actual electricity consumed, the production and product costing teams can monitor the variances between standard consumption and actual consumption.

You can assign up to six parameters to an SVK. In other words, you can monitor and record up to six important parameters that have direct cost implications on a given work center. You can also use SVKs in scheduling and capacity calculations.

To define a parameter, use configuration (Transaction SPRO) menu path, PRODUCTION • BASIC DATA • WORK CENTER • GENERAL DATA • STANDARD VALUE • DEFINE PARAMETERS, or use Transaction OP7B. You'll see the initial screen consisting of standard and user-defined parameters. Double-click on SAP_02, and the screen shown in Figure 3.9 appears. You can see the standard parameter with TIME as a DIMENSION and STANDARD VALUE UNIT in MIN (minutes). If you've created a self-defined parameter such as steam or electricity, you can give the dimension and the unit of measure in which you want to record the consumption value.

Next, to create the SVK, follow the configuration (Transaction SPRO) menu path, PRODUCTION • BASIC DATA • WORK CENTER • GENERAL DATA • STANDARD VALUE • DEFINE STANDARD VALUE KEY, or use Transaction OP19 or Transaction OPCM.

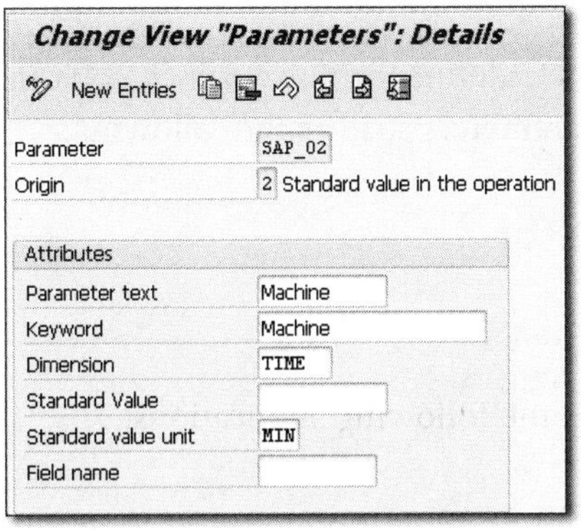

Figure 3.9 Machine Standard Parameter with Unit of Measure

Figure 3.10 shows the STD VAL. KEY SAP1, which consists of the standard parameters SAP_01, SAP_02, and SAP_03. If you have any self-defined parameters that you want to be part of the SVK, you can enter them here. As noted previously, you can enter up to six parameters in SVK. Make sure to select the GENERATE checkbox when defining SVK because then the system automatically performs the calculations defined in the formulas. If not selected, then it does the calculation for scheduling and capacity planning during production order creation, which often leads to system performance issues.

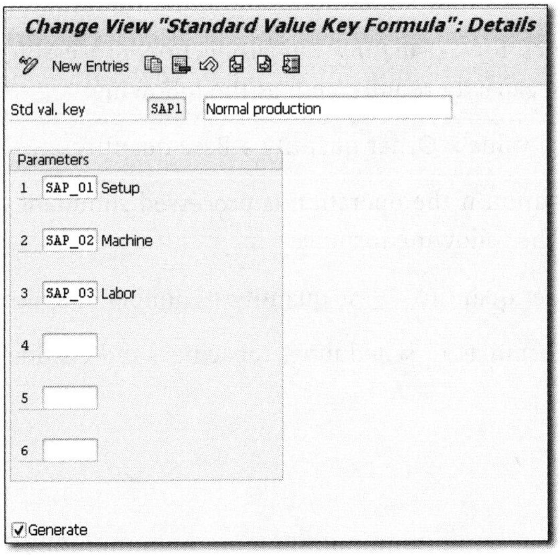

Figure 3.10 Standard Value Key Formula

3.3.4 Formulas for the Work Center

The system uses previously defined parameters to define formulas, which you can then use in CRP or scheduling. You can use parameters such as the following:

- SAP_08 Base quantity
- SAP_09 Operation quantity
- SAP_11 Number of operation splits

A formula definition also holds the control for the following applications:

- CRP
- Scheduling
- Costing

To define the formula parameter, if it's different from the ones already available, use the configuration (Transaction SPRO) menu path, Production • Basic Data • Work Center • Costing • Work Center Formulas • Define Formula Parameters for Work Centers, or use Transaction OP51.

To define the formula that you can use in the work center for costing, CRP, and scheduling, follow the configuration (Transaction SPRO) menu path, Production • Basic Data • Work Center • Costing • Work Center Formulas • Define Formula for Work Centers, or use Transaction OP54.

In Figure 3.11, notice the formulas for calculating the production processing duration. The system calculates the capacity requirement as the following:

Capacity requirement = Standard value × Order quantity ÷ Base quantity

You can reduce the processing duration if the operation is processed simultaneously at several work centers, per the following formula:

Duration = Standard value × Order quantity ÷ Base quantity ÷ Number of splits

In Chapter 6, you'll assign these formulas to scheduling, capacities, and costing views of the work center.

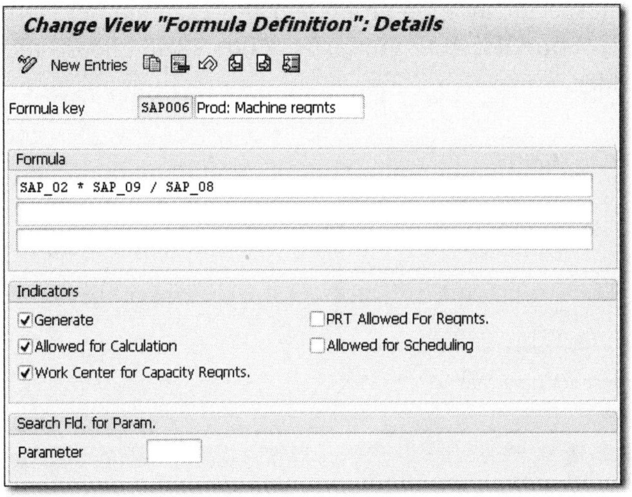

Figure 3.11 Formula Definition in the Work Center

3.3.5 Location Groups

A *location group* consists of a physical location where each work center is located. You can combine several work centers into one location group if they are in close proximity to one another. You can use the move time matrix to provide standardized values to different transitions times (also known as interoperation times) such as queue time, wait time, and move time. If you don't use the move time matrix, then you have to maintain this transition's time information either in the work center or in the routing.

You assign the location group to the work center, and in the move time matrix, you define the normal and the minimum move times between locations. The system uses the working times that you maintain in the move time matrix for scheduling the move times. Basically, you're performing two steps:

- Maintain the location groups.
- Maintain the matrix of move times from one location group to another.

To set up location groups, follow the configuration (Transaction SPRO) menu path, Production • Basic Data • Work Center • General Data • Define Move Time Matrix, or use Transaction OP30 (or Transaction OPJR).

Figure 3.12 shows the plant-specific location group. For this example, set PLNT (plant) as "3000", set LOC. GROUP (location group) as "0001", and choose the TRANSPORT TIME MATRIX folder.

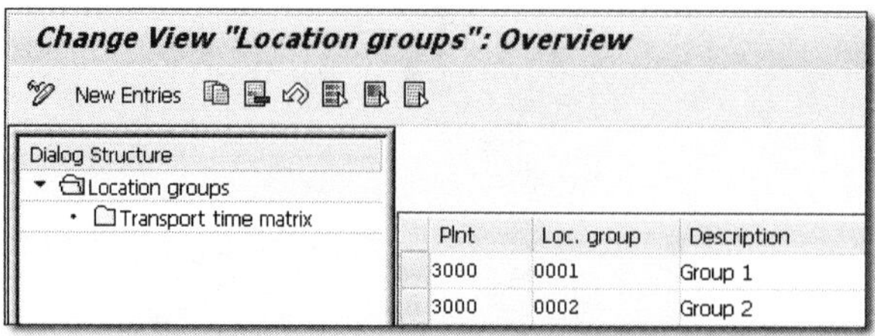

Figure 3.12 Location Groups/Transport Time Matrix

[»] **Note**

Move time matrix and transport time matrix are synonymous.

In the screen that appears as shown in Figure 3.13, assign the standard and the minimum move times for the plant–location group combination. In the CALENDAR field, you can choose whether the move times defined are applicable every day of the week, according to the factory calendar, or according to the operating time and calendar given in the work center.

When the necessary configuration settings are in place, assign the location group in the SCHEDULING area of the work center.

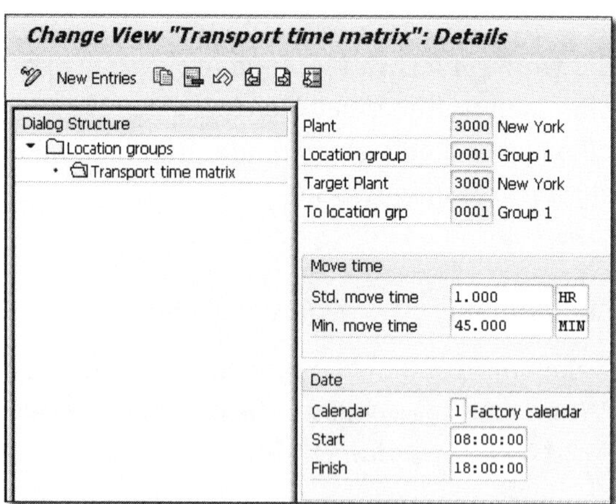

Figure 3.13 Transport Time Matrix

3.3.6 Control Key for Operations

A *control key* acts as a central function to control which actions a user or system can or can't perform on a specific work center. You have to select the relevant checkboxes and make the relevant settings in the control key. You then assign this control key in the DEFAULT VALUES area of the work center (or resource). When you assign the work center to the material's routing, the control key becomes an integral part of the operation and acts as a control function.

In a control key, you can define (among other things), whether some or all of the following functions are possible:

- Scheduling
- CRP
- Inspection characteristics required for in-process (during production) inspection
- Automatic goods receipt
- Scheduling for external operations

You can also control the confirmation type required for a work center. You can control whether confirmation is required or optional for a given operation during the order confirmation process. Further, you can also control whether an operation is a milestone confirmation. In milestone confirmation, all of the preceding operations are automatically confirmed, as soon as you confirm the operation as having a milestone confirmation.

To create a new control key, follow configuration (Transaction SPRO) menu path, PRODUCTION • BASIC DATA • WORK CENTER • ROUTING DATA • DEFINE CONTROL KEY, or use Transaction OP00.

Note [«]

In Chapter 6, we'll show how you can assign this control key in the DEFAULT VALUES area of the work center.

3.4 Routing

A *routing* is a set of operations that the material will undergo during the entire production process, and the routing operations are performed at relevant work centers. For example, the transformation of a raw material into a finished product is due to undergoing various operations (defined in the routing) that are performed at relevant work centers.

Most of the important parameters for routing selection are made in the order type-dependent parameters (see Section 3.6), in which you incorporate controls such as routing selection as a mandatory or an optional requirement. Order type-dependent parameters also control whether routing selection is an automatic process or manual.

To set up the routing parameters, follow the configuration (Transaction SPRO) menu path, PRODUCTION • BASIC DATA • ROUTING • ROUTING SELECTION • SELECT AUTOMATICALLY, or use Transaction OPEB. In Figure 3.14, the selection ID specifies the selection sequence according to which the system will select the planned type, usage, and status of the routing. The following list explains the first few columns:

- ID
  Combines all of the different priorities listed in the next column, SP.
- SP
  Lists the sequence in which the system will look for relevant routing during production order creation. In other words, it's the priority setting for routing selection.
- TASK LIST TYPE
  Refers to the task list type that the system will consider, such as N is for routing, R is for rate routing, and S is for reference operation set.
- PLAN USAGE
  It controls if the routing only be used for production, or is available for inspection plan usage too. The value '3' denotes that this routing will be available for all types of usages.
- DESCRIPTION
  The description that appears after you select relevant usage from the PLAN USAGE field.
- STAT
  Denotes the status that the system should look for in a routing or in a reference operation set and, if found, incorporate it in the production order.

- DESCRIPTION OF THE STATUS
  The description that appears after you select relevant usage from the DESCRIPTION OF THE STATUS field.

Change View "Automatic Selection": Overview

New Entries

ID	SP	Task List Type	Plan Usage	Description	Stat	Description of the status
01	1	N	1	Production	4	Released (general)
01	2	S	1	Production	4	Released (general)
01	3	N	1	Production	2	Released for order
01	4	N	3	Universal	4	Released (general)
01	5	2	1	Production	4	Released (general)
01	6	N	1	Production	3	Released for costing
01	7	R	1	Production	4	Released (general)
01	8	2	3	Universal	4	Released (general)
02	1	R	1	Production	4	Released (general)
02	2	R	3	Universal	2	Released for order
03	1	3	1	Production	4	Released (general)
03	2	3	3	Universal	4	Released (general)
04	1	2	1	Production	4	Released (general)
04	2	2	1	Production	2	Released for order
04	3	2	3	Universal	4	Released (general)

Figure 3.14 Automatic Routing Selection

Note [«]

You can use the selection IDs during scheduling of orders (planned orders or production orders). You can use Transactions OPU3, OPU4, or OPU5 to assign the relevant selection IDs to the order.

See Section 3.11 to see how you can incorporate selection IDs into production orders (Transaction OPU3).

3.5 Production Order Creation

During an SAP ERP system implementation, the PP consultant conducts and finalizes several configuration objects, including order types and their number ranges. An *order type* is a unique identification and control function that enables the business process owner to quickly perform several functions, such as creating an order, automatically releasing an order, or scheduling an order.

The decision to create and make available different order types depends on how the company wants to see data and information and the depth of the planning and controls required. For example, a company might want to sort all orders by separating all of the rework production orders from the normal production orders. This logical segregation of two different order types enables the planner to not only quickly segregate the available data in the system in the form of reports but also put in place the necessary system authorizations. Normally, companies also prefer to have production order types in direct relation to work centers. For example, in the cold-rolled steel sheets industry, if there are five different work centers, such as cold-roll sheets uncoiling, pickling, annealing, recoiling, and packing, then it makes sense to have five different production order types to enable every work center to have clear visibility of the activities and controls involved. Similarly, in the automotive assembly, the main work centers may be press shop, paint shop, and final assembly. In the textile industry, the work centers might be spinning, weaving, bleaching and dyeing, stitching, and finishing.

However, we suggest not overdoing it in this exercise because creating too many production order types can be confusing and also demands greater data entry and maintenance work. You can use operations and other work center criteria to meet your reporting needs.

In the following sections, we'll show you how to create a new production order type in discrete manufacturing and assign it a new number range.

3.5.1 Maintain Order Types

An order type controls the following functions directly or through additional profiles:

- Internal and external number assignment
- Classification system
- Commitments management
- Status management for the different production statuses
- Settlement
- Scheduling

When defining an order type, the system automatically assigns the relevant order category. An order category distinguishes among various order types, not just in the production processes, but also in SAP ERP Plant Maintenance (PM), QM, or CO orders. Whenever you create a new order type for a particular application, the system automatically assigns the correct order category.

Figure 3.15 shows the list of order categories available in the SAP ERP system when you hover over ORDER CATEGORY and press F4. A PP PRODUCTION ORDER has ORDER CATEGORY 10, whereas the ORDER CATEGORY for a PROCESS ORDER is 40.

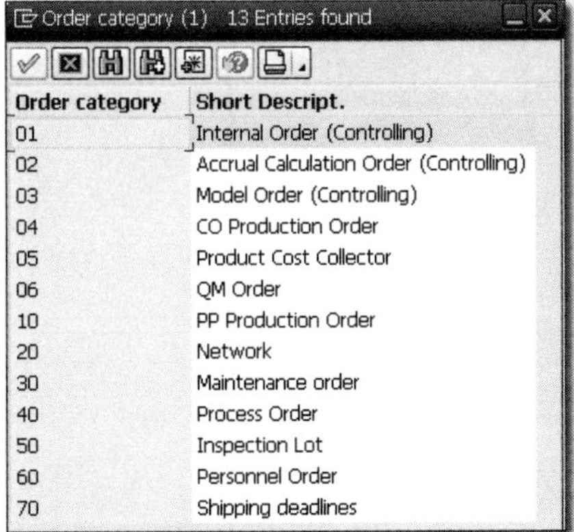

Order category	Short Descript.
01	Internal Order (Controlling)
02	Accrual Calculation Order (Controlling)
03	Model Order (Controlling)
04	CO Production Order
05	Product Cost Collector
06	QM Order
10	PP Production Order
20	Network
30	Maintenance order
40	Process Order
50	Inspection Lot
60	Personnel Order
70	Shipping deadlines

Figure 3.15 Order Categories

To create a new order type, follow the configuration (Transaction SPRO) menu path, PRODUCTION • SHOP FLOOR CONTROL • MASTER DATA • ORDER • DEFINE ORDER TYPES, or use Transaction OPJH.

In Figure 3.16, notice that the system automatically assigns ORDER CATEGORY 10 to the production order. Create ORDER TYPE "PP10", and give a short description. For the SETTLEMENT PROFILE, you need to coordinate with the product costing team to assign the appropriate settlement profile to the production order.

Tips & Tricks [+]

To save you the effort of defining all of the parameters in any configuration object and also lessen the chance of errors, we suggest that you copy it from one of the standard objects available, and then make the desired changes to the newly created object. The copy function is available in all configuration objects. Look for the COPY AS icon.

For this example, copy the production order PP01 to create the new production order type PP10.

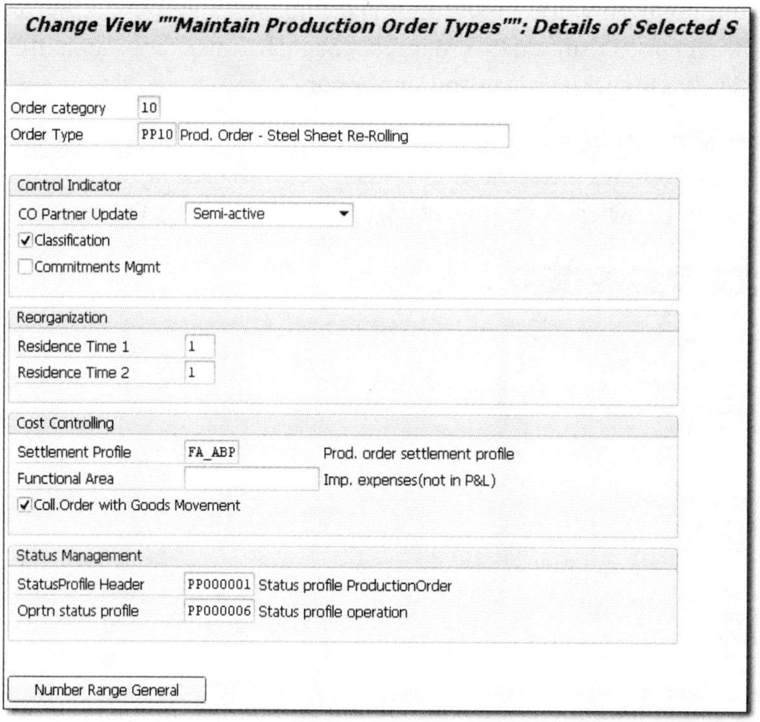

Change View ""Maintain Production Order Types"": Details of Selected S

Order category 10
Order Type PP10 Prod. Order - Steel Sheet Re-Rolling

Control Indicator
CO Partner Update Semi-active
☑ Classification
☐ Commitments Mgmt

Reorganization
Residence Time 1 1
Residence Time 2 1

Cost Controlling
Settlement Profile FA_ABP Prod. order settlement profile
Functional Area Imp. expenses(not in P&L)
☑ Coll.Order with Goods Movement

Status Management
StatusProfile Header PP000001 Status profile ProductionOrder
Oprtn status profile PP000006 Status profile operation

Number Range General

Figure 3.16 Maintain Production Order Types

Click on the Number Range General button in Figure 3.16 to define the new number range to the newly created production order type PP10.

[»]

Note

You can also use Transaction CO82 to create a new number range group and define the interval for the production order. On this initial screen, click on the Change Groups icon Groups, and then perform all of the steps listed in Section 3.5.2.

3.5.2 Number Ranges

On the initial screen that appears, choose Group • Insert. In the screen that appears as shown in Figure 3.17, enter the new number range group. All production orders assigned to this number range group will use this number range. For this example, enter a short Text "Sheets Processing", and define the new number range interval as "80000001" to "84999999". To facilitate your work, the system displays the existing number ranges available in the lower half of the screen. The newly defined number range can't overlap with the existing number ranges. Click on the INSERT icon, and the system enables you to define a new number range interval.

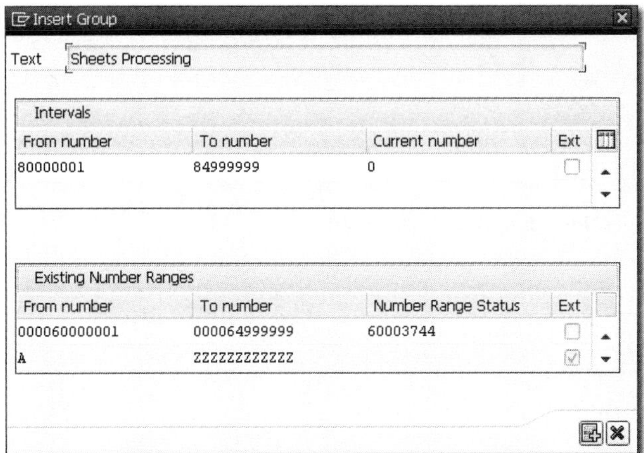

Figure 3.17 Insert Group Screen to Enter a New Number Range Interval

Next, perform the following steps:

1. Select the SHEETS PROCESSING checkbox, which represents the new number range group.
2. At the bottom of the same screen, look for NOT ASSIGNED, including the new created order type PP10. Click on order type PP10 (see Figure 3.18), and then click on the SELECT icon (not shown).
3. Click on the ELEMENT/GROUP icon.

If you've successfully followed these steps in order, the final screen will look like the screen shown in Figure 3.19. Notice that we had given a short text to order type PP10, and it's still not appearing in this figure. Rather, the text TEXT DOES NOT EXIST appears. This is due to the fact that you haven't saved the settings yet. The moment you save it and come back to this screen, you'll see the self-defined order type text right next to the order type.

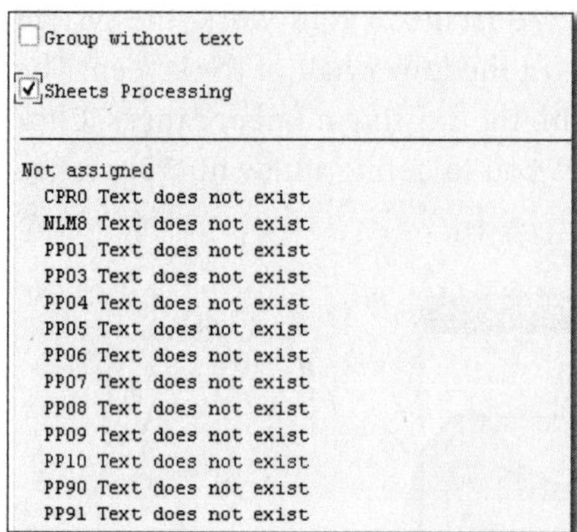

Figure 3.18 Unassigned Order Types, Including Order Type PP10

Sheets Processing
PP10 Text does not exist

Figure 3.19 Number Range Group Assigned to Order Type PP10

Save your settings.

In Chapter 6, you'll see the production order for order type PP10 has the number range you've assigned here.

3.6 Order Type-Dependent Plant Parameters

After you've defined the new order type, the next step is to define order type-dependent plant parameters. In this step, you make several settings to enable the system to control different functions during production order creation. You can, for example, control whether the system should automatically select the production version or provide the user with the option to select it manually in a pop-up that appears during production order creation. Similarly, you can assign a batch search strategy for batch-managed materials. You can also control which information the system should update to make it a part of the order information system and Logistics Information System (LIS). Finally, you coordinate with the product costing team to enable them to make relevant settings from a Cost Center Accounting (within SAP ERP CO) perspective.

To set up order type-dependent plant parameters, follow the configuration (Transaction SPRO) menu path, PRODUCTION • SHOP FLOOR CONTROL • MASTER DATA • ORDER • DEFINE ORDER TYPE-DEPENDENT PLANT PARAMETERS, or use Transaction OPL8. In Figure 3.20, notice the plant and order type combinations already available. For this example, select the PLANT 3000 and ORDER TYPE PP01, and choose the COPY AS icon to copy all settings from order type PP01 to order type PP10. The system creates a new entry for plant 3000 and order type PP10.

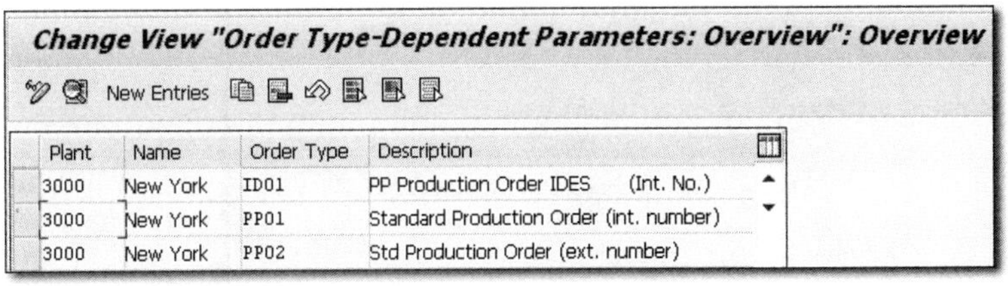

Change View "Order Type-Dependent Parameters: Overview": Overview

New Entries

Plant	Name	Order Type	Description
3000	New York	ID01	PP Production Order IDES (Int. No.)
3000	New York	PP01	Standard Production Order (int. number)
3000	New York	PP02	Std Production Order (ext. number)

Figure 3.20 Available Order Type-Dependent Parameters

We'll now walk through the different tabs that are available for order type-dependent parameters.

3.6.1 Planning

In the PLANNING tab area shown in Figure 3.21, you can choose whether the system should bring up a pop-up window for you to manually select the PRODUCTION VERSION or automatically assign the first available production version to the production order. In the ROUTING area, the SELECTION ID is the same as that covered in Section 3.4. You can choose from SELECTION ID for automatic or manual routing selection (or reference operation set) during production order creation, and you can also choose whether it's mandatory or optional for the system to have routing (or reference operation set) available during production order creation. You can also select the CHECK OP. DETAILS checkbox to enable the system to check and automatically copy the work center details in the routing. You can even copy the routing header text into the production order.

In the BILL OF MATERIAL area, you can either choose to use the standard BOM application or enter a custom-defined BOM application, if you've previously created one.

In the BATCH DETERMINATION area, you can enter the previously created batch-search strategy (primarily in the process industry but also in discrete manufacturing for the batch-managed materials). You can coordinate with the MM team to define the batch search strategy and then assign it in the SEARCH PROCEDURE field.

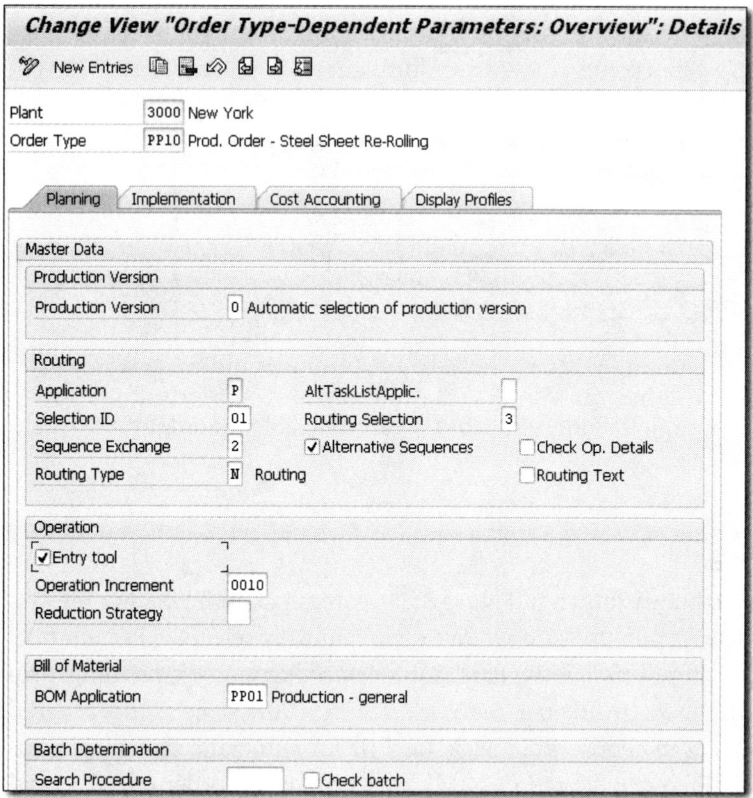

Figure 3.21 Planning Tab of Order Type-Dependent Plant Parameters

Scroll down on the same screen to see the area shown in Figure 3.22, where you can assign the SUBSTITUTE MRP CTRLLER or SUBSTITUTE SCHEDULER, if the system is unable to find it in the material master or when you create a production order without reference to a material. You assign the MRP controller in the MRP 1 view of the material master and the production scheduler (supervisor) in the WORK SCHEDULING view of the material master.

When you create a production order, you can choose when the system should create reservations for in-house or stock-based materials and purchase requisitions for externally procured materials. Finally, on this screen, you can enter INSPECTION TYPE "03" to enable the system to perform in-process (during production) inspection of the material. This is an integration point between PP and QM.

Batch Determination

Search Procedure		☐ Check batch

General

Assignment

Substitute MRP Ctrller	
Substitute Scheduler	

Purchase Requisitions

Reservation/Purch.Req.	3	☐ Collective PR

Quality Inspection

Inspection Type	

Figure 3.22 Planning Subarea of Order Type-Dependent Plant Parameters

Note [«]

Chapter 20 shows you the integration aspect of PP with QM, including an end-to-end in-process (during production) inspection.

3.6.2 Implementation

In the IMPLEMENTATION tab (shown in Figure 3.23), make sure to select all of the relevant checkboxes, especially for the SHOP FLOOR INFORMATION SYSTEM and DOCUMENTATION OF GOODS MOVEMENT areas.

If you select the checkboxes, the system automatically updates the LIS, and you have the option to view documented goods movements (goods issuance or receipt) for the production orders. The documented goods movements in the production order information system (Transaction COOIS) also show up in the complete goods issuances and receipts details.

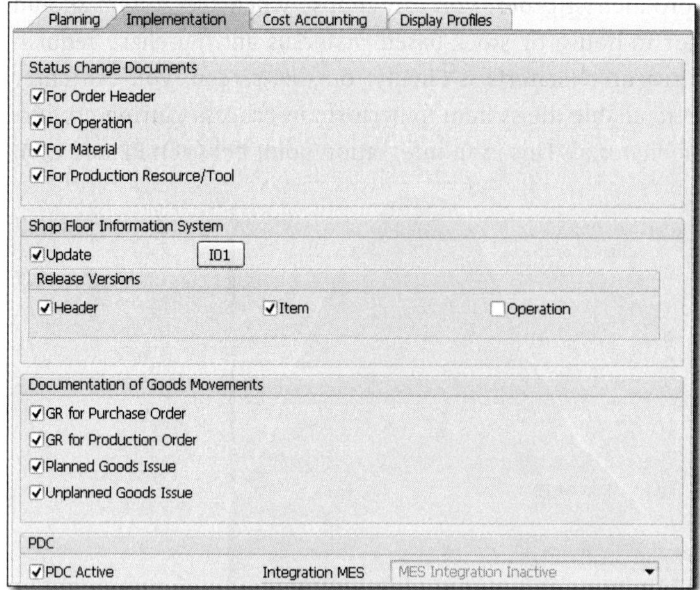

Figure 3.23 Implementation Tab of the Order Type-Dependent Plant Parameters

3.6.3 Cost Accounting

Figure 3.24 shows the Cost Accounting tab. We suggest that you coordinate with the CO resource (who handles SAP ERP Controlling's Cost Object Controlling) to ensure that the necessary configuration settings are in place.

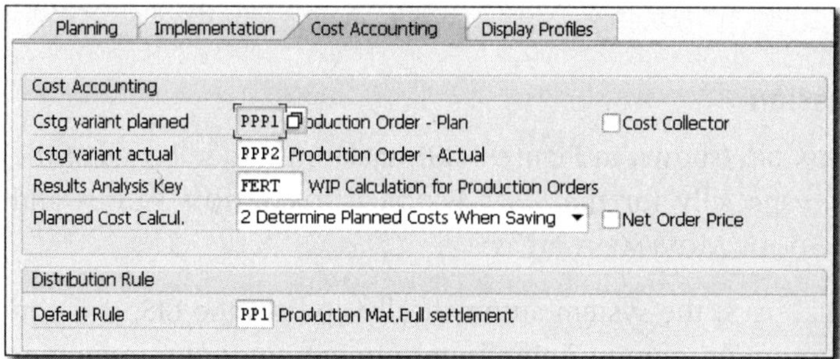

Figure 3.24 Cost Accounting Tab of the Order Type-Dependent Plant Parameters

3.7 Production Scheduling Profile

You use the production scheduling profile to define the tasks of the production order that the system automatically executes during creation or release of a production order. For example, the system can automatically release the production order as soon as you create it, or the system can automatically send the shop floor papers for printing after you release an order. This saves you the time and effort to perform individual functions, when in fact you can combine a few of them. Further, you can assign a specific production order type that is used in make-to-stock (MTS) production, make-to-order (MTO) production, project planning, or production without material reference. You assign the production scheduling profile in the WORK SCHEDULING view of the material master.

Additional control functions available in the production scheduling profile include the following:

- Automatic actions on creating a production order
- Automatic actions on releasing a production order
- Material availability
- Capacity availability
- Automatic goods receipt (along with the control key; see Section 3.3.6)
- Confirmation controls
- Batch Management settings

To create a new production scheduling profile, follow the configuration (Transaction SPRO) menu path, PRODUCTION • SHOP FLOOR CONTROL • MASTER DATA • DEFINE PRODUCTION SCHEDULING PROFILE, or use Transaction OPKP. In the AUTOMATIC ACTIONS area shown in Figure 3.25, you can control whether the system should automatically release a production order on its creation. This is applicable in those business scenarios when a formal or separate approval on the release of the production order isn't required; hence, immediate creation and its release serve the purpose. Similarly, on release of the production order, you can control whether the system is to automatically execute the printing function, schedule the order, or even create control instructions (see the information on Process Management/XSteps in Chapter 4, Section 4.4 and Chapter 7, Section 7.6).

For capacity planning, you can assign an overall profile, and then in the production order screen, you can navigate to the capacity planning screen, in which the system uses the profile you assign here. You can control whether the system should confirm the capacity availability as well as perform finite scheduling during production order creation.

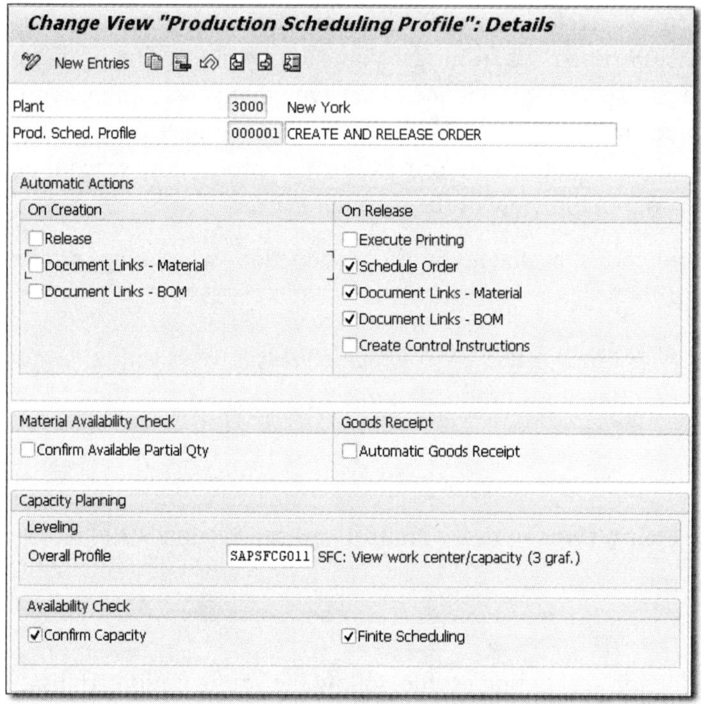

Figure 3.25 Production Scheduling Profile (Selection)

Scroll down the screen to see the options shown in Figure 3.26. Here we recommend that you don't set the CONFIRMATION area checkboxes. This ensures that the system automatically updates the production deficit or surplus (excess) in the production order as soon as you perform the necessary business processes such as confirmation and goods receipt. The only exception to selecting these checkboxes is when your production process is generally stable with few operational problems; then, a bit of a deficit or excess production doesn't make much difference to your reporting or monitoring needs.

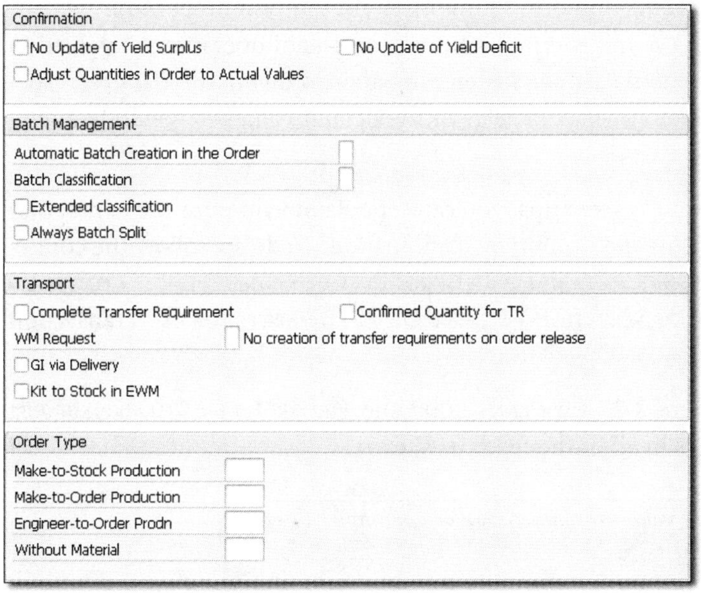

Confirmation

No Update of Yield Surplus
No Update of Yield Deficit
Adjust Quantities in Order to Actual Values

Batch Management

Automatic Batch Creation in the Order
Batch Classification
Extended classification
Always Batch Split

Transport

Complete Transfer Requirement
Confirmed Quantity for TR
WM Request
No creation of transfer requirements on order release
GI via Delivery
Kit to Stock in EWM

Order Type

Make-to-Stock Production
Make-to-Order Production
Engineer-to-Order Prodn
Without Material

Figure 3.26 Production Scheduling Profile (Selection)

You can use the Adjust Quantities in Order to Actual Values parameter to adjust the operation durations and component requirements of subsequent operations on the basis of a scrap-adjusted quantity calculation. The system reschedules the relevant operations.

You can control how the system handles the batch-managed materials, and if it should automatically create the batch during production order creation, during its release, or not at all. If your company plans to integrate SAP ERP Warehouse Management (WM) into the production process, then you can coordinate with the relevant WM resource to make the necessary settings on this screen.

3.8 Default Values for the Generation of Operations

During the production order creation, if the system is unable to find routing or a reference operation set, it no longer allows the user to create the production order. Further, if you create a production order without reference to a material, the system still needs a routing or reference operation set.

When you create a production order for a material, the system looks for its master data, such as a BOM and routing. Routing consists of operations involved in the production of a material. If you haven't maintained the master data yet, especially routing, then the system can refer to the default values to generate operations.

To manage these business scenarios, you can set default values for the generation of operations with plant and order type combination. To do so, follow the configuration (Transaction SPRO) menu path, PRODUCTION • SHOP FLOOR CONTROL • OPERATIONS • TASK LIST SELECTION • DEFINE DEFAULT VALUES, or use Transaction OPJG.

In the screen in Figure 3.27 for PLANT 3000 and ORDER TYPE PP10, we suggest that you make entries in all of the fields as shown.

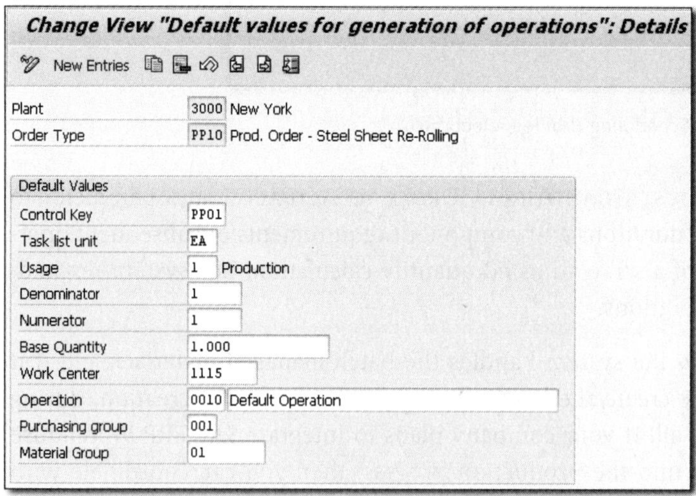

Figure 3.27 Default Values for Generation of Operations

3.9 Availability Check

An *availability check* enables the system to check and validate whether one or all of the required resources, whether it's material, capacity, or production resources/tools (PRT), are available in the production process. As a production planner and scheduler, you want to have a comprehensive overview of material and capacity availability or shortage to take necessary action, if required. In case of missing component availability, the system maintains an entry in the missing parts information system. Not only this, you can also control whether the system should allow creation or release of the production order if it finds a shortage of material or capacity needed to produce the goods.

Several options are available for you to perform material or capacity availability checks. You can manually check these within the production order creation screen, or you can even use a separate transaction to perform the material availability check. Alternatively, you can make the settings in the availability check for the system to perform these functions automatically.

There are three different types of availability checks in the SAP ERP system:

- **Material availability check**
  Checks the availability of the components for the production order, either against actual stocks (and optionally receipts) or against the planning.
- **Capacity availability check**
  Checks whether sufficient free capacity is available for the order's operations.
- **PRT check**
  Checks whether the required PRT is free via a status in the master data.

Note [«]

See Chapter 14 for more information on CRP.

The system makes availability checks during two levels:

- Production order creation
- Production order release

The following sequence of steps is involved in the availability checks:

1. Define the checking group.
2. Define the checking rule.
3. Define the scope of check.
4. Define the checking control.

We cover each of these steps in the following sections.

3.9.1 Define Checking Group

To set the availability checking control, follow the configuration (Transaction SPRO) menu path, PRODUCTION • SHOP FLOOR CONTROL • OPERATIONS • AVAILABILITY CHECK • DEFINE CHECKING GROUP, or use Transaction OVZ2. In the screen shown in Figure 3.28, the availability (AV) 01 for DAILY REQUIREMENTS has options to enter TOTALSALES and TOTALDLVREQS on a daily, weekly, or monthly basis. If you select the NO CHECK checkbox, then the system switches off the available-to-promise (ATP) check.

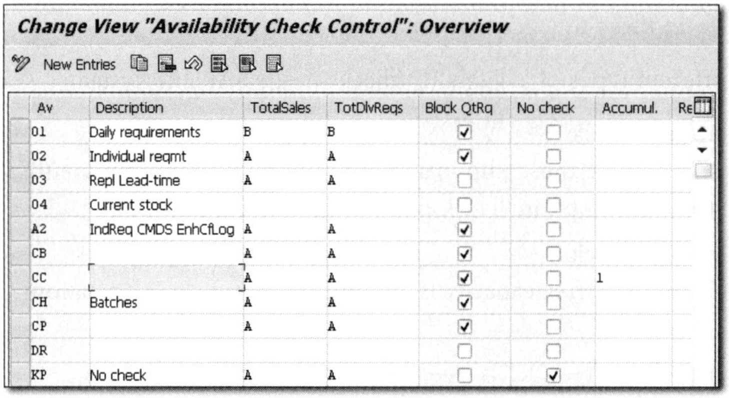

Change View "Availability Check Control": Overview

New Entries

Av	Description	TotalSales	TotDlvReqs	Block QtRq	No check	Accumul.	Re
01	Daily requirements	B	B	☑	☐		
02	Individual reqmt	A	A	☑	☐		
03	Repl Lead-time	A	A	☐	☐		
04	Current stock			☐	☐		
A2	IndReq CMDS EnhCfLog	A	A	☑	☐		
CB		A	A	☑	☐		
CC		A	A	☑	☐	1	
CH	Batches	A	A	☑	☐		
CP		A	A	☑	☐		
DR				☐	☐		
KP	No check	A	A	☐	☑		

Figure 3.28 Availability Checking Control

3.9.2 Define Checking Rule

You can define a checking rule to cater to various applications such as SAP ERP Sales and Distribution (SD), MRP, production order processing, or Inventory Management. This enables you to configure and integrate different checking rules for different applications to meet your business needs. To define the checking rule, follow the configuration (Transaction SPRO) menu path, PRODUCTION • SHOP FLOOR CONTROL • OPERATIONS • AVAILABILITY CHECK • DEFINE CHECKING RULE.

3.9.3 Define Scope of Check

From the combination of the previously defined availability checking control as 01 (DAILY REQUIREMENTS) and CHECKING RULE as PP, we proceed to define the scope of check. In a *scope of check*, you primarily define which inward and outward movements and stock types the system should consider during the availability check. For example, you can choose if the system should consider purchase orders whose goods receipt on the delivery dates will ensure material availability. Similarly, you can set the relevant settings for the system to include (or exclude) firmed planned orders in its quantity and material availability calculation.

In the standard settings, the system only considers available unrestricted stock for the availability check; however, you may have business needs in which quality stock may also be considered as available stock during an availability check. For example, in ice cream manufacturing, the stock is kept in quality inspection even during transportation from one plant to another, until it reaches its destination. If you want to include this quality stock in the material availability check, set the relevant indicator.

To define the scope of check, follow the configuration (Transaction SPRO) menu path, Production • Shop Floor Control • Operations • Availability Check • Define Scope of Check, or use Transaction OPJJ.

In the screen shown in Figure 3.29, you can choose the stock types and inward/outward movements that best meet the business scenarios. Notice that in the Receipts in past field, you can control whether the system should consider all of the past and future receipts or future receipts only.

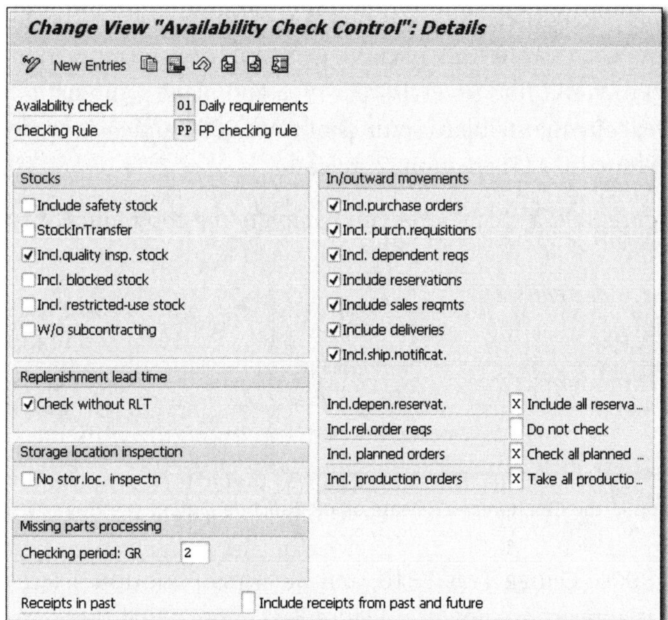

Figure 3.29 Scope of Check (Availability Checking Control)

You can now proceed to assign the checking group in the MRP 3 view of the material masters for all of the components in the AVAILABILITY CHECK field. For this example, use availability checking group 01, and assign it in the MRP 3 view of the material masters for all four components.

3.9.4 Define the Checking Control

As mentioned previously, you can maintain comprehensive availability checks of material, PRT, and capacity during production order creation and release. Because creation and release of production orders are two important business functions, you can define the level of controls for each business function. For example, if you have a long lead time between production order creation and release, you can skip some of the availability checks such as for capacity or PRT until you release the order. Similarly, instead of a stringent control that the system doesn't allow the release of a production order until the material is available to produce it, you can select the option in which the user decides on releasing the production order, which meets business needs more closely so that at least partial production can commence.

To set up a checking control for production order creation and release, follow the configuration (Transaction SPRO) menu path, PRODUCTION • SHOP FLOOR CONTROL • OPERATIONS • AVAILABILITY CHECK • DEFINE CHECKING CONTROL, or use Transaction OPJK. In the initial screen, you'll find several order type and plant combination entries. For this example, copy the standard settings of plant 0001 and order type PP01 and both business functions (1: creation, 2: release).

The screen shown in Figure 3.30 appears after you complete the copy function.

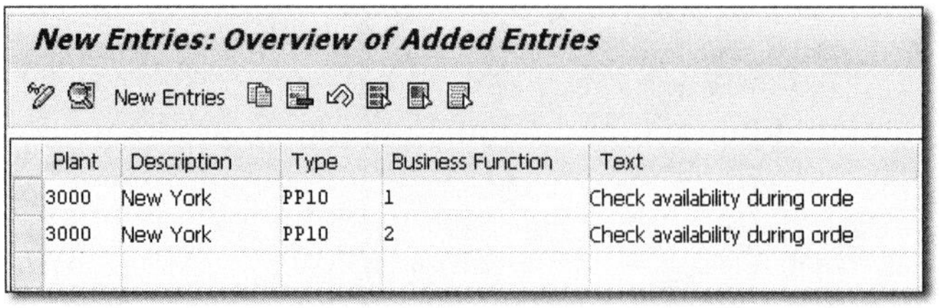

New Entries: Overview of Added Entries

New Entries

Plant	Description	Type	Business Function	Text
3000	New York	PP10	1	Check availability during orde
3000	New York	PP10	2	Check availability during orde

Figure 3.30 Checking Controls in the Creation and Release of a Production Order

Double-click on PLANT 3000, ORDER TYPE PP10, and BUSINESS FUNCTION 1 (creation). In the screen that appears as shown in Figure 3.31, notice that there are three distinct areas: MATERIAL AVAILABILITY, PRT AVAILABILITY, and CAPACITY AVAILABILITY. In the CHECKING RULE field, assign the rule PP that was discussed previously. For the PRT AVAILABILITY, select the NO CHECK checkbox, and the system won't check for PRT availability during production order creation. You can have different settings when you're ready to release the production order (BUSINESS FUNCTION 2).

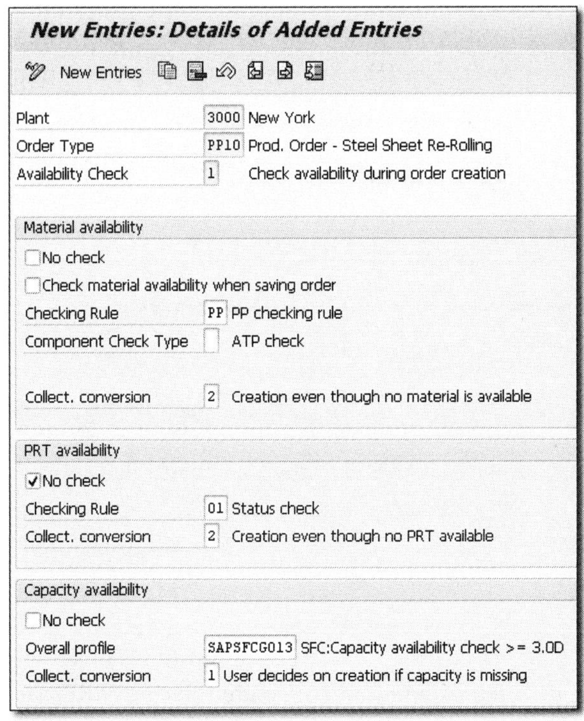

Figure 3.31 Availability Checks during Production Order Creation

Finally, for CAPACITY AVAILABILITY, assign an overall profile, and also select the option 1 for the COLLECT. CONVERSION field. This option enables you to decide on a runtime basis whether you can create a production order despite a capacity shortage.

Save your settings, and repeat the same steps for BUSINESS FUNCTION 2 (release production order).

3.10 Stock and Batch Determination

If you have materials that you manage in batches, then you can make the relevant settings to enable the system to automatically consider stock or batch determination (or both) for a given order type. To do so, follow the configuration (Transaction SPRO) menu path, Production • Shop Floor Control • Operations • Stock and Batch Determination of Goods Movement, or use Transaction OPJ2.

In the screen shown in Figure 3.32, you can set the plant and order type combination and select the no stock determination (No SD) and batch determination (BD) checkboxes.

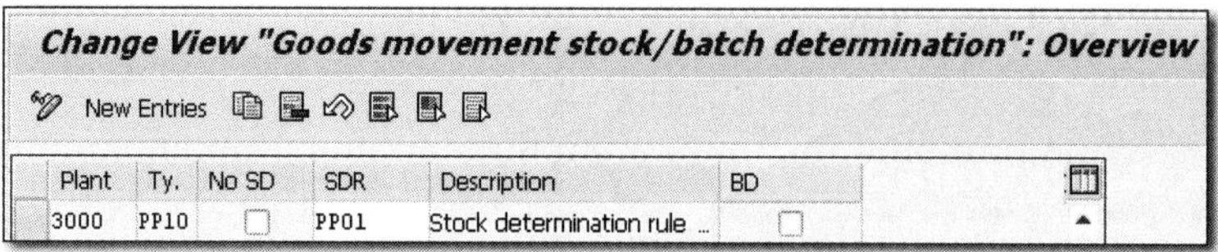

Figure 3.32 Stock and Batch Determination

[»]

> **Note**
>
> We suggest that you engage a MM resource to help with stock and batch determinations.

3.11 Scheduling

If you compare the planning of planned orders that are created during MRP, the system needs to incorporate dates of production for the individual operations in the production order. Apart from specific production times (setup and machine), you also need to add transition times, such as wait time, queue time, and move time, as well as safety times, such as opening period, float before production, or float after production.

Before we cover the scheduling in detail, it's important to note how the system takes the various dates into account during scheduling:

- There are two scheduling types in SAP ERP: basic dates scheduling and lead-time scheduling.
- For planned orders, you can have only basic date scheduling (only basic dates). For production orders, you can have both (basic and production dates).
- Whenever the system carries out lead-time scheduling, it automatically carries out basic dates scheduling beforehand.

In finite scheduling and reduction of lead-time scheduling, the first step the system takes is to define the start and end dates of the production order without taking capacity constraints into account. The objective is to evaluate whether the dates are synchronized for timely and complete production. Primarily, there are two types of scheduling:

- **Backward scheduling**
  Determines the start date of the production order based on the latest finish date that you define in the production order.
- **Forward scheduling**
  Determines the end date of the production order based on the earliest start that you define in the production order.

In finite scheduling, the system carries out individual planning (scheduling) for the production order and doesn't take interdependencies of various orders into account. If the system can't manage the production within the stipulated dates, then you can reduce the lead time with overlapping, splitting, or reducing the transition times. If you want the system to consider reduction during the planned orders creation stage (during an MRP run), then you can use Transaction OPU5 to configure it. Here are some of the steps you can take to reduce the lead time as well as optimize the production process:

- If you've defined the float before production as three days, then you can reduce or eliminate this time to enable the system to bring back the production schedule to meet the target finish date. It also applies to the opening period and float after production. You can use Transaction OPU3 to achieve this.
- Similarly, with splitting, you can divide a large production lot into several smaller lots, provided you have several individual capacities to manage it in parallel.

- With overlapping, the system passes the processed parts of the total lot directly on to the next machine.
- You can reduce other times, such as queue time and wait time, where possible.
- You can add another workday or shift to the factory calendar to increase the number of available production hours.

In the following sections, we explain the different scheduling types available for production orders and how the system adjusts the scheduled dates if these aren't within the time frame of the basic dates. We also discuss how the scheduling margin key enables various buffers in production to account for any unforeseen or unexpected delays.

3.11.1 Scheduling Types for Production Orders

As previously covered, there are two primary types of scheduling: backward scheduling and forward scheduling. The other available options are backward or forward scheduling with time and current date. In CURRENT DATE, the system performs forward scheduling and uses the current date (today's date) as the start basic date in the planning.

With the scheduling type ONLY CAPACITY REQUIREMENTS, the system calculates the capacity requirements only. The individual operations aren't scheduled, and the basic dates are copied as operation dates.

To set up the scheduling type, follow the configuration (Transaction SPRO) menu path, PRODUCTION • SHOP FLOOR CONTROL • OPERATIONS • SCHEDULING • SPECIFY SCHEDULING TYPE, or use Transaction OPJN.

3.11.2 Scheduling Parameters for Production Orders

You assign the scheduling type in the scheduling parameters of the production order. You can make other settings in the scheduling parameters for the production order, and you can even control it up to the production scheduler level.

To set the scheduling parameters for the production order, follow the configuration (Transaction SPRO) menu path, PRODUCTION • SHOP FLOOR CONTROL • OPERATIONS • SCHEDULING • DEFINE SCHEDULING PARAMETERS FOR PRODUCTION ORDERS, or use Transaction OPU3. The screen in Figure 3.33 shows PLANT 3000 and ORDER TYPE PP10 with no specific production scheduler.

Select the GENERATE CAPACITY REQS. checkbox to enable the system to generate capacity requirements during scheduling.

The ADJUST DATES field describes whether the system determines the basic start and finish dates during scheduling. In backward scheduling, this applies to the basic start dates. Further, this only applies if lead-time scheduling creates production dates so that the time frame given by the basic dates isn't sufficient to include the production dates from lead-time scheduling anymore. In this case, one of the two basic dates is adjusted automatically.

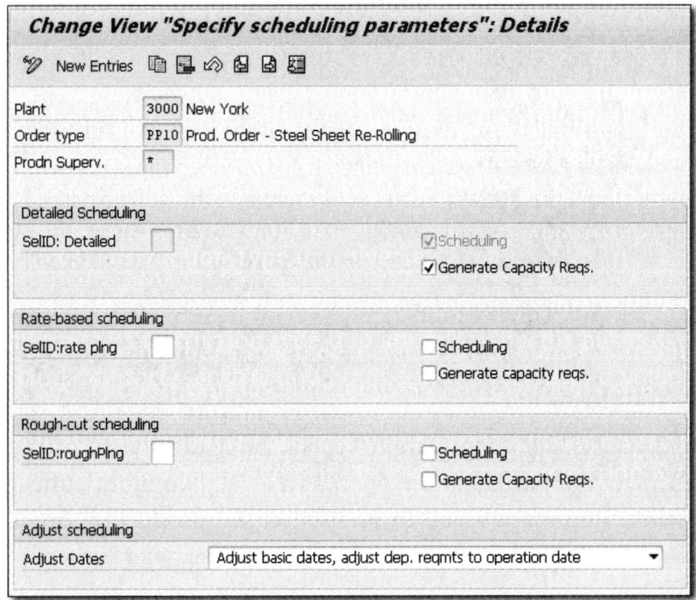

Figure 3.33 Scheduling Parameters for the Production Order

Figure 3.34 shows the lower half of the screen. Here you first set the SCHEDULING TYPE that the system will consider in production order scheduling.

The START IN THE PAST parameter specifies how many days in the past the basic start date may be before the scheduling type is overridden and the today scheduling (forward scheduling from today) is used.

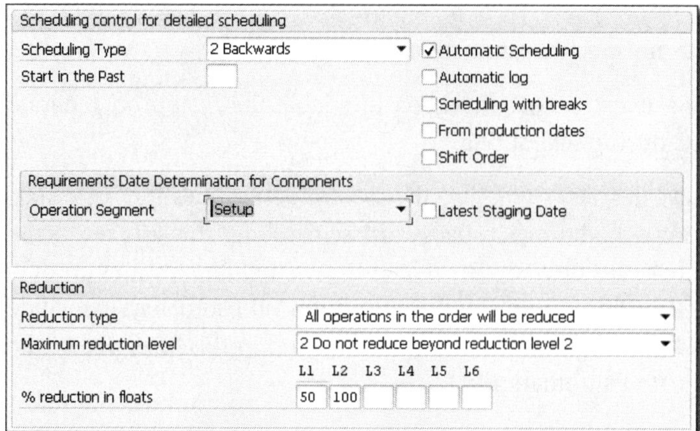

Figure 3.34 Scheduling Parameters for the Production Order

[+]

Tips & Tricks

This is especially useful during an SAP ERP system implementation when the user needs to enter the backlog entries from the cutover dates. For example, if the go-live date of an SAP ERP system implementation project is July 8, 2016, and the user has to enter backlog from July 1, 2016, then an entry of 10 days in the START IN THE PAST field will facilitate that requirement.

3.11.3 Scheduling Margin Key

You assign the plant-specific scheduling margin key in the MRP 1 view of the material master. Apart from the buffers at the operation level, two more buffers are available in the scheduling margin key to be used at the production order level:

- **Float before production**
  In forward scheduling, the system adds up the number of days to the basic start date of production.
- **Float after production**
  In forward scheduling, the system adds up the additional days after production ends to calculate the basic end date.

These two buffers are available to account for any unforeseen delay in the production process. For example, late arrival of raw materials needed to produce the good or delayed end of previous production orders. In other words, it acts a safety margin in planning. Similarly, the float after production is in place to account for any production or unforeseen machine breakdown delays that may affect the overall planning and scheduling. Further, these floats provide a certain degree of freedom from the capacity-planning step to move operations on resources and also consider dependencies of other operations of the order to maintain the sequence of the operations.

We suggest that you evaluate the actual production realities, including average or normal delays faced from the historical data, experience, and trends, and make the necessary settings in the configuration for scheduling the margin key. Later, when you have significant actual and real data built in the SAP ERP system, you can use several statistics and other evaluation reports and information systems available in the PP component to revisit and update the data, which forms the basis for planning.

Finally, apart from floats before and after production, the scheduling margin key also contains the opening period and the release period. The *opening period* is a time interval at the beginning of the planned order and is used for the (collective) conversion of planned orders into production orders. The opening date is the beginning of the opening period and is prior to the basic start date of the planned order. The *release date* is the beginning of the release period and is prior to the start date of the production order. The released period may also be considered as the time it takes from creating a production order to its release, due to internal company processes and controls in place.

To set up the scheduling margin key, follow the configuration (Transaction SPRO) menu path, Production • Shop Floor Control • Operations • Scheduling • Define Scheduling Margin Key.

For scheduling margin Key 001 for Plnt 3000, the Opening Period is 10 days, the fLoat After Production is 1 day, the Float Before Production is 2 days, and the Release Period is 5 days (Figure 3.35).

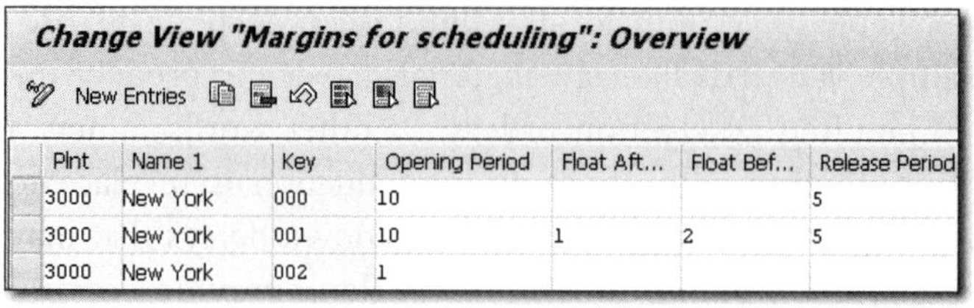

Change View "Margins for scheduling": Overview

New Entries

Plnt	Name 1	Key	Opening Period	Float Aft...	Float Bef...	Release Period
3000	New York	000	10			5
3000	New York	001	10	1	2	5
3000	New York	002	1			

Figure 3.35 Scheduling the Margin Key

3.12 Reduction Strategy

It's only logical and practical to maintain buffers during the production processes to account for any unforeseen or other business contingencies. However, when needed, you can reduce or eliminate several of these buffers, especially when the system is unable to schedule a feasible and practical production plan.

To cancel the buffers, you can use a reduction strategy, which consists of reduction types and reduction levels. In a reduction type, you define whether the reduction is applicable to all operations of the production order or only to the critical part. In a reduction level, you define the levels, available up to six levels, to which you can reduce the floats. Depending on the reduction level that you select, you define the percentage reduction at each level. For this example, set the REDUCTION LEVEL to level 2 and then define the percentage reduction in floats as 50% and 100%, respectively. Although you can set up different reduction levels, the system always tries to make it without reduction in the first place. If unable to do without reduction, and if the system is able to meet the scheduling needs during the first reduction in floats by 50%, no further reduction will take place; otherwise, it will move to the next reduction level.

[!]

Warning!

We suggest that you avoid using all reduction levels for system performance reasons. The system first schedules the planned order with standard scheduling, but with each reduction level, the required system's resources to achieve the reduction increase greatly.

The reduction levels only reflect the reduction of the operations by shortening the queue, wait, and move times, as well as through splitting and overlapping.

If you haven't maintained the reduction strategy in the routing, then the system uses order type-dependent parameters at the plant level to perform reduction. A reduction strategy refers to the percentages by which the system reduces the interoperation times, such as transport time, and performs further reduction with overlapping and splitting.

To set the reduction strategies, follow the configuration (Transaction SPRO) menu path, Production • Shop Floor Control • Operations • Define Reduction Strategies, or use Transaction OPJS. In the screen shown in Figure 3.36, notice that there are three control elements—Transport (move time), Overlap, and Split—that you can use for reduction by the defined percentages given in the Red. Queue column. Be sure to select the relevant checkboxes for which you want to perform the reductions.

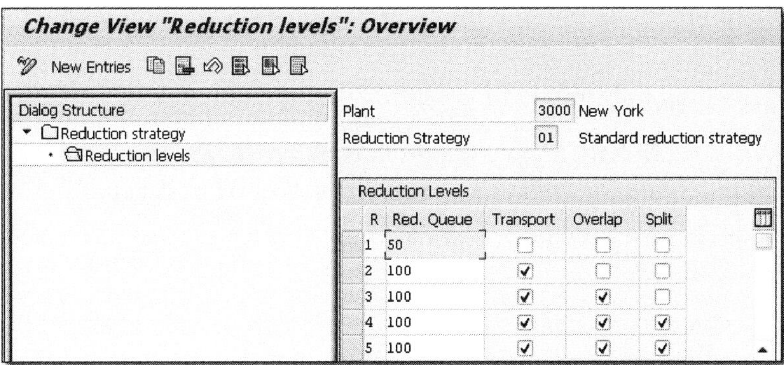

Figure 3.36 Reduction Levels

3.13 Confirmation

So far, all of the work that you've undertaken is merely at the various planning stages. From initial material forecast to running MRP to generation of planned orders, their conversions to production orders are all planning activities with no financial implications. For example, the production order creation signals to the production department to start production of the material. The production order quantity denotes the target or the planned quantity for production. The release of a production order enables the system to activate certain business functions such as printing shop floor papers, allowing issuance of components, or enabling confirmation of operations. When the production department has actually produced some or all of the target/planned quantity, it proceeds to confirm this produced quantity in the SAP ERP system.

It's imperative that the user performs the confirmation as soon as the actual production data is available and ready for entry in the system. The data entry confirmation can also be an automatic process via bar code using a conveyor belt. Alternatively, the business normally agrees on the frequency with which it will perform confirmation, for example, once at the end of every 8-hour shift or once in 24 hours. The confirmation process also enables several other business processes such as automatic goods issuance (backflush), automatic goods receipt, capacity reduction of the relevant work center, and updating costs.

In the following sections, we'll show you the various options and controls available for confirmation of an order.

3.13.1 Process Overview

When you perform confirmation against a production order, whether it's for an entire order or a milestone confirmation, the system makes actual data updates of the following objects:

- Yield
- Scrap
- Rework

Depending on the settings that you've already made, when confirming an order, the system updates the following data in the production order:

- Quantities (yield, scrap, rework), activities, dates, and order status
- Backflush of the components
- Automatic goods receipt
- Capacity reduction of the relevant work centers
- Costs based on actual data entered during confirmation
- Excess/shortage quantities for MRP-relevant materials

On the configuration side of confirmation, you can incorporate several checks and controls during the confirmation, including (but not limited) to the following:

- Time when the system should backflush components
- Check operations sequence
- Check underdelivery and overdelivery quantity tolerances
- Check if QM results of in-process (during production) are entered for an operation
- Goods receipt control for co-products
- Actual cost of error handling during confirmation
- Quantity (goods movements) of error handling during confirmation

Further, you can control the screen layout of confirmation to suit your business needs. You can even control the fields in which the data entry is required, optional, or hidden. See Section 3.13.3 to see how to manage this.

Let's move on to discuss the specifics of confirmation.

3.13.2 Parameters for Order Confirmation

To set up parameters for order confirmation, follow the configuration (Transaction SPRO) menu path, PRODUCTION • SHOP FLOOR CONTROL • OPERATIONS • CONFIRMATION • DEFINE CONFIRMATION PARAMETERS.

For PLANT 3000 and ORDER TYPE PP10 (Figure 3.37), the PROCESS CONTROL field controls when the system should perform some or all of the following steps:

- Goods movements such as backflush
- Automatic goods receipt
- Actual costs calculation

If you leave the PROCESS CONTROL field blank, then the system performs all activities online and immediately. Refer to Section 3.13.4 to see how you can control the settings as either online, immediately, or in the background.

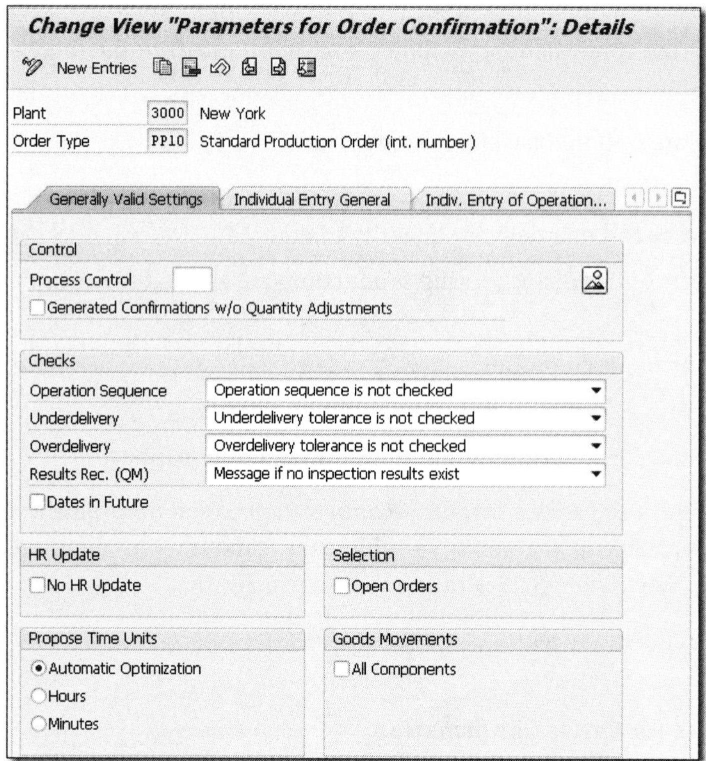

Figure 3.37 Generally Valid Settings Tab of Confirmation Parameters

In the CHECKS area, you can control whether the system should ensure that the user adheres to the OPERATION SEQUENCE, with varying degree of controls from no check to warning to error. Similarly, if you've maintained the minimum and maximum tolerance limits in the UNDERDELIVERY and OVERDELIVERY fields in the WORK SCHEDULING view of the material master, the system copies this information during production order creation. When you perform confirmation of the order and enter the yield, the system checks to see if it's within the tolerance limits and acts according to the level of control set—no message, warning, or error. In case of an error message, the system won't allow you to proceed further or save the confirmation until you correct the error. The system considers the sum of yield and rework quantities for underdelivery/overdelivery calculation.

[«]

Note

During the SAP ERP system implementation, you discuss and agree with the client on the level of control required and then make the necessary settings. Generally, the control should neither be too loose nor too stringent to compromise the day-to-day working of business process owners. The level of control also depends on the stability of production processes and business maturity to reflect that optimum production should be within the defined tolerance limits. If there are large variations and deviations in yield, then a less stringent control will suffice.

[+]

Tips & Tricks

If you've set the underdelivery or overdelivery control as an error setting, the system won't allow you to exceed the defined tolerance limits. To attend to this, you can make changes in the Goods receipt tab of the production order and perform the confirmation process again. This enables the user to deviate from one-time or limited deviation to the standard tolerance limits.

In the Results Rec. (QM) field and for in-process (during production) inspection, the system creates a quality inspection lot for one or more operations. This is an integration point between PP and QM. You can control whether the system should check to see if the inspection results of an operation for the inspection lot are already entered, before it allows you to perform the confirmation of the next operation.

In the Goods Movements area of the screen, you can control whether the system should perform goods movement for all of the components by selecting the checkbox.

Click on the Individual Entry General tab, and the screen shown in Figure 3.38 appears. Here, you select the Confirmation Type that the system makes available during the confirmation process. It can be a Partial confirmation, Final confirmation, or an Automatic final confirmation.

The settings depend on how frequently the business performs the confirmation process. If there are large and frequent numbers of confirmations against a production order, then it makes sense to set Partial confirmation as the default. When you perform the final confirmation against the production order, you can manually set the Confirmation Type as Final confirmation. For Automatic final confirmation, the system checks whether the confirmed quantity is equal or less than the production order quantity. If it's less, then it automatically sets the status as PARTIAL CONFIRMATION, whereas if the confirmation quantity is equal to or exceeds the production order quantity, then it automatically sets the status as FINAL CONFIRMATION.

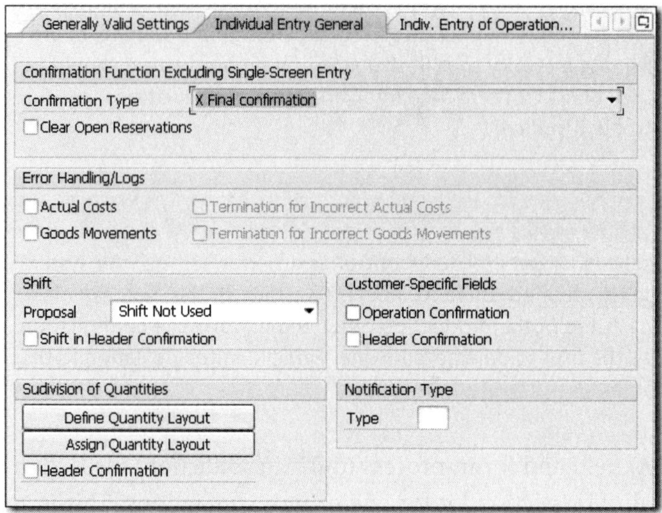

Figure 3.38 Individual Entry General Tab of Confirmation Parameters

For ACTUAL COSTS and GOODS MOVEMENTS in the ERROR HANDLING/LOGS area, you can set whether the system should terminate the confirmation processing until you correct the error.

Click on the INDIV. ENTRY OF OPERATION W. INIT. SCREEN tab, and the screen shown in Figure 3.39 appears. The system offers you a large number of checkboxes to facilitate your data entry work. For example, if you select the PROPOSE checkbox in the QUANTITIES area, the system proposes the quantity that you need to confirm for an operation or an order. You can also select a DISPLAY checkbox to display any information for your reference purposes.

We suggest that you select the ACTUAL DATA radio button in the SCREEN CONTROL TIME TICKET area, so that the system automatically proposes the data (quantity and activities) that you need to confirm. If you don't, then on the actual confirmation screen, you can manually click the ACTUAL DATA icon to bring the data that you need to confirm.

Save the confirmation settings.

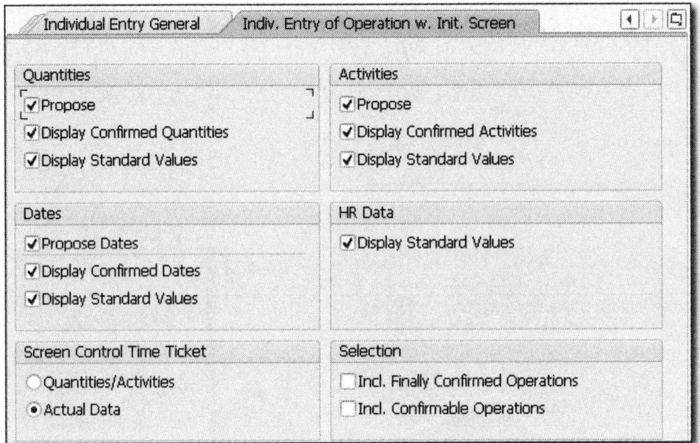

Figure 3.39 Individual Entry of Operation Tab Confirmation Parameters

3.13.3 Single Entry Screen for Confirmation

You can control how the system displays the information on the confirmation screen (Transaction CO11N for discrete manufacturing or Transaction COR6N for process industries). In this step, you set up a profile and assign that profile to an individual user in the user parameter CORUPROF.

A profile for a single entry screen controls the user-defined layout for confirmation. In the layout, you can define the fields and the sequence in which they appear on the confirmation screen. For example, if you'd like the provision to enter long text in the confirmation screen to enable the process operator to write extensive shift details, you can select the relevant option. Similarly, you can control whether the system should list the activities to be confirmed as individual fields or in a tabular form.

To define a single entry screen for confirmation, follow the configuration (Transaction SPRO) menu path, Production • Shop Floor Control • Operations • Confirmation • Single Screen Entry • Define Single Screen Entry, or use Transaction OPK0. In Figure 3.40, you define the Header area so that during the actual confirmation, the system brings up the relevant fields for you to enter the details. For example, you can choose whether the system should only allow you to enter the confirmation number at the header level or allow you to enter the production order number together with its operation.

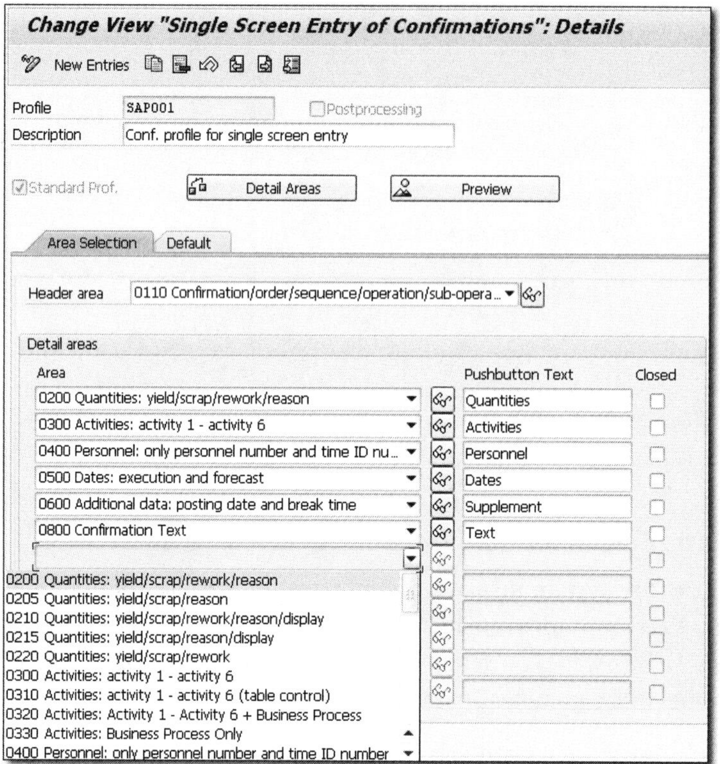

Figure 3.40 Single Entry Screen of Confirmation

[+] **Tips & Tricks**

At any time, if you want to see how your chosen option will look on the confirmation screen, click on the Display Header (or Detail) Area icon, and the system will show you a preview.

In Figure 3.40 and in the Detail Areas, you can select from the large number of options, such as quantities, activities, HR data, short and long text entry provision, dates, and even external IDs if you use a third-party product to integrate the confirmation process with the production process. You can choose the sequence in which these details appear on the confirmation screen.

In the Pushbutton Text fields, you can define the text of the pushbuttons to suit your business needs. The Closed checkbox enables you to have a minimized (closed) view of the information on the confirmation screen, which you can open when you're ready for data entry.

[+]

Tips & Tricks

To check that your settings are correct with no duplication or other layout error, you can use the DETAIL AREAS button for a consistency check. Further, you can use the PREVIEW button to simulate the confirmation screen.

[«]

Note

You have to activate enhancement CONFPP07 if you want more than one reason for variances. A practical workaround (without activating the enhancement) is that in the screen in Figure 3.40 and the DETAIL AREAS, you can incorporate ADDITIONAL DATA: REASON as many times as you want, from the dropdown list.

3.13.4 Time of Confirmation

If you have a large number of confirmations that you need to perform on a daily basis, then it's practical to perform confirmations in the background, such as quantity update or backflush, when there is less of a workload on the system. However, if you want an immediate update of all quantity and activities data posted, you can set the necessary settings. These are the same settings used for the PROCESS CONTROL field at the beginning of Section 3.13. To set up time of confirmation, follow the configuration (Transaction SPRO) menu path, PRODUCTION • SHOP FLOOR CONTROL • OPERATIONS • CONFIRMATION • SINGLE SCREEN ENTRY • DEFINE TIME FOR CONFIRMATION PROCESSES, or use Transaction OPKC. In Figure 3.41, you can choose when the system should update the following:

- AUTO GR (automatic goods receipt)
- BACKFLUSH
- ACTUAL COSTS

The available update options are listed here:

- IMMEDIATELY ONLINE
- IMMEDIATELY IN UPDATE PROGRAM
- LATER IN BACKGROUND JOB

We cover how you can set up background jobs in Section 3.17.

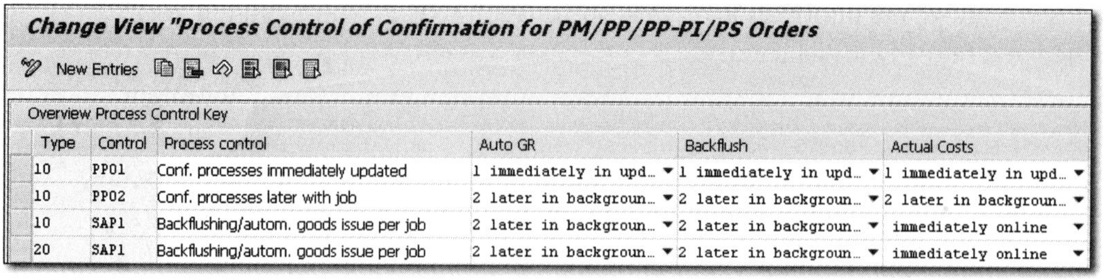

Change View "Process Control of Confirmation for PM/PP/PP-PI/PS Orders

New Entries

Overview Process Control Key

Type	Control	Process control	Auto GR	Backflush	Actual Costs
10	PP01	Conf. processes immediately updated	1 immediately in upd...	1 immediately in upd...	1 immediately in upd...
10	PP02	Conf. processes later with job	2 later in backgroun...	2 later in backgroun...	2 later in backgroun...
10	SAP1	Backflushing/autom. goods issue per job	2 later in backgroun...	2 later in backgroun...	immediately online
20	SAP1	Backflushing/autom. goods issue per job	2 later in backgroun...	2 later in backgroun...	immediately online

Figure 3.41 Process Control for Confirmation

3.14 Reason for Variances

During confirmation against the production order, when you record the yield, scrap, and rework, you can assign a reason for variance between the planned and actual production. For example, assume that, due to machine malfunction or defective components, your actual production was less than planned. You can assign the specific reason for this deviation or variance. Similarly, when entering the scrap quantity, you can give the reason for variance. Later, when you execute the production order information system (Transaction COOIS), you can filter out the information based on the reason for variance. This helps you get a better understanding of the reasons behind the problems and helps you take corrective or preventive actions.

As a PP consultant during an SAP ERP system implementation, you make a complete list of plant-related reasons for variances and configure them accordingly in the system. You can perform multiple partial confirmations and enter different reasons for variance for each deviation.

[»] **Note**

If you have the business need to enter multiple reasons for variances, you can engage an ABAP resource and activate enhancement `CONFPP07`.

You can also assign a user status to a reason for variance to add further control in the confirmation.

[»] **Note**

To set up a new status profile, use Transaction BS02.

To configure a new reason for the variance for a given plant, follow the configuration (Transaction SPRO) menu path, PRODUCTION • SHOP FLOOR CONTROL • OPERATIONS • CONFIRMATION • DEFINE REASONS FOR VARIANCES, or use Transaction OPK5.

Figure 3.42 lists the reason for variance for PLANT 3000. If you select a specific reason for variance and click on the USER STATUS folder, the system prompts you to enter the status profile and then the user status.

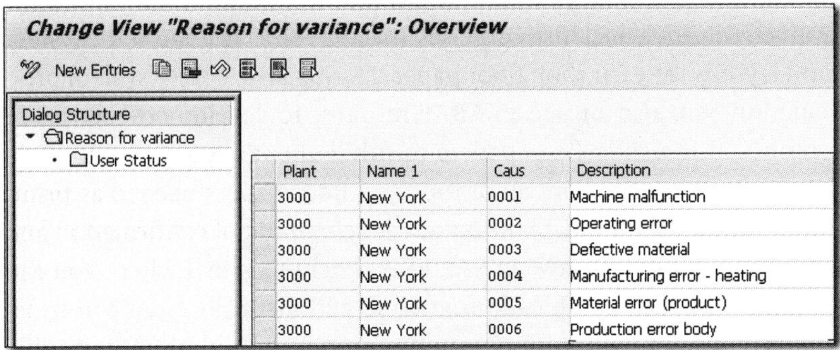

Figure 3.42 Reasons for Variance

3.15 Trigger Points

Trigger points are automatic functions that the system can perform when it's able to meet the user-defined condition. For example, when a user releases an operation, the system can automatically release all preceding operations via a trigger point. Minimal configuration steps are involved in trigger points: you define the trigger point usage and then define a group for the trigger points.

The TRIGGER POINT USAGE field provides a unique identification to list trigger points that will have common usage. This enables you to quickly select the relevant trigger point based on its usage. To define the (standard) trigger points usage, follow the menu path, PRODUCTION • SHOP FLOOR CONTROL • MASTER DATA • TRIGGER POINT • DEFINE (STANDARD) TRIGGER POINT USAGE. On the initial screen, enter an alphanumeric key, and give a short description of the trigger point usage.

You can group all of the trigger points together, which you can then assign to an operation. In doing so, you save the time and effort involved in individually creating and assigning trigger points.

To define a group for standard trigger points, follow the menu path, PRODUCTION • SHOP FLOOR CONTROL • MASTER DATA • TRIGGER POINT • DEFINE GROUP FOR STANDARD TRIGGER POINTS. On the initial screen, enter an alphanumeric key, and give a short description of the trigger points group.

3.16 Define Print Control

During the production process, you need to print several shop floor papers, for example, production order, pick list, or goods issuance slip. The SAP ERP system offers standard layouts for each shop floor paper. During an SAP ERP system project implementation, you also engage an ABAP resource to develop new scripts or modify the scripts of the shop floor papers to fulfill client's requirements. You discuss and evaluate with the client all of the shop floor papers needed as printouts. For example, you can eliminate the layout development of confirmation and its printout requirement if you have an SAP ERP system terminal where you can make live data entries or use an upload program to automatically enter confirmation. Further, if you've installed Adobe Document Services, then it's also possible to have PDF printouts of the shop floor papers.

To make necessary settings to print shop floor papers, follow the configuration (Transaction SPRO) menu path, PRODUCTION • SHOP FLOOR CONTROL • OPERATIONS • DEFINE PRINT CONTROLS, or use Transaction OPK8. The resulting screen in Figure 3.43 lists the shop floor papers that you can individually work on and make the desired settings.

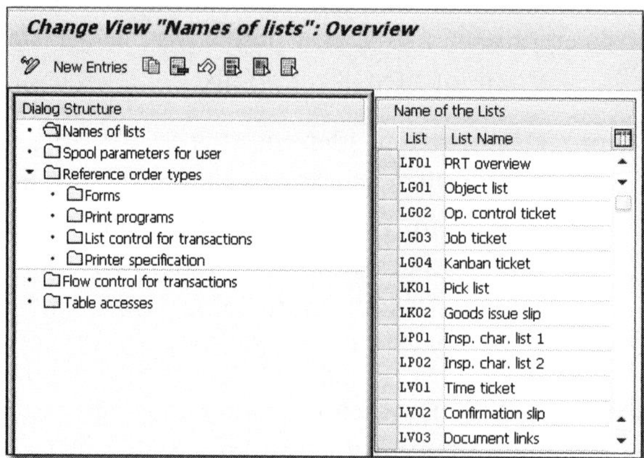

Figure 3.43 Printing Shop Floor Papers

[«]

Note

Chapter 6 shows how we used list LG01 (OBJECT LIST) shown on the right-hand side of Figure 3.43 to get printouts with the SAP ERP system standard layout.

3.17 Background Jobs

To optimize your SAP ERP system's resources, which will help to improve performance and minimize undue burdens, there are several processes you can set up as background jobs. A *background job* enables you to define the frequency with which the system automatically performs the desired function without manual intervention. For example, if you run MRP on a weekly basis and are generally satisfied with the results, you can set up a background job to automatically convert planned orders into production orders.

To set up a background job, follow the configuration (Transaction SPRO) menu path, PRODUCTION • SHOP FLOOR CONTROL • OPERATIONS • SCHEDULE BACKGROUND JOBS, or use Transaction SM36.

We suggest that you use the JOB WIZARD icon because it provides a step-by-step approach of setting up a background job. Table 3.1 lists the program names of selected background jobs that you can set up, and the system asks for this information. Before setting up the background job, you should also set up the variant of the background job, so that the system processes only the relevant data.

Function	Program Name
Convert planned orders into production orders	PPBICO40
Print production orders	PPPRBTCH
Print shop floor papers	PPPRBSEL
Mass processing	COWORKDISPATCHNEW
Confirmation: Componentized processes (GR, GI, HR, actual)	CORUPROC
Confirmation: Fast entry records	CORUPROC1

Table 3.1 Background Jobs

[+]

Tips & Tricks

On the initial screen of Transaction SE38, you can enter the program name listed in Table 3.1 and choose EXECUTE. The system takes you to the screen in which you can set screen parameters and save them by pressing Ctrl+S or by choosing GOTO • VARIANTS • SAVE AS VARIANT.

You can read additional documentation on the background program by clicking on the information icon, where available.

3.18 Process Integration

The process integration is a relatively new offering for discrete manufacturing, including creation of control recipes. Refer to Chapter 4 and Chapter 7 for in-depth coverage of Process Management, as the underlying working and concepts are the same for discrete manufacturing.

3.19 Summary

The logical and sequential configuration steps that we undertook in this chapter will enable you to perform the same set of steps and activities in configuration steps in process manufacturing. If you follow all of the steps covered in the chapter in a sequential and logical order, you'll be able to cover all of the major business processes and scenarios of production with production orders (shop floor control). See Chapter 6 to find more details that will align with your business processes.

The next chapter covers the configuration basics of process manufacturing.

Configuration basics in process manufacturing are generally the same as in discrete manufacturing. However, the Process Management functionality demands in-depth coverage because it's unique to process manufacturing only.

4 Configuration Basics of Process Manufacturing

The configuration basics of process manufacturing are similar to those in discrete manufacturing. For example, the steps to define order types, assign number ranges, and define the screen layout for confirmation are the same in both production types. We recommend that you also read Chapter 3 and Chapter 6 for an in-depth understanding of the configuration settings and business processes, respectively.

The focus of this chapter is primarily on the configuration that is unique only to process manufacturing. For example, the master recipe profile is unique to process manufacturing only. At the same time, this chapter gives extensive coverage to the configuration area of Process Management.

We'll start by covering the standard settings and tools that you can use to optimize most of your configuration work related to Process Management. We then cover process messages and process message categories, followed by control recipes and process instruction sheets. Next, we'll discuss the process instructions, types, and categories, as well as how the system can facilitate several configuration steps using wizards. Using the scope of generation functionality, you can further reduce the data maintenance efforts involved in Process Management. We'll also cover background jobs for sending control recipes and how you can send process messages back to the SAP ERP system as scheduled (background) jobs.

The concepts and the basics that you'll learn will enable you to implement a newer and user-friendlier functionality of Process Management known as Execution

Steps (XSteps). You can also transition from Process Management to XSteps and still use all of the configuration steps that you'll learn in this chapter.

We'll also show how the entire configuration undertaken so far facilitates the quick definition, simulation, and activation of process manufacturing cockpits. We present a simple example of how you can use a self-defined process instruction type and incorporate its details in the process instruction sheet. When the system sends process messages back to its predefined destination, it also sends the details of the self-defined process instruction characteristic.

[»]

Note

Refer to the appendix for features comparison among discrete manufacturing, process manufacturing, and repetitive manufacturing, as well as how to decide which manufacturing type is most relevant for a given industry.

Let's get started.

4.1 Master Data in Process Industries

The master data in process manufacturing is similar to the master data of discrete manufacturing. For detailed configuration settings and its business processes, refer to Chapter 3 and Chapter 6, respectively.

The following master data are specific to process industries and will be the main topics covered in this section:

- Resource
- Materials list (bill of materials [BOM])
- Master recipe
- Production version

The *resource* in process manufacturing is similar to the work center in discrete manufacturing. The creation of the *materials list* in process manufacturing is similar to the BOM in discrete manufacturing, so we won't cover these master data items here (see Chapter 3 and Chapter 6). For process industries, it's mandatory to have a master recipe and production version in the system. While the configuration of the master recipe has some uniqueness and is covered in this chapter, the creation of the production version is once again the same for discrete manufacturing and is therefore not covered here.

4.1.1 Master Recipe Profile

Although most of the details in the master recipe profile are the same as in routing, the one exception is the option available for Process Management. Depending on the master recipe profile that you use during the creation of the master recipe, the system offers the following options for Process Management:

- PROCESS INSTRUCTIONS
- XSTEPS (Execution Steps)
- XSTEPS OPTIONAL

If you choose XSTEPS OPTIONAL during the master recipe creation, the system provides you the option to either select PROCESS INSTRUCTIONS or XSTEPS. You can't use both of them at the same time. Also, if you previously used process instructions in the master recipe and now want to transition to XSteps, the system issues a warning message that you can't revert back to process instructions after you save the master recipe with XSteps activated.

To set up the master recipe profile, follow the configuration (Transaction SPRO) menu path, PRODUCTION PLANNING FOR PROCESS INDUSTRIES • MASTER DATA • MASTER RECIPE • DEFINE PROFILE WITH DEFAULT VALUES, or use Transaction OPN1. Choose PROFILE/RECIPE on the initial screen, which takes you to the overview of the master recipe profile with three entries as shown in Figure 4.1.

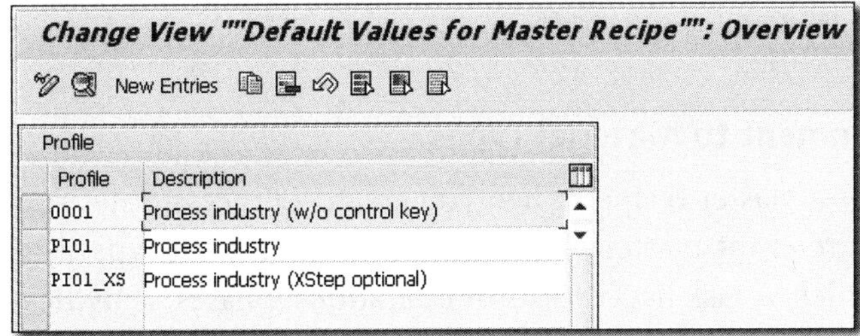

Figure 4.1 Overview Screen for Default Master Recipe Values

Select PROFILE PI01_XS, and choose the DETAILS icon, which leads to the screen shown in Figure 4.2. In this screen, you can update the fields, which then apply when you create the master recipe. Assigning a profile is a mandatory requirement when creating a master recipe. Notice the PROCESS INSTRUCTION MAINTENANCE dropdown in the PROCESS MANAGEMENT area and the available options, including XSTEPS OPTIONAL.

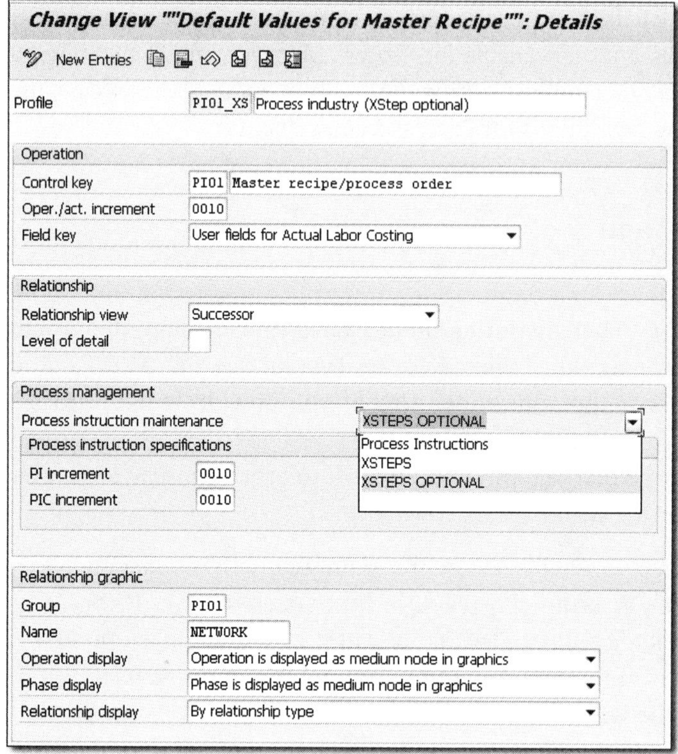

Figure 4.2 Detailed Screen for the Master Recipe Profile

4.1.2 Task List Assignment to Material Types

Task list type 2 is used for a master recipe. When you create a process order and the system looks for the relevant master recipe, it looks for task list type 2 to assign it to the process order. A task list consists of operations, phases, activities, control keys, and the relationship to the BOM. If your company has material types other than those available in a standard SAP ERP system, then you need to maintain the task list relationship for each material type.

To define the task list assignment to the material type, follow the configuration (Transaction SPRO) menu path, Production Planning for Process Industries • Master Data • Master Recipe • Assign Material Types. In Figure 4.3, you can choose New Entries to define the task list type (TLType) as 2 and the company-specific material types used in process manufacturing.

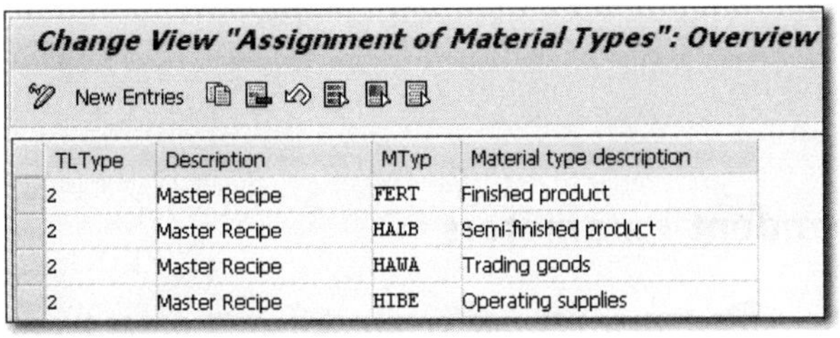

Change View "Assignment of Material Types": Overview

New Entries

TLType	Description	MTyp	Material type description
2	Master Recipe	FERT	Finished product
2	Master Recipe	HALB	Semi-finished product
2	Master Recipe	HAWA	Trading goods
2	Master Recipe	HIBE	Operating supplies

Figure 4.3 Task List Assignment with Material Type

4.1.3 Task List Status

The standard SAP ERP system offers task list statuses to cater to various business scenarios. For example, when you're in the initial stages of defining the master recipe of a material, you can set the status to 1, which is for Creation. The system won't make this task list available for selection during process order creation or in production version creation. You can also have a task list that's specific for Product Cost Controlling (CO-PC), and in that case, it becomes available for costing purposes but not for production.

To define a task list status at the header level, follow the configuration (Transaction SPRO) menu path, Production Planning for Process Industries • Master Data • Master Recipe • Data for the Recipe Header • Define Recipe Status, or use Transaction OP46.

Figure 4.4 shows that the task list status (Stat) 4 enables you to perform all of the steps, such as using it in production (RelInd), using it for costing (Cstng), and performing consistency checks (Cons.chk.). You can perform a consistency check during master recipe creation to enable the system to point out any deviations or shortcomings in the master recipe that you can deal with before saving. You can even save a master recipe if the system issues warning or error messages that occur during consistency checks, and then you can correct them later.

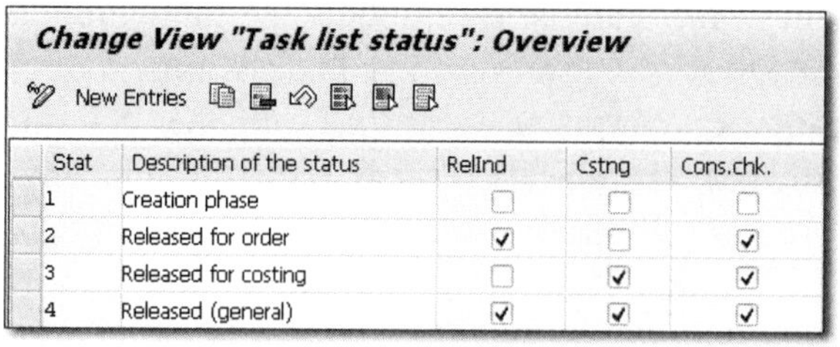

Change View "Task list status": Overview

New Entries

Stat	Description of the status	RelInd	Cstng	Cons.chk.
1	Creation phase	☐	☐	☐
2	Released for order	☑	☐	☑
3	Released for costing	☐	☑	☑
4	Released (general)	☑	☑	☑

Figure 4.4 Task List Status

4.2 Order Type-Dependent Parameters

To have better control of your order types, you can create different order types within a plant. For example, you may have one order type for finished goods and another for semifinished goods or a third one to capture rework details from normal production. Each order type can have a separate number range and its own order type settings. Further, you can control authorization to users based on the order type and plant combination. In order type-dependent parameters, you make settings that are applicable each time you create a process order for that specific order type and plant combination.

To set up order type-dependent parameters, follow the configuration (Transaction SPRO) menu path, PRODUCTION PLANNING FOR PROCESS INDUSTRIES • PROCESS ORDER • MASTER DATA • ORDER • DEFINE ORDER TYPE-DEPENDENT PARAMETERS, or use Transaction COR4. In the initial screen, select PLANT 3000 and ORDER TYPE PI01, which leads to the MASTER DATA tab. We'll discuss each of the tabs for configuring parameters in the following sections.

4.2.1 Master Data

Figure 4.5 shows the MASTER DATA tab of order type-dependent parameters. Here you select AUTOMATIC SELECTION OF PRODUCTION VERSION from the PRODUCTION VERSION dropdown, and the system assigns the first available production version, during planning and production execution steps. Alternatively, you can also use the manual selection option, and the system opens a pop-up during process order creation to enable you to select a production version manually.

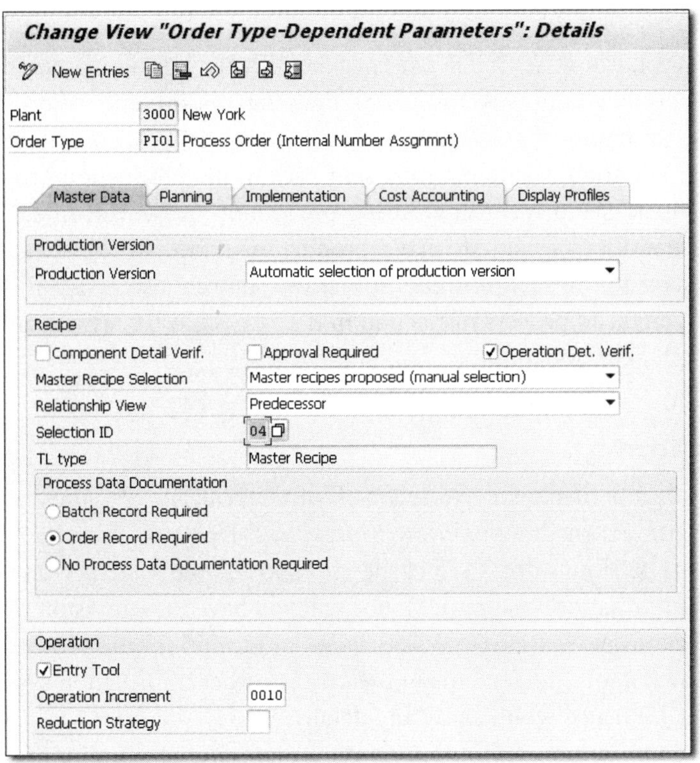

Figure 4.5 Order Type-Dependent Parameters: Master Data Tab

If you select Approval Required in the Recipe area, then your Master Recipe Selection dropdown should consist of either a mandatory or an optional option. When a user creates a process order, the system requires the order to be approved before allowing the user to release it. Similarly, if you set up automatic process order release during creation (in the production scheduling profile), the system overrules the automatic release settings if you've set up the approval process (see Section 4.3).

Note [«]

To grant approval to a process order during its creation, follow the menu path, Process Order • Functions • Approval • Individual Approval.

4.2.2 Planning

In the PLANNING tab, you can select the SCRAP checkbox if your business process has any such need. Normally in process industries, the scrap option or its recording finds less use. For example, in the fertilizer industry, any visible or overflow material (spilled-over material) is automatically sent back in the production process chain, so there is no scrap recording during production. However, in the steel rerolling industry, it's common to have production scrap, so the Scrap option needs to be activated in such business scenarios. You can also assign the relevant inspection type for in-process inspection in the PLANNING tab, as well as make settings for batch determination.

4.2.3 Implementation

In the IMPLEMENTATION tab, make sure to select all of the relevant checkboxes, especially for the SHOP FLOOR INFORMATION SYSTEM and DOCUMENTATION OF GOODS MOVEMENT areas. If you don't select the checkboxes, the system won't update the Logistics Information System (LIS), nor will you have the provision to see the documented goods movements (goods issuance or receipt) for the process orders. The documented goods movements in the process order information system (Transaction COOISPI) also won't show any details.

4.2.4 Cost Accounting

In the COST ACCOUNTING tab, we suggest that you coordinate with a Controlling resource (responsible for Cost Object Controlling) to ensure that the necessary configuration settings are in place.

4.3 Production Scheduling Profile

The production scheduling profile in process manufacturing is similar to the one in discrete manufacturing. However, in process manufacturing, it provides the option to automatically generate the control recipe when you release a process order. A control recipe enables the process operator to record important plant parameters. We cover the details concerning control recipes in Section 4.4.1.

To set up the production scheduling profile, follow the configuration (Transaction SPRO) menu path, PRODUCTION PLANNING FOR PROCESS INDUSTRIES • PROCESS

Order • Master Data • Define Production Scheduling Profile, or use Transaction CORY (see Figure 4.6).

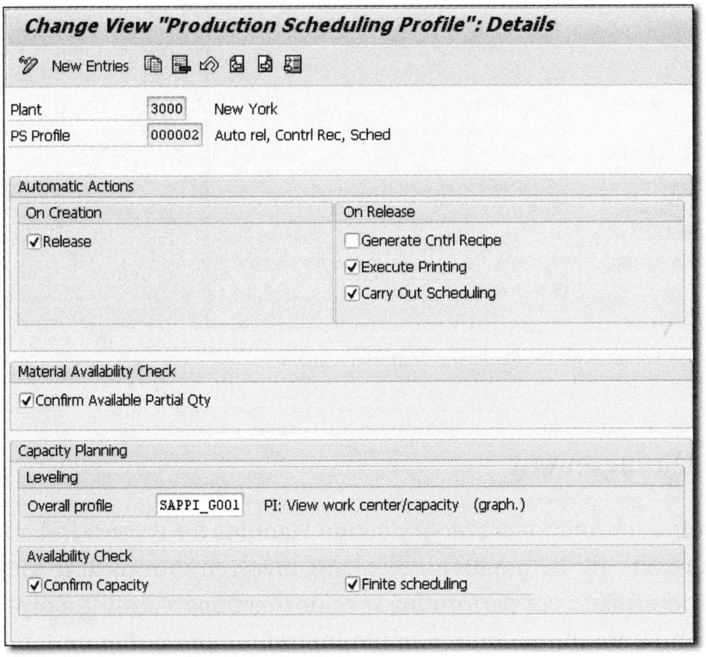

Figure 4.6 Production Scheduling Profile

On the initial screen, select Plant 3000 and PS Profile (production scheduling profile) 000002.

In this example, we haven't selected the Generate Cntrl Recipe checkbox in the On Release area (within the Automatic Actions area). In Chapter 7, we'll show you how to manually generate a control recipe when you release the process order.

Using Transaction MM02 to change the material master, you can assign the production scheduling profile in the Work scheduling tab (see Figure 4.7). Whenever you create a process order of this material and plant combination, the system automatically triggers the settings you made in the production scheduling profile.

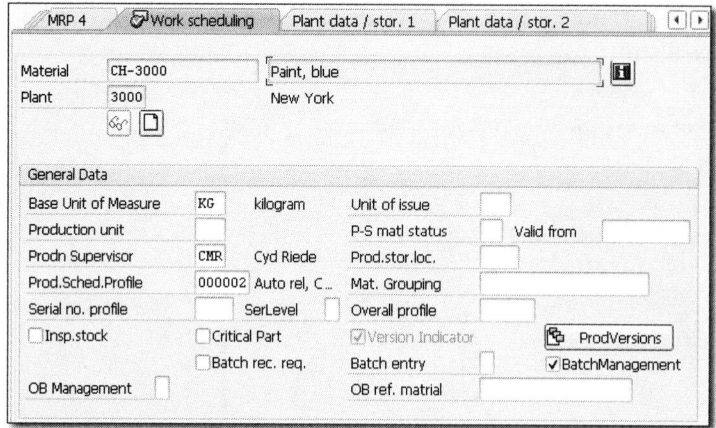

Figure 4.7 The Production Scheduling Profile in the Material Master

4.4 Process Management

Process Management is a subcomponent of Production Planning for Process Industries (PP-PI), which covers all the important elements involved in recording the process parameters' information or performing specific functions. Creating a process order forms the basis for the generation of the control recipe (see Figure 4.8). The control recipe is sent to the process control in the form of a process instruction sheet. The process operator not only follows the instructions given in the process instruction sheet but also fills the process instruction sheet with relevant plant parameters and other important data and returns it as a process message back to either the SAP ERP system or to an external system.

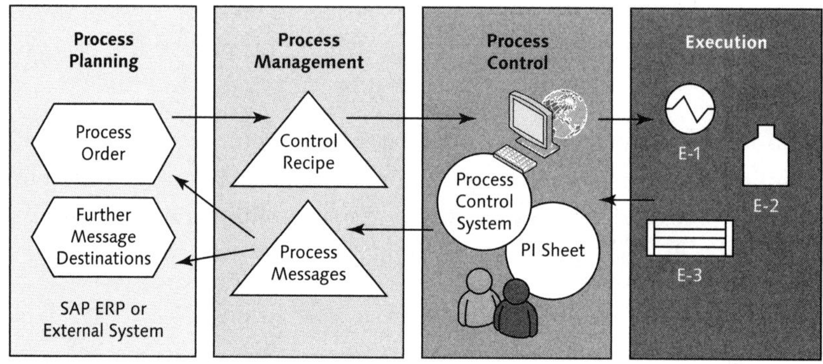

Figure 4.8 Process Management Overview

Although this section applies more to the business process side of Process Management, it's important to have a basic understanding of the concepts and terminologies used throughout the remaining chapter. The important terms used in Process Management are discussed in the following sections.

4.4.1 Control Recipe Destination

The *control recipe destination* specifies the technical address to which a control recipe is transferred. An operation in a master recipe can have several phases, and each phase can have a different control recipe destination. In the process order, you can create one control for each control recipe destination. Therefore, if there are three different phases, it's possible to have three different control recipe destinations (e.g., mixing, blending, and liquid ammonia sections). Typically, at the process order release, the system creates control recipes for each control recipe destination in the process order. All phases of a process order with the same control recipe destination are combined into the same control recipe. A control recipe destination defines whether the process instruction of a phase is transferred to an external system (process control system, PCS) or is converted to a process instruction sheet for a line operator. It also specifies which system or line operator executes the process instructions of a phase.

4.4.2 Process Instruction Characteristic

A released characteristic within a characteristic group is available for use in the process instruction. Process instruction characteristics are allocated to process instructions either directly in the master recipe or by process instruction category. Process instruction, together with the corresponding characteristic value, determines the information transferred or requested, that is, the status of the control recipe. It also determines how the requested data is processed (e.g., the message category to be used to report the data).

4.4.3 Process Instruction Category

The process instruction category specifies the type of process instruction (i.e., process parameter or process data request). It also specifies the characteristics of this process instruction, that is, information to be sent via this instruction.

4.4.4 Process Message Characteristic

A released characteristic within the characteristic group is available for use in the process messages. Process message characteristics are allocated to process message categories. Process message characteristics, together with the corresponding characteristic values, determine the content of the process message. The process message can be sent to an external system, SAP Office Mail recipients, or custom-defined ABAP tables.

4.4.5 Process Message Category

The process message category specifies the characteristics to be allocated to a process message, that is, the information to be sent. It also specifies the message destinations for the process message. Every process message sent via Process Management must be assigned to a process message category.

4.4.6 Process Instruction Sheet

A process instruction sheet is used as an input and output form for plant parameters and is customizable according to individual production processes. This customized process instruction sheet is part of either configuration activity or master data management. A process instruction sheet, due to its intuitive features, can be used for many functions such as calling up a transaction while remaining in a process instruction sheet, displaying documents, recording shift highlights, posting a goods issue or goods receipt, undertaking confirmation of a phase, calculating results, and jumping to record results for quality inspections.

To create a custom process message or a custom process instruction, you can leverage your knowledge and experience in creating characteristics in the classification system. The only difference between classification and process message/instruction creation is that the former doesn't offer the option to incorporate details specific to Process Management. In our example, we show how you can use process instruction to fetch a field, Process Order Finish Date, from the process order automatically and integrate it in the process instruction sheet.

Now let's discuss the important one-time settings that you need to make in your SAP ERP system to activate Process Management. This entails sequence steps, in which you can copy the standard settings delivered by the SAP ERP system to your specific plants.

4.4.7 Standard Settings and Tools

The system offers a large number of standard process messages, process message categories, process instruction, and process instruction categories. These are all available in the SAP ERP system's plant 0001. You can copy all of these standard settings from plant 0001, except process instruction categories, which you need to copy manually. An easier way to perform this copy function is to do it with reference to plant 0001 to your desired plants.

You need to perform a few important steps in the right sequence to ensure that minimal work and effort are involved in integrating Process Management in your business processes. Use configuration (Transaction SPRO) menu path, Production Planning for Process Industries • Process Management • Standard Settings, as shown in Figure 4.9.

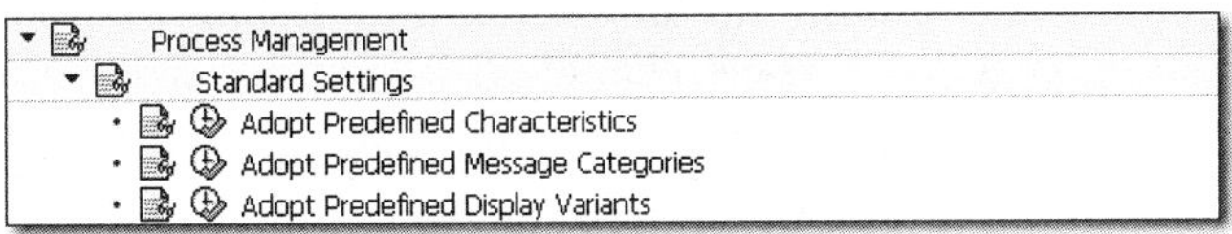

Figure 4.9 Standard Settings in Process Management

Next, follow these steps:

1. Use Transaction O23C to transport predefined characteristics from the SAP ERP system reference client, which is 0001, to your SAP ERP system client. The Copy Object Lists screen appears when you successfully perform the first step.
2. Access Transaction O22C to transport preconfigured process message categories from the SAP ERP system reference plant 0001 to your plants. The screen shown in Figure 4.10 appears after you enter the target plant.

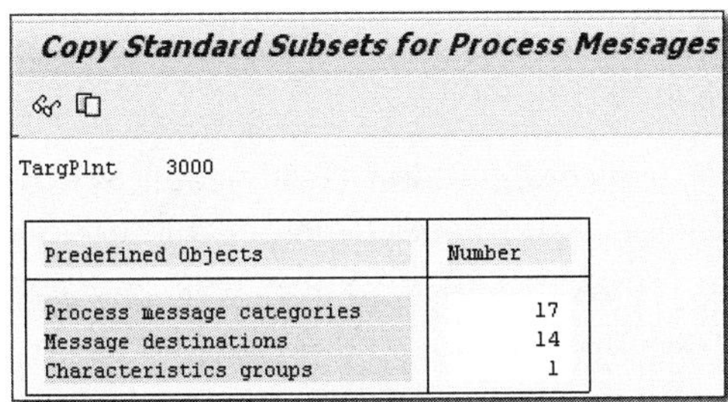

Predefined Objects	Number
Process message categories	17
Message destinations	14
Characteristics groups	1

Figure 4.10 Predefined Objects of Process Messages

3. In this example, enter the target plant as "3000". Use the COPY icon to copy it to the target plant. The system issues a warning message: THE SAP STANDARD SETTINGS ARE COPIED TO YOUR TARGET PLANT. OBJECTS WITH IDENTICAL NAMES ARE OVERWRITTEN.
4. When you see the DO YOU WANT TO COPY? message, click YES.
5. Access Transaction CO60_VM to transport the process instruction sheet's display variants from the SAP reference plant to your login client. Make sure to select all of the available display variants, and choose SAVE before you exit the screen.

The SAP ERP system also offers standard characteristics groups in Process Management, both for process instructions and for process messages. The system groups the process instruction characteristics in group PPPI_01, whereas it groups the process message characteristics in group PPPI_02. When you create a self-defined process instruction or process message characteristic, ensure that you assign them to the relevant characteristics group to enable them to work correctly.

[+]

Tips & Tricks

To ensure that you can see the characteristics groups (PPPI_01 or PPPI_02) in the dropdown, click on CUSTOMIZE LOCAL LAYOUT (or press [Alt]+[F12]) on the main screen, select OPTIONS • INTERACTION DESIGN • VISUALIZATION AND INTERACTION, and then choose SHOW KEYS WITHIN THE DROPDOWN LISTS.

When you implement Process Management, you need to ensure that you release the process instruction and process message groups to the requisite plants. To release process messages group, use configuration (Transaction SPRO) menu path, PRODUCTION PLANNING FOR PROCESS INDUSTRIES • PROCESS MANAGEMENT • PROCESS MESSAGES • DEFINE CHARACTERISTICS GROUPS FOR PROCESS MESSAGES • RELEASE FOR PROCESS MESSAGES, or use Transaction O08C.

To release a process instructions group, use configuration (Transaction SPRO) menu path, PRODUCTION PLANNING FOR PROCESS INDUSTRIES • PROCESS MANAGEMENT • PROCESS INSTRUCTIONS • DEFINE CHARACTERISTICS GROUPS FOR PROCESS INSTRUCTIONS • RELEASE FOR PROCESS INSTRUCTIONS, or use Transaction O09C.

If the process messages or process instructions groups aren't defined, available, or released for your desired plant, you can copy them with reference to another plant, where these are available and released.

To copy all settings from reference plant 0001 to your desired plant, use configuration (Transaction SPRO) menu path, Production Planning for Process Industries • Process Management • Tools • Copy Settings between Plants, or use Transaction O20C. In the ensuing screen, enter the reference plant, which is "0001", and the target plant to which you want to copy the standard system's settings (which is "3000" in this example).

Tips & Tricks [+]

If you're using a new SAP client for the first time (e.g., production system [PRD]), then there's a possibility that the Process Management won't work. If that happens, just run Transaction O23C and then Transaction O22C once (and in the same order), and this should solve the problem.

Alternatively, you can also make it a part of your SAP go-live checklist to ensure that you run these two transactions in the PRD system before starting any PP-related data entry in the system.

4.5 Process Messages

Process messages enable the process operator to send the requested data back to the SAP ERP system, which then performs the necessary functions. For example, you want the process operator to inform you (through process messages) about the consumption (goods issuance) of all of the materials used in the production process (against process order). The process operator fills in the process instruction sheet, which contains the provision to record actual material consumption details and mark the status of the process instruction sheet as complete. The process operator then sends the process message containing the consumption details to the process message destination in the SAP ERP system, after which the system will perform the actual goods issuance posting against the process order in the system.

The system offers a large number of standard process messages and process message categories, but you can still define your own process messages to attend to specific business needs. For example, you might want specific information that the system should send back to the SAP ERP system, based on the process operator's input in the process instruction sheet.

In the following sections, we'll go through the details of how to create or define and set up a process message. Your first step is to create a process message characteristic.

4.5.1 Create a Process Message Characteristic

Your first step is to use configuration (Transaction SPRO) menu path, Production Planning for Process Industries • Process Management • Process Messages • Define Characteristics for Process Messages • Create Characteristics for Process Messages, or you can use Transaction O25C.

For this example, create process message characteristic ZPI_Creation_Date1, and define the characteristics group (Chars Group field) as Process Message Characteristic in the Basic Data tab. Define the Data Type as Date. In Chapter 7, we show how you can use this self-defined process message to send details back to the SAP ERP system.

4.5.2 Process Message Destination

To create a new process message destination, use the configuration (Transaction SPRO) menu path, Production Planning for Process Industries • Process Management • Process Messages • Process Messages Destinations • Define and Set up Process Messages Destinations, or use Transaction O03C.

For this example, create a new process message Destination PP10 for Plant 3000, and then choose Target fields/Message destinations on the left-hand side of the screen. The Change View screen shown in Figure 4.11 appears. Here the system allows you to enter the process message characteristic, "ZPI_CREATION_DATE1", which you created in the previous step.

You can set your self-defined process message destination so that the system is able to send the process messages to the defined destination. When the system receives information from process messages, it processes them in the SAP ERP system using function modules. For external systems, the process messages with information are sent back to the SAP ERP system using remote function call (RFC) connections. You can select the Individual checkbox (not shown) if you want the system to process each process message individually. This option is generally used in complex process messages involving function module destinations.

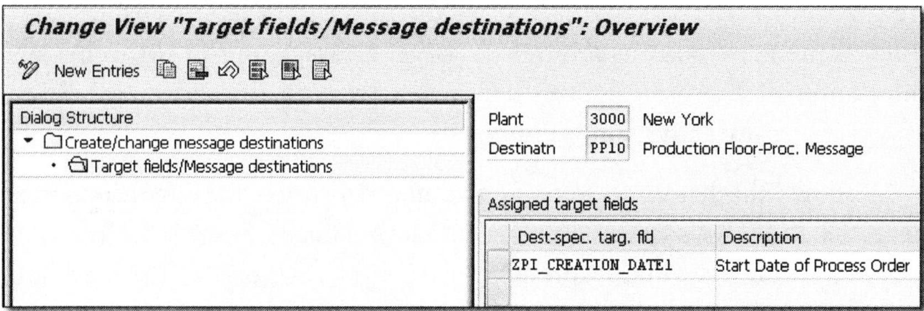

Figure 4.11 Target Field in the Process Message Destination

[«]

Note

We suggest that you engage a Basis (SAP NetWeaver) resource to manage the configuration of RFC destinations.

4.5.3 Process Message Categories

In Process Management, the system first sends the control recipe to the control recipe destination as a manual or automatic (background) job. The plant operator receives the control recipe in the form of a process instruction sheet and fills in the requisite data. When the data entry is complete, the process operator sets the status of the process instruction sheet as complete. This enables the system to communicate the data back into the SAP ERP system in the form of *process messages*. The process messages are the intimation or instructions to proceed with performing the necessary functions such as data entry or data updates.

The system embeds the process message category within the process instruction category. For example, the system embeds the process message category PI_PROD within the process instruction category APROD_1. The process instruction category APROD_1 enables you to record the production and also perform goods receipt against the process order through the process message category PI_PROD. When you create a process order, generate a control recipe, send it to the control recipe destination, maintain the process instruction sheet, and set the status to COMPLETE, the system processes the process messages by sending the completed data such as quantity produced or goods issuance to the SAP ERP system, which then process the data using function modules or Business Application Programming Interfaces (BAPIs).

There are several process message categories that cater to different business scenarios, such as the material consumption message, message on control recipe status, confirmation of a phase, or even new batch creation (provided that Batch Management is already active for the material).

Figure 4.12 shows the list of selected and standard process message categories available in the SAP ERP system. The system automatically assigns the relevant process message category in the process instruction category in the standard offering. For self-defined process message and process instruction categories, you need to ensure that you incorporate the relevant process message category within the process instruction category.

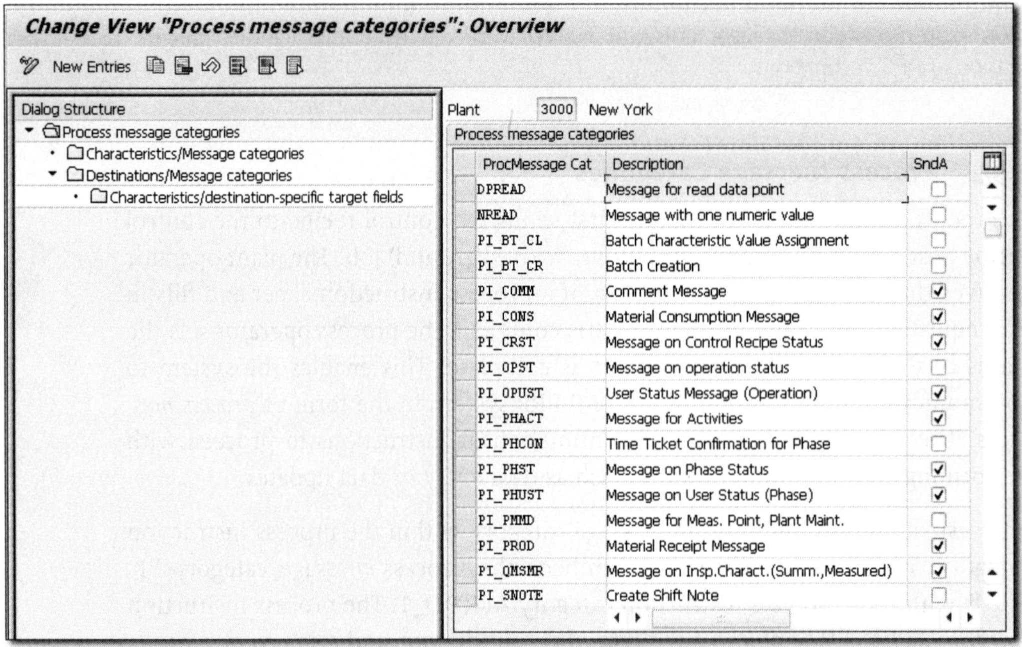

Figure 4.12 Overview of Process Message Categories

The screen shown in Figure 4.13 appears when you select process message category PI_CONS and double-click on DESTINATIONS/MESSAGE CATEGORIES. On the right-hand side of the figure, the system assigns two different destinations

(DEST)—PI01 and PI04—to which it will send the process messages. You can further map source and target characteristics that you want to send to specific destinations.

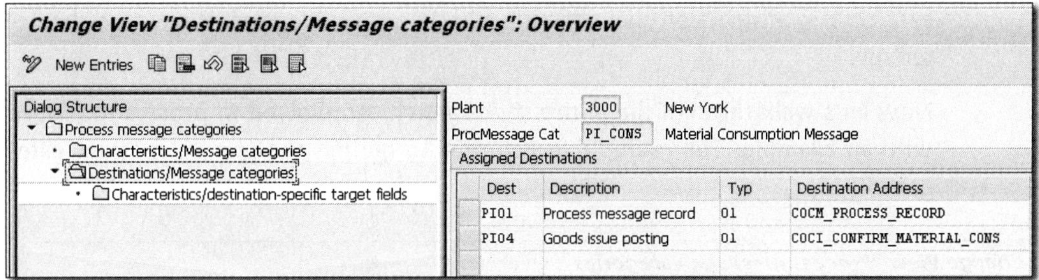

Figure 4.13 Process Message Categories

To view the list of standard process message destinations available in the SAP ERP system, place the cursor on the DEST field and press F4.

The screen shown in Figure 4.14 appears when you select PI_CONS in the PROCMESSAGE CAT field and select PI04 GOODS ISSUE POSTING in the DESTINATION field.

Change View "Characteristics/destination-specific target fields": Over

New Entries

Dialog Structure
- Process message categories
  - Characteristics/Message categories
  - Destinations/Message categories
    - Characteristics/destination-specific target fields

Plant 3000 New York
ProcMessage Cat PI_CONS Material Consumption Message
Destination PI04 Goods issue posting

Assignment of message characteristics to dest.-spec. target fields

Characteristic	Target Field
PPPI_BATCH	BATCH
PPPI_EVENT_DATE	EVENT_DATE
PPPI_EVENT_TIME	EVENT_TIME
PPPI_FINAL_ISSUE	FINAL_ISSUE
PPPI_MATERIAL	MATERIAL
PPPI_MATERIAL_CONSUMED	QUANTITY_CONSUMED
PPPI_OPERATION	OPERATION
PPPI_PHASE	PHASE
PPPI_PROCESS_ORDER	PROCESS_ORDER
PPPI_RESERVATION	RESERVATION_NUMBER
PPPI_RESERVATION_ITEM	RESERVATION_ITEM
PPPI_SPECIAL_STOCK	SPECIAL_STOCK
PPPI_STORAGE_LOCATION	STORAGE_LOCATION
PPPI_UNIT_OF_MEASURE	UNIT_OF_MEASURE
PPPI_VENDOR	VENDOR

Figure 4.14 Process Message Category PI_CONS and Destination PI04

On the right-hand side of the screen, the system maps each process instruction characteristic (source characteristic) with the process message characteristics (target characteristic/field). The CHARACTERISTIC column lists all of the process instruction characteristics, whereas the TARGET FIELD column enables you to map each process instruction characteristic with its respective process message characteristic.

Now let's walk through the entire set of activities involved in process messages with an example. The example in Figure 4.15 defines the process message category PP10.

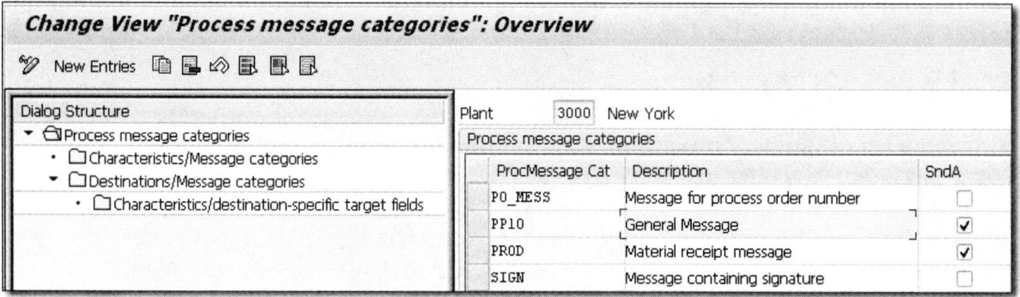

Figure 4.15 Process Message Category PP10 for Plant 3000

Follow these steps:

1. Select PP10, and click on CHARACTERISTICS/MESSAGE CATEGORIES to open the screen shown in Figure 4.16. Now you see the PP10 in the PROCMESSAGE CAT field and 3000 in the PLANT field. Choose CHARACTERISTICS/MESSAGE CATEGORIES on the left-hand side of the screen, and then on the right-hand side, enter the process instruction CHARACTERISTIC "ZPI_CREATION_DATE". Select the REQ checkbox to denote that it's required (mandatory) to enter details in this field.
2. Double-click on DESTINATIONS/MESSAGE CATEGORIES on the left-hand side, and the screen shown in Figure 4.17 appears. Incorporate the process message destination PP10 (PRODUCTION FLOOR-PROC. MESSAGE), and also assign the destination type (TYP), which in this example is 01 or COCM_PROCESS_RECORD.

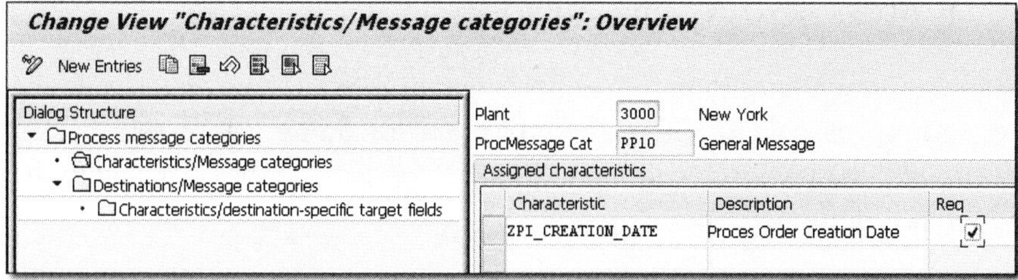

Figure 4.16 Process Instructions Characteristic in the Process Message Category

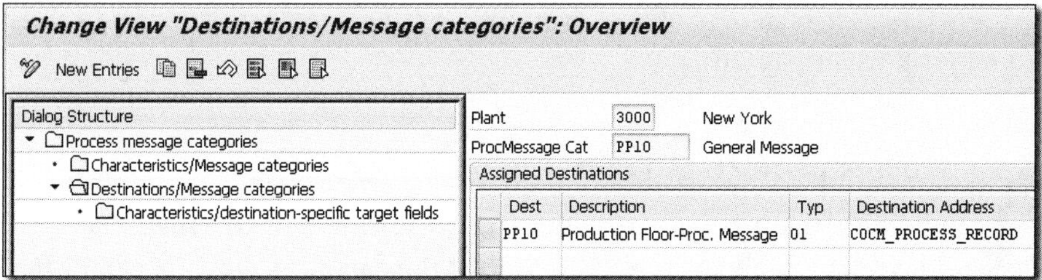

Figure 4.17 Process Message Destination in the Process Message Category

3. Select the PROCESS MESSAGE DESTINATION PP10, and choose CHARACTERISTICS/DESTINATION-SPECIFIC TARGET FIELDS on the left-hand side. The screen shown in Figure 4.18 appears. On the right-hand side of this figure, perform one-to-one mapping of the process instruction characteristic with the process message characteristic at the process message DESTINATION PP10. For this example, map process instruction characteristic ZPI_CREATION_DATE with the process message characteristic ZPI_CREATION_DATE1.

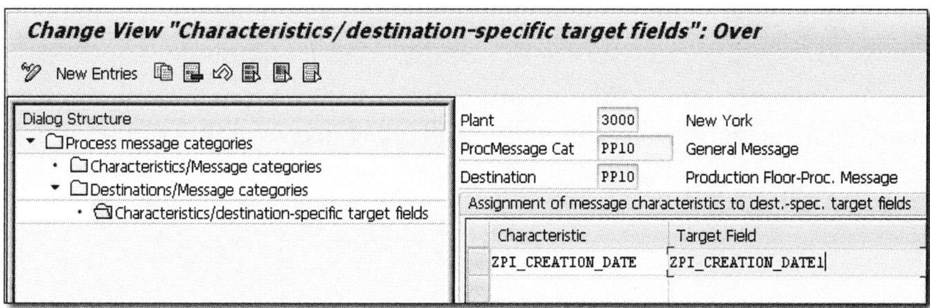

Figure 4.18 Fields Mapping of Process Message Characteristics with Destination-Specific Target Fields

4.6 Process Instruction Category

A *process instruction category* consists of different process instruction types. In the master recipe, you have to assign the process instruction category, which then automatically incorporates the different process instruction types assigned to the process instruction category.

In the following section, we deliberately took a longer route to explain the various process instruction types to cover the concepts and the fundamentals involved in each type. Toward the end of this section, we also show you how to use the wizard to create any of the seven different process instruction categories for various business scenarios.

4.6.1 Process Instruction Types

When you define the process instruction category, you have to assign a process instruction type to enable the system to process the data accordingly. Eight different process instruction types are available:

- Process parameters
- Process data requests
- Process message subscriptions
- Calculations
- Inspection results requests
- Dynamic function calls
- Sequence definitions
- Universal

Table 4.1 summarizes the different process instruction types, which are also explained in detail later. Universal isn't included because it applies to all types.

We discuss each of these in the following sections.

Process Instruction Type	Process Instruction Type Description	Usage in Process Instruction Sheets (Types 1 and 4)	Usage in External Process Control System (Types 2 and 3)	Sending Process Parameter Details	Process Message Requests to Process Message Destinations
1	Process parameters	✓	✓	✓	
2	Process data requests	✓	✓		✓
3	Process message subscriptions		✓		✓
4	Process data calculation formulas	✓			✓
5	Inspection results requests	✓			✓
6	Dynamic function calls	✓			✓
7	Sequence definitions	✓		✓	✓

Table 4.1 Process Instruction Categories

Process Instruction Type 1: Process Parameters

Process instruction type 1 is the simplest of all because it only requires the user to enter parameter values. In the process instruction sheet, you'll see a label for process parameter, into which you enter the parameter value. Figure 4.19 shows the screen for the process instruction category AMAT_1 for process instruction type 1. If you use this process instruction type in the process order (or in the master recipe), the system automatically fills in most of the details, such as material number, short text, issuance quantity for the operation, and phase. In column A (for automatic), the system automatically fills in all of the requisite fields.

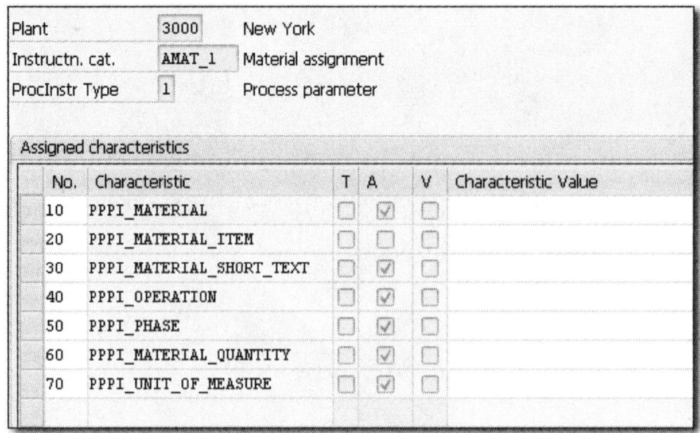

Figure 4.19 Process Instruction Type 1: Material Assignment

Process Instruction Type 2: Process Data Requests

Process instruction type 2 requires the user on the shop floor to enter the required data and send it back to the SAP ERP system through process messages, so that the system is able to perform the necessary data entry or updates. Either the system requires the user to enter the process parameter value manually, or the system can fill in the value automatically, depending on how you've set up the process instruction characteristics. We'll cover both examples of process parameter value entry (manual input as well as automatic).

Figure 4.20 shows the screen for the process instruction category AQMSMR_1 and process instruction type 2. It contains the process message category PI_QMSMR to enable the system to send back the parameter values that the shop floor operator has entered.

This example is for quality inspection results for in-process (during production) inspection. The system automatically fills in the CHARACTERISTIC VALUES such as operation number and inspection lot number, but you have to maintain the characteristic value for the characteristic PPPI_INSPECTION_CHARACTERISTIC. This value is basically the master inspection characteristic (in SAP ERP Quality Management [QM]), against which you're requesting the process operator to enter the values. The process operator also enters values (PPPI_REQUESTED_VALUES) in the process instruction sheet. During process messages sending, the system sends the entered information back to the SAP ERP system for further processing.

Change View "Characteristics/Process instruction category": Overview

New entries

Dialog Structure
- Process instruction categories
  - Characteristics/Process instruction category

Plant: 3000 New York
Instructn. cat.: AQMSMR_1 Request inspection results
ProcInstr Type: 2 Process data request

Assigned characteristics

No.	Characteristic	T	A	V	Characteristic Value
20	PPPI_MESSAGE_CATEGORY			x	PI_QMSMR
30	PPPI_PROCESS_ORDER		x		
40	PPPI_OPERATION		x		
50	PPPI_INSPECTION_LOT		x		
60	PPPI_INSPECTION_CHARACT...				
70	PPPI_REQUESTED_VALUE			x	PPPI_NUMBER_OF_INSPECTIONS
80	PPPI_REQUESTED_VALUE			x	PPPI_INSPECTION_SHORT_TEXT
90	PPPI_REQUESTED_VALUE			x	PPPPI_INSPECTION_RESULT
100	PPPI_REQUESTED_VALUE			x	PPPI_EVENT_DATE
110	PPPI_REQUESTED_VALUE			x	PPPI_EVENT_TIME

Figure 4.20 Overview of the Control Instruction Category for AQMSMR_1

Another example for process instruction type 2 is the process instruction category AREAD1 shown in Figure 4.21.

Plant: 3000 New York
Instructn. cat.: AREAD1 Request a measured value
ProcInstr Type: 2 Process data request

Assigned characteristics

No.	Characteristic	T	A	V	Characteristic Value
10	PPPI_DATA_REQUEST_TYPE			x	Simple Data Request
20	PPPI_MESSAGE_CATEGORY			x	READ
30	PPPI_PROCESS_ORDER		x		
40	PPPI_CONTROL_RECIPE		x		
40	PPPI_DATA_POINT_NAME				
45	PPPI_SOURCE				
50	PPPI_REQUESTED_VALUE			x	PPPI_EVENT_DATE
60	PPPI_REQUESTED_VALUE			x	PPPI_EVENT_TIME
80	PPPI_REQUESTED_VALUE			x	PPPI_MATERIAL_QUANTITY
90	PPPI_UNIT_OF_MEASURE		x		
100	PPPI_OPERATION		x		
110	PPPI_PHASE		x		

Figure 4.21 Process Instruction Type 2: Request a Measured Value

It contains the process message category READ to enable the system to send back the parameter values that the shop floor operator has entered. In the process instruction sheet, when the process operator enters the parameter value for the characteristic value PPPI_MATERIAL_QUANTITY, the system sends this value during the process messages sending. The system automatically fills in the characteristic values for the PPPI_EVENT_DATE and PPPI_EVENT_TIME, with the date and time the user enters the requested process parameter values.

Later, we'll also show how the system automatically fills in the characteristic value for a self-defined process instruction characteristic.

Process Instruction Type 3: Process Message Subscriptions

Figure 4.22 shows the screen for the process instruction category APROD_1 and process instruction type 3.

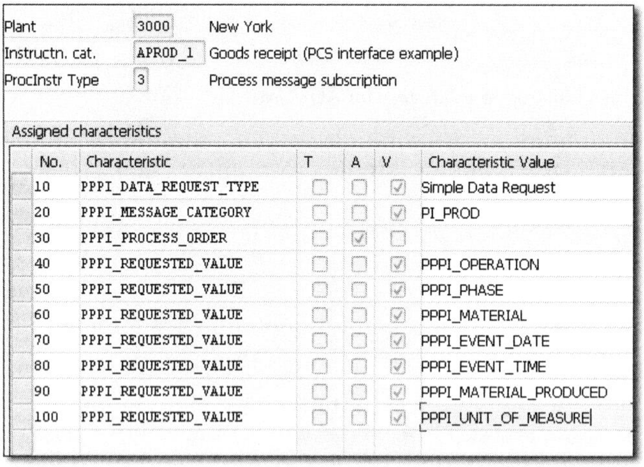

Plant 3000 New York
Instructn. cat. APROD_1 Goods receipt (PCS interface example)
ProcInstr Type 3 Process message subscription

Assigned characteristics

No.	Characteristic	T	A	V	Characteristic Value
10	PPPI_DATA_REQUEST_TYPE	☐	☐	☑	Simple Data Request
20	PPPI_MESSAGE_CATEGORY	☐	☐	☑	PI_PROD
30	PPPI_PROCESS_ORDER	☐	☑	☐	
40	PPPI_REQUESTED_VALUE	☐	☐	☑	PPPI_OPERATION
50	PPPI_REQUESTED_VALUE	☐	☐	☑	PPPI_PHASE
60	PPPI_REQUESTED_VALUE	☐	☐	☑	PPPI_MATERIAL
70	PPPI_REQUESTED_VALUE	☐	☐	☑	PPPI_EVENT_DATE
80	PPPI_REQUESTED_VALUE	☐	☐	☑	PPPI_EVENT_TIME
90	PPPI_REQUESTED_VALUE	☐	☐	☑	PPPI_MATERIAL_PRODUCED
100	PPPI_REQUESTED_VALUE	☐	☐	☑	PPPI_UNIT_OF_MEASURE

Figure 4.22 Process Instruction Type 3: Process Message Subscription

It contains the process message category PI_PROD to enable the system to send back the parameter values that the shop floor operator has entered. In this specific example, the process operator sends the quantity produced against the process order for which the operator wants to record goods receipt against the process order. The process operator records the goods produced information in the process instruction sheet and sets the status of the process instruction sheet as COMPLETE. Upon processing the process message, that is, sending it back to the

SAP ERP system, the system updates the goods produced information in the process order and also creates the material and accounting documents.

Process Instruction Type 4: Calculations

In this process instruction type, the system can perform calculations in the process instruction sheet based on the calculation formula that you've defined. The system makes use of the process message category to send back the calculated value to the predefined destination. The standard process message category is NREAD.

Table 4.2 provides the details of some of the important process instruction characteristics that you can use for calculation. You may also refer to Figure 4.23.

Characteristic	Characteristic Value	Details
PPPI_INPUT_REQUEST	KS PRODUCTION	You can define any characteristic value, and the system displays it accordingly in the process instruction sheet.
PPPI_VARIABLE	B	This is the variable that you can define for the calculated value.
PPPI_EVENT	PARAMETER_CHANGED	If you use this characteristic and its value, the system automatically does the calculation. If not, you have to manually choose the CALCULATOR icon to trigger calculation.
PPPI_CALCULATED_VALUE	NH3_02	This characteristic value was already created using Transaction CT04 and is assigned here. It governs and controls several parameters such as whether the field input is mandatory, whether a negative data entry is possible, or whether you can select a value from the dropdown option in the process instruction sheet.

Table 4.2 Process Instruction Characteristics for Calculation

Characteristic	Characteristic Value	Details
PPPI_CALCULATION_FORMULA	A*20	This is the calculation formula. The variable A needs to be previously defined in the process instruction sheet. You can use up to eight lines of characteristic to define a long or complex formula.
PPPI_UNIT_OF_MEASURE	KG	This is the unit of measure of calculated value.

Table 4.2 Process Instruction Characteristics for Calculation (Cont.)

In the screen shown in Figure 4.23, you can give a field description to the CHARACTERISTIC PPPI_INPUT_REQUEST, and the system will display it in the process instruction sheet. Further, you can define the formula that the system will use in the calculation in the value field for CHARACTERISTIC PPPI_CALCULATION_FORMULA. If the formula is too long to fit on one line, you can enter the formula in up to eight characteristics. All you have to do is add a new line, enter the characteristic "PPPI_CALCULATION_FORMULA", and continue entering formula details in the value field until the formula is complete.

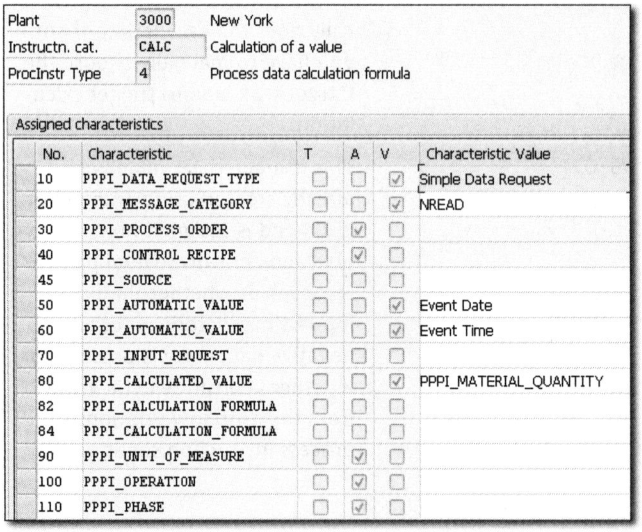

Plant 3000 New York
Instructn. cat. CALC Calculation of a value
ProcInstr Type 4 Process data calculation formula

Assigned characteristics

No.	Characteristic	T	A	V	Characteristic Value
10	PPPI_DATA_REQUEST_TYPE	☐	☐	☑	Simple Data Request
20	PPPI_MESSAGE_CATEGORY	☐	☐	☑	NREAD
30	PPPI_PROCESS_ORDER	☐	☑	☐	
40	PPPI_CONTROL_RECIPE	☐	☑	☐	
45	PPPI_SOURCE	☐	☐	☐	
50	PPPI_AUTOMATIC_VALUE	☐	☐	☑	Event Date
60	PPPI_AUTOMATIC_VALUE	☐	☐	☑	Event Time
70	PPPI_INPUT_REQUEST	☐	☐	☐	
80	PPPI_CALCULATED_VALUE	☐	☐	☑	PPPI_MATERIAL_QUANTITY
82	PPPI_CALCULATION_FORMULA	☐	☐	☐	
84	PPPI_CALCULATION_FORMULA	☐	☐	☐	
90	PPPI_UNIT_OF_MEASURE	☐	☑	☐	
100	PPPI_OPERATION	☐	☑	☐	
110	PPPI_PHASE	☐	☑	☐	

Figure 4.23 Process Instruction Type 4: Calculation

Process Instruction Type 5: Inspection Results Requests

In this process instruction type, you can navigate to the screen or transaction within the process instruction sheet, in which you can record inspection results of an operation or a phase. In other words, it provides the option to record the quality results. The standard process instruction category available for operations is QMJUM, and for phase, it's QMJUM_PH. Depending on the process instruction category that you select, you can then define the operation (see Table 4.3) or the phase (see Table 4.4).

Characteristic	Characteristic Value	Details
PPPI_OPERATION	–	The operation for which you want to record quality inspection results

Table 4.3 Process Instruction Type 5 for Operation

Characteristic	Characteristic Value	Details
PPPI_PHASE	–	The phase for which you want to record quality inspection results

Table 4.4 Process Instruction Type 5 for Phase

Process Instruction Type 6: Dynamic Function Calls

Process instruction type 6 makes use of the dynamic function call, wherein you can call any SAP ERP system transaction in the process instruction sheet to facilitate your business process. For example, before you initiate the process of goods issuance against a process order within the process instruction sheet, you want to quickly check the stock overview of the materials. Similarly, while remaining in the process instruction sheet, you also want to perform the phase confirmation of the process order. In the first example, you can call Transaction MMBE for a stock overview. In the second example, you can call or navigate to the phase confirmation Transaction COR6N.

Table 4.5 provides an overview of each process instruction characteristic and its function.

Characteristic	Characteristic Value	Details
PPPI_FUNCTION_NAME	COPF_CALL_TRANSACTION	The system uses this function module to call up a transaction.
PPPI_BUTTON_TEXT	Display Process Order	The system displays the text DISPLAY PROCESS ORDER (or any other text) on the pushbutton in the process instruction sheet.
PPPI_FUNCTION_DURING_DISPLAY	Allowed	This control function determines whether the system should show or forbid display of the desired information.
PPPI_EXPORT_PARAMETER	NEW_SESSION	–
PPPI_INSTRUCTION	–	–
PPPI_EXPORT_PARAMETER	TCODE	–
PPPI_TRANSACTION_CODE	COR3	This is the transaction code that we want to display in the process order.

Table 4.5 Process Instruction of Type 6

In Figure 4.24, the process instruction category C_DURAT enables you to calculate the duration.

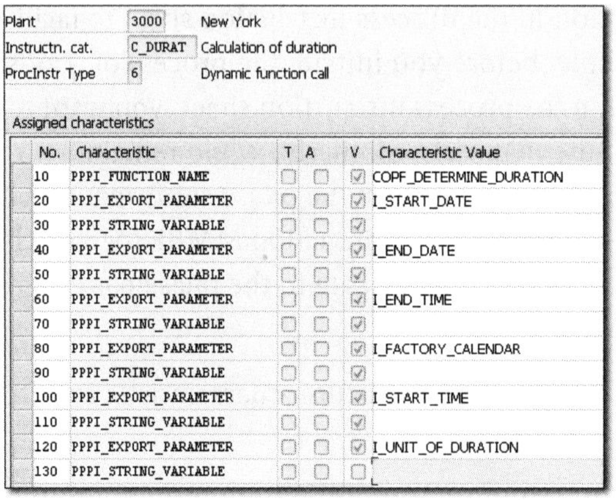

Plant 3000 New York
Instructn. cat. C_DURAT Calculation of duration
ProcInstr Type 6 Dynamic function call

Assigned characteristics

No.	Characteristic	T	A	V	Characteristic Value
10	PPPI_FUNCTION_NAME	☐	☐	☑	COPF_DETERMINE_DURATION
20	PPPI_EXPORT_PARAMETER	☐	☐	☑	I_START_DATE
30	PPPI_STRING_VARIABLE	☐	☐	☑	
40	PPPI_EXPORT_PARAMETER	☐	☐	☑	I_END_DATE
50	PPPI_STRING_VARIABLE	☐	☐	☑	
60	PPPI_EXPORT_PARAMETER	☐	☐	☑	I_END_TIME
70	PPPI_STRING_VARIABLE	☐	☐	☑	
80	PPPI_EXPORT_PARAMETER	☐	☐	☑	I_FACTORY_CALENDAR
90	PPPI_STRING_VARIABLE	☐	☐	☑	
100	PPPI_EXPORT_PARAMETER	☐	☐	☑	I_START_TIME
110	PPPI_STRING_VARIABLE	☐	☐	☑	
120	PPPI_EXPORT_PARAMETER	☐	☐	☑	I_UNIT_OF_DURATION
130	PPPI_STRING_VARIABLE	☐	☐	☐	

Figure 4.24 Process Instruction Type 6: Calculation of Duration

For each export parameter (PPPI_EXPORT_PARAMETER), such as I_START_DATE or I_START_TIME, you can enter the values in PPPI_STRING_VARIABLE, and the system automatically calculates the duration. You may use this duration for calculating machine hours or production duration. The system makes use of function module `COPF_DETERMINE_DURATION`.

Process Instruction Type 7: Sequence Definitions

When you incorporate process instruction type 7 in the process instruction of a master recipe, the system prompts you to enter the phases' relationships with each other. This enables the system to provide the business process owner with the necessary control so that unless a predecessor phase is complete, the successor phase can't start, or its data entry in the process instruction sheet can't be made. When you send the process message of a predecessor phase, only then will the system enable you to enter the details of operation of the next (successor) phase. In this process instruction type, you assign the following process instruction character, along with its value.

Process Instruction Type 0: Universal

Now that you know how each type of process instruction type differs, you can use process instruction type 0 to attend to specific business processes, which the standard process instruction types don't cover. Also, if you plan to implement the process manufacturing cockpit, you need to define all of your process instruction categories of type 0 to enable the system to use it. See Section 4.11 on configuring the process manufacturing cockpit in the SAP ERP system.

4.6.2 Using a Wizard or Process Instruction Assistant

With the exception of process instruction type 0, you can use a wizard to help you define the category of each process instruction type. For example, if you want to create a new process instruction type 4 (calculation) for a new process instruction category, you can use the PI ASSISTANT button to open this step-by-step instruction tool, which eliminates the need to know the complexities involved, such as characteristics PPPI_INPUT_VALUE. Alternatively, look for the icons [Assistant] or [icon] in the master recipe or in the process order, which serve the same purpose.

4.6.3 Creating a Self-Defined Process Instruction Category

Now that we've covered the basics of the process instruction category, let's create a self-defined process instruction category of type 0 as an example. In Chapter 7, we'll show how you make use of this self-defined process instruction category PP10.

To create a new process instruction category, use configuration (Transaction SPRO) menu path, Production Planning for Process Industries • Process Management • Process Instructions • Define Process Instruction Categories (General), or use Transaction O12C.

Figure 4.25 shows the initial screen in which you enter "3000" in the Plant field and choose the Continue icon to open the screen shown in Figure 4.26.

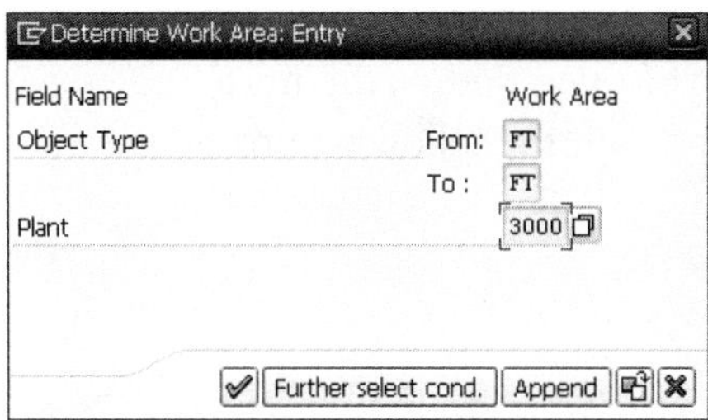

Figure 4.25 Initial Screen for the Process Instruction Category

Figure 4.26 shows the list of available process instruction categories for Plant 3000. Notice the ProcInstr.type column on the far right-hand side of the screen. If you place the cursor on any field of that column and press F4, the system shows a dropdown list of process instruction types. These are the same process instructions that we covered in Section 4.6.1. Choose New Entries and you'll see the screen as shown in Figure 4.26.

On this screen, define the process instruction category PP10 of type 0 (universal) and give a short Description. Next, select the process instruction category, and double-click on the Characteristics/Process instruction category field on the left-hand side of the screen to enter specific process instruction characteristics. For this example, enter the characteristics as shown in Table 4.6.

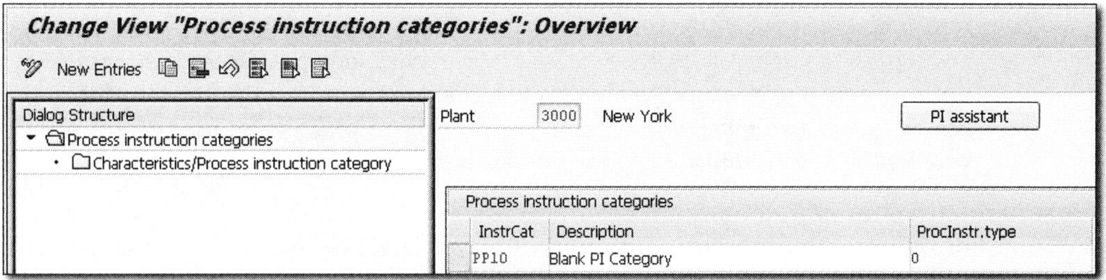

Figure 4.26 Overview Screen of Process Instruction Categories

Characteristic	Characteristic Value
PPPI_DATA_REQUEST_TYPE	Simple Data Request
PPPI_MESSAGE_CATEGORY	PP10
ZPI_CREATION_DATE	
PPPI_OUTPUT_TEXT	P.O. Basic Finish date
PPPI_OUTPUT_CHARACTERISTIC	ZPI_CREATION_DATE
PPPI_PROCESS_ORDER	
PPPI_CONTROL_RECIPE	

Table 4.6 Process Instruction Characteristics in Self-Defined Process Instruction Category PP10

Table 4.6 consolidates the configuration that you've made so far. It incorporates the process message category PP10 as well as the self-defined process instruction characteristic ZPI_CREATION_DATE. See Section 4.6.4 to create a self-defined process instruction characteristic. Also note that the system automatically fills in the characteristic values that are blank in the preceding table.

4.6.4 Creating a Self-Defined Process Instruction Characteristic

Although the SAP ERP system offers a large number of standard process instruction (and process message) characteristics, unique business processes or business needs often require custom-defined or self-defined characteristics.

Example [Ex]

The process operator wants to refer to specific information for a process order or its operations in the process instruction sheet. You notice that this isn't available in the standard offering. Similarly, there are some standard fields for which you want the process operator to send back the requisite information through process messages. The option to self-define the process instruction characteristic offers the flexibility that you can choose a large number of fields in the production process and use them to bring together different information in the process instruction sheet.

For our example, create a simple self-defined process instruction characteristic that automatically fetches the process order end date when the user creates the process order. You can use this characteristic to display this information (process order end date) in the process instruction sheet. Although this example uses automatic value assignment—that is, the system automatically fills in the date field—you can also use it to request that the process operator send the requisite information in the form of a process message. You also use process instruction characteristics to define the RFC destination to which you want to send process instructions to an external control system.

To create a new process instruction characteristic, follow the configuration (Transaction SPRO) menu path, Production Planning for Process Industries • Process Management • Process Instructions • Define Characteristics for Process Instructions • Create Characteristics for Process Instructions, or use Transaction O25C.

Figure 4.27 displays the initial screen to create the process instruction characteristic. Notice the similarities that you find in creating a standard characteristic (Transaction CT04) when compared with creating a process instruction characteristic.

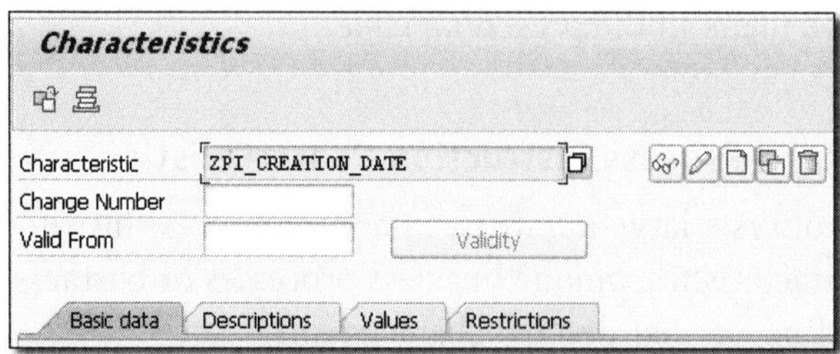

Figure 4.27 Initial Screen to Create a Process Instruction

For this example, create ZPI_CREATION_DATE, and click on the Basic data tab. In this tab, it's important to ensure that the Chars (characteristics) Group is defined as a process instruction characteristic (PPPI_01). This is the same characteristic group that we covered in Section 4.4.7.

Define the DATA TYPE as DATE, and click on the PROC. MGMT. icon. The system displays a pop-up that allows you to incorporate additional process instruction and process messages details. If you place the cursor on the TABLE field and press F4 or click on the dropdown option, the dialog box shown in Figure 4.28 appears. This screen provides you the option to select any of the five tables available. These are PROCESS ORDER (HEADER), OPERATION, PHASE, RESOURCE, and MATERIAL (HEADER). Depending on the table that you select, the system displays the fields associated with it.

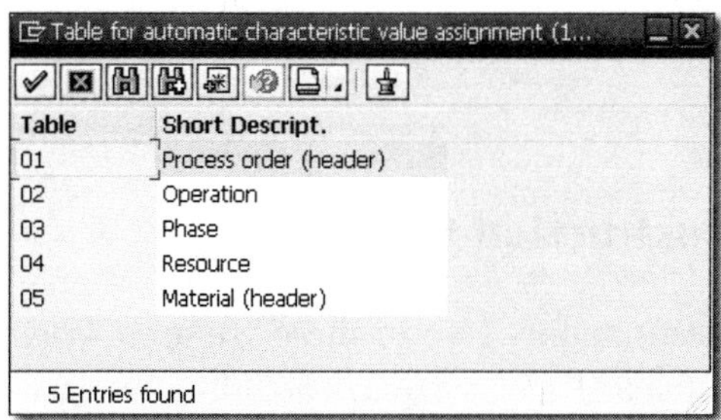

Figure 4.28 Automatic Value Assignment for Process Order Tables

For this example, select 01 PROCESS ORDER (HEADER), and the system brings you back to the previous screen. This time place the cursor in the FIELD field, and again press F4 to see a limited selection of fields that are available in the process order table. Select the FIELD GLTRP, which is the technical name of the process order finish date (see Figure 4.29).

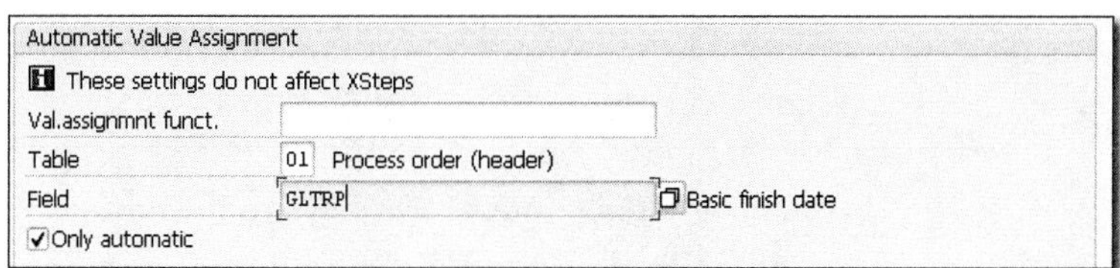

Figure 4.29 Automatic Value Assignment of the Basic Finish Date in the Process Order

Apart from entering the TABLE and the FIELD, select the ONLY AUTOMATIC checkbox to enable the system to automatically fill in the requisite information in the process instruction characteristic and save it. When the user creates the process order and the system assigns start and finish basic dates of the process order, the system fetches the basic finish date of the process order and makes it an integral part of the process instruction sheet.

[+]

Tips & Tricks

If your business process involves the creation of a simple process instruction characteristic, then you can use Transaction CT04 (which doesn't have a Process Management option) to create characteristics and ensure selection of the correct characteristic group.

[»]

Note

See Chapter 15 on the classification system for information on how you can create characteristics and classes.

4.7 Control Recipe/Process Instruction Sheets

There are four types of control recipe destinations. Two of these are used when you want to transfer information within the SAP ERP system. The other two facilitate the transfer of information with external systems. The system creates a control recipe when you release a process order. This creation of the control recipe can be automatic (controlled in order type-dependent parameters, see Section 4.2), or you can generate it manually. Creating the control recipe and then sending the process instruction sheet to its destination enables the process operator to begin recording important plant parameters.

Table 4.7 provides the necessary details on the different types of control recipes.

Control Recipe Destination Type	Destination Location	Output Type
1	SAP ERP system	ABAP list-based process instruction sheet
2	External control system	External control system
3	External control system	External system
4	SAP ERP system	Browser-based process instruction sheet

Table 4.7 Control Recipe Destination Types

If the control recipe destination is of either type 1 or 4, then the system sends the recipes to different destinations within the SAP ERP system. The system creates a control recipe, either user-driven (manually) or automatically, and consolidates the process instructions for a given order and destination in the form of the process instruction sheet. The process instruction sheet contains user-defined details, for example, process notes, process steps, input fields, calculations, options to sign the phases of the process instruction sheet, details to enter goods issuance (consumption), and navigation to the confirmation screen or to record in-process quality results. When the process operator sets the status of the process instruction sheet to COMPLETE, the system sends the data back to the SAP ERP system in the form of process messages. The system then uploads the information from the process messages into the SAP ERP system tables using function modules and BAPIs.

If the control recipe destination is of either type 2 or 3, then the system consolidates the process instructions in the control recipe of the process order and sends them to external PCSs.

An internal SAP ERP system destination can be any logical or virtual destination where the process operator receives the process instruction sheet and fills in the data. For example, internal destinations can be mixing unit, packing unit, blending, ammonia section, urea section, or quality section. For external destinations, that is, external PCSs, these are the RFC destinations of the external system.

You can also integrate a digital signature using an SAP ERP system login and password. The options available for when to use digital signatures are the following:

- Activating or deactivating the process instruction sheet
- Locking or unlocking process instruction sheet
- Changing the status of the process instruction sheet to COMPLETE
- Entering each parameter value in the process instruction sheet

Digital signature options are also available in case a parameter value deviates from predefined limits, as well as for completion of each phase in the process instruction sheet. If you want to automatically generate process instructions to minimize maintaining them in the master recipe, you can do so when creating the control recipe destination. See Section 4.7.2 for information on the scope of generation.

[»] **Note**

See Chapter 17 for more on the digital signature functionality.

In the following sections, we'll explain how to set up a control recipe destination, as well as make the necessary settings for the scope of generation for process instructions.

4.7.1 Create a Control Recipe Destination

To create a new control recipe destination, follow the configuration (Transaction SPRO) menu path, Production Planning for Process Industries • Process Management • Control Recipe/PI Sheets • Control Recipe Destinations • Define and Set Up Control Recipe Destinations, or use Transaction O25C.

In the screen shown in Figure 4.30, create a new control recipe destination 10 of Type 4 (browser-based process instruction sheet). Give a short description as well as the Destination address as PRODUCTION FLOOR. Select the SortMat checkbox to enable the system to display materials in the same order they appear in the master recipe or in the process order. If you place the cursor on the Type field and press F4 or use the dropdown option, you'll see the four different types of control recipe destinations available to choose from (this is the same as in Table 4.7).

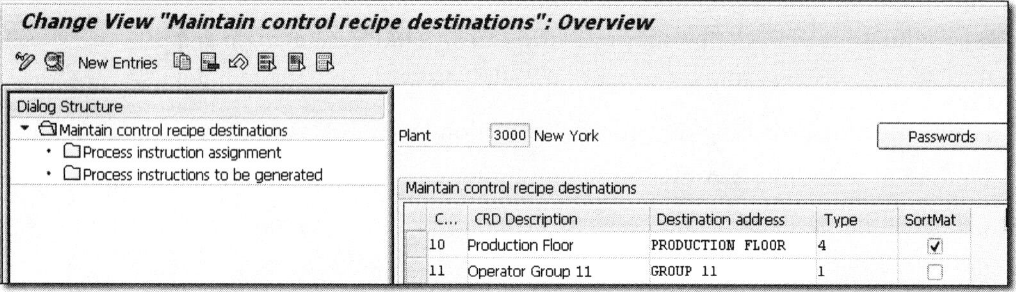

Figure 4.30 Control Recipe Destination: Overview

Figure 4.31 shows the details screen to create the control recipe destination. The options to incorporate digital signatures in some or all of the processes are shown.

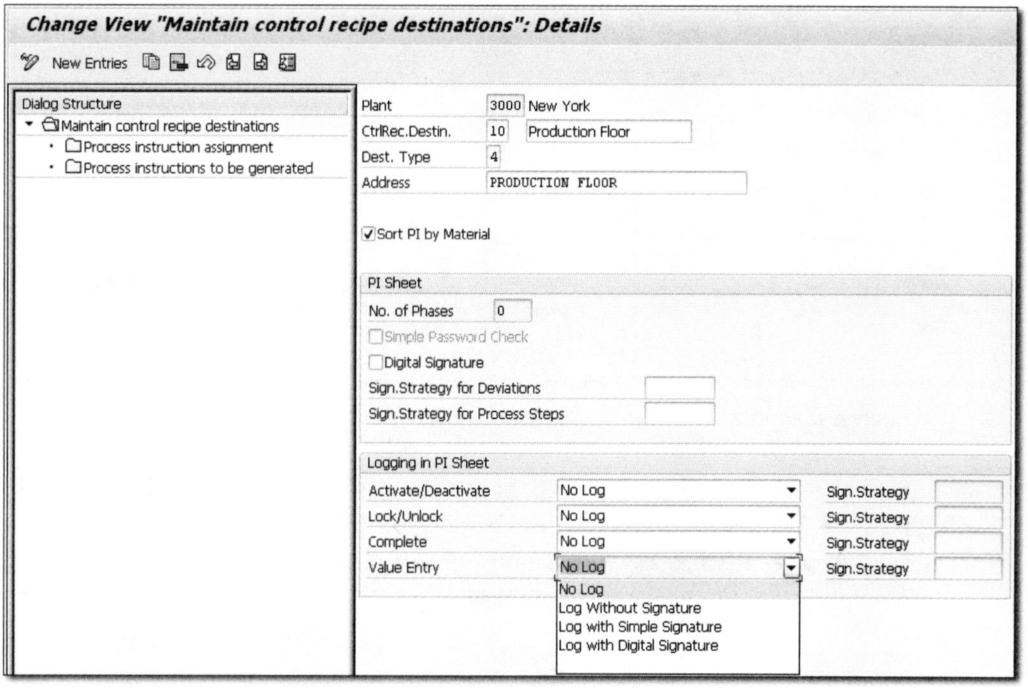

Figure 4.31 Maintaining Control Recipe Destination Detail Screen

4.7.2 Scope of Generation

In some situations, the process operator enters the same data. You can automate the entry of this data so that it's the same every time with the use of scope of generation. This is helpful in the following situations:

- To confirm the quantity produced or activities consumed at the end of each phase
- To enter shift details at the end of every phase or operation
- To ensure that each master recipe created for a specific destination should have standard instructions and notes, which then become visible in the process instruction sheet

Figure 4.32 shows the New Entries: Overview of Added Entries screen that appears during the creation of control recipe destination 10 for Plant 3000 and when you choose Process instructions to be generated on the left-hand side of the figure.

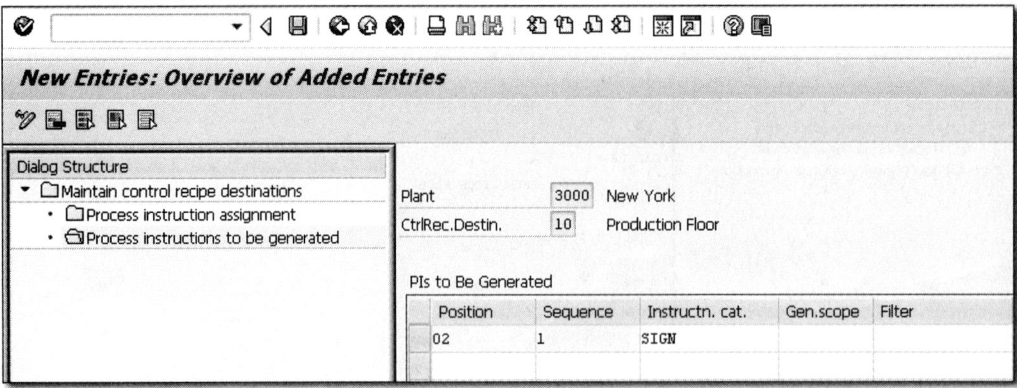

Figure 4.32 Process Instructions to Be Generated during Control Recipe Creation

Place the cursor on the POSITION field and press F4, and the dialog box shown in Figure 4.33 appears. Here you see the position in the control recipe where you want to place the process instruction category. It can be at the start or end of a control recipe, an operation, or a phase.

Position of generated proc. instructions in...

Position in ctrl.rec	Short Descript.
01	At the start of the control recipe
02	At the finish of the control recipe
03	At the start of an operation
04	At the finish of an operation
05	At the start of a phase
06	At the finish of a phase

Figure 4.33 Position of Automatic Generation of Process Instructions

When you select the position, the next step is to assign a process instruction category, place the cursor on the GEN.SCOPE field, and press F4. This opens the screen shown in Figure 4.34, where you see the scope of generation to enable the system to generate process instructions based on the given details. For example, you can select the 01 option for the goods issuance process instruction category. When the system creates a process order and incorporates the materials list based on the BOM, it also simultaneously creates a reservation with item numbers. If you want to record consumption (goods issuance) against the process order using the process instruction sheet, you can select the relevant process instruction category (CH_CONS or CH_CONS2) and define the scope of generation as 01 (For all reservations).

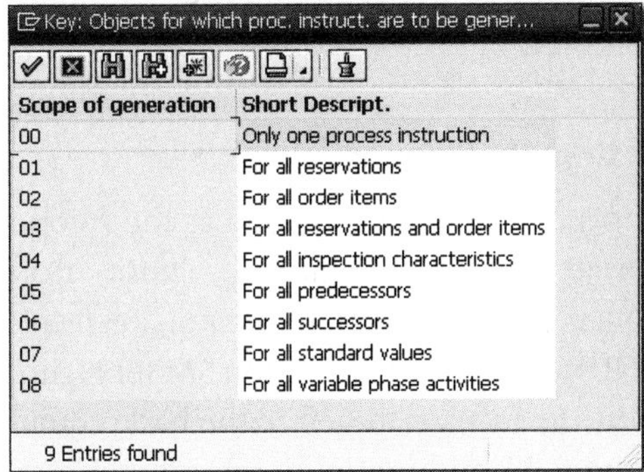
Key: Objects for which proc. instruct. are to be gener...

Scope of generation	Short Descript.
00	Only one process instruction
01	For all reservations
02	For all order items
03	For all reservations and order items
04	For all inspection characteristics
05	For all predecessors
06	For all successors
07	For all standard values
08	For all variable phase activities

9 Entries found

Figure 4.34 Objects for Process Instructions Generation

For this example, assign the instruction category SIGN that the system will generate once and at the end of the process instruction sheet (refer to Figure 4.32).

Tips & Tricks [+]

When you create the process instruction category SIGN, make sure that you assign a characteristic value (e.g., PI Completion Signoff) to the process instruction characteristic PPPI_INPUT_VALUE.

4.8 Background Jobs

Background jobs help to automate processes that otherwise require manual intervention. For example, when you create a process order and generate its control recipe, you have to manually send it to the control recipe destination. You can configure the system to automatically send the control recipe as soon as the system generates it or even periodically. Similarly, sending back process messages to the SAP ERP system can be a manual effort that you can manage as a background (automatic) job.

Apart from sending control recipe and process messages, you can also delete them periodically to save the system's precious resources. For example, you may not process messages older than six months and can delete them.

In this example, we show how to set up a background job for sending a control recipe, and the same steps apply in sending process messages and deleting process messages as background jobs. You need to know the program that is responsible for the background job.

4.8.1 Background Job for Sending Control Recipes

To set up the background job for sending a control recipe as soon as the system creates it, follow the configuration (Transaction SPRO) menu path, PRODUCTION PLANNING FOR PROCESS INDUSTRIES • PROCESS MANAGEMENT • CONTROL RECIPE/PI SHEET • DEFINE BACKGROUND JOB FOR SENDING, or use Transaction SM36. Figure 4.35 shows the initial DEFINE BACKGROUND JOB screen to define the background job. Enter "ZPP_PI_SEND" in the JOB NAME field, and press Enter.

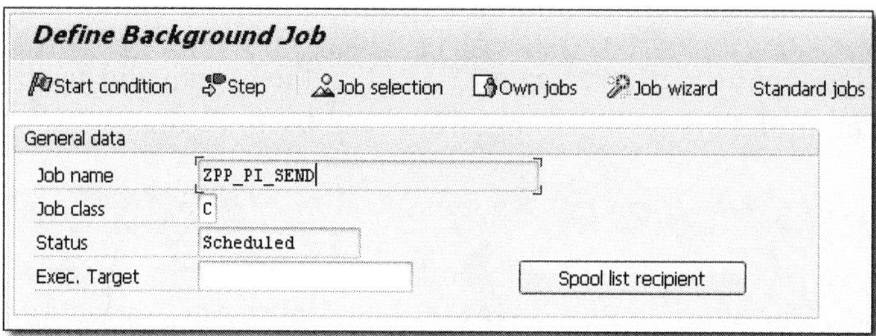

Figure 4.35 Initial Screen for Defining a Background Job to Send a Control Recipe

On the next screen, enter "SAP_NEW_CONTROL_RECIPES" in the EVENT field, which causes the system to send the control recipe as soon as the system creates it. If you then click on the START CONDITION icon, you can define the schedule for sending control recipes to various destinations. You can also assign the program name RCOCB006 to set up the background job.

[+] **Tips & Tricks**

You can use the JOB WIZARD icon to enable the system to guide you in a step-by-step process of setting up a background job, not just for sending a control recipe to a destination but for any background jobs. It's important to know the program name for which you want to define the background job.

4.8.2 Background Job for Sending Process Messages

In Transaction SM36, you use Program RCOCB004 to send newly created process messages as a background job. The system also processes the process messages that have statuses of To be submitted and To be submitted with Warning. This program is for sending process messages within the plant. If you want to send process messages across different plants, then use Program RCOCB002. The corresponding event is SAP_NEW_PROCESS_MESSAGES.

[+]

Tips & Tricks

Whenever you define a background job using Transaction SM36, you also have to assign a variant to it. You can use Transaction SE38 to create a new variant for any background job (and its associated program). On the initial screen, enter the program name, click on the Variants radio button, and then click the Change button. On the next screen, give the variant that you want to create a name, and click on Create.

4.8.3 Background Job for Deleting Process Messages

The background program enables you to delete process messages as a background job. You can use Transaction SM36 to set up the background job.

4.9 Process Management Configuration: At a Glance

You can significantly reduce the complexities and the steps involved in implementing Process Management if you follow a step-by-step, logical approach. For example, creating a process instruction category requires you to assign a process message category; therefore, it makes sense to define a process message category before you create a process instruction category. Similarly, defining a process message category requires both process instructions and a process message destination for field mapping. You need to ensure that these are already available before you create a process message destination.

Table 4.8 provides the checklist that you can follow in the correct order to facilitate Process Management implementation in your company. For each step, cross-references to the section in which we covered the details for that step are provided. We also list the dependency of one step on another.

Step	Step Details	Additional Details	Reference Section	Dependency (if any)
01	Copy standard settings and tools		Section 4.4.7	Independent
02	Release characteristic groups		Section 4.4.7	Independent
03	Create process message characteristics		Section 4.5.1	Independent
04	Create process instruction characteristics		Section 4.6.4	Independent
05	Create process message destination	Assign process message characteristics	Section 4.5.2	Section 4.5.1
06	Create process message categories	Map process instruction characteristics with process message characteristics	Section 4.5.3	Section 4.6.4 and Section 4.5.2
07	Create process instruction categories	Assign process instruction characteristics and also process message category	Section 4.5.2	Section 4.6.4 and Section 4.5.3
08	Create control recipe destination	Assign process instruction categories	Section 4.7.1	Section 4.5.2
09	Define background jobs (if any)		Section 4.8	Independent

Table 4.8 Process Management Implementation Checklist

4.10 Process Management: Configuration and Implementation Roadmap

Regardless of whether you use process instructions or XSteps in Process Management, you need to take a series of logical steps or ask the company's business process owners a series of questions to enable you to successfully implement Process Management.

Following are some of the most relevant points for consideration:

- **Process message characteristics**
  You should create process messages only in those scenarios in which you want the system to report data back to SAP ERP and perform any action. Also, if the process messages are simply sending data back to SAP ERP without the use of Process Management, then you can use Transaction CT04 to create process message characteristics. In fact, with Transaction CT04, you can also use a data upload program to quickly create large numbers of process message characteristics.
- **Process instruction characteristics**
  For complex requirements only, you should create process instruction characteristics using Transaction O25C. Otherwise, using Transaction CT04 suffices. In fact, with Transaction CT04, you can also use a data upload program to quickly create a large number of process instruction characteristics. If several process instruction characteristics have the same properties, then you can create one characteristic (using Transaction CT04) and assign it to all of them. For example, if there are six process parameters that use the same unit of measure and same format (number of decimal places), then you can use this one characteristic and assign it to all six process parameters. This saves you the effort to create six different characteristics for six process parameters.
- **Process instruction categories**
  You should be able to use the available process instruction categories in most business scenarios. If you need to create new ones, then use one of the existing process instruction categories as a reference and modify it. Alternatively, you can also use wizards to facilitate the creation process. We suggest that you create a blank process instruction category of type 0 (universal) because you can then use it in different scenarios such as process parameter input, calculations, or even dynamic function calls.
- **Process message categories**
  Create process message categories only for those business scenarios in which you want the system to report back the data or perform any function. If there is no such need, then the details that you enter in the process instruction sheet are generally sufficient to fetch data from ABAP tables and quickly develop

custom-reports using SAP Query. (See Chapter 19 on reporting, including specifically Section 19.7.3 on creating an SAP Query.)

- **Control recipe destinations**
  When deciding on creating control recipe destinations, evaluate all of the physical and virtual locations from where you need the process operators (or concerned person) to fill in the process instruction sheets and send back the data.

[+]

Tips & Tricks

You can create a virtual control recipe destination named, for example, Coordination. The process operator (or the coordinator) reviews all of the parameters values entered in the process instruction sheet, performs some more tasks such as adding shift notes, signs the process instruction sheet, and sets the status of the process instruction sheet to COMPLETE.

4.11 Process Manufacturing Cockpit

The process manufacturing cockpits find diverse application in areas where the activities performed, actions taken, or results recorded aren't referencing any specific process order. In the cockpit, you can make use of function calls, for example, to call up a transaction or to display a document. You can also create a cockpit if you have a predefined set of activities that you perform on a specific resource or work center. The cockpit also facilitates input or output data entry as well as calculations. The system creates a process instruction sheet with reference to a process order, and the system reflects any changes or updates made to it. In a cockpit, you don't have to give reference to a process order. Here are some examples in which you can create process manufacturing cockpits to meet your business needs:

- You need to record values of five specific process parameters of the plant and report these back to the main control room. The reporting for the parameters doesn't reference a specific process order but is used to ensure that there is regular recording and consistent transmission of this information.
- The production process generates an ancillary product (a by-product of low value), but you don't want to record it with reference to any specific process order, nor do you want to book the cost of the by-product generation to a process order. Further, you want to record its quantity after every three days along

with other details such as date of recording, person who recorded this information, and signature.

- Your company has installed a new production line for which it wants to ensure that the latest technical drawings and manuals are available for process operators to refer to. These drawings are for each work center or work station. In the process manufacturing cockpit, the process operator only has to click on the relevant drawing's icon, and all details become visible.

You can create process manufacturing cockpits for as many applications or activities as needed. Because the cockpit makes use of a process instruction category, you can also incorporate a process message category into it. When you've made entries in the cockpit, the system generates process messages that you can send to the predefined destinations.

To define a process manufacturing cockpit, follow the configuration (Transaction SPRO) menu path, Production Planning for Process Industries • Process Management • Define Process Manufacturing Cockpits (see Figure 4.36).

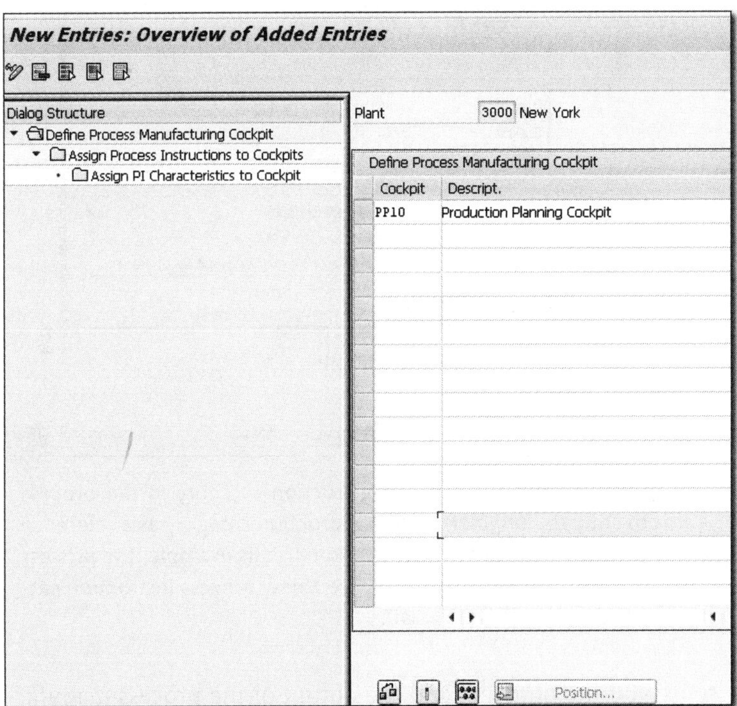

Figure 4.36 Initial Screen for a Process Manufacturing Cockpit

Here you define Cockpit PP10 for Plant 3000. Notice that you can check the newly created cockpit for correctness by choosing the Check icon, and you can also simulate it by choosing the Simulate icon. Finally, you have to ensure that you activate a process manufacturing cockpit before it becomes available or before changes made to the cockpit become effective. To activate the cockpit, choose the Activate icon.

For this example, select Cockpit PP10, and double-click on Assign Process Instructions to Cockpits. This takes you to the screen shown in Figure 4.37, where you enter four different process instruction categories. The system only allows process instruction categories of type 0 (universal). The Line no. column defines the sequence in which the process instruction categories will appear in the process manufacturing cockpit. You can create a new process instruction category with reference to the existing ones and assign it as type 0.

The Ref (reference) checkbox ensures that the system automatically reflects any changes made in the specific process instruction category.

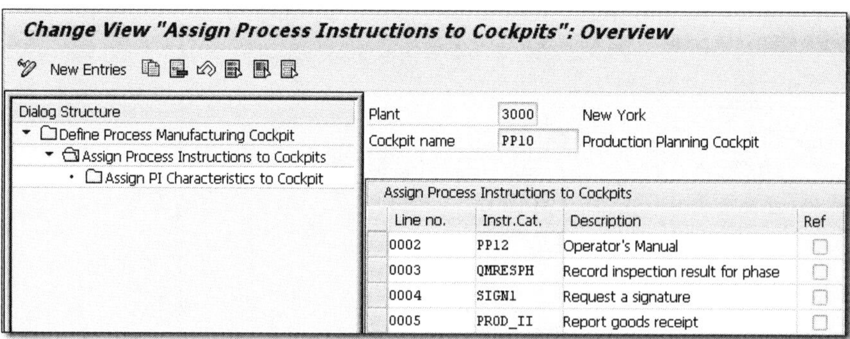

Figure 4.37 Assigning Process Instruction Categories to Cockpits

[+] **Tips & Tricks**

A quick and easy way to use the desired process instruction category in the process manufacturing cockpit is to copy the original process instruction category as a reference and define it as type 0 (universal) during the copy function. In this example, the process instruction category SIGN is used as a reference to create a new process instruction category SIGN1, which is defined as type 0 (universal).

In Chapter 7, we show you the outcome and the output of the process manufacturing cockpit, PP10, which you created in this chapter.

4.12 Summary

Most of the configuration settings that are applicable in discrete manufacturing also apply to process manufacturing. However, the Process Management area in PP for process industries requires detailed coverage. The concepts and fundamentals covered in the chapter equally apply if you plan to implement XSteps in discrete manufacturing or use process manufacturing cockpits.

The next chapter covers the configuration basics of the last main manufacturing type we'll discuss in this book: repetitive manufacturing.

Configuration steps in repetitive manufacturing include creating a repetitive manufacturing profile, setting scheduling parameters for a run schedule quantity (planned orders), and selecting backflush settings for use online or in the background.

5 Configuration Basics of Repetitive Manufacturing

The configuration steps in repetitive manufacturing (REM) are relatively straightforward, without many complexities when compared to discrete or process manufacturing types. In fact, the very purpose of REM is to enable lean manufacturing in actual business scenarios with correspondingly fewer entries in the SAP ERP system. REM may not be able to manage complex manufacturing processes, but it can be used to bring about significant process optimization with a decreased data entry workload and improved system performance.

This chapter begins by explaining how to set up the REM profile using the REM assistant. The assistant guides the user with a step-by-step approach to ensure that the correct settings are made during the REM profile creation stage. The assistant also provides the detailed function of each step of configuration and gives recommendations where necessary. You can even choose the desired production method such as make-to-stock (MTS) or make-to-order (MTO). We deliberately took the long route to REM profile creation by using the assistant because we wanted to focus on explaining each and every available option in detail and with business examples. At the end of REM profile creation by using the assistant, we explain the "normal" or the prevalent method of creating an REM profile with a single screen (and multiple tabs). Here, we also explain the options that aren't available in the REM profile assistant. If you're configuring an REM profile for the first time, we suggest that you use the REM profile assistant. If you're already familiar with the REM profile and the functionalities it offers, then you can directly proceed to using the single-screen REM profile creation option.

We cover scheduling REM planned orders and running schedule quantity and then proceed to discuss the layout and display settings available on various screens in REM processing. You'll get to see the effects of the REM configuration in Chapter 8, in which we cover the production planning for REM in detail. Finally, we also cover the settings available to optimize day-to-day business transactions, such as backflushing.

[»] **Note**

Refer to the appendix for features comparison among discrete manufacturing, process manufacturing, and repetitive manufacturing, as well as how to decide which manufacturing type is most relevant for a given industry.

Let's get started.

5.1 Repetitive Manufacturing Profile

The REM profile enables you to control several important functions that form the basis of regular or periodic data recording in the system. This includes the options to post activities online or at a later date, automatic goods movements of all or some of the materials, and stock and/or automatic batch determination, among others. To create an REM profile using the REM profile assistant, follow the SAP configuration (Transaction SPRO) menu path, Logistics • Production • Repetitive Manufacturing • Control • Create Repetitive Manufacturing Profile Using Assistant, or use Transaction OSPT.

In the following sections, we'll explain the steps involved in setting up an REM profile.

5.1.1 Repetitive Manufacturing Production Type

While creating an REM profile, you can decide whether you want to use the profile for the MTO or MTS production method.

[»] **Note**

Chapter 10 covers MTO and MTS production methods.

Depending on the option you select, the system correspondingly displays the relevant screens and options to choose from. If you choose MTO production, you create a profile suited to REM, which references the sales order. If you choose MTS production, you create a profile suited to REM that doesn't reference the sales order.

You can also use an REM profile template to copy previously created and available REM profiles to create a new REM profile. This step copies all of the settings of the REM template profile into the new REM profile, and you can then make the desired changes where needed.

Tips & Tricks [+]

Choose the MEANING icon available on each screen of the REM profile creation to get information on the underlying concept of that functionality or area.

Choose the RECOMMENDATION icon available in each screen of the REM profile creation to see system advice on recommended settings.

For this example, you'll create an REM profile for an MTS production method. Figure 5.1 shows the initial screen to create an REM profile using the assistant. For this example, select the MAKE-TO-STOCK - REM radio button, and then choose NEXT.

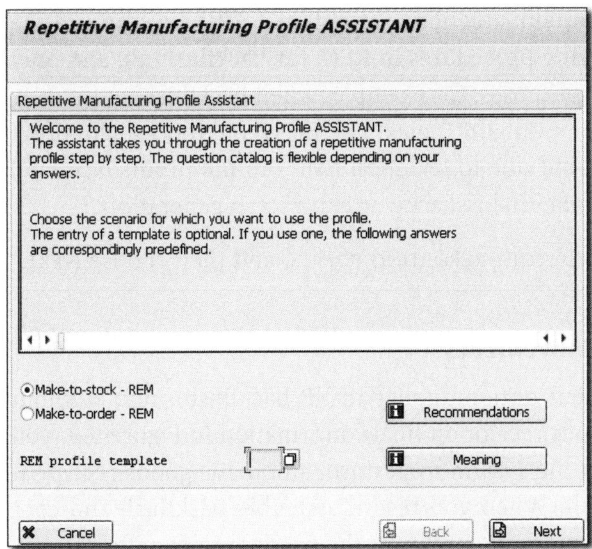

Figure 5.1 Initial Screen of the Repetitive Manufacturing Profile Assistant

5.1.2 Reporting Points

A *reporting point* (RP) enables the system to record consumption or other details, such as identifying the work in process (WIP) and better stock visibility in Inventory Management, which are specific to actual information in time and equally helpful in reporting. As shown in Figure 5.2, you can select whether you want the system to include no RPs or make the RP mandatory or optional. The RP serves the same purpose as a milestone does in routing for discrete manufacturing. If you select the MANDATORY REPORTING POINTS option, then you have to define a RP at the time of assembly or component backflush.

Figure 5.2 Reporting Points

You can use RPs in one of the following situations:

- If you use *mandatory* RPs, then you must backflush at every RP. It makes sense to backflush at RPs if the goods issue is to be posted as close as possible to the actual issuance of the materials or if WIP is to be calculated.
- If you use *optional* RPs, you can carry out the normal final backflush, and then use RP backflush only in certain situations. At optional RPs, you can use the standard goods receipt posting procedures in REM for backflushing, and only backflush at RPs for special purposes, for example, when backflushing scrap or when calculating WIP. The optional RP makes sense when you usually don't have a need for a RP at all, but due to technical issues in the production line, you want to evaluate extra material issuance or extra scrap generation.

For this example, choose MANDATORY REPORTING POINTS, and then choose NEXT.

5.1.3 Automatic Goods Movements

The automatic goods movement option applies to RP backflush only. Without RPs, you can only post the goods receipt on final confirmation In Figure 5.3, you can choose whether you want the system to perform automatic goods receipt at the time of assembly backflush. When you record assembly backflush and confirm the yield, the system automatically performs the goods receipt at the last RP.

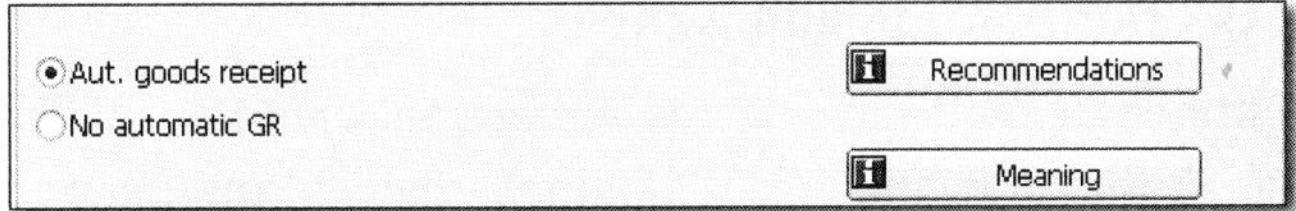

Figure 5.3 Automatic Goods Receipt

When using optional RPs, you choose the setting in automatic goods receipt because you usually backflush the components along with the goods receipt posting.

When using mandatory RPs, you can only backflush at RPs. The normal goods receipt posting isn't available here. Therefore, for mandatory RPs, the recommendation is to define the last workplace on your production line as a RP with the automatic goods receipt setting. If you set the automatic goods receipt option, you have to post activities and goods issues manually for operations that lie after the last RP. This is normally an exception to the preceding rule/recommendation

For this example, choose AUT. GOODS RECEIPT, and then choose NEXT.

[«]

Note

While configuring the REM profile using the assistant, your system displays either the screen shown in the preceding figure or the screen shown in the following figure. This is due to different SAP ERP versions and the options available.

In Figure 5.4, you can choose whether you want the system to simultaneously post goods receipt and goods issuance in the same transaction of assembly confirmation. In REM, the final confirmation also encompasses goods receipt. You can choose whether you also want to post goods issuance during the confirmation process. If you choose the POST GR ONLY option, you can post the goods issues for the components collectively, for example, using the goods issue transaction in REM.

For this example, choose POST GR AND GI in the assembly confirmation.

Figure 5.4 Automatic Goods Receipt and Goods Issuance

5.1.4 Reporting Points Confirmation and Kanban

Depending on your system release, you can select whether you want to use RPs with or without Kanban as shown in Figure 5.5. RP Kanban enables you to monitor the production process at the intermediate level by using Kanban for processing products in individual operations. The RPs record confirmation of the completion of operations. For this example, choose RP CONFIRMATION W/O KANBAN, and then choose NEXT.

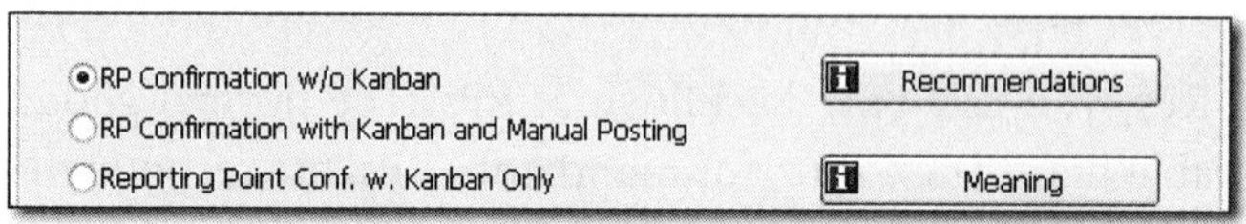

Figure 5.5 Reporting Point Confirmation for Kanban

5.1.5 Activities Posting

In the next screen (Figure 5.6), you can choose whether you want the system to also post the activities to a product cost collector at the time of assembly backflush. Activities may be machine hours, labor hours, or processing time, and they are defined in rate routing. If you follow the lean manufacturing approach, in which even the data entry effort is minimal, activities posting involves additional use of the system's resources such as material calculation and price updates. Activities posting should only be considered if it's significant when compared with cost. We suggest that you engage a Product Cost Controlling (CO-PC) resource to help with the decision-making process on activities positing.

Figure 5.6 Activities Posting

If you instruct the system to post production activities, it calculates these activities either on the basis of details from the standard cost estimate or from preliminary costing in the cost collector for material. Additionally, the system posts the activities during the confirmation process. For this example, choose POST ACTIVITIES, and then choose NEXT.

5.1.6 Separated Backflush

As shown in Figure 5.7, you can choose whether you want the system to perform separated backflushes or not. This option is primarily used to improve a system's performance, especially if it gets slow due to a large number of confirmations. With separated backflush, you can uncouple certain aspects of backflushing during the confirmation process by scheduling them as a separate background job offline.

Figure 5.7 Backflush Posting

You use the separated backflush in the following cases, for example:

- To improve system performance
- If you're dealing with large bills of materials (BOMs)
- For sales orders with many schedule lines for small quantities

For this example, choose SEPARATED BACKFLUSH, and then choose NEXT.

5.1.7 Process Control

In Figure 5.8, you can choose when goods issue and calculation of costs are performed and whether goods issue and calculation of costs are aggregated during confirmation. While you have the option to let the system perform several steps/functions together, you can use the PROCESS CONTROL option to customize the process according to your business needs. You can click on the CUSTOMIZING icon on this screen to make the necessary changes. However, we cover the customization of process controls later in Section 5.5.

Figure 5.8 Process Control in Backflushing

It's important to note that you'll only get this option if you chose SEPARATED BACKFLUSH in the previous step. In that case, you have to define the process control for this separated backflush, and there are four options available to maintain.

For this example, select the PROCESS CONTROL SAP1, and choose NEXT.

5.1.8 Firming Planned Orders

It makes sense to firm the planned orders when you create them in the REM planning table because a firmed planned order won't change during the MRP run. You also have the options to not firm the planned order at the time of its creation and to firm planned orders in the planning time fence only.

For this example, choose FIRM PLANNED ORDERS IN THE PLANNING TIME FENCE, and then choose NEXT.

5.1.9 Automatic Stock Determination

In the screen shown in Figure 5.9, you can choose to take advantage of the system performing the automatic stock determination at the time of component or assembly backflush. The prerequisite is that you've already set up the automatic stock determination procedure in the system. Stock determination enables the system to suggest available stock for consumption based on the defined criteria.

For this example, choose DO NOT USE STOCK DETERMINATION, and then choose NEXT.

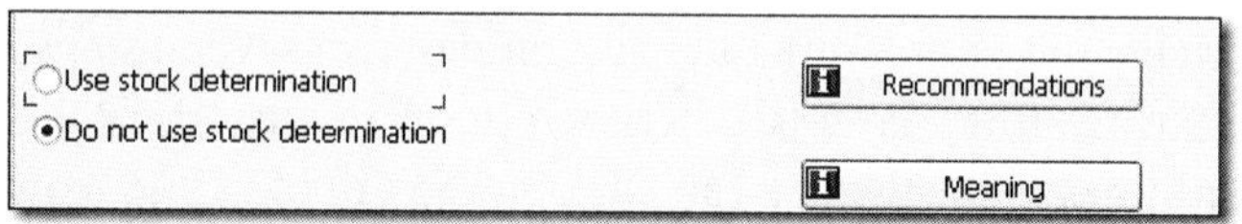

Figure 5.9 Stock Determination

5.1.10 Batch Determination Procedure

Similar to the previous automatic stock determination procedure, you can also choose whether the system should perform the batch determination procedure in REM. The material, in this case, must be batch-managed to use this functionality. Batch determination enables the system to suggest available batches for consumption based on the defined criteria. In Figure 5.10, choose No separate batch search proc., and then choose Next.

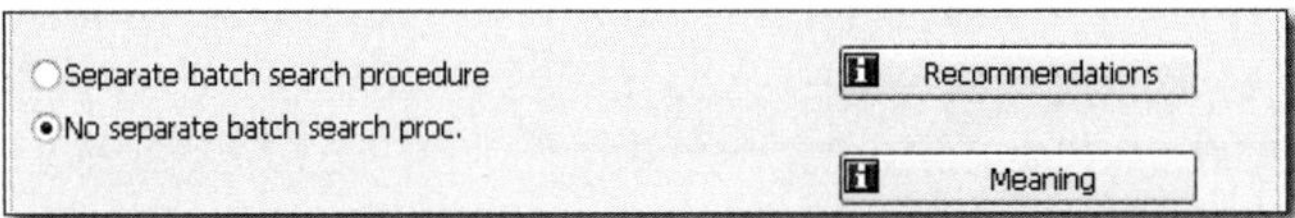

Figure 5.10 Batch Determination

5.1.11 Reduction in Planned Order Quantities

In the screen shown in Figure 5.11, you can choose whether you want the system to reduce the production quantities in the planned order or in the run schedule quantity (RSQ) when you perform assembly confirmation of a material. You can also choose whether you want the system to reduce the planned order quantity only if it pertains to a specific production version. As each production version reflects a production line, this option enables you to reduce planned order quantities as soon as the goods on the production line (in direct relation to the production version) are produced.

For this example, choose All, which causes the system to reduce planned order quantities in all cases. Choose Next.

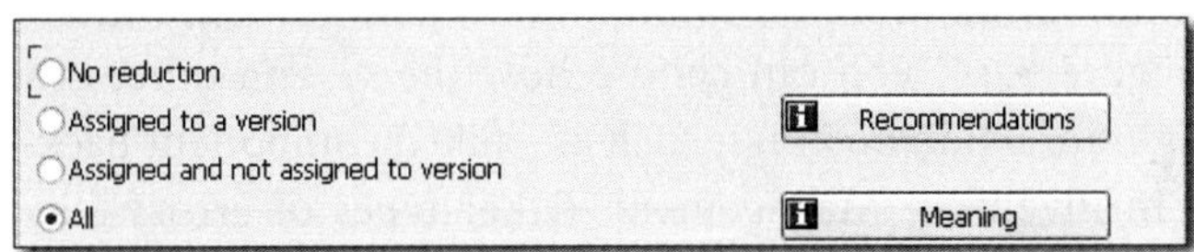

Figure 5.11 Reduction in Planned Order Quantities

5.1.12 Reduction Period

In the screen shown in Figure 5.12, you can choose the period in days during which the system should reduce the planned order quantities. The system considers the reduction period from today and expands to consider the defined dates in the future. For example, if you're consistently faced with overproduction and want to avoid additional production for the defined days in the future, then you can enter the reduction period here. Also important to note is that the system reduces the planned order quantities dates of the past in any case and accords them higher priority than the current date or the dates in future.

For this example, enter "3" in the REDUCTION PERIOD field, and then choose NEXT.

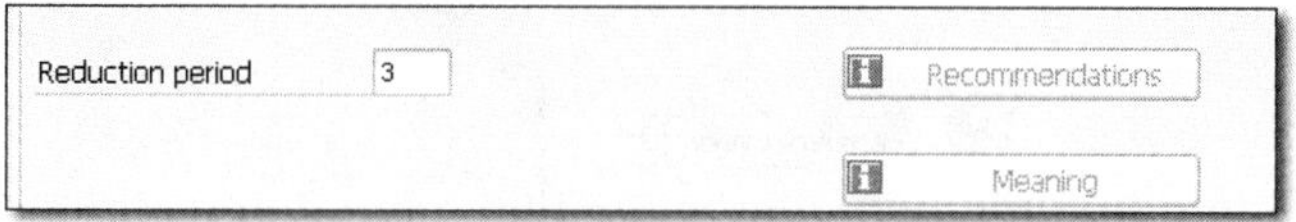

Figure 5.12 Reduction Period for Planned Orders

5.1.13 Create New Planned Orders on Goods Receipts Reversals

In the screen shown in Figure 5.13, you can choose whether the system should automatically create planned orders for reversed quantities. If you choose not to, then the system automatically creates planned orders during the MRP run if necessary. For this example, choose NO PLANNED ORDER CREATION, and then choose NEXT.

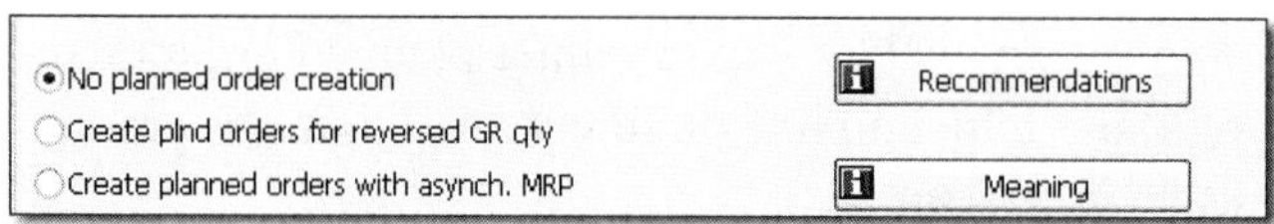

Figure 5.13 Planned Order Creation on Goods Receipt Reversals

5.1.14 Online Error Correction

In the screen shown in Figure 5.14, you can choose how the system processes errors encountered during transaction postings, such as a failed component backflush due to a mismatch in Inventory Management. Other types of errors may include an insufficient quantity of a component's stock in a component backflush, a missing storage location for goods issuance or goods receipt, and so on. If you opt for optional or no online correction, you can use a separate transaction to attend errors separately, provided that reprocessing records are created (see the next step). For this example, choose OPTIONAL ONLINE CORRECTION, and then choose NEXT.

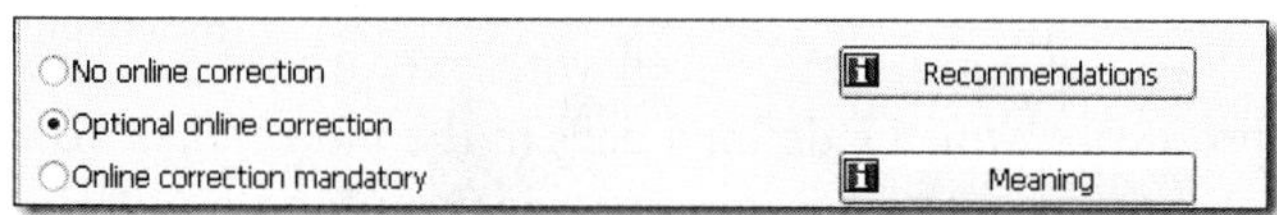

Figure 5.14 Online Error Correction

5.1.15 Reprocessing Errors Log Maintenance

During automatic goods movement processing such as backflushing, sometimes the processing isn't successful. This may be due to insufficient stock, an incorrect storage location, or missing or nonupdated rates for activity types. All such unsuccessful records end up in errors log maintenance for reprocessing and to note that they haven't been resolved yet.

In the screen shown in Figure 5.15, you can choose how the system should manage all of the reprocessing records of all items that caused the errors. The available options are that the system won't maintain any reprocessing record, the system will maintain collective reprocessing records, or the system will maintain individual and collective reprocessing records of errors. For this example, choose INDIV.AND CUMUL. REPROCESSING RECORDS, and then choose NEXT.

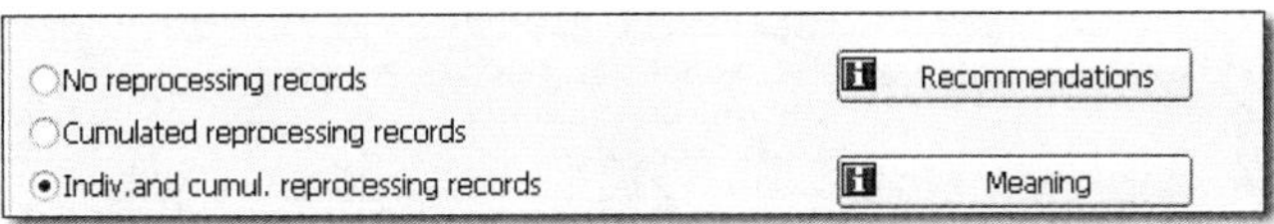

Figure 5.15 Reprocessing Incorrect Records

5.1.16 Movement Types for Stock Postings

On the next screen (Figure 5.16), the system provides standard movement types that the system uses to perform various Inventory Management transactions. For example, for goods issuance against an REM planned order, the movement type is 261, and its reversal is reflected in movement type 262.

Goods issue	261	Goods issue/reversal	262	
Goods receipt	131	Goods receipt/reversal	132	Recommendations
Scrap	551	Scrap/reversal	552	
By-product	531	By-product/reversal	532	Meaning

Figure 5.16 Movement Types for Make-to-Stock Production

If you have any customized movement type to record an inventory transaction, you can enter it on this screen. The prerequisite for using a customized movement type is that you've maintained it in the system previously and then assign it on this screen.

Choose NEXT.

5.1.17 Naming the Repetitive Manufacturing Profile

You'll now see the last screen of the REM profile creation process using the assistant. In this screen, enter the REM PROFILE "PP15", and give it a short description. You can go back and make changes to any of the settings you've made so far.

5.1.18 Summary of Repetitive Manufacturing Profile Settings

Figure 5.17 shows the overview/summary screen of the REM profile PP15. Choose CONTINUE, and then save the REM profile PP15.

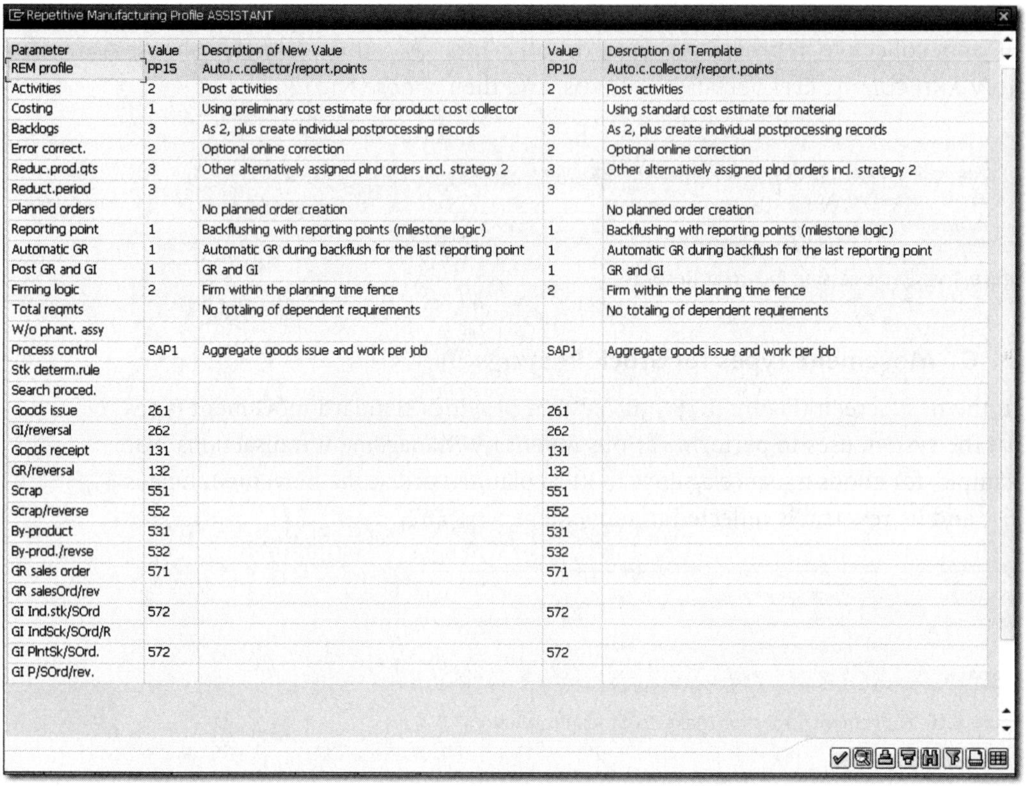

Repetitive Manufacturing Profile ASSISTANT

Parameter	Value	Description of New Value	Value	Desrciption of Template
REM profile	PP15	Auto.c.collector/report.points	PP10	Auto.c.collector/report.points
Activities	2	Post activities	2	Post activities
Costing	1	Using preliminary cost estimate for product cost collector		Using standard cost estimate for material
Backlogs	3	As 2, plus create individual postprocessing records	3	As 2, plus create individual postprocessing records
Error correct.	2	Optional online correction	2	Optional online correction
Reduc.prod.qts	3	Other alternatively assigned plnd orders incl. strategy 2	3	Other alternatively assigned plnd orders incl. strategy 2
Reduct.period	3		3	
Planned orders		No planned order creation		No planned order creation
Reporting point	1	Backflushing with reporting points (milestone logic)	1	Backflushing with reporting points (milestone logic)
Automatic GR	1	Automatic GR during backflush for the last reporting point	1	Automatic GR during backflush for the last reporting point
Post GR and GI	1	GR and GI	1	GR and GI
Firming logic	2	Firm within the planning time fence	2	Firm within the planning time fence
Total reqmts		No totaling of dependent requirements		No totaling of dependent requirements
W/o phant. assy				
Process control	SAP1	Aggregate goods issue and work per job	SAP1	Aggregate goods issue and work per job
Stk determ.rule				
Search proced.				
Goods issue	261		261	
GI/reversal	262		262	
Goods receipt	131		131	
GR/reversal	132		132	
Scrap	551		551	
Scrap/reverse	552		552	
By-product	531		531	
By-prod./revse	532		532	
GR sales order	571		571	
GR salesOrd/rev				
GI Ind.stk/SOrd	572		572	
GI IndSck/SOrd/R				
GI PlntSk/SOrd.	572		572	
GI P/SOrd/rev.				

Figure 5.17 Overview of the REM Profile after All Settings Are Made

After you've created the REM profile, you can't use the same transaction to make changes to it because it's only for the creation of the REM profile. At the same time, when you become familiar with the REM profile, you can proceed to directly create it without using the REM profile assistant. This saves time and effort because all of the selection options are available on one screen and in different tabs.

To make changes or to update any field in the already-created REM profile, follow the SAP configuration (Transaction SPRO) menu path, PRODUCTION • REPETITIVE MANUFACTURING • CONTROL • DEFINE REPETITIVE MANUFACTURING PROFILES, or use Transaction OSP2. Here, you can make changes to REM PROFILE PP15. For this example, choose BACKFLUSH USING STANDARD COST ESTIMATE FOR MATERIAL, and then save the REM profile (see Figure 5.18).

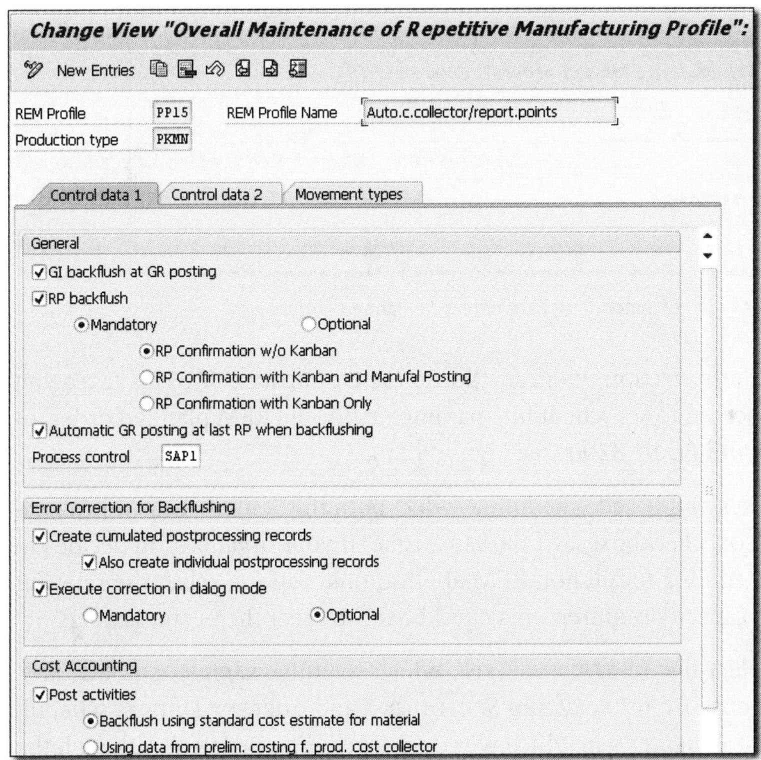

Figure 5.18 Overall Maintenance of the REM Profile

5.2 Scheduling Planned Orders

The next basic configuration topic we'll discuss is scheduling planned orders. As in the discrete manufacturing or process manufacturing types, you can take advantage of the available scheduling options. You can set scheduling parameters for planned orders in REM by following the configuration (Transaction SPRO) menu path, PRODUCTION • REPETITIVE MANUFACTURING • PLANNING • DEFINE SCHEDULING PARAMETERS FOR PLANNED ORDERS. Figure 5.19 shows the initial screen for maintaining the scheduling parameters for a planned order. Because the order type for the REM planned order is PE, select PLANT 3200 and ORDER TYPE PE, and double-click the PLANT field (3200). It's important to note here that the ORDER TYPE PE is identical to the RSQ.

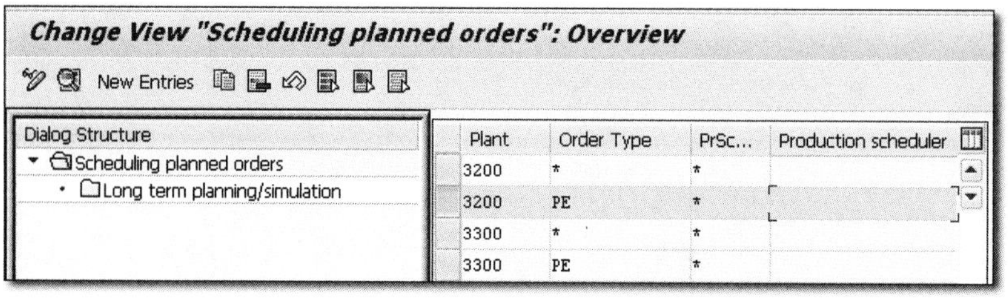

Change View "Scheduling planned orders": Overview

New Entries

Dialog Structure
- Scheduling planned orders
  - Long term planning/simulation

Plant	Order Type	PrSc...	Production scheduler
3200	*	*	
3200	PE	*	
3300	*	*	
3300	PE	*	

Figure 5.19 Initial Screen of Scheduling Parameters of Planned Orders

Alternatively, after selection, you can choose the DETAIL icon. This takes you to the detailed screen to set scheduling parameters for the REM planned order, of ORDER TYPE PE and PLANT 3200 (see Figure 5.20).

Because REM uses rate-based scheduling, select both the SCHEDULING and GENERATE CAPACITY REQS. checkboxes. At the same time, if your organization decides to use detailed scheduling for all non-REM applications, you can select the relevant scheduling and capacity requirements checkboxes to meet the business need.

You can also select the SCHEDULING TYPE, which for this example is FORWARDS. You can also choose from BACKWARD SCHEDULING and CURRENT DATE as scheduling types. Refer to Chapter 4 in which we cover scheduling in detail. Although the details covered in Chapter 4 are applicable to scheduling production orders, the same equally applies to planned orders.

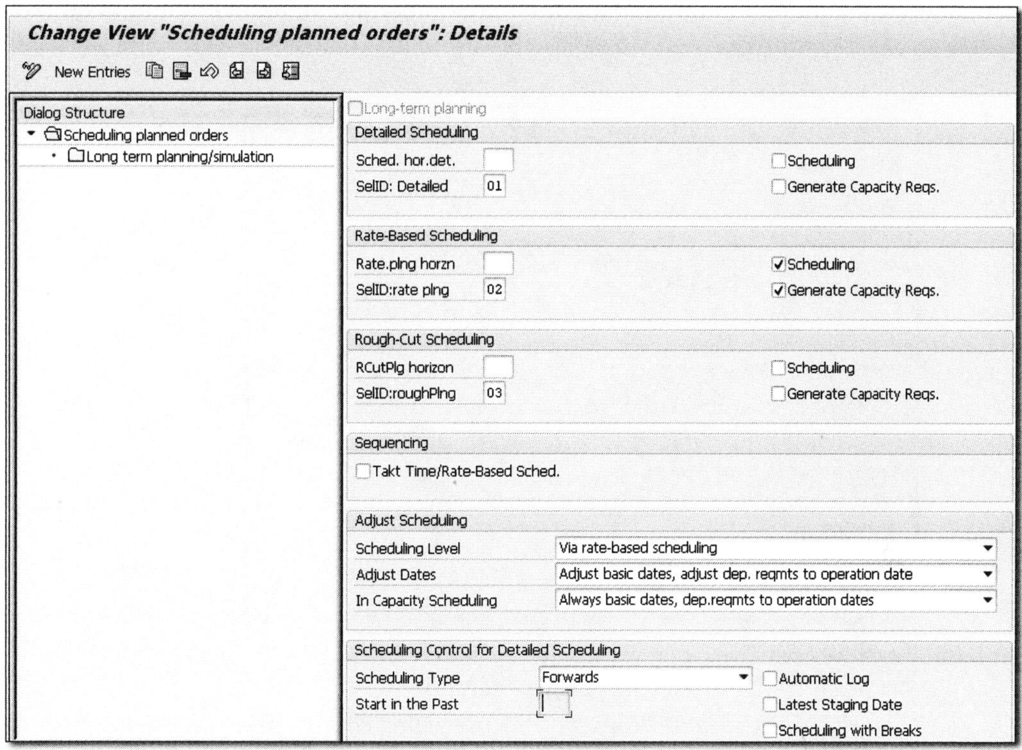

Figure 5.20 Detailed Screen of Scheduling Parameters for Planned Orders

5.3 Display

You have a few different options to control how different settings will appear in your system. The follow sections cover two display options.

5.3.1 Entry Parameters for a Planning Table

You can choose how the display entry parameters of the REM planning table should look or how to present the requisite information. The settings you make in this section will eventually be available on the initial parameters selection screen of the REM planning table (Transaction MF50). To maintain or change the entry parameters of the planning table, follow configuration (Transaction SPRO) menu path PRODUCTION • REPETITIVE MANUFACTURING • PLANNING TABLE • MAINTAIN ENTRY PARAMETERS (see Figure 5.21).

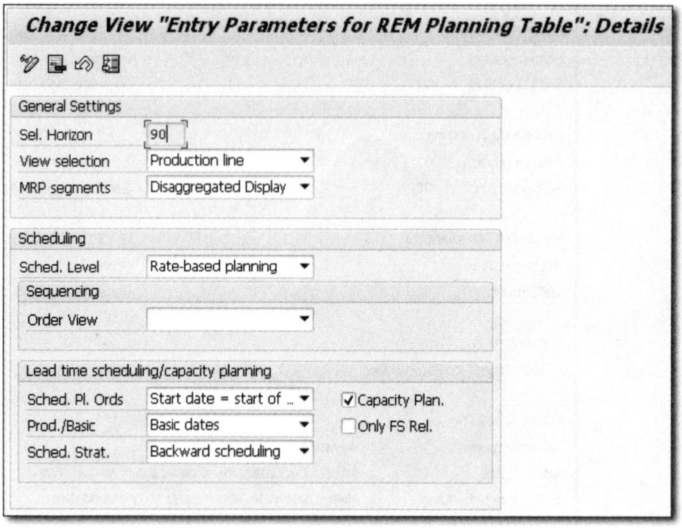

Figure 5.21 Entry Parameters of the REM Planning Table

You'll be able to see the impact of the preceding configuration on the parameters selection screen of the planning table (Transaction MF50), which we cover in Chapter 8.

5.3.2 Maintain Rows Selection

You can control the display of information within the planning table, including receipts, requirements, and stock situations. To configure rows selection in the planning table, follow the configuration (Transaction SPRO) menu path PRODUCTION • REPETITIVE MANUFACTURING • PLANNING TABLE • MAINTAIN ROWS SELECTION. For this example, select all of the checkboxes (see Figure 5.22); however, you should select only the checkboxes relevant to your business needs to keep things lean when dealing with your company's unique situation.

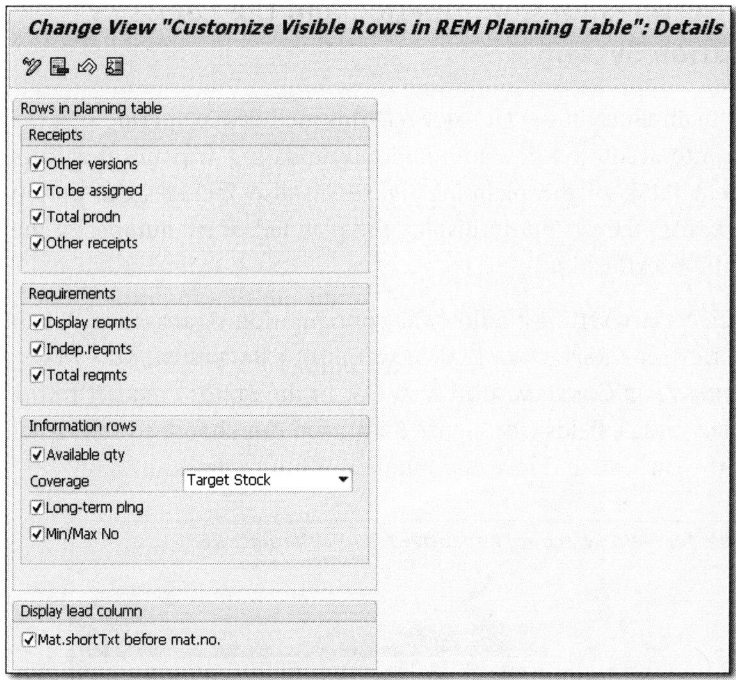

Figure 5.22 Visible Rows in the REM Planning Table

5.4 Material Staging

A *pull list* is basically a movement of stock (stock transfer) from storage location to storage location that uses inventory movement type 311. It's also known as *material staging*. For the necessary settings, follow the configuration (Transaction SPRO) menu path, PRODUCTION • REPETITIVE MANUFACTURING • MATERIAL STAGING • DEFINE CONTROL DATA FOR PULL LIST (see Figure 5.23).

Change View "Pull List Control": Overview

New Entries

Plnt	MoveType	MTO	StDR	Search Pr.
3200	311		PP01	
5000	311		PP01	

Figure 5.23 Pull List Control

5.5 Global Settings for Confirmation and the Logistics Information System

Some settings are maintained at the global level. This includes the period that the system is to take into account when automatically updating statistics (Logistics Information System [LIS]) of the planning figures. It also includes the option allowing you to require the system to display the planned order number of the RSQ number during backflushing.

To configure the necessary settings, follow the configuration (Transaction SPRO) menu path, PRODUCTION • REPETITIVE MANUFACTURING • BACKFLUSHING • MAINTAIN GLOBAL SETTINGS FOR CONFIRMATION AND LIS. In the START: CURRENT DATE+ and FINISH: CURRENT DATE+ fields (see Figure 5.24), you can choose to start from the current date (the earliest) and give any number of future dates.

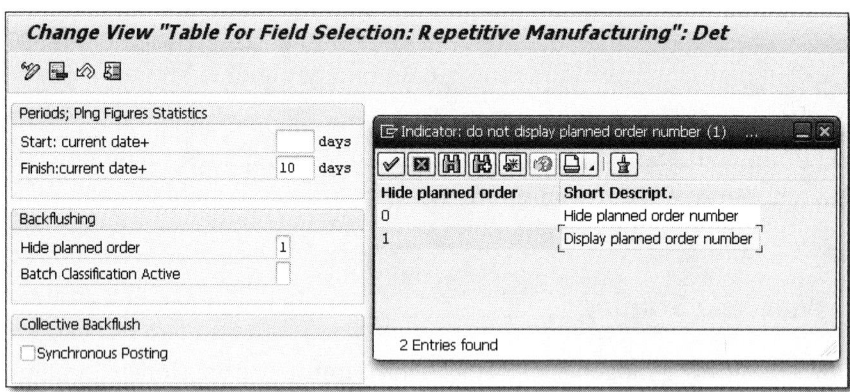

Figure 5.24 Field Selection Table in REM

When you maintain the process control for confirmation, you can control when the system should update the transactions, such as backflush or activities costs calculation. The system can update the information immediately and online by using an update program or later in a background program. It's therefore important to ensure that you select the process control when you're creating the REM profile.

In Figure 5.25, if you place the curser in the BACKFLUSH field and press [F4], the dropdown on the right side of the screen appears. From the dropdown, you can choose one of the three backflushing options available. To set the relevant parameters, use configuration (Transaction SPRO) menu path, PRODUCTION • REPETITIVE MANUFACTURING • BACKFLUSHING • SEPARATE BACKFLUSHING PROCESSES • DEFINE CONFIRMATION PROCESSES.

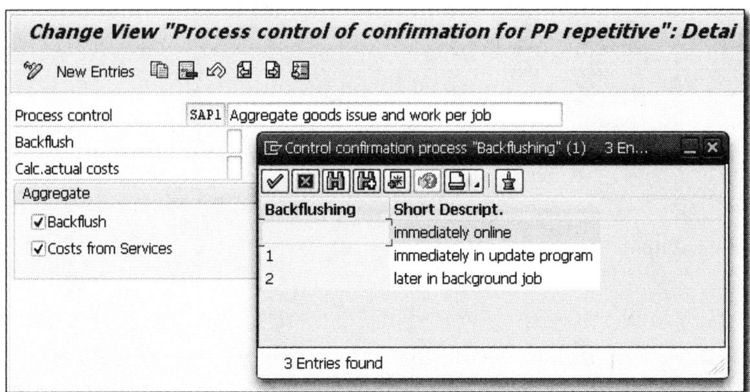

Figure 5.25 Process Control Confirmation Parameters

5.6 Operational Methods Sheet

The operational methods sheet (OMS) supports business tasks in the entire REM cycle. An OMS shows the production details that a plant operator needs to produce the goods. You can make several layout and configuration settings to customize the display layout of an OMS. For example, you can specifically select the desired fields from the work center, operation, component, production resources/tools (PRT), or Document Management System (DMS) that you want to appear in the OMS. You can separately customize the header and table (item) details.

To create an OMS, follow the configuration (Transaction SPRO) menu path, PRODUCTION • BASIC DATA • LINE DESIGN • OPERATIONAL METHOD SHEET, or use Transaction OLDPS. In the TABLE ASSIGNMENTS area shown in Figure 5.26, you can select COMPONENT, MATERIAL, PRT, or OPERATION, for example, from the TABLE ID column. Next, you can use the ASSIGN FIELDS folder available on the left-hand side of the screen to choose the field that you want to display in the OMS.

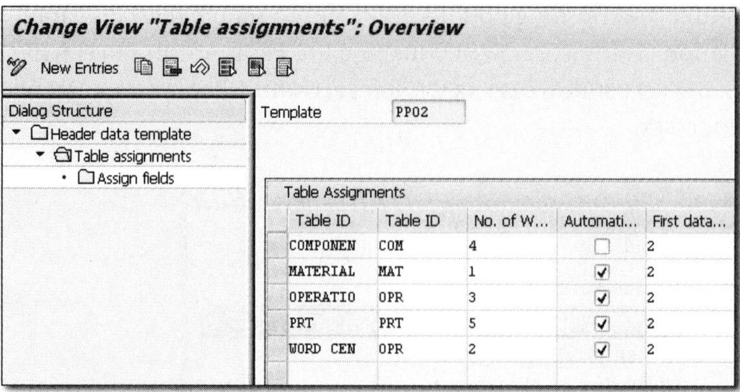

Figure 5.26 Operational Method Sheet

In Chapter 8, we show how the settings you made here enable you to display (and print) the OMS in REM. You have to ensure that integration with Microsoft Word is in place as well. The prerequisite is to have at least Word for Windows 95 installed with macros enabled. The template file is placed within the SAP ERP system, which the system refers to while creating the OMS.

5.7 Summary

The REM profile is the main configuration activity, and it controls the various functions that you can perform in REM. You assign the configured REM profile in the material master. You can configure the scheduling parameters of RSQs as well as alter the display layout of the planning table. You can control when the system performs automated functions, such as backflush, either online or in the background.

This concludes Part II of this book. In the next part, we'll jump into business processes for each of the manufacturing types, as well as further settings you can make.

PART III
Production Planning Workflow by Production Type

This chapter covers the important business processes and functions that are used in discrete manufacturing. You'll find the vital and logical links to understand how the configurations you made in Chapter 3 match with your business processes when dealing with products or materials that are subject to high changeability.

6 Production Planning for Discrete Manufacturing

Discrete manufacturing, which is also interchangeably referred to as shop floor control (SFC) in the SAP ERP system, is characterized by frequent change in products on the production lines. The demand for the product is generally irregular, and often the entire production process is complex, including the routing (the sequence of work steps). The assemblies are often placed in interim storage, and the components are staged with reference to production order.

Discrete manufacturing finds extensive implementations in the high-tech, steel sheets recoiling, pumps, textile, steel rerolling, and automobile industries. SAP ERP Production Planning (PP) for discrete manufacturing helps with production order creation, material/capacity availability checks, variance calculation, and much more. It also helps manage and smooth out some of the related business processes that are involved with this type of manufacturing, including scheduling, costing, and goods issue/receipt. To help manage the business processes seamlessly, PP for discrete manufacturing integrates with most components in SAP ERP.

Note [«]

The master data and processes in discrete manufacturing and process manufacturing have significant similarities. You can find the specific details of process manufacturing in Chapter 4, but this chapter covers the general master data and process specifics in detail.

We'll begin with an overview of discrete manufacturing and explain how it fits in the planning and production perspectives. Important discrete manufacturing master data is covered next, with extensive focus on creating material, the bill of materials (BOM), work center and routing, and finally the production version. The concepts and the fundamentals that we cover in this chapter equally apply to process manufacturing and repetitive manufacturing (REM). After covering the master data creation, we use the same master data to cover end-to-end business processes of discrete manufacturing, from production order creation to completion and settlement.

Finally, we cover some of the additional functions available in discrete manufacturing to optimize the business processes. Let's get started with the overview of discrete manufacturing.

6.1 Process Overview

Figure 6.1 provides an understanding of the end-to-end process flow in discrete manufacturing. The process begins when requirements from manually created planned orders, material requirements planning (MRP), or the SAP ERP Sales and Distribution (SD) component flow in, and for which the production process must initiate. To initiate the production process, you need a production order ❶. The system schedules the order ❷, as well as checks for material and production resources/tools (PRT) requirements, to see if it's able to produce the required quantities or not ❸. It also checks for capacity ❹ needed to produce the material. If it's unable to find the requisite capacity, it reflects an overload.

The next step is to release the production order ❺ to enable the system to perform additional functions. You can't perform these functions until you release the order, including printing shop floor papers ❻. The production execution step is the actual and physical production of material, and the first step toward production is material withdrawal from the warehouse ❼.

The plant operator then enters the operation or order confirmation to record the actual processing or production, including scrap produced or rework required ❽. The production department delivers the produced material to the warehouse, which performs the goods receipt against the production order ❾. The last two steps reflect the SAP ERP Controlling (CO) component's integration with PP, in which you perform order settlement ⑩ and proceed to close the production order ⑪.

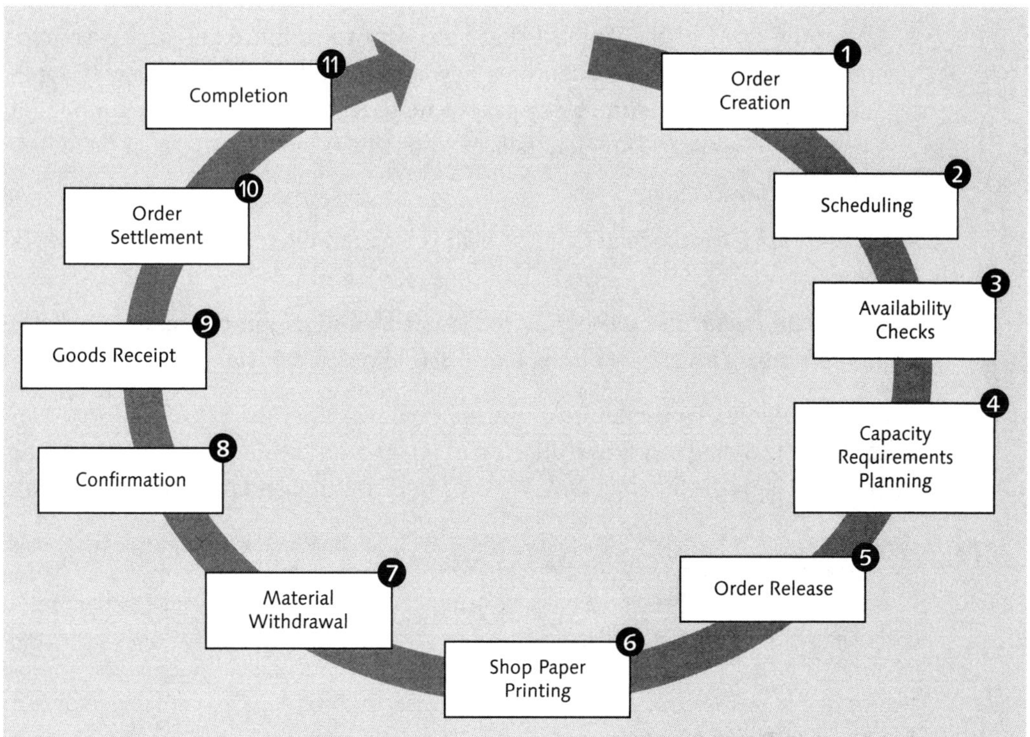

Figure 6.1 Process Flow in Discrete Manufacturing

6.2 Master Data

Before running processes and transactions in the SAP ERP system, it's imperative that you set up complete and comprehensive master data. In discrete manufacturing (and in other manufacturing types), if you set up master data in the right sequence, then it's much easier to connect with other types of master data, as you've already taken care of the predecessor-successor relationship.

The following steps outline the procedure you need to follow to set up your master data correctly, which we discuss in more detail in the following sections:

1. Create the *material master* of the product (a finished good or an assembly) that you want to produce. You can also create the material master for the raw material that will eventually become part of the production process.
2. Create the *BOM* of the product that you want to produce and assign components, together with the quantities needed to produce the product. If applicable, you can also define a scrap percentage at the operation or component levels. You can also maintain details of co-products or by-products in the BOM.
3. Create the *work center.*
4. Create *routing* for the material, in which you assign the previously created work center.
5. Create the *production version* for the material, and assign the material's BOM and routing. (This step is optional in discrete manufacturing.)

Due to the highly diverse nature of the material master creation process, we cover it briefly in this chapter, while offering much greater coverage of the PP-specific master data (specifically the BOM, work center, routing, and production version).

[»]

Note

You can refer to Chapter 3 for a more in-depth understanding of the configuration specifics for discrete manufacturing.

6.2.1 Material Master

A material master is whatever the company wants to maintain as an inventory item. Examples are far ranging and include the following: a finished product that the company first produces in-house and then sells, a raw material it procures and consumes in the production process, consumables (e.g., greases, lubricants, or gloves), trading goods that it buys and then sells, or spare parts for maintenance on machines.

Creating and maintaining information on a material master is the collective and the combined responsibility of a company's various departments and divisions. A material master is a central element; therefore, everyone contributes their share of maintaining different information. In the SAP ERP system, maintaining logically structured information absolutely helps the relevant department efficiently perform day-to-business functions.

Example [Ex]

A company produces a finished good (a product). For this product, the company needs to maintain information for producing the product, selling it, maintaining planning and design data, quality control and life span, costing, and more.

More than likely, one person or department won't know all of the information related to many business functions. For example, the sales personnel may not know the quality tests a product goes through. Similarly, the production department won't necessarily care about the selling price of the product because they are just concerned or focused on producing it. However, the salesperson needs instant access to selling price and the applicable taxes or discount on creation of a sales order. Similarly, the production supervisor wants better control on underdelivery and overdelivery tolerances allowed for a material, whereas the warehouse in-charge needs instant information on storing the product.

To add greater visibility to the entire material master creation process, the SAP ERP system provides *views*, in which the relevant departments of the company (depending on which SAP ERP component they're working with) can fill in their share of information. The BASIC DATA view contains basic information that is applicable to the entire material master, such as base unit of measure, gross and net weights, volume, product hierarchy, and design/drawing information. We cover the details of these views later in this section.

When you create a material master, you also have to define the *material type*, that is, whether it's a raw material, a semifinished product, a trading good, a spare part, or a consumable item. The material type controls the views that it makes available for the user to maintain information for later use. The views contain either organization-independent or organization-dependent details. The organization-independent information is, for example, the material description or its weight and volume. The organization-dependent information, that is, plant or storage location-specific information, may be the planning or the forecasting data that is unique to a plant only. For a different plant (organization-dependent), the planning or forecasting data may be different.

Table 6.1 provides an overview of plant-independent and plant-dependent views of the material master. The appearance of a checkmark in a column denotes the data that you need to maintain for a given function.

Material View	Organization-Independent	Plant	Storage Location	Storage Number	Sales and Distribution
Basic Data 1 and 2	✓				
Classification	✓				
Sales: Sales Org.	✓				✓
Sales: General/Plant	✓	✓			
Foreign Trade/Export Data		✓			✓
Sales Text		✓			✓
Purchasing		✓			
Foreign Trade/Import Data		✓			
Purchase Order Text	✓				
MRP 1–4		✓			
Forecast		✓	✓		
Work Scheduling		✓			
Plant Data/Storage 1 and 2	✓	✓	✓		
Warehouse Management 1 and 2		✓		✓	
Quality Management		✓			
Accounting 1 and 2		✓			
Costing 1 and 2		✓			

Table 6.1 Material Master Views

Following is a broad explanation of the data in some of the views of the material master:

- Basic Data
  These views consist of information such as base unit of measure, which is the smallest unit of measure in which you maintain the product's inventory. It consists of gross weight, net weight, volume, product hierarchy, size/dimension,

and design/drawing details by enabling you to use the Document Management System (DMS).

- CLASSIFICATION
  This view enables you to classify specific materials with identical or similar characteristics into groups and to categorize those groups based on specific characteristics. Due to its diverse and practical nature of application, it's available in several components of the SAP ERP system. You can classify materials, as well as customers and equipment. Further, when there isn't any other option to do so, you can use classification to maintain material-specific information.

Note [«]

We cover the classification system in Chapter 15.

- SALES
  These views contain details such as delivering plant, selling unit of measure, taxes, conditions (e.g., gross price and taxes), sales order text, transportation group, loading group, and item category. You have to maintain these views for materials that you want to sell.
- PURCHASING
  This view contains the purchase unit of measure, purchase group, material group that groups together materials with similar attributes, purchasing value key that incorporates underdelivery and overdelivery tolerances, goods receipt processing time, and Batch Management (BM). For externally procured material, you have to maintain this view.
- MRP
  These views enable you to define how to procure a material and how to plan and control this material within the production process. The definition of the procurement type specifies whether a company procures the material externally or produces it in-house, or if both scenarios are possible. It enables you to specify whether the material follows reorder point planning or is demand driven. Using a strategy group, you can also control whether material is make-to-stock (MTS), make-to-order (MTO), or assemble-to-order (ATO) production. Material planning for a procured or produced material requires that you maintain some or all of the MRP views.

[»] **Note**

We cover MRP in Chapter 11.

- Forecast
  This view enables you to define the forecasting parameters and models such as constant, seasonal, moving average, or trend. If you already know the forecasting data properties, then you should make a selection from one of the available forecasting models. If you're unsure of which model to select, the system provides you the option for automatic model selection. You can also define the number of historical and forecast periods based on which the system runs forecasting.

[»] **Note**

We cover forecasting in Chapter 9.

- Work Scheduling
  This view contains details for underdelivery and overdelivery tolerances for in-house production, the production scheduler, the production scheduling profile that controls several business functions, the production unit of measure, and the production time in days. For material produced in-house, you have to maintain this view. It's possible to have a different unit of measure, which in this case is the production unit of measure that may be different from the base unit of measure.
- Plant data
  This view contains information such as storage bin, temperature conditions, and shelf-life data for batch-managed materials, including maximum storage period, minimum remaining shelf life, and total shelf life.
- Quality Management
  This view provides the option to select different inspection types that a material may go through, for example, finished goods inspection, raw material inspection, and in-process (during production) inspection. The SAP ERP Quality Management (QM) procurement key denotes whether raw material must be quality-managed and the relevant certificate type.

- ACCOUNTING
  These views contain the price control indicator, that is, whether the material will be valued at a standard price or moving average price, and the valuation class.
- COSTING
  These views contain details such as overhead group, origin group, and variance key.

[«]

Note

The creation of material types and also the material master is largely driven by SAP ERP Materials Management (MM), with other components supporting and facilitating it.

Now let's create a new material, in which you can maintain some of the important views. To create a new material master, follow the menu path, LOGISTICS • PRODUCTION • MATERIAL MASTER • CREATE (GENERAL) • IMMEDIATELY, or use Transaction MM01 (see Figure 6.2). On the initial screen, select the INDUSTRY SECTOR as M MECHANICAL ENGINEERING and the MATERIAL TYPE as FERT FINISHED PRODUCT.

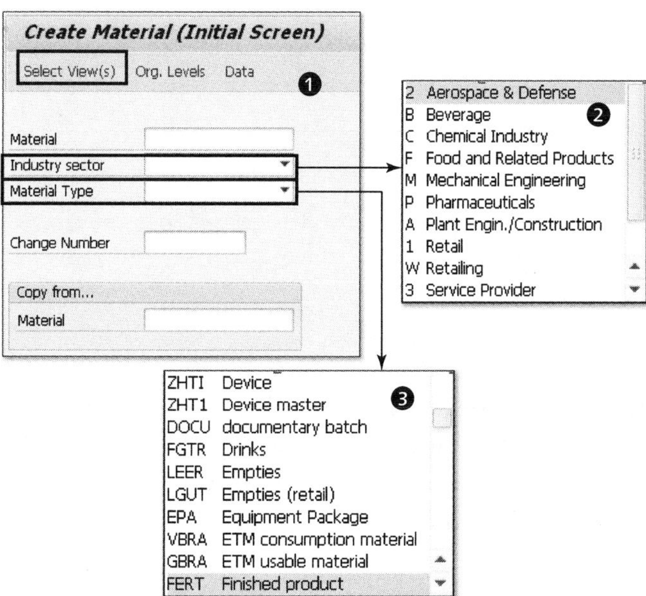

Figure 6.2 Initial Screen for Material Master Creation

Next, click on the SELECT VIEW(S) icon, and the screen shown in Figure 6.3 appears, showing the list of views that were briefly covered earlier. You don't have to select and maintain all of the views, only those that you'll eventually use in your business processes. For example, if you don't intend to include a material in planning, then you don't have to maintain MRP views of the material.

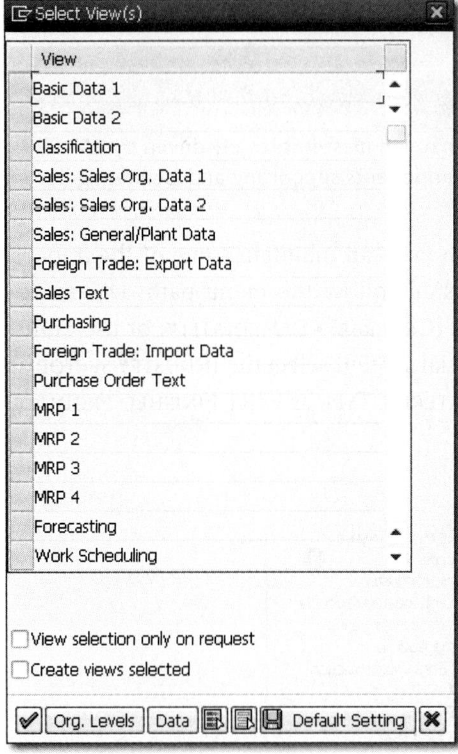

Figure 6.3 Select the Views Option in Material Master Creation

Select the BASIC DATA views, and the screen shown in Figure 6.4 appears, showing the BASIC DATA 1 tab of the material master. Here, the system automatically assigns it an internal material number, 1991. In this view, you can maintain information such as BASE UNIT OF MEASURE, GROSS WEIGHT, and NET WEIGHT.

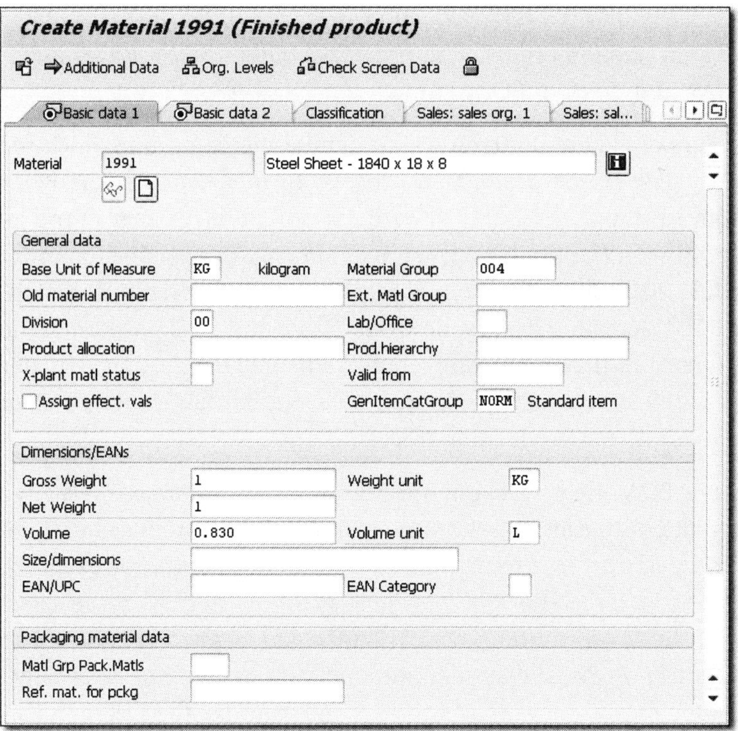

Figure 6.4 Basic Data View of the Material Master

6.2.2 Bill of Materials

A material's BOM consists of components grouped together with the relevant quantities that are needed to produce the material in the production process. For example, if vanilla ice cream is the main product, then its BOM will consist of all the ingredients (components) needed to produce it. Similarly, to produce an automobile, you need an engine, a chassis, a body, and a large number of other components.

To create the BOM of a material, follow the menu path, Logistics • Production • Master Data • Bills of Material • Bill of Material • Material BOM • Create, or use Transaction CS01. On the initial screen, enter Material "1991" and Plant "3000". This is the material number for which you want to create the BOM. In the BOM Usage field, enter "1" to specify the purpose of creating and using a BOM.

BOM USAGE 1 denotes that it's for production purposes and will be available when you create a production order for this material. Other BOM usages are in SD BOM, PM BOM, and COSTING BOM.

Enter the VALID FROM field as "04/01/2013" to denote the date from which the BOM becomes available in the production process. In this example, the BOM validity starts from April 01, 2013 (04/01/2013), when a user creates a production order. The BOM won't be available when the user creates a production order in the month of July 2012, for example. If no date is given, the system automatically assigns the BOM's creation date as the VALID FROM date. The purpose of the BOM validity is to ensure that if the company has a new BOM starting at a later date, then the validity of the new BOM can be set accordingly, which then becomes available during the production process.

In the ALTERNATIVE BOM field, you can specify up to 99 different alternative BOMs for a material. This means that the same material can be produced in up to 99 different ways by having an alternative BOM option available. For example, when one component of a material's BOM is unavailable in stock, you can select another alternative BOM to continue with the production process. When you first create a material's BOM, the system automatically assigns the alternative BOM as 1. When you create a new BOM for the same material, the system issues a warning message and internally assigns the next alternative BOM number to it.

Click on the HEADER icon to open the CREATE MATERIAL BOM: HEADER OVERVIEW screen shown in Figure 6.5. The base quantity of a material denotes the quantity based on which you define the material's BOM components. The system then automatically divides the component's quantity with the base quantity during production order creation. This is especially useful when component quantities are very small and defined, for example, in milligrams for vanilla flavor.

For this example, assign a base quantity of "10KG" in the BOM header. Hence, the components and their quantities that you define in the BOM denote the requirements to produce 10 kg of material 1991.

The BOM STATUS field controls whether the BOM is active and is available for usage. The value 1 denotes that it's active. If you want to deactivate a BOM, you can change the status of the BOM to INACTIVE at any time.

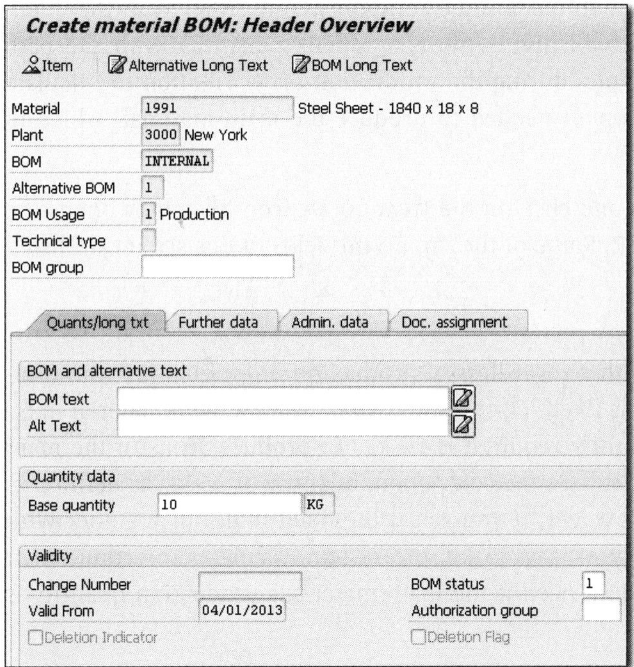

Figure 6.5 BOM Header

Click on the ITEM OVERVIEW icon Item to go to the screen shown in Figure 6.6, which shows the item overview of the BOM.

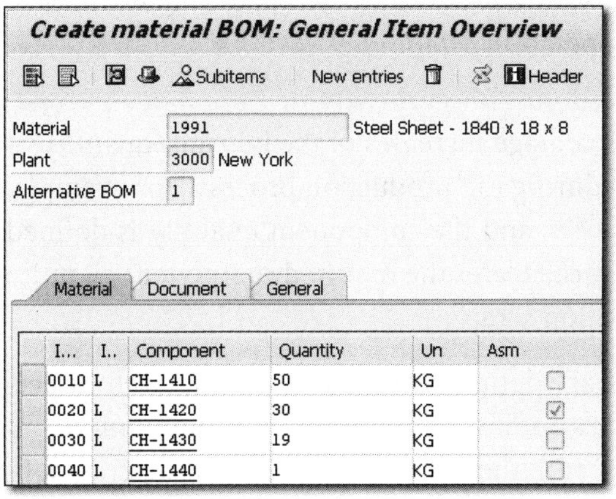

Create material BOM: General Item Overview

Subitems New entries Header

Material 1991 Steel Sheet - 1840 x 18 x 8

Plant 3000 New York

Alternative BOM 1

Material | Document | General

I...	I..	Component	Quantity	Un	Asm
0010	L	CH-1410	50	KG	☐
0020	L	CH-1420	30	KG	☑
0030	L	CH-1430	19	KG	☐
0040	L	CH-1440	1	KG	☐

Figure 6.6 BOM Item Overview

Enter material numbers for the individual components, together with the quantity for each. The system will divide the component's quantity with the base quantity (10, in this example) during the production order creation to calculate the quantity of the component needed to produce the main material, which in this case is 1991.

Select the line item 0010, and click on the ITEM DETAIL icon, which opens the screen shown in Figure 6.7. Some of the important fields in this screen are listed here:

- FIXED QUANTITY
  This checkbox denotes that regardless of production order quantity, the component quantity remains fixed. For example, to produce a main material in 50 kg, the component quantity required is 50 kg. To produce 35 kg of the main material, the system calculates the component quantity of 35 kg, as defined in the material's BOM. However, if you select the fixed indicator together with the component quantity of, say 35 kg, the system calculates the component quantity required as 35 kg even if the main material's production quantity is 50 kg.
- OPERATION SCRAP IN %
  Operation scrap defined in percentage increases the issuance quantity of a component to account for scrap during the production process. For example, if operation scrap is defined as 5%, and the component quantity is defined as 100 kg, the system automatically calculates the required component quantity as 105 kg to account for the operation (during-production) scrap. It's mandatory to select the NET ID when defining operation scrap.
- COMPONENT SCRAP (%)
  Component scrap is defined as percentage increases of the issuance quantity of a component to account for scrap during the production process. For example, if component scrap is defined as 10%, and the component quantity is defined as 100 kg, the system automatically calculates the required component quantity as 110 kg to account for the operation scrap.

If both scrap types that are selected are defined (operation scrap as well as percentage scrap), the system considers both scrap types during the component quantity calculation. During the MRP run, the system takes all of these scrap definitions into account to suggest planning proposals.

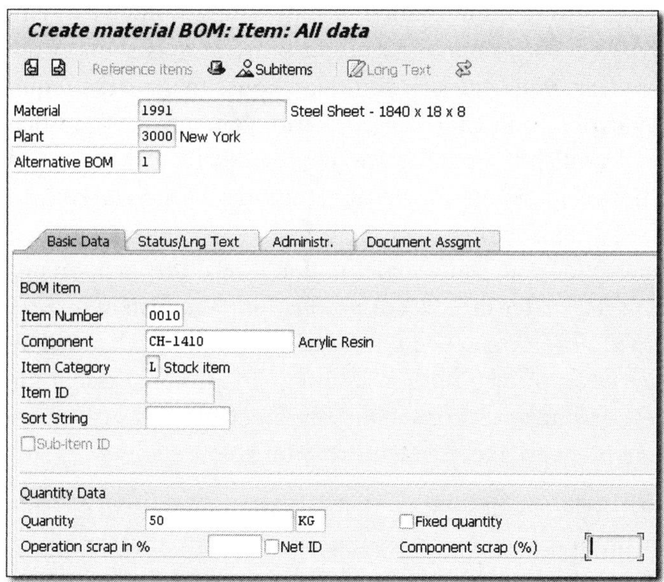

Figure 6.7 Detailed View of the BOM Item

Scroll down the screen (see Figure 6.8). If the normal production of a product also generates a co-product, then you can select the CO-PRODUCT checkbox.

General Data
Co-product
Recurs. allowed
AltItemGroup
Recursive
CAD Indicator
Discontin. data
ALE indicator
Reference point
MRP Data
Lead-time offset
Oper. LT offset
Distribution key
Phantom item
Explosion type
Special procurement

Figure 6.8 Additional Details in the BOM Item

Note [«]

See Chapter 16 for a discussion of co-products and by-products in the production processes.

If you have business scenarios in which the production process entails that the main material can also be one of its BOM components, then this is only possible if you select the RECURS. (recursive) ALLOWED checkbox.

In the MRP DATA area of the screen, the LEAD-TIME OFFSET field helps in a planning component's availability. For example, to produce material as soon as the first shift starts in the morning, you approach your warehouse to issue the requisite components one day earlier. To handle this scenario, you enter "–1" so that the component's required availability date is one day earlier than the planned start date of the production. You denote the negative number of the workday to reflect that you need components earlier than the start date of production, and you use a positive number of days to show that you don't need the components for the defined number of days after the start of production. The system ignores this option if you work with lead-time scheduling (and not just basic date scheduling). To enable this behavior for lead-time scheduling, you need to assign the components to the corresponding operations in the routing. Note that such a positive or negative lead time is always independent from lot size.

The SPECIAL PROCUREMENT field enables you to define a component-specific special procurement-like phantom assembly or procurement from another plant. Maintaining a special procurement type at the component level has higher precedence than at the material level. During an MRP run, the system explodes the material BOM and evaluates parameters maintained at the component level for planning purposes before it looks for the information at the material level.l

[»] **Note**

See Chapter 13 for information on special procurement types.

Click on the STATUS/LNG TEXT tab in Figure 6.7 to see the see the fields in Figure 6.9.

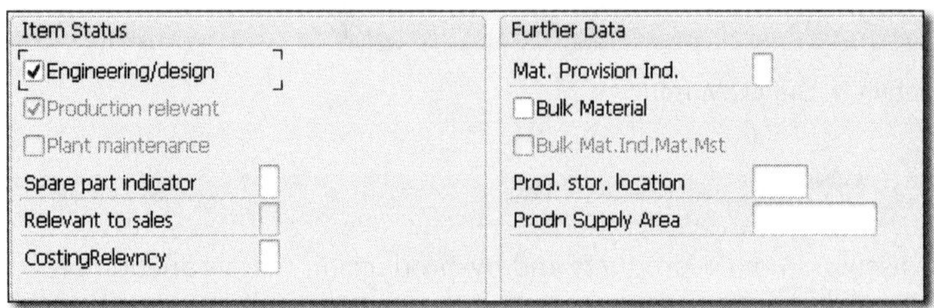

Figure 6.9 Status Tab of BOM Item Details

Enter the PROD. STOR. LOCATION. During the backflush process in SFC, the system looks for information in this field because it acts as an issue storage location. If the customer or vendor supplies the material instead of the company using its own material in production, then you can select the relevant option from the MAT. PROVISION IND. dropdown field. Finally, you can also control whether the component is considered during product cost calculation or the system excludes it by selecting the relevant option in the COSTINGRELEVNCY field.

[+]

Tips & Tricks

You can make mass changes to the BOMs to attend to any new business requirements. For example, you might want to replace an old and obsolete component number in all of the materials' BOMs in which it's currently in use, or you might want to change the component's quantity in all of the materials' BOMs in which it's in use. You can use Transaction CS20 to make mass changes to the BOM. On the upper half of the screen, define the selection criteria, and then in the lower half of the screen, define the changes you want to make. While making the desired changes, be sure to select from one of the several radio buttons available.

6.2.3 Work Center

A work center can be a production facility, a processing unit, a machine or group of machines, or even a laborer or group of laborers directly involved in the production process. You can also group together machines and laborers in a single work center. To create a work center, follow the menu path, LOGISTICS • PRODUCTION • MASTER DATA • WORK CENTERS • WORK CENTER • CREATE, or use Transaction CR01.

[«]

Note

For process industries, you create a resource and follow the menu path, LOGISTICS • PRODUCTION – PROCESS • MASTER DATA • RESOURCES • RESOURCE • RESOURCE • CREATE, or use Transaction CRC1.

On the initial screen, enter PLANT "3000", for which you want to create a new work center. It's mandatory to assign a plant while creating a work center. Enter the work center as "BAF". It's an alphanumeric identification to denote a work center. Enter "0001" in the WORK CENTER CATEGORY field; this is a control function available to ensure only the relevant screens are eventually available during work center creation. Work center category 0001 denotes a machine, whereas 0007 denotes a production line.

Other important functions on the screen are COPY FROM PLANT and COPY FROM WORK CENTER. You can select the checkboxes of the reference work center views that you want to copy to the new work center. These functions help in reducing the data entry efforts during the new work center creation by automatically making available all of the necessary information from the reference work center to the new work center, which you can change, as needed.

In the following sections, we'll cover the various tabs available in a work center and the data you need to maintain, as well as this data's impact on business processes.

Basic Data

Click on the BASIC DATA tab of the work center (see Figure 6.10), and enter a short description of the work center. In the USAGE field, enter "009". A *usage* is a control function that controls whether this work center eventually becomes available for various other tasks, such as during the creation of routing or in the production version. USAGE 009 denotes that you can use this work center in all task list types.

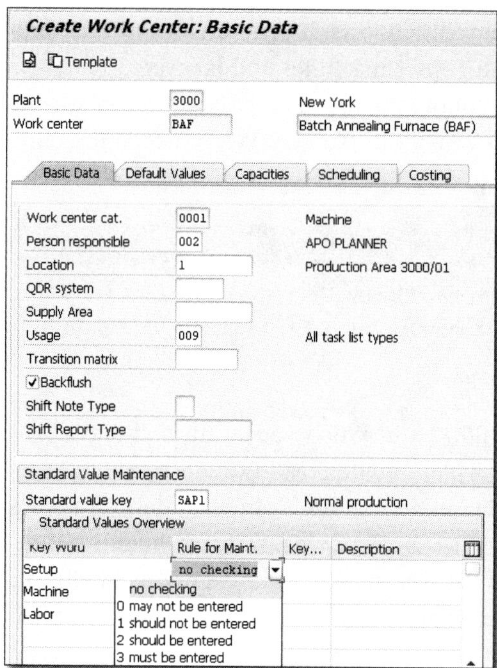

Figure 6.10 Basic Data View of the Work Center

The two important fields on this screen are the following:

- SHIFT NOTE TYPE
  A *shift note* is an option to record the important shift-related details of a work center. These can be general details of the shift as well as any specific problem encountered at this specific work center during the production process.
- SHIFT REPORT TYPE
  A *shift report* compiles and consolidates all of the information from a single or multiple shift notes and is available either in Adobe Acrobat (PDF) or other formats, as well as on the layout defined for a company. The user can also digitally sign a shift report before a printout is possible. The user can also send a shift report by email.

In the STANDARD VALUE KEY field, enter "SAP1". A standard value key denotes activities performed on the work center as well as the option to record them or otherwise. Some of the activities are production time, setup time, processing time, or labor time. When you enter "SAP1", the system automatically brings up three activities (SETUP, MACHINE, and LABOR) for which you'll record the actual details in the confirmation process.

Enter "Machine" under KEY WORD, and select 3 MUST BE ENTERED under RULE FOR MAINT. If you mark this activity as MUST BE ENTERED, then during the creation of a routing, you're required to enter the actual duration of the activity; otherwise, the system won't allow you to save the routing. Other maintenance options are MAY NOT BE ENTERED, SHOULD NOT BE ENTERED, and SHOULD BE ENTERED, which offers various degrees of control from an information message to a warning message, if the requisite information (duration in this case) isn't recorded during confirmation.

Also, be sure to select the BACKFLUSH checkbox on this screen, so that the system automatically issues components at the work center (in this example, BAF) during the confirmation process.

[«]

Note

In Chapter 3, we made the field settings that require the selection of the BACKFLUSH field whenever you create a work center or category 001.

Default Values

Click on the DEFAULT VALUES tab to see the options shown in Figure 6.11.

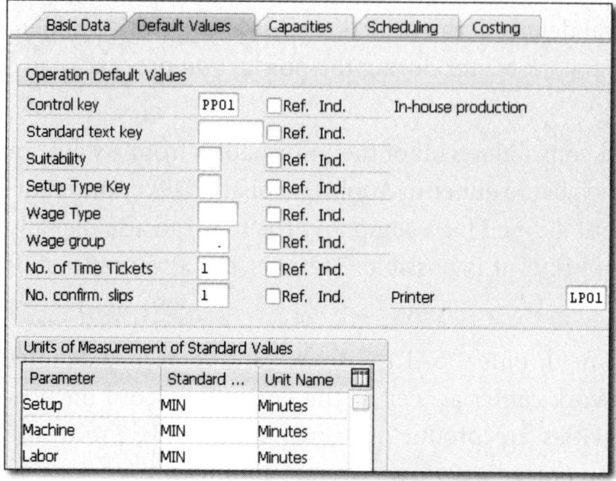

Figure 6.11 Default Values View of the Work Center

Assign the CONTROL KEY "PP01" to the work center. A *control key* is a control function that determines whether you can perform an activity or a task associated with the work center or not. A control key controls several functions, such as scheduling, cost, automatic goods receipt, print confirmation, and print. If any of the checkboxes of the control key aren't selected, then you can't perform that step or function on this work center (BAF). For example, if the SCHEDULING checkbox isn't selected in the control key, then the system won't perform scheduling on this work center.

[»] **Note**

We show how you can configure the control key in Chapter 3, Section 3.3.6.

Capacities

Click on the CAPACITIES tab to see the options shown in Figure 6.12. Enter the SETUP FORMULA "SAP005", which denotes the setup formula that the system will use during capacity calculation. For the PROCESSING FORMULA, enter "SAP006". For the TEARDOWN FORMULA, enter "SAP007". These formula entries are for CAPACITY CATEGORY 001, which is for machine.

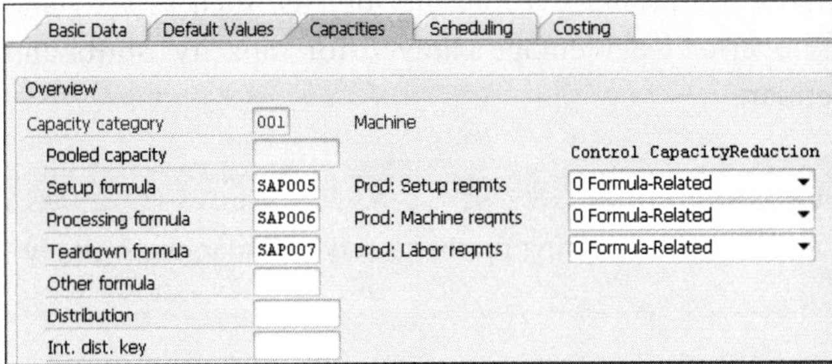

Figure 6.12 Capacities View of the Work Center

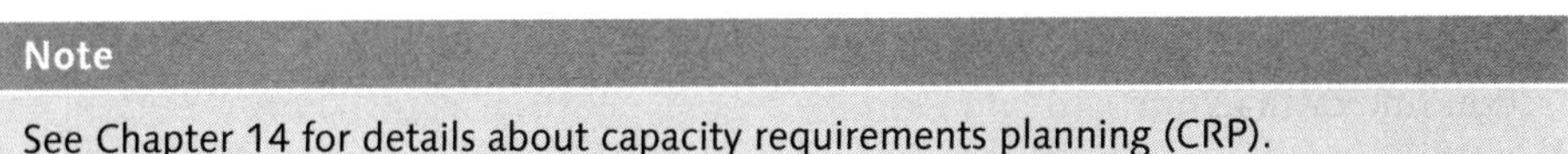

[«]

Note

See Chapter 14 for details about capacity requirements planning (CRP).

Click on the CAPACITY HEADER icon in the toolbar at the bottom of the CAPACITIES tab to open the CHANGE WORK CENTER CAPACITY: HEADER screen for work center BAF (Figure 6.13).

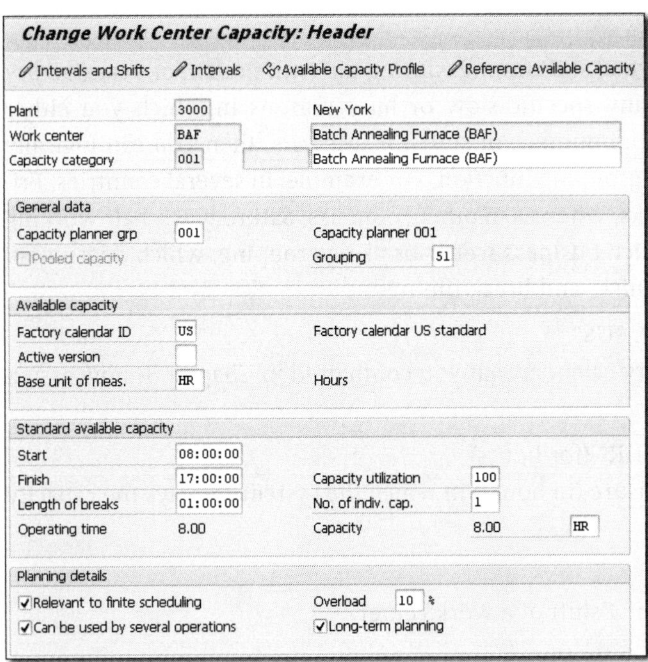

Figure 6.13 Capacity Header

The capacity header contains comprehensive capacity details, such as start and end time of shift, breaks, overload percentage allowed for capacity, shifts and intervals, and factory calendar.

[»] **Note**

Refer to Chapter 2 for information about setting up the factory calendar and capacity planner group.

For this example, maintain the following in each field:

- Short Text for Capacity: "Batch Annealing Furnace (BAF)"
  This short text describes the capacity category, for example, a specific production line or machine.
- Capacity planner grp: "001"
  It's important to assign a capacity planner group to a work center, so that during evaluation of capacities, the capacity planner group can be one of the evaluation criteria. The capacity planner group shouldn't be the name of the person but instead the name of the area or domain of responsibilities.
- Grouping: "51"
  If apart from your normal shift timings during the entire calendar year (factory calendar), you have any specific slow or busy seasons in which you either increase or decrease the number of shifts or intervals, then you can logically group them using the grouping function. For example, in several countries, Fridays have longer breaks, whereas in other countries, Saturday is a half-working day. You can attend such business scenarios using grouping, which doesn't fall under normal start, finish, and break timings.
- Factory calendar ID: "US"
  This is the same factory calendar that you configured in Chapter 3; now assign it to the work center.
- Base unit of meas.: "HR" (for hours)
  This is the unit of measure (in hours) in which the system displays the capacity of a work center.
- Start: "08:00:00"
  This is the start time of a shift of a work center.

- FINISH: "17:00:00"
  This is the end time of a shift of a work center.
- LENGTH OF BREAKS: "01:00:00"
  This is the total length of break during the start and end of shifts. Break timing deducts the availability time of the capacity of a production line. For example, if a machine runs 12 hours a day, with 2 hours of daily breaks, then the available capacity of the production line is 10 hours.
- CAPACITY UTILIZATION: "100" (percent)
  This is the actual capacity utilization factor. This factor has to be as realistic and as close to actual capacity of a production line as possible to enable the system to calculate the available capacity of the production line. You can use it to model for statistically unforeseen downtimes. When your SAP ERP system has built up a significant amount of data, you can use one of many available statistics reports on capacities to update the information in this field. For example, you've maintained a value of 100% in this field, but the statistics reports reveal that capacity is never utilized by more than 90%. In this case, you can update the information to 90% here so that the system ensures realistic and practical capacity planning, and you have greater confidence in the system-generated results. (Refer to Chapter 19 on reporting.)
- NO. OF INDIV. CAP.: "1"
  This is the number of individual capacities associated with a production line. For example, four individual ice cream mixers are created as one production line. Thus, the individual capacities are 4 in number and directly multiply with the timings of the capacity of the production line. For example, if the available capacity of a production line is 10 hours in a day, and there are 4 individual capacities, then the total available capacity of the production line is 40 hours. If you're going to use splitting in routing and also define the number of splits, then it's important to also define the number of capacities because the system checks for this information during scheduling.
- RELEVANT TO FINITE SCHEDULING
  This checkbox causes the work center to consider capacity overload during capacity evaluation.
- OVERLOAD: "10%"
  A capacity overload in percentage directly increases the capacity availability of the production line by the defined percentage. If, for example, the total capacity

of a production line is 40 hours with 10% overload, then the available capacity is 44 hours. A 10% overload of 40 hours is 4 hours, and it adds up to 40 hours to show the total available capacity at 44 hours.

- LONG-TERM PLANNING (checkbox)
  This checkbox enables the system to consider the production line availability during Long-Term Planning (LTP). LTP typically is a simulative planning of material and production lines over a longer period.

[»]

Note

See Chapter 12 for detailed information on LTP.

Click on the INTERVALS button Intervals, and then in the screen shown in Figure 6.14, you can assign specific intervals or shifts that are applicable for the defined duration. For this example, define that from 06/01/2013 until 06/10/2013, there will be a shift SEQUENCE "PP", consisting of a 7-day cycle in which there will be 3 shifts.

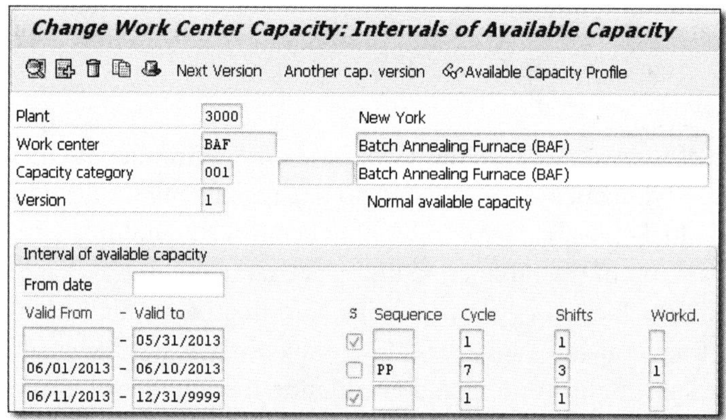

Figure 6.14 Interval in Capacity Header

Scheduling

Click on the BACK icon, and then click on the SCHEDULING tab to see the options shown in Figure 6.15. Enter the standard formulas available for discrete manufacturing, as covered in detail in Chapter 3.

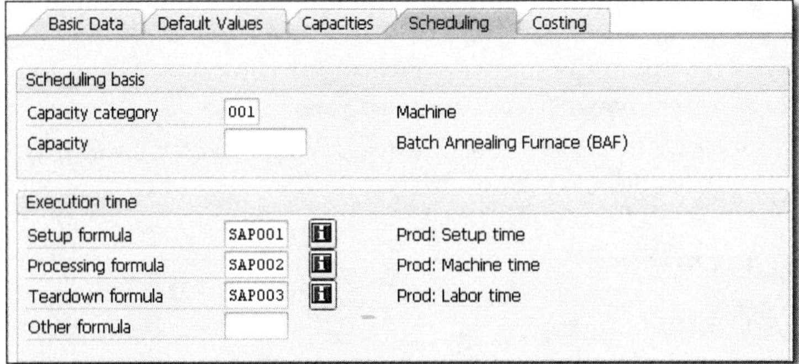

Figure 6.15 Scheduling View of the Work Center

Costing

Click on the COSTING tab to see the options shown in Figure 6.16. Because PP comprehensively integrates with Product Cost Controlling's Cost Object Controlling, each work center needs to have a cost center. You can assign one cost center to multiple work centers, but you can't have multiple cost centers assigned to a work center.

Basic Data | Default Values | Capacities | Scheduling | Costing

Validity

Start date 01/01/2012 End Date 12/31/9999

Link to cost center/activity types

Controlling Area 2000 CO N. America

Cost Center 4240 Sheet Production I

Activities Overview

Alt. activity descr.	Activity Type	Activity Unit	Refer...	Formula...	Formula descriptic
Setup	1421	HR		SAP001	Prod: Setup time
Machine	1421	HR		SAP002	Prod: Machine time
Labor	1421	HR		SAP003	Prod: Labor time

Figure 6.16 Costing View of the Work Center

[»] **Note**

You need to coordinate with the CO person to ensure that each cost center and associated activity type for the given controlling area has price planning.

In the same step, you also have to ensure that activity types and cost centers are already in place so that you can assign them in the work center.

Enter the following details:

- START DATE: "01/01/2012"
  This is the start date of the availability of the cost center for the work center.
- END DATE: "12/31/9999"
  This is the end date of the availability of the cost center for the production line.
- COST CENTER: "4240"
  Every work center requires direct assignment of a cost center so that when you perform work using the work center, the system is able to accumulate all of the costs on the designated cost center.
- ACTIVITY TYPE (SETUP TIME): "1421"
  Activity type denotes the predefined rate with which the system calculates the cost incurred to perform machine setup.
- FORMULA (SETUP TIME): "SAP001"
  A formula denotes the parameters that the system considers to calculate the cost incurred to perform machine setup.

Repeat the same process of entering activity types and formulas for the other remaining two activities, MACHINE and LABOR, and then save the work center.

6.2.4 Routing

The routing of a material can contain a series of operations. The operation details are defined for each operation based on the production quantity, production time, and associated unit of measure. You assign work centers to the routing, which should also contain a standard value key, in which the standard values for setup time, production time, and teardown time are defined. The routing is the sequence of logical production processes through which a product will, for example, transform from raw material into a finished product.

To create the routing of a material, follow the menu path, LOGISTICS • PRODUCTION • MASTER DATA • ROUTINGS • ROUTINGS • STANDARD ROUTINGS • CREATE, or use Transaction CA01.

Figure 6.17 shows the initial screen to create routing. Enter the MATERIAL as "1991" and PLANT as "3000". In the KEY DATE field, enter "04/01/2013" to denote the date when the routing becomes available in the production process. Because the routing validity starts on April 01, 2013 (04/01/2013), when you create a production order, for example, for the month of July 2012, this routing won't be available.

If no date is given during routing creation, then the system automatically assigns the routing's creation date as the valid from date.

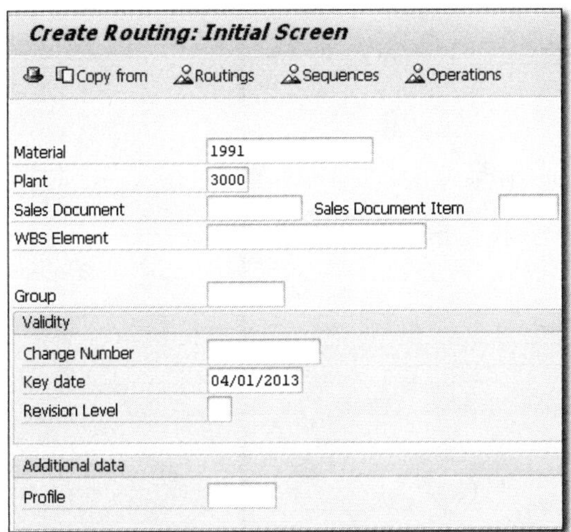

Figure 6.17 Initial Screen for Routing

In the following sections, we'll cover the various elements and options available in routing.

Header Details

Click on the HEADER icon. In the header details of the routing shown in Figure 6.18, enter the following:

- USAGE: "1" (which denotes PRODUCTION)
  This makes the routing available during subsequent production activities, in which the system uses routing during creation of the production order.
- STATUS: "4" (which denotes RELEASED)
  This makes the routing available during subsequent production activities, in which the system uses routing during creation of the production order. If the status of the routing is set to BEING CREATED (STATUS 1), it won't be available when you create a production order for the material.
- PLANNING WORK CENTER: "BAF"
  This is the work center that you previously created and is now used in the planning and production of the material.

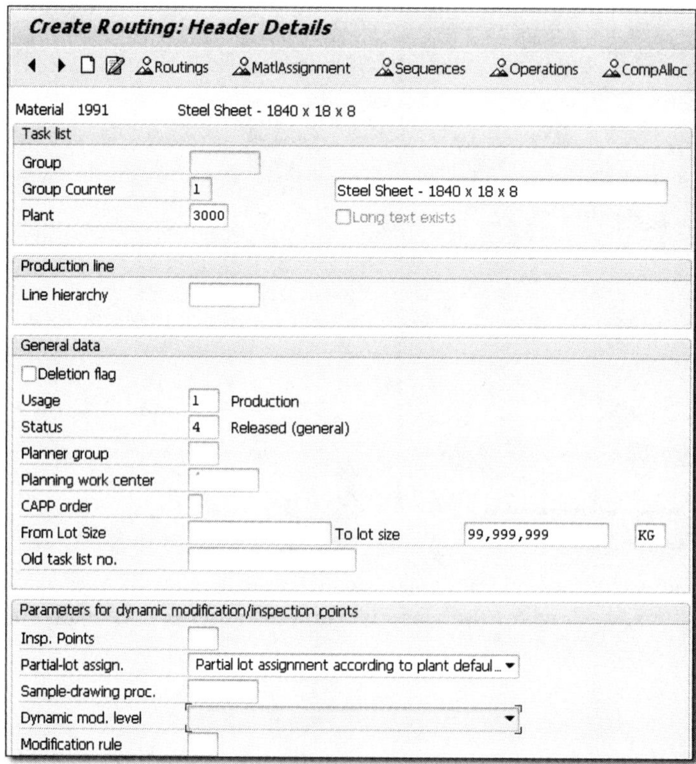

Figure 6.18 Header Details in Routing

> **Note** [«]
>
> Normally, you should have several work centers created before creating a routing. The PLANNING WORK CENTER field should either be left open/blank or made to include the "bottleneck" work center that you want to evaluate during the planning process.

The FROM LOT SIZE field denotes that the smallest quantity of the production order must be at least 1 PC for the system to consider using this routing. The field TO LOT SIZE denotes the maximum quantity of the production/order.

The PARAMETERS FOR DYNAMIC MODIFICATION/INSPECTION POINTS area of the screen is the integration point between PP and QM components in SAP ERP. The parameters you enter here determine whether the system carries out an inspection for the inspection type, for the individual inspection lot, or for the master inspection characteristics.

> **Note** [«]
>
> Because QM also integrates with discrete manufacturing, you can engage the QM resource for this activity.
>
> See Chapter 20 on integration aspects of PP with the SAP logistics components, including QM.

Click on the OPERATIONS button, which takes you to the screen shown in Figure 6.19 where you enter the WORK CENTER as "BAF". This denotes assigning a previously created work center to an operation, which is used in the production of material and for which you're creating the routing. An operation defines, for example, a single-level detail of a production step. Select the first line item, OPERATION 0010, and double-click on the OPERATION 0010, which takes you to the screen shown in Figure 6.20 in the next section.

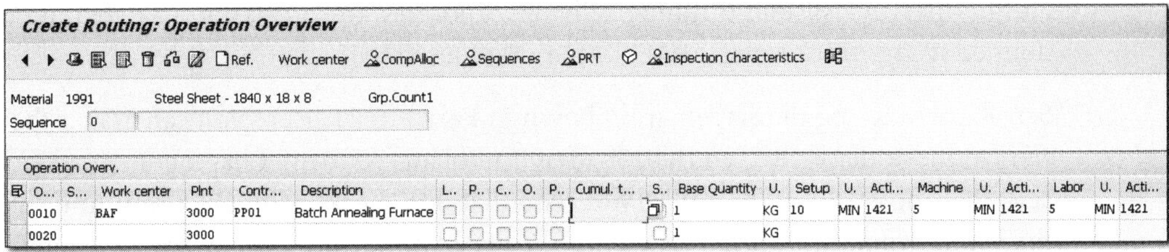

Figure 6.19 Operations Overview of Routing

Operations

Figure 6.20 shows the operations detailed view in which you can enter the following:

- SETUP: "10 MIN"
  You define the time it takes to set up the machine for this task. Any deviation in setup time, either positive or negative, has direct impact on the costs of goods manufactured (COGM).
- ACT. TYPE: "1421"
  An *activity type* is a result of the direct integration of PP with the Product Costing subcomponent of CO (CO-PC). In fact, PP completely integrates with CO-PC. An activity type contains the financial rate (value-based) to perform an activity during production. So, five hours of production time has an associated cost of producing the material. Any deviation in production time, either positive or negative, has direct implication on the COGM.
- MACHINE: "5 MIN"
  You've defined in the work center that recording duration information in routing is a mandatory field (in the rule for maintenance when you selected the MUST BE ENTERED option), so you have to define the time it will take in this work center's operation of the manufacturing process.
- LABOR: "5 MIN"
  You define the time it takes to dismantle or tear down the setup of the machine or the production line. Any deviation in teardown time, either positive or negative, has a direct implication on the COGM.

Enter ACT. TYPE as "1421" for each of the three activities: SETUP, MACHINE, and LABOR. These durations are for the BASE QUANTITY of 1 KG of material (1991) produced.

Notice that the system automatically incorporates the CONTROL KEY that you defined in the work center, BAF, in the previous step.

Scroll down to see the screen area shown in Figure 6.21. Here you can assign the reduction strategy that you configured for plant 3000 in Chapter 3. Further, selecting the TEARDOWN/WAIT SIMUL. (simulation) checkbox enables you to have the simulated details during scheduling. You can define the normal and minimum interoperation time. The interoperation times consist of wait time, queue time, and move (transport) time. During reduction, that is, during scheduling conflict in basic dates, the system switches to minimum interoperation times, as well as further reduction based on the reduction strategy and reduction levels defined.

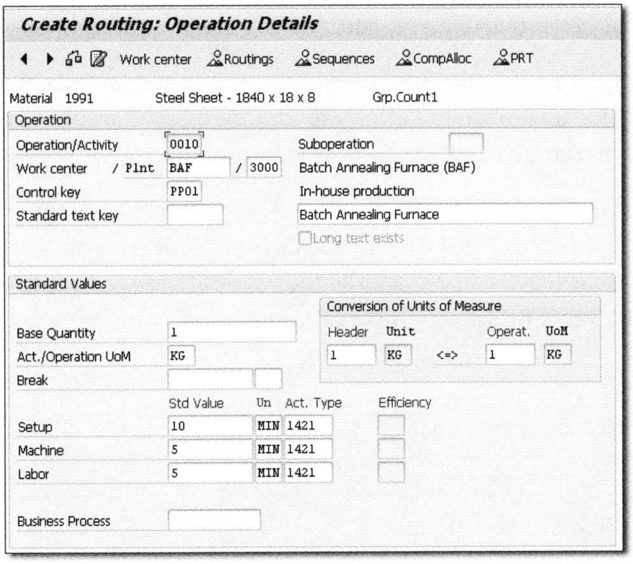

Create Routing: Operation Details

Work center | Routings | Sequences | CompAlloc | PRT

Material 1991 Steel Sheet - 1840 x 18 x 8 Grp.Count1

Operation

Operation/Activity 0010 Suboperation

Work center / Plnt BAF / 3000 Batch Annealing Furnace (BAF)

Control key PP01 In-house production

Standard text key Batch Annealing Furnace

Long text exists

Standard Values

Conversion of Units of Measure

Header	Unit		Operat.	UoM
1	KG	<=>	1	KG

Base Quantity 1

Act./Operation UoM KG

Break

	Std Value	Un	Act. Type	Efficiency
Setup	10	MIN	1421	
Machine	5	MIN	1421	
Labor	5	MIN	1421	

Business Process

Figure 6.20 Operation Details in Routing

Transfer to orders

Cumul. type

Suboper. do not apply to order

Determination of Standard Values

Calculation type

Calculation date

Calculation basis

Standard value code

CAPP

CAPP prod. order

CAPP text created

Interoperation times

Reduction strategy

Teardown/wait simul.

Maximum wait time | Min. wait time

Std. queue time | Min. queue time

Std. move time | Min. move time

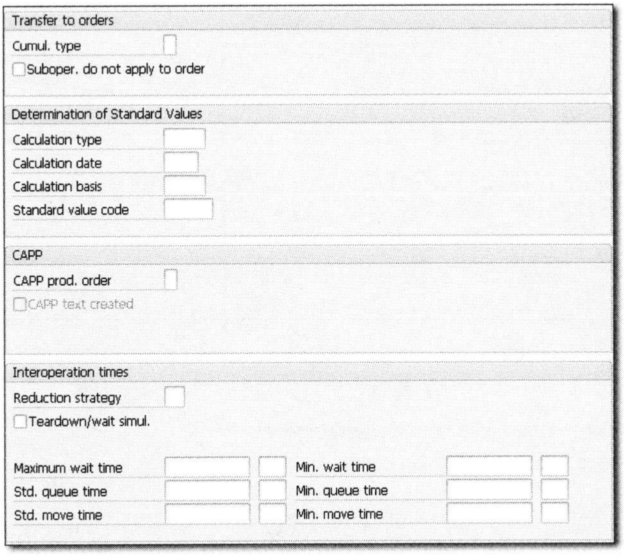

Figure 6.21 Operation Details in Routing: Interoperation Times

Scroll down the same screen to see the screen areas shown in Figure 6.22. With SPLITTING, you can divide a large production lot into several smaller lots provided you have several individual capacities to manage it. The system checks the number of splits with individual capacities that you had defined in the work center.

With OVERLAPPING, the system passes the processed parts of the total lot directly on to the next machine. You can choose to define the minimum overlap time as well as the minimum send-ahead quantity to the next operation.

The information in the GENERAL DATA tab relates to the percentage scrap (planned) for the operation and the SAP ERP Human Capital Management (SAP ERP HCM) data. You can also integrate, for example, the person doing the confirmation at the shop floor to SAP ERP HCM. This is an integration point between PP and SAP ERP HCM.

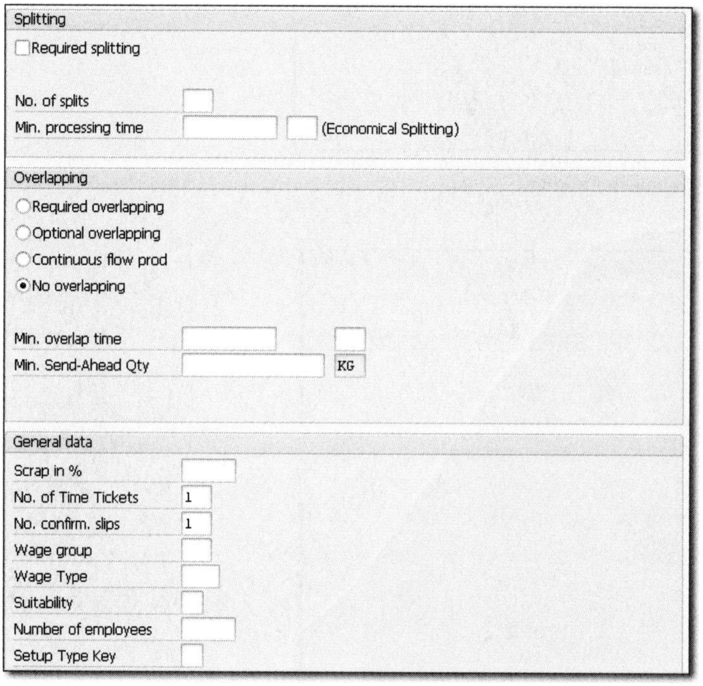

Figure 6.22 Item Details in Routing: Splitting and Overlapping

Scroll down further on the same screen to see the screen area shown in Figure 6.23. Here, the SUBCONTRACTING checkbox reflects the details that you can enter in case the material processing involves subcontracting. A subcontracting process entails that you'll engage services of a vendor to perform some of the steps in the production process, for which you either don't have the in-house expertise or find it cost-effective to outsource it. You select the SUBCONTRACTING checkbox and enter the procurement details in the remaining fields. When the system creates a production order for the material, it also creates a purchase requisition for the subcontracting operations.

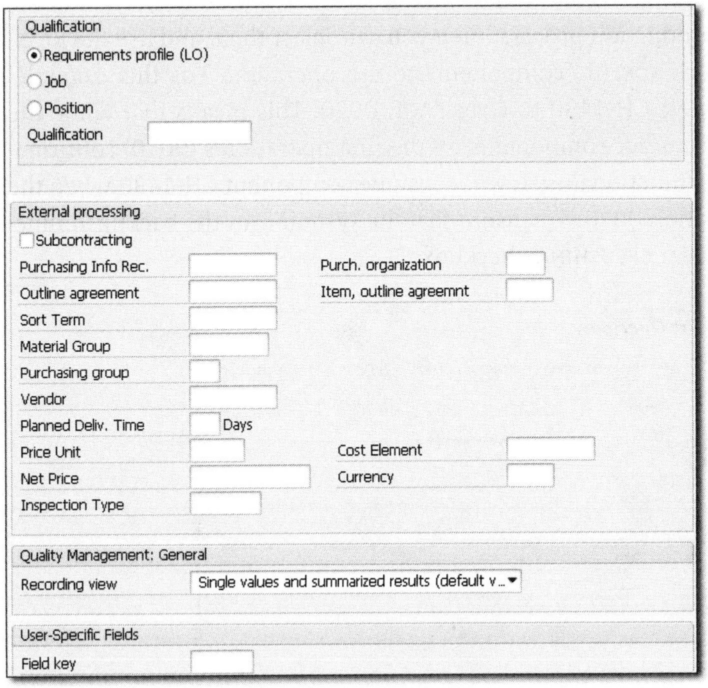

Qualification
(•) Requirements profile (LO)
( ) Job
( ) Position
Qualification

External processing
[ ] Subcontracting
Purchasing Info Rec.
Purch. organization
Outline agreement
Item, outline agreemnt
Sort Term
Material Group
Purchasing group
Vendor
Planned Deliv. Time Days
Price Unit
Cost Element
Net Price
Currency
Inspection Type

Quality Management: General
Recording view: Single values and summarized results (default v...

User-Specific Fields
Field key

Figure 6.23 Item Details in Routing: Subcontracting

[«]

> **Note**
>
> See Chapter 13 on special procurement types, including subcontracting.

Components Assignment

Next, press F6, or follow the menu path, GOTO • OPERATIONS OVERVIEW, and the system brings you back to the OPERATIONS OVERVIEW screen. In this screen, click on the COMPONENTS ALLOCATION icon CompAlloc to open the screen shown in

Figure 6.24. Here you find the component allocation in routing. These components are directly taken from a material's BOM, which you've already created in the previous step. If you don't assign specific component to the specific operation, the system assigns all components to the first operation of the routing during lead-time scheduling. This means that all components must be available at the first operation, so that during backflush, the system is able to consume all components in the first operation during its confirmation.

If your business process entails that a few components are used or needed in the later stages of the production process, then you can select them and use the REASSIGN button to assign specific components to the operation. For this example, assign the COMPONENT CH-1430 to OPER./ACT. 0020. This means that while the system will consume other components on the first operation's (0010) confirmation using backflush functionality, it will consume component CH-1430 when the user confirms operation 0020. To ensure that the system uses the backflush functionality, check the BACKFLUSHING checkbox.

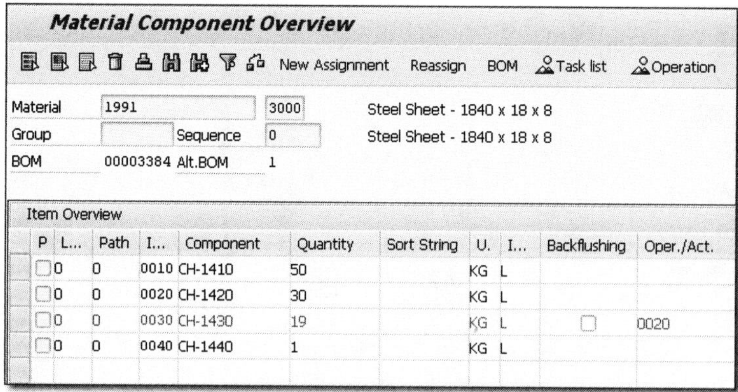

Figure 6.24 Material Component Overview

Material Assignment

Go back to the header screen of routing, click on the MATLASSIGNMENT icon, and the pop-up shown in Figure 6.25 appears. Material 1991 in PLANT 3000 is assigned an internal group as 10022 with the group counter (GRC) as 2 (not shown). Later, we'll use this group (10022) and the group counter (2) in creating the next PP master data, which is the production version. It's important to add that every combination of group and group counter reflects a certain planned working process.

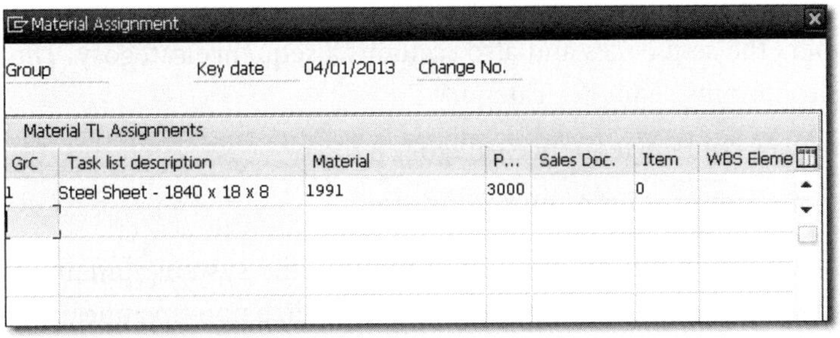

Material Assignment

Group Key date 04/01/2013 Change No.

Material TL Assignments

GrC	Task list description	Material	P...	Sales Doc.	Item	WBS Eleme
1	Steel Sheet - 1840 x 18 x 8	1991	3000		0	

Figure 6.25 Material Assignment

You can assign several materials, together with their respective plants, to this routing group. All such materials belonging to this routing group will have the same group but will have ascending group counter numbers.

[+]

Tips & Tricks

If you have several materials undergoing the same production process, you no longer need to create separate routings for each. Just create one routing, and assign relevant materials (with plants) to it. This saves time and effort in creating an important master data (routing) and also eliminates or reduces redundant data entry efforts.

Sequences

Whenever we refer to routing, we imply that it will be a sequence of operations in a logical order. For example, when we define operations 10, 20, and 30 for a routing, we concur that operation 10 will be the first step of the production process, then operation 20, and finally operation 30. It also means that we confirm the operations in the same sequence. However, you can also maintain parallel and alternate sequences in routing.

Parallel sequences enable you to split up the production process if you can carry out specific production steps of an order simultaneously. The *alternative sequences* represent an alternative in being able to produce a product in the production process in different ways.

You can call the sequence overview by clicking the SEQUENCES icon in the initial routing screen. The sequence overview displays all parallel and alternative sequences contained in the routing:

- The system numbers the sequences and also contains a sequence category. The standard sequence contains sequence category 0.
- The parallel sequences contain sequence category 1.
- The alternative sequences contain sequence category 2.

When you first create a routing and define an operation, the system contains a standard sequence. When you click on the NEW ENTRIES icon, a pop-up appears in which you can select either ALTERNATIVE SEQUENCE or PARALLEL SEQUENCE. For this example, choose PARALLEL SEQUENCE (see Figure 6.26), and the system navigates to the screen shown in Figure 6.27.

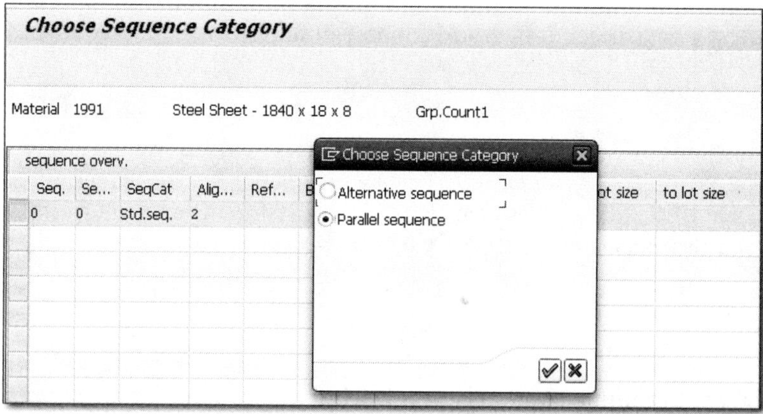

Figure 6.26 Choosing a Sequence Category

In Figure 6.27, SEQUENCE 1 refers to the first sequence with reference to the original operation's sequence, which is 0. The BRANCH OPERATION field stipulates that operation when the system should proceed with the parallel sequence. The RETURN OPERATION field denotes the operation number that the system must return to or branch back to after the parallel operation. It's mandatory to define the return operation.

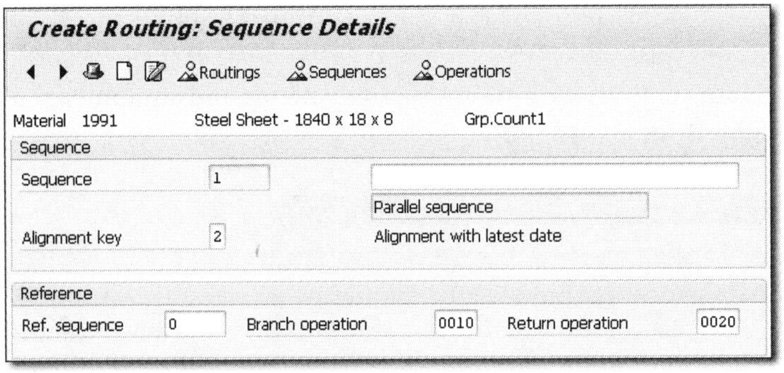

Figure 6.27 Sequence Details

Click again on the SEQUENCES OVERVIEW icon, and you'll see the screen shown in Figure 6.28. Here is a sequence overview for MATERIAL 1991. It stipulates that the system will branch off to a parallel sequence on operation 0010 and will return to a standard sequence in operation 0020.

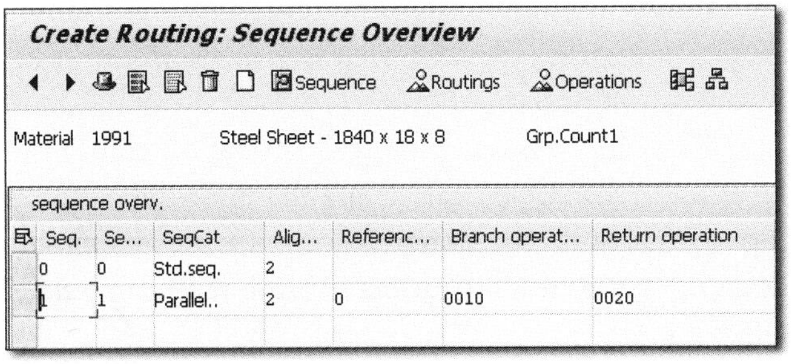

Figure 6.28 Sequence Overview

For ALTERNATIVE SEQUENCE, you can define the minimum and maximum lot sizes that are applicable for the alternative sequence.

Production Resources and Tools

Production resources/tools (PRTs) are movable operating resources that you can assign to internal and external operations in the routing. Examples of PRTs include tools, materials, documents or drawings, and test equipment. PRTs play an important role in facilitating the production process. For example, to move heavy metal containers from one production line to the next, you need a pair of large metal hooks, which is a PRT.

You can assign a PRT to an operation of a routing by clicking on the PRT icon. In Figure 6.29, assign a MATERIAL PRT "1988", and assign a QUANTITY of "1.000". You can use this screen to create more PRTs such as documents and equipment by using the corresponding DOCUMENTS and EQUIPMENT buttons. Save the routing.

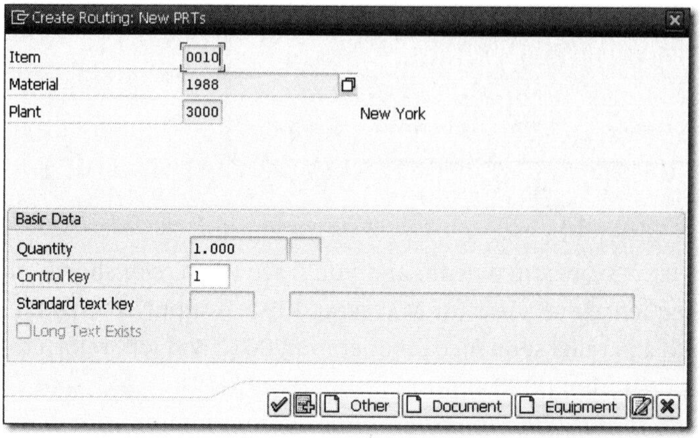

Create Routing: New PRTs

Item 0010
Material 1988
Plant 3000 New York

Basic Data
Quantity 1.000
Control key 1
Standard text key
Long Text Exists

Other | Document | Equipment

Figure 6.29 PRT Assignment

Delete Task List

One of the advantages of PP is the fact that you can delete the master data from the system and avoid having to archive it. You delete master data, for example, when you wrongly create a work center or an incorrect routing. You can delete any PP master data that isn't in use.

To delete routing or, in fact, any task lists (e.g., rate routing, master recipe, or inspection plan), use Transaction CA98. On the initial screen, enter the selection parameters, and choose the relevant deletion options.

Engineering Workbench

You can use the Engineering Workbench to create or make changes to PP master data, such as BOMs or routing. The Engineering Workbench is an interactive and intuitive tool to enable you to make collective changes to, for example, BOM headers of several materials at once.

You use Transaction CEWB to access the Engineering Workbench, and then on the initial pop-up, select the work area that you want to work on. The work area can be the BOM header, BOM item, routing header, or routing item. On the parameters selections screen, enter the parameters, and then choose LOAD BOMS AND TASK LISTS. In the ensuing worklist, you can make the desired changes to PP master data or create a new worklist.

6.2.5 Production Version

Production versions denote different ways by which a company produces a material. This includes, for example, a material having one routing but multiple BOMs. Hence, for each BOM and the same routing, the system needs a separate production version to denote uniqueness during production order creation. The requirement to create a separate production version also applies when there is one BOM for a material but multiple routings. In process manufacturing and in REM, it's mandatory to create and make available a production version. Creating a production version isn't a mandatory requirement in discrete manufacturing, however. The salient features of the production version include the following:

- Production versions can also be quantity range-specific, thus acting as a control function. For example, you can create one production version that has a quantity limitation of 1 to 1,000 PC and another production version for quantities greater than 1,000 PC.
- Production versions can also be validity period-specific, thus acting as a control function. For example, you can have a production version that is valid from April 01, 2000, until March 31, 2013. From April 01, 2013, onward, you may have a new production version. When the system has to assign a production version to the production order, it searches for specific validity dates. When the system finds the relevant production version, it automatically assigns it.
- If the system is able to find multiple production versions fulfilling the search and application criteria, it assigns the first available production version to the production order.

To create a new production version, follow the SAP menu path, LOGISTICS • PRODUCTION • MASTER DATA • PRODUCTION VERSIONS, or use Transaction C223.

In Figure 6.30, and in the SELECTION CONDITIONS area, enter the PLANT as "3000", enter the MATERIAL as "1991", and then press Enter. If the system is unable to find any production version of the material, then the lower half of the screen is blank. If it does, then the system displays the relevant production versions.

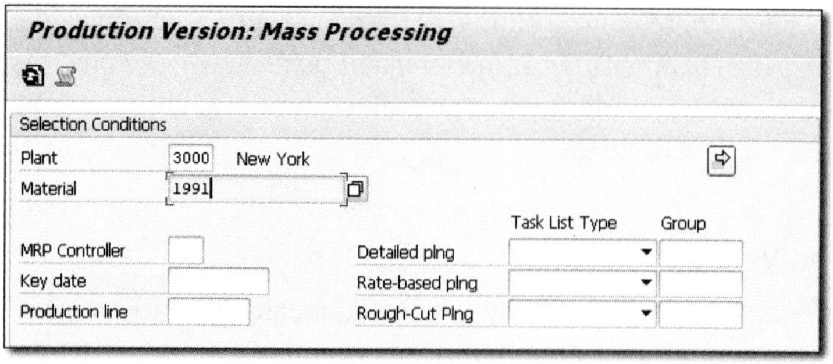

Figure 6.30 Production Version Initial Screen

From the menu bar available at the bottom of the production version creation screen, you have just about all of the options normally needed while creating production version master data. This includes copying, deleting, copy from a template, and so on. Choose the CREATE PRODUCTION VERSION icon to open the screen shown in Figure 6.31. Enter the basic details such as PLANT "3000", PRODUCTION VERSION "0001", a TEXT, VALID FROM "04/01/2013", and VALID TO "12/31/9999".

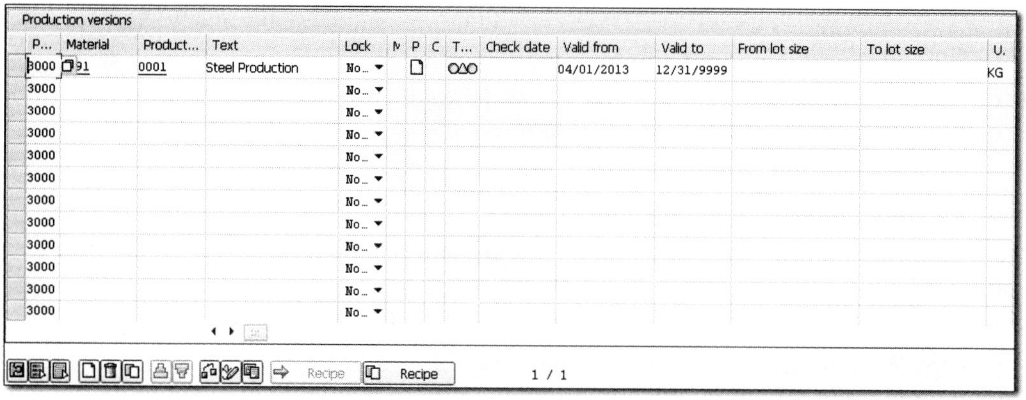

Figure 6.31 Overview Screen of the Production Version

Select the first line item, and double-click on it, or click on the DETAIL icon, which brings up the screen shown in Figure 6.32.

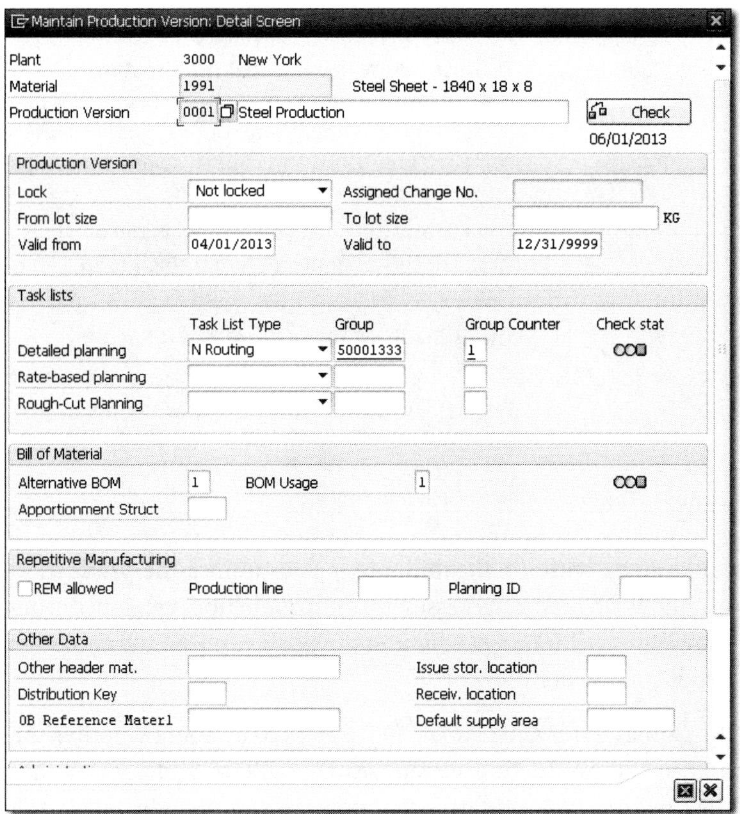

Figure 6.32 Detailed Screen of the Production Version

Enter the details as noted in Table 6.2.

Field	Value	Additional Details/Remarks
Material	1991	–
Production Version	0001	–
Valid from	04/01/2013	This field is the start of the validity date of the production version.
Valid to	31/12/9999	This field is the end of the validity date of the production version.

Table 6.2 Parameters Entry in the Production Version

Field	Value	Additional Details/Remarks
DETAILED PLANNING	Routing (N)	Task list type N denotes that it's a routing. Each task list type represents a different function. Type list type R represents a rate routing, whereas task list type Q represents a quality inspection task list. Enter the routing that you created in the previous step. For this example, enter routing group "50001333" and a group counter of "1".
ALTERNATIVE BOM	1	This is the BOM that you created in the previous step.
BOM USAGE	1	This field is the BOM usage, where 1 refers to the usage production.

Table 6.2 Parameters Entry in the Production Version (Cont.)

The LOCK field provides you with the flexibility to lock or unlock the production version anytime to allow or restrict accesses to the production version during production order creation. A locked production version won't be available for selection during production order creation.

Finally, click on the CONSISTENCY CHECK icon Check to evaluate whether the details incorporated in the production version, such as BOM and routing, are correct and valid. It's important to add here that you can also include data in the production version that doesn't make sense, such as selecting an incorrect work center as a production line.

The system confirms that it's able to find a relevant task list (routing) as well as the BOM, which meets the other criteria of the production version. This includes having validity dates of the routing and BOM, which falls within the validity specified in the production version as well as the lot sizes mentioned in the production version. Click on the CANCEL icon.

The green traffic light against the task list as well as the BOM denotes that the consistency check was successful, and this production version will be available to users during production order creation. If a warning or error is observed during consistency check, the system denotes it with a yellow or a red sign, respectively. In the production version, it's possible to check if previously created and nonchecked production versions are still are valid or not.

Click on the CLOSE DETAILED SCREEN AND ADOPT CHANGES icon, and save the production version.

6.3 Production Order Management

The *production order* is the formal element that informs the production department on how to proceed with producing the product. It contains details on production quantities, the components required to produce the material, the operations (consisting of work centers) that are involved in production, and the scheduling details of production, among many other details. Due to the highly integrated nature of production with CO-PC, all planned costs are in place when you first create the production order.

The master data of a material that we covered in the previous sections forms the core and the central information for the production order. In other words, when you create a production order for a material, the system looks for its BOM, routing, and production version (if applicable). The system copies all of the information from the master data, so that minimal efforts are involved in production order creation. Regardless of whether you create the production order by converting planned orders from MRP or manually, the system fetches the master data in the production order. You can individually or collectively convert planned orders of MRP into production orders. You can also create the production order without reference to any planned order.

The production order is also the central component used to record and update all of the activities or functions that the users perform. For example, you can manually or automatically enable the system to check and perform the following functions or activities in the production order (depending on the configuration settings): checking material or capacity availability, generating the control instruction, calculating planned costs, and more.

Later, when you perform functions with reference to the production order such as goods issuance, confirmation, or yield/scrap updates, the system keeps updating the information.

The production order consists of the following elements, which we'll discuss in the following sections:

- **Header data**
  Contains organizational information such as the order number and costs, quantities, and dates for the order as a whole.
- **Operations data**
  Contains detailed operations dates, default values, and the confirmed quantities and dates.
- **Components data**
  Contains detailed information about components, such as the requirements quantity and reservation number.

Figure 6.33 provides an overview of some of the elements, features, and functionalities available in the production order.

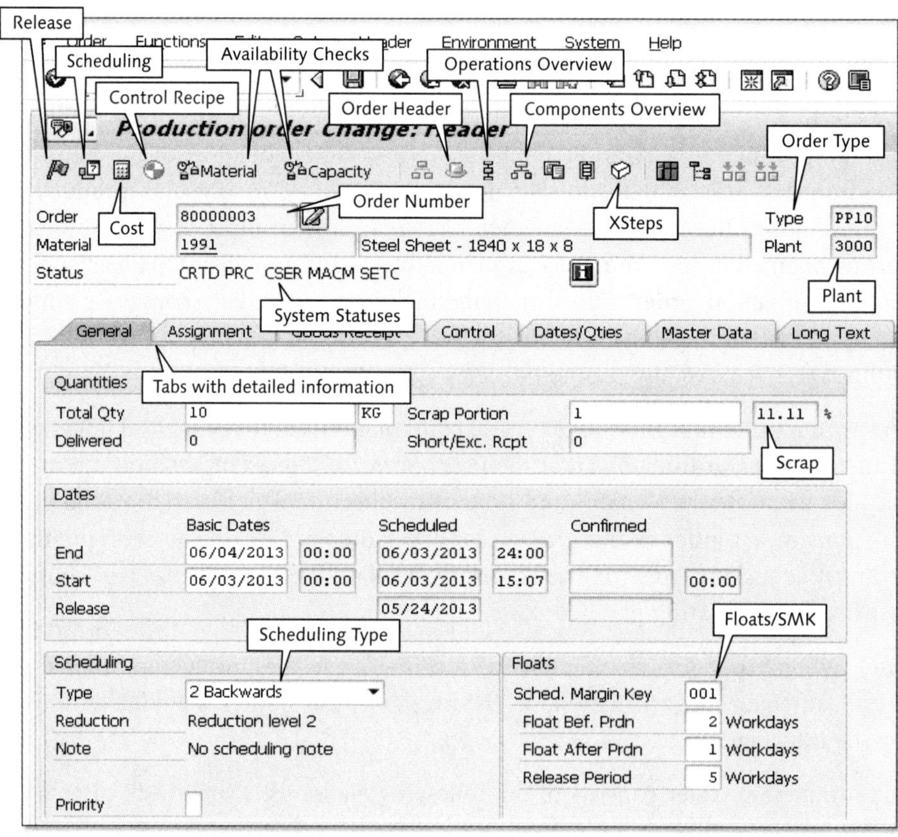

Figure 6.33 Elements of a Production Order

To create a production order, follow the menu path, LOGISTICS • PRODUCTION • SHOP FLOOR CONTROL • ORDER • CREATE • WITH MATERIAL, or use Transaction CO01.

On the initial production order screen, enter MATERIAL "1991", production PLANT "3000", and order TYPE "PP10", as configured in Chapter 3. Choose CONTINUE.

6.3.1 Header Data

Figure 6.34 shows the header data screen of the production order. It consists of information that is applicable to the entire order, such as production order number, the material number, the plant, and the order type. The GENERAL tab includes the planned production quantity, the scrap quantity, the scheduling details, and the reduction details. If you've converted a planned order into a production order, the system automatically copies the details from the planned order, such as quantity, scrap percentage (and its equivalent quantity), BOM, routing (if available), scheduling, and capacity details.

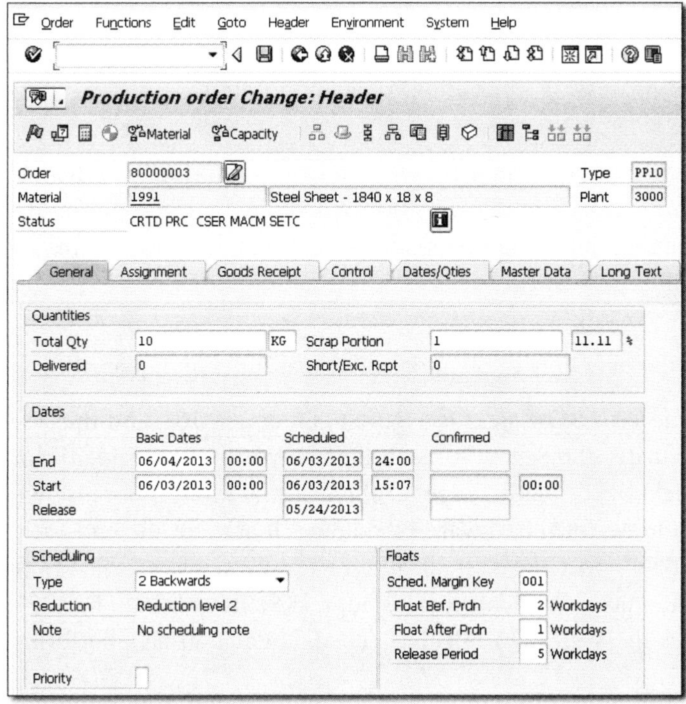

Figure 6.34 Production Order Header

When you perform subsequent business functions, such as confirmation, the system automatically updates the relevant fields, at both the production order level and the operation level.

You'll notice a large number of system statuses, which the system keeps on updating as you continue to perform business functions. These business functions can include activities such as releasing or printing the production order, providing confirmation, and closing or locking the production order.

Further, if you've set up User Status at the production order level, then it also becomes visible just below the System Status in the production order header.

The following briefly covers some of the information available in various tabs of the production order header:

- Assignment
  This view contains the responsible MRP controller, the production scheduler, the sales order number and its item number (for MTO production), the work breakdown structure (WBS) element (for engineer-to-order [ETO]).
- Goods Receipt
  This view contains control parameters for goods movements, deviation tolerances, goods receipt processing time, and production location.
- Control
  This view contains the parameters for the calculation, the production scheduling profile, and the control parameters for scheduling: creating capacity requirements, scheduling allowing for breaks, and automatic rescheduling in the event of scheduling-relevant changes.
- Dates/Qties
  This view contains the overview of planned and actual times and quantities.

For this example, in the Total Qty field in Figure 6.34, enter "10", and the system automatically calculates the Scrap Portion field, if you've maintained it in the Assembly scrap field of the MRP 1 view of the material master. Because we configured the system to perform backward scheduling, it asks for the end basic date. Enter "06/04/2013", and the system automatically calculates the Start Basic Date. Notice that it takes the floats into account, as Sched. Margin Key 001 is in place and enables the system to calculate schedule dates. It also performs reduction up to level 2.

Let's now evaluate the information maintained in the operations view of a production order. Click on the OPERATIONS OVERVIEW icon.

6.3.2 Operations Overview

The OPERATION OVERVIEW screen details the sequence with which a product will undergo a production process. It uses information from the routing and the operations that are scheduled for the order. The system denotes each operation with an operation number, together with the associated work center and the control key that you defined in the work center. The control key determines the functions that the user can perform for an operation, such as printing and whether confirmation is required or not allowed. It also controls scheduling and CRP. Other details that are available in this screen are trigger points, the PRTs, and an assignment flag for material components.

Figure 6.35 shows the OPERATION OVERVIEW screen of a production order. Double-click on 0010 in the OP. column, and the system navigates to the operation detail screen.

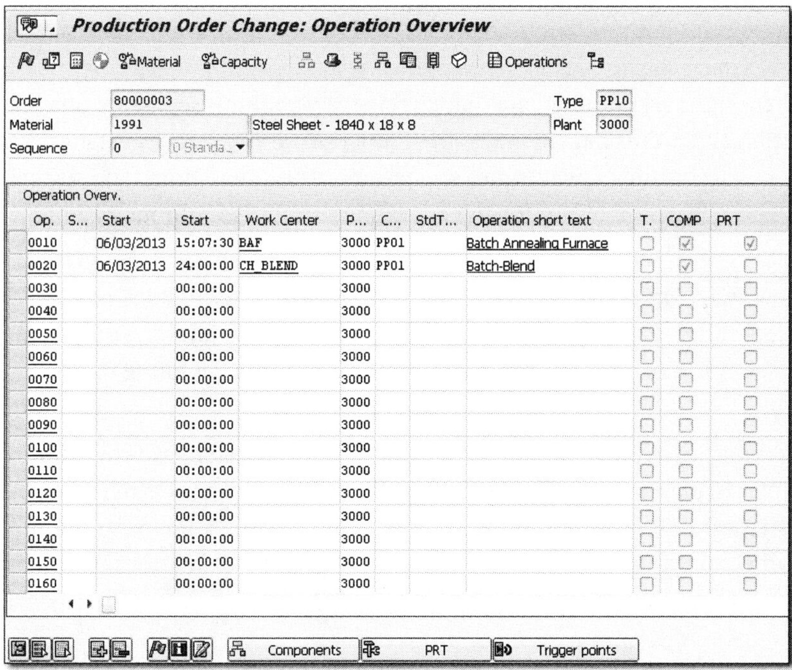

Figure 6.35 Operation Overview in the Production Order

The detail operation screen has several tabs, of which some of the important ones are described here:

- General
  This view contains general information on the operation, such as the percentage scrap, number of shop floor papers to print, and setup type key. It also contains the cost calculation relevance indicator, which specifies whether the operation is included in the cost calculation.
- Default Values
  This view contains the default and standard values copied from the routing and directly relates to the operation quantity, that is, the planned values for performing the operation and the corresponding units of measure.
- External Processing
  This view contains information on externally processed operations such as subcontracting, which involves operations processed at the supplier's end. The material–supplier relationship known as the purchasing info record is entered in this view. The system checks to see if the control key for the operations allows external processing or not. If it doesn't, then you can't enter the external processing details here.
- Transition Times
  This view contains scheduling data that affects the operation segments outside the execution time. This data includes the reduction strategy and level and the pertinent transition times: wait time, queue time, and move time. The system checks to see if the control key for the operations allows scheduling or not.
- Splitting
  In this view, you can maintain splitting data that controls whether, how, and when an operation may or must be split during the scheduling. The system checks to see if the control key for the operations allows scheduling or not.
- Overlapping
  In this tab, you can maintain data on operations that the system can consider in overlapping during production order scheduling. The system verifies if overlapping is required, and, if so, the minimum overlap time and the minimum send-ahead quantity. The system checks to see if the control key for the operations allows scheduling or not.
- Dates
  This view contains the results of the infinite scheduling for the operation. The

scheduling determines dates for the operation segments such as setup time, processing time, teardown time, and wait time. For each of these operation segments, the system determines the duration and the earliest and/or latest start date with time. In determining the earliest dates and times, the system makes an assumption that the operation will use minimal queue and wait times. In determining the latest dates and times, the system uses normal queue and wait times in scheduling. The system checks to see if the control key for the operations allows scheduling or not.

- QUANTITIES/ACTIVITIES
  In this view, the system displays the confirmed quantities and activities for the operation.
- DATES CONFIRMED
  This view contains the actual processing dates for the operation.
- CAPACITY REQUIREMENT ASSIGNMENT
  In this view, you can split the total requirement of the operation into subrequirements.

6.3.3 Standard Trigger Points

As an SAP ERP system consultant or a business process owner, you're always looking for ways to optimize existing business processes and functions, especially when the processes have matured, and you can transition from manual intervention to automation.

You'll use a *trigger point* to enable the system to perform predefined functions when an event occurs in a production order. Trigger points automatically trigger functions, for example, releasing all subsequent operations, when you release the first operation. You can assign trigger points in the routing or to individual production orders.

You can perform the following functions with trigger points:

- **Release of directly following operations**
  An event occurrence triggers the automatic release of the subsequent operations.
- **Release up to release point**
  An event occurrence triggers the release of all subsequent operations up to and including the next operation, which is given the RELEASE STOP indicator.

- **Release of previous operations**
  An event occurrence triggers the release of all previous operations.
- **Create order with template**
  An event occurrence (e.g., material is faulty) means that the system creates a new production order to post-process the material. The system creates an order using standard routing without reference to a material.
- **Include standard routing**
  An event can incorporate a standard routing in an existing production order. In this case, you have to specify the operations between which the system inserts standard routing.
- **Trigger workflow**
  An event occurrence, such as material unavailability, triggers a workflow task, such as sending an email.

You can create standard trigger points, which then serve as a template that you can use to create new trigger points. You can also group together different trigger points into a trigger point group. This group then copies all of the trigger points when you assign it to the routing or the production order.

[»] **Note**

We cover creating trigger point groups in Chapter 3.

You first have to create a trigger point and then assign it to an operation in the routing (or production order). Figure 6.36 appears after you enter STDTRIGGER-POINT "PP10" on the initial screen. Give a short text to the trigger point, and then assign this trigger point to the TRIGPOINTUSAGE "FERT" and TRIGGERPTGROUP "FERT".

Next, select the TRIGPOINTFUNCTS checkbox in the USAGE area, and then select one of the trigger point functions, such as RELEASE SUCCEEDING OPERATIONS, when the SYST.STS (system status) is CRTD (created). Click on the PARAMETERS icon, and the PARAMETERS FOR FUNCTION pop-up shown in Figure 6.37 appears. Here you specify an event so that when the system status of the operation of an order is CRTD (created), this triggers an automatic release of succeeding operations. The STATUS CHANGE field enables you to define if a particular event sets or resets the status.

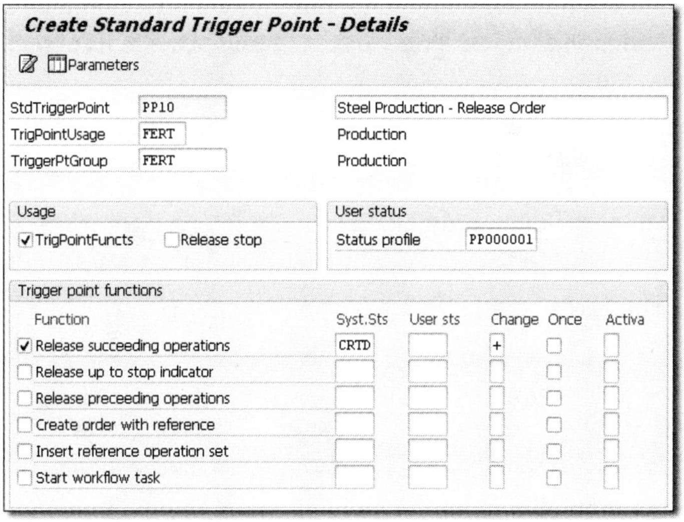

Figure 6.36 Trigger Point Function

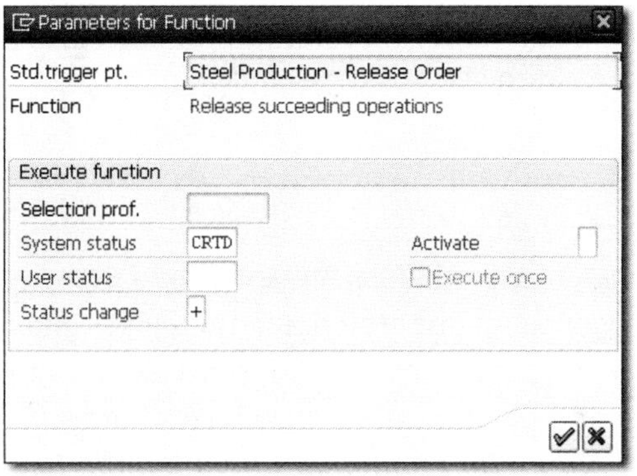

Figure 6.37 Parameters for a Trigger Point Function

Depending on which trigger point function you chose to implement, the system then brings up the corresponding parameter's entry options. For example, if you select the CREATE ORDER WITH REFERENCE function, the system asks you to enter the status (e.g., partially confirmed or scheduled), which it will check to activate the trigger function. It then asks you to define parameters such as routing group number, group counter, and production order type.

Figure 6.38 shows the routing operations screen for operation 0010 for MATERIAL 1991 and plant 3000. Assign the TRIGGER POINT USAGE "FERT". One of its trigger points is PP10, which you just created.

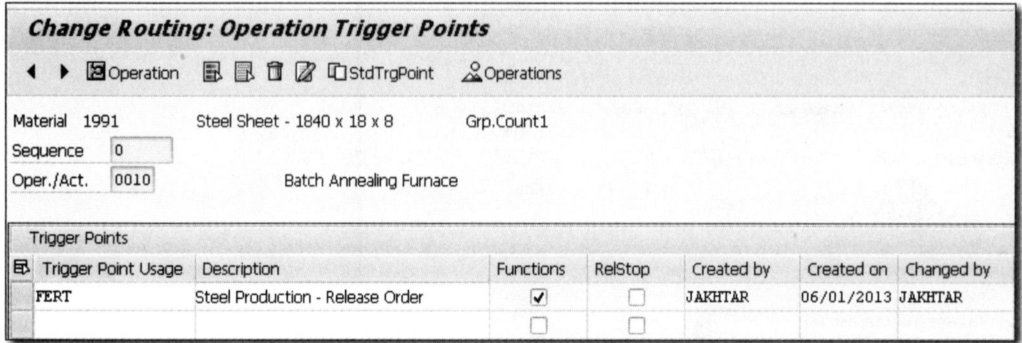

Figure 6.38 Operation Trigger Points

[»] **Note**

If you refer to Figure 6.36, you'll notice the TRIGGER POINTS icon, which you can use to assign TRIGGER POINT USAGE in the operation screen of a production order.

6.3.4 Components Overview

At any time in the production order creation screen, you can navigate to COMPONENTS OVERVIEW by clicking on the relevant icon. The system reads the BOM data to bring up the components that will be used in the production process. You can assign each component of the BOM to an operation in the production order. If you don't do this, the system automatically assigns all components to the first operation. You can also change this assignment by using the REASSIGN option. For nonstock components, the system creates a purchase requisition when the production order is created and saved. This is an integration point between PP and MM. During the entire procurement process, the system maintains the link with the production order.

Figure 6.39 shows the components overview screen of the production order. The details covered here are the same that we previously covered in the routing master data creation.

General Data	Purch. Data	V-S Itm Data	Co-Product	Text Item	Long Txt

Component

Material	CH-1410	Acrylic Resin	
Item	0010 ItemCat L	Acrylic Resin	
Stor. Location	0001	Status	CRTD
Batch		Item ID	00000001
Plant	3000		
Goods Recipient		Unloading Point	
Reservation	72552 1	Movement Type	261
Revision Level		Sort String	
D/C Indicator	H Credit		

Quantities

Requirement Qty	50	Unit of Measure	KG	Fixed Qty
Committed Qty	0	Operation scrap	0.00	Net Indic.
Withdrawal Qty	0	Component Scrap	0.00	Final Issue

Figure 6.39 Detailed View of a Component in a Production Order

6.3.5 Reread Master Data

When you create a production order, the system reads the relevant PP master data and makes it an integral part of the production order. In the SAP ERP system, you first create the master data and then use this master data in day-to-day business transactions, such as production order creation. The system doesn't automatically read any changes you make to the master data later on. For example, you've already created a production order of a material that reads the PP master data, including its BOM. Now, if you make changes in a material's BOM, the system doesn't automatically reflect these changes in the newly created production order. If you have a business need to reread the master data, you can do this by choosing FUNCTIONS • READ PP MASTER DATA in the production order screen. The READ PP MASTER DATA pop-up screen appears as shown in Figure 6.40.

If you've made any interactive or manual changes in the production order, reading PP master data again will overwrite all changes made.

Further, there is a limit on when you can read the PP master data again. For example, if you've already confirmed or performed a goods receipt against a production order, then it's no longer possible to read the PP master data again.

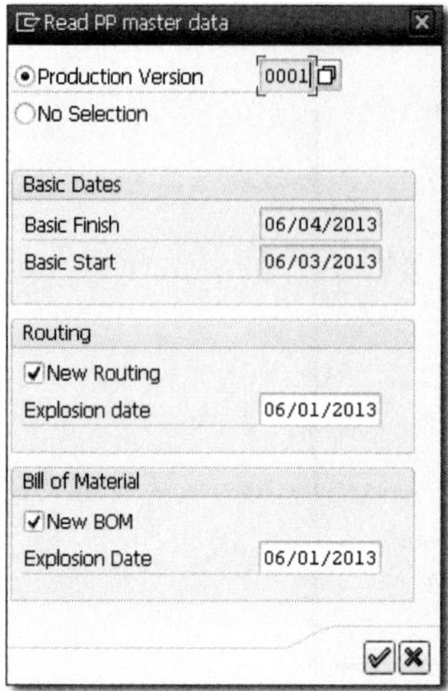

Figure 6.40 Read PP Master Data

6.3.6 Statuses

Whenever the business process owner performs any function, such as creating a production order, performing scheduling, checking capacity requirements or material availability, or generating a control instruction, the system keeps updating the status. These updates in status can be at the header level or operations level of the production order. Similarly, when you perform partial or complete confirmation, perform goods movement, or receive goods, the system correspondingly updates the status after every activity. In addition to system status, there is also user status. The system status works according to preconfigured rules and logic, whereas you're free to define your own rules in the user status.

When you run the production order information system (Transaction COOIS), you can choose status or user status as one of the selection parameters. For example, you list all of the production orders that have partial confirmation status (PCNF).

Figure 6.41 appears when you click on the STATUS icon in the production order header.

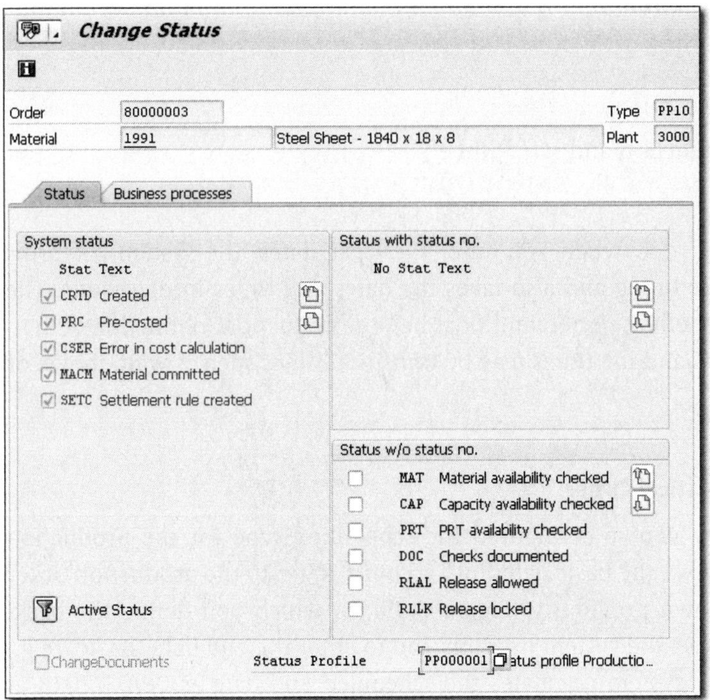

Figure 6.41 Status View of the Production Order

6.3.7 Scheduling

Building on the information that we covered in the work center discussion (see Section 6.2.3) and in the routing discussion (see Section 6.2.4), this section provides more in-depth information on the factors that the system considers when you're working with scheduling. For example, the system takes the following into account from the work center: factory calendar, shift timings, maximum capacity utilization (in percentage), overload percentage, and number of individual capacities. From the routing, it considers information such as setup, processing, transition times, splitting, and overlapping.

When the system performs scheduling for the production order, it takes the following factors into consideration (we'll discuss these in more detail in the following sections):

- Dates in the production order
- Finite scheduling

- Transition times
- Splitting
- Overlapping
- Float before production and safety time
- Reduction

In the production order, when you enter the basic dates, the system performs comprehensive scheduling and also takes the dates and times into account. The dates may be the opening period and float before production (defined in SCHEDULING MARGIN KEY), and the times may be transition times such as wait, move, or queue times.

Dates of the Production Order

Depending on how you've configured the scheduling type for the production order, the system takes the basic dates into account. Refer to the production order creation screen shown previously (Figure 6.35), in which you define the backward scheduling type, the system prompts you to enter the end date (BASIC) (e.g., "06/04/2013"). The system calculates the scheduled end date as 06/03/2013 at 24:00. The system arrives at this schedule based on the following factors (and in the same sequence by performing forward scheduling):

- Float before production
- Operations times
- Safety time (float after production)
- Goods receipt processing time

In MRP, if you use backward scheduling, the system starts with the planned availability date. Therefore, the first object that the system considers is the goods receipt processing time to get the basic order finish date. Then, based on the in-house production time, the system calculates the basic order start date. Afterwards, starting from the basic order finish date, the float after production is considered to get the production finish date. From there (due to backwards scheduling), the operations are considered for calculating the production start date. Finally, the float before production is applied, and, if necessary and allowed by customizing Transaction OPU5, the initial basic start data is adjusted. The durations of the operations are made up of waiting, setup, edit, teardown, and transport. In the production order, the system displays the operation dates and timings on the DATES tab. Figure 6.42 shows the operation details screen of OPERATION 0010. Starting from the queue time, it takes the setup, processing, teardown, wait, and finally move times into account for the operation.

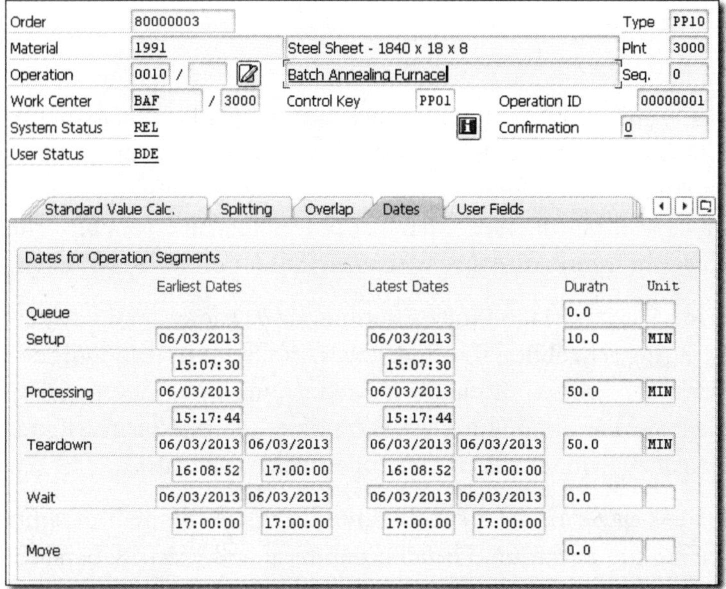

Figure 6.42 Dates and Times Details in the Operation View of the Production Order

Finite Scheduling

The objective of finite scheduling is an organization type in capacity planning that takes account of current capacity loads. *Finite scheduling* calculates the start and finish dates for operations in an order. In infinite scheduling, the capacity requirement isn't given due consideration. It also doesn't take into account the interdependencies of production orders.

The system uses backward and forward scheduling. In backward scheduling, the system determines the start date of the production order based on the latest finish date. In forward scheduling, the system determines the finish date of the production order based on the earliest start date. It's equally important to ensure that apart from scheduling, the material availability is another very important factor that you need to take into account. If it isn't possible to manage production within the given dates, then you can minimize the lead time by overlapping, splitting, and reducing the transition times, float before production, and safety time. During finite scheduling (capacity planning), you can also use the float times as degrees of freedom to move operations there to get a feasible plan for your resource-critical operations; that is, the production dates can be variable within the limits of the basic order dates.

Splitting

You can use *splitting* to split an operation into several suboperations that you can process in parallel with each other. This reduces the overall lead time of the operation. A prerequisite for splitting is that the work center capacity must contain multiple individual capacities; otherwise, parallel processing isn't possible. The system looks for this split information in the capacity header of the work center.

Figure 6.43 shows the Splitting tab in the operations view. The system only performs the splitting during scheduling if it finds the Required splitting indicator checked; otherwise, it considers splitting during reduction. If you don't specify the number of splits and the minimum processing time, the system distributes the processing time equally among the number of available capacities.

The Maximum Number of Splits field specifies the maximum number of splits that may take place for the operation. For an economical and practical perspective, a Minimum Processing Time must also be set.

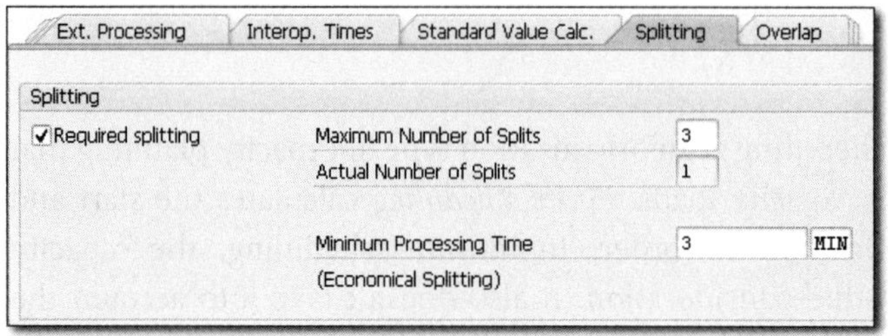

Figure 6.43 Splitting in the Operation View of the Production Order

[»] **Note**

Refer to Section 6.2.4, in which we cover splitting and overlapping, explained next.

Overlapping

You maintain the OVERLAPPING parameters, the MINIMUM SEND-AHEAD QTY, and the MINIMUM OVERLAP TIME in the operation data of the routing. These parameters relate to the overlapping with the follow-up operation.

During scheduling, if the minimum overlapping time isn't reached, no overlapping takes place. This is the case, for example, when the remaining processing time after the processing of the minimum send-ahead quantity is too small. The minimum overlapping relates to the processing time. Transition times such as wait time or move time don't contribute to the minimum overlapping time.

Depending on the settings made in the routing or production order for overlapping, when the system schedules the order, it performs overlapping during reduction (OPTIONAL OVERLAPPING), always (REQUIRED OVERLAPPING), or never (NO OVERLAPPING).

Figure 6.44 shows the OVERLAP tab in the production order.

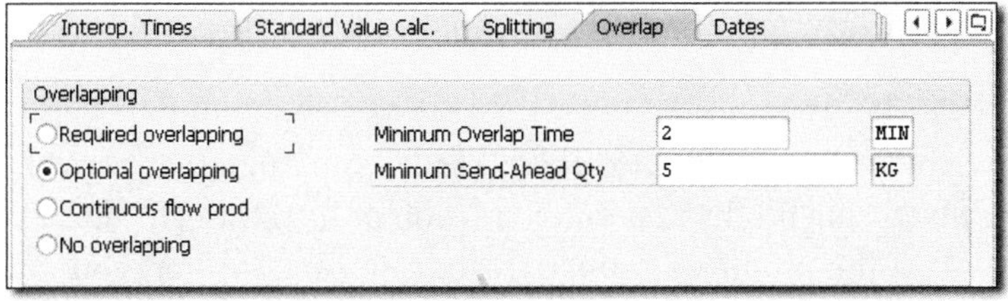

Figure 6.44 Overlapping in the Operation View of the Production Order

Float before Production

In Chapter 3, we covered the scheduling margin key, which provides additional buffers to the production process from unforeseen delays. If you refer to Figure 6.34 again, you'll notice the FLOATS area on the lower-right side of the screen. The float with SCHED. MARGIN KEY stipulates float before production, float after production, and the release time.

Reduction

While the buffers help in maintaining overall stability for the production process, you can reduce or eliminate them when needed. For example, if the system is unable to schedule production within the defined basic dates, it implements reduction up to the levels that you defined in Customizing. If you refer back to Figure 6.34, notice that the system performs REDUCTION LEVEL 2 on the lower-left side. This is the maximum level that we allowed for reduction in Chapter 3. The system first tries to resolve the scheduling issue without reduction, then with reduction level 1, and only if necessary finally with reduction level 2. Before making reduction, the system considers if the production dates (after scheduling) are within the basic dates, and if so, it makes no reduction. If not, the system continues the reduction process until it either meets the basic production dates or reaches the last reduction level, whichever comes first.

For example, you've configured the system to reduce up to two levels; however, if the system is able to meet the basic production date after first-level reduction, it secures the production dates. In this case, it doesn't proceed to the next reduction level. For such cases, there is also the chance to finally let the system adjust the basic dates to the new lead-time scheduling situation in Customizing Transactions OPU5/OPU3 (this is recommended). Otherwise, there might be production dates outside the basic dates, and the corresponding exception message occurs.

6.3.8 Availability Checks

Depending on how you've configured *availability checks*, you can either perform this step automatically during production order creation or its release, or interactively (manually) within the production order creation screen. Further, you can perform the material availability check for an individual production order or for a large number of production orders collectively.

Three different types of availability checks are available for a production order:

- **Material availability check**
  Checks the availability of the components for the production order, either against actual stocks (and optionally receipts) or against the planning.
- **Capacity availability check**
  Checks whether sufficient free capacity is available for the order's operations.
- **PRT check**
  Through a status in the master data, checks whether the required PRT is free.

For automatic availability checks, you can make the relevant configuration settings to enable the system to check for one or all of the availability checks during an order creation or order release (refer to Chapter 3 for these). For example, if there is a significant lead time between order creation and its release, you can choose not to perform a capacity check on order creation but instead during order release.

You can also set how the system should proceed if it finds missing capacity, missing PRTs, or a material shortage (due to missing parts) during the availability check. The following options are available for material, capacity, and PRT checks during order creation or release:

- Disallow order creation or release when there is a shortage.
- Allow order creation or release when there is a shortage.
- Allow the user to decide (interactive) to allow creation or release of an order when there is a shortage.

Material Availability Check

Despite all of the planning, checks, and balances in place, and even keeping the minimum stock levels, it's imperative for a production planner to consistently check material availability to ensure there is no disruption in the entire production process. The immediate and visible purpose of the material availability check is to ensure that the requisite components are available for the production order. If not, the system creates an entry in the missing parts list, which the production and the procurement planners can work on together to procure the components. So for in-house produced parts, the production planner can coordinate internally to ensure the requisite component's availability. For externally procured material, the procurement planner can take advantage of the missing parts list to coordinate and attend to procurement priorities.

With every availability check that you perform, the system correspondingly updates its status. In the material availability check, for example, the system assigns a status NMVP, if it makes no material availability check. A production order whose component requirements are checked and confirmed attain the status MABS (material confirmed). If the system can't confirm the availability of one or several components, the order receives the status FMAT (missing material), and the system creates a missing parts list.

In the production order, click on the COMPONENT AVAILABILITY button [Component]. If all components are available, then the system issues an information message to confirm material availability; otherwise, the pop-up shown in Figure 6.45 appears. Click on the MISSING PARTS OVERVI (overview) button.

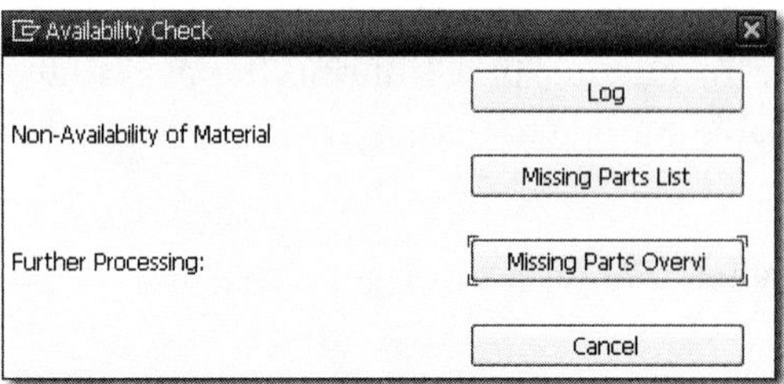

Figure 6.45 Material Availability Check

Figure 6.46 shows the missing parts overview, missing parts list, and other detailed information on the production order, such as the requirements quantity and the component's availability situation.

Availability Check

No. of Components Checked: 4
Missing Parts: 3
Overall Commitment Date: 06/26/2013

Material	Plnt	Sto...	Requirement quantity	Reqmt Date	Conf./Allocated qty	Comm. Date	ATP/avail.qty	Missing Part
CH-1410	3000	0001	2,500	06/10/2013	0	06/12/2013	0	X
CH-1420		0001	1,500	06/10/2013	3	06/12/2013	3	X
CH-1440		0001	50	06/10/2013	3.100	06/12/2013	3.100	X

Figure 6.46 List of Missing Components

While this example shows how you can interactively check for material availability within the production order change screen (Transaction CO02), the system also offers a choice called the *missing parts information system*. You should first perform a mass availability check on production orders, and then if this activity finds missing parts, you can use the missing parts information system to further work on producing or procuring the components.

To access the missing parts information system, follow the menu path, LOGISTICS • PRODUCTION • SHOP FLOOR CONTROL • INFORMATION SYSTEM • MISSING PARTS INFO

SYSTEM, or use Transaction CO24. Figure 6.47 shows the initial screen in which you assign the PLANT as "3000". In the SELECTION FROM ORDERS area, enter the MATERIAL as "1991", for which you want the system to check for missing components.

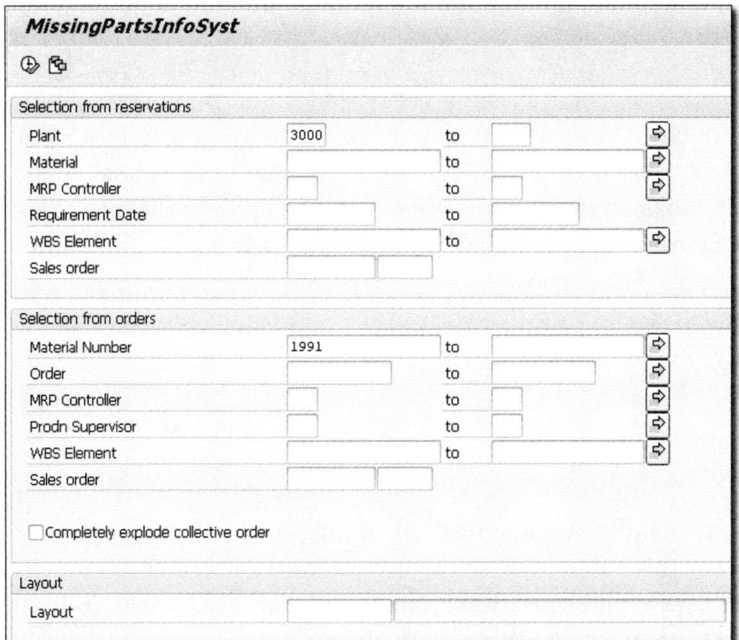

Figure 6.47 Initial Screen of the Missing Parts Information System

Press [F8] or choose EXECUTE, and the screen shown in Figure 6.48 appears, listing the missing components against the production ORDER 80000012.

Missing Parts Info System

Material	ReqmtDate	Order	Reqmt Qty	Comm. qty	Unit	Plnt	StLc	MRPC	ReservNo	Item
CH-1410	06/17/2013	80000012	20,000	0	KG	3000	0001	001	72571	1
CH-1420			12,000	3	KG		0001	001	72571	2
CH-1440			400	3.100	KG		0001	001	72571	4

Figure 6.48 Missing Components in the Missing Parts Info System Screen

[+]

Tips & Tricks

If you run the missing parts information system for the first time in a newly created SAP ERP system client, the system may produce an error stating that you need to run a specific program first. Use Transaction SE38 and Program PPCOXPR1, and choose EXECUTE (or press F8). You need to perform this activity only once for a given SAP ERP system client.

To collectively check material availability for several production orders, follow the menu path, LOGISTICS • PRODUCTION • SHOP FLOOR CONTROL • CONTROL • COLLECTIVE AVAILABILITY CHECK, or use Transaction COMAC. On the initial screen, define the selection parameters for which you want to perform the availability check, and then choose EXECUTE. In the ensuing screen, select all of the production orders and press F8 (or choose EXECUTE) to perform the collective availability check. The system creates a log that you can check to see the results of the collective availability check.

Capacity Availability Check

Similar to the material availability check, you can also enable the system to perform a capacity availability check automatically during the production order creation or release. For this example, when you create a production order, the system automatically checks the capacity and informs you of the shortage. You can choose the DETAILED INFO icon to gain a comprehensive overview of the capacity shortage, or you can choose the FINITE SCHEDULING icon and choose CONFIRM ALL in the next pop-up. This enables the system to perform finite scheduling and propose the next available date when available capacity can ensure production.

Further, when you're in the production order creation or change screen, you can choose the CAPACITY AVAILABILITY CHECK icon Capacity. You'll see a screen offering you the option to release a production order despite capacity shortage or cancel an order release. (This is the same setting that you configured in Chapter 3.)

Finally, the screen shown in Figure 6.49 appears when you use the following menu path within the production order: GOTO • CAPACITY PLANNING TABLE.

[»]

Note

Chapter 14 describes CRP in detail.

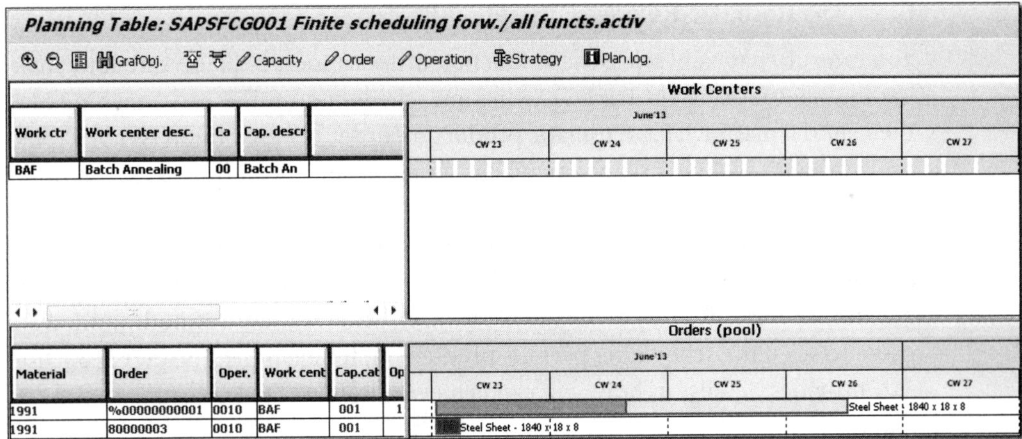

Figure 6.49 Capacity Requirements Planning Table

6.4 Release Production Order

When you create a production order, the system ensures that it contains the complete information you need to proceed with actual production. However, you can't initiate the actual production until you release the production order. Only a released production order enables you to do the following:

- Print shop floor papers.
- Withdraw materials from the warehouse.
- Confirm the operations for the order.
- Record goods receipt for the order.

As with other business functions in the production execution, there are several options which you can use to efficiently manage the release of individual or multiple production orders:

- Automatic release on creation of production order
- Individual release
- Collective release

We'll discuss these options in the following sections.

6.4.1 Automatic Release

You can automatically release a production order as soon as you create it by making the relevant settings in the production scheduling profile and assigning it to the material master. The automatic production order release option is more applicable in those business scenarios when there is a short lead time between order creation and release or in scenarios when the business and production processes are mature enough to use this option. While using this option enables you to eliminate a step of creating and then releasing an order and thus optimizes the business process, it also offers a loose control, wherein there is no intermediate check to see if the scheduling permits production, for example. Chapter 3, Section 3.7, provides details on how you can achieve automatic order release on its creation.

6.4.2 Individual Release

When you create a production order, the system assigns it a CRTD status, which denotes that it's in the creation stage. You can release it by choosing FUNCTIONS • RELEASE, or by clicking on the RELEASE icon. On its release, the system updates the status of the order from CRTD to REL. While this option will release the entire production order and thus releases all of its operations, you can also control the release of an individual operation within the production order. For example, if there is a long production or lead time for operations, then you can maintain greater control with a release at the operation's level. To do so, choose OPERATION • RELEASE in the operation overview screen of the production order.

To avoid having to release individual operations, you can also use TRIGGER POINT, which we've already covered in this chapter.

6.4.3 Collective Release

To release several production orders simultaneously, you can use the collective release option. To do so, follow the SAP menu path, LOGISTICS • PRODUCTION • SHOP FLOOR CONTROL • CONTROL • COLLECTIVE RELEASE, or use Transaction CO05N.

Figure 6.50 shows the parameters selection screen in which you can define the parameters such as material number, plant, order type, MRP controller, or production supervisor to enable the system to bring up relevant orders for release.

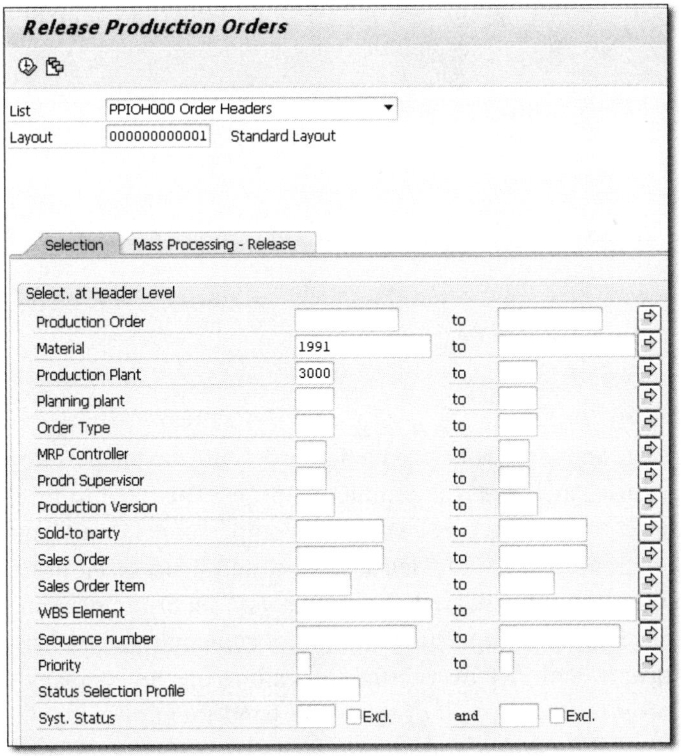

Figure 6.50 Mass Release of Production Orders

For this example, define MATERIAL as "1991" and PRODUCTION PLANT as "3000", and choose EXECUTE or press F8. The screen shown in Figure 6.51 appears, which consists of two production orders that have the CRTD (created) status. Select both the orders, and then select the RELEASE option from the MASS PROCESSING icon dropdown.

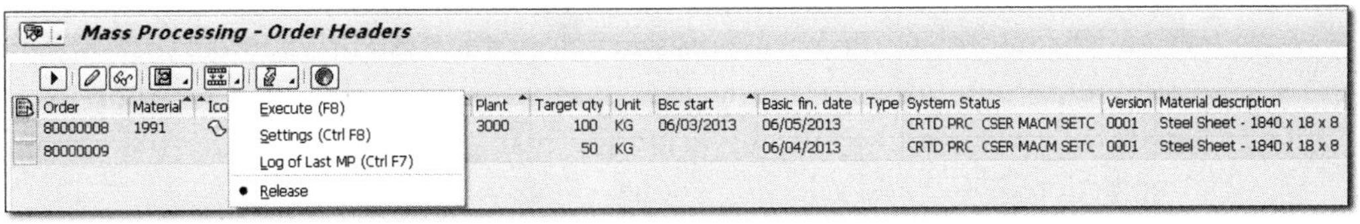

Figure 6.51 Mass Processing of Production Orders: Detailed Screen

The mass processing log appears showing that the collective release of production orders was successful (Figure 6.52). During the production order's release, the system also carried out the material availability check, and on finding no missing part (material shortage), successfully executed the release function.

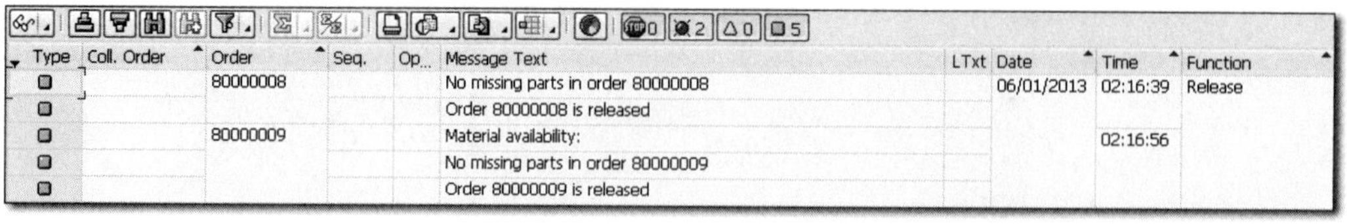

Type	Coll. Order	Order	Seq.	Op...	Message Text	LTxt	Date	Time	Function
		80000008			No missing parts in order 80000008		06/01/2013	02:16:39	Release
					Order 80000008 is released				
		80000009			Material availability:			02:16:56	
					No missing parts in order 80000009				
					Order 80000009 is released				

Figure 6.52 Log of Mass Processing Activity (Production Orders Release)

6.5 Printing

Upon the creation, release, and saving of a production order, the next logical step is to print shop floor papers, including the production order. The need to print shop floor papers is more prevalent where the production personnel don't have access to the SAP ERP system to check the details online, and hence need printouts. The shop floor papers consist of several objects that you can print, including production orders, goods issue slips, confirmation slips, document lists, pick lists, time tickets, Kanban tickets, and PRT overviews. Depending on the company's business needs, you can print some or all of them and hand them over to concerned production personnel for their information. If required, shop floor papers can be saved as PDFs using the Adobe Document Server.

The layout, printer settings, and other configuration settings were covered in Chapter 3. For this example, use the standard form (layout) for production orders that the SAP ERP system provides for various shop floor papers.

It's also important to note that you can only print a production order (or other shop floor papers), after you release it. You can use the manual option to print the shop floor papers or control the printing from the production schedule profile, in which you can select the checkbox that causes the system to automatically perform the printing function on release. You can individually print the shop floor papers of a production order by using the change production order transaction (Transaction CO02) and selecting Order • Settings • List Control.

To print shop floor papers, follow the menu path, Logistics • Production • Shop Floor Control • Control • Print, or use Transaction CO04N. On the resulting screen shown in Figure 6.53, enter these parameters: PLANT "3000", ORDER TYPE "PP10", and ORDER "80000009". Choose the ORIGINAL PRINTOUT radio button.

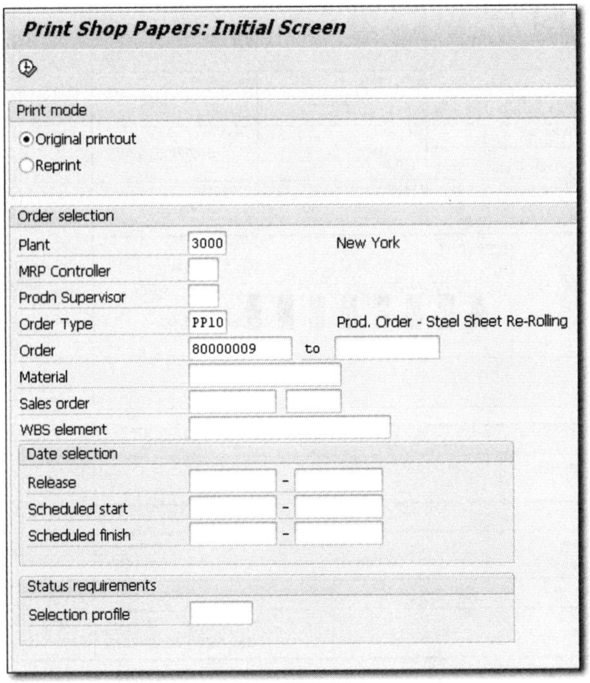

Figure 6.53 Initial Screen to Print Shop Floor Papers

Note [«]

You can only print an original shop floor paper once. For any subsequent print requirement of the same shop floor paper, select the REPRINT radio button.

Choose EXECUTE or press F8, and Figure 6.54 appears showing the OBJECT LIST (production order) print preview. It contains just about all of the information you need on a production order, such as order start and end dates, the operations involved in production, and the components required at each operation.

If you define more generic selection criteria on the initial parameters selection screen (refer back to Figure 6.53), such as material number or plant only, then Figure 6.55 appears consisting of a large number of production orders that you can print selectively.

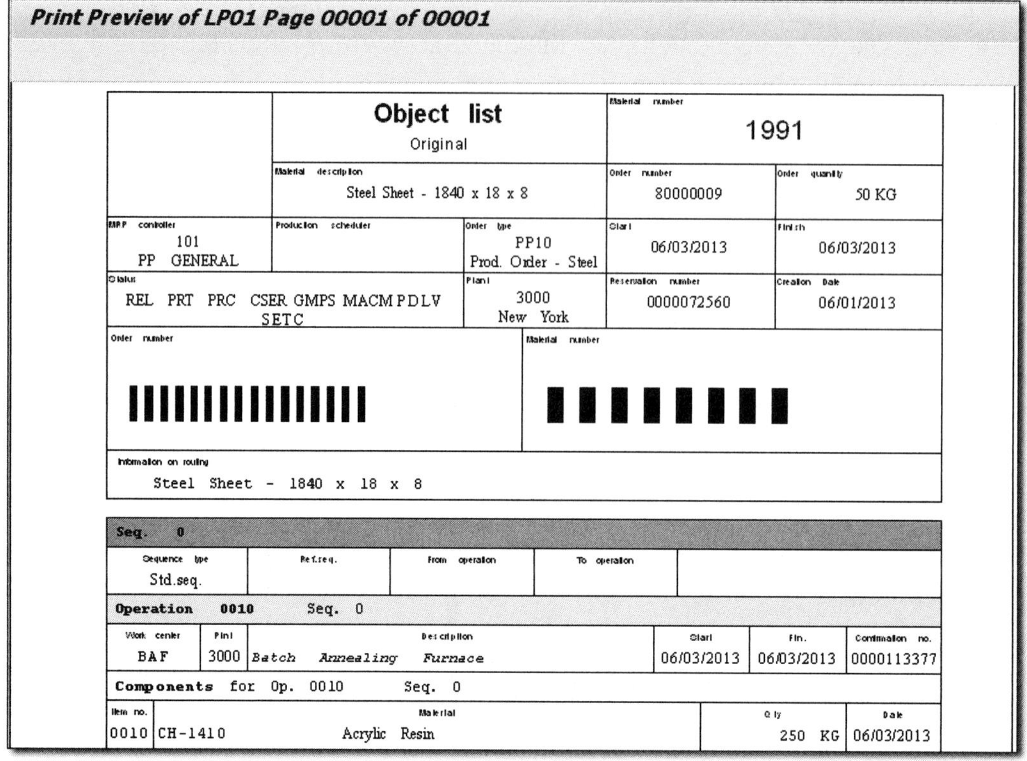
Print Preview of LP01 Page 00001 of 00001

Object list
Original
Material number: 1991
Material description: Steel Sheet - 1840 x 18 x 8
Order number: 80000009
Order quantity: 50 KG
MRP controller: 101 PP GENERAL
Production scheduler:
Order type: PP10 Prod. Order - Steel
Start: 06/03/2013
Finish: 06/03/2013
Status: REL PRT PRC CSER GMPS MACM PDLV SETC
Plant: 3000 New York
Reservation number: 0000072560
Creation Date: 06/01/2013
Order number
Material number
Information on routing: Steel Sheet - 1840 x 18 x 8

Seq. 0
Sequence type: Std.seq.
Ref.seq.
From operation
To operation

Operation 0010 Seq. 0

Work center	Plnt	Description	Start	Fin.	Confirmation no.
BAF	3000	Batch Annealing Furnace	06/03/2013	06/03/2013	0000113377

Components for Op. 0010 Seq. 0

Item no.		Material	Qty	Date
0010	CH-1410	Acrylic Resin	250 KG	06/03/2013

Figure 6.54 Print Preview of Object List (Production Order)

Print Shop Papers: List

Plant	3000 New York	Print Type	Original Printout
MRP Cont.		Print proc.	Online
Order Type	PP10 Prod. Order - Steel Sheet Re-Rolling		

Order Selection

Order	Material	Qty	U.	Errors	Short Text	Order Ty...	Status	Start	Finish
80000001	1988	1	KG	☐	Paint - Dull White	PP10	REL PRC BCRQ MACM SETC	05/30/2013	05/30/2013
80000002	1988	1	KG	☐	Paint - Dull White	PP10	REL CNF DLV PRC BASC BCRQ GMPS MACM*	05/30/2013	05/30/2013
80000003	1991	100	KG	☐	Steel Sheet - 1840 x 18 x 8	PP10	REL PRC CSER MACM SETC	06/03/2013	06/04/2013
80000005	1991	500	KG	☐	Steel Sheet - 1840 x 18 x 8	PP10	REL PRC CSER MACM MSCP SETC	06/03/2013	06/14/2013
80000006	1991	1,000	KG	☐	Steel Sheet - 1840 x 18 x 8	PP10	REL PRC CSER MACM SETC	06/03/2013	06/27/2013
80000007	1991	1,000	KG	☐	Steel Sheet - 1840 x 18 x 8	PP10	REL PRC CSER GMPS MACM SETC	06/03/2013	06/27/2013
80000008	1991	100	KG	☐	Steel Sheet - 1840 x 18 x 8	PP10	REL PCNF PRC CSER GMPS MACM SETC	06/03/2013	06/04/2013
80000010	1991	500	KG	☐	Steel Sheet - 1840 x 18 x 8	PP10	REL PRC MACM SETC	06/03/2013	06/14/2013
80000004	1991	1,000	KG	☐	Steel Sheet - 1840 x 18 x 8	PP10	REL PRC CSER MACM SETC	06/10/2013	07/05/2013

Figure 6.55 Item Overview of Shop Floor Papers for Printing

6.6 Material Withdrawal

After the system creates a production order (even before the production order is released), it automatically creates a material reservation. You'll pull from this reserve in the next actual production process. The next business function is to withdraw material from the warehouse. During the SAP ERP system implementation, the production and the product costing teams decide which components will form BOMs and which will be charged off to the production cost center. For example, low-value items such as greases, scrubbers, cleaning solutions, and lubricants are used in the production process but aren't the components used in manufacturing a product. Similarly, bulk material isn't part of the BOM. There are three ways in which you can perform material withdrawal against the production order, which we'll discuss in detail in the following sections:

- Goods issuance against production order
- Picking list
- Backflush

On goods issue posting, the system enables several activities in the background:

- Reduces the stock quantity of the material by the withdrawn quantity at both the plant and storage location levels
- Updates the consumption statistics, as these consumption figures eventually facilitate forecasting
- Reduces the reservation with the quantity withdrawn
- Debits the production order with the actual costs
- Creates a material document showing the goods issue and an accounting document, consisting of all postings in SAP ERP Financials (material stock and consumption accounts)

6.6.1 Goods Issuance against the Production Order

The system uses movement type 261 for goods issuance against the production order. This means that you can withdraw the material directly from the warehouse against the production order or first transfer it from your main warehouse to the virtual production floor storage location and then post goods issue as you consume it. If you use the second option, you must first create a transfer posting (movement type 311) to transfer material from the main storage location to your shop floor storage location. There is an issuing and a receiving storage location for this type of transfer posting. Once at the shop floor storage location, you can issue the material as and when consumed, and the system reflects the material stock at the shop floor storage location also. This business scenario is more applicable when you can issue smaller quantities of materials that are used frequently.

To perform goods issuance against the production order, follow the SAP menu path, Logistics • Materials Management • Inventory Management • Goods Movement • Goods Movement, or use Transaction MIGO. Figure 6.56 shows the initial screen in which you select Goods Issue from the first dropdown list and then select Order from the next dropdown list. Next, enter the production order number as "80000009". It's also important to note that the system should use movement type 261, which we discussed earlier. This is shown on the far-right side of the screen in the GI for Order field.

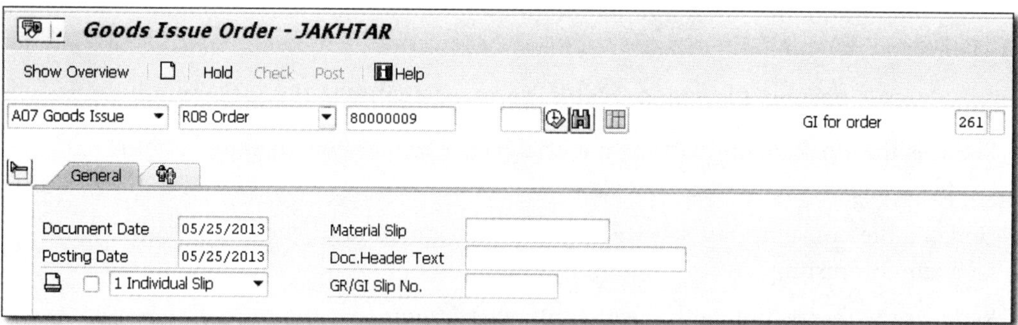

Figure 6.56 Header Screen of Transaction MIGO

When you press Enter, the system brings up the list of components directly from the production order (see Figure 6.57). This screen shows the same components that the system reads from the production order's components overview screen and suggests for goods issuance. You can make changes to the quantities, if desired, and also enter the storage location (SLoc), from where you want to issue the components. Finally, select the OK checkbox for each component that you want to issue. Save your entries, and the system creates a material document number (and a corresponding accounting document).

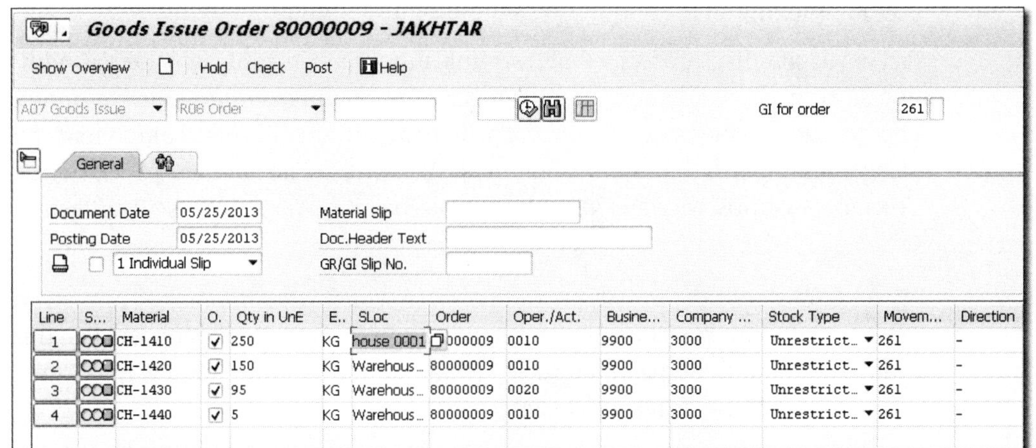

Figure 6.57 Goods Issuance against the Production Order

Figure 6.58 is the display production order screen (Transaction CO03) in which you navigate to the COMPONENTS DETAILS screen. It shows RESERVATION 72560 and item 1 for the production order, along with the REQUIREMENT QTY, and the WITHDRAWAL QTY (after the goods issuance step).

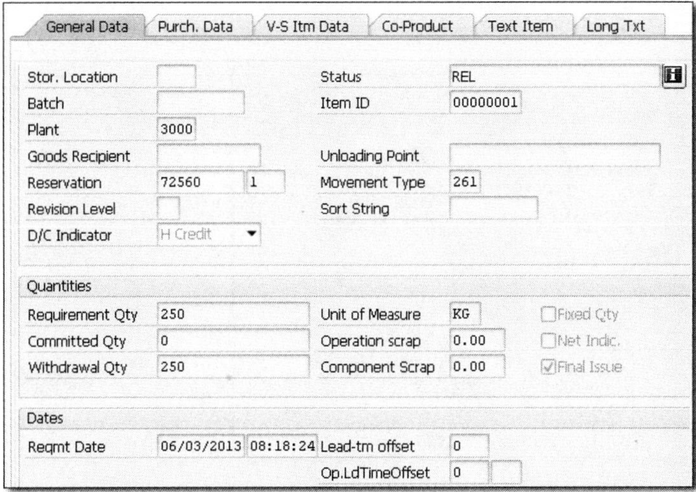

Figure 6.58 Material Withdrawal in the Component View of the Production Order

6.6.2 Picking List

Picking is another process of performing the same function of material withdrawal from the warehouse. A *picking list* is an intuitive and interactive option you can use to issue materials against production orders. To issue components by using a picking list, follow the menu path, LOGISTICS • PRODUCTION • SHOP FLOOR CONTROL • GOODS MOVEMENTS • MATERIAL STAGING • PICK, or use Transaction CO27.

Figure 6.59 shows the initial screen that offers you a large number of options to enter parameters, based on which the ensuing picking list will appear. For this example, define these parameters: MATERIAL "1991", PLANT "3000", WORK CENTER "BAF", and PLANT (for work center) as "3000".

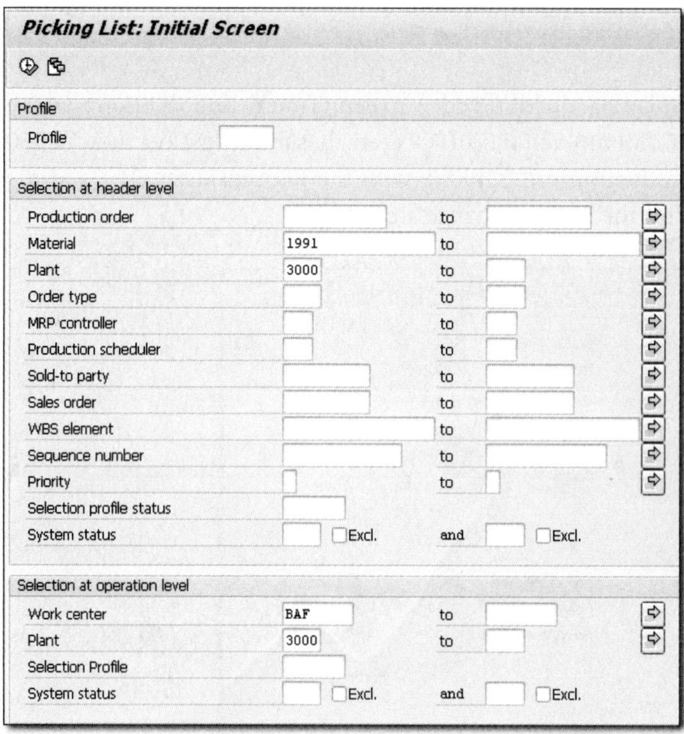

Figure 6.59 Initial Screen of Picking List

Choose EXECUTE or press F8 to open the screen shown in Figure 6.60. This screen consists of several production orders against which you can initiate the picking process. Select the four components that pertain to the ORDER 80000007, and click on the PICKING button in the toolbar.

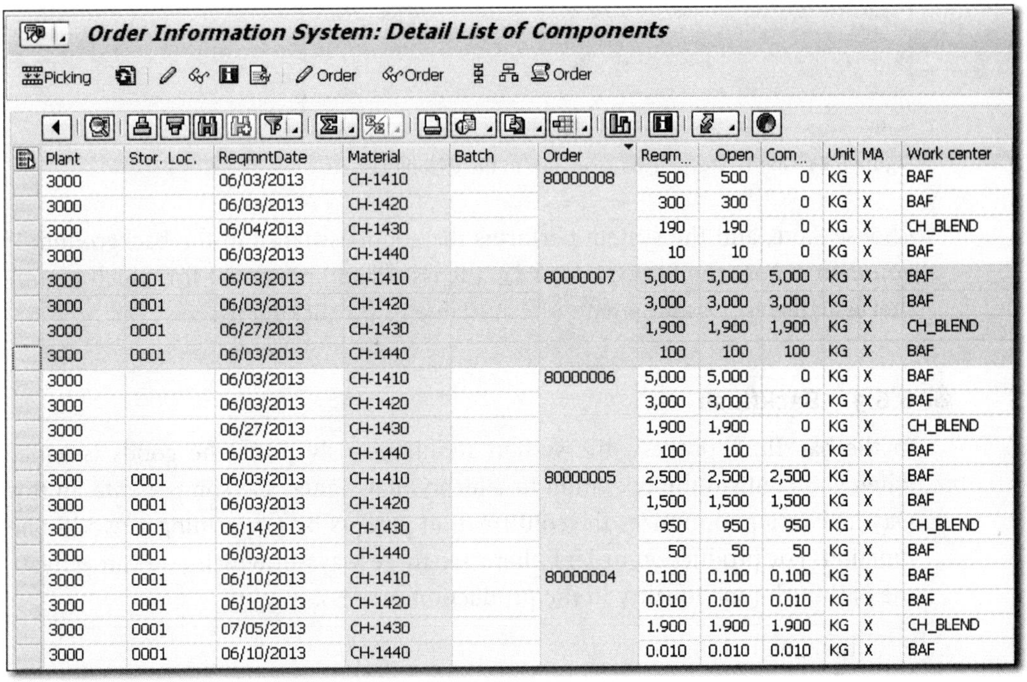

Order Information System: Detail List of Components

Picking | Order | Order | Order

Plant	Stor. Loc.	ReqmntDate	Material	Batch	Order	Reqm...	Open	Com...	Unit	MA	Work center
3000		06/03/2013	CH-1410		80000008	500	500	0	KG	X	BAF
3000		06/03/2013	CH-1420			300	300	0	KG	X	BAF
3000		06/04/2013	CH-1430			190	190	0	KG	X	CH_BLEND
3000		06/03/2013	CH-1440			10	10	0	KG	X	BAF
3000	0001	06/03/2013	CH-1410		80000007	5,000	5,000	5,000	KG	X	BAF
3000	0001	06/03/2013	CH-1420			3,000	3,000	3,000	KG	X	BAF
3000	0001	06/27/2013	CH-1430			1,900	1,900	1,900	KG	X	CH_BLEND
3000	0001	06/03/2013	CH-1440			100	100	100	KG	X	BAF
3000		06/03/2013	CH-1410		80000006	5,000	5,000	0	KG	X	BAF
3000		06/03/2013	CH-1420			3,000	3,000	0	KG	X	BAF
3000		06/27/2013	CH-1430			1,900	1,900	0	KG	X	CH_BLEND
3000		06/03/2013	CH-1440			100	100	0	KG	X	BAF
3000	0001	06/03/2013	CH-1410		80000005	2,500	2,500	2,500	KG	X	BAF
3000	0001	06/03/2013	CH-1420			1,500	1,500	1,500	KG	X	BAF
3000	0001	06/14/2013	CH-1430			950	950	950	KG	X	CH_BLEND
3000	0001	06/03/2013	CH-1440			50	50	50	KG	X	BAF
3000	0001	06/10/2013	CH-1410		80000004	0.100	0.100	0.100	KG	X	BAF
3000	0001	06/10/2013	CH-1420			0.010	0.010	0.010	KG	X	BAF
3000	0001	07/05/2013	CH-1430			1.900	1.900	1.900	KG	X	CH_BLEND
3000	0001	06/10/2013	CH-1440			0.010	0.010	0.010	KG	X	BAF

Figure 6.60 Overview Screen of the Picking List

Figure 6.61 appears in which you can make any last-minute changes to the issuance quantities. If you have missing issuance storage locations, then you can enter them on this screen. Notice that the MOVEMENT TYPE is 261, which is for goods issuance against the order. In this step, the system also performs the stock determination to ensure that the components required for issuance are available in stock. If you've configured any specific stock determination rule and assigned it in the STOCK DETERM. GROUP field in the PLANT DATA/STORAGE 2 of the material master, then the system accordingly suggests the components during the picking function.

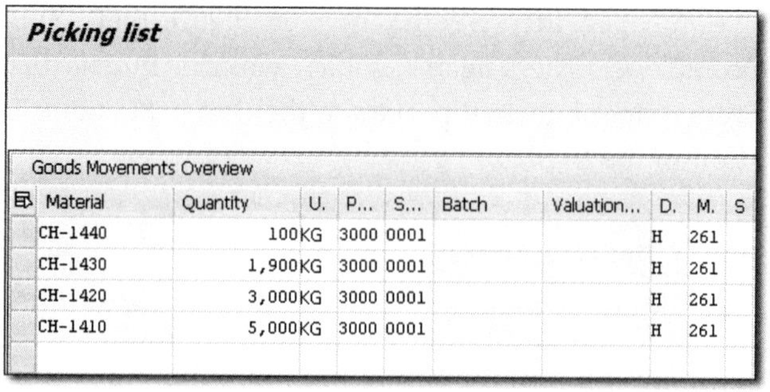

Picking list

Goods Movements Overview

Material	Quantity	U.	P...	S...	Batch	Valuation...	D.	M.	S
CH-1440	100	KG	3000	0001			H	261	
CH-1430	1,900	KG	3000	0001			H	261	
CH-1420	3,000	KG	3000	0001			H	261	
CH-1410	5,000	KG	3000	0001			H	261	

Figure 6.61 Goods Movements Overview in the Picking List

Choose SAVE, and the system performs the goods issuance in the background. If you then choose the REFRESH icon, the issued components—for which no further issuance is possible—won't be available in the picking list.

6.6.3 Backflush

In the backflush process, the system simultaneously posts the goods issuance when you confirm the operation to which you assigned the components. In this way, the system optimizes the confirmation process by performing a two-in-one function. For production orders, there are three ways in which you can activate the backflush functionality in the production process:

- In the material master
- In the work center
- In the material component assignment

We'll discuss these in further detail in the following sections.

Activate Backflushing in the Material Master

Figure 6.62 shows the MRP 2 view of the material master (component) with the BACKFLUSH field. If you place the cursor on the field and press F4, the system brings up the two available options shown on the right-hand side of the screen. The first option is that the system always backflushes (1) this component during the confirmation process. The second option is that it will look for the work center to see if the backflush checkbox (2) is selected to enable it to execute the backflush function.

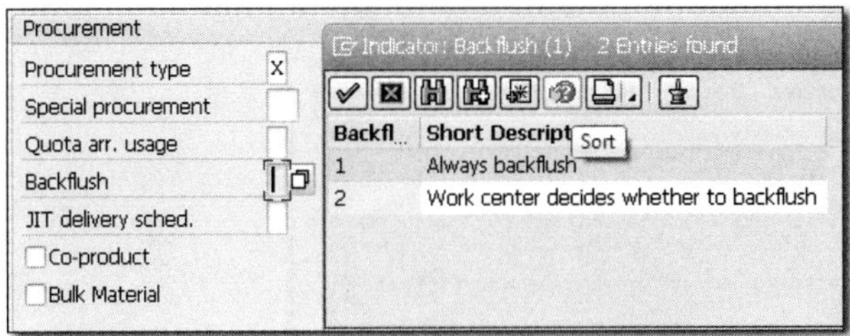

Figure 6.62 Backflush Option in the Material Master

Activate Backflushing in the Work Center

Figure 6.63 shows the Basic Data tab of work center BAF for plant 3000 (Transaction CR02), in which you can select the Backflush checkbox.

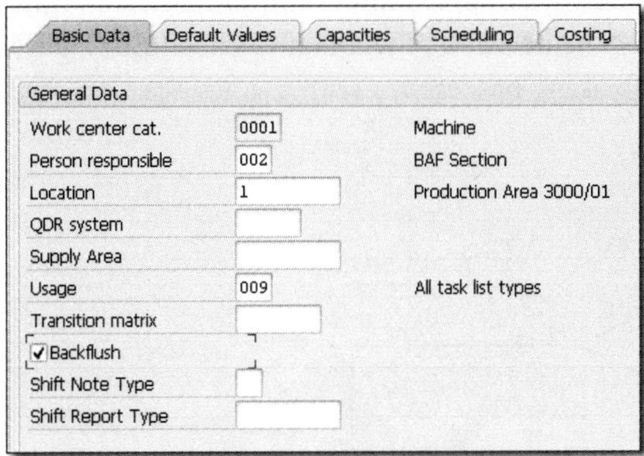

Figure 6.63 Backflush Option in the Work Center

Activate Backflushing in the Material Component Assignment

Figure 6.64 shows the Material Component Overview screen of routing (Transaction CA02), in which Component CH-1430 includes the Backflushing option when the user confirms operation 0020 (see the last column Oper./Act.).

Tips & Tricks [+]

If you haven't maintained the Backflush indicator in any of the master data (material master, BOM, or routing), you can still select the Backflush checkbox in the Material Component Overview screen of the production order. However, this is applicable to the specific production order only and will in no way update the master data.

Material Component Overview

New Assignment Reassign BOM Task list Operation

Material	1991	3000	Steel Sheet - 1840 x 18 x 8
Group		Sequence	0	Steel Sheet - 1840 x 18 x 8
BOM	00003384	Alt.BOM	1	

Item Overview

P	L...	Path	I...	Component	Quantity	Sort String	U.	I...	Backflushing	Oper./Act.
☐	0	0	0010	CH-1410	50		KG	L		
☐	0	0	0020	CH-1420	30		KG	L		
☐	0	0	0030	CH-1430	19		KG	L	☐	0020
☐	0	0	0040	CH-1440	1		KG	L		

Figure 6.64 Backflush Option in Routing

With the three backflush options just discussed, it's important to understand the sequence in which the system makes the decision of whether to backflush. Figure 6.65 shows the backflush decision-making process.

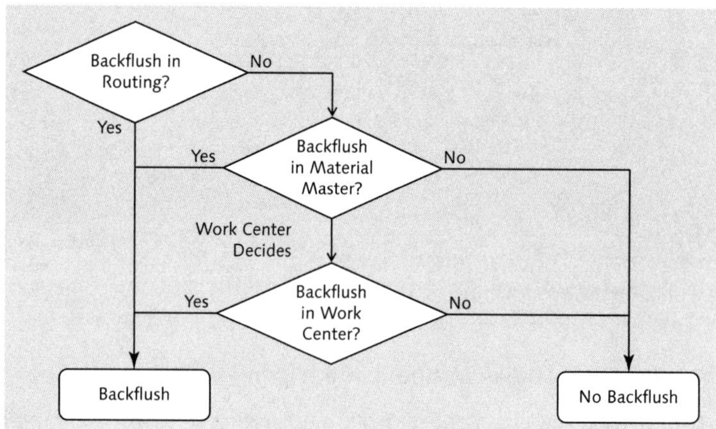

Figure 6.65 Decision Making on Backflush

6.7 Confirmation

To bring comprehensive visibility to the entire production process, the confirmation process provides the production scheduler and the MRP controller with important information: the quantities produced and confirmed for each operation or order. The types of confirmation that you can perform are listed here:

- Quantities/subsets
- Activities (setup time, processing time, etc.)
- Dates (start and end dates for setup, processing, etc.)
- Personnel data (personnel number, number of employees)
- Work centers used

As previously mentioned, a confirmation process automatically triggers some or all of the following business functions, depending on how you've configured the system or set up the master data:

- Updates the actual costs
- Reduces the capacities of the work centers involved
- Performs goods issuance for backflush components
- Performs goods receipt for the automatic goods receipt material

The system calculates the remaining order quantity from the four quantity fields in the production order based on the following formula:

> Remaining quantity = Total quantity – scrap – delivered + expected yield variation (due to underdelivery/overdelivery)

The available options in confirmation are listed here:

- **Time events**
  If you want the confirmations to be based on specific points in time, you can have *time events* confirmed. For example, you can specify the date and time the process starts and the date and time the process ends. The system automatically calculates the processing time based on this data.
- **Milestones**
  If you've defined an operation as a *milestone*, the confirmation of the milestone operation includes all previous operations. For example, during in-process (during production) quality inspection, you can only confirm scrap quantity after you perform the quality inspection at an operation.
- **Progress confirmations**
  Like the milestone confirmation, the *progress confirmation* of an operation automatically includes all previous operations. However, in contrast to the milestone confirmation, you don't have to specifically mark an operation as a progress confirmation. A special characteristic of the progress confirmation is

that each confirmation always considers the entire quantity that has been produced until the confirmation is triggered. A progress confirmation is useful if several partial confirmations are required and if it's easier to determine the total quantity rather than only the newly added quantity.

You can also confirm an entire production order. However, you won't be able to confirm the details of the actual activities in this case. The system uses standard activities details given in the production order.

Next, let's discuss the various options available to perform confirmation for a production order.

6.7.1 Confirmation at the Operations Level

You can opt to confirm an operation either with a time ticket or a time event. For a confirmation with a time ticket, the system confirms the quantities, activities, dates, and personnel data of an operation. The activities that you can confirm depend on the standard value key, whereas the quantities that you can confirm are yield, scrap, and rework. For dates, you can specify the start and end of the execution, including time and calendar date.

To bring greater visibility to the production process, you can periodically confirm the produced quantity at a work center. The frequencies with which you confirm an operation or an order depend entirely on business needs. For example, in 24 hours, you can confirm the quantities produced at the end of an 8-hour shift. This means you have to perform two partial confirmations and one final confirmation. When you perform the final confirmation, the system no longer expects any further confirmation against the operation or the order. The partial confirmation option is especially useful and practical for operations with long lead times. When you partially confirm an operation, the system assigns the production order a status of PCNF, whereas for final confirmation, it assigns the status as CONF.

To proceed with confirmation of a time ticket, follow the menu path, Logistics • Production • Shop Floor Control • Confirmation • Enter • For Operation • Time Ticket, or use Transaction CO11N.

Figure 6.66 shows the initial screen to confirm the time ticket. Enter the Order number as "80000008", enter Operation as "0010", and press Enter. The system automatically brings up the data that you can confirm. In this screen, you enter the actual quantities, activities, and duration details to enable the system to perform automatic functions as discussed earlier (such as updating costs, releasing capacities, etc.).

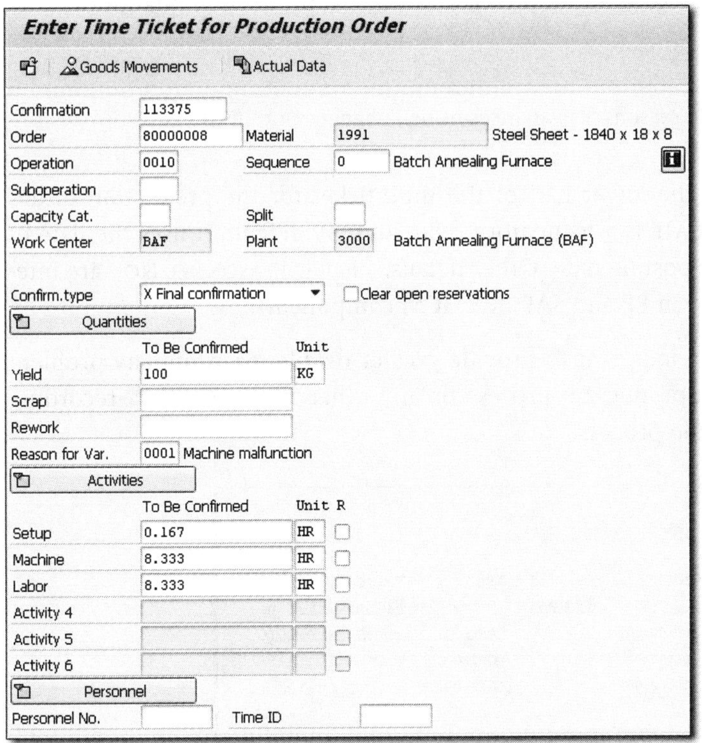

Figure 6.66 Operation Confirmation of the Production Order

Tips & Tricks [+]

If the system doesn't automatically bring up the data that you want to confirm, click on the ACTUAL DATA button.

Notice the CONFIRM.TYPE field, which in this case has the value of FINAL CONFIRMATION. In partial confirmation, you can select the relevant option from the dropdown list. In the AUTOMATIC FINAL CONFIRMATION option, the system checks whether the confirmed quantity is equal to, higher than, or less than the production order quantity. If it's less, then it automatically sets the status as PARTIAL CONFIRMATION, whereas if the confirmation quantity is equal to or exceeds the production order quantity, then it automatically sets the status as FINAL CONFIRMATION.

Further down this screen, you can also assign a REASON FOR VAR. (variance) in case there is a deviation between planned and actual production quantities.

[+] **Tips & Tricks**

You can add multiple reasons for variances using enhancement CONFPP07.

Figure 6.67 shows the lower half of the time ticket for the production order. Here, the POSTING DATE is a mandatory field, and, by default, it uses the date of confirmation as the posting date. Other details, such as PERSONNEL NO., are integration points between PP and SAP ERP HCM components.

You can also enter a long text to provide greater details regarding any problem encountered in the production process or any other findings worth recording during the production process.

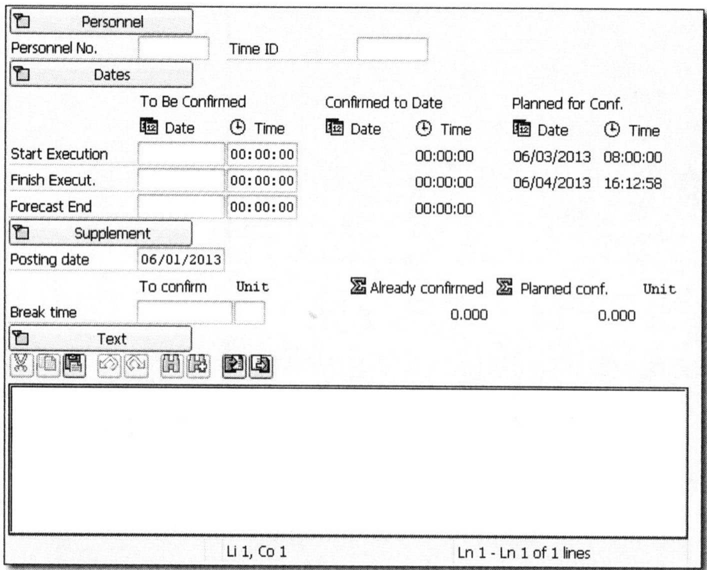

Figure 6.67 Additional Input Options in Operation Confirmation

Click on the GOODS MOVEMENTS icon, and the system navigates to the screen shown in Figure 6.68. This screen shows the components that the system will backflush when you save the confirmation. Notice that out of the four components that you want to backflush, only three are available on this screen. This occurs because you allocated the second component to operation 0020, whereas in this case, you're performing the confirmation process for operation 0010.

Enter Confirmation for Production Order: Goods Movements

Order 80000008 Status: REL
Material Number 1991 Steel Sheet - 1840 x 18 x 8
Oper./Act. 0010 Sequence 0

Goods Movements Overview

Material	Quantity	U.	P...	S...	Batch	Valuation...	D.	M.	S	Vendor	Customer	Compltd
CH-1410	500	KG	3000	0001			H	261				☑
CH-1420	300	KG	3000	0001			H	261				☑
CH-1440	10	KG	3000	0001			H	261				☑

Figure 6.68 Goods Movements (Backflush) in Confirmation

Save the confirmation, and you'll see a message indicating the successful save of the confirmation (time ticket) and also that the system was able to successfully perform goods movement (which in this case is components backflush) for the three components. (In Section 6.9.1, we cover what to do if the goods movement fails.)

6.7.2 Progress Confirmation

Each progress confirmation specifies the total quantity. For example, you create a production order of 50 PC of a material. The production process is distributed across two days. On the first day, 30 PC are produced, while the remaining 20 PC are produced on the second day. If you use the regular confirmation, the system confirms 30 PC in a partial confirmation on the first day and includes the remaining 20 PC in the final confirmation on the second day. For progress confirmation, the system confirms 30 PC on the first day and 50 PC on the second day.

You can enter progress confirmations using Transaction CO1F.

6.7.3 Confirmation for Order

You can also enter a confirmation for the entire order. In this case, the system confirms the operations based on the confirmed quantity. To proceed with the confirmation of an order, follow the menu path, LOGISTICS • PRODUCTION • SHOP FLOOR CONTROL • CONFIRMATION • ENTER • FOR ORDER, or use Transaction CO15.

Figure 6.69 displays just about all of the options available for confirmation, with the exception that you can't enter details of the actual activities consumed. When you confirm the quantity (YIELD TO CONF.), the system uses standard activities details. Notice that the option to backflush components, using the GOODS MOVEMENTS button, is also available in this screen.

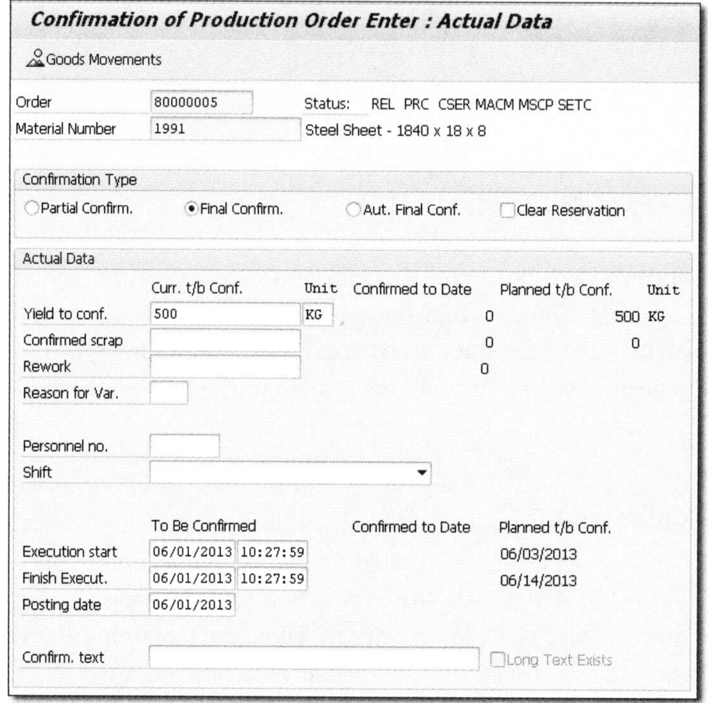

Figure 6.69 Production Order Confirmation

6.7.4 Confirmation Cancellation

If you've entered incorrect confirmation details, such as quantity or activities, you can cancel the confirmation using Transaction CO13. On the initial screen, enter the order number together with the operation number. On the next screen, choose SAVE to save the confirmation cancellation. There's no provision to make changes to a confirmation. You'll have to first cancel the confirmation and then do the confirmation again.

6.7.5 Display Confirmation

You can display the entered confirmation using Transaction CO14.

6.8 Goods Receipt

As soon as the last step or the operation in the production process is complete, you'll place the produced material in stock for future sales or for consumption in the next assembly. This step increases the physical inventory of the material. When you put the material in stock, you can post it in one of the following stock types:

- Unrestricted (freely usable) stock
- Quality inspection stock
- Blocked stock

If you integrate QM with PP in the production process, then the stock will go into quality inspection. The QM user performs the results recording and usage decision and also performs the stock posting of quality stock into unrestricted (or blocked) stock.

[«]

Note

Chapter 20 covers the integration of PP with QM.

When you post goods receipt against the production order, the system carries out many activities in the background:

- The system increases the relevant stock type and also updates the information in the Delivered Quantity field of the production order.
- If you post the material to consumption, then the system updates consumption statistics, which facilitate the forecasting process.
- The system credits the production order, and the stock value increases.
- If you post material to consumption, the system debits the recipient.
- The system updates the SAP General Ledger (G/L) accounts in FI, such as the material stock account and the factory activity account.

- The system creates a material document showing the goods receipt as well as an accounting document, consisting of all postings in FI.

You can use either a manual or automatic process to perform the goods receipt, which we'll discuss in the following sections.

6.8.1 Goods Receipt: Manual Process

To perform goods receipt against the production order, follow the menu path, Logistics • Materials Management • Inventory Management • Goods Movement • Goods Movement, or use Transaction MIGO. Figure 6.70 shows the initial screen in which you first select the Goods Receipt dropdown option, and then from the next dropdown, select Order. Enter the production order number as "80000009".

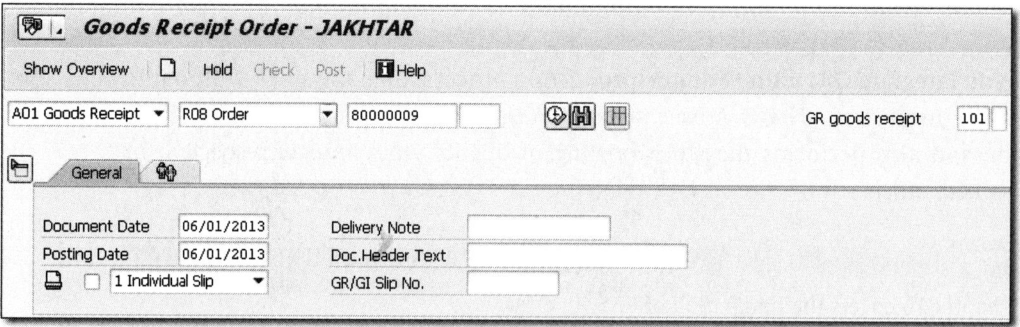

Figure 6.70 Initial Screen for Goods Receipt of the Production Order

It's important to note that the system uses movement type 101, as shown on the far-right side of the screen in the GR goods receipt field. When you press Enter, the system brings up the details of the material for which you can perform goods receipt against the production order (see Figure 6.71).

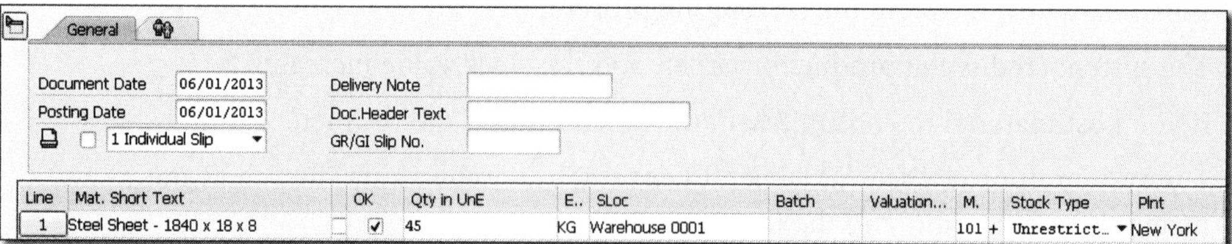

Figure 6.71 Goods Receipt of the Production Order

You can make changes to the quantity to reflect the actual goods receipt quantity and enter the storage location (SLoc) of the receiving storage location. Finally, select the OK checkbox. Save your entries, and the system creates a material document number (and a corresponding accounting document).

[«]

Note

You can perform the goods receipt of a co-product in the same way. Read Chapter 16 on handling co-products and by-products.

For this example, we performed the goods receipt of 45 kg of the MATERIAL 1991 against the production order.

If you go back to the display production order screen (Transaction CO03), the system updates the DELIVERED field to reflect the goods receipt against the production order. Because no operation or production order scrap was recorded, the system expects another delivery of 5 kg against the order (see Figure 6.72).

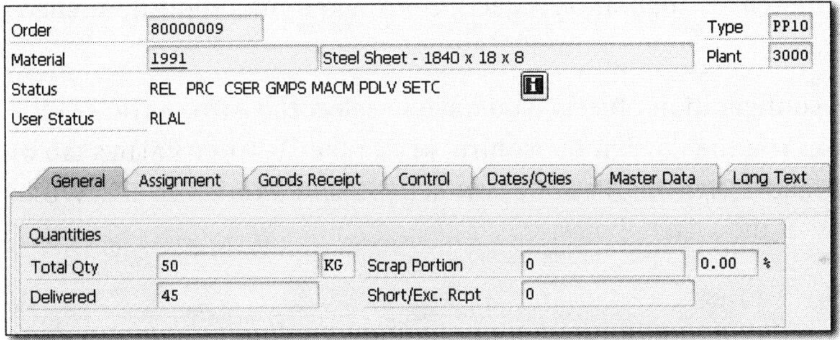

Figure 6.72 Production Order Header with Updated Quantities

Figure 6.73 shows that a production scrap of quantity 5 kg was first recorded and then a goods receipt was performed against the production order.

General | Assignment | Goods Receipt | Control | Dates/Qties | Master Data | Long Text

Quantities

Total Qty 50 KG Scrap Portion 0 0.00 %

Delivered 45 Short/Exc. Rcpt 5-

Figure 6.73 Shortage/Excess Production Updates in the Production Order

The system updates two fields, that is, DELIVERED and SHORT/EXC. RCPT. of 5– kg. A negative quantity in this field reflects a shortage in production and its eventual goods receipt.

Depending on how you've set up the confirmation parameters of the production order, the system either issues a warning or an error message if it finds confirmation that the goods receipt quantity deviates from the predefined tolerance limits. You can maintain the upper and lower tolerance limits in the WORK SCHEDULING view of the material master. If you don't maintain any tolerance limits, then the system infers this to be 0% tolerance and doesn't allow you any underdelivery or overdelivery.

6.8.2 Goods Receipt: Automatic Process

To optimize the business process and eliminate manual data entry efforts, you can also perform automatic goods receipt against the production order when you confirm the last operation. To activate automatic goods receipt, you need to make settings at two different configuration objects: control key and production scheduling profile.

In each of these configuration objects, you have to select the AUTOMATIC GOODS RECEIPT checkbox. You then assign the control key in the DEFAULT VALUES tab of the work center and assign the PRODUCTION SCHEDULING PROFILE in the WORK SCHEDULING view of the material master.

[»] **Note**

We cover the automatic goods receipt settings in Chapter 3.

6.9 Postprocessing

When you opt for automated business functions, such as backflush or automatic goods receipt, sometimes the system faces errors in posting these automated transactions. For example, during a scheduled backflush, if the system is unable to find the quantity to be backflushed in stock or if the storage location (to backflush from) is missing, it leads to errors in goods movements.

As a business process owner, you need to make sure that you frequently check the log for errors in goods movements and attend to the issues in a timely fashion.

In the following sections, we'll cover how to resolve errors due to goods movement and errors in cost calculations.

6.9.1 Reprocessing Goods Movements

To correct or reprocess the errors in goods movements, follow the menu path, Logistics • Production • Shop Floor Control • Confirmation • Reprocessing • Goods Movement, or use Transaction COGI.

Figure 6.74 shows the initial parameters selection screen, in which you can enter several relevant parameters, so that the system only brings up the relevant records for you to process. For this example, enter Plant "3000" and Material "1991", and choose Execute.

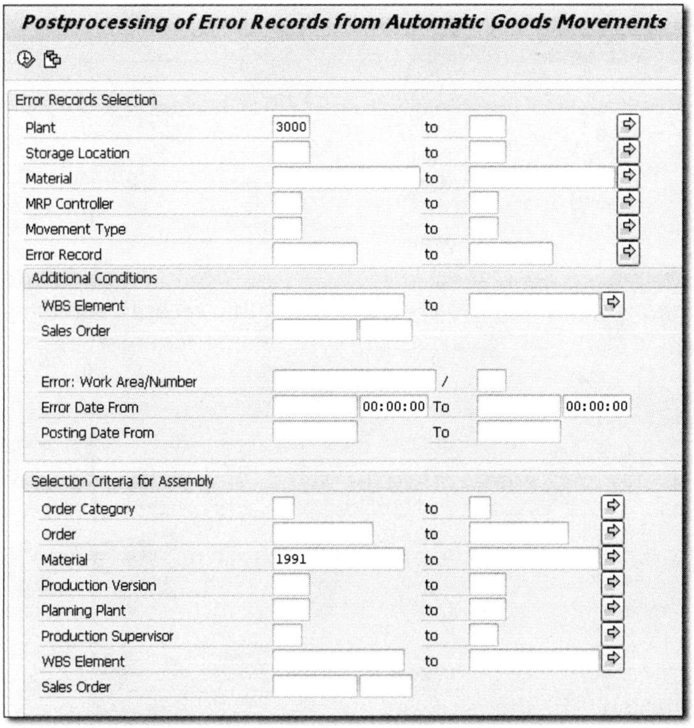

Postprocessing of Error Records from Automatic Goods Movements

Error Records Selection

Plant	3000	to	
Storage Location		to	
Material		to	
MRP Controller		to	
Movement Type		to	
Error Record		to	

Additional Conditions

WBS Element		to	
Sales Order			
Error: Work Area/Number		/	
Error Date From	00:00:00	To	00:00:00
Posting Date From		To	

Selection Criteria for Assembly

Order Category		to	
Order		to	
Material	1991	to	
Production Version		to	
Planning Plant		to	
Production Supervisor		to	
WBS Element		to	
Sales Order			

Figure 6.74 Error Handling in Automatic Goods Movements

The screen in Figure 6.75 consists of a list of four records that need to be reprocessed. You can select a record and click on the DISPLAY ERRORS icon to know the nature of the error, as shown in Figure 6.76.

Postprocessing of Error Records from Automatic Goods Movements

Individual Records   Stock   Material

06/01/2013 Goods Movements with Errors: Summarized Records

Status	Material	Plant	SLoc	Batch	M...	Qty in UnE	EUn	ID	No.	Created On	Error date	Counter
	CH-1410	3000			261	2,500	KG	M7	018	06/01/2013	06/01/2013	1
	CH-1420	3000			261	1,500	KG	M7	018	06/01/2013	06/01/2013	1
	CH-1430	3000			261	950	KG	M7	018	06/01/2013	06/01/2013	1
	CH-1440	3000			261	50	KG	M7	018	06/01/2013	06/01/2013	1

Figure 6.75 Postprocessing Records of Automatic Goods Movements

Figure 6.76 shows that the system found missing storage location information, which leads the system to terminate automatic goods movement processing. Select all four records, and click on the DETAILS icon.

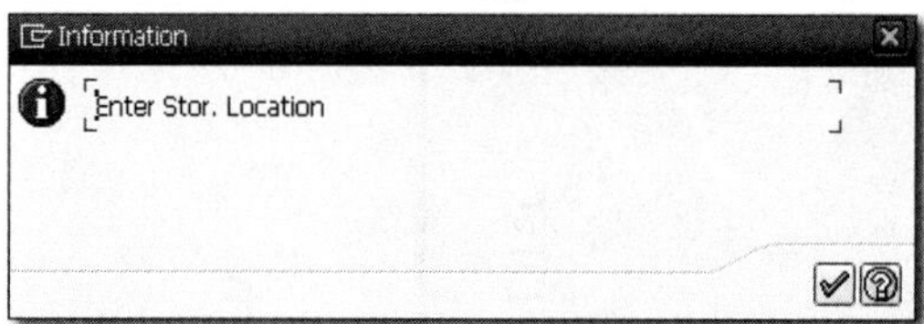

Figure 6.76 Error Details in Automatic Goods Movements

In Figure 6.77, enter the STORAGE LOCATION for each of the four records, and then click on the BACK icon.

Goods Movements with Errors

Goods Movements Overview

Material	Quantity	U.	P...	Storage Lo...	Batch		Valuation...	Pr...	Item	D.	M...
CH-1410	2,500	KG	3000	0001					10	H	261
CH-1420	1,500	KG	3000	0001					20	H	261
CH-1430	950	KG	3000	0001					30	H	261
CH-1440	50	KG	3000	0001					40	H	261

Figure 6.77 Goods Movements with Errors

The screen in Figure 6.78 shows the successful reprocessing status of all four records with a green traffic light. Save the records, and the system reprocesses them and posts goods issuance.

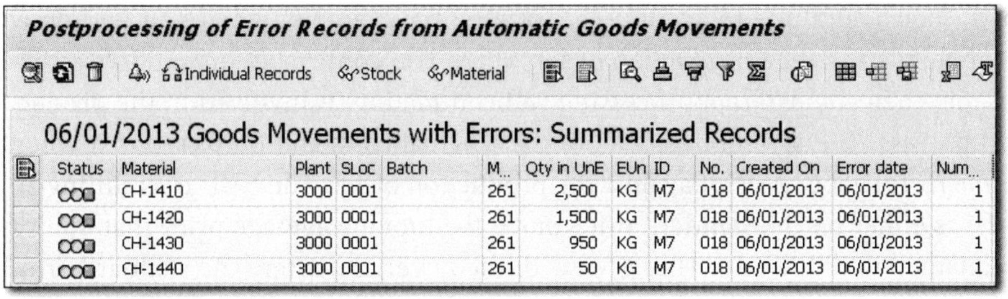

Postprocessing of Error Records from Automatic Goods Movements

Individual Records | Stock | Material

06/01/2013 Goods Movements with Errors: Summarized Records

Status	Material	Plant	SLoc	Batch	M...	Qty in UnE	EUn	ID	No.	Created On	Error date	Num...
OOO	CH-1410	3000	0001		261	2,500	KG	M7	018	06/01/2013	06/01/2013	1
OOO	CH-1420	3000	0001		261	1,500	KG	M7	018	06/01/2013	06/01/2013	1
OOO	CH-1430	3000	0001		261	950	KG	M7	018	06/01/2013	06/01/2013	1
OOO	CH-1440	3000	0001		261	50	KG	M7	018	06/01/2013	06/01/2013	1

Figure 6.78 Log on Automatic Goods Movements before Posting

6.9.2 Cost Calculation

Similar to automatic goods movement errors, the cost calculation errors arise during the confirmation process when the system updates the costs based on the entered data such as activities. Depending on how you've configured the system, errors occurring during the confirmation process and the cost calculation errors can be warnings or errors. For warnings, the system allows you to save the confirmation so that you can process the actual costs later on. For errors during costs calculation in confirmation, the system doesn't allow you to save the confirmation.

To reprocess the actual costs of the production order, follow the menu path, LOGISTICS • PRODUCTION • SHOP FLOOR CONTROL • CONFIRMATION • REPROCESSING • ACTUAL COSTS, or use Transaction COFC. Figure 6.79 shows the postprocessing error log that you need to engage and coordinate with the CO-PC resource to manage.

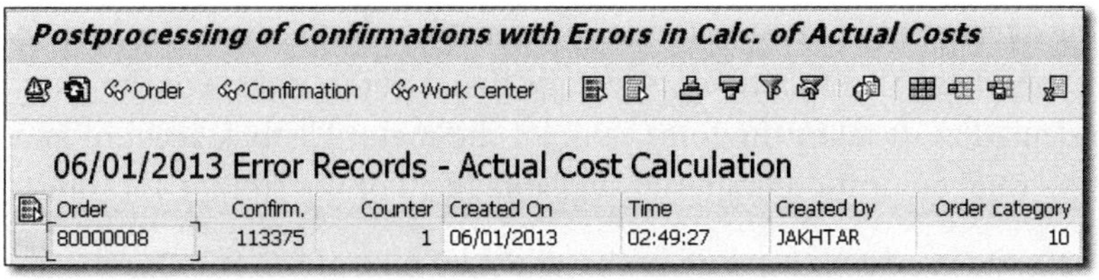

Postprocessing of Confirmations with Errors in Calc. of Actual Costs

Order | Confirmation | Work Center

06/01/2013 Error Records - Actual Cost Calculation

Order	Confirm.	Counter	Created On	Time	Created by	Order category
80000008	113375	1	06/01/2013	02:49:27	JAKHTAR	10

Figure 6.79 Postprocessing Confirmations with Errors in Cost Calculations

6.10 Settlement and Completion

When you create a production order, the system calculates the planned cost of the order. Thereafter, every business function that you perform will directly reflect as costs, which continue to accrue. This occurs, for example, when you withdraw components against a production order, when you confirm an operation's activities, or when any external processing (subcontracting) is involved in the production process. The system debits all of these actual costs and only credits it when you perform goods receipt against the production order. The basis of crediting an order is either on the standard price or on the moving average price that you've maintained in the ACCOUNTING 1 view of the material master. The standard price control may not reflect the actual costs of production and so must be passed on to settle the production order.

The CO-PC team normally proceeds with the settlement process after all production steps are ensured and the order has the status of DELIVERY COMPLETED. With the right settings in place, the system assigns a settlement rule to the production order as soon as you create the order. A settlement rule defines the recipient to which the system will settle the order and distributes the percentages of costs to various settlement receivers. Some of the settlement receivers can be a material, a cost center, a project, or a sales order. Normally, the CO-PC team assigns the settlement receiver of the production order to the material account. So, the settlement receiver is the material, and the settlement share is 100%. You can display the settlement rule of a production order by choosing HEADER • SETTLEMENT RULE.

When you perform goods receipt against the production order, the system credits the production order and debits the material stock account. If the price control in the material master is standard price, then the system valuates the material at the standard price. Because the actual production costs often vary, the system posts the difference to the price difference account during the settlement process.

For the moving average price control in the material master, the system automatically valuates the material at the moving average that it determines. For the moving average calculation, it takes the total value of the material stock account and divides it by the total quantity of material available in all of the storage locations.

You can evaluate the *overheads* separately that must be part of CO-PC. Overheads can be management or employee salaries, administration, and even utilities, such as electricity.

You can also evaluate the *variances* separately between planned and actual costs that occur due to the actual components and activities consumed that are different (either more or less) from the plan according to the BOM and routing.

Similarly, as soon as the goods issuance for the production order takes place, the system reflects this as *work in process (WIP)*, and it remains so until the goods receipt is posted and the stocks are updated. You can individually or collectively view the WIP details of production orders.

The completion of an order is logically divided into two distinct areas to distinguish logistical and accounting-based completion. When you technically complete the order and assign it the status TECO, the system updates the MRP-relevant details and deletes all reservations of components and capacity requirements. You can technically complete an order any time by choosing FUNCTIONS • RESTRICT PROCESSING • COMPLETE TECHNICALLY in the order screen. You'll then no longer be able to make changes in the production order. Further, with a technically completed order, you can no longer post costs to the order or perform any further confirmation.

While in the production order screen, you can set the completion status by choosing FUNCTIONS • RESTRICT PROCESSING • CLOSE. The system denotes the completion by the status CLSD.

You can also carry out both the completion and the technical completion in mass processing steps. We explain the mass processing function in Section 6.12.4.

6.11 Production Order Batch Traceability Using Work in Process Batches

Batch traceability is a key functionality used not only in the chemical and food industries but also in almost all industries, including in discrete manufacturing. SAP delivers the batch where-used list (Transaction MB56) in the SAP ERP system. This where-used list enables business users to trace the batch from the finished product to the raw material (top down) or from the raw material to the finished product (bottom up). The WIP batch is used to keep the batch traceability on a production order when several batches of raw material are consumed and several batches of finished product are generated.

Transaction MB56 works well in most cases. However, there is a limitation on a production order with N:N batches. In a standard production order, there is no link between the raw material batch consumed and the finished product's batch received in the warehouse; therefore, the batch where-used list determines all the batches of raw material for each batch of the finished product.

Consider, for example, that a problem was detected in one of the raw material batches after production. In this case, it isn't possible to identify exactly which batch of finished product was produced with this problematic batch of raw material. On previous releases, the only solution for this scenario was to create separate production orders for each raw material batch, but that required more data entry, inefficient business processes, and a significant amount of manual work.

In the example in this section, we focus on the WIP batch functionality that is available in the system for production and process orders in SAP ERP 6.0 EHP 4 and later releases. More specifically, we explain how to enhance the batch where-used list to show a link between each raw material batch consumed and the respective finished product batch produced on a production order. Our example involves a simple production order with two operations and a single component.

In the following sections, we start out with the necessary configuration that you'll need to set up WIP batches, followed by the required master data setup. Next, we transition to the business processes involved in WIP batches, and finally conclude by using the Batch Information Cockpit reporting tool to trace WIP batches.

6.11.1 Configuration Settings

Following are the three configuration steps that you'll need to perform, before creating the confirmation profile.

Activating Work in Process Batches Business Functions

Two business functions are related to the WIP batch functionality that you'll need to activate: LOG_PP_WIP_BATCH and LOG_PP_WIP_BATCH_02. For the basic use of the WIP batch and for the scenario that covered here, it's necessary to only activate the first business function. However, you should activate the second business function as well because it brings additional functionalities for the WIP batch such as inventory valuation, use with documentary batches, and use on an order split.

To activate the business function using the Switch Framework (Transaction SFW5), select the business function, and choose ACTIVATE CHANGES button.

[!]

Warning!

You can't deactivate these business functions after they are activated.

Activate Work in Process Batch

After the business function activation, it's also necessary to activate the WIP batch functionality in configuration. To complete this step, use Transaction OMCWB (see Figure 6.80), choose the ACTIVE radio button, and then save.

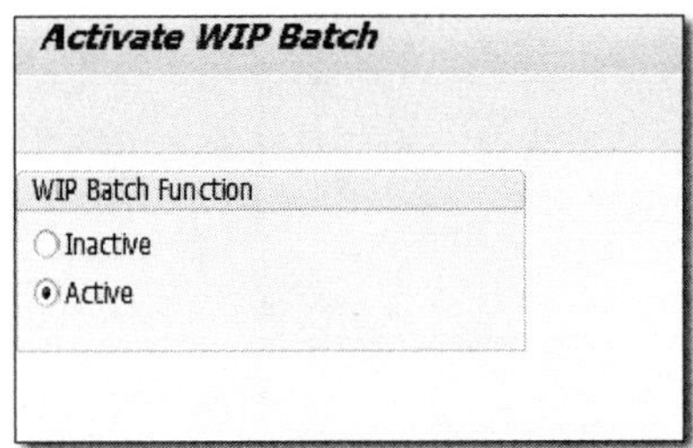

Figure 6.80 WIP Batch Activation

[!]

Warning!

Once activated, this configuration setting also can't be reversed, In other words, when you allow the use of the WIP batch functionality in your SAP ERP system, it isn't possible to get back to the original state in which WIP batches can't be used.

Activating the Original Batch

You must also activate ORIGINAL BATCH REFERENCE MATERIAL. To complete this step, execute Transaction OMTO. In the initial screen that appears (see Figure 6.81), select the ORIGINAL BATCH REFERENCE MATERIAL indicator and save. Doing so enables the system to use this functionality in master data, which is covered later in Section 6.12.2.

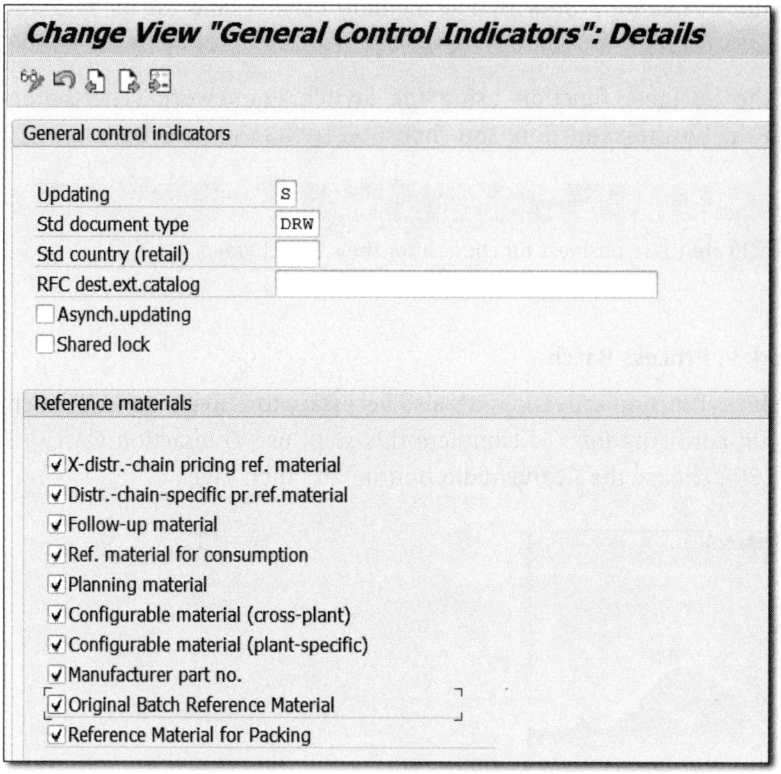

Figure 6.81 Activating Original Batch Reference Material

Confirmation

Now that you've already activated WIP batch and original BM functionalities, you need to create a new confirmation profile. Because the WIP batch information must be posted during the production order confirmation, your confirmation profile will include the WIP batch-specific subscreens.

For necessary configuration settings for the confirmation profile, use Transaction OPK0. In the screen that appears, copy the standard confirmation profile SAP001 by selecting it and choosing Copy. In the next screen (see Figure 6.82), you need to add Areas 8285 WIP Batch: Goods Receipts, 8290 WIP Batch: Goods Issues, and 8295 WIP Batch: WIP Batch Entry to the newly created Profile WIP-BATCH01.

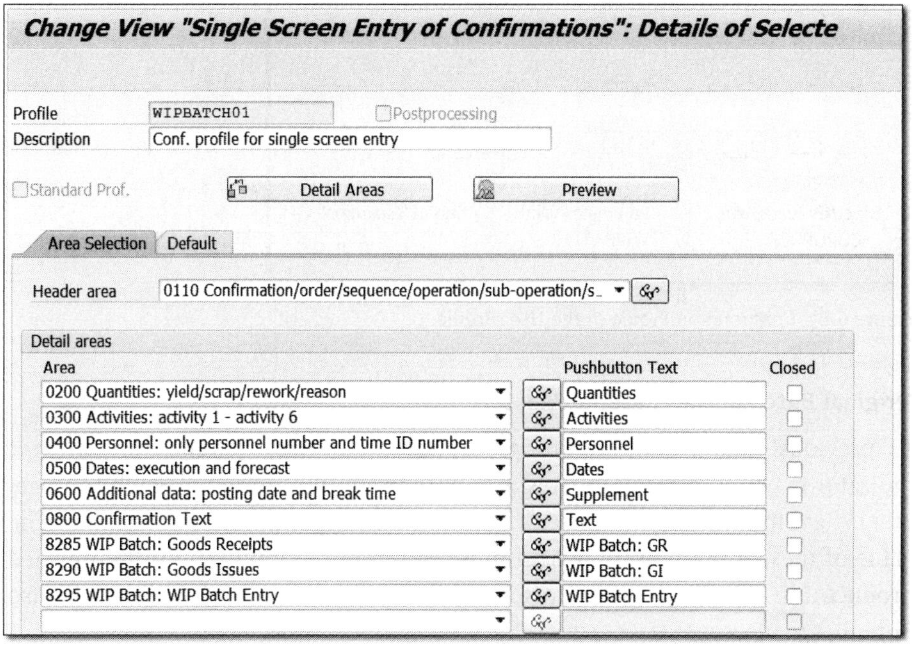

Figure 6.82 Confirmation Profile

6.11.2 Master Data Setup

Let's now delve into the requisite master data that you need to set up for WIP batches.

Confirmation Profile in the User Profile

To set the confirmation profile created in the previous step to default while posting a confirmation using Transaction CO11N, you need to choose the STANDARD PROF. (standard profile) checkbox on the initial screen of Transaction OPKO. However, because WIP batch is a very specific functionality that may not be used in all scenarios, here's an alternative. For each business user who posts a confirmation using WIP batch, you can add this confirmation profile to the user parameters. User parameters are used to store user-specific settings. To add a new user parameter, use Transaction SU3, add the SET/GET PARAMETER ID as "CORUPROF", and enter the previously created confirmation profile ("WIPBATCH01") as the PARAMETER VALUE (see Figure 6.83). You can either do this step yourself or request your Basis team member for help.

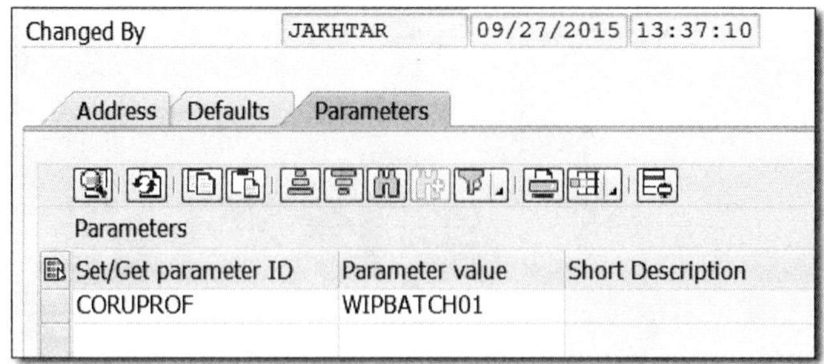

Figure 6.83 Confirmation Profile in the User Profile

Original Batch in the Material Master

As previously explained, activating the Original Batch Reference Material indicator in configuration activates the OB Management (original Batch Management) field in the material master's Work scheduling tab (see Figure 6.84). The value of the OB Management field must be set to 1 (Allowed) for the finished product, but it isn't necessary to set any value for the raw material. You'll also note that the BatchManagement checkbox has been selected.

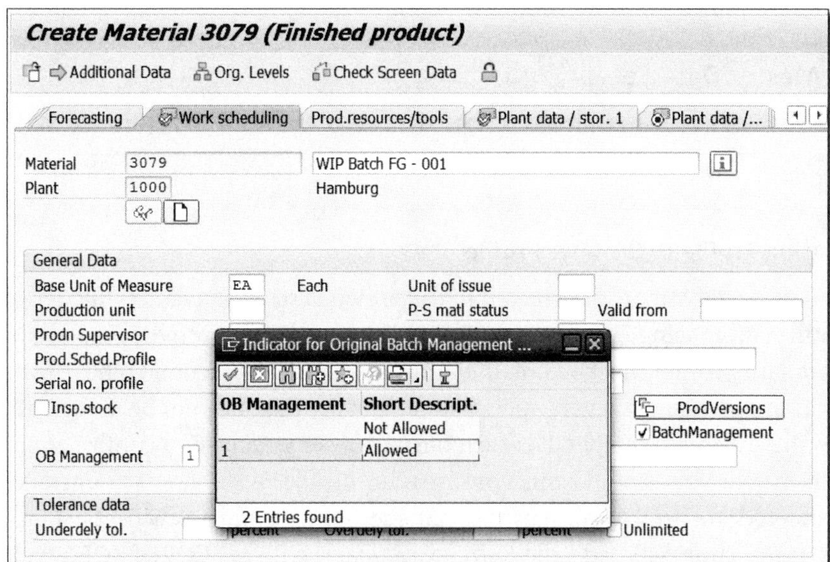

Figure 6.84 Allowing Original Batch Management in the Material Master

Let's now move to the business processes of WIP batches.

6.11.3 Business Processes for Work in Process Batches

The business process in WIP batches entails creating a production order as you normally would. This is then followed by confirmation in which you create the WIP batches. The system automatically posts the goods issuance and goods receipt.

Production Order

Consider the example shown in Figure 6.85 of a regular production order that appears after you execute Transaction CO02. This screen shows two operations in the OPERATION OVERV. (operation overview) section and a single component in the OBJECTS section.

To post the confirmation using WIP batches, you need to create a link between the raw material and the finished product. It isn't necessary to make any change on the production order header or components. You can go directly to the order confirmation.

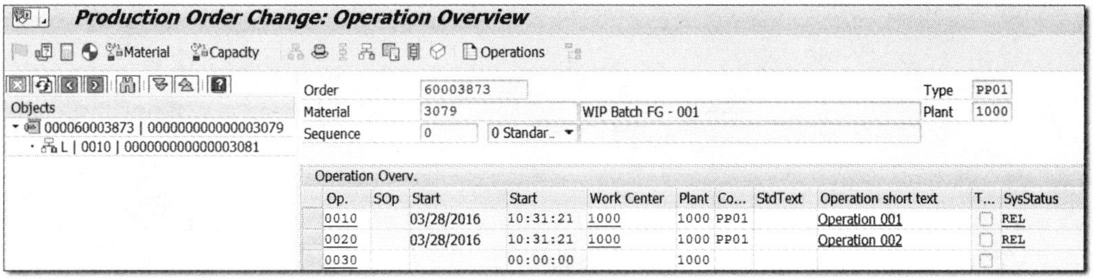

Figure 6.85 Production Order with Component and Operations

Confirmation and Goods Issuance

When positing a confirmation using the WIP batch confirmation profile, you'll notice some of the following details:

- The button to display the GOODS MOVEMENTS overview isn't displayed.
- The goods receipt goods movement is always proposed, even if automatic goods receipt isn't set for the production order.
- The goods issues goods movements is always proposed, even if the BACKFLUSH indicator isn't set for the components.
- All the goods movements are proposed without quantity.

Access Figure 6.86 using Transaction CO11N, and enter the production ORDER "60003873" and its first OPERATION, which, in this example, is "0010". Scroll down the screen to evaluate the information provided here.

The same three areas—WIP BATCH: GR, WIP BATCH: GI, and WIP BATCH ENTRY—that were set in the confirmation profile are displayed here. In the WIP BATCH: GI area ❶, the system proposes the raw MATERIAL 3081 together with its BATCH WIP-RM-01 for consumption (using MVMNT TYPE 261). Select this line item, and then choose the WIP BATCH button ❷.

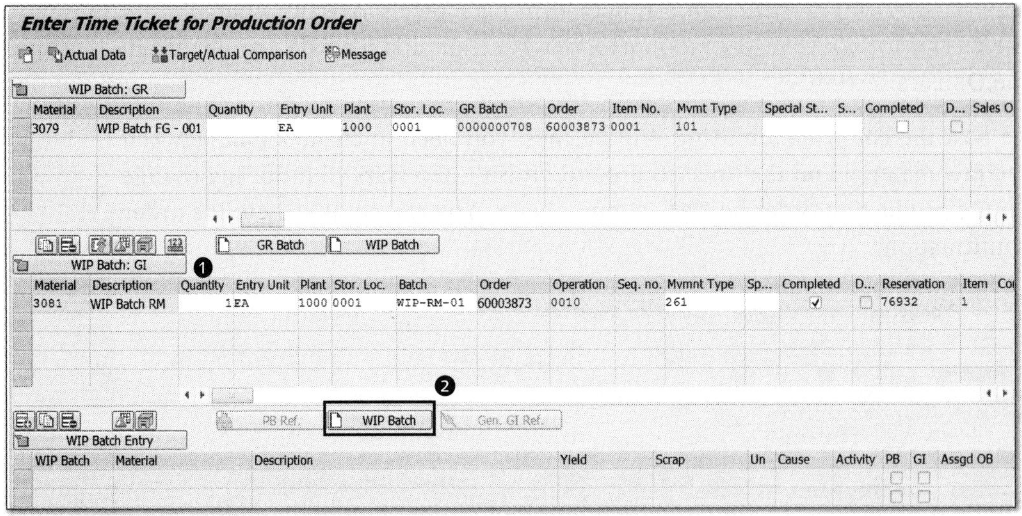

Figure 6.86 Confirmation with WIP Batches

In the pop-up screen that appears (Figure 6.87), enter the name of your WIP batch in the WIP BATCH field (e.g., "WIP_B_001"). With this process, you're creating a link between the raw material batch consumed and the WIP batch. Click the CONTINUE button (green checkmark).

Figure 6.87 WIP Batch

The screen that appears shows the WIP Batch WIP_B_001 (Figure 6.88), for which you'll have to manually enter the Yield, Scrap, or any Rework (scroll to the right to see the Rework field). Save your confirmation for this operation 0010.

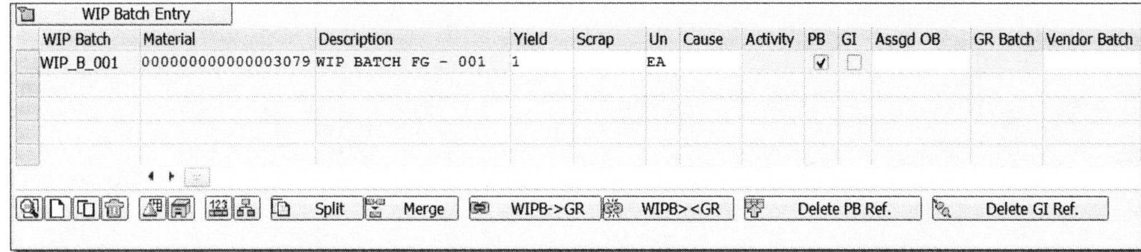

Figure 6.88 Yield Entry for the WIP Batch

The next step is to post a confirmation for operation 0020. Because it's the second and last operation of the production order, you consume the WIP batch created previously and post a goods receipt of the finished product batch, which is shown in Figure 6.89 in the next section.

Work in Process Batch Goods Receipt

As explained previously in this section, the goods receipt movement is proposed by default in the WIP Batch: GR section; however, it's proposed without quantity, so you need to input the quantity. After that, you must select the goods receipt item ❶ and the respective WIP Batch ❷ as shown in Figure 6.89. Now choose the WIPB->GR button ❸.

Notice that the in the WIP Batch: GR section, the system has created the GR Batch number 0000000708 for finished good Material 3079. This confirmation can now be saved, and because the WIP batch was already linked to the raw material, you now have a link between the raw material and the finished product material.

The following summarizes how the system links up a raw material's batch to a WIP batch and then to a finished good's batch for the production order 60003873:

> Raw Material 3081 (batch number WIP_RM_01) → WIP Batch (WIP_B_001) → Finished Good 3079 (batch number 0000000708).

You'll see the same link-up when we cover reporting in Section 6.11.4.

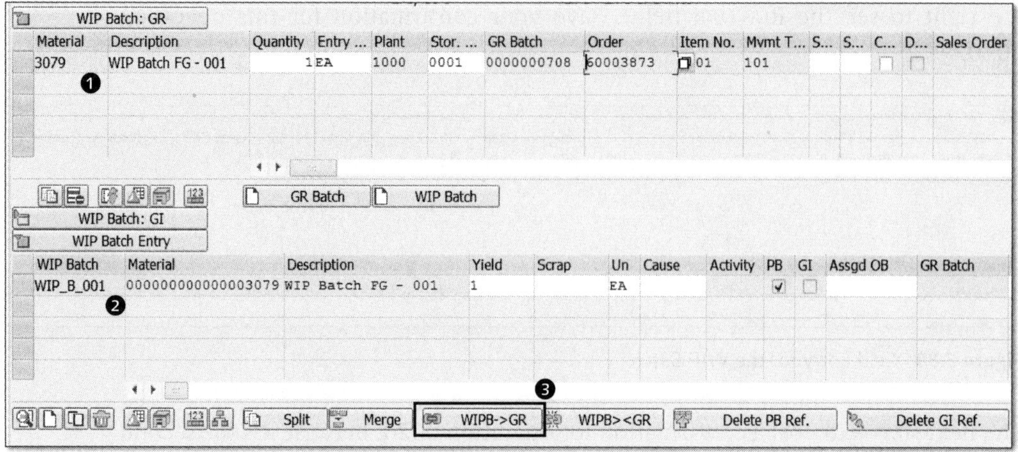

Figure 6.89 WIP Batch Goods Receipt

[»]

Note

Now that you have a basic understanding of WIP batches and how they work, you should explore several additional features and functionality available to handle all business processes.

For example, in Figure 6.89 ❸, we show how to link up WIPB->GR (linking WIP batch with goods receipt), but what if you mistakenly created a wrong link? To undo or delink this linkage, you can use the WIPB><GR icon to correct the error.

6.11.4 Work in Process Batches Reporting

Access the Batch Information Cockpit screen (see Figure 6.90) using Transaction BMBC to report on the WIP batches. Enter the Material "3079", which is the finished good, and click Execute or press F8.

In the screen that appears as shown in Figure 6.91, the system brings up the one batch (0000000708) ❶ for material 3079 that meets the selection criteria. Click on this batch 0000000708, choose Execute ❷, and then select Usage from the Details dropdown icon ❸.

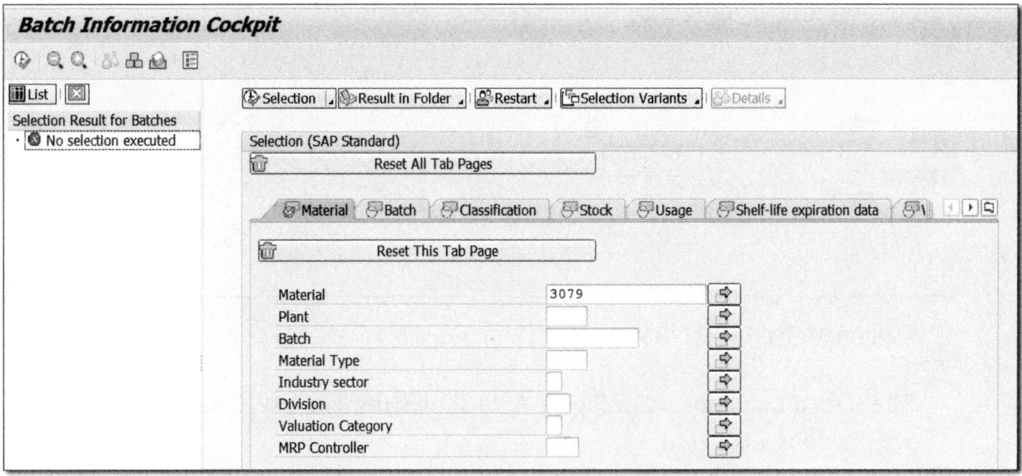

Figure 6.90 Batch Information Cockpit

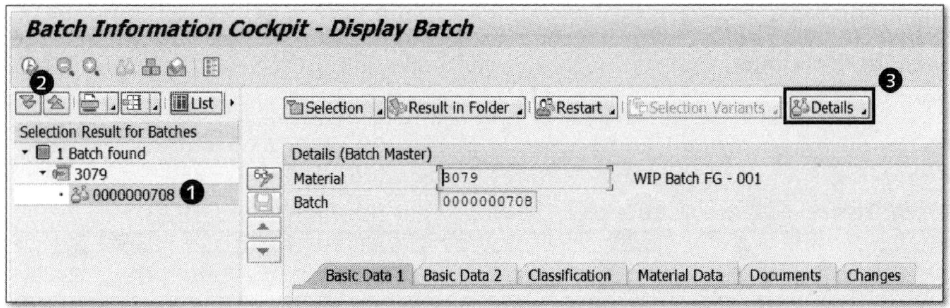

Figure 6.91 Batch usage

For batch 0000000708 ❶ for the finished good 3079, Figure 6.92 shows the TOP-DOWN view ❷ of the production order 60003873, the raw material 3079, and its associated batch WIP-RM-01. However, these screen details still don't show the associated WIP batch that created during confirmation.

To view the details of the WIP batch, click on production order 60003873, and then choose WIP BATCH ❸.

Tips & Tricks [+]

WIP BATCH ❸ won't be immediately visible in Figure 6.92. You'll have to scroll to the right of the menu bar to get to this icon.

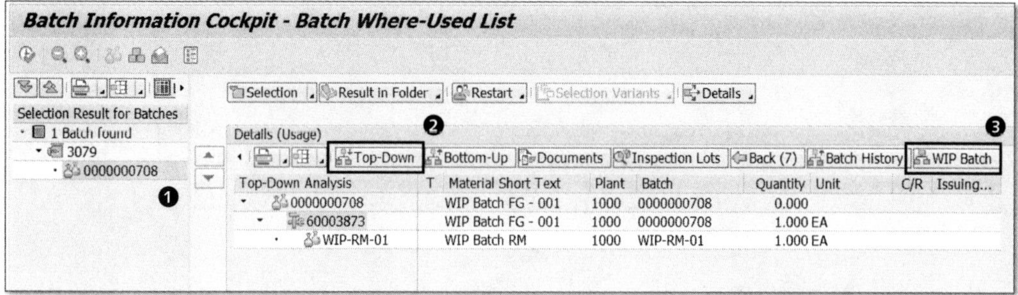

Figure 6.92 Top-Down Analysis

The screen that appears in Figure 6.93 shows the same WIP batch's linkage that we previously covered:

Raw Material 3081 (batch number WIP_RM_01) → WIP Batch (WIP_B_001) → Finished Good 3079 (batch number 0000000708)

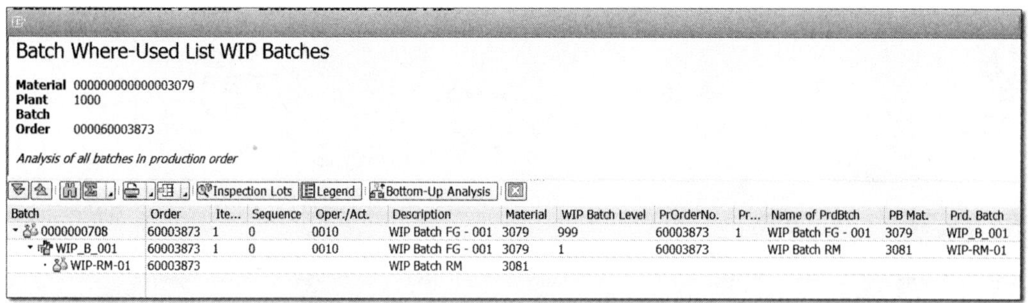

Figure 6.93 WIP Batch Traceability

[»]

Note

See Chapter 19 for reporting in the SAP ERP system.

See Chapter 15 for more information on classification or SAP Note 1473025, which answers frequently asked questions about WIP batches.

6.12 Additional Functions and Information Systems

In this section, we cover several of the functions you can use to optimize your business processes. We suggest that you check out the features and functionalities available in each area of working, such as master data, transactional data, or reports. You can do so by navigating the menu path and checking out the available options in relevant folders (e.g., routing or work center).

6.12.1 From Planned Order: Individual Conversion

Although this entire chapter is primarily focused on covering the master data and the transactional data in the production processes, you can also effectively use the planning data, which the system generates when you run MRP. The output of MRP is the generation of planned orders, which you convert into production orders for in-house production and purchase requisitions for external procurement. For in-house produced materials, it's important that the system is able to find the PP master data such as BOM, routing, and production version.

There are two options for you to convert planned orders into production orders: interactive conversion in Transaction MD04 or individual conversion using Transaction CO40. With production order creation, all of the remaining functions are the same as covered in this chapter, such as its release, cost calculation, control instruction generation, and scheduling.

[«]

Note

See Chapter 11 on MRP for more information.

6.12.2 From Planned Orders: Collective Conversion

You can convert several planned orders into production orders collectively using Transaction CO41. On the initial screen, enter the selection parameters that the system will consider in bringing up the relevant planned orders. You also assign an Order Type on the initial screen, so that the collective conversion of planned orders to production orders is all of the same order type.

[«]

Note

You can use Report PPBICO40 for the automatic conversion of planned orders into production orders in a batch job.

6.12.3 Production Order Creation without Material

You can create a production order without reference to a material; for example, you may want to create a capacity placeholder for postprocessing or for maintenance works without using the maintenance orders. You can also use this option to record extra production costs incurred during rework and can give reference to the main (parent) production order for settlement.

You use Transaction CO07 to create production orders without a material. On the initial screen, you need to enter the short text and assign a routing. If no routing is available, the system generates the operation from the default value that you defined in Customizing.

6.12.4 Mass Processing

To process large volume of transactions data or to simply optimize the routine business processes, you can use the mass processing functionality to perform a number of functions:

- Performing confirmation
- Creating capacity requirements
- Picking
- Costing
- Checking material availability
- Generating control instructions
- Printing shop floor papers

For this example, you can perform the confirmation function using mass processing by following the SAP menu path, Logistics • Production • Shop Floor Control • Control • Mass Reprocessing, or by using Transaction COHV.

Figure 6.94 shows the initial parameters entry screen, in which the system offers a large number of parameters to select from. Enter Production Plant "3000", enter Order Type "PP10", and choose the Mass Processing tab.

In the next screen, select Confirmation from the Function dropdown, and choose Execute (see Figure 6.95).

Mass Processing Production Orders

List: PPIOH000 Order Headers

Layout: 000000000001 Standard Layout

☑ Production Orders

☐ Planned Orders

Selection | Mass Processing

Select. at Header Level

Field	From		To
Production Order		to	
Material		to	
Production Plant	3000	to	
Planning plant		to	
Order Type	PP10	to	
MRP Controller		to	
Prodn Supervisor		to	
Production Version		to	
Sold-to party		to	
Sales Order		to	
Sales Order Item		to	
WBS Element		to	
Sequence number		to	
Priority		to	
Status Selection Profile			
Syst. Status	☐ Excl.	and	☐ Excl.

Figure 6.94 Initial Screen of the Mass Processing Functions

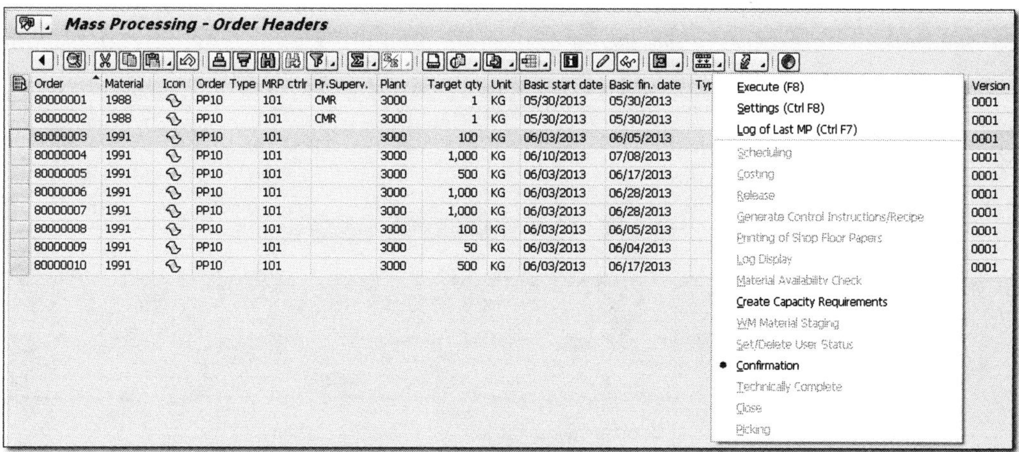

Mass Processing - Order Headers

Order	Material	Icon	Order Type	MRP ctrlr	Pr.Superv.	Plant	Target qty	Unit	Basic start date	Basic fin. date	Typ	Version
80000001	1988		PP10	101	CMR	3000	1	KG	05/30/2013	05/30/2013		0001
80000002	1988		PP10	101	CMR	3000	1	KG	05/30/2013	05/30/2013		0001
80000003	1991		PP10	101		3000	100	KG	06/03/2013	06/05/2013		0001
80000004	1991		PP10	101		3000	1,000	KG	06/10/2013	07/08/2013		0001
80000005	1991		PP10	101		3000	500	KG	06/03/2013	06/17/2013		0001
80000006	1991		PP10	101		3000	1,000	KG	06/03/2013	06/28/2013		0001
80000007	1991		PP10	101		3000	1,000	KG	06/03/2013	06/28/2013		0001
80000008	1991		PP10	101		3000	100	KG	06/03/2013	06/05/2013		0001
80000009	1991		PP10	101		3000	50	KG	06/03/2013	06/04/2013		0001
80000010	1991		PP10	101		3000	500	KG	06/03/2013	06/17/2013		0001

Figure 6.95 Confirmation through the Mass Processing Function

The system opens the worklist, where you select the PRODUCTION ORDER 80000003 and choose EXECUTE. The system then displays the log of mass processing in Figure 6.96.

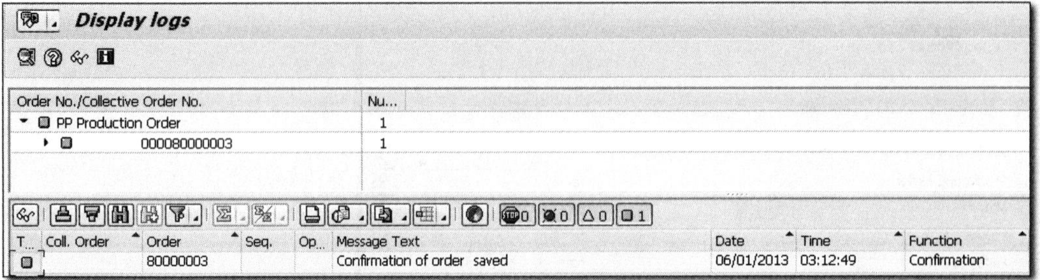

Figure 6.96 Log of Mass Processing

6.12.5 Information Systems

Due to the highly integrated nature of the SAP ERP system, any activity or business function that you perform with reference to the production order automatically updates the relevant information.

Figure 6.97 shows the PRODUCTION ORDER - DOCUMENTED GOODS MOVEMENTS screen within the production order display option (Transaction CO03), in which all of the goods issuances and receipts performed in MM are automatically updated.

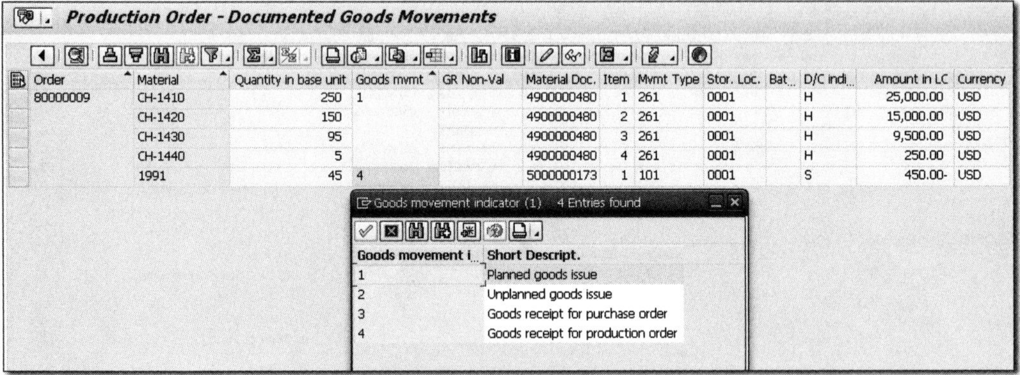

Figure 6.97 Documented Goods Movement in the Production Order

Figure 6.98 shows the itemized costs of the production order, providing a detailed costs breakdown. It also keeps on updating the planned costs with actual costs to provide a quick evaluation. This is an integration point between PP and the CO-PC team.

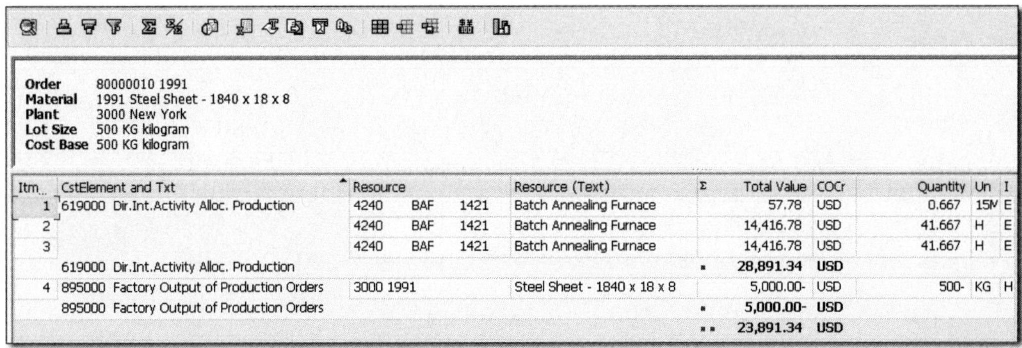

Order 80000010 1991
Material 1991 Steel Sheet - 1840 x 18 x 8
Plant 3000 New York
Lot Size 500 KG kilogram
Cost Base 500 KG kilogram

Itm	CstElement and Txt	Resource	Resource (Text)	Σ	Total Value	COCr	Quantity	Un	I
1	619000 Dir.Int.Activity Alloc. Production	4240 BAF 1421	Batch Annealing Furnace		57.78	USD	0.667	15M	E
2		4240 BAF 1421	Batch Annealing Furnace		14,416.78	USD	41.667	H	E
3		4240 BAF 1421	Batch Annealing Furnace		14,416.78	USD	41.667	H	E
	619000 Dir.Int.Activity Alloc. Production			▪	**28,891.34**	**USD**			
4	895000 Factory Output of Production Orders	3000 1991	Steel Sheet - 1840 x 18 x 8		5,000.00-	USD	500-	KG	H
	895000 Factory Output of Production Orders			▪	**5,000.00-**	**USD**			
				▪▪	**23,891.34**	**USD**			

Figure 6.98 Item Cost Overview of the Production Order

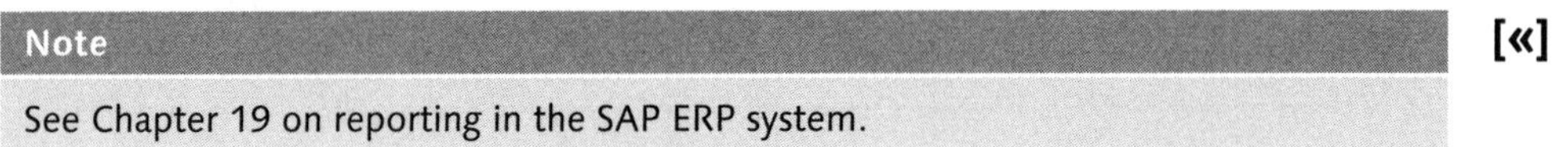

Note

See Chapter 19 on reporting in the SAP ERP system.

[«]

6.13 Summary

The production process in discrete manufacturing entails the creation of important master data followed by production execution that itself consists of order creation, goods issuance, confirmation, and goods receipt. Handling scrap, co-products, by-products, completion, and settlement processes occurs in both manufacturing types (discrete and process). You can print shop floor papers, such as object lists. You can also use a large number of tools and functionalities to optimize your business processes, such as using backflush, automatic goods receipt, mass processing, and postprocessing. You can also implement WIP batches to have better control over the entire production process.

The next chapter discusses the business processes in Production Planning for Process Industries (PP-PI).

Building on the important configuration of process manufacturing that you undertook in Chapter 4, this chapter covers the important business processes and functions and also provides the vital and logical links of configuration with business processes. Greater focus is placed on Process Management in the master recipe, which is unique to process industries only.

7 Production Planning for Process Industries

Production Planning for Process Industries (PP-PI) is characterized by product complexity. For many industries, it is necessary to integrate Batch Management (BM) and SAP ERP Quality Management (QM) with process manufacturing in PP-PI. Some of the industries in which process manufacturing finds extensive implementation include chemicals, edible oil refining, pharmaceuticals, fertilizers, beverages, food, and food processing. Any manufacturing industry that deals with liquids, where the product flows in a liquid or semisolid form, or where the processed material can't be brought back to its original state or disassembled, characterizes process manufacturing.

The chapter begins with an overview of process manufacturing and how it fits into the planning and production perspectives. The process manufacturing process flow provides a comprehensive and step-by-step explanation of each stage involved. Important process manufacturing master data is covered next, with extensive focus on the master recipe, in which the system not only facilitates material quantity calculation but also Process Management. We cover some of the standard features available in Process Management such as input and calculated values, integration with the Document Management System (DMS), and digital signature. We then cover the end-to-end business processes involved from the creation of the process order to how Process Management integrates with it.

Next, we cover the highly versatile and intuitive functionality of Execution Steps (XSteps) when you either want to implement it or simply transition from process instructions to XSteps. More features and functionalities of XSteps are shown, as well at their correlations to the configuration made in Chapter 4.

We then cover the process manufacturing cockpit that you've already configured in Chapter 4 to see how it helps and facilitates the business processes. We also cover process messages evaluation.

Finally, the remaining sections provide brief coverage of the rest of the standard processes of PP-PI, such as goods issuance, confirmation, and goods receipt. Because these processes are all similar in discrete manufacturing, we suggest that you visit the relevant sections of those chapters (Chapter 3 and Chapter 6).

7.1 Process Manufacturing Overview

Figure 7.1 shows an overview of the end-to-end process involved in process manufacturing. The business processes involved can broadly be divided into the following areas:

- ▸ Process planning
- ▸ Process order execution
- ▸ Process Management
- ▸ Order closure

Production planning in PP-PI begins when you convert the output of material requirements planning (MRP), which in this case is a planned order, into a process order. This is then followed by a material availability check to ensure that the required quantities of components needed to produce the material are available. If you've enabled material quantity calculation in the master recipe of the material, the system calculates the components' quantities. If not, it reads off the information from the material's bill of materials (BOM). At this stage, you can also enable the system to perform batch determination of the components that you want to use in production.

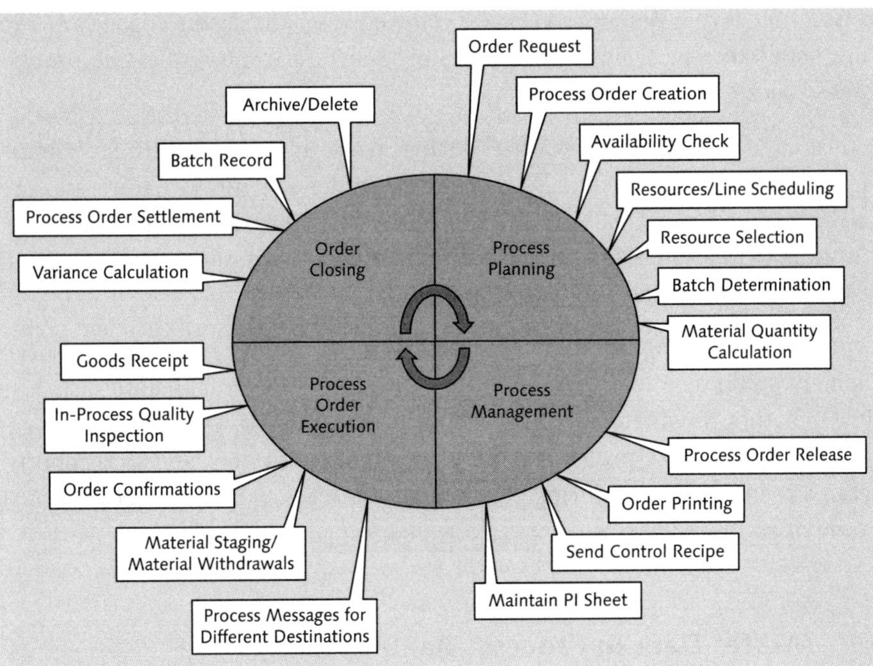

Figure 7.1 Production Planning and Execution in Process Industries

You proceed with releasing the process order as well as printing the process order. With a released process order, you can generate a control recipe. A generated control recipe takes the form of a process instruction sheet. You can run several of these process order management activities automatically or in the background to minimize managing them manually. For example, you can determine that on creation of the process order, the system can automatically release it too. If not, you have to manually release the process order. Alternatively, you can use a separate transaction to release a large number of process orders (mass processing), which again can be carried out as a manual task.

SAP ERP Materials Management (MM) plays an important role when you want to issue raw materials and components against a process order. QM (if integrated with SAP ERP Production Planning [PP]) enables extensive in-process (during production) quality inspection checks. During this time, you also maintain the process instruction sheet and assign it a COMPLETE status. You then perform confirmation of the process order, either at the individual phase level or at the entire process order level. When goods are produced, you can again engage MM to ensure goods receipt against the process order. You can now send the process messages back to the SAP ERP system.

The Cost Object Controlling activities such as work in process (WIP) determination, variance calculation, and settlement are order-specific in nature and are usually processed in the background. PP completely integrates with Cost Object Controlling in the SAP ERP system, so it's imperative that extensive coordination is ensured for comprehensive business processes mapping.

To optimize and bring greater visibility to your business processes, you can implement and integrate several additional processes and functionalities, such as digital signature, Engineering Change Management (ECM), Document Management System (DMS), co-products and by-products, shift notes, and shift reports. You can also integrate QM during production (in-process quality inspection) or at the time of goods receipt.

7.2 Master Data in Process Manufacturing

Process manufacturing has its own unique and often overlapping master data with other production types, such as discrete manufacturing or repetitive manufacturing (REM). If you set up master data in the right sequence, it's much easier and logical to interconnect them because you've already taken care of the predecessor-successor relationship.

The creation of master data for process manufacturing begins with the material master of the product (a finished good or an assembly). You create the BOM of the product that you want to produce and assign components, together with the quantities needed to produce the product. If needed, you can also define the scrap percentage at the operation or component levels. You then create the resource and create the master recipe for the material, in which you also assign the previously created resource. Finally, you create the production version for the material and assign the material's BOM, that is, the master recipe. When all of the logistical master data is in place, your Controlling (CO) team can create a product cost estimate of the material and also release it.

[«]

Note

You need to maintain a close coordination and liaison with the CO team to ensure that when working in PP, you're completely aligned with their working and reporting needs.

For example, for each resource, you need to assign a cost center, which your CO team should provide you with. They may provide you with one cost center for multiple resources or one cost center for an individual resource, depending on how they want to see the cost center reporting and evaluation.

The following make up the important master data in PP-PI:

- Material master
- BOM
- Resource
- Production version
- Master recipe

We'll discuss each in detail in the following sections.

7.2.1 Material Master

The material master is the central master record in logistics and the supply chain. The system defines a *material* as a substance or commodity that you can buy or sell on commercial basis. You can also relate a material to either being consumed or produced. A few examples of material are raw material, packing material, consumables, semifinished goods, and finished goods. The material isn't just restricted to production-based processes but all those for which the company wants to maintain inventory (stock items). So, you may also have materials that are used in SAP ERP Plant Maintenance (PM) processes, or you can even have nonvaluated materials.

For PP-PI, there is an extensive use of *Batch Management (BM)*. A *batch* is a uniquely identifiable partial quantity of a material. The batches of a material are managed in separate stocks. In a production process, a batch is a quantity of a specific material produced during a standardized production run. This quantity therefore represents a nonreproducible unit with unique specifications. The key properties of a batch are homogeneity and nonreproducibility.

A batch can be traced across the entire supply chain, that is, from the receipt of the raw material to processing in production and the creation of the final product, all of the way to sales and delivery to the customer. Complete batch traceability, batch determination, and batch derivation functionalities are available. You can use the batch information cockpit (Transaction BMBC) for complete top-down or bottom-up evaluation of material batches.

The system creates batches for a material, and the data of the material master are valid for all batches assigned to it. In contrast to the material master, a batch master record contains data that uniquely identify the corresponding batch and characterize the unit as one that can't be reproduced. The characteristic batch specifications are assigned using characteristics from the classification system in the material master and are inherited by the corresponding batch master records.

[»]

Note

Chapter 15 describes how to create classes and characteristics that you can eventually use in BM. We suggest that you extensively coordinate with the MM consultant for activation as well as complete business process mapping of BM in production processes.

7.2.2 Bill of Materials

The BOM in PP-PI is the same as in discrete manufacturing. Refer to Chapter 3 and Chapter 6 for a detailed understanding of the configuration and business processes involved in BOMs.

The material quantity calculation is unique only to the PP-PI and uses components of the material defined in its BOM. When calculating the components' quantities that the system should use in reference to each other, it refers to the information in the BOM. See Section 7.2.5 concerning the master recipe for a detailed understanding of material quantity calculation.

To create a BOM, you use Transaction CS01.

7.2.3 Resource

The resource in process manufacturing is the same as the work center is in discrete manufacturing. Refer to Chapter 3 and Chapter 6 for a detailed understanding of the configuration and business processes involved in work centers (resource in PP-PI).

To create a resource, you use Transaction CRC1.

The system offers and makes available standard configuration for PP-PI, which you can use if your business processes aren't too complex. For example, you can set USAGE as "008" (for Master Recipe + Process Order) and STANDARD VALUE KEY as "SAP4" (Process Manufacturing), in which only DURATION is listed as an activity. The available control key that you can use is PI01 (Master Recipe/Process Order).

7.2.4 Production Version

A production version determines which alternative BOM the system should use in combination with the master recipe for process manufacturing. In PP-PI, it's mandatory to define a production version. The system uses the production version during the creation of a master recipe to identify the BOM for the material and pull the BOM details from the master recipe.

When you create the master recipe for a material and plant combination, you should also enter the production version for the material on the initial screen. The production version should be created prior to the creation of the master recipe and then be used for creation of the master recipe.

To create a new production version, use Transaction C223. You can also create a production version in the MRP 4 view of the material master or even in the WORK SCHEDULING view. In this view (Transaction MM02), make sure that SELECTION METHOD is set as either "2" (SELECTION BY PRODUCTION VERSION) or "3" (SELECTION ONLY BY PRODUCTION VERSION). Refer to Chapter 6 for a detailed understanding of the business process of a production version and how to create one in the SAP ERP system. It's mandatory to create a production version for process manufacturing (and also in REM), but it's optional in discrete manufacturing.

Note [«]

Creating a production version directly from Transaction MM02 should be an exception because there may still be some incomplete data at this stage. Use Transaction C223 to achieve this objective instead.

7.2.5 Master Recipe Creation

Before you create the master recipe, you can create a production version and include BOM details only (and not the master recipe details because you don't have them at that time). Next, you'll create the master recipe and give reference to the production version because it's a mandatory requirement to enter a production version during master recipe creation. You can then go back to the production version and incorporate the master recipe details, including the group number and group counter that the system generated when you saved the master recipe. The system suggests the master recipe group number and the group counter when you again go back to production version. This approach in creating the master recipe helps in having a materials list (BOM) in the master recipe, which you can then also use in material quantity calculation.

A second approach that you can use in creating the master recipe is to first create a master recipe group without reference to a material and plant combination. When the system generates the recipe group number, create a production version of the material, and enter the BOM and master recipe details. Finally, when you assign the header material number in the change master recipe option for the master recipe group, the system prompts you to enter a production version to enable it to explode the BOM.

To create a master recipe for which the production version already exists, follow the SAP menu path, Logistics • Production – Process • Master Data • Master Recipes • Recipe and Material List • Create, or use Transaction C201. On the initial screen of the master recipe, enter the material, the plant, and the production version, and the header screen appears. We'll discuss the different screen elements of this screen in the following sections.

Recipe Header

Figure 7.2 shows the header details screen of the master recipe.

The Charge Quantity Range area is valid as the lot size quantities in the master recipe. It contains the default values for the operation, phase, and secondary resources. A proportional relationship exists between the default values for operation quantities and their unit of measure versus the recipe quantities and their unit of measure. Compared with master recipes, you enter this relationship directly in the operation details in routing and rate routings.

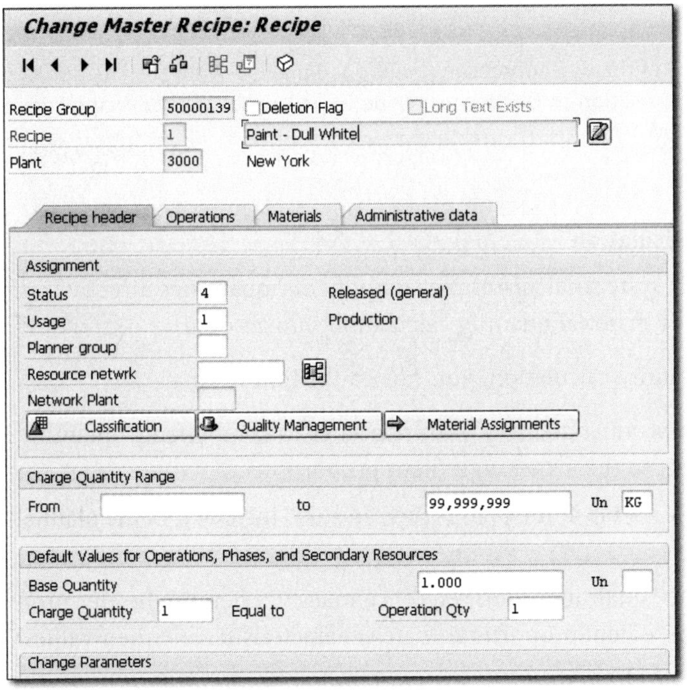

Figure 7.2 Change Master Recipe Screen

As an example, when the master recipe unit of measure is pieces, and the operation unit of measure is kilogram (kg), then for every 7 kg of the operation, there are 4 pieces (PC) of the master recipe, the quotient is 4/7. The charge quantity is 4 PC, and the operation quantity is 7 kg. The system also provides the option to maintain a base quantity for detailed working.

Materials

The master recipe integrates the details of the operations and BOM together as one master data by using the production version. The system explodes the BOM in the master recipe to bring up the details of the material BOM. The material BOM details in the task list (master recipe) help enable a unique feature to process manufacturing known as *material quantity calculation*.

[»]

Note

You need to adopt one of the two approaches mentioned previously for the creation of a production version in relation to the master recipe to ensure the master recipe contains the materials (BOM).

Material Quantity Calculation

In a process order, the system calculates the components' quantities directly from the BOM and takes the material quantity calculation into account.

With the material quantity calculation, you can do the following:

- Change the header product quantity with reference to components' quantities or even with respect to the active ingredient proportions.
- Calculate the planned scrap at the phase level, and also include it in the planned production cost.
- Change components' quantities with reference to each other, the header product, or the active ingredient proportions (batch characteristics and their values).
- Change operation or phase quantities when these aren't in proportion to the product quantity.

For material quantity calculation to work effectively, you need to make sure that you create the master recipe with reference to the BOM and consisting of components and quantities.

Because the planned scrap of the component is entered either in the material master or in the BOM, the system automatically increases the component quantity during planned order or process order creation. You can use the planned scrap of a component as a variable to calculate the other component's quantity using the material quantity calculation formula.

When you create the process order, the system automatically calculates the quantities based on the formulas. For a formula that is processed at the batch level and also uses active ingredient proportions (batch characteristics values), you need to manually trigger the material quantity calculation in the process order and after batch determination.

Note [«]

Note that the system only considers batch characteristics with numeric values.

When the system explodes the BOM in the master recipe, you can go to the MATERIAL QUANTITY CALCULATION screen shown in Figure 7.3 by choosing GOTO • MATERIAL QUANTITY CALCULATION or by clicking on the MATERIAL QUANTITY CALC. (calculation) icon in the MATERIALS tab of master recipe.

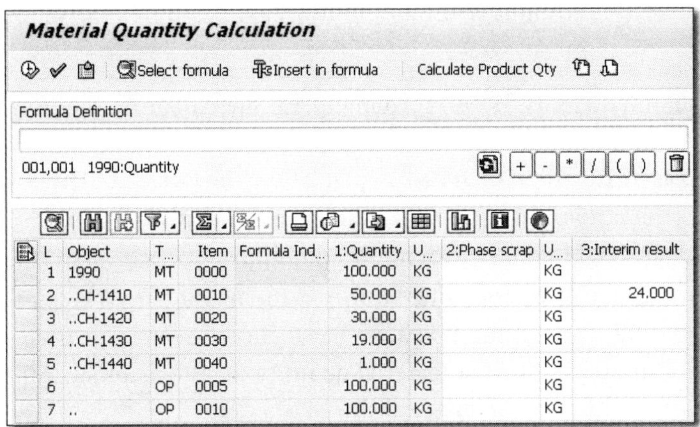

Figure 7.3 Material Quantity Calculation in the Master Recipe

Generally, the following steps are involved in entering the formula for the material quantity calculation:

1. On the screen shown in Figure 7.3, place the cursor on the field for which you want to change the quantity using a formula, and click on the SELECT FORMULA button in the menu bar.
2. In the FORMULA DEFINITION box, enter the formula or equation, which derives the output field value.
3. While creating a formula, you can also double-click on the variables that you want to include in the formula or place the cursor on the variable, and click on the INSERT IN FORMULA button in the menu bar.

   You can use formula operators such as +, –, *, /, DIV, and MOD. You can also use exponential, rounding (ROUND), absolute values (ABS), truncation

(TRUNC), EXP, LOG, SIN, COS, TAN, square root (SQRT), IF THEN ELSE conditions, and IF THEN NOT conditions.

We'll now show two examples to demonstrate how you can use the material quantity calculation to calculate product quantity and to show the interdependency of one component on another in calculations.

Example 1

In our first example, enter a formula using the following steps (refer to Figure 7.3):

1. For the header material quantity (1990) formula, place the cursor on the FORMULA INDICATOR field, and click on the SELECT FORMULA button in the menu bar. This shows up as 001,001 1990:QUANTITY just below the FORMULA DEFINITION bar.
2. Place the cursor on the field with the quantity 50.000 KG for MATERIAL CH-1410, and click on the INSERT IN FORMULA button in the menu bar. This automatically brings up [002, 001] in the FORMULA DEFINITION bar, in which you then enter "* 1.9". This means that the material quantity for the material 1900 will be 1.9 times the quantity of the material CH-1410.
3. Click on the REFRESH icon, and the system denotes the row containing the material 1990 with the FORMULA icon.
4. If you then click on the CALCULATE PRODUCT QTY button, the system updates the product quantity of material 1990 from 100 kg to 95 kg (50 kg for material CH-1410 × 1.90 = 95 kg).
5. Figure 7.4 shows the updated product quantity for material 1990. This compares with 100 kg as shown in Figure 7.3.

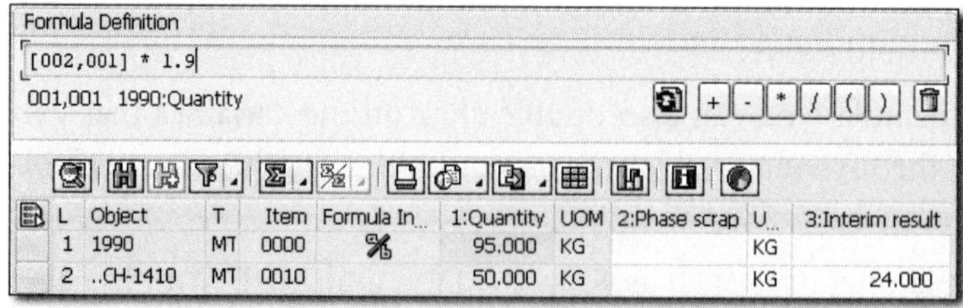

Figure 7.4 Updated Product Quantity after the Material Quantity Calculation

Example 2

In the second example of material quantity calculation, the system calculates one component's quantity based on the calculation that is associated with another component. Perform the following steps (refer to Figure 7.3 again):

1. To enter the formula for the component quantity (CH-1430), place the cursor on the FORMULA INDICATOR field, and choose the SELECT FORMULA button in the menu bar. This shows up as 004,001 CH-1430:QUANTITY just below the FORMULA DEFINITION bar.
2. Place the cursor on the field with quantity 30KG for MATERIAL CH-1430, and choose the INSERT IN FORMULA button. This automatically brings up [002, 003] in the FORMULA DEFINITION bar, in which you then manually enter "– 8". This means that the material quantity for the material CH-1430 will be subtracted by 8 kg from the quantity of material CH-1420.
3. Click on the REFRESH icon, and the system denotes the row containing the material 1430 with the FORMULA icon.
4. Because the quantity for material CH-1420 is 30 kg, the system subtracts it by 8 kg to update the quantity for material CH-1430 as 22 kg. If you refer to Figure 7.3, the original quantity (before the material quantity calculation) for this material, CH-1430, was 19 kg.

Figure 7.5 shows the updated product quantity for material CH-1430.

Formula Definition

[002,001] * 1.90

001,001 1990:Quantity

L	Object	T	Item	Formula Ind...	1:Quantity	UOM	2:Phase scrap	U...	3:Interim result
1	1990	MT	0000	%	95.000	KG		KG	
2	..CH-1410	MT	0010		50.000	KG		KG	24.000
3	..CH-1420	MT	0020		30.000	KG		KG	
4	..CH-1430	MT	0030	%	22.000	KG		KG	
5	..CH-1440	MT	0040		1.000	KG		KG	
6		OF	0005		100.000	KG		KG	
7	..	OF	0010		100.000	KG		KG	

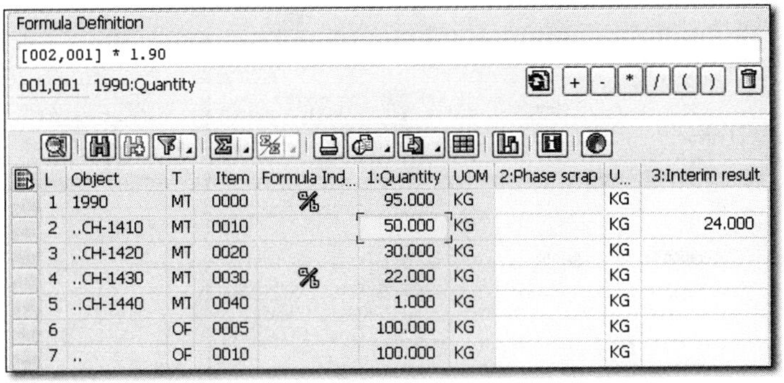

Figure 7.5 Component Quantity Calculation for Material CH-1430

Figure 7.6 appears when you click on the FORMULA OVERVIEW icon and contains comprehensive details of all the formulas and calculations involved.

Material 000000000000001990 Plant 3000 Prod. Version 0001

Line	Object: OP = Operation / Phase MT ...	+	Value	Unit	Formula (According To Column Descr...
1	MT 1990	-	95.000	KG	Quantity
	MT 0010 CH-1410		50.000	KG	Quantity
					* 1.90
2	MT 0010 CH-1410		50.000	KG	Quantity
3	MT 0020 CH-1420		30.000	KG	Quantity
4	MT 0030 CH-1430		22.000	KG	Quantity
	MT 0020 CH-1420		30.000	KG	Quantity
					- 8
5	MT 0040 CH-1440		1.000	KG	Quantity
6	OP 0005		100.000	KG	Quantity
7	OP 0010		100.000	KG	Quantity
2	MT 0010 CH-1410		24.000		Interim result
	MT 0020 CH-1420		30.000	KG	Quantity
					* 0.8

Figure 7.6 Overview of the Material Quantity Calculation

When you click on the BACK icon twice, the system takes you to the screen shown in Figure 7.7, which now has updated quantity details of all components, including a base quantity of 95 kg for material 1990.

If you create a process order for material 1990 for a quantity of 100 kg, the system will divide the components' quantities by 95 kg (the new base quantity) and then multiply each quantity with 100 kg (the process order quantity) to arrive at the individual component quantity. For example, for component CH-1410, the quantity calculation for 100 kg of process order is the following: 50 kg ÷ 95 kg × 100 kg = 52.63 kg.

Recipe header | Operations | Materials | Administrative data

Material 1990 — Plant 3000 — BOM

Prod. Version 0001 Paint-Dull White

Base quantity 95 KG

Material Component Assignments

Material	Oper...	P	S...	Operation Desc.	Quantity	Co...	B	Item Text	Item ...
CH-1410		☐			50	KG	☐	Acrylic Resin	0010
CH-1420		☐			30	KG	☐	Additive BG99	0020
CH-1430		☐			22	KG	☐	Solvent Mix A	0030
CH-1440		☐			1	KG	☐	Pigment, blue	0040

Figure 7.7 Updated Material Quantities in the Master Recipe

Operations and Phases

Master recipes use something called a *phase*, which works in the same manner as operations do in routing for discrete manufacturing. It's easier to maintain detailed levels working at the phase level in the master recipe because you can manage and incorporate more production details, including Process Management.

In the master recipe, you assign activities such as production duration or labor hours at the phase level and not at the operation level. Hence, the confirmation of a process order is recorded for a phase and not an operation. You also assign a resource (work center) at the operation level. The phases below the operation then adopt the resource that you assigned at the operation level. The system assigns the standard values and activities (controlled by a control key in the resource) as active at the phase level and not at the operation level. The sum total of standard values at a phase is in fact the total time required to process the operation. The system assigns the components of the BOM (materials list) to phases and not to operations. You can, however, integrate in-process quality inspections of QM either at the operation level or the phase level.

To create a phase below an operation, you need to select the PHASE checkbox in the OPERATIONS tab, which then automatically copies the resource from the operation. At the same time, when defining a phase, you also have to assign the *superior* operation so that the system knows which specific phase relates to which operation.

You can maintain the relationships among various phases as start-finish, finish-start, finish-finish, or start-start. The phases can either work in parallel or in overlapping sequences. In the OPERATIONS tab of the master recipe, you can access the phase relationship screen for phases by selecting the phases and choosing GOTO • RELATIONSHIPS.

You assign individual control recipe destinations at the phase level and assign the process instructions in the respective phases of the master recipe. If you've defined the scope of generation in the configuration of the process instructions, it reduces the data maintenance efforts at the master recipe level. Alternatively, you can maintain the desired process instruction details either in the master recipe or in the process order. For process instructions that have characteristic values based on a material, you need to assign them at the master recipe level. To assign process instructions to the phases in the OPERATIONS tab of the master recipe, use the menu path, GOTO • PROCESS MANAGEMENT • PROCESS INSTRUCTIONS.

Figure 7.8 shows the OPERATIONS tab of the master recipe, in which the OPERATION is 0005. Enter the RESOURCE "CH_BLEND" at the operation level, and the system automatically copies it in all of the phase below it. The phase is 0010 and is denoted by the PHASE checkbox. When you define an operation as a phase, you also have to define the SUPERIOR OPERATION, which, for this example, is 0005 (the operation).

Recipe header | Operations | Materials | Administrative data

Ops

Op...	P..	Sup. O...	Destinatn	Resource	C...	L..	Stan...	Description	L...	R...	C...	O...	Base Qty	Ac...	1st Std V...	S..
0005	☐			CH_BLEND	PI01	☐					☐	☐	1	KG		
0010	☑	0005	10	CH_BLEND	PI01	☐		Paint Blending		X	☐	☐	1	KG	1.0	HR

Figure 7.8 Operations Overview in the Master Recipe

The control recipe destination is 10. This is the same control recipe destination that you configured in Chapter 4.

Notice that the system automatically copies the CONTROL KEY, PI01, from the RESOURCE CH_BLEND. Select the phase 0010, and double-click the line item 0010 (the operation), and the system takes you to the screen shown in Figure 7.9.

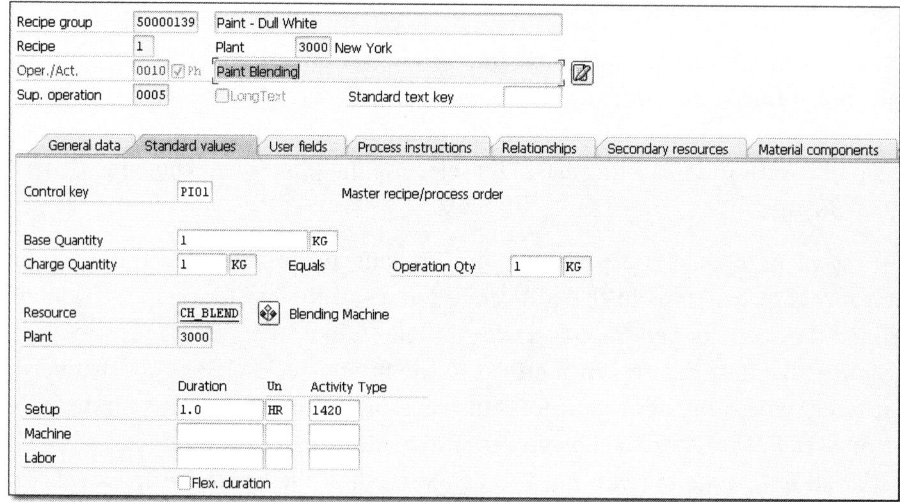

Figure 7.9 Standard Values in the Master Recipe

This shows the STANDARD VALUES tab of the master recipe in which you can enter the duration of the activities, such as SETUP, MACHINE, or LABOR hours required to produce the material at each operation. From here, you can click on the PROCESS INSTRUCTIONS tab to configure the process instructions that are an integral part of Process Management. We'll visit this screen a bit later in the upcoming section.

7.3 Process Management

Because a large number of features and functionalities of process instructions exist within an operation's phase of the master recipe, it warrants a separate section in this chapter. This section deals with process instructions (which is a part of Process Management) that you need to define in the PROCESS INSTRUCTIONS tab shown previously in Figure 7.9.

If you're working with a manufacturing organization, a permanent requirement is to monitor system performance and plant parameters. For example, when the production of a certain item is scheduled, the plant operator needs to have a series of clear and comprehensive instructions to follow. Similarly, the plant operator is required to record and report back data, such as steam temperature twice a shift or an abnormal vibration in the suction pump, so that it will be available for future reference or corrective action.

Therefore, a functionality is needed in the SAP ERP system that can transfer and communicate all such information in a timely manner from plant operator back to the process control system (PCS). This has been made possible by the Process Management functionality.

Note [«]

Transmitting information between an SAP ERP system and a PCS is possible by defining the type of the control recipe destination. We focus on the transfer to process instruction sheets to show you that implementing Process Management can still yield significant added value without integrating SAP with a PCS.

Process Management completely integrates with core SAP ERP system components such as MM, QM, and the cross-application DMS. It offers functionality such as goods issues and goods receipts, process order confirmations, and results recording of quality inspection data. All of this information helps in analysis and report generation functions, not to mention benefiting the business process owners who are directly using the information.

In the following sections, we'll cover some of the elements of process management, and how you can intuitively create a process instruction sheet.

7.3.1 Functions in Process Management

The following summarizes the functions supported by Process Management in PP-PI:

- Receiving control recipes from released process orders
- Sending control recipes to process operators or PCSs
- Preparing process instructions as texts so that the process operators can display them on their computer screens
- Receiving, checking, and sending process messages with actual process data
- Monitoring process messages and control recipes
- Manually creating process messages

7.3.2 Elements in Process Management

Figure 7.10 shows an illustration of the various elements involved in Process Management for data flow. Starting from the top left, creating a process order forms the basis for the generation of the control recipe.

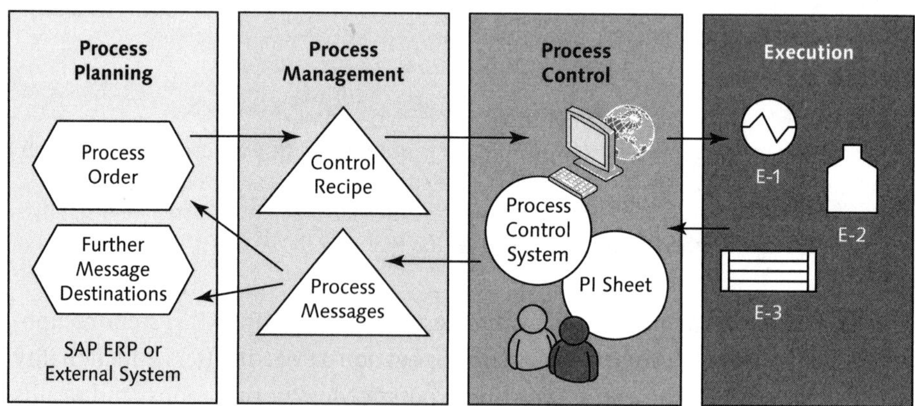

Figure 7.10 Overview of Process Management

The system sends the control recipe in the form of a process instruction sheet to the predefined control recipe destinations. The process operator follows the instructions given in the process instruction sheet and also fills the process instruction sheet with relevant plant parameters and other important data, and then returns it as a process message either back to the SAP ERP system or to an external system.

7.3.3 Integrating Process Management with External Systems

In an automated environment, Object Linking and Embedding (OLE) for Process Control (OPC) and OPC Data Access (ODA) enables the system to read and write data points and events using the OPC server for the SAP ERP system. This function is also available in production orders (discrete manufacturing).

[«]

Note

The OPC is a standard that uses Component Object Model/Distributed Component Object Model (COM/DCOM) technology to define interfaces independent of the manufacturer for use in an industry. The SAP ERP system designed the OPC standard especially for the process control level. OPC servers allow access to various data sources, such as PCSs, programmable logic controllers, and temperature sensors, and thus provide process data that can be requested by OPC clients.

7.3.4 Process Management and Manufacturing Integration and Intelligence

With SAP Manufacturing Integration and Intelligence (SAP MII), the SAP ERP system offers an adaptive manufacturing solution for production. SAP MII provides manufacturing companies increased flexibility through improved linking of the SAP ERP system to the production process level and by making real-time information available. You can use SAP MII both in the process order and production order environment. SAP MII provides standardized, preconfigured connectors to enable real-time data integration in the manufacturing execution systems (MES) and Supervisory Control and Data Acquisition (SCADA) systems.

[«]

Note

You can find more information on SAP MII/ME in Chapter 20.

You can run real-time analyses and display the results in browser- and role-based dashboards. These analyses provide important information for checking and supporting decision making such as warnings, job lists, analyses, reports, and real-time messages about production variance.

7.3.5 Process Instructions

An operation in the master recipe may have several phases, and each phase requires a control recipe destination. After a control recipe destination is defined for an operation, it automatically applies to all of the phases of that operation. Within each phase, Process Management-related information is incorporated, including process message categories, process instruction characteristics, and control recipe destinations.

You can assign process instructions to the phases in the OPERATIONS tab of the master recipe. To do this, select the specific phase, and use the following path in the OPERATIONS tab: MASTER RECIPE – GOTO • PROCESS MANAGEMENT • PROCESS INSTRUCTIONS.

In the resulting screen shown in Figure 7.11, there are two process instruction categories, AREAD1 and PP10. AREAD1 relates to the request to the shop floor to get the measured value of the process parameter. The second process instruction category, PP10, is the same one that you configured in Chapter 4. The CONTROL RECIPE DESTINATION 10 (PRODUCTION FLOOR) is also the same one that you configured in Chapter 4.

General data | Standard values | User fields | Process instructions

Control Recipe Destination 10 Production Floor

Process Instr.

Proc. Instr.	Proc.instr. category	Typ	Description
0010	AREAD1	2	Request a measured value
0020	PP10	0	Blank PI Category
0030			

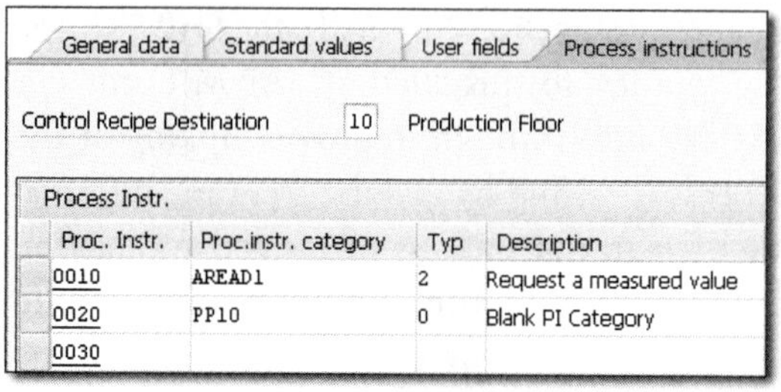

Figure 7.11 Process Instructions for Control Recipe Destination PP10

Double-click on the PROC. INSTR. (process instruction) 0010 (with PROC.INSTR. CATEGORY AREAD1) to go to the screen shown in Figure 7.12.

A major benefit that Process Management offers is that its results can be checked for consistency and simulated to ensure completeness and correctness. Click the CHECK PROCESS INSTRUCTION icon in Figure 7.12 to check the consistency of the sequence of process instruction characteristics and the value of each characteristic defined. Then click the SIMULATE PROCESS INSTRUCTION SHEET icon to show the simulated version of what the field and other information will eventually look like in a process instruction sheet.

Figure 7.12 contains the MESSAGE CATEGORY PP10 that you configured in Chapter 4. It also contains the process instruction OUTPUT CHARACTERISTIC ZPI_CREATION_DATE that you created earlier in Chapter 4. In the process instruction sheet, this field should show the BASIC FINISH DATE of the process order. The output characteristic also has the same value (ZPI_CREATION_DATE) assigned.

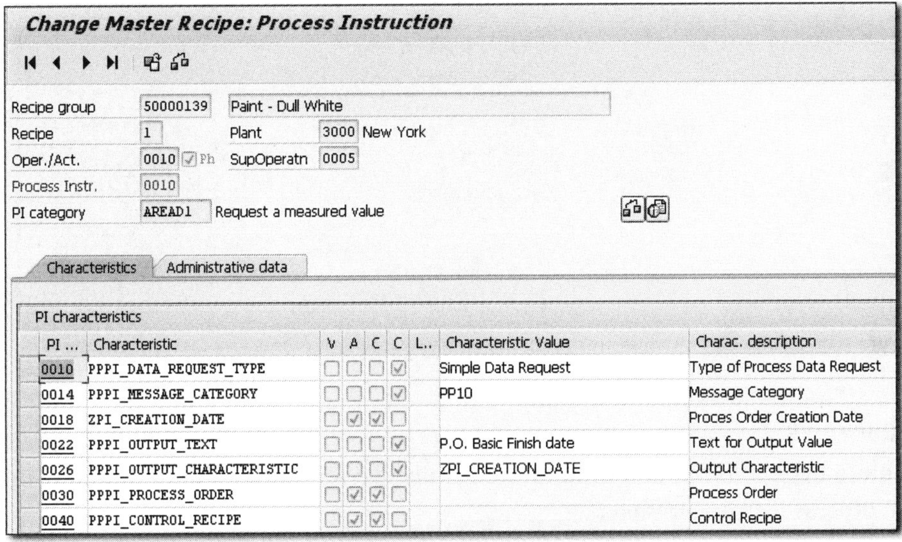

Figure 7.12 Process Instruction Characteristics for PI Category AREAD1

7.3.6 Process Instruction Sheet

Figure 7.13 shows a general example of a process instruction sheet.

The following sections explain some of the options available in the process instruction sheet and the data or other information that you need to maintain for using a specific function/option.

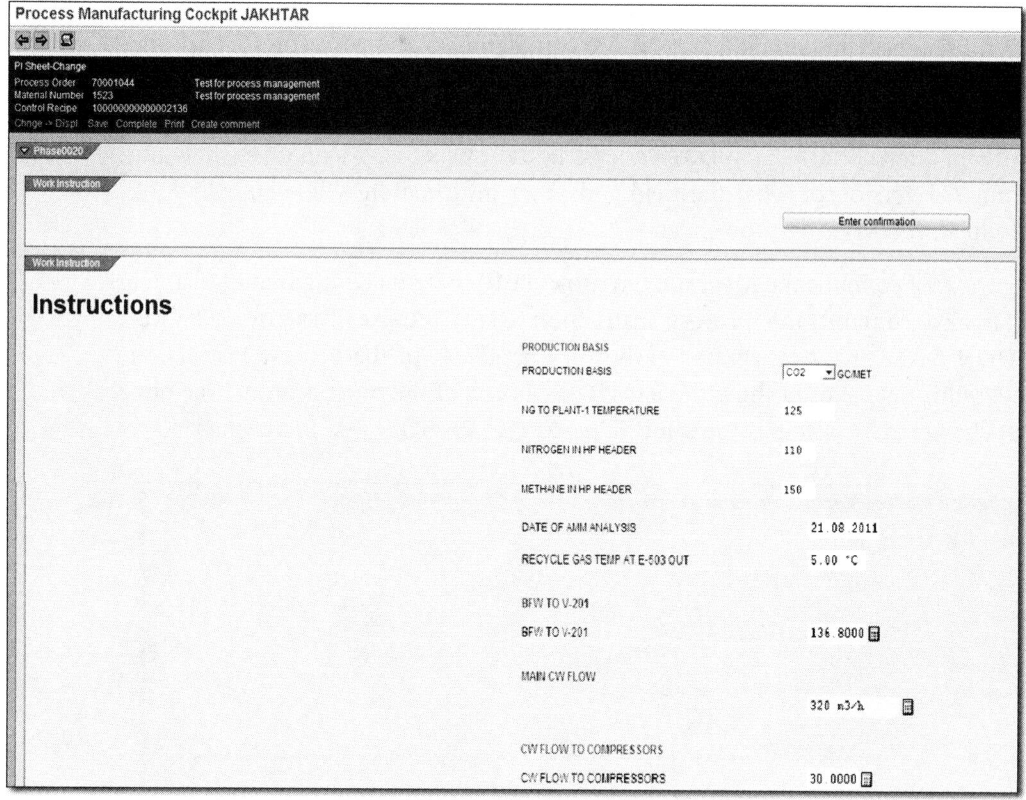

Figure 7.13 Example of a Process Instruction Sheet

Input Value

Table 7.1 contains the PP-PI characteristics needed for input field functionality in the process instruction sheet.

Characteristic	Characteristic Value	Description
PPPI_INPUT_REQUEST	CARBON DI OXIDE IN HP HEADER	This is the standard PP-PI characteristic whenever it's required that a field has an input value. The field value corresponding to this characteristic given as CARBON DI OXIDE IN HP HEADER will be the display name of the field in the process instruction sheet.

Table 7.1 PP-PI Characteristics and Their Values as Defined in Process Management

Characteristic	Characteristic Value	Description
PPPI_VARIABLE	F	Each field can be tagged as a variable whose value can subsequently be used in calculations, if needed. For this example, give the variable tag as "F" as defined in the characteristic value.
PPPI_REQUESTED_VALUE	NH3_CO2_HPH	This PP-PI characteristic is the output value of a field. However, what governs the format and other details actually comes from characteristic value NH3_CO2_HPH.

Table 7.1 PP-PI Characteristics and Their Values as Defined in Process Management (Cont.)

Figure 7.14 shows the simulation of the CARBON DI OXIDE IN HP HEADER field and how it will look in the process instruction sheet. Hence, the PP-PI characteristic PPPI_INPUT_REQUEST is the display field in the process instruction sheet. The value (any numeric value) will be given a tag of PPPI_VARIABLE as "F", and the output format of the numeric value will be governed by characteristic NH3_CO2_HPH. For example, characteristic NH3_CO2_HPH stipulates having a field length of 5 with two decimal places and no negative values. In such a case, values such as 45.35 or 15.88 are acceptable, but –15.88 isn't acceptable in the process instruction sheet.

CARBON DI OXIDE IN HP HEADER

Figure 7.14 Input Value in the Process Instruction Sheet

Note [«]

See Chapter 15 on the classification system, including classes and characteristics, for further information on creating characteristics that you can use in Process Management.

Tips & Tricks [+]

If you're not going to use the PP-PI characteristic in any subsequent calculation and if you're using it only for data entry purposes, you can eliminate the entire PPPI_VARIABLE row and its value F.

Calculated Value

You can extensively use the process instruction sheet for all kinds of calculations, as long as all of the relevant parameters required for calculation are available in the same process instruction sheet.

Table 7.2 contains the PP-PI characteristics needed for the calculation field functionality in the process instruction sheet. It also shows that if the calculation formula is too long for a single line, it can be continued on the next line (up to eight lines can be used for the calculation formula). Also, for the calculation formula, the variables AA1, AA2, and AA3 must previously be defined in the same process instruction sheet.

Characteristic	Characteristic Value
PPPI_INPUT_REQUEST	KS PRODUCTION
PPPI_VARIABLE	AA
PPPI_EVENT	PARAMETER_CHANGED
PPPI_CALCULATED_VALUE	NH3_02_FR_9
PPPI_CALCULATION_FORMULA	AA1*SQRT(((AA2+1.03*783)/((
PPPI_CALCULATION_FORMULA	AA3+273)*106))*3*24)

Table 7.2 Example of PP-PI Characteristics and Their Values for Calculated Fields

The simulated version of the calculated field will appear as shown in Figure 7.15.

Figure 7.15 Calculation Field in a Process Instruction Sheet

Input Group and Dropdown Selection

Table 7.3 contains the PP-PI characteristics needed for the input field functionality in the process instruction sheet.

Characteristic	Characteristic Value
PPPI_INPUT_GROUP	PRODUCTION BASIS
PPPI_INPUT_REQUEST	PRODUCTION BASIS

Table 7.3 PP-PI Characteristics and Their Values as Defined in Process Management

Characteristic	Characteristic Value
PPPI_VARIABLE	A
PPPI_REQUESTED_VALUE	NH3_PR_201
PPPI_UNIT_OF_MEASURE	GC/MET

Table 7.3 PP-PI Characteristics and Their Values as Defined in Process Management (Cont.)

They will result in a display as shown in the screen in Figure 7.16.

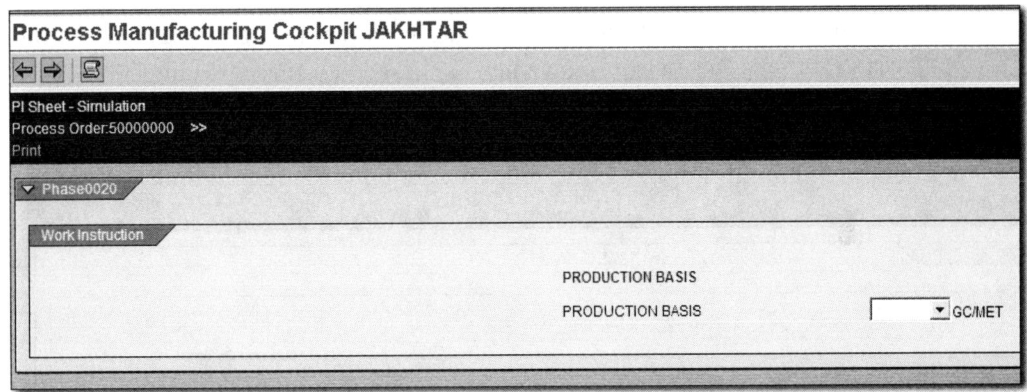

Figure 7.16 PP-PI Characteristics for the Input Field Functionality in a Process Instruction Sheet

Call Function

As explained in Table 7.4, you can use the process instruction sheet to call up a transaction while remaining on the process instruction sheet screen. The process instruction characteristics together with their values call up the Display Process Order transaction while remaining in the process instruction sheet. The PPPI_BUTTON_TEXT enables you to define a meaningful description of the icon while remaining in the process instruction sheet. Set the icon text as DISPLAY PROCESS ORDER, and set PPPI_TRANSACTION_CODE as COR3 for this example, but note that these fields are flexible and can be set to whatever you need.

Characteristic	Characteristic Value
PPPI_FUNCTION_NAME	COPF_CALL_TRANSACTION
PPPI_BUTTON_TEXT	DISPLAY PROCESS ORDER

Table 7.4 PP-PI Characteristics and Their Values as Defined in Process Management

Characteristic	Characteristic Value
PPPI_FUNCTION_DURING_DISPLAY	ALLOWED
PPPI_EXPORT_PARAMETER	NEW_SESSION
PPPI_INSTRUCTION	
PPPI_EXPORT_PARAMETER	TCODE
PPPI_TRANSACTION_CODE	COR3

Table 7.4 PP-PI Characteristics and Their Values as Defined in Process Management (Cont.)

The simulated version of the characteristics is shown in Figure 7.17. When the DISPLAY PROCESS ORDER icon shown here is clicked, it brings up Transaction COR3 (Display Process Order).

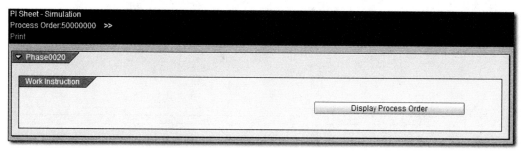

Figure 7.17 Call Function in the Process Instruction Sheet

Table Entry

Often there is a business need to enter multiple values in a tabular form for a single value or multiple values of parameters. Table 7.5 lists all of the PP-PI characteristics needed to use the table-entry format in the process instruction sheet. Notice that you can control the table size (minimum four values, maximum six values in this example).

Characteristic	Characteristic Value
PPPI_DATA_REQUEST_TYPE	REPEATED DATA REQUEST
PPPI_MINIMUM_TABLE_SIZE	4
PPPI_MAXIMUM_TABLE_SIZE	6

Table 7.5 PP-PI Characteristics for Activating the Format with Multiple Values

Characteristic	Characteristic Value
PPPI_INPUT_REQUEST	Hourly Flow Meter Readings
PPPI_VARIABLE	ABC
PPPI_REQUESTED_VALUE	NH3_02_FR_9

Table 7.5 PP-PI Characteristics for Activating the Format with Multiple Values (Cont.)

Figure 7.18 illustrates the simulated version table entry format and shows six values being entered. Also note that up to four decimal places are allowed for each value (this is controlled via characteristic NH3_02_FR_9).

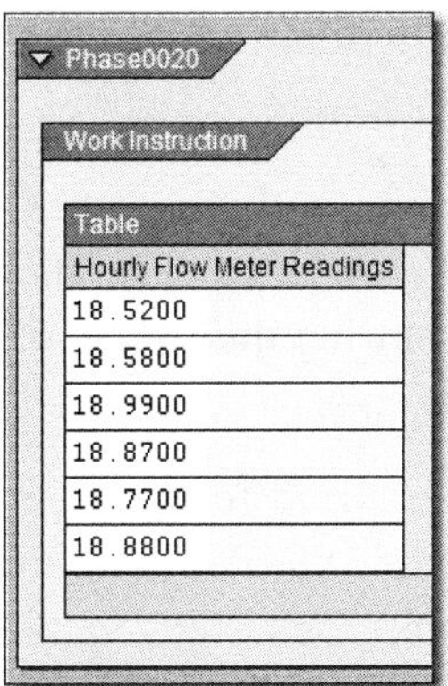

Figure 7.18 Table in Process Instruction Sheet

Long Text Input

Generally, at the end of every shift, the company requires the process operator to prepare shift highlights and/or other important details. To do this, you can use the long text functionality in the process instruction sheet. Table 7.6 lists the PP-PI characteristics needed for the long text functionality in a process instruction sheet.

Characteristic	Characteristic Value
PPPI_INPUT_REQUEST	Shift Highlights
PPPI_REQUESTED_VALUE	PPPI_MESSAGE_TEXT

Table 7.6 Required PP-PI Fields for Activating the Text Box in a Process Instruction Sheet

Figure 7.19 shows the text box that becomes available in the process instruction sheet, in which the process operator has entered shift details.

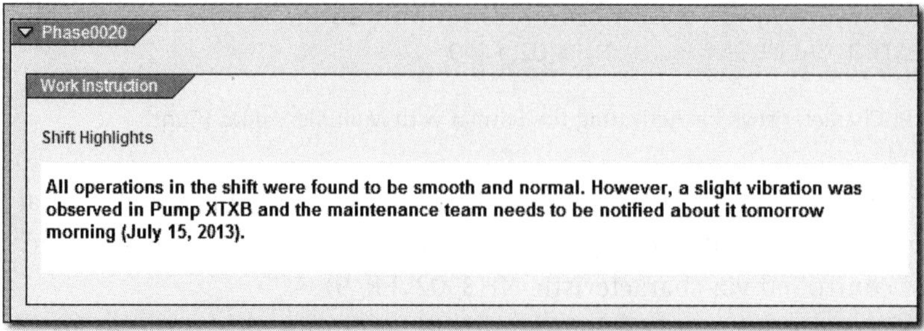

Phase0020
Work Instruction
Shift Highlights
All operations in the shift were found to be smooth and normal. However, a slight vibration was observed in Pump XTXB and the maintenance team needs to be notified about it tomorrow morning (July 15, 2013).

Figure 7.19 Shift Details (Notes) in the Process Instruction Sheet

Instructions and Notes

Generally, businesses want to make sure that the line operator has easy and instant accessibility to various instructions and other notes during the course of the shift. The functionality of instructions and notes is useful in such scenarios.

Table 7.7 provides a list of the PP-PI characteristics that you need to maintain instructions and notes in a process instruction sheet. To enter the desired text that you want to display on the process instruction sheet, select the characteristic PPPI_INSTRUCTION (or PPPI_NOTE), and click on the VALUE icon at the bottom of the screen. You can enter the text that you want to see being displayed in the process instruction sheet, as well as use some of the formatting options such as bold, italic, large fonts, or text alignment center/right.

Characteristic	Characteristic Value
PPPI_INSTRUCTION	–
PPPI_NOTE	–

Table 7.7 PP-PI Fields for Activating the Instructions and Notes in a Process Instruction Sheet

Figure 7.20 shows the simulated version of the instructions and long text.

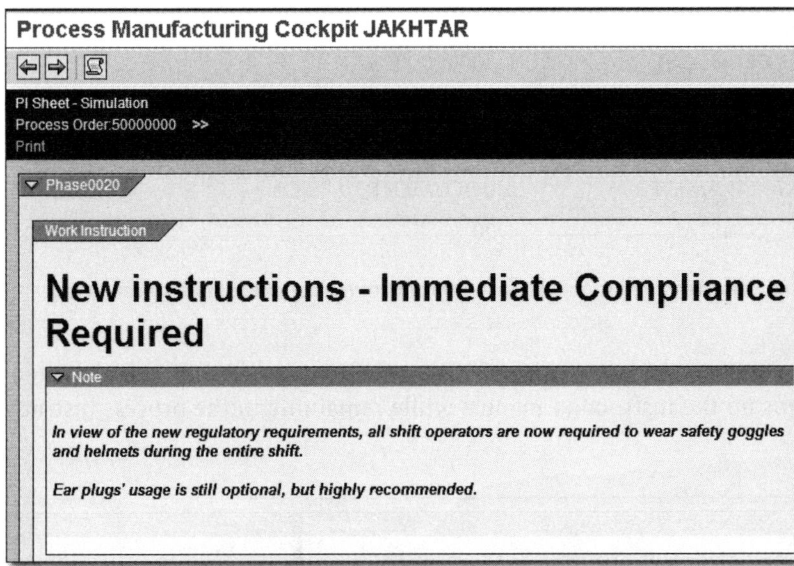

Figure 7.20 Work Instructions in the Process Instruction Sheet

Document Management System (DMS) in the Process Instruction Sheet

Sometimes the line operator is required to refer to a specific user manual or machine instruction manuals to perform a task or to rectify a production problem during operation. For this purpose, Process Management completely integrates with the Document Management System (DMS) in SAP, which is a central repository used to store important documents and other references for use.

Table 7.8 contains the PP-PI characteristics that you need to maintain to activate DMS in a process instruction sheet.

Characteristic	Characteristic Value
PPPI_FUNCTION_NAME	COPF_DOCUMENT_SHOW_DIRECT
PPPI_BUTTON_TEXT	INSTRUCTION MANUAL – COMPLETE
PPPI_FUNCTION_DURING_DISPLAY	ALLOWED
PPPI_EXPORT_PARAMETER	DOCUMENT_TYPE
PPPI_STRING_CONSTANT	Z01
PPPI_EXPORT_PARAMETER	DOCUMENT

Table 7.8 PP-PI Characteristics for Activating the Document Management System

Characteristic	Characteristic Value
PPPI_STRING_CONSTANT	10000000332
PPPI_EXPORT_PARAMETER	DOCUMENT_PART
PPPI_STRING_CONSTANT	000
PPPI_EXPORT_PARAMETER	DOCUMENT_VERSION
PPPI_STRING_CONSTANT	00

Table 7.8 PP-PI Characteristics for Activating the Document Management System (Cont.)

Figure 7.21 points to the INSTRUCTION MANUAL – COMPLETE button, which, when clicked, opens up the instruction manual while remaining in the process instruction sheet.

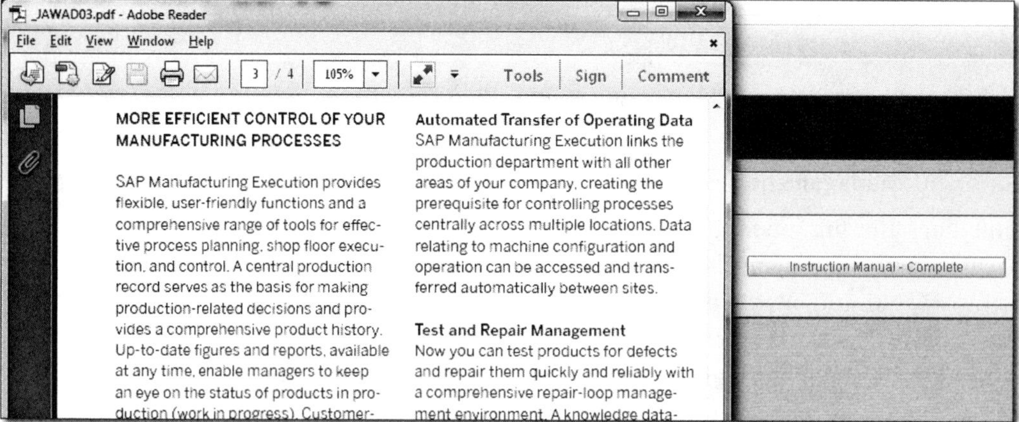

Figure 7.21 Document Management System in the Process Instruction Sheet

Digital Signature on Completion of the Process Instruction Sheet

Digital signature in the SAP ERP system is the process of authenticating information and requires the user to enter his name along with a password before the system can accept an entry or perform a function. For example, when a function is executed, the user is required to digitally sign the process instruction sheet in the pop-up box that appears (Figure 7.22). Refer to Chapter 4 to see how to activate digital signature at the process instruction sheet level.

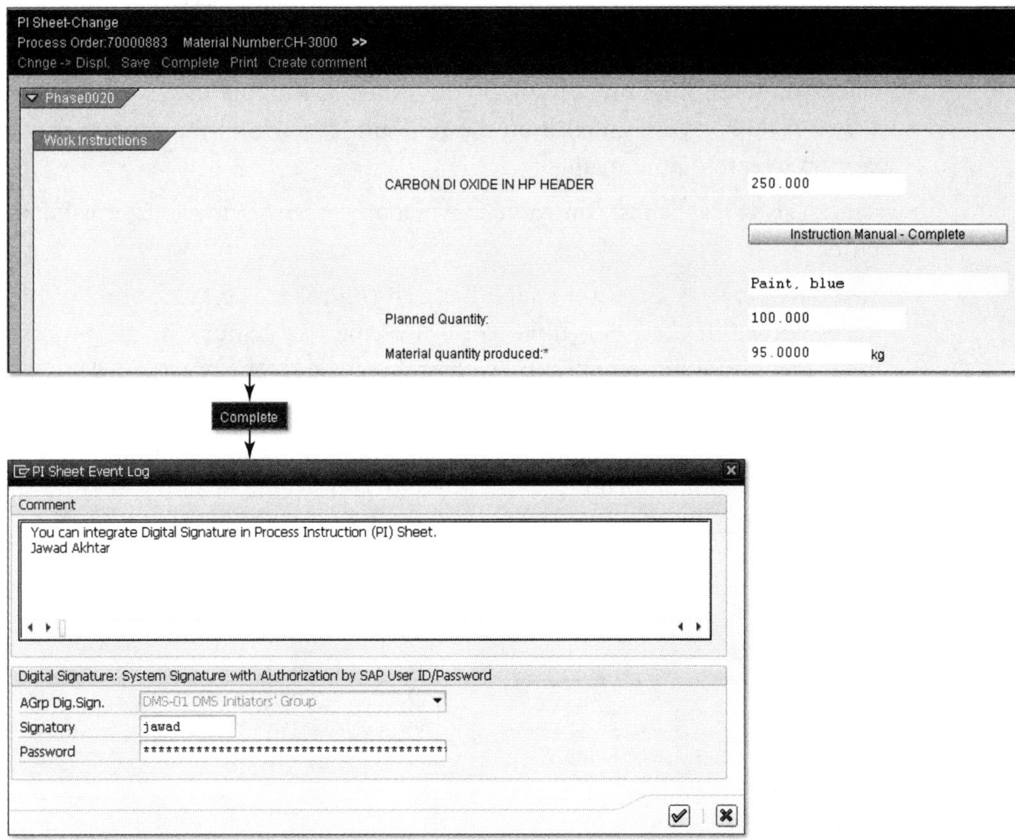

Figure 7.22 Digital Signature in a Process Instruction Sheet

[«]

> **Note**
>
> See Chapter 17 on how to set up the digital signature functionality.

7.4 Process Order Execution

When compared with production order execution in discrete manufacturing, process order execution has differences as well as similarities:

- You can use batch allocation, either automatically or manually, within the process order.

- The system automatically executes the formulas of the material quantity calculation that is defined in the master recipe using the batch allocations.
- When you release the process order, the system can create the control recipe automatically, or you can click on the relevant icon within the process order creation screen to do it manually.
- You can also create a test control recipe in a process order to check or validate settings.
- The system creates a control recipe for each control recipe destination of the process order. If there are three control recipe destinations in the process order, then the system creates three control recipes for the process order.
- Integration points such as in-process inspection, finished goods inspection, capacity requirements planning (CRP), material requirements planning (MRP), sales and operations planning (S&OP), and Long-Term Planning (LTP) also remains the same.

Figure 7.23 illustrates the important elements in a process order.

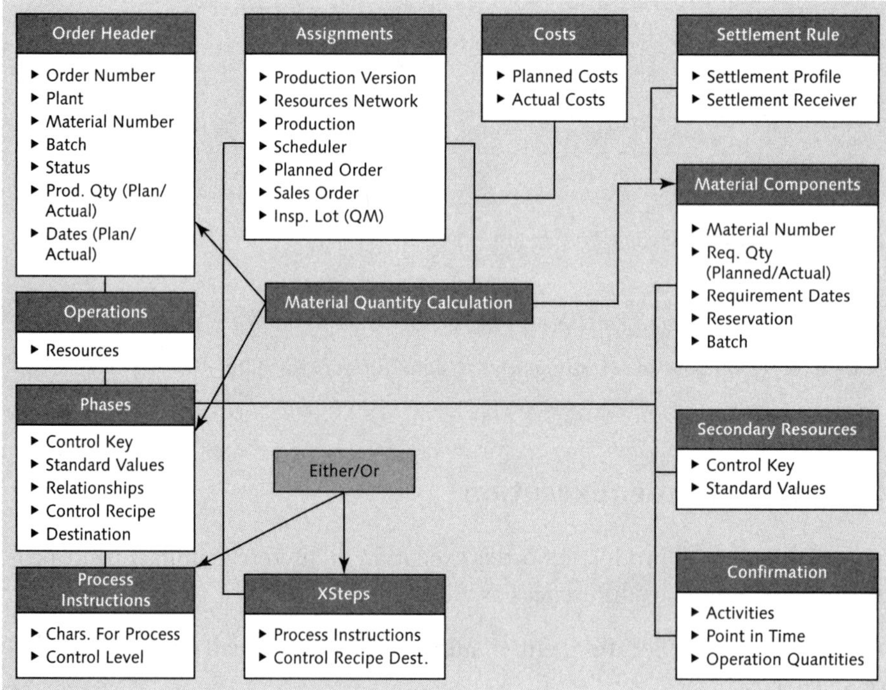

Figure 7.23 Elements of the Process Order

7.5 Process Management in Action

The main steps involved in Process Management are listed here with numbers corresponding to each step as shown in Figure 7.24:

1. The first step is the process order creation. The process order reads the material master data, including the master recipe and materials list (BOM). It also takes the material quantity calculation into account. The master recipe contains all of the information related to Process Management and is used for the generation of a control recipe.
2. When you release the process order and generate the control recipe, the system creates a process instruction sheet.
3. The next step is to send the generated control recipe to the predefined destination in the master recipe (in this example, it's 10 – PRODUCTION FLOOR).
4. The process instruction sheet is now available at the destination, and you maintain the process instruction sheet.
5. You update the status of the process instruction sheet as COMPLETE. Once completed, no further changes to the process instruction sheet are possible.
6. The last step is to send the process messages to their required locations so that the system can perform the necessary predefined actions.

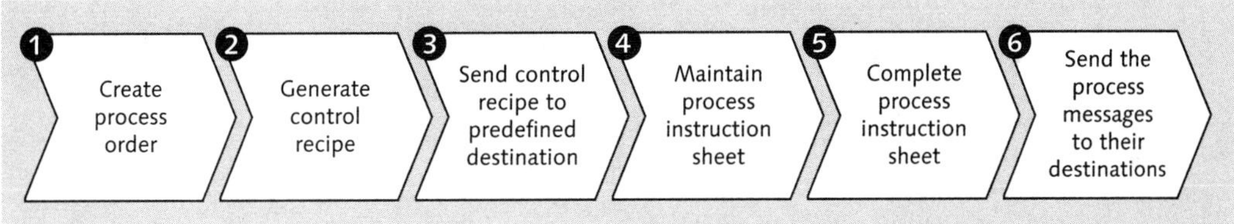

Figure 7.24 The Main Steps in Process Management

You'll learn how to accomplish each of these steps in the following sections.

7.5.1 Creating and Releasing a Process Order

To create a process order, follow the menu path, LOGISTICS • PRODUCTION – PROCESS • PROCESS ORDER • PROCESS ORDER • CREATE • CREATE WITH MATERIAL, or use Transaction COR1. Enter MATERIAL number "1988", PLANT "3000", and order TYPE "PI01", and then press Enter. Figure 7.25 shows the detailed process order creation screen. On this screen, enter the production quantity in the TOTAL QTY field as well as the order finish date in the END field. Notice that the end date of the process order is 05/15/2013. In the process instruction sheet, the system should automatically be able to fetch this date in the P.O. BASIC FINISH DATE field. Release the process order by clicking on the RELEASE icon.

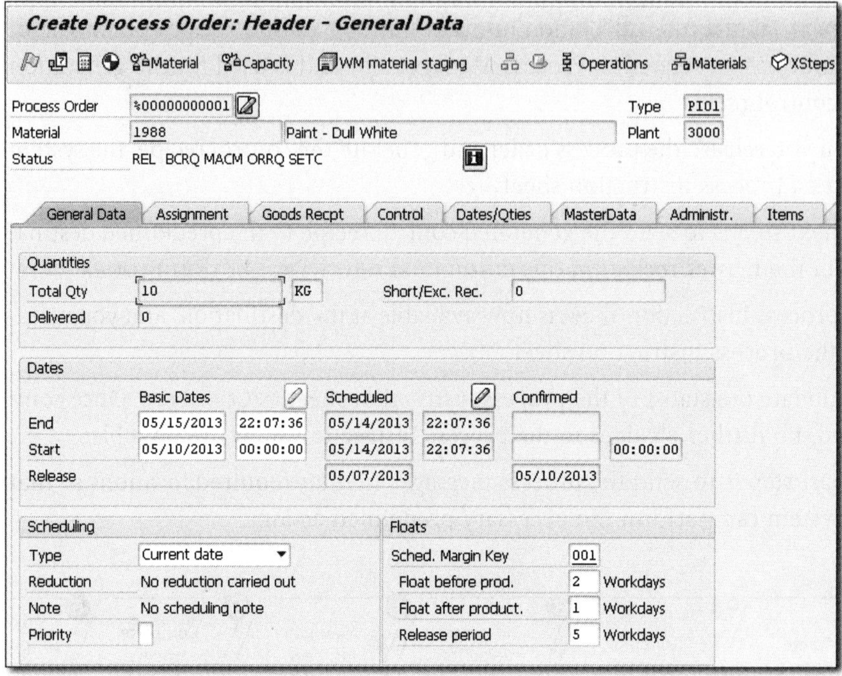

Figure 7.25 Header Screen for Process Order Creation

7.5.2 Generating a Control Recipe

After you release the process order, choose the GENERATE CONTROL RECIPE icon at the top of the process order creation screen to generate a control recipe. An information message appears to inform you that the system created a control recipe and also generated a log. The system has now created a process instruction. With the control recipe being generated, save the process order by choosing SAVE in the process order screen. The system generates a process order number.

Tips & Tricks [+]

You can automate the control recipe creation on process order release by making the relevant settings in the production scheduling profile (refer to Chapter 4) and then assigning this production scheduling profile in the WORK SCHEDULING view of the material master.

7.5.3 Downloading and Sending a Control Recipe

You can manually send a control recipe to a destination in a nonautomated environment (as a background job). To do so, follow the menu path, LOGISTICS • PRODUCTION – PROCESS • PROCESS MANAGEMENT • CONTROL RECIPE • CONTROL RECIPE MONITOR, or use Transaction CO53XT. Figure 7.26 shows the initial screen with selection parameters for the control recipe. For this example, use the same PROCESS ORDER number 70000888, and assign the DESTINATION ADDRESS as PRODUCTION FLOOR (10), which you configured in Chapter 4. Click on the DISPLAY button to view the generated control recipe.

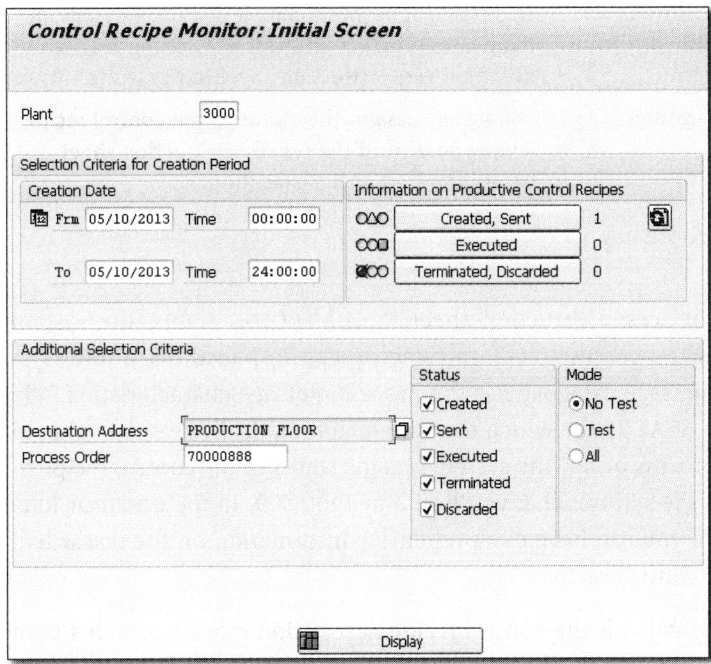

Figure 7.26 Initial Screen of the Control Recipe Monitor

In the screen shown in Figure 7.27, select the line item, and click the SEND button to send the control recipe to destination 10. The status is updated to SENT.

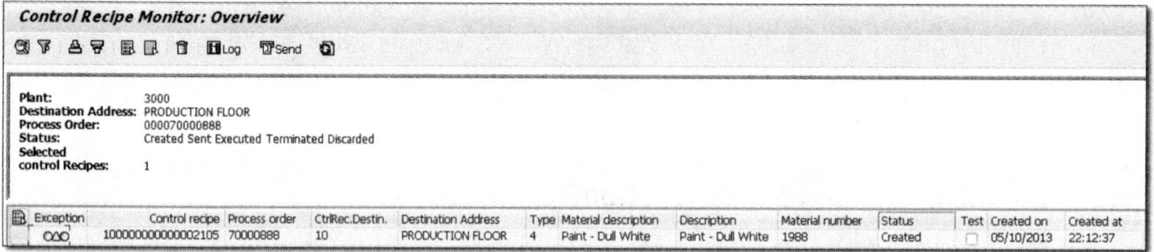

Figure 7.27 Control Recipe Monitor

The control recipe has the following statuses as detailed in Table 7.9.

Control Recipe Status	Control Recipe Description	Details
CRCR	Created	This status appears in the process order, when the system generates a control recipe.
CRFI	Processed/ finished	This status appears on completion of all details in the process instruction sheet and setting the status of the process instruction sheet to COMPLETE.
CRAB	Discarded	The system assigns this status to the control recipe when you discard the process instruction sheet before the start of the execution.

Table 7.9 Control Recipe Statuses

When you set the process instruction sheet to COMPLETION status, the system automatically sends the process message category PI_CRST to the SAP ERP system. This process message category has the process message characteristic PPPI_CONTROL_RECIPE_STATUS for which the system automatically sets the value to PROCESSED. In the process order, the system sets the status of the control recipe to CRFI. This is one of the statuses that you'll find in Table 7.9. In the CONTROL RECIPE MONITOR screen, you can have comprehensive information on the discarded, terminated, or sent control recipes.

When the system downloads the control recipe destination types 1 or 4, it's then known as a process instruction sheet. When the system downloads control recipes of destination types 2 or 3, it reaches an external system or external PCS.

[«]

Note

You can schedule a background job for downloading and sending control recipes with Program RCOCB006 (batch job name: SAP_NEW_CONTROL_RECIPE).

Refer to the detailed settings of the background job covered in Chapter 4.

7.5.4 Maintaining Process Instruction Sheets

To maintain the process instruction sheet, follow menu path, Logistics • Production – Process • Process Management • Process Instruction Sheet • Worklist – Maintain, or use Transaction CO55. You can click on the All Selections icon so that the system can offer you more selection options on the initial screen that aren't visible when you first execute Transaction CO55. For example, apart from new and in-process process instruction sheets, you can also select process instruction sheets that you discarded and terminated (see Figure 7.28). You can use this transaction to also display completed process instruction sheets by selecting the relevant checkbox.

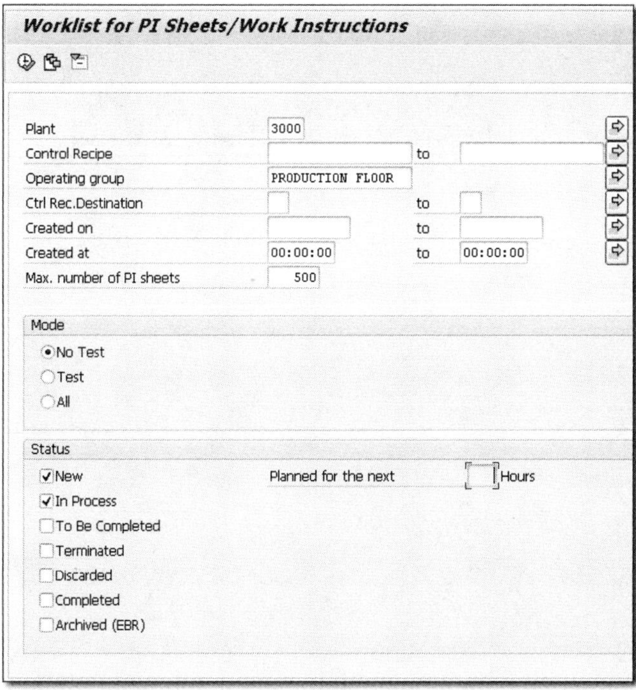

Figure 7.28 Initial Screen for Process Instruction Sheets

Enter PLANT "3000" and OPERATING GROUP (which is the description of the control recipe destination) as "PRODUCTION FLOOR", and then choose EXECUTE or press F8. The corresponding process instruction sheet's worklist is now available for maintenance (see Figure 7.29).

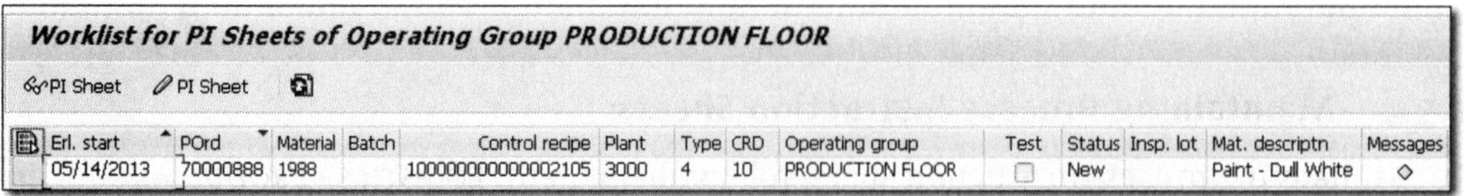

Erl. start	POrd	Material	Batch	Control recipe	Plant	Type	CRD	Operating group	Test	Status	Insp. lot	Mat. descriptn	Messages
05/14/2013	70000888	1988		100000000000002105	3000	4	10	PRODUCTION FLOOR		New		Paint - Dull White	

Figure 7.29 Worklist of Process Instruction Sheets for Operating Group Production Floor

This is the process instruction worklist, which meets the selection criteria. For this example, there is only one process instruction sheet in the worklist. Select the relevant process instruction sheet, and click the MAINTAIN PI SHEET icon PI Sheet to start data input (data maintenance).

Figure 7.30 shows the process instruction sheet with all of the input fields available for maintenance.

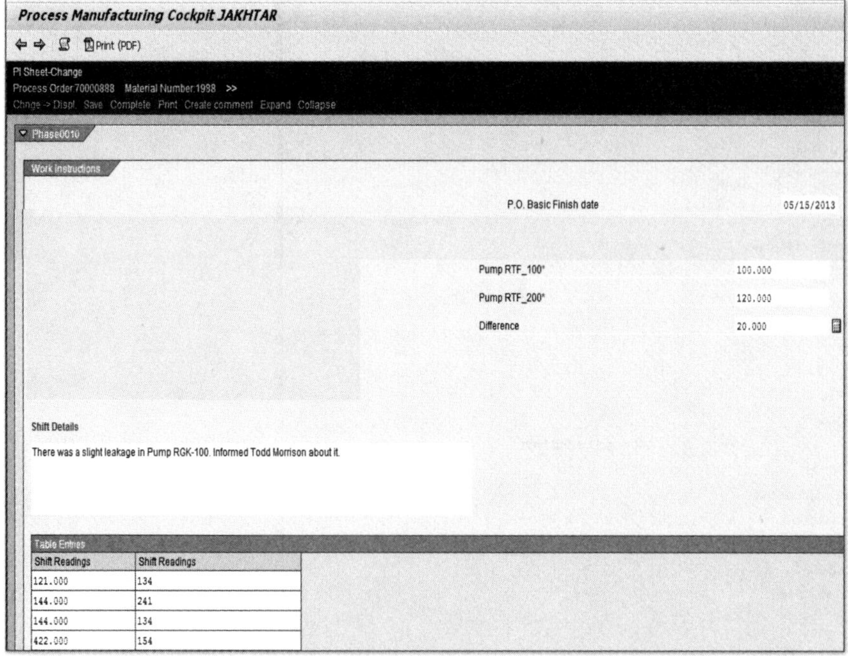

Figure 7.30 Maintaining the Process Instruction Sheet

The process instruction sheet may contain several fields and other important functions already filled in, such as default views or calculation fields.

Notice that the system has automatically filled in the P.O. Basic Finish date field as 05/15/2013 and is also noneditable. Enter the values for the Pump RTF_100 and Pump RTF_200 parameters. There is a red asterisk next to these two parameters to denote that these are mandatory fields you need to maintain.

Tips & Tricks [+]

You can control whether an entry in the field is optional or mandatory by selecting the relevant checkbox in the characteristic (Transaction CT04).

The system automatically calculates the Difference field by using the formula that you defined in the master recipe for this field. Notice a small calculator icon right next to the Difference field to denote that it's a calculated field. This field is also noneditable. If you make changes to the parameter values, the system automatically updates the calculated values.

The Shift Details text box enables you to enter detailed text. You can also make entries in the two tables at the bottom of the screen.

7.5.5 Completing a Process Instruction Sheet

When you've made all entries in the process instruction sheet, you can either save it or mark the status as Complete by using the icons at the top of the process instruction sheet (refer to Figure 7.30). You can use the Save function as many times as needed to save partial and incomplete information in the process instruction sheet. You use the Complete function when all details in the process instruction sheet have been incorporated. Note that after using the Complete function, it's no longer possible to make changes to the process instruction sheet.

Note [«]

If you've made the relevant settings in the configuration (Chapter 4) of the control recipe destination, then the system shows a pop-up message to digitally sign the process instruction sheet on completion (refer to Figure 7.22).

7.5.6 Sending Process Messages

The process instruction sheet automatically creates a process message with all of the information that the system needs to send to specific destinations. This information may be a goods issuance against a process order, a goods receipt against a process order, or a confirmation of a process order. Sending the process message executes the function module with the gathered information.

Process messages in a productive system are typically sent via a background job of Program RCOCB002 (cross-plant) or Program RCOCB004 (plant-specific).

[!]

Warning!

Be sure to monitor process messages because sometimes a process message sending fails if, for example, data is missing from the process instruction sheet. You can monitor and delete process messages with Transaction CO54XT and manually create them with Transaction CO57.

You can also create, test, check, and monitor process messages, although the system doesn't process these process test messages at the destination as it does for normal process messages. You create test process messages to check whether the system is able to transfer all of the relevant information from various systems and destinations.

To send a process message, follow the menu path, LOGISTICS • PRODUCTION – PROCESS • PROCESS MANAGEMENT • MESSAGE • MESSAGE MONITOR, or use Transaction CO54XT. Enter the PLANT and the CREATION DATE range for the process message (see Figure 7.31), and then click the MESSAGE LIST button to send the messages.

In the process messages worklist shown in Figure 7.32, there are two process messages categories, PP10 and PI_CRST, that the system needs to send. Notice that the sender of these process messages is PRODUCTION FLOOR (configured in Chapter 4). Select the first process message, PP10, and click on the CHANGE icon .

In Figure 7.33, the system automatically incorporates the CHAR. VALUE, which is the date 05/15/2013 for the CHARACTERISTIC ZPI_CREATION_DATE.

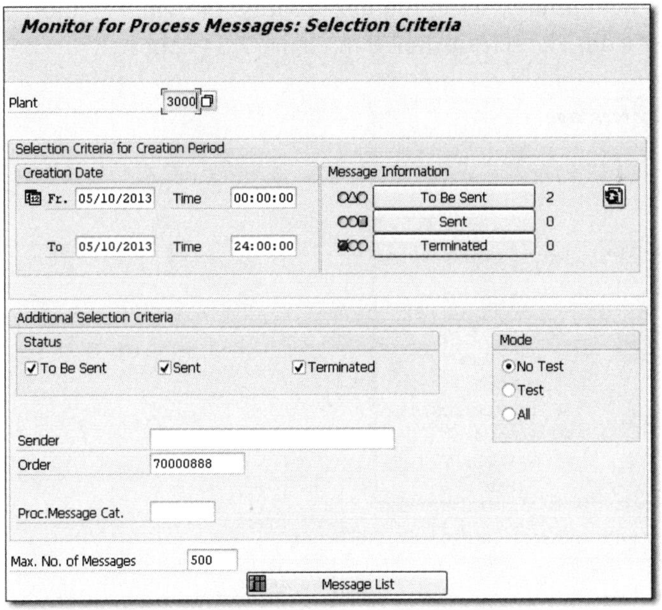

Figure 7.31 Process Messages Monitor

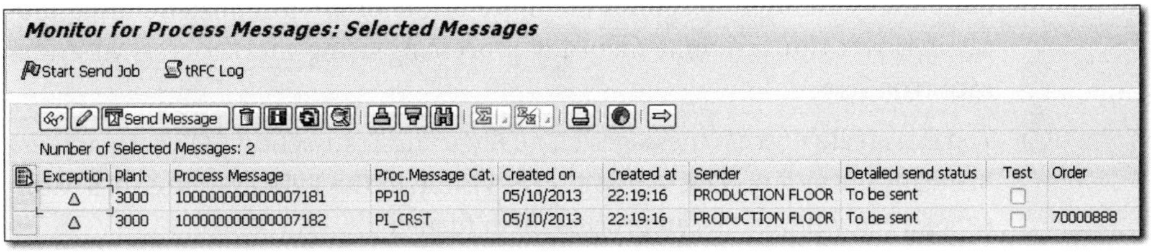

Figure 7.32 Process Messages from the Production Floor Sender

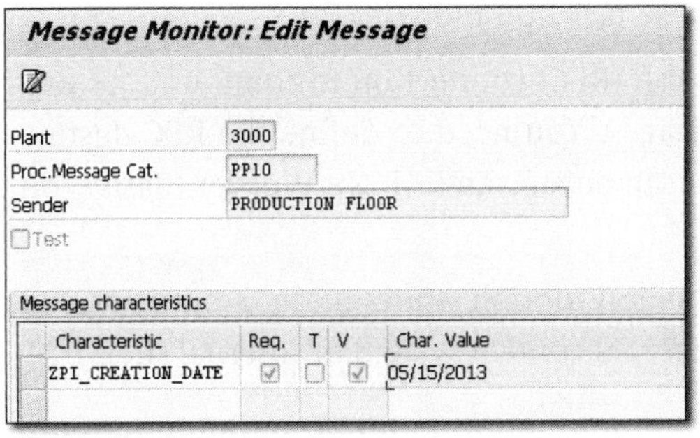

Figure 7.33 Process Message with Message Category PP10

Click on the Back icon (not shown), select the next process message (PI_CRST), and click on the Display icon. The screen shown in Figure 7.34 appears.

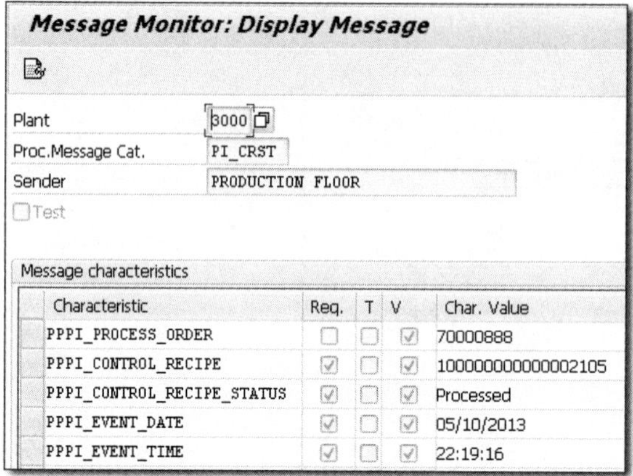

Figure 7.34 Process Message with Message Category PI_CRST

Some of the other salient features of process messages are described in the following list:

- For process message categories that have errors or are terminated during processing, you can resubmit them for processing using the Process Messages Monitor screen.
- You can create process messages manually using Transaction CO57. Then these manually created process messages are visible in the Process Messages Monitor screen where you can process them further.
- The system uses Business Application Programming Interfaces (BAPIs) or function modules to update the process messages. The system processes the received process messages using these function modules.
- The system uses a remote function call (RFC) connection to communicate with an external PCS using process messages. You need to define the RFC destination for every external system; you can engage an SAP NetWeaver (Basis) consultant for this task.
- The system uses alerts of critical situations or milestones in the production run. The alerts can be either from the external control system or internal shop floor operator.

- You can use a special destination type in the process message, SAP ERP Office User, to inform specific SAP ERP system users about the status of process messages (and also control recipes).

7.5.7 Generating a New Control Recipe

Generating a control recipe creates a process instruction sheet. You've maintained the necessary checks and balances such as defining tolerance limits to parameters values, requiring sign-off at the end of each operation or phase, or integrating digital signatures in Process Management. If you realize that you've sent incorrect information in the process instruction sheet, you can create a new process instruction sheet (by generating a new control recipe) to attend to this. This situation arises after you've already performed all of the necessary steps in Process Management.

To generate a new process instruction sheet, follow these steps:

1. Use Transaction CO67, and in the WORKLIST FOR PI SHEETS screen containing the process instruction sheet with the COMPLETE status, select the process instruction sheet, and then choose PI SHEET • DELETE.

   The system issues a message stating that the selected entry (process instruction sheet) was deleted.

2. Use Transaction CO57 to manually create a process message. In this option, enter the PLANT, and use PROCESS MESSAGE CATEGORY PI_CRST. Enter the process order number of the deleted control recipe, enter the number of the deleted control recipe (from the previous step), and set the CONTROL RECIPE STATUS as 00007 (DISCARDED).
3. Save the process message, and the system states that a new process message has been created, including the process message number.
4. Use Transaction CO54XT, and on the initial parameters selection screen, enter the process order number and the process message category as "PI_CRST". Press F8, or click on the MESSAGE LIST icon.
5. Select the process message number created in the previous step, and click on the SEND icon to send the process message. The system issues an information message to inform that the process message has been processed.

6. Use Transaction CO53XT, and on the initial screen, enter the PLANT, the process order number, and the destination address (in this example, it's "Production Floor"). Press F8, or click on the DISPLAY icon.
7. In the CONTROL RECIPE MONITOR screen, select the relevant control recipe number, and click on the DELETE icon. The system issues an information message that the control recipe has been deleted, and at the same time assigns the status of the control recipe as DISCARDED.
8. Use Transaction COR2 for the change in process order, and click on the GENERATE CONTROL RECIPE icon to create a new process instruction sheet for the same process order.

7.6 Execution Steps (XSteps)

Execution Steps (XSteps) are extensively used not only in PP-PI but also in discrete manufacturing. In fact, all of the concepts, fundamentals, features, and functionalities covered in Chapter 4 and in this chapter apply if you need to integrate XSteps with discrete manufacturing. In the following sections, we'll discuss where you can find a repository of standard XSteps that you can use as a reference to create your own XSteps and how to set up a specific function or parameter in the process instruction sheet. We then address the transition that you can make from Process Management to XSteps. The remaining sections then focus on how you can leverage the configuration you made in Chapter 4 and the Process Management concepts you developed earlier in this chapter for XSteps.

7.6.1 Repository for Standard XSteps

The standard XSteps repository is an invaluable resource to gain a better understanding of how to use XSteps. It contains detailed and step-by-step instructions on how you can effectively use all of the features and functionalities of XSteps.

To access the repository of standard XSteps, follow the menu path, LOGISTICS • PRODUCTION – PROCESS • MASTER DATA • STANDARD XSTEP REPOSITORY, or use Transaction CMXSV. For this example, enter PLANT "1100", and choose EXECUTE. The screen shown in Figure 7.35 appears, containing comprehensive details on XSteps.

Repository for Standard XSteps with Versions, Plant 1100

Change Display

Object	Description
SXS Library	Standard XStep Library
SXS library BASICS	
00 General Remarks Concerning Simulation	Limitations for the simulation of PI Sheets
SXS: Calculations (manual/automatic)	Executes calculations manually/automatically
SXS: Changing the 'Look and Feel'	Usage of different style sheets or CSS modification
SXS: Checklists (SAP stand.)	Examples for checklists
SXS: Defining Tables	Grouped instructions, parameterized tables etc.
SXS: Grouped Instructions (Table)	Grouping of process instructions (value input)
0001	Version 0001: Released; Validity Period: 31.10.2007 - 31.12.9999
SXS: Grouped Instructions	XStep Tree
PI Sheet	Plant , control recipe destination , address
PI: SXS Documentation	
<Grouping>	
Read this first (SXS docu)!	
Instruction	
<Grouping>	
Example:	
Line: Machine Type	
Input of Machine Parameters	
Parameter:	
Value:	
Line: Machine Type	
SXS: Parameterized Table	Build a flexible table and calculate average value
SXS: Dynamic Function Calls (SAP stand.)	Usage of function modules of the SAP standard
SXS: Execution of Commands via Triggers	Triggers: Formula, functions or events
SXS: Input and Output of Data	Set and get values of different formats
SXS: Output of Actual Date/Time	Output of actual date and time
SXS: SAP Process Messages	Building blocks for SAP process messages
SXS: Using MDA Data Points	Working with Manufacturing Data Access (MDA)

Figure 7.35 Repository in XSteps

7.6.2 Switching from Process Instructions to XSteps

Whenever you transition from process instructions to XSteps in the master recipe, the system issues the warning message that all previously created process instructions become invalid when you save the changes made in the master recipe. Additionally, the profile in the master recipe (see Chapter 4, Section 4.1.1) should allow for the creation of XSteps. You'll see a warning message that appears when you first click on the XSteps icon (in the master recipe). Click Yes to proceed further.

7.6.3 XSteps: General Information

On the XSteps creation screen in the master recipe, click on the XStep Tree icon, and the system creates an XSteps icon on the left-hand side of the screen. Select the XStep Tree icon again, right-click, and select Properties from the dropdown.

[+]

Tips & Tricks

You can also select the relevant field in XSteps, such as process instruction or calculation, and press Enter, and the system will bring up the PROPERTIES option, wherein you can make the desired changes.

In the NODE ATTRIBUTES screen that appears as shown in Figure 7.36, enter a short DESCRIPTION, which then becomes visible throughout the XSteps tree. Click on the PARAMETERS tab to bring up the next screen, which is described in the next section.

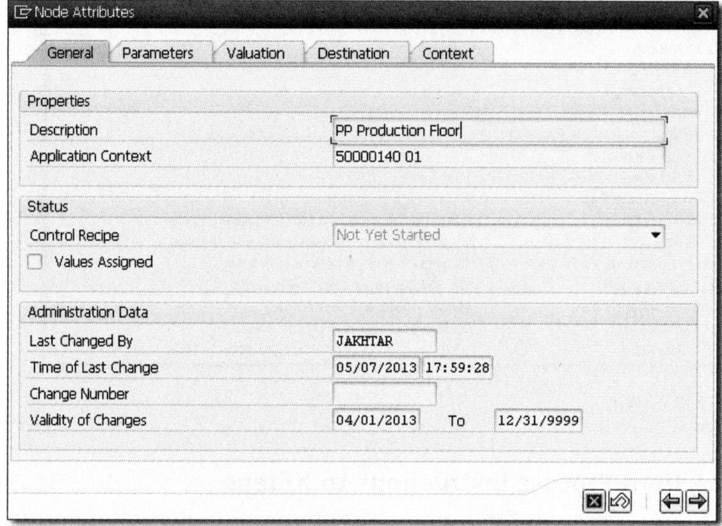

Figure 7.36 Node Attributes in XSteps

7.6.4 Parameters in XSteps

Figure 7.37 shows the screen in which you define all of the parameters you'll use in the XSteps for data entry or for calculation. Unless you define parameters on this screen, you won't be able to select them when you want to; for example, you won't be able to create an input value in the process instructions. If you only have a numeric value, which you won't subsequently use in calculation, then you don't need to define the parameter.

For this example, define two parameters, RFT_100 and RFT_200, and give short DESCRIPTIONS to each. It's important to enter the CHARACTERISTIC ("PPPI_

PARAMETER_VALUE", in this example). You can also enter the characteristics that you had previously created using Transaction CT04 (in the classification system). However, in this case, use characteristic value "PPPI_PARAMETER_VALUE" delivered by the SAP ERP system. If you need to enter parameter values in a table, then you can select the TABULAR VALUE checkbox.

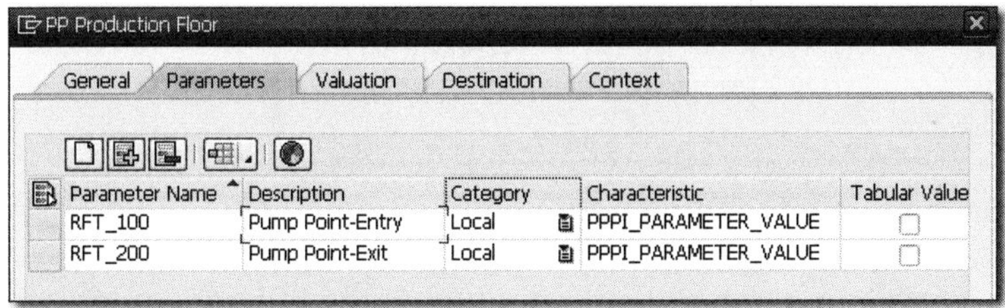

Figure 7.37 Parameters in XSteps

7.6.5 Valuation in XSteps

In the same screen, click on the VALUATION tab to open the screen shown in Figure 7.38, which provides the option for how you want the system to evaluate the parameters. You can even define a fixed value in the VALUE field, if a value remains constant. Additionally, if the value refers to another parameter or value, you can refer to the relevant parameter.

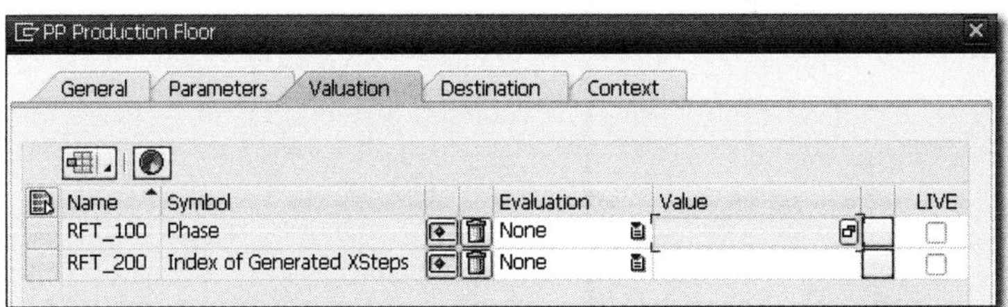

Figure 7.38 Valuation in XSteps

7.6.6 Control Recipe Destination in XSteps

Click on the DESTINATION tab, and the screen shown in Figure 7.39 appears, in which you assign the control recipe destination defined in the configuration section in Chapter 4. For this example, double-click the control recipe destination PP10, which was defined in Chapter 4 and is now assigned here.

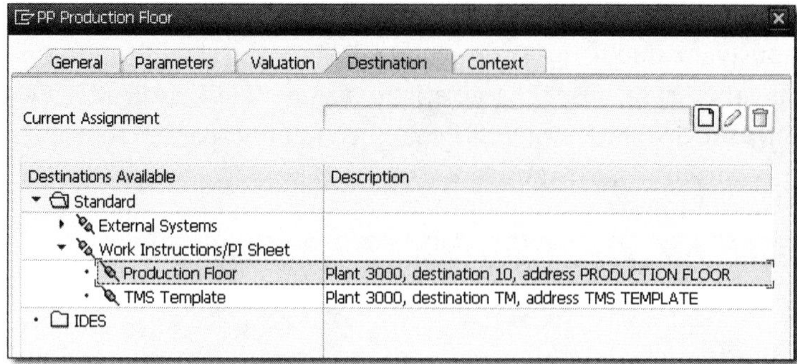

Figure 7.39 Destination in XSteps

Click on the CLOSE icon in Figure 7.39.

After you've successfully entered the details in the previous four figures, the screen appears as shown in Figure 7.40, and you can perform the next step: the creation of process instructions in the XSteps tree. To do so, place the cursor on the PP PRODUCTION FLOOR field (in the XSteps tree), and right-click to bring up the options. From the available options, choose CREATE • PROCESS INSTRUCTION.

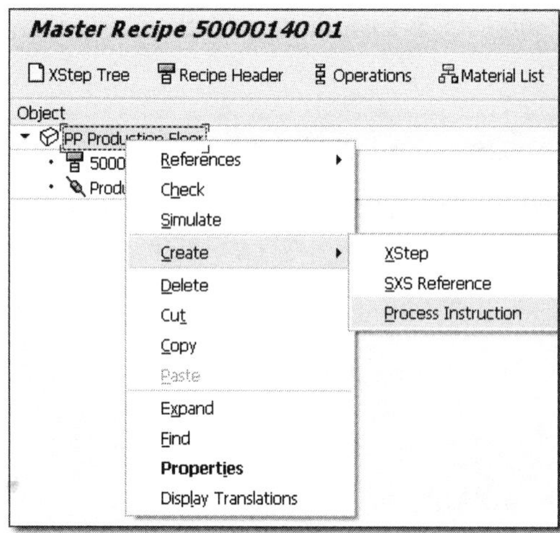

Figure 7.40 Process Instruction Creation in XSteps

7.6.7 Process Instructions in XSteps

You'll now be able to create a new process instruction by right-clicking on NEW PROCESS INSTRUCTION, choosing CREATE • CONTROL DATA, and then choosing your option. You can now freely define the input parameters, output parameters, and function modules, or you can manage control data such as calculations, process messages, commands, or function calls.

Figure 7.41 appears when you select the process instruction and bring up the properties (by either pressing Enter or right-clicking to select the properties option). For this example, enter the description to identify the process instruction.

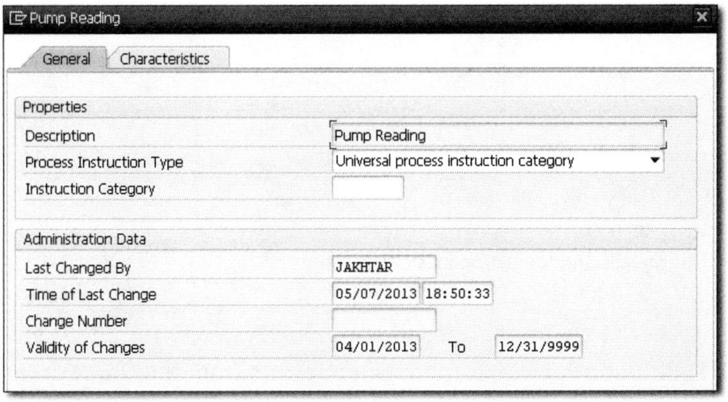

Figure 7.41 General Tab in Calculations

[«]

> **Note**
>
> Now that you know how to access the properties option of any element in XSteps, we'll only cover the details of the element such as input value, output value, or calculation without repeatedly referring to the properties option.

The following sections covers some XSteps features that we've previously touched upon to help you understand the similarities between Process Management and XSteps. We also show additional features in XSteps.

Parameter Values in XSteps

Figure 7.42 appears when you create the first parameter value by choosing Create • Entry • Parameter Value. Enter the short Description, "Pump RTF_100", and from the Parameters area, select RFT_100 Pump Point-Entry from the Name dropdown list. Notice that this is one of the same parameters that you defined when you first created the XSteps tree. The remaining three tabs provide in-depth options on how you want to validate the values entered for a parameter. You can choose to always accept a value, accept it with a digital signature, or never accept an out-of-range value. You can even create your own text that the system will display in case of a failed check.

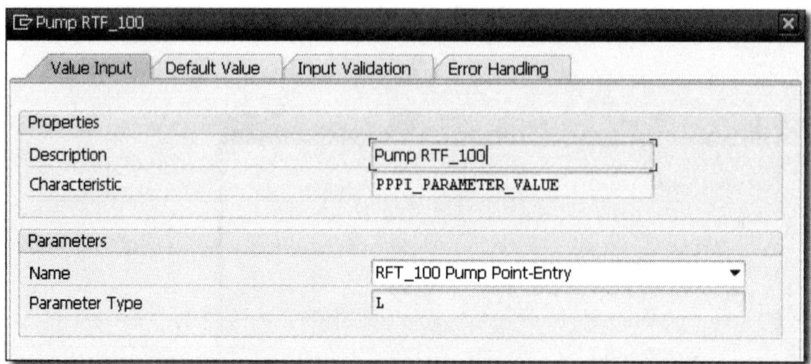

Figure 7.42 Characteristic Details and Validation in XSteps

Repeat the same process to create the second parameter value, with a DESCRIPTION of "Pump_RTF_200", and select the parameter's name from the NAME dropdown list.

[+]

Tips & Tricks

You can also choose CREATE • ENTRY • NUMERIC VALUE to achieve the same results as just described, but you need to select the PARAMETERS option from the dropdown list in the DEFAULT VALUE tab.

Calculation in XSteps

Now that you've created two parameter values in the XSteps tree, you can use them in calculations. You can perform any kind of calculation, as long as the parameters are already defined in the XSteps tree. For this example, use the simple calculation of subtracting the value of one parameter from the other to display the result as a calculated field.

To perform calculation, choose CREATE • CONTROL DATA • CALCULATION. Figure 7.43 appears in which you enter a short DESCRIPTION: "Difference". In the lower half of the screen, enter "RTF_200 – RTF_100" in the CALCULATION FORMULAS box.

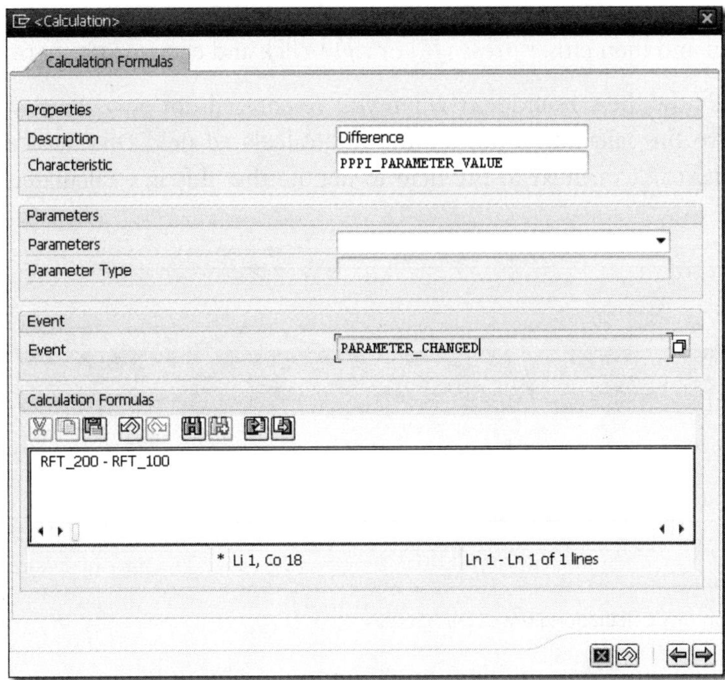

Figure 7.43 Calculation Formula in XSteps

An important aspect to note is the EVENT field, in which you enter "PARAMETER_CHANGED". An event is a control function that triggers the necessary action when the user or the system fulfills a certain and predefined event. For this example, stipulate that whenever there is a change (an event) in the DIFFERENCE field, the system should automatically calculate and update the value in the field. If you don't define an event, you'll then manually have to click on the CALCULATOR icon to update the calculation in the DIFFERENCE field, based on any changes made to the parameter values.

Other examples of events include the system unlocking the table fields when the user maintains the process instruction sheet or locking the tables when all entries in the table are made, or the process instruction sheet being activated or deactivated.

Tips & Tricks [+]

To view all of the available options in events, place the cursor on the EVENT field and press [F4], or click on the dropdown option.

To simulate the fields, you can place the cursor on either the XSteps tree or on the process instruction and then either press F8 or right-click and choose SIMULATE.

Figure 7.44 shows how the screen looks when you've entered not only the two parameters but also the calculation. Notice that the calculated field DIFFERENCE has a small CALCULATOR icon next to the field to denote that this is a calculated value.

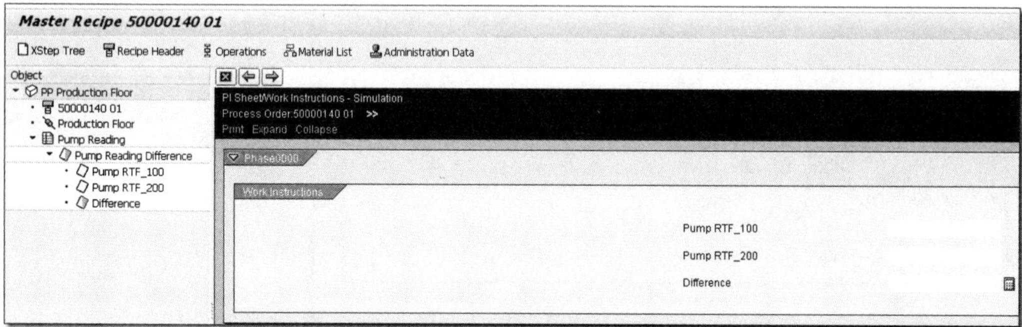

Figure 7.44 Simulation of a Calculation Formula in XSteps

Many input, output, and control options are available in the XSteps tree, so we briefly discuss the most relevant ones in the following sections.

[»] **Note**

We encourage you to explore the large number of options available in XSteps, including the option to incorporate website addresses and images/figures (such as hazard or gloves signs), the option to enter long text or a short note, and so on. You can immediately simulate the results to see if your settings are correct. If they aren't correct, the system provides you the guidance to correct them or points out the shortcoming or deviation.

Output Characteristics and Values in XSteps

If there is a fixed value that you want to display in the XSteps or subsequently in the process instruction sheet, you can use the OUTPUT CHARACTERISTIC option. Enter the DESCRIPTION "Value Output (Fixed)" and use CHARACTERISTIC PPPI_PARAMETER_VALUE with a fixed value of 40, as shown on the right-hand side of the screen in Figure 7.45.

On the left-hand side of the same figure, you can again use the OUTPUT CHARACTERISTIC option, but this time select the CHARACTERISTIC as NOTE, and in the LONG TEXT, enter the text shown. This is a fixed text that will display when you process the process instruction sheet.

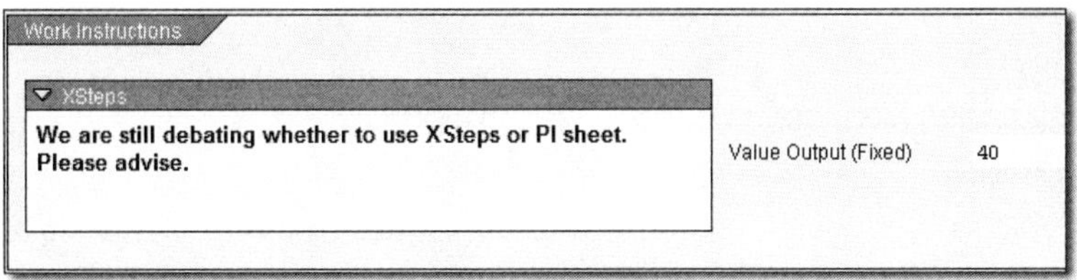

Figure 7.45 Simulation of Output Values

Control Instruction

In the screen shown in Figure 7.46, you can change the description of how each control instruction in the process instruction sheet appears. This option helps in combining the relevant process instructions in one visual tab, so that it's easier for the line operator to either enter the details of the process parameters or simply use the available functions.

To ensure that the system displays control instructions as a tab, first enter the desired description of the control instruction, and then select the CONTROL INSTRUCTION characteristic from the dropdown menu.

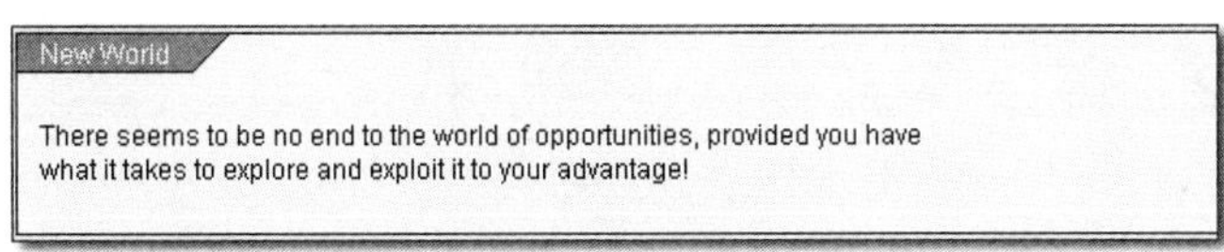

Figure 7.46 Simulation of Output Characteristic in XSteps

Tables in XSteps

To generate a table in the process instruction sheet using XSteps, you need to first create a process instruction and then choose CREATE • CONTROL DATA • TABLE.

Figure 7.47 provides the option to enter the minimum and the maximum number of lines (rows) that must appear in the table. When you've created the table, you can create the input value of the parameters by choosing CREATE • ENTRY • NUMERIC VALUE (OR PARAMETER VALUE).

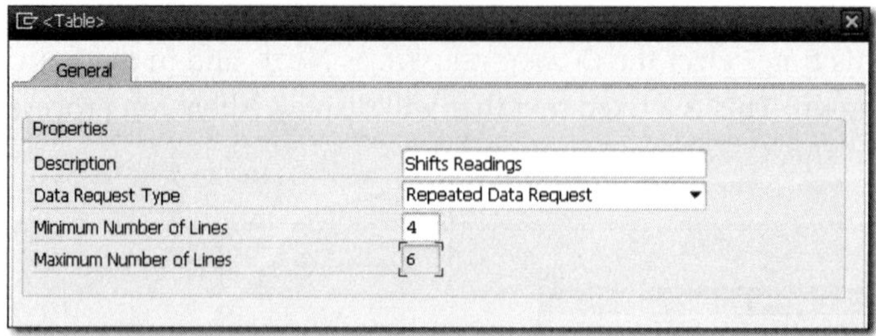

Figure 7.47 Tables in XSteps

Scope of Generation in XSteps

Whenever you create an XStep, the system provides the option to define the scope of generation, just as it does for process instructions. You can define the scope of generation for materials, resources/work centers, documents, or orders. The generation options in XSteps are far greater when compared to process instructions.

Figure 7.48 lists the available options for scope of generation.

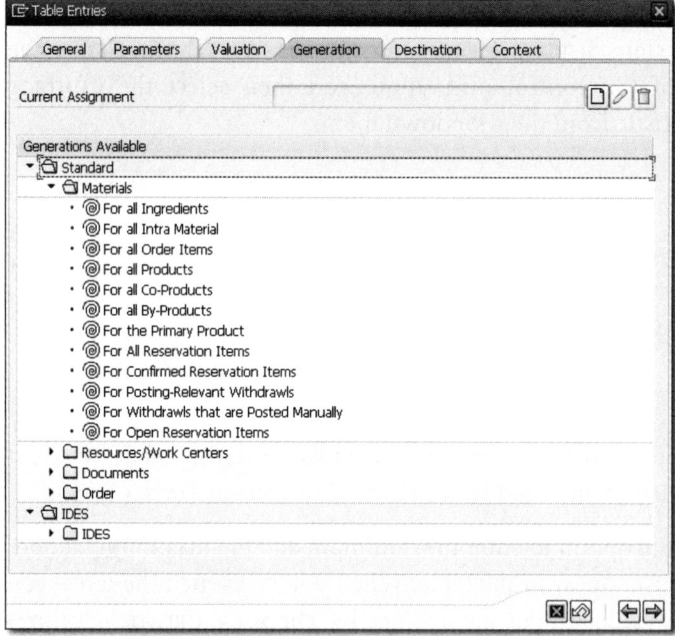

Figure 7.48 Scope of Generation in XSteps

Signature in XSteps

At the process instruction level in XSteps, you can incorporate a signature at various steps of data entry (or other input) to validate the information entered. You can even enter details of a digital signature, if you've already configured it in the system.

The screen shown in Figure 7.49 appears when you choose CREATE • CONTROL DATA • SIGNATURE.

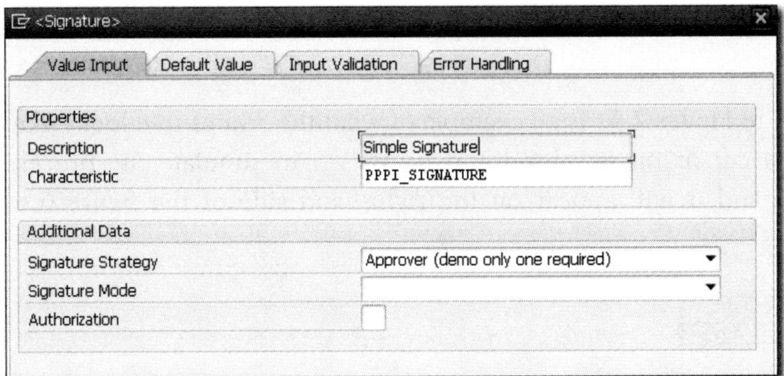

Figure 7.49 Signature in XSteps

[«]

> **Note**
>
> To integrate the digital signature functionality in the production processes, see Chapter 17.

Process Messages in XSteps

At the same level where you can enter a signature in the process instruction, you can also choose to select the associated process message.

Figure 7.50 appears when you choose CREATE • CONTROL DATA • PROCESS MESSAGE. We use the same process message example here that we created in Chapter 4. Enter the PROCESS MESSAGE CATEGORY "PP10", and the relevant message characteristic automatically appears.

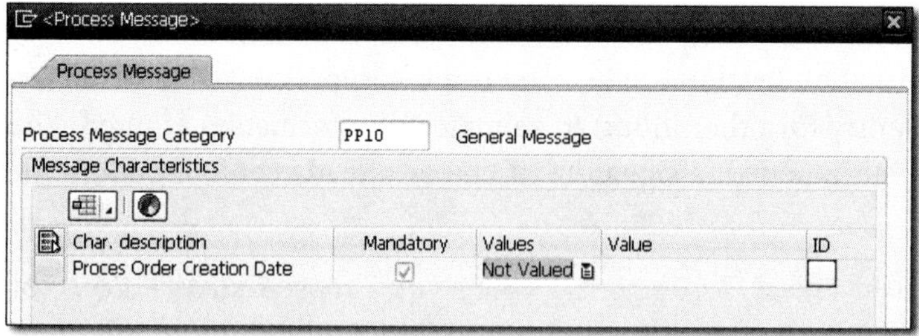

Figure 7.50 Process Messages in XSteps

XSteps Overview

The screen shown in Figure 7.51 is an example of what the XSteps tree looks like. You can place the cursor on any level and press F8 to simulate the process instruction sheet, and it will appear on the right-hand side of the figure (not shown).

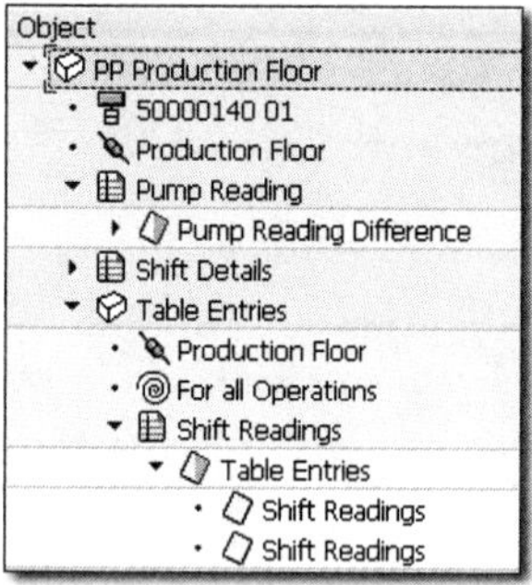

Figure 7.51 Overview of XSteps

[+] **Tips & Tricks**

You can drag and drop any element in the XSteps to your desired location. For example, if you want the Table Entries to appear earlier in the XSteps tree, simply select it and drag it to your desired new location.

Actual Process Instruction Sheet Using XSteps

Figure 7.52 shows the actual process instruction sheet in the maintenance mode (Transaction CO55) that the system created using XSteps. You can also incorporate graphics, such as hazard signs or even process flow charts.

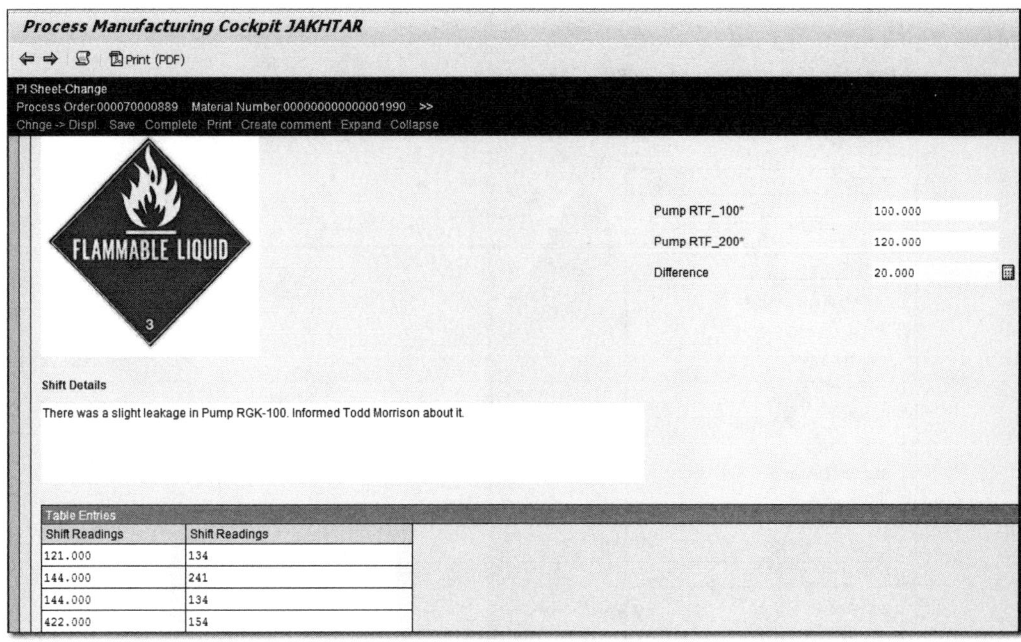

Figure 7.52 Process Instruction Sheet Created by XSteps

7.7 Process Manufacturing Cockpit

Building on the configuration made in Chapter 4 for the process manufacturing cockpit, PP10, we now show how it appears when you want to use it for data entry and reporting purposes. To access the process manufacturing cockpit, use the menu path, LOGISTICS • PRODUCTION – PROCESS • PROCESS MANAGEMENT • MANUFACTURING COCKPIT • START, or use Transaction COPOC.

On the initial screen, enter the PLANT as "3000" and the cockpit NAME as "PP10". Choose EXECUTE.

Figure 7.53 shows the manufacturing cockpit PP10, in which you enter the required details. All details are optional, except for the RESULTS field because it's a mandatory field denoted by the red asterisk. When you exit the manufacturing cockpit, the process messages automatically become available in the message monitor (Transaction CO54XT). From there, you can send them to the predefined destinations.

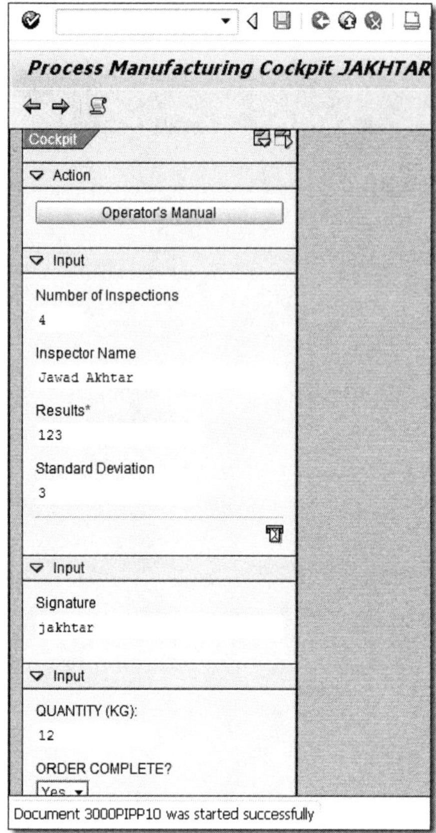

Figure 7.53 Process Manufacturing Cockpit: PP10

7.8 Process Messages Evaluation

The system offers process messages evaluation as a report that you can use to access information on process messages. The process messages evaluation integrates with Microsoft Excel and offers you the option to save the evaluated information either as a pivot table or simply as an Excel worksheet.

You can use all of the standard or self-defined process message characteristics in evaluation. You can self-define the layout for evaluation; that is, on how the system should display the information during evaluation. You need to create an evaluation version before you can use it to evaluate process messages.

For process messages evaluation, use the menu path, LOGISTICS • PRODUCTION – PROCESS • PROCESS MANAGEMENT • EVALUATION • PROCESS MESSAGES, or use Transaction CO52.

On the initial screen, first enter the PLANT as "3000" and the EVALUATION VERSION as "PP10", and then choose the CREATE icon. You need to create an evaluation version first before you can use it in process messages evaluation. You can also create a new evaluation version with reference to an existing version. To create an evaluation version, simply enter the plant, and click on the SAP XXL icon. The system takes you to the screen shown in Figure 7.54, wherein you can select the relevant checkboxes and save. The system asks you to enter the EVALUATION VERSION and its DESCRIPTION.

Process Message Evaluation: Layout for SAP XXL

Evaluate Preview Selection criteria

Plant 3000 New York
Evaluation version PP10 PRODUCTION FLOOR

XXL layout

Description	Line char.	Column grp	Col. char.	Sort no.
PPPI_BLOCK_REF_ID	☐	☐	☐	
PPPI_BLOCK_REF_PARAM	☐	☐	☐	
PPPI_BLOCK_SUCCESSOR	☐	☐	☐	
PPPI_BUTTON_TEXT	☐	☐	☐	
PPPI_CALCULATED_VALUE	☐	☐	☐	
PPPI_CALCULATION_FORMULA	☐	☐	☐	
PPPI_CHANGING_PARAMETER	☐	☐	☐	
PPPI_CLEAR_RESERVATIONS	☐	☐	☐	
PPPI_CODE_CATALOGUE	☐	☐	☐	
PPPI_CODE_GROUP	☐	☐	☐	
PPPI_COMMAND	☐	☐	☐	
PPPI_CONFIRMATION_SHORT_TEXT	☐	☐	☐	
PPPI_CONTROL_RECIPE	☐	☐	☑	4
PPPI_CONTROL_RECIPE_DEST	☐	☐	☑	3
PPPI_CONTROL_RECIPE_STATUS	☑	☐	☐	5
PPPI_DATA_ACCESS	☐	☐	☐	

Figure 7.54 Selection Screen of Process Messages Evaluation

For this example screen in Figure 7.54, use the previously created EVALUATION VERSION PP10, and select the checkboxes in the line (LINE CHAR.) and column characteristics (COL. CHAR.) columns for the characteristics (process messages) that you want to evaluate. For the LINE CHAR. option, select the requisite checkboxes so that the data appear in horizontal lines (in rows). For the COL. CHAR. option, select the requisite checkboxes so that the data appear in columns. For this example, the control recipe status appears in a column, whereas the control recipe destination appears in a row.

If you select the EVALUATE option, the system downloads the information either in the Excel spreadsheet or in pivot table format. Alternatively, you can also use the PREVIEW option, as we did here, and the screen shown in Figure 7.55 appears. The first column and the related rows show the control message ID. In the second column, the first row pertains to the control recipe status, whereas the second row reflects the control recipe number.

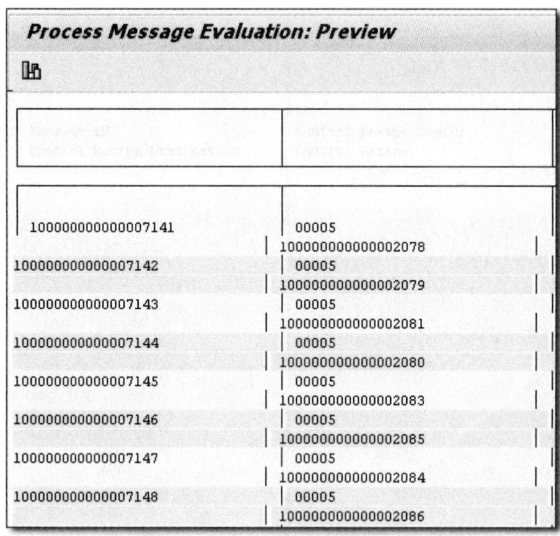

Figure 7.55 Preview of Process Messages Evaluation

7.9 Miscellaneous Cross-Manufacturing Topics

While you can certainly use Process Management to record the consumption of components, here we cover the standard business processes of material staging (goods issuance), confirmation of process order, backflushing, goods receipt, and order settlement. All of the following business processes in PP-PI are the same as in discrete manufacturing:

- **Material staging**
  The material staging in process manufacturing is no different than it is in discrete manufacturing (production order). However, in process manufacturing, you can also use Process Management (process instruction sheet) to record the consumption of components.
- **Confirmation and backflushing**
  In process manufacturing, you can enter confirmations for phases and for secondary resources but not for operations. You can also confirm an entire process order, in which the system takes the standard values into account. You can also backflush components during confirmation.

  In process manufacturing, you can also use Process Management (process instruction sheet) to record confirmation. The system delivers standard process instructions categories and process messages to facilitate this activity.

[«]

Note

Refer to Chapter 3 (discrete manufacturing) and Chapter 4 (process manufacturing) for a detailed understanding of configuration and business processes for confirmation.

- **Goods receipt**
  The goods receipt in process manufacturing is no different than it is in discrete manufacturing. You can perform just about the same functions of goods receipt, including automatic goods receipt upon confirmation, as those available in discrete manufacturing.
- **Settlement**
  The overhead calculation, the variance calculation, WIP, and the process order settlement remain the same as in discrete manufacturing.
- **Reporting**
  Refer to Chapter 19 for an extensive coverage of the reporting options available, including that for process manufacturing.

7.10 Summary

Generally, the production processes in process industries is no different from that of discrete manufacturing, except that it offers options to record greater details of the processes. However, it offers functionalities such as Process Management, which is unique to PP-PI only. With the introduction of XSteps, the Process Management functionality not only becomes easier to implement in PP-PI but also finds its usage in the discrete manufacturing industries.

The next chapter covers production planning in repetitive manufacturing.

This chapter covers the important business processes and functions in repetitive manufacturing, along with the vital and logical links of configuration required.

8 Production Planning for Repetitive Manufacturing

Repetitive manufacturing (REM) is mainly used for production scenarios with high repetition rates, high product stability, and low product complexity. REM is characterized by quantity-based and period-based production (uninterrupted production over a longer period of time). The REM type is often mixed up with takt-based flow manufacturing, which is usually used whenever the production process is based on a takt time; for example, in the automobile industry. Both production types (REM and takt-based flow manufacturing) have different business background and properties, and use different SAP technology, different master data, and even different planning tools.

REM is also often used in the Japanese automobile industry-specific production planning and is known as *heijunka*. Heijunka is used effectively for production smoothing, and also uses sales and operations planning (S&OP), SAP Demand Management, material requirements planning (MRP), capacity requirements planning (CRP), and sequencing.

Note [«]

Refer to the appendix for comparisons between discrete manufacturing, process manufacturing, and repetitive manufacturing, as well as how to decide which manufacturing type is most relevant for a given industry.

In this chapter, we explain what REM is and how it fits into the planning and production perspectives. We walk you through the REM process flow and provide a comprehensive, step-by-step understanding of each process involved. We describe the important REM master data and walk you through how to run the end-to-end MRP process, to show how you can use planning to optimize your production and procurement processes. We also discuss performing the material availability check for missing components.

Then, we cover actual and daily business processes: planning tables and material staging, based on the current material situation; the creation of a production list; REM confirmation at the assembly, components, and activity levels; and CRP. Next, the chapter moves toward handling scraps at various levels, cancellations, and reversals. Finally, the chapter concludes with a few important evaluation reports available in REM.

We've linked this chapter's discussion to the configuration you made in Chapter 5 as often as possible. If deemed necessary, the pointers to necessary configurations are given in this chapter.

8.1 Overview

Companies that work with high-volume and mass-produced materials (pens, pencils, or batteries) find that REM best reflects their production processes, rather than discrete or process manufacturing. If the actual business process is quite simple and straightforward, consider implementing the lean REM production type as long as it's able to fulfill the business needs from a functional perspective.

The SAP ERP system supports and offers make-to-order (MTO) REM, which enables you to handle configurable materials. REM supports handling unit management, in which goods and packaging are mapped as a logistic unit. Serial number assignment for unique identification is also possible in REM.

In the following sections, we'll cover how REM plays a central role in production planning and production execution, as well as discussing the REM process flow.

8.1.1 Roles of Repetitive Manufacturing in Planning and Production

Figure 8.1 provides an overview of the role of REM in the planning and production areas of the SAP ERP system. The planning process starts with requirements coming from SAP ERP Sales and Distribution (SD) or from the material forecast, and becomes a part of SAP Demand Management. The system runs the MRP based on the figures given in the demand program, and the outcome of the MRP run (planning run) is the creation of procurement proposals. These procurement proposals can be either planned orders (order proposals), purchase requisitions for external procurement, or delivery schedules for scheduling agreements with vendors. In REM, you can't convert the planned orders because these are sufficient to trigger the production process. In fact, the REM planned orders obtain the new order type of the *run schedule quantity (RSQ)* when the system assigns a production version to them.

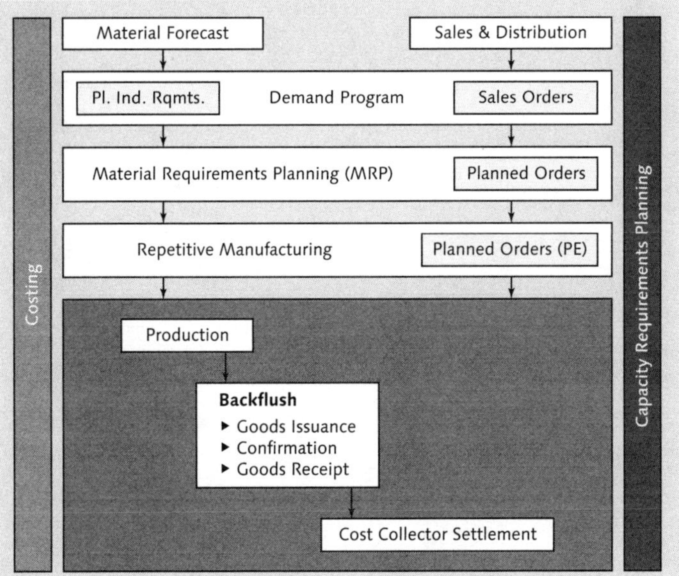

Figure 8.1 Repetitive Manufacturing in Production Planning

After actual production execution, you record the relevant information in the system, such as components consumptions, activities consumptions, and production yield. Because REM is a period-based manufacturing type, you finally proceed to the settlement process, in which variances and overheads are also settled during the cost collector settlement.

8.1.2 Repetitive Manufacturing Process Flow

Figure 8.2 illustrates the end-to-end process flow in REM.

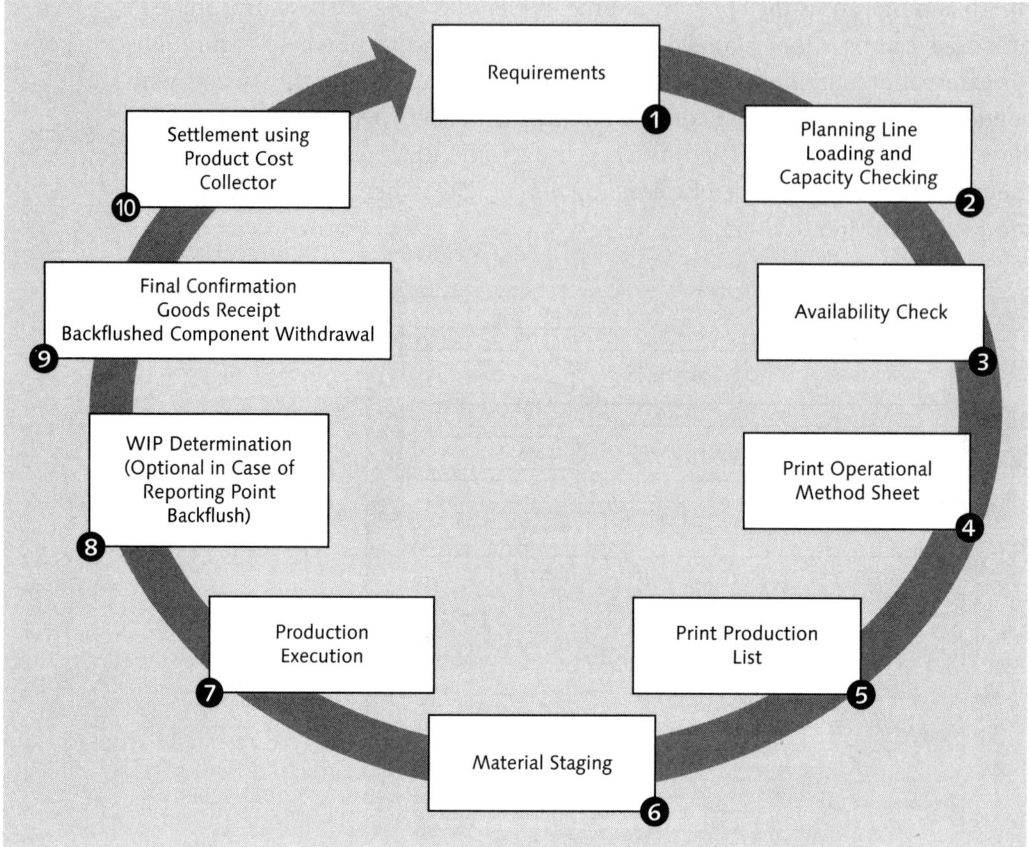

Figure 8.2 Process Flow in Repetitive Manufacturing

The process flow sequence is described in the following list:

❶ The process begins when requirements from MRP (through planned independent requirements [PIRs]) or the SD component flows in, and for which the production must initiate.

The system covers the infinite requirements with planned orders during the MRP run.

❷ The system performs the line loading and checks the capacity requirements to see if it's able to produce the required quantities. If not, it's overloaded.

❸ The system checks for the availability of the components needed to produce the material.

❹ You can print the operational methods sheet (OMS), containing detailed instructions for producing the material.

❺ You can also print the production list, which contains quantities to produce and important dates. You can print a production list specific to a material or to a production line. Following the lean manufacturing principle, these are the only two shop floor papers.

❻ You need to stage the material by undertaking transfer posting from the main store to the production floor, from where you issue the components to the production lines.

❼ The production execution step is the actual and physical production of the material.

❽ If your business process entails working and reporting with reporting point (RP) backflush (milestone confirmation), then you can calculate the work in process (WIP).

❾ You record the production yield, record issuance, consumption of components, and postproduction activities. You can also record excess component consumption, excess activities such as machine hours or labor hours consumed (as compared to standard durations), and record scrap. You can also record any deviation in consumption.

❿ The last step reflects SAP ERP Controlling (CO)'s integration with PP, in which you perform settlement using the product cost collector (PCC).

Note [«]

You can also integrate SAP ERP Quality Management (QM) into the REM production process to ensure that all product produced in-house will go through the quality checks before being dispatched to customers.

See Chapter 20 in which we cover the integration of PP with QM.

8.2 Repetitive Manufacturing Master Data

Before running processes and jobs in the SAP ERP system, it's imperative that you set up complete and comprehensive master data. REM has its own unique master data that often overlaps with other production types, such as discrete manufacturing or process manufacturing. If master data is set up in the right sequence, it's much easier and logical to interconnect, because you've already taken care of the predecessor–successor relationship.

Table 8.1 provides an overview of the master data that you need in REM. It begins with the material master of the product (a finished good or an assembly) in which you specifically mark the REM type in the MRP 4 view of the material master by selecting the relevant checkbox, and also assign the REM profile, which you previously configured (in Chapter 5). You also assign the selection method (entry 2) for the production version in the material master.

Material
▸ Selection method of the production version ▸ Activating REM indicator ▸ Assigning REM profile
Bill of materials (BOM)
▸ Material BOM
Work center (production line)
▸ Production line as work center
Routing
▸ Assign work center (production line)—usually one operation ▸ Production rate (quantity per time)
Production version
▸ Activating REM indicator ▸ Rate-based planning ▸ BOM assignment ▸ Production line ▸ Receipt storage location
Cost object controlling
▸ Product cost collector

Table 8.1 Master Data of Repetitive Manufacturing

You create the BOM of the product that you want to produce and assign components together with the quantities needed to produce that product. If needed, you can also define the scrap percentage at the operation or component level.

You then create the work center or production line and create routing for the material, in which you also assign the previously created work center.

Finally, you create the production version for the material and assign the material's BOM and routing. You also select the REM checkbox to denote that the material produced will follow the REM process.

When all of the logistics master data is in place, your CO team can create PCCs of the material. This can be specific to the production version or can be independent of the production version.

[«]

Note

You need to maintain a close coordination and liaison with the CO team to ensure that when working in PP, you're completely aligned with their working and reporting needs.

For example, for each work center (production line), you need to assign a cost center, which your CO team should provide you with. They may provide you with one cost center for multiple work centers or one cost center for an individual work center, depending on how they want to see the cost center reporting and evaluation.

One possible approach to having several cost centers within one production line is by incorporating additional cost-relevant operations (in the routing) and still keeping the principle of just one work center for production (and PP).

We'll now discuss how to set up each of the REM-specific master data areas.

8.2.1 Material Master

The first step in creating REM specific master data begins by creating the necessary settings in the material master of the product (an assembly or a finished good). To change the material master, follow the menu path, Logistics • Materials Management • Material Master • Material • Change • Immediately, or use Transaction MM02. On the resulting screen, enter Material number "C-1112" and press Enter to bring up the organizational level, where you enter Plant "3200" and Stor. Loc. "0001". Choose Continue to bring up Select Views, in which you select MRP 4 and press Enter again.

Figure 8.3 shows the screen where you can maintain the details related to REM production processes in the MRP 4 view of the material master.

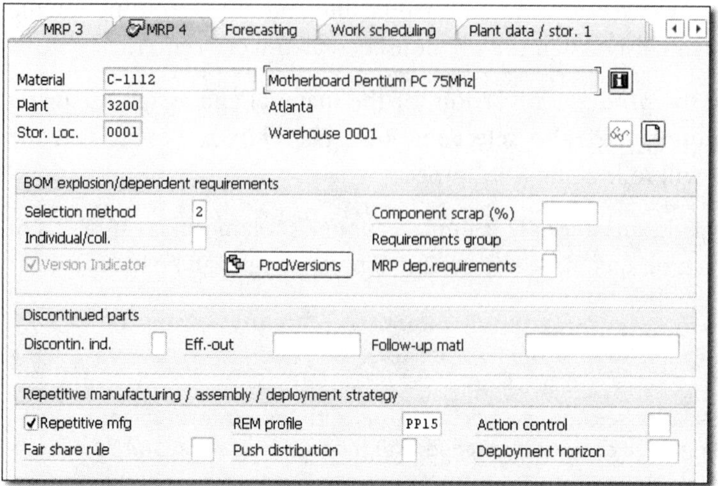

Figure 8.3 REM Indicator in the MRP 4 View of the Material Master

The SELECTION METHOD field provides you with the option to explode the material's BOM with several options. Select 2 in this field, which is always the choice for REM.

Next, select the REPETITIVE MFG checkbox, which denotes that the material will undergo the REM production process. Select PP15 in the REM PROFILE field.

[»]

Note

This is the same REM profile that you configured in Chapter 5.

An REM profile is a four-digit code that determines the following:

- The movement type that the system uses to post goods receipts and goods issues.
- How the system reduces planned order and RSQs in the confirmation process.
- How the system makes the correction in the assembly's components (according to BOM). This enables you to define whether goods issuance posting isn't possible in the system due to forgotten goods receipt postings, or whether the system should process the same, later on in a separate backflush processing transaction.

- How the system deals with production activities during confirmation, such as setup or machine hours.

Save the settings in the material master.

8.2.2 Bill of Materials

The process used to create a material BOM for REM is no different from the process in discrete manufacturing. Refer to Chapter 6, Section 6.2.2, in which we cover the details of material BOM creation.

8.2.3 Work Center (Production Line)

The alternative term for a work center is *production line* in the SAP ERP system. *Work center* is specific to discrete manufacturing, whereas *production line* is specific to REM. Production lines are usually created as simple work centers in the system (and because the SAP system refers to these as work centers, we'll do the same here). In the work center, you define the availability of the production line. In REM, following the lean manufacturing principle, all work stations in the production line aren't modeled as individual work centers. Rather, the entire production line (with all of its work centers) is modeled as "one" work center. The production line created as the work center is entered in the PRODUCTION LINE field in the production version. This same work center is specified in the single operation of the routing. Production lines that need to include or model more than one work center can be represented in a line hierarchy.

To create a work center (production line), follow the menu path, LOGISTICS • PRODUCTION • MASTER DATA • WORK CENTERS • WORK CENTER • CREATE, or use Transaction CR01 (see Figure 8.4 ❶, which shows the initial screen to create the work center). Enter PLANT "3200", for which you want to create a production line. It's mandatory to assign a plant while creating a production line.

Enter the WORK CENTER as "LINE3". It's an alphanumeric identification to denote a production line. A production line can be a production facility, a processing unit, a machine or group of machines, or a laborer or group of laborers, directly involved in the production process.

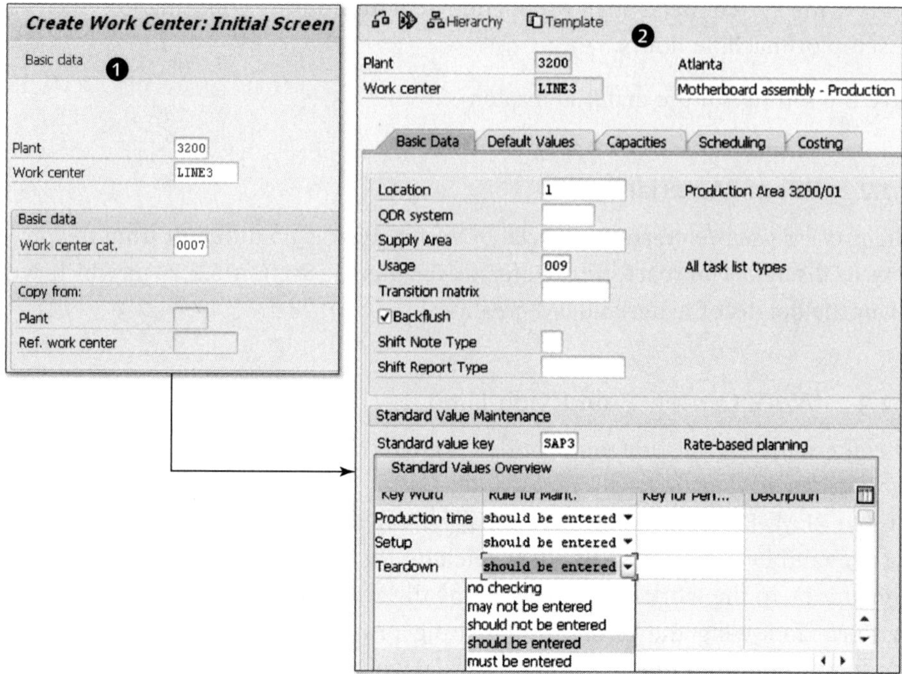

Figure 8.4 Work Center Basic Data

For this example, set the WORK CENTER CAT. field as "0007". This control function ensures that only the relevant screens are eventually available during production line creation. A work center category 0007 denotes a production line.

Other important functions in this screen are COPY FROM PLANT and COPY FROM REF. WORK CENTER. These functions help reduce the data entry efforts during the new production line creation by automatically making available all of the necessary information from the reference production line to the new production line, which you can change as needed.

Click on the BASIC DATA tab of a production line to see the BASIC DATA options described next ❷.

Basic Data

Enter a short description of the production line in the text box at the top-right of this screen. In the USAGE field, enter "009". A *usage* is a control function that controls whether this production line eventually becomes available for various other tasks, such as during the creation of routing or the production version. A usage of 009 denotes that this production line can subsequently be used in all task list types.

The two important fields on this screen are described here:

- SHIFT NOTE TYPE
  A shift note is an option to record important shift-related details of a production line. These can be general details of the shift as well as any specific problem encountered at this specific production line during the production process.
- SHIFT REPORT TYPE
  A shift report compiles and consolidates all of the information from a single shift note or multiple shift notes. These are available as PDFs and other formats, as well as on the layout defined for a company. The user can also digitally sign a shift report before a printout is possible, and send a shift report by email.

In the STANDARD VALUE KEY field, enter "SAP3". A standard value key (SVK) denotes activities performed on a specific production line as well as the option to record them or otherwise. Some of the activities include production time, setup time, processing time, and labor time. For REM, the SAP ERP system already provides the SVK SAP3, which you can use.

You must enter the production time as a rule for maintenance. If the PRODUCTION TIME is assigned as MUST BE ENTERED from the dropdown list, you're required to enter the actual duration of the activity during the confirmation of the REM planned order. Otherwise, the system won't allow you to save the confirmation. Other maintenance options are MAY NOT BE ENTERED, MAY BE ENTERED, or SHOULD BE ENTERED, offering various degree of control from an information message to a warning message, if the requisite information (duration in this case) isn't recorded during confirmation.

Default Values Tab

Click on the DEFAULT VALUES tab to see the screen shown in Figure 8.5 ❶, where you assign the CONTROL KEY PP97 to the work center. A *control key* is a control function that manages whether you can perform an activity or a task associated with the work center. Place the cursor on the CONTROL KEY field, and press F4, which leads to the CONTROL KEY screen ❷.

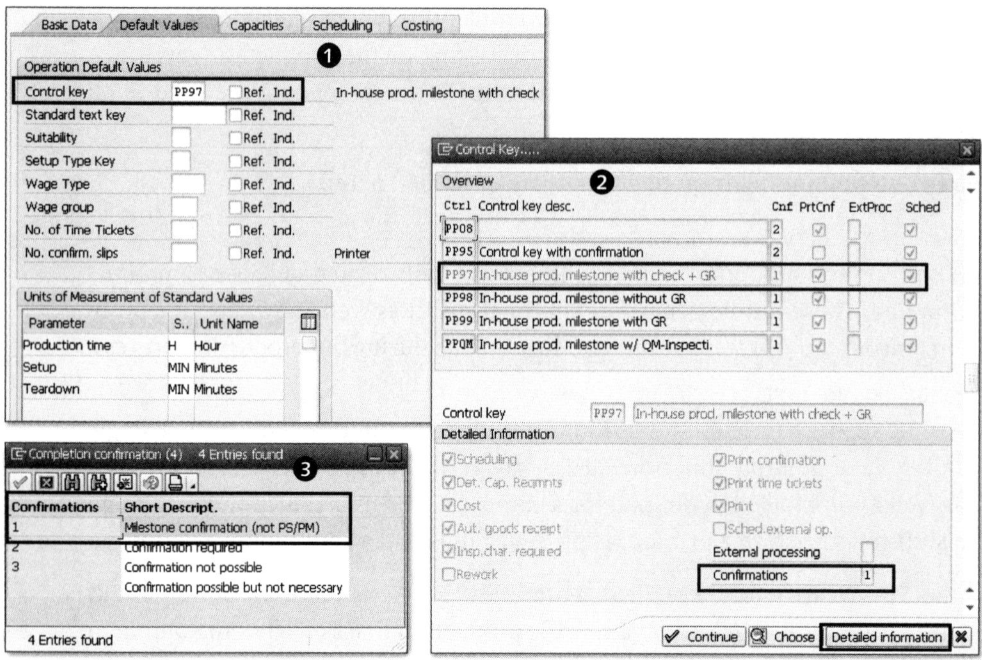

Figure 8.5 Work Center: Default Values

While placing the cursor on CONTROL KEY PP97, if you choose the DETAILED INFORMATION button, the system shows the details in the lower half of the same screen. Notice that in REM, only the following checkboxes are included: SCHEDULING, DET. CAP. REQMNTS., COST, INSP.CHAR. REQUIRED, CONFIRMATIONS. All other checkboxes don't work for REM. If the checkbox isn't selected, then you can't perform that step or function on this work center (Line3). For example, if the SCHEDULING checkbox isn't selected in the control key, then the system won't perform scheduling on this work center.

Also, notice that for the CONFIRMATIONS field, you have to select option 1 (milestone confirmation) if you want to use RP backflush during confirmation.

[»] **Note**

You can set up the control key by using Transaction OPJ8.

If you place the cursor on the CONFIRMATIONS field and press F4, a screen will appear that provides several Confirmations options ❸.

Capacities Tab

Click on the CAPACITIES tab in the work center (production line) creation screen to see the options screen shown in Figure 8.6 ❶. In the CAPACITIES tab, enter the SETUP FORMULA "SAPR05", which denotes the setup formula that the system will use during capacity calculation. For PROCESSING FORMULA, enter the formula "SAPR06". This is for CAPACITY CATEGORY 001 for MACHINE. Repeat the same process of entering formulas for capacity category 002 for labor.

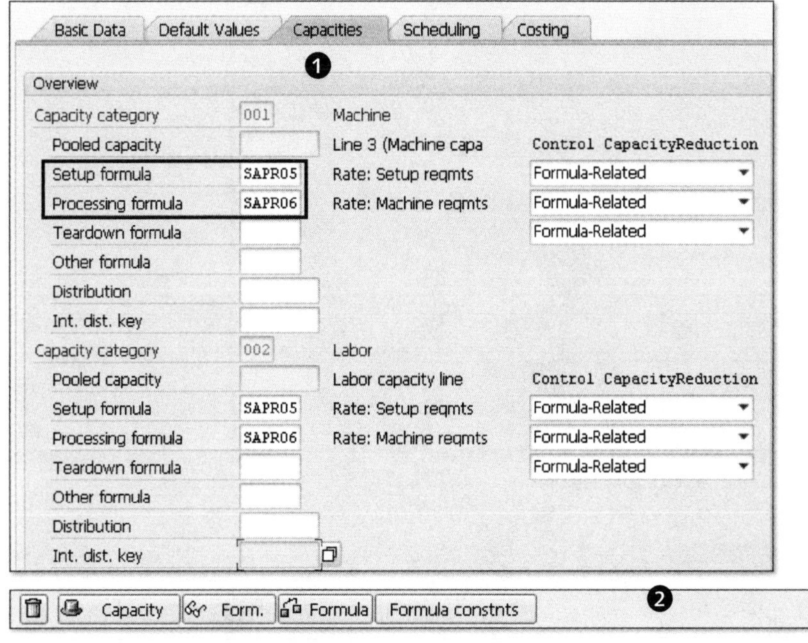

Figure 8.6 Work Center: Capacities

[«]

> **Note**
>
> See Chapter 14 for information on capacity requirements planning (CRP).

Click on the CAPACITY HEADER button [Capacity] at the bottom of the CAPACITIES tab ❷. This brings up the header screen of capacity for work center Line3 as shown in Figure 8.7. The capacity header contains comprehensive details for the capacity: start and end time of the shift, length of breaks, overload percentage allowed for capacity, shifts and intervals, and factory calendar.

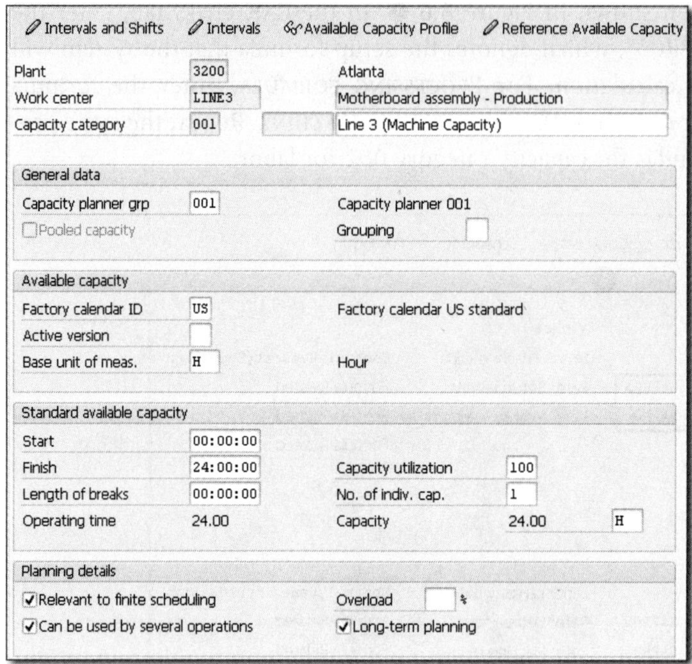

Figure 8.7 Work Center: Capacity Header

[»] **Note**

Refer to Chapter 2 for information on how to set up the factory calendar and capacity planner group.

For this example, enter the following in each field (or select the relevant checkbox):

- CAPACITY CATEGORY short text: "Line3 (Machine Capacity)"
  This short text describes the capacity category, for example, the specific production line or machine.

- Capacity planner grp: "001"
  It's important to assign a capacity planner group to a production line, so that during evaluation of capacities, the capacity planner group can be one of the evaluation criteria. Capacity planner group shouldn't be the name of the person but instead the area or domain of responsibilities.
- Base unit of meas.: "H" (for hours)
  This is the unit of measure (in hours) in which the system displays the capacity of a production line.
- Start: "00:00:00"
  This is the start time of a production line.
- Finish: "24:00:00"
  This is the end time of a production line.
- Length of breaks: "00:00:00"
  This is the total lengths of break during the start and end of shifts. Break timing deducts the availability time of the capacity of a production line. For example, if a machine runs 12 hours a day, with 2 hours of daily breaks, then the available capacity of the production line is 10 hours. Note that this fact is considered in lead-time scheduling of orders.
- Capacity utilization: "100" (percent)
  This is the actual capacity utilization factor. This factor has to be as realistic and as close to actual capacity of a production line as possible to enable the system to calculate the available capacity of the production line. The capacity utilization is reduced accordingly if the production line can't be used for a certain percentage of time, for example, due to machine failure.
- No. of indiv. cap.: "1"
  This is the number of individual capacities associated with a production line. For example, four individual ice cream mixers are created as one production line. Thus, the individual capacities are 4 in number and directly multiply with the timings of the capacity of the production line. For example, if the available capacity of a production line is 10 hours in a day, and there are 4 individual capacities, then the total available capacity of the production line is 40 hours.
- Relevant to finite scheduling
  This checkbox requires the system to consider capacity overload during capacity evaluation.

- OVERLOAD %:
  A capacity overload in percentage directly increases the capacity availability of the production line by the defined percentage. If, for example, the total capacity of a production line is 40 hours with 10% overload, then the available capacity is 44 hours. A 10% overload of 40 hours is 4 hours, and it adds up to 40 hours to show the total available capacity at 44 hours. Since we are not considering any overload, you don't need to maintain any information here.
- LONG-TERM PLANNING
  This checkbox enables the system to consider the production line availability during Long-Term Planning (LTP). LTP refers to a simulative planning of material and production lines over a longer period.

[»] **Note**

Refer to Chapter 12 for more information about LTP.

Scheduling

Click on the BACK icon, and then click on the SCHEDULING tab (see Figure 8.8). Enter the standard formulas available for REM (these are covered in detail in Chapter 6).

Basic Data | Default Values | Capacities | Scheduling | Costing

Scheduling basis

Capacity category	001	Machine
Capacity		Line 3 (Machine capacity)

Execution time

Setup formula	SAPR01	Rate: Setup time
Processing formula	SAPR02	Rate: Machine time
Teardown formula		
Other formula		

Interoperation times

Location group		
Std. queue time		Min. queue time

Dimension and unit of measure of work

Work dimension	
Work unit	

Figure 8.8 Work Center: Scheduling

[«]

Note

See Chapter 14 to learn more about CRP.

Costing

Because PP comprehensively integrates with Product Cost Controlling's Cost Object Controlling, each work center needs to have a cost center. You can assign multiple cost centers or just one cost center to multiple work centers in the COSTING tab, but you can't have multiple cost centers assigned to a single work center.

[«]

Note

You need to coordinate with the CO person to ensure that each cost center and associated activity type for the given controlling area has price planning.

In the same step, you have to ensure that activity types and cost centers are also in place, so that you can assign them in the work center.

In the screen shown in Figure 8.9, enter the following details:

- START DATE
  This is the start date of the availability of the cost center for the production line.
- END DATE
  This is the end date of the availability of the cost center for the production line.
- COST CENTER
  Every production line requires the direct assignment of a cost center, so that when you perform work using a production line, the system is able to accumulate all of the costs on the designated cost center.
- ACTIVITY TYPE (Production time)
  Activity type denotes the predefined rate with which the system calculates the work performed by a production line.
- FORMULA (Production time)
  A formula denotes the parameters that the system considers while calculating the cost incurred in performing work at a production line.

Save the work center.

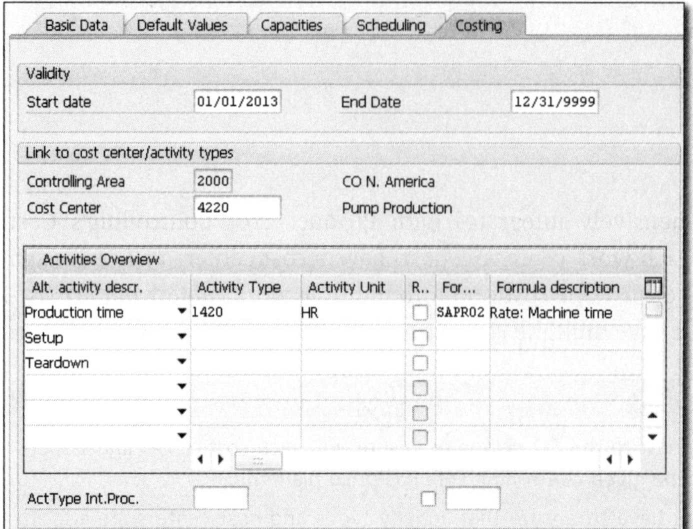

Figure 8.9 Work Center: Costing

8.2.4 Routing

Routings are designed for the maintenance of production rate-based task lists and are used in REM, among other business scenarios. In standard routing, the base quantity (e.g., pieces) generally remains constant, and the time-related data such as the processing time is maintained using standard values. Rate routing works exactly the opposite. The production quantities are maintained on the basis of individual items, and the reference time is constant. Thus, a rate routing describes the quantity of an item that must be produced within a given period of time.

[»] **Note**

There's no need to switch from standard routing to rate routing when starting to model with REM. Both work with REM, and the information included is the same.

A routing can contain a series of operations. The production rate is defined for each operation based on the production quantity, production time, and the associated unit of measure. The production rate is determined as the quotient of the production quantity and production time. This enables you to determine the production quantity in terms of tons per shift, for example. Usually, only one operation is production relevant, and the others may be controlling or print relevant. Using line hierarchies, you can always have several production-relevant operations to support capacity leveling on different work centers along the production line.

All other aspects of the functionality and maintenance of standard routings and rate routings are identical. You assign work centers to the routing, which should also contain the SVK in which the standard values for production time, setup time, and teardown time are predefined.

Reference routings are used as a reference or template in routings. A reference routing contains a sequence of operations. As is the case with the routing, the production rate for an operation is defined on the basis of the production quantity and reference time. The functionality and maintenance of reference routings are similar to those of reference operation sets.

To create a rate routing, follow the menu path, LOGISTICS • PRODUCTION • MASTER DATA • LINE DESIGN • RATE ROUTING • CREATE, or use Transaction CA21. Figure 8.10 shows the initial screen to create routing. Enter the MATERIAL "C-1112" and PLANT "3200". In the KEY DATE field (not shown), enter "04/01/2013", and denote the date when the routing becomes available in the production process. Because the routing validity starts from April 01, 2013, when you create a REM planned order for the month of July 2012, for example, this routing won't be available.

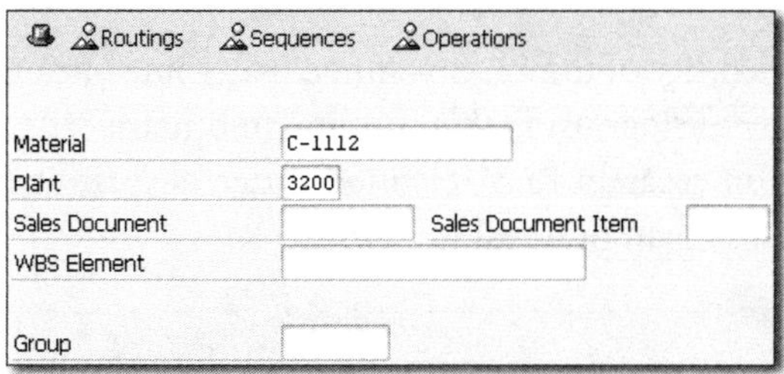

Figure 8.10 Routing: Initial Screen

If no date is given during routing creation, then the system automatically assigns the routing's creation date as the valid from date.

In the following sections, we'll cover the various elements of routing.

Header Details

Click on the HEADER icon. In the header details of the routing, enter the following:

- USAGE: "1" (for production)
  This makes the routing available during subsequent production activities, in which the system will use routing during the creation of the REM planned order.
- STATUS: "4" (released)
  This makes the routing available during subsequent production activities, in which the system will use routing during the creation of the REM planned order. If the status of the routing is set to BEING CREATED (STATUS 1), it won't be available when you create the REM planned order for the material.
- PLANNING WORK CENTER: "Line3"
  This is the production line you previously created, which is now used in the planning and production of the material.
- FROM LOT SIZE: "1" PC
  This denotes that the smallest quantity of the REM planned order must be at least 1 PC for the system to consider using this routing. In other words, this routing is applicable when you create the REM planned order of quantity greater than or equal to 1 PC.
- TO: "100,000" PC
  This denotes that the maximum quantity of the REM planned order must be at least 100,000 PC for the system to consider using this routing. In other words, this routing is applicable when you create a REM planned order of quantity greater than or equal to 1 PC but less than or equal to 100,000 PC.

[»] **Note**

Because the QM also integrates with REM, you can engage the QM resource for this activity.

See Chapter 20 for the integration of PP with SAP logistics components, including QM.

Click on the OPERATIONS button. The resulting screen is shown in Figure 8.11, in which LINE3 is the work center. This denotes assigning a previously created work center (production line) to an operation, which is used in the production of material and for which you're creating the routing. An operation defines a single level of detail for a production step.

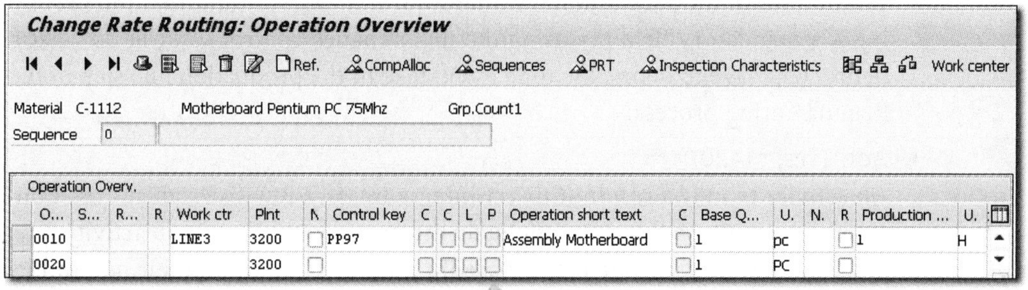

Figure 8.11 Routing: Operation Overview

Operations

Select and double-click on the first line item, OPERATION 0010, which takes you to the screen shown in Figure 8.12.

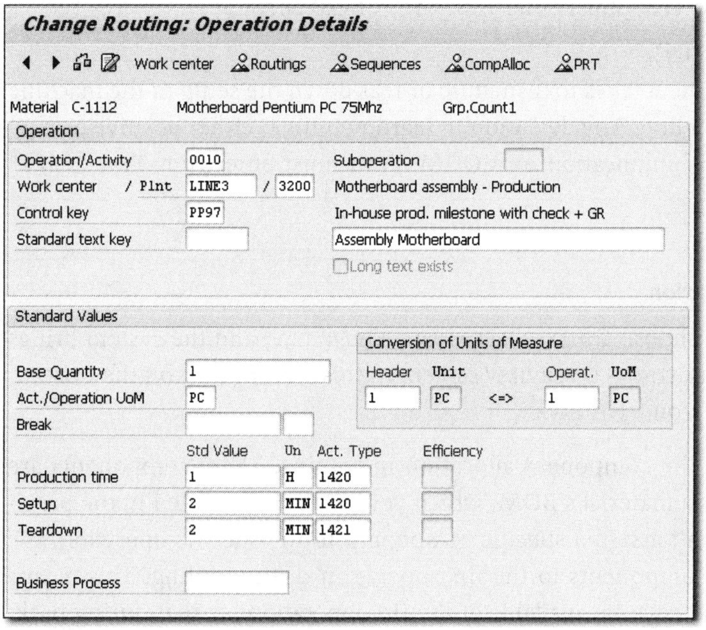

Figure 8.12 Routing: Operation Details

Here you see the operation's detailed view in which you can enter the following:

- Production time: "1 H" (hour)
  You defined in the production line that recording duration information in routing is a mandatory field (by using the maintenance control must be entered); hence, you have to define the time it will take in the production line step of the manufacturing process.
- Act. Type: "1420"
  An activity type is a result of direct integration of PP with Product Cost Controlling (CO-PC). In fact, PP completely integrates with CO-PC. An activity type contains the financial rate (value-based) to perform an activity during production. So, 5 hours of production time has an associated cost of producing the material. Any deviation in production time, either positive or negative, has a direct implication on the cost of goods manufactured (COGM).
- Setup: "2 MIN"
  You define the time it takes to set up the machine or the production line for this task. Any deviation in setup time, either positive or negative, has a direct implication on COGM. (You must enter Act. Type "1420" here.)
- Teardown: "2 MIN"
  You define the time it takes to dismantle or tear down the setup of the machine or the production line. Any deviation in teardown time, either positive or negative, has a direct implication on COGM. (You must enter Act. Type "1421" here.)

Components Allocation

Next, press F6 or choose Goto • Operations Overview, and the system brings you back to the Operations Overview screen. In this screen, click on the Components Allocation icon CompAlloc .

Figure 8.13 shows the component allocation in routing. These components are directly taken from a material's BOM, which you've already created in the previous step. If you don't assign a specific component to the specific operation, the system assigns all components to the first operation of the routing. This means that all components must be available at the first operation, so that during backflush, the system is able to consume all components in the first operation.

Rate Routing Change: Material Component Overview

New Assignment Reassign BOM Task list Operation Variant

Material	C-1112	3200	Motherboard Pentium PC 75Mhz
Group	10022 Sequence	0	CPU-100 production line 3
BOM	00000234 Alt.BOM	1	

Item Overview

P	L..	Path	I...	Component	Quantity	Sort String	U.	I.	B	O...	Seq.	Material Description
	0	0	0010	C-1212	1		PC	L				PROCESSOR CHIP, 75 MHZ
	0	0	0020	R-1220	1		PC	L				Memory, 128 MB
	0	0	0030	R-1230	1		PC	L				BIOS

Figure 8.13 Material Component Overview in Routing

If your business process entails that a few components are used or needed in the later stages of the production process, then you can select them and use the REASSIGN button shown in Figure 8.13 to assign specific components to the operation.

Because there is only one operation in REM, the preceding scenario of material assignment is broadly more applicable in discrete manufacturing than in REM.

Material Assignment

Going back to the routing header screen, click on the MATL (material) ASSIGNMENT icon to see the screen shown in Figure 8.14. Here, MATERIAL C-1112 in PLNT 3200 is assigned an internal GROUP as 10022 with the group counter (GRC) as 2. Later, you'll use this group (10022) and the group counter (2) in creating the next part of the REM master data, the production version.

Material Assignment

Group 10022 Key date 04/01/2013 Change No.

Material TL Assignments

GrC	Task list description	Material	Plnt	Sales Doc.	Item	WBS Element
2	CPU-100 production line 3	C-1112	3200		0	

Figure 8.14 Material Assignment to Routing

You can assign several materials, together with their respective plants, to this routing group (10022). All such materials belonging to this routing group will have the same group but with an ascending group counter number.

In other words, if you have several materials undergoing the same production process, you no longer need to create separate routings for each. Just create one routing, and assign relevant materials (with plants) to it. This saves time and effort in creating an important part of the master data (routing) and also eliminates or reduces redundant data entry efforts.

Save the routing.

8.2.5 Production Version

Production versions denote different ways by which a company produces a material. For example, a material with one routing but multiple BOMs. Hence, for each BOM and the same routing, the system needs a separate production version to denote uniqueness during the REM planned order creation. The requirement to create a separate production version also applies when there is one BOM for a material but multiple routings. In the REM process, it's mandatory to create and make available a production version.

The production version can also be specific to the quantity range, thus acting as a control function. For example, you can create one production version that has a quantity limitation of 1 to 1,000 PC and another production version for a quantity greater than 1,000 PC.

The production version can also be specific to the validity period, thus acting as a control function. For example, you can have a production version that is valid from April 01, 2000, to March 31, 2013. From April 01, 2013 onward, you may have a new production version. When the system has to assign the production version to the REM planned order, it searches for specific validity dates, and on finding the relevant production version, the system automatically assigns it.

If the system is able to find multiple production versions fulfilling the search and application criteria, it assigns the first available production version to the REM planned order. The first available production version refers to the list that you can call from the material master (in the MRP 4 or WORK SCHEDULING views). To create a new production version, follow the SAP menu path, LOGISTICS • PRODUCTION • MASTER DATA • PRODUCTION VERSIONS, or use Transaction C223.

On the SELECTION CONDITIONS area of the screen shown in Figure 8.15, enter PLANT "3200" and MATERIAL "C-1112", and then press [Enter]. If the system is unable to find a production version of the material, then the lower half of the screen appears blank. If it does, then the system displays the relevant production versions.

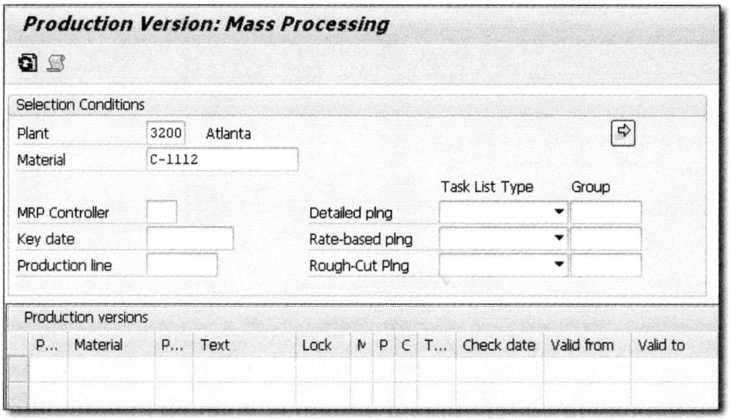

Figure 8.15 Production Version: Initial Screen

A menu bar is available at the bottom of the production version creation screen. You have just about all of the options normally needed while creating production version master data. This includes copying, deleting, copying from a template, and so on.

Click on the CREATE PRODUCTION VERSION icon , and an almost empty screen appears, in which you maintain details as shown in Figure 8.16.

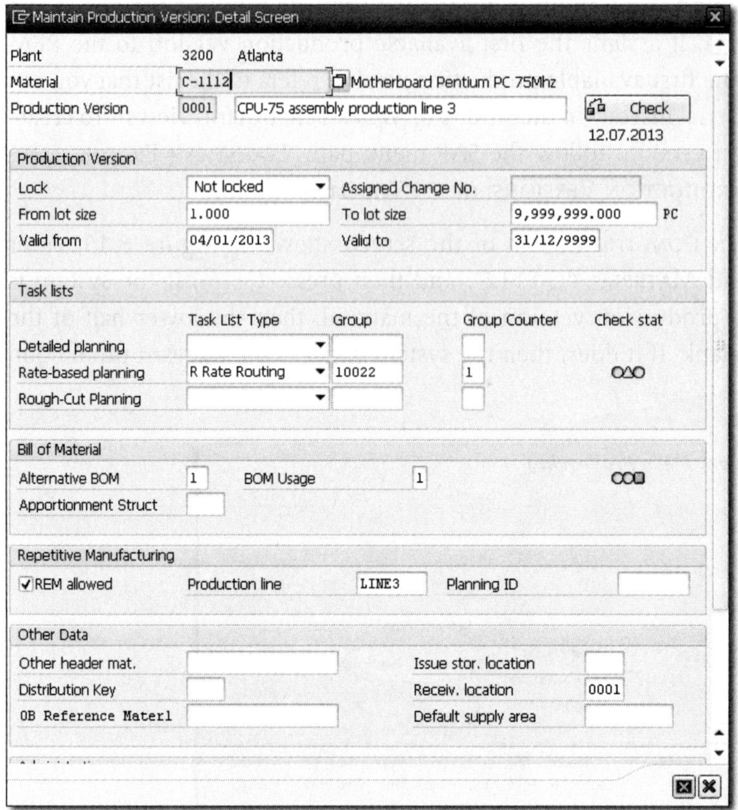

Figure 8.16 Production Version of Material C-1112

Enter the details as noted in Table 8.2.

Field	Value	Additional Details/Remarks
MATERIAL	C-1112	–
PRODUCTION VERSION	0001	–
FROM LOT SIZE	1.000	This field sets the minimum REM planned order quantity for which the system offers this production version for selection. Leaving this field blank entails that there is no minimum quantity limit.

Table 8.2 Parameters Entry in the Production Version

Field	Value	Additional Details/Remarks
To lot size	9,999,999.000	This field sets the maximum planned order quantity for which the production version is applicable.
Valid from	04/01/2013	This field denotes that when you create the REM planned order starting on or after 04/01/2013, the system offers this production version for selection. If the start date of the REM planned order is before 04/01/2013, the system won't make this production version available for selection. If you leave this field blank, the system automatically incorporates the current date as the validity start date of the production version.
Valid to	31/12/9999	This field sets the end validity date of the production version.
Detailed planning	Routing (R)	Task list type R denotes that it's a routing. Each task list type represents a different function; for example, task list type Q represents quality inspection. For this example, enter routing Group "10022" and Group Counter "2".
Rate-based planning	Rate Routing (R)	Enter routing Group "10022" and Group Counter "2".
Alternative BOM	1	This is the BOM that you created in the previous step.
BOM Usage	1	In this field, 1 refers to the usage production.
REM allowed	Select checkbox	This checkbox selection is mandatory to denote that the production version is applicable and becomes available during the REM process.
Production line	Line3	Because REM is generally a simpler production process, only one production line (work center) is involved.
Receiv. location	0001	Entering a value makes it a default setting, and the system automatically proposes it during goods receipt or confirmation of the REM planned order.

Table 8.2 Parameters Entry in the Production Version (Cont.)

The LOCK field provides you with the flexibility to lock or unlock the production version anytime to allow or restrict accesses to the production version during the REM planned order creation. A locked production version won't be available for selection during the REM planned order creation.

Finally, click on the CONSISTENCY CHECK button Check to evaluate if the details incorporated in the production version are correct and valid.

The system confirms that it's able to find a relevant task list (routing) as well as the BOM, which meets the other criteria of the production version. This includes having validity dates for routing and the BOM, which falls within the validity specified in the production version as well as the lot sizes mentioned in the production version. Click on the CANCEL icon.

The green traffic light against the task list as well as the BOM indicates that the consistency check was successful, and this production version will be available to the user during REM planned order creation. If a warning or error is observed during the consistency check, the system indicates that with a yellow or a red traffic light, respectively.

[»] **Note**

The consistency check isn't a prerequisite to using a correct production version. But if you use this option, it will give you the needed confidence that you'll eventually have a valid production version to work with.

Click on the CLOSE DETAILED SCREEN AND ADOPT CHANGES icon, and save the production version.

8.3 Material Requirements Planning in Repetitive Manufacturing

We provide extensive coverage of MRP in Chapter 11. In this section, however, we show how you can use MRP for preplanning of REM-based manufacturing types. The preplanning process entails that you've already planned to produce predetermined quantities of a product and use MRP for comprehensive planning, including the components (BOM) of the product. It's important to note that there are no REM-specific details of MRP. REM uses the standard MRP. For example, preplanning isn't a prerequisite for REM. You can also use REM based on sales orders only (MTO). For this example, we use strategy 40 (make-to-stock [MTS] production method), which is a typical case in REM. You can use any planning strategy in conjunction with REM (MTO and MTS). We'll focus on a typical real-world example in the consumer products industry.

For normal and standard planning, the planning process starts with either a sales order forming the basis of planning or quantities calculated from forecasting. Any or all of these quantities are forwarded to SAP Demand Management, which ends up forming the PIRs. The MRP considers PIRs and generates procurement proposals, including planned orders. You evaluate the planning results and either convert them into procurement elements for external procurement, you don't need to convert them any further or for REM types. We'll discuss this further in the following sections.

8.3.1 Planned Independent Requirements

Use Transaction MD61 to enter the PIR into the screen shown in Figure 8.17 ❶.

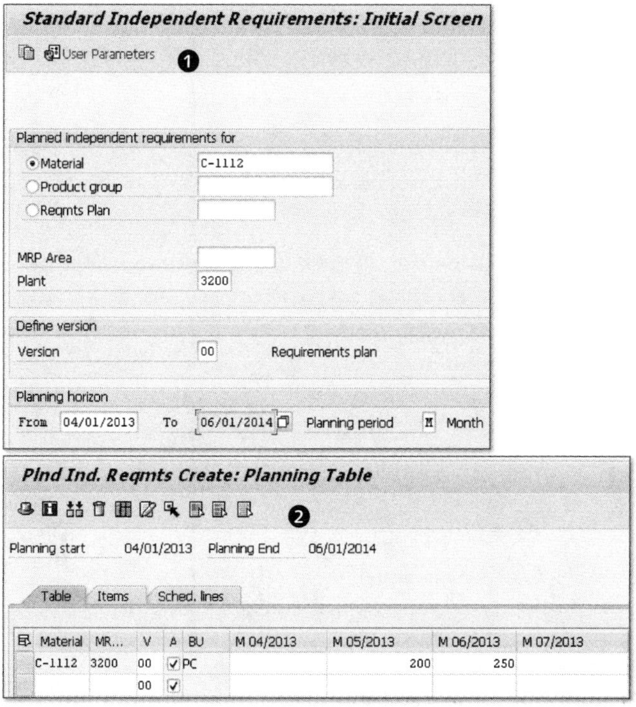

Figure 8.17 Planned Independent Requirements

Enter the MATERIAL "C-1112", PLANT "3200", and VERSION "00", and then define the planning horizon from "04/01/2013" to "06/01/2004". Press Enter. Next, enter planning quantities of "200" and "250" PC ❷ for the month of May 2013 and June 2013, respectively, and save the PIR. The next step is to run the MRP.

[»] **Note**

The material C-1112 in plant 3200 is available in the SAP Internet Demonstration and Evaluation System.

8.3.2 Run Material Requirements Planning

In this step, you run the single-item, multi-level MRP by using Transaction MD02. Figure 8.18 shows the initial screen to run the MRP. Set the parameters as shown, and press Enter twice to begin the process of running MRP. As soon as the system completes the planning process (MRP run), it displays the planning results.

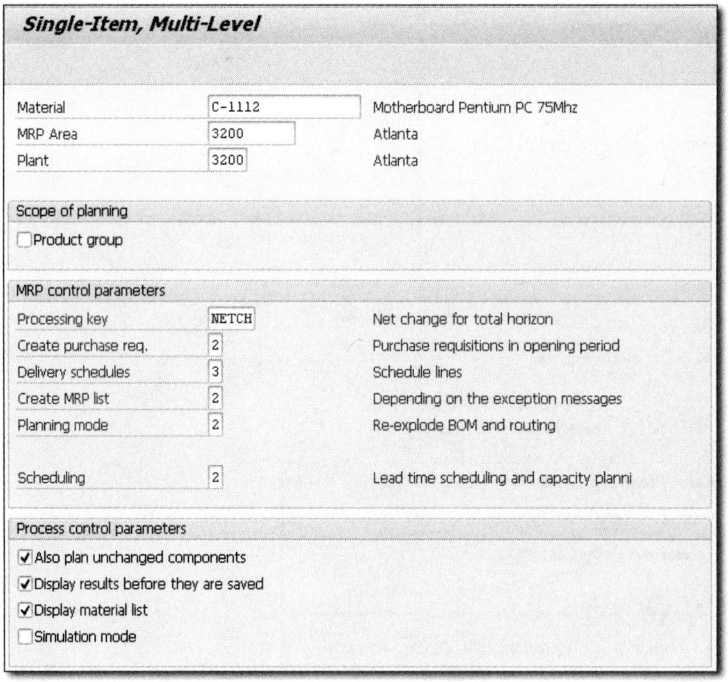

Single-Item, Multi-Level

Material	C-1112	Motherboard Pentium PC 75Mhz
MRP Area	3200	Atlanta
Plant	3200	Atlanta

Scope of planning

☐ Product group

MRP control parameters

Processing key	NETCH	Net change for total horizon
Create purchase req.	2	Purchase requisitions in opening period
Delivery schedules	3	Schedule lines
Create MRP list	2	Depending on the exception messages
Planning mode	2	Re-explode BOM and routing
Scheduling	2	Lead time scheduling and capacity planni

Process control parameters

☑ Also plan unchanged components
☑ Display results before they are saved
☑ Display material list
☐ Simulation mode

Figure 8.18 Initial Screen of the Single-Item, Multi-Level MRP Run

8.3.3 Planning Results

Figure 8.19 shows the planning results for both the material C-1112 and the components associated with it. The next step is to evaluate the planning results of material C-1112 by either double-clicking directly on the material or first saving the planning results (in simulation mode only) and then using Transaction MD04 for material and plant combination to see the stock/requirements list. For this example, use the first option by directly double-clicking on the planning results of material C-1112.

Planning Result: Material List

Selected Results   1st Material

Material C-1112 Motherboard Pentium PC 75Mhz
MRP Area 3200 Atlanta
Plant 3200 Atlanta

Valid from ...	Material	Ext. Manuf	MRP Area	Material Description	M...		1	2	3	4	5	6	7	8	Stc...
	C-1112		3200	Motherboard Pentium PC 75Mhz	001	✓			1	1			1		2.4
	R-1220		3200	Memory, 128 MB	001	☐									999.9
	R-1230		3200	BIOS	001	✓				1					20.0

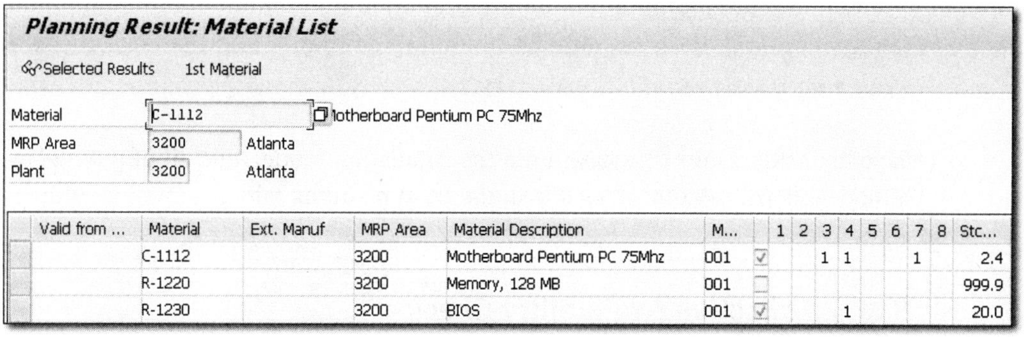

Figure 8.19 Planning Results of the MRP Run

8.3.4 Evaluate Planning Results (Material Level)

Double-click MATERIAL C-1112 to open the screen shown in Figure 8.20. Here you see the planning results for material C-1112, which contains an initial stock of 121 PC. It subtracts a quantity of 50 PC to account for safety stock to arrive at the available quantity of 71 PC. Next, the system takes the existing and firmed RSQ (REM planned order) of 125 PC into account and arrives at a cumulative quantity of 196 as the available quantity. The needed or the preplanned quantity for May 2013, is 200 PC, so the system creates a new REM planned order for a quantity of 4 PC.

Finally, the system creates another REM planned order for 250 PC, which we manually changed to 270, against the second PIR quantity 250 PC.

Note [«]

In principle, you should only make manual changes to the REM planned orders in Transaction MD04.

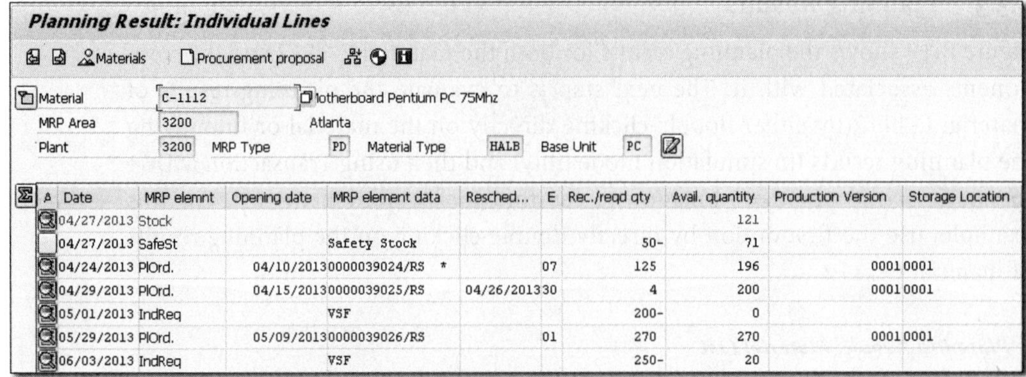

Figure 8.20 Planning Results of Material C-1112

Save the MRP results (if you've used the simulation mode), and all of these planning details will become available in the REM planning table.

8.4 Collective Availability Check

As a production planner, it's important that you frequently and consistently monitor the availability of the components that are needed to produce the material. While you can always perform an availability check of an individual planned order in REM, it's more efficient if you perform a collective availability check on several materials simultaneously to gain a better understanding of the material availability situation and then plan or take action accordingly.

In the collective availability check option, you can use a large number of selection parameters to choose the information that the system should display.

To perform a collective availability check, use menu path, Production • Repetitive Manufacturing • Planning • Collective Availability Check, or use Transaction MDVP. On the initial screen, enter the desired parameters, and choose Execute or press F8. If the system finds a materials shortage, it identifies the shortage with a red traffic light. You can then click on the Missing Parts icon for an in-depth analysis and evaluation because it contains details of the requirements versus available quantities as well as the dates when the required and shortage material should be in store to timely initiate the production process. The availability check function in SAP ERP only checks the availability but isn't able to trigger replenishment.

8.5 Operational Method Sheet

The operational method sheet (OMS) is primarily used for production personnel to have a clear understanding and step-by-step directions involved in producing a material. You can configure the layout and the information that you want to see on the OMS. It integrates with Microsoft Word, and you can make any last-minute changes and print it. Further, there is an option to print the sheet at the header level or at the operation level.

To print an OMS, follow the menu path, Logistics • Production • Master Data • Line Design • Rate Routing • Print Operational Method Sheets, or use Transaction LDE1. On the initial parameters selection screen, enter Material "C-1112" and Plant "3200". It's also important to select a template for either the cover sheet or individual operation, or both. For this example, select the template for the individual operation, and choose Execute or press F8. Word opens, and when the system asks you to choose to enable or disable the macros, choose Enable.

Figure 8.21 shows the print preview of the OMS. The layout is the same as configured in Chapter 5.

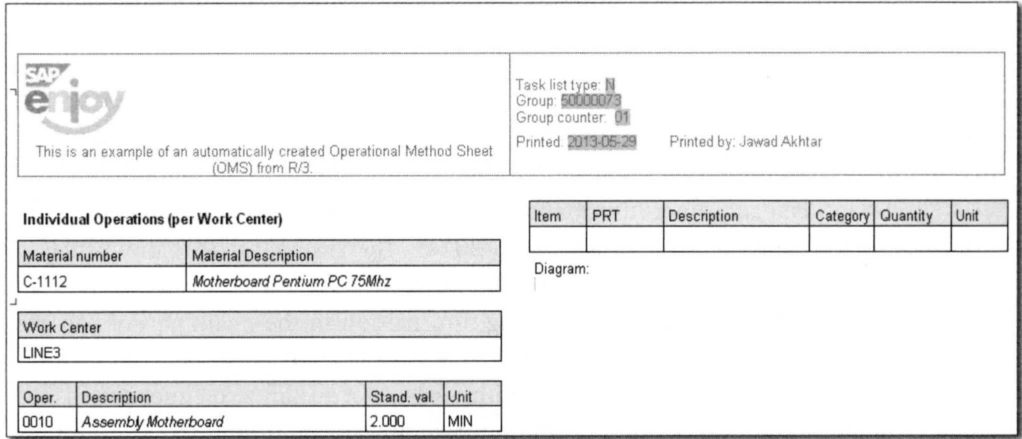

This is an example of an automatically created Operational Method Sheet (OMS) from R/3.

Task list type: N
Group: 50000073
Group counter: 01
Printed: 2013-05-29 Printed by: Jawad Akhtar

Individual Operations (per Work Center)

Material number	Material Description
C-1112	*Motherboard Pentium PC 75Mhz*

Work Center
LINE3

Oper.	Description	Stand. val.	Unit
0010	*Assembly Motherboard*	2.000	MIN

Item	PRT	Description	Category	Quantity	Unit

Diagram:

Figure 8.21 Print Preview of the Operational Method Sheet

8.6 Planning Table in Repetitive Manufacturing

The planning table in REM is highly interactive and enables the planner to perform a large number of functions and activities with all details automatically updated. For example, if you create a new planned order in the planning table, the system automatically updates the capacity requirements in the upper half of the planning table screen. The planning table also provides comprehensive visibility on material and capacity availabilities. In fact, important planning in REM, such as line loading planning, is generally undertaken in the planning table.

In the following sections, we'll discuss the planning table parameters and how to create orders.

8.6.1 Parameters Selection for the Planning Table

To access the planning table in REM, follow the menu path, Logistics • Production • Repetitive Manufacturing • Planning • Planning Table • Change Mode, or use Transaction MF50. Figure 8.22 shows the initial screen for the selection criteria, which allows you to enter important selection parameters. Based on the input parameters you define in the parameters selection screen, the ensuing planning table then displays the relevant information.

In REM, the system refers to the main information with reference to the production version of the material; therefore, it allows you to have the option to enter data specific to the production version on the selection screen. Some of the other selection options are Production line, Material, Product group, MRP controller, and so on.

For this example, enter Plant "3200" (entering either the MRP area or Plant on the parameter selection of the planning table is mandatory). Select the Material radio button, and enter the material "C-1112", for which the system brings up the relevant data and other information in the planning table. For Start, enter "04/27/2013", and for End, enter "07/26/2014". The system restricts the display of information in the planning table based on these start and finish dates. Additionally, the system won't show planning information in the planning horizon that lies outside the planning horizon. Assign the Period field by Month, and the system shows planning details in a monthly basket. You can also choose to display planning information in the planning table, either in days or weeks.

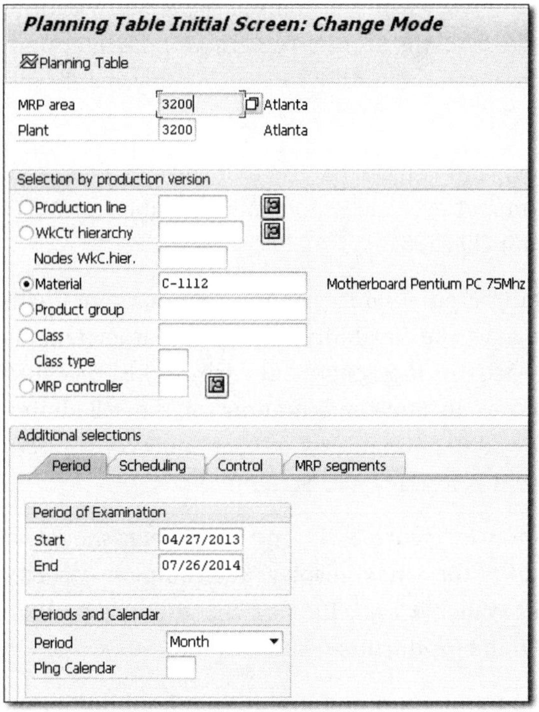

Figure 8.22 Initial Screen of the Planning Table

Click on the Scheduling tab shown in Figure 8.22. The scheduling options available in this screen enable you to select options to make them available in the planning table. You can choose whether you want the system to perform rate-based planning and select whether it should consider basic dates or production dates during planning. You can also enable the system to perform capacity planning by selecting the relevant checkbox.

Note [«]

We suggest that you make the same selection for the planning table as you did when you maintained the routing in the production version. Except, in this case, the capacity requirements can be evaluated properly during planning. We also suggest that you check on the configuration in Transaction OPU5 to see if the relevant settings are in place. We cover the settings for Transaction OPU5 in Chapter 3.

Click on the Control tab, in which you can choose whether you want the system to show the Material overview, line overview, or both. You can also limit your information display in the planning table further if you select the SEL. PRODUCTION ONLY checkbox and then choose from the SELECTION RULE dropdown option. In the selection rule, you can choose to display only receipts, requirements and stocks, or PIRs.

Click on the MRP SEGMENTS tab next, and you can choose to display various MRP elements, such as direct production and even sales document number for MTO production, by selecting the relevant checkboxes.

All of the parameter selection options are the same that you configured in Chapter 5, Section 5.3. However, you have the flexibility to change parameters, as we've done with SCHEDULING STRATEGY in the SCHEDULING tab of the PLANNING TABLE INITIAL SCREEN: CHANGE MODE. In our configuration for the scheduling strategy, we defined the scheduling strategy as BACKWARD SCHEDULING, whereas in the SCHEDULING tab, we changed this to FORWARD SCHEDULING.

Click on the PLANNING TABLE button in Figure 8.22, and the screen shown in Figure 8.23 appears. The upper half of the screen displays the complete capacity data, including the required and available capacities of the production line, LINE3, which you've maintained in the production version before.

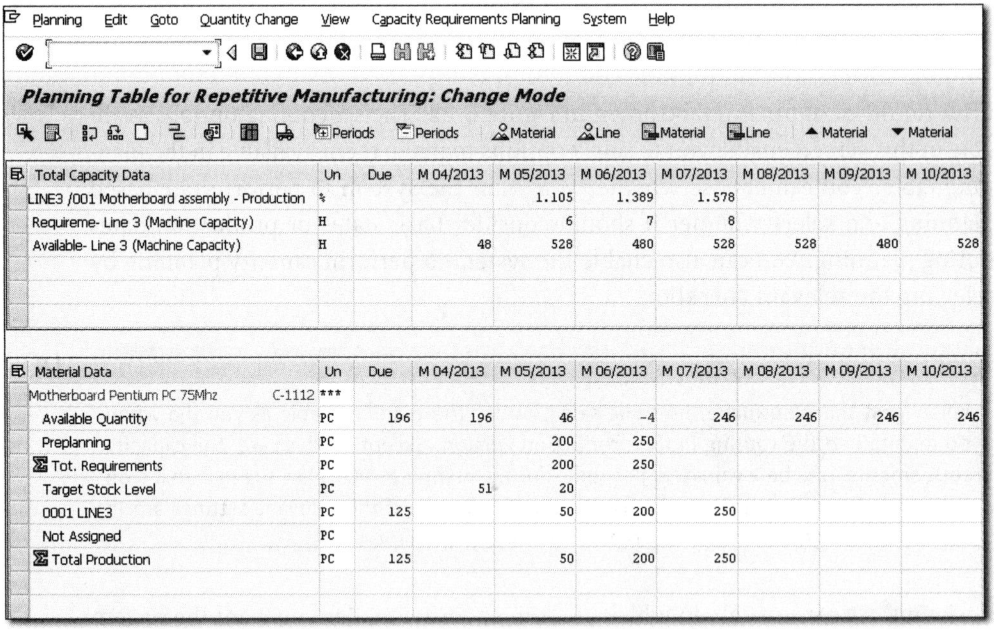

Planning Edit Goto Quantity Change View Capacity Requirements Planning System Help

Planning Table for Repetitive Manufacturing: Change Mode

Periods Periods Material Line Material Line ▲ Material ▼ Material

Total Capacity Data	Un	Due	M 04/2013	M 05/2013	M 06/2013	M 07/2013	M 08/2013	M 09/2013	M 10/2013
LINE3 /001 Motherboard assembly - Production	%			1.105	1.389	1.578			
Requireme- Line 3 (Machine Capacity)	H			6	7	8			
Available- Line 3 (Machine Capacity)	H		48	528	480	528	528	480	528

Material Data	Un	Due	M 04/2013	M 05/2013	M 06/2013	M 07/2013	M 08/2013	M 09/2013	M 10/2013
Motherboard Pentium PC 75Mhz C-1112	***								
Available Quantity	PC	196	196	46	-4	246	246	246	246
Preplanning	PC			200	250				
Σ Tot. Requirements	PC			200	250				
Target Stock Level	PC		51	20					
0001 LINE3	PC	125		50	200	250			
Not Assigned	PC								
Σ Total Production	PC	125		50	200	250			

Figure 8.23 Planning Table

The lower half of the screen displays material-specific information. Notice that in the PREPLANNING field, the system displays the PIRs of 200 and 250 for May 2013, and June 2013, respectively, that you created in the previous section (see Section 8.3.1).

The planning table also displays the settings you made in Chapter 5, Section 5.3, including the display of available quantity, total production, total requirements, and the coverage as target stock.

Tips & Tricks [+]

If you're unable to see editable fields such as LINE3 or NOT ASSIGNED, as shown in the planning table in Figure 8.23, double-click on the NOT ASSIGNED field under MATERIAL DATA, and these fields will appear.

8.6.2 Creating a Repetitive Manufacturing Planned Order in the Planning Table

As a planner, you often have to make changes to the existing planning elements, such as planned orders, or create new planned orders to account for production constraints or other practical realities such as an unplanned plant or machine shutdown. Because the REM planning table is interactive, you can perform all of the important planning functions and much more within the REM planning table. On exiting the REM planning table, you have to save your planning to enable the system to update the information accordingly.

For this example, create a new planned order by placing the cursor in the LINE3 field and entering a quantity of "50" for May 2013 (column M 05/2013). Then double-click the field in which you entered the quantity 50 to open the screen shown in Figure 8.24.

Here, the system has given a temporary planned order number (9999999003) and enables you to make changes to the production version, the production line, or to the production start or finish dates. You can also choose the FIRM checkbox, which will ensure that no changes are made to the planned order during the MRP run. Click on the ADOPT button, and the details are copied to the planning table.

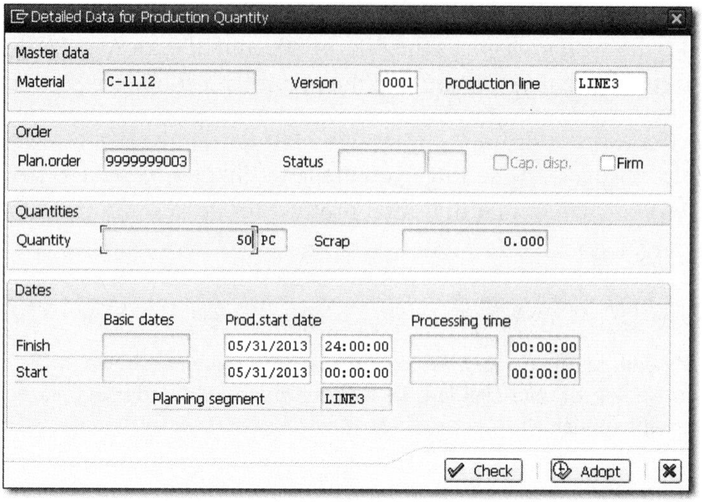

Figure 8.24 Planned Order Creation in the Planning Table

You can double-click the QUANTITY field, and this time the system assigns a permanent planned order number to which you can make a large number of requisite changes, if desired. To make changes to the planned order, click on the CHANGE icon, and the screen shown in Figure 8.25 appears. This screen shows that the REM planned order is referred to as RUN SCHEDULE QUANTITY, which is due to the fact that you created the new quantity directly on the production line. You can change the ORDER QUANTITY and add the SCRAP QUANTITY to account for any production losses. The system automatically copies all other details, such as storage location data and the production version number from the REM master data. The system also automatically selects the REPETITIVE MFG checkbox from the PRODUCTION VERSION 0001.

The PLND ORDER checkbox in the FIRMING area indicates that during the MRP run, the system won't make automatic changes to this REM planned order. The system automatically firms all manually created REM planned orders, according to the setting you made in the REM profile in Chapter 5.

Click on the SCHEDULING icon at the top of the screen, and the system shows the DET. SCHEDULING tab. When you've performed detailed scheduling for the material, you can choose GOTO • CAPACITY LEVELING in the planned order, and the screen in Figure 8.26 appears showing the capacity overload (reflected by the OVERLOAD button).

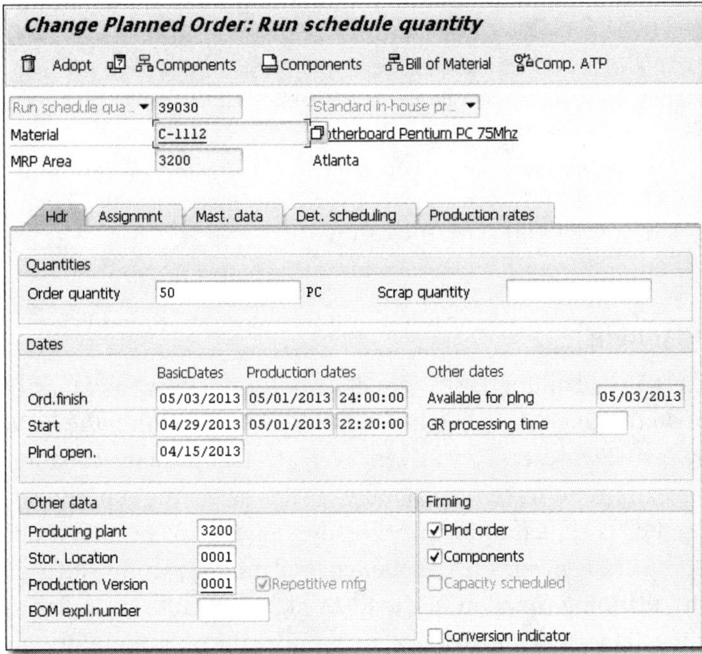

Figure 8.25 Header Screen of the Planned Order (Run Schedule Quantity)

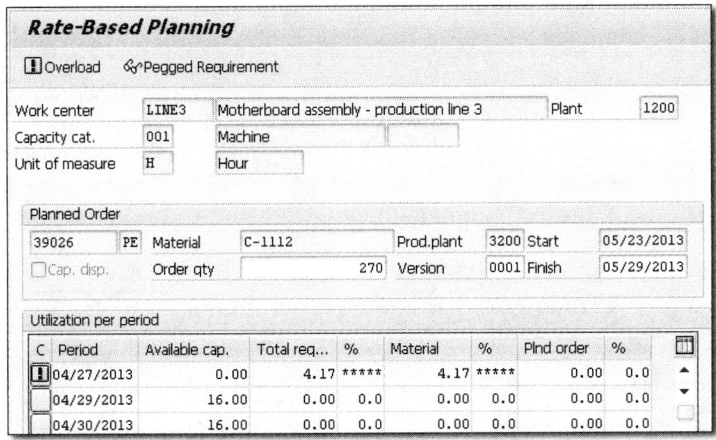

Figure 8.26 Overload in Planning

[+]

Tips & Tricks

Because you're in the Change Planned Order screen, you can perform a large number of activities and functions, for example, components availability check, capacity check, or reexploding the BOM for the material.

As an MRP planner, you can decide whether you want to use Transaction MD04 or Transaction MF50 for daily work. In Transaction MF50, you can work on a period basis, whereas in Transaction MD04, all work is order based.

8.6.3 Capacity Planning

If you go back to the REM planning table, you can set up graphics for capacity planning by clicking on the Graphical Planning Table icon. Within the REM planning table shown in Figure 8.27, you can evaluate the capacity situation (located in the upper half of the screen). Generally, at this stage, the capacity situation at the work center is agreed upon and already known, except in cases of urgent or immediate business need, such as sequence planning (within shifts or in days). The capacity planning function in the REM planning table is similar to the graphical planning table used in discrete manufacturing. It's therefore an exception rather than a norm to plan capacity in the planning table.

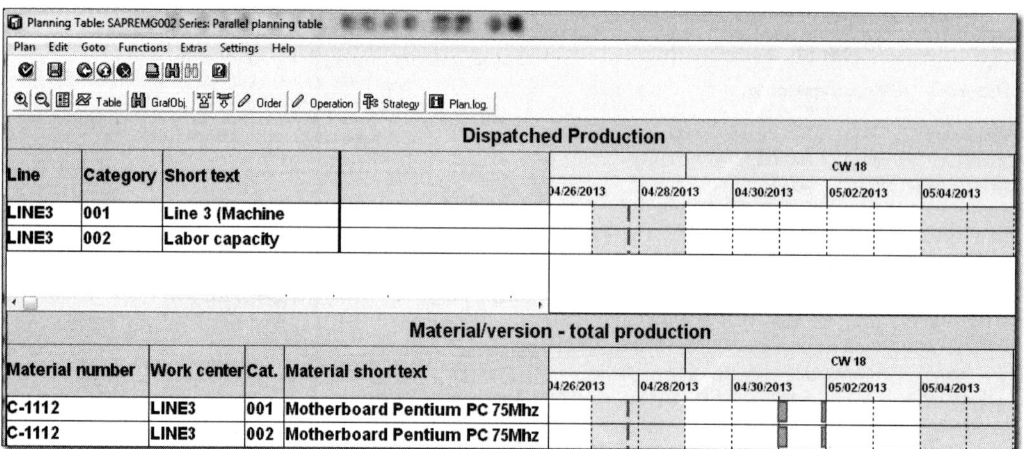

Figure 8.27 Capacity Planning

You can perform all of the CRP and capacity leveling functions in the REM planning table. Some of these functions include the following:

- Rescheduling
- Dispatching
- Performing deallocation
- Making changes to the REM planned order
- Making changes to the routing operations

[«]

Note

See Chapter 14 for more about CRP.

8.6.4 Functions in the Planning Table

You can make any changes or evaluations in the planning table, and the system updates all of the information on a real-time basis. Some of these functions are listed in Table 8.3.

Icon	Function	Function Detail
	Create new planned order	Place the cursor on the NOT ASSIGNED field in the planning table, and click on this icon to create a new planned order.
Periods	Expand periods	If you had initially selected the planning periods in months and now want to see the planning figures in weeks, you can use this icon.
Periods	Collapse periods	If you had initially selected the planning periods in days and now want to see the planning figures in weeks, you can use this icon. The shift is the smallest period, whereas the month is the largest period.
Material	Select material	You can use this icon to display more planning materials, associated with the production line.
Line	Select (production) line	You can use this icon to display production lines from the routing.
Material	Hide material	You use this icon to hide the already-displayed materials.

Table 8.3 Selected Functions in the REM Planning Table

Icon	Function	Function Detail
Line	Hide (production) line	You use this icon to hide the already-displayed production lines.
	Graphical planning table	You can use this icon to display a graphical version of capacity planning.

Table 8.3 Selected Functions in the REM Planning Table (Cont.)

8.6.5 Range of Coverage

If you've already undertaken the necessary configuration for the range of coverage profile in the MRP 2 view of the material master, then the option to evaluate the coverage becomes available in the planning table. You have to first activate it from the VIEW menu. The range of coverage profile contains the parameters for calculating the dynamic safety stock. This is a statistical calculation on the basis of average daily requirements. The range of coverage profile includes, for example, maximum stock level, minimum stock level, target stock, statistical, and actual range of coverage.

Because we've configured the TARGET STOCK display in the COVERAGE (refer to Chapter 5, Section 5.3, Figure 5.22 showing the COVERAGE field and the TARGET STOCK value), the system displays the TARGET STOCK as a default setting in the REM planning table. You can change this display (TARGET STOCK) in the REM planning table by choosing VIEW • RANGE OF COVERAGE PROFILE. You can select the maximum or minimum stock level displays in the planning table.

[»]

Note

We encourage you to explore the features and functionalities in the planning table for REM.

8.7 Material Staging

Material staging refers to the step in which the components from the replenishment storage location are transferred to the issuing storage location. These components are consumed during the actual production process. This step involves the stock transfer transaction with standard movement type 311, and it transfers stock from the replenishment storage location to the issuing storage location. An issuing storage location is the storage location from where the system issues or backflushes the components of the material.

[+]

Tips & Tricks

The alternative term used in material staging is *pull list*. A pull list provides details on which material's components are faced with shortages. At the same time, it also triggers stock transfers. You can also use storage location MRP or Kanban as the pull list.

You can view the current or existing material situation of the planned orders before proceeding to stage the materials. Evaluating the current material situation gives you a better idea, as well as helps in deciding which components you should stage first. Material staging in REM is typically carried out with reference to a certain time horizon or a certain time interval, for example, a shift or until end of the week. Except for MTO, you don't stage material with reference to any specific planned order. Similarly, the goods receipt of the material produced is period based, with or without reference to planned orders.

In the following sections, we'll explain how to view the current situation, trigger replenishment, and evaluate the material documents that the system created for material staging.

8.7.1 Material Staging: Current Situation

To view the current situation in the pull list, follow the menu path, LOGISTICS • PRODUCTION • REPETITIVE MANUFACTURING • MATERIAL STAGING • PULL LIST – CURRENT SITUATION, or use Transaction MF63 (see Figure 8.28).

In the initial parameters selection screen, you can choose from a large number of selection options and also define the end date for requirements in the SELECTION HORIZON FOR RQMTS field. This is the date until which the system will consider requirements of materials. There are also selection options in case your company uses the Kanban production process or has integrated with SAP ERP Warehouse Management (WM).

For this example, define the PLANT as "3200" (this is a mandatory entry requirement), and set the SELECTION HORIZON FOR RQMTS as "04/28/2014"—the end date until which the system will consider all materials requirements. If the materials requirements lie outside the horizon, they won't show up in the current list of material staging.

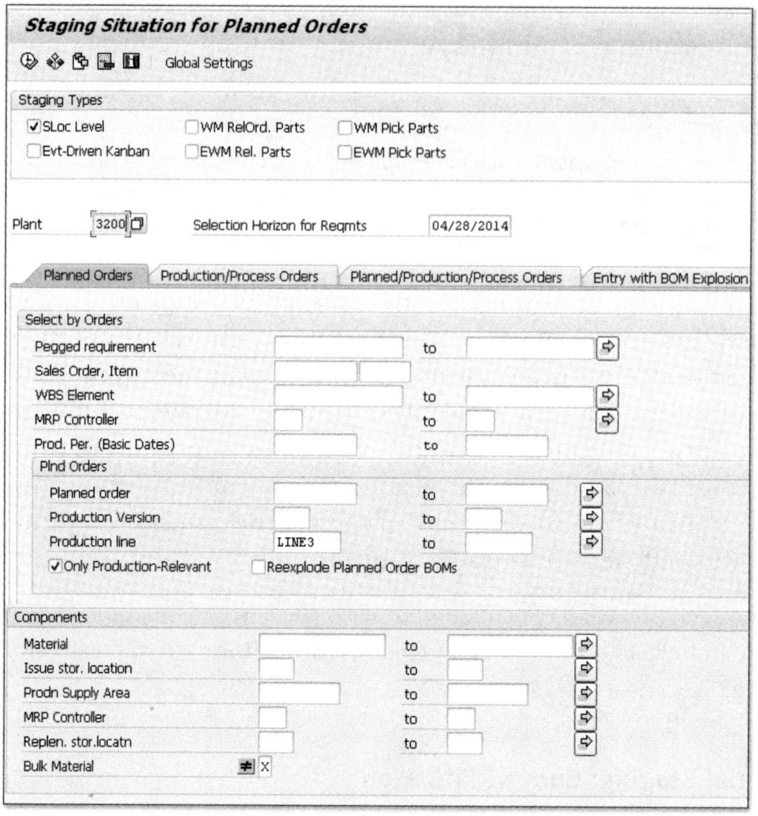

Figure 8.28 Staging Situation of Planned Orders

Further limit your selection to PRODUCTION LINE (work center) "Line3", and choose EXECUTE or press [F8]. This brings up the screen shown in Figure 8.29. The system shows the current stock and requirements situation of the planned orders in detail. It also shows the available stock, as well as the shortage quantities for the period under evaluation. In this figure, the system shows no shortage quantity for the first two planned orders, whereas for the third planned order (last one), the system shows the available quantity of 238 PC against the requirement of 270 PC, thus having a shortage of 32 PC.

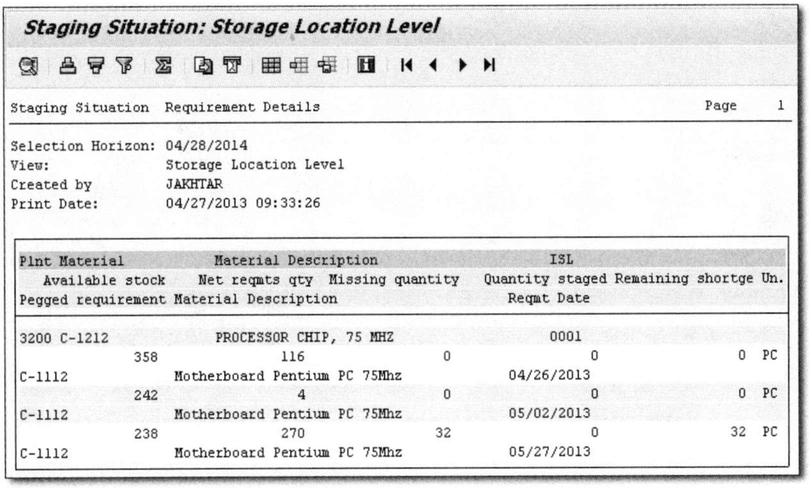

```
Staging Situation: Storage Location Level

Staging Situation  Requirement Details                                    Page    1

Selection Horizon: 04/28/2014
View:              Storage Location Level
Created by         JAKHTAR
Print Date:        04/27/2013 09:33:26

Plnt Material              Material Description                  ISL
   Available stock    Net reqmts qty  Missing quantity   Quantity staged Remaining shortge Un.
Pegged requirement Material Description                     Reqmt Date

3200 C-1212                PROCESSOR CHIP, 75 MHZ                 0001
                358            116                0                0                  0  PC
C-1112            Motherboard Pentium PC 75Mhz             04/26/2013
                242              4                0                0                  0  PC
C-1112            Motherboard Pentium PC 75Mhz             05/02/2013
                238            270               32                0                 32  PC
C-1112            Motherboard Pentium PC 75Mhz             05/27/2013
```

Figure 8.29 Staging Situation: Storage Location Level

8.7.2 Material Staging: Trigger Replenishment

To trigger replenishment from one storage location (replenishment storage location) to another (issuing storage location within a plan), follow the menu path, LOGISTICS • PRODUCTION • REPETITIVE MANUFACTURING • MATERIAL STAGING • PULL LIST – TRIGGER REPLENISHMENT, or use Transaction MF60.

In the initial screen, which is similar to the one shown earlier in Figure 8.26, enter the same parameters that you entered on the initial parameters selection screen shown previously in Figure 8.29. This takes you to the screen shown in Figure 8.30. It's important to highlight that the subsequent screen appearance will greatly depend on the parameters that you defined on the initial parameters selection screen shown earlier in Figure 8.29.

In the upper half of Figure 8.30, the system shows a summarized view of MATERIAL C-1212; it needs a total quantity of 390 PC against various planned orders, whereas a quantity of 358 PC is available in stock, thus leaving a shortage of 32 PC. This is the same quantity shortage you've seen in the previous Section 8.7.1.

You can click on the REPLENISHMENT ELEMENTS button in Figure 8.30, and the details in the lower half of the screen appear, including details of various requirements for the material C-1212.

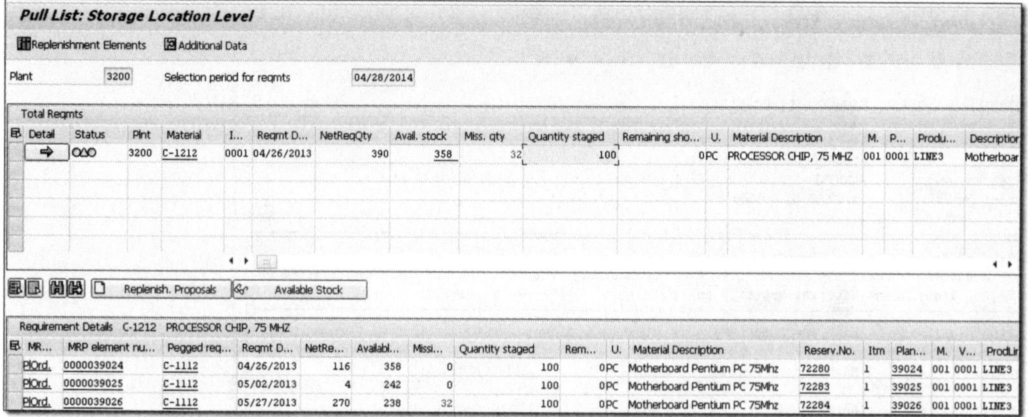

Figure 8.30 Pull List at the Storage Location Level

Click on the ADDITIONAL DATA button to see the detailed information on the requirements for material C-1212. Enter the QUANTITY STAGED as "100", as shown in Figure 8.31, and in the lower half of the same screen, enter the REPLOC (replenishment storage location) as "0002".

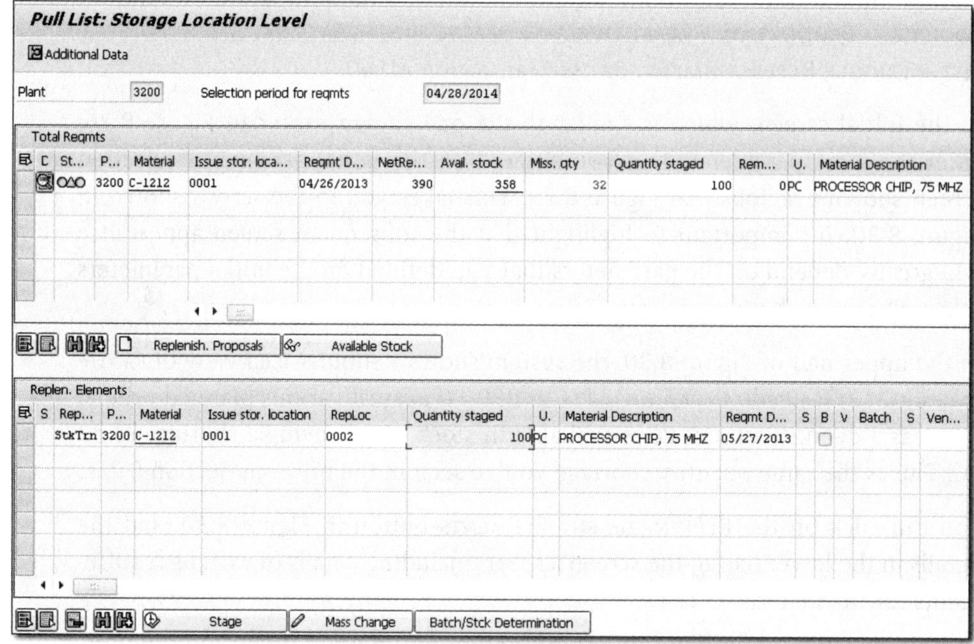

Figure 8.31 Material Staging

The system is able to detect the replenishment storage location automatically if you've activated it in the global settings of the transaction. Select the line item, and click on the STAGE button.

If the staging is successful, the system displays the relevant message at the bottom of the screen and also turns the status traffic light green. Because this is a transfer posting between two storage locations, the system creates a material document. A material document records the details of inventory movements.

8.7.3 Material Document of Material Staging

You can use Transaction MD03 to view the material document created after material staging. Figure 8.32 shows the DISPLAY MATERIAL DOCUMENT 4900000456: OVERVIEW screen that reflects the storage location transfer movement type 311. Material staging transferred (staged) a 100 PC of MATERIAL C-1212 from storage location 0002 to storage location 0001. The two signs, negative (for issuance) and positive (for receipt) of material from storage location 0001 to 0002 also reflect the same details.

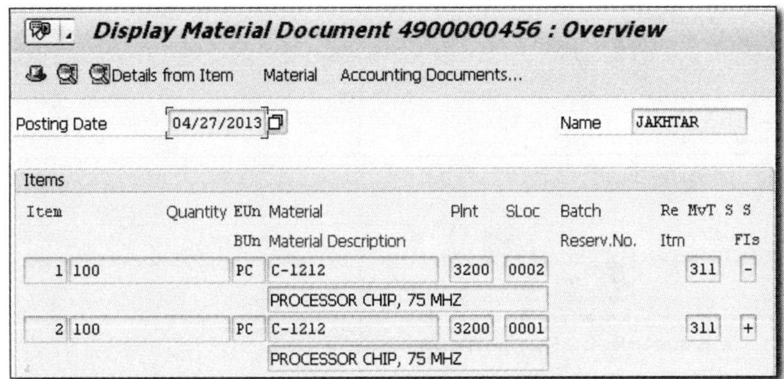

Figure 8.32 Material Document after Material Staging

8.8 Production List

A *production list* is a list of REM planned orders that you can print for your information and record. You can create and print the production list for a specific production line or work center hierarchy. This is helpful for the production supervisor or line operator (shop floor), who don't have access to the SAP ERP system, for their information as well as for necessary planning and action, where needed. On the selection screen, you can even choose how the ensuing production list should appear as an output.

To create a production list, follow the menu path, LOGISTICS • PRODUCTION • REPETITIVE MANUFACTURING • PLANNING • PRODUCTION LIST, or use Transaction MF51 (see Figure 8.33).

Figure 8.33 Initial Screen of the Production List

In the initial selection parameters screen, you can choose to print the production list based on the defined period, and you can choose if you want the system to consider only firmed planned orders, scheduled planned orders, or dispatched planned orders.

For this example, define the period of examination as 04/27/2013 to 07/26/2013, and enter the PRODUCTION LINE as "LINE3". Choose EXECUTE or press F8 to open the screen shown in Figure 8.34. This screen shows the production list of REM planned orders. In the period under evaluation, there are two REM planned orders in the production list. These are RSQs, which means that you can proceed with production because they are already assigned to the production line LINE3.

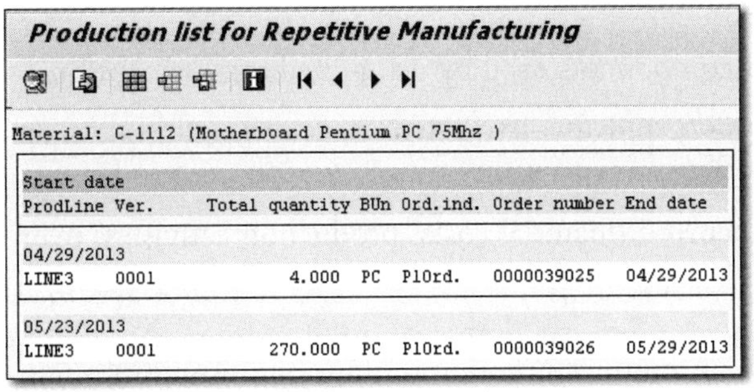

Production list for Repetitive Manufacturing

Material: C-1112 (Motherboard Pentium PC 75Mhz)

Start date						
ProdLine	Ver.	Total quantity	BUn	Ord.ind.	Order number	End date
04/29/2013						
LINE3	0001	4.000	PC	PlOrd.	0000039025	04/29/2013
05/23/2013						
LINE3	0001	270.000	PC	PlOrd.	0000039026	05/29/2013

Figure 8.34 Production List

8.9 Confirmation

Assume that your company manufactures products on a long production line with a long lead time. With long lead times, the goods issues for the components may be posted much later in the system than when they are actually physically withdrawn. All components used along the production line are usually backflushed during final confirmation. However, the goods issues for the first components at work centers are posted on time and not by backflushing with final confirmation at the end of the production line.

An REM confirmation indicates that you've performed actual production and are now entering the details of the goods produced in the system. It's important to note that during the assembly backflush, the system automatically posts the component consumptions (goods issuance).

Depending on how you've set up the REM profile and assigned it to the material, the system accordingly performs the necessary functions. In this case, it makes sense to use RP confirmation to post the withdrawal of components (and production activities) at an earlier stage after operations are already completed.

In the following sections, we cover how you can perform confirmation for REM assembly, REM components, and REM activities. We'll give you an overview of when and how backflushing should be used in REM situations, as well as the different types of backflushing that are available in REM.

8.9.1 Overview

When you post the goods receipt for the finished parts during confirmation, a number of different actions are performed by default (these are controlled to a large extent using the REM profile):

- Posting goods issue for the components used by using the backflush option
- Reducing RSQs/planned orders and the associated capacity requirements on the production lines
- Posting production costs incurred (material costs, production activities, overhead costs) to a PCC
- Creating a data entry document (material and/or financial document)
- Updating statistics in the Logistics Information System (LIS)

Other optional actions (see the DETAILS icon in the confirmation transaction) include the following:

- Taking account of the setup costs for material quantities that are independent of their lot sizes
- Posting component/assembly scrap
- Performing collective entry for different materials in a line
- Archiving confirmation documents
- Performing goods issue confirmation only or activity confirmation only (more/less activity)
- Using RP confirmations
- Decoupling confirmation processes
- Performing aggregation
- Using postprocessing and reversal options

In addition, production based on sales orders enables confirmation with reference to sales orders.

[»] **Note**

We configured the REM profile PP15 in Chapter 5, Section 5.1; please revisit that chapter to view which settings were made.

Although the system-recorded components' consumptions are based on a BOM calculation, you have the option to make changes to consumption quantities during the REM assembly backflush. Separately, you also use this component's backflush option to record excess component consumption, either against a specific REM planned order or for general consumption during the production process.

The confirmation can be any of the following types:

- **Assembly confirmation**
  The production of goods is complete, and you want to record the yield of the produced material. At the same time, you're also recording the consumption of the components against the produced goods based on the standard BOM. The user can also record the assembly scrap on the same screen with a different posting.
- **Component confirmation**
  You can record the *extra consumption* of the components against the produced goods based on the standard BOM. You can also record the component scrap on the same screen. Further, you can also issue components if backflush isn't applicable (this scenario isn't used often).
- **Activity confirmation**:
  You can record *excess consumption* of the activities, such as machine hours, labor hours, or setup hours, against the produced goods based on the routing. You can also record the activity scrap on the same screen.

As with assembly backflush, several options are available on the component backflush screen:

- Record corresponding information during REM confirmation and backflush. This flexibility is available because the REM production type caters to all MTS or MTO productions.
- Record components' scrap, either with reference to the assembly material or as standalone. The movement type for scrap is 551.
- Reverse the material document for an already-posted consumption or scrap.
- Perform a reversal function, without reference to any material document.

8.9.2 Repetitive Manufacturing Assembly Confirmation

To access the REM confirmation for an assembly backflush, follow the menu path, Logistics • Production • Repetitive Manufacturing • Data Entry • Repetitive Manufacturing Confirmation, or use Transaction MFBF. In the initial screen of REM confirmation, select the Assembly backflush radio button (see Figure 8.35). You use the screen to confirm the production yield as well as record components' consumption and activities (if configured in the REM profile).

You can also record assembly scrap by clicking on the Scrap button. We cover how to record assembly scrap later in this chapter in Section 8.10.3.

If you click on the Details button, the system brings up the REM profile (in this example, PP15), in which you can still change the following options:

- Post activity backflush
- Stock posting to unrestricted, quality, or blocked stock
- Backflushing options, for example, from online to background
- Post ordering costs

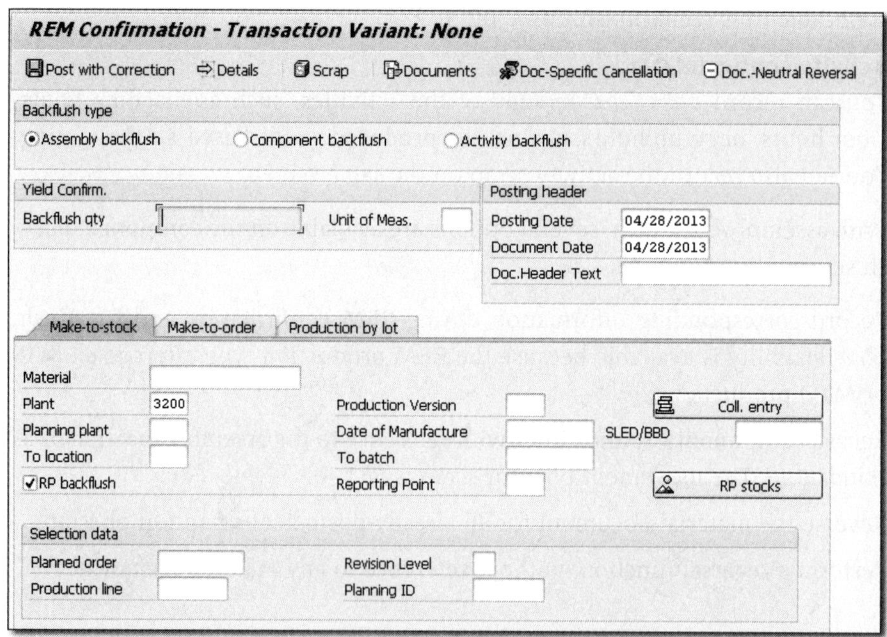

Figure 8.35 Assembly Backflush in REM Confirmation

You can assign the posting date as the document date on which you want to post the REM assembly backflush. Further, you can also use this screen for document-specific or document-neutral reversals of the entered values. We cover document-specific reversal in Section 8.10.

It's also important to note that you have the flexibility to record the yield of the assembly without reference to the REM planned order because then the system is able to account for period-based confirmation. However, doing so won't reduce the quantity specific to the planned order for the MTS scenario (if such settings are made in the REM profile). For MTO, the system reduces the planned order quantity.

The three tabs MAKE-TO-STOCK, MAKE-TO-ORDER, and PRODUCTION BY LOT cater to various production scenarios.

8.9.3 Repetitive Manufacturing Component Confirmation

As soon as you select the COMPONENT BACKFLUSH radio button shown previously in Figure 8.35, the system offers the following two options:

- NO BOM EXPLOSION (MANUAL COMPONENT ENTRY)
  In this option, you have to enter the assembly in the MATERIAL field so that the system is able to find the link with the PCC. If you then select NO BOM EXPLOSION, the PROCESS COMPONENT LIST is empty.
- PROPOSE COMPONENTS ACCORDING TO BOM
  This option enables you to enter a quantity of the assembly material, and the system explodes the assembly BOM and proposes the component quantities for backflush or excess consumption. While selecting this option, you need to provide the reference quantity of the assembly material.

You can also select if you want excess component consumption for a specific RP or after the last RP, by selecting the relevant radio button. This radio button only appears if you enter a valid RP.

8.9.4 Repetitive Manufacturing Activities Confirmation

When you select the ACTIVITY BACKFLUSH radio button in the REM confirmation screen (Transaction MFBF) shown earlier in Figure 8.35, the system displays the following two options:

- NO PLANNED ACTIVITIES FROM ROUTING
  In this option, you have to enter the material number together with the activity types as well as the corresponding duration of each activity. The system doesn't make any suggestions.
- FETCH PLANNED ACTIVITIES FROM ROUTING
  With this option, when you enter the quantity of the assembly material, the system refers to the routing of the assembly material and proposes the details. When selecting this option, you need to give the reference quantity of the assembly material.

8.9.5 Repetitive Manufacturing Actual Assembly Confirmation

In the screen shown in Figure 8.36, you now perform the actual assembly backflush posting for MATERIAL C-1112, PLANT 3200, and PRODUCTION VERSION 0001. The yield confirmation for the material in this REM assembly backflush is 10 PC.

Because the configured and assigned REM profile PP15 stipulates that there should be mandatory RP backflushing, select the RP BACKFLUSH checkbox, and also assign REPORTING POINT 0010.

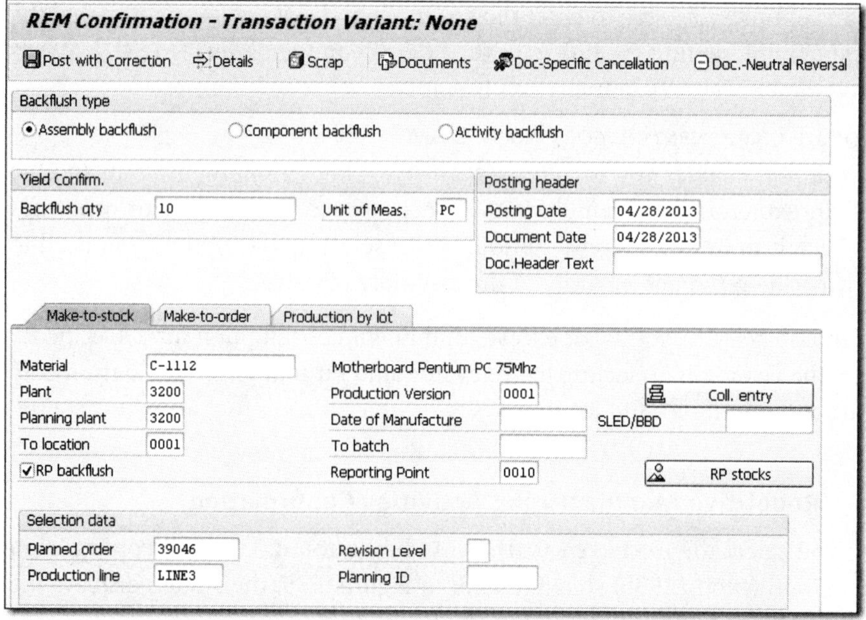

Figure 8.36 Assembly Backflush against the Planned Order

RP backflush in REM occurs when the system is able to automatically backflush several operations in the processing sequence. Using the RP backflush procedure, you can backflush components synchronized with actual consumption as much as possible. From the controlling perspective, RP backflush displays WIP for assemblies that haven't completely been backflushed. The REM profile controls whether you can use the RP backflush for a material; further, you use the control key to determine whether an operation is an RP operation (milestone). If an operation is an RP operation, then all previous operations that haven't been confirmed yet or all previous operations that aren't mandatory RPs are automatically backflushed together. If several operations are marked as RPs, the system should backflush components according to the processing sequence of the RP operations.

Save your entries at this stage, and the system performs the following three functions:

- Performs goods receipt (yield confirmation) of 10 PC for material C-1112
- Performs backflush (goods issuance) of components according to the BOM of C1112
- Records and updates activities according to the routing of material C-1112

Click on the Post with Correction button to open the screen in Figure 8.37, which shows that the system is performing two functions simultaneously. First, it's confirming the yield of 10 PC for the assembly Material C-1112; then, in the second step, it's recording the consumption of all of the components that are used in the production of the assembly. Notice the Movement Type 261, which indicates consumption or goods issuance.

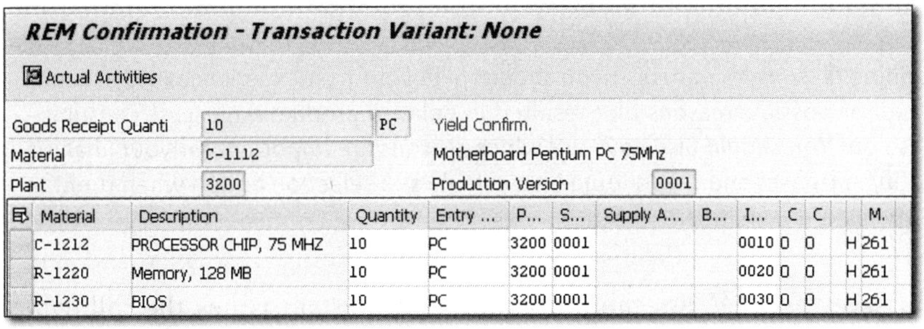

REM Confirmation - Transaction Variant: None

Actual Activities

Goods Receipt Quanti 10 PC Yield Confirm.
Material C-1112 Motherboard Pentium PC 75Mhz
Plant 3200 Production Version 0001

Material	Description	Quantity	Entry ...	P...	S...	Supply A...	B...	I...	C	C		M.
C-1212	PROCESSOR CHIP, 75 MHZ	10	PC	3200	0001			0010	0	0	H	261
R-1220	Memory, 128 MB	10	PC	3200	0001			0020	0	0	H	261
R-1230	BIOS	10	PC	3200	0001			0030	0	0	H	261

Figure 8.37 Components Overview of Assembly Backflush

You can change the quantity of any component, if needed, and add or delete components, as necessary. For example, to overwrite (change) the quantity of component C-1212, you need to overwrite the quantity given in the QUANTITY column.

Similarly, to delete a component from this figure, you need to first select the component, and then click on the DELETE icon. In the same screen, notice that the first field, GOOD RECEIPT QUANTITY (YIELD CONFIRM.), shows that you're recording a yield of 10 PC. Click on the ACTUAL ACTIVITIES button.

In the ensuing screen, POSTING ACTUAL ACTIVITIES, you can make changes to the activities such as machine hours or labor hours. It takes the standard duration details from the routing of material C-1112. You can make changes to the durations to reflect actual durations (whether on the higher or lower side), and the system correspondingly and automatically updates the cost details when you save the confirmation.

In the REASON field, you can also define the reason for the excess consumption of activities. This is also known as reasons for variance. When you run the standard SAP ERP system reports, you can give reasons for variance as one of the selection criteria to generate a report, which eventually helps in analysis and evaluation. Examples of reasons for variance can be machine malfunctions, electrical faults, labor shortages, emergency shutdowns, and so on.

[»] **Note**

Reasons for variances in confirmation are plant-specific, and you can define them using configuration Transaction OPK5.

[+] **Tips & Tricks**

When defining reasons for variance, you should take your past experience into account on the major or possible reasons that resulted in delayed production, excess consumption, and so on. You should also consider which reasons are important for your analysis and reporting purposes and thus should be available as a selection option when running standard or custom-developed reports.

When you save the REM assembly backflush, the system issues the following message at the bottom of the screen confirming the goods movements (goods issuance and goods receipt) and activities posting: GR AND GI WITH DOCUMENT XXX AND ACTIVITIES POSTED. This means that the system successfully posted goods receipt (production yield), goods issuance (components consumption), and recording of the actual production activities durations. It's important to add that the business process depends on whether the planned activities, per the routing, have been adjusted manually during the confirmation according to the real production efforts or not.

8.9.6 Separated Backflush

If you have a large number of backflush processes to manage—that is, a high volume of data entry for goods issuance as well as to post activities—you can use the separated backflush functionality. In separated backflush, you can let the system post goods issuance as well as post activities during the slower time of the day (or even schedule it at night). You can control whether you want separate backflush in the REM profile creation (configuration).

To perform separate backflush, follow the SAP menu path, Logistics • Production • Repetitive Manufacturing • Data Entry • Separated Backflush, or use Transaction MF70.

Enter the appropriate parameters on the initial screen, and select the appropriate checkboxes, such as Post Goods Issue and/or Post Production Activities.

8.9.7 Postprocessing of Components

If you've defined that you don't want the system to perform online error correction during REM confirmation, then you can correct these errors later. The nature of the error can be that there were insufficient component quantities for backflush (goods issuance) during confirmation, for instance, due to incorrect or delayed postings in SAP ERP Materials Management (MM). Alternatively, there can be errors when the system is unable to find the storage location to backflush (goods issue) components from.

It's important to ensure that you do attend to these errors later because these continue to accumulate in the incorrect goods movement log. To postprocess components (or other errors), follow the SAP menu path, Logistics • Production • Repetitive Manufacturing • Data Entry • Postprocess • Postprocessing Individual Components, or use Transaction COGI.

[»] **Note**

To use Transaction COGI, you also need to ensure that you've selected the ALSO CREATE INDIVIDUAL POSTPROCESSING RECORDS option in the REM profile.

Enter the appropriate parameters on the initial screen, and choose EXECUTE or press F8. The ensuing screen lists all of the confirmations in which the system found errors and wasn't able to process them successfully. You can even display the details of the errors to help you understand why they occurred, for example, a missing storage location.

[»] **Note**

The standard tool for REM postprocessing is Transaction MF47.

8.10 Reversals and Scrap

In a normal business scenario, the process operator occasionally faces a situation in which there was a data entry error or the confirmation details posted in the system differ from the actual production figures. In such cases, reversal becomes inevitable. In REM, you can perform document-specific reversal or document-neutral reversal. You can perform the reversal while remaining in the REM confirmation screen (Transaction MFBF) and selecting the DOC-SPECIFIC CANCELLATION icon. When you do this, the system takes you to Transaction MF41. Alternatively, you can also use a separate transaction code (Transaction MF41) to reverse a specific document (we use this option in this example).

In the following sections, we'll discuss how you can perform reversal to the documents already posted in the system, such as during RP backflush. As already stated, there are two types of document reversals. In *document-specific reversal*, you can select the specific document that you want to reverse. On the other hand, in *document-neutral reversal*, you can correct the total confirmed quantity, for example, for a day. In this case, you just have to reverse the difference between the previously posted quantity with the correct quantity, which reduces the overall posting efforts. Later in this section, we also cover how you can manage REM actual scrap for component, assembly, and activity.

8.10.1 Document-Specific Reversal

To proceed with a document-specific reversal of confirmation, follow the menu path, LOGISTICS • PRODUCTION • REPETITIVE MANUFACTURING • DATA ENTRY • REVERSE, or use Transaction MF41. In the initial screen for parameters selection (see Figure 8.38), you can give as much detail as possible or available to enable the system to bring up only the relevant documents for reversal.

Document-Specific Cancellation of Confirmation

Posting Date

Date of Reversal	04/28/2013

Selection of Reversal

(•) Make-to-Stock				
( ) Make-to-order	Sales Order		Sales Order Item	0
( ) Kanban	Control cycle number		Kanban item no.	0
( ) Production Lot	Production lot			
( ) Material Document	Material Document		Material Doc. Year	

Further Selection Data

Plant	3200
Material	C-1112
Production Version	0001
Posting Date	
Entry Date	
User Name	
Reporting Point Operation Number	
Document Log	

Dialog if no. of docs greater than 800 [ ] No note dialog

Figure 8.38 Initial Screen for Document-Specific Reversal

For this example, define the posting date or DATE OF REVERSAL as "04/28/2013", which means the system will post the reversal on that posting date.

To limit the selection further, enter PLANT "3200", MATERIAL "C-1112", and PRODUCTION VERSION "0001". Choose EXECUTE or press F8 to open the screen in Figure 8.39, which shows a selective list of material documents and the corresponding RP documents. Place the cursor on the first material document, which is 4900000465, and click on the REVERSE button. Upon a successful reversal (of both material documents and activity posting), a message appears at the bottom of the screen: MATERIAL MOVEMENT WITH DOCUMENT AND POSTING REVERSED.

Document-Specific Cancellation of Confirmation

Choose Save Reverse

```
Documt.log Material          Plnt Ver.   Created on Time      P Document   Item
User Name        PrTp   RPnt  Pl.plant    Pstng Date GI Ac Pos St      Quantity UM
Documt.log Item MatDoc.No. Act.Postin Counter  RP Documen Doc.GMvtEr

        2017 C-1112           3200 0001   04/28/2013 08:22:43  01
JAKHTAR         Z      0010   3200        04/28/2013                        10  PC
        2017    1 4900000465
        2017    2               26684       12
        2017    3                             0000000097
        2017    4                             0000000098
        2016 C-1112           3200 0001   04/28/2013 08:08:46  01
JAKHTAR         Z      0010   3200        04/28/2013                        10  PC
        2016    1 4900000464
        2016    2               26684       11
        2016    3                             0000000095
        2016    4                             0000000096
        2015 C-1112           3200 0001   04/28/2013 08:07:06  01
JAKHTAR         Z      0010   3200        04/28/2013                         5  PC
        2015    1 4900000463
        2015    2               26684       10
        2015    3                             0000000093
        2015    4                             0000000094
        2013 C-1112           3200 0001   04/28/2013 08:04:39  01
JAKHTAR         Z      0010   3200        04/28/2013                        10  PC
        2013    1 4900000462
        2013    2               26684        9
        2013    3                             0000000091
        2013    4                             0000000092
```

Figure 8.39 Detailed Screen for Document-Specific Reversal

On reversal, the system reverses the production yield confirmation with movement type 132, whereas the component consumption is reversed with movement type 262. The document number of the reversal document posted is 4900000466. These are the standard reversal movement types for REM. The material document reversal and the associated reversal movement types also validate the settings you made in Chapter 5, Section 5.1.16.

Figure 8.40 shows the Material Document List, accessed using Transaction MB51. It contains the reversal details, including the movement types 132 and 262 for assembly and components reversals.

Material Document List

Material	Material Description	Plant	SLoc	Mvmt Type	Material Doc.	Item	Posting Date	Qty in Un. of Entry	Entry Unit
C-1112	Motherboard Pentium PC 75Mhz	3200	0001	132	4900000466	1	04/28/2013	10-	PC
C-1212	PROCESSOR CHIP, 75 MHZ	3200	0001	262	4900000466	2	04/28/2013	10	PC
R-1220	Memory, 128 MB	3200	0001	262	4900000466	3	04/28/2013	10	PC
R-1230	BIOS	3200	0001	262	4900000466	4	04/28/2013	10	PC

Figure 8.40 Material Document with Reversal Movement Types

8.10.2 Document-Neutral Reversal

To proceed with the document-neutral reversal of confirmed quantities, use Transaction MFBF (the REM backflush screen). Enter the PLANT and MATERIAL, click on the DOC.-NEUTRAL REVERSAL icon, and enter the quantity that you want to reverse. Finally, click on the POST icon to post the reversal in the system.

8.10.3 Repetitive Manufacturing Actual Assembly Scrap

At any time during the REM confirmation screen for assembly, component, or activity backflush (Transaction MFBF), you can switch the screen layout to record scrap of any of these three backflush types. To do so, click on the SCRAP icon available in the REM confirmation screen. For this example, show the assembly backflush for the material by using the reference quantity.

The system consumes the components per the BOM of the material (assembly) along with activities. It uses movement type 261 to consume the components without recording any assembly yield.

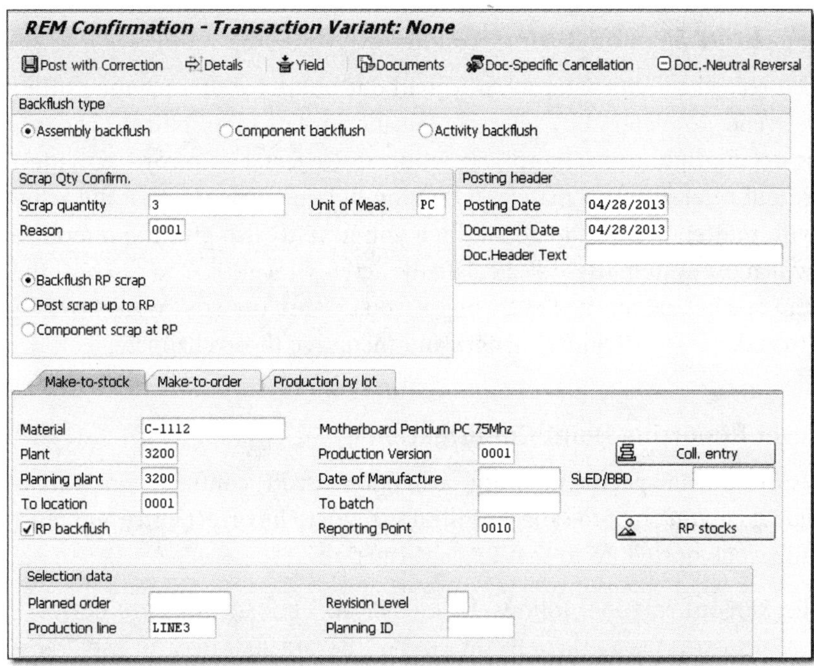

Figure 8.41 Scrap Posting for Assembly

After you click on the SCRAP icon, select the BACKFLUSH RP (REPORTING POINT) SCRAP radio button, enter the reference scrap quantity of "3", and the screen shown in Figure 8.41 appears. As usual, you also give reference to MATERIAL C-1112 and PLANT 3200.

8.10.4 Repetitive Manufacturing Actual Component Scrap

If you need to post scrap with reference to component scrap, select the appropriate radio button, and then the system will post the goods issuance to scrapping while booking the cost of scrapped components to the PCC. The movement type for goods issuance against scrapping is 551.

If you choose to do actual components reversal with reference to the material (assembly), then you have to also give the reference quantity, which the system uses to calculate the components reversal. You can post the component scrap by clicking on the SAVE icon, or you can use the POST WITH CORRECTION icon to make any individual changes and then save those changes.

8.10.5 Repetitive Manufacturing Actual Activity Scrap

You can also record activity scrap with or without reference to a planned order but with reference to material by selecting the appropriate radio button. The system brings up the option to record activity scrap (reversal of record activities) with or without reference to a material's routing. If you choose to post the activity scrap with reference to the material, then you have to also give the reference quantity, which the system uses to calculate the activity reversal. You can post the activity scrap by clicking on the SAVE icon, or you can use the POST WITH CORRECTION icon to make any individual changes and then save those changes.

8.10.6 Reset Reporting Point Confirmation

You can also reset the specific quantity posted against RP confirmation. This is necessitated when you want to cancel a milestone due to incorrect entry and must also reset the stock posted up to that RP confirmation.

To reset the RP confirmation, follow the menu path, LOGISTICS • PRODUCTION • REPETITIVE MANUFACTURING • DATA ENTRY • RESET RP CONFIRMATION, or use Transaction MF4R. On the initial screen, the system provides you the option to enter a document and the posting dates and make further selections, which are specific to the material or production line. You can also click on the RP OVERVIEW icon to display RP information. When you've made the desired parameters selection, click on the SAVE icon (on the initial screen), and the system resets the RP confirmation.

8.11 Collective Confirmation

The SAP ERP system provides an option to collectively enter a large volume of data that you don't need to enter and post individually. However, this option is limited to recording assembly backflush and assembly scrap. As already covered, with assembly backflush, the system records production yield and consumes components according to the BOM. For assembly scrap, it consumes additional components of the assembly with reference to the assembly's BOM.

To use the collective confirmation functionality, follow the menu path, LOGISTICS • PRODUCTION • REPETITIVE MANUFACTURING • DATA ENTRY • COLLECTIVE ENTRY, or use Transaction MF42N (see Figure 8.42).

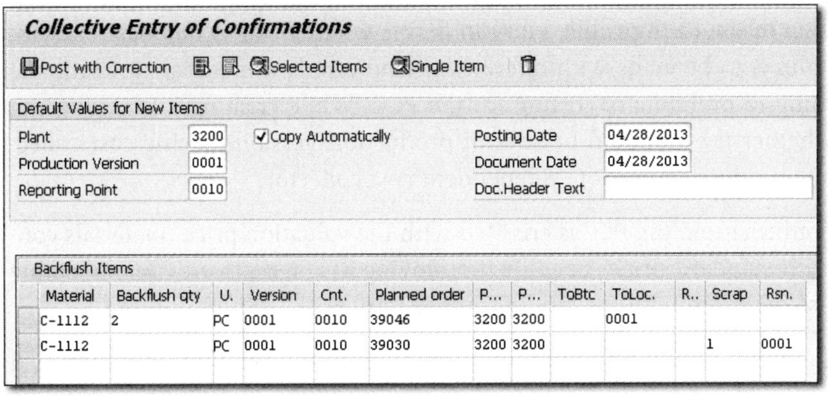

Material	Backflush qty	U.	Version	Cnt.	Planned order	P...	P...	ToBtc	ToLoc.	R..	Scrap	Rsn.
C-1112	2	PC	0001	0010	39046	3200	3200		0001			
C-1112		PC	0001	0010	39030	3200	3200				1	0001

Figure 8.42 Collective Confirmations

For this example, enter the details shown, including assembly MATERIAL "C-1112" BACKFLUSH QTY of "2" PC, and in the second line, enter the MATERIAL "C-1112" and the SCRAP quantity of "1" PC. The system creates two material documents, the first for the assembly backflush and the second one for assembly scrap (excess components consumption, without any yield).

8.12 Costing Activities (Cost Object Controlling)

Before recording the actual yield confirmation, components backflush, or activities recording, all of which have financial implications, you need to coordinate with the Cost Object Controlling person to ensure that the necessary and preliminary work on the costing side is taken care of. Some of these activities include the following:

- Creating a PCC
- Creating a material cost estimate
- Releasing the material cost estimate
- Performing settlement

During confirmation in REM, the material costs for the components and the production activities (except if a material cost estimate with price update has been carried out in advance) are automatically posted to the PCC during backflushing. The option to post activities is enabled in the REM profile before you can actually post activities. The PCC is debited with the material costs, production activities, and overhead. A PCC such as an order (a cost controlling object) may have an eternal validity. In the REM profile, you can define whether the actual data entry of the activities is to be made with reference to the standard material cost estimate or according to preliminary costing for the PCC. While creating a PCC, you can decide whether there should be several production version-specific cost collectors or one production version-independent cost collector.

During confirmation, the PCC is credited with the valuation price (materials controlled with standard price) or with the moving average price of the assembly (materials controlled with the moving average price).

Overheads aren't determined with reference to an operation, that is, not for each goods movement or activity confirmation. Rather, they are determined periodically during period-end closing in CO using an overhead structure and are posted to the PCC.

Components different from the planned components may be used in production, or the overhead cost rates may change. This results in variances in the PCC, and a settlement must be carried out. The aim of the settlement is to credit the PCC in full. During settlement, the costs not yet credited to the PCC are transferred. For a material that is valuated using the standard price, the variant costs are posted to the price difference account.

Settlement in REM is always period-based and based on the costs in the PCC (product-related Cost Object Controlling).

8.13 Reporting

We'll be covering standard reporting extensively in Chapter 19, including several reports available specifically for REM. In the following sections, we briefly cover two of them:

- Document log information
- RPs statistics

8.13.1 Document Log Information

The SAP ERP system maintains a comprehensive log of each and every material document posted or cancelled, which is known as the *document principle*. As a planner, it's often important to have access to the material document information for evaluation purposes, as well as for identifying individual documents against which incorrect or incomplete entries were made and must now be reversed or cancelled.

To access the document log information in REM, follow the menu path, LOGISTICS • PRODUCTION • REPETITIVE MANUFACTURING • EVALUATIONS • DISPLAY DATA ENTRY DOCUMENTS, or use Transaction MF12. On the initial parameters selection screen, enter the parameters such as PLANT, MATERIAL, PRODUCTION VERSION, or POSTING DATE, and choose EXECUTE or press [F8]. Not only does the system bring up relevant documents for display, it also provides you with the option to reverse any document, if needed. We've already covered document-specific reversal in Section 8.10.1.

8.13.2 Reporting Point Statistics

RP statistics is one of the several statistics reports available to facilitate evaluation. To access RP statistics, follow the menu path, LOGISTICS • PRODUCTION • REPETITIVE

MANUFACTURING • EVALUATIONS • LIS STATISTICS • REPORTING POINT STATISTICS, or use Transaction MCRM. On the initial parameters selection screen, enter the relevant parameters such as MATERIAL, PLANT, PRODUCTION VERSION, or RP. You can also provide a period of evaluation, so that the system only brings up relevant information. Choose EXECUTE or press F8, and this brings up the screen shown in Figure 8.43.

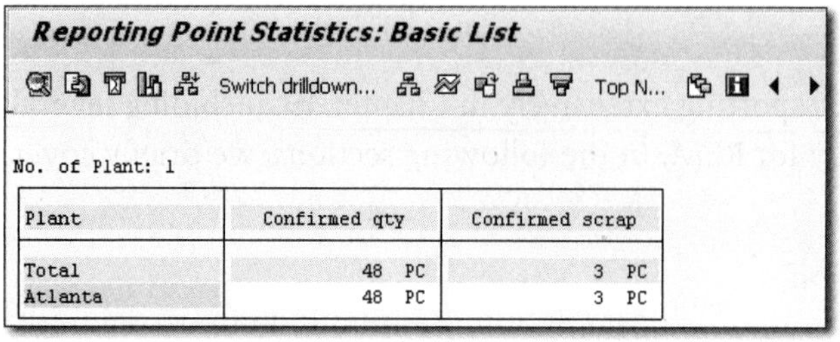

Reporting Point Statistics: Basic List

Switch drilldown... Top N...

No. of Plant: 1

Plant	Confirmed qty	Confirmed scrap
Total	48 PC	3 PC
Atlanta	48 PC	3 PC

Figure 8.43 REM Reporting: Reporting Point Statistics

In this statistics report, you can get details on the confirmed quantity at each RP, as well as the confirmed scrap recorded. You can also switch to different views and details, for example, date-wise evaluation, RP-wise evaluation, production version-wise evaluation, or production line-wise evaluation, by clicking on the SWITCH DRILLDOWN button and then selecting the desired radio button. You can also use these RP evaluations as a valuable reporting tool to evaluate the WIP.

8.14 Summary

For material produced in bulk quantity and with relatively fewer production steps and complexity when compared with production types such as discrete or process, you can implement the REM production type in a company. You can plan line loading, material staging, backflush assembly, components, and activities as frequently as deemed necessary and practically from the business perspective. All of the costs associated with REM production of a material continue to get accumulated in the PCC, and you can periodically settle it. Due to the highly integrated nature of PP with CO-PC, you maintain extensive coordination to ensure that the inventory, financial, and business results are in line with the company's business processes.

With this, we come to an end on the configuration and the business processes sides of each of the three production types (discrete, process, and repetitive). The next part of the book covers PP workflow tools to optimize your business processes.

The next chapter discusses sales and operations planning (S&OP) as a start to the next part of the book.

PART IV
Production Planning Workflow Tools

Sales and operations planning strives to maintain a balance between SAP Demand Planning and operations planning, which takes initial stocks, machine capacities, and constraints into account to finalize a realistic production plan. You can use standard sales and operations planning to come up with a feasible production plan or use Flexible Planning if you have complex and diverse planning needs.

9 Sales and Operations Planning

Sales and operations planning (S&OP) is an iterative form of business process management, in which you use several planning scenarios and versions until you arrive at a production plan. You can then confidently use it in production, procurement, and capital investment processes.

With the complexities involved in global supply chain management processes, there has never been a greater appreciation for ensuring effective and efficient use of S&OP. S&OP is mainly about sales-driven forecasting and consolidating forecasts and uses only neutral key figures (numbers) for that. In contrast to this, production planning (material requirements planning [MRP], capacity requirements planning [CRP]) is all about production and takes into account requirements instead of key figures. Still, MRP in the SAP ERP system doesn't consider capacities, but CRP does. For a value-based plan, you need to transfer the sales or production figures to SAP ERP Controlling's Profitability Analysis (CO-PA).

The biggest incentive for implementing S&OP in a company comes from the ability of a planner to evaluate various what-if models and to perform scenario planning in simulative modes, before passing on the finalized operations plan to SAP Demand Management in the form of PIRs. Forecasting plays a major role in helping the planner arrive at a plausible operations plan.

[»] **Note**

At the same time, the Long-Term Planning (LTP) option is also available in SAP ERP as a planning and simulation tool. LTP offers several planning options, including simulating the components' requirements quantities, inventory controlling, activities, and capacity requirements. LTP has its limitations, however. For example, LTP can't take into account a product's demand fluctuation (which is possible in Flexible Planning) or the effect of changes in one key figure such as sales quantity or production quantity (possible in Flexible Planning). Refer to Chapter 12 for more information on LTP.

In this chapter, we'll start with an overview of what S&OP is and also introduce Flexible Planning, which you can use as an alternative. You'll find an explanation of all functions and tools that are a part of these planning types and learn how to work with them and interpret their results. We also cover a lesser-known and used application of Flexible Planning known as standard analysis reporting. The standard analysis in Flexible Planning makes use of all the concepts and fundamentals that you'll learn in this chapter. Let's get started with an overview.

9.1 Standard Sales and Operations Planning

The following sections begin with an overview of S&OP and the similarities and differences between standard S&OP and Flexible Planning. It then moves to cover forecasting, and concludes with coverage of standard analysis in Flexible Planning.

9.1.1 Overview

In standard S&OP, you can plan individual materials or a group of materials (known as a product group). The product group consists of individual materials or other product groups and enables you to define the proportion factor (percentage) for each material in the overall product group. It also offers the option to aggregate and disaggregate at various planning levels. If the planning processes in your company are relatively simple and straightforward, that is, if they are restricted to individual materials or a group of materials, then standard S&OP can fulfill your business needs.

Example [Ex]

Consider the following business scenario: Before each financial year begins, the sales and production teams spend countless hours working on planning figures that are acceptable to both. The sales team uses forecasting tools to arrive at next year's sale figures (targets) based on historical data. The production team looks at things differently. They evaluate whether they are able to meet the demand of sales with the existing production capacities or not. The procurement planner needs a better understanding of how the production figures will impact the procurement process; that is, do the vendors even have the capabilities to meet the supply requirements of the company? The inventory controller (warehouse) is concerned about whether there is enough space in the warehouse to manage and store the produced quantities. The management of the company is interested in not only increasing profitability on the product (and at the same time reducing cost) but also gaining a broader understanding of the capital tie-up involved and whether new investment (such as capacity enhancement or increasing the number of working shifts) is warranted. Further, knowing which product (or group of products), regions, or markets can bring in greater revenue for the company also helps in the decision-making process.

This is where S&OP can help. In general, the entire "planning" exercise entails the creation of various planning versions in the system, adjusting sales or production figures until there is a mutual consensus. The "finalized" planning figures, which usually reflect a production plan, are sent forward to SAP Demand Management where they appear as planned independent requirements (PIRs) with an "active" version. When the system runs the MRP, it considers the active PIRs to arrive at procurement proposals (in-house production and external procurement).

In the planning process, a planning table is needed in the SAP ERP system that can take all of the important planning considerations into account and at the same time account for dependencies of one factor on another.

The Flexible Planning functionality, while very sophisticated, is also a slightly complex tool to manage. With better comprehension, the dividends that Flexible Planning offers are far higher and bring forth much more realistic planning figures, which the company can use to reap greater financial benefits. For example, with Flexible Planning, you can configure your own planning layout (known as a planning table) of important key figures, include self-defined macros to manage complex calculations, perform forecasting (also possible in standard S&OP), take special events such as trade shows or the Olympics into account for increases in sales or natural calamities such as drought or flood to factor in decreases in sales (can also be an increase, if your company manufactures relevant products), and have a broader understanding of "commitments" (known as pegged requirements) such as capacities, materials, or production resources/tools (PRT).

When deciding which S&OP options (standard S&OP or Flexible Planning) to use, it makes sense to evaluate the business requirements and at the same time strive to maintain simplicity and a straightforward approach to business processes. As mentioned already, standard S&OP offers a very limited set of functions, but being almost completely predefined, it enables you to immediately start taking advantage of the functionalities without any configuration. A general recommendation is to first try and cover the business process with standard S&OP, and if important functions are still missing, then consider using the rather complex and sophisticated Flexible Planning tool instead. Flexible Planning offers full functionality but requires configuration, many presettings, and definitions.

SAP ERP offers several standard analyses reports in all logistics components, consisting of characteristics and key figures for a period. A characteristic values combination (CVC) is the combination of characteristic values with which you want to plan. Characteristics can be materials, plants, sales organizations, distribution channels, or purchase organizations. Key figures (values or quantities) can be quantity produced, quantity procured, operations quantity confirmed, production scrap (quantity), invoiced value, or purchasing value. The period can be an interval, such as six months.

For example, standard analyses bring forth information such as the purchase and invoiced values (key figures) of all the materials (characteristics) during the past six months (period). The standard analyses reports that SAP ERP offers for each Logistic component has its predefined information structures with no option to add new key figures. However, when you create your own self-defined information structure in Flexible Planning, you can choose your desired key figures from several available catalogs. A catalog consists of a large number of characteristics and key figures of the specific application area, such as SAP ERP Production Planning (PP) or SAP ERP Quality Management (QM). You can even define how frequently you want to update the values of key figures in Flexible Planning standard analysis. Hence, standard analysis in Flexible Planning isn't just applicable to the PP component, but concepts and details covered in this chapter are equally applicable to other SAP ERP components such as QM, Plant Maintenance (PM), Materials Management (MM), or Sales and Distribution (SD).

[«]

Note

See Chapter 19 in which we cover reporting, including standard analyses in the PP component.

This example of Flexible Planning from both perspectives (Flexible Planning and standard analysis in Flexible Planning) in this chapter remains primarily focused on the SD component and its integral correlation with the PP component.

[«]

Note

Regardless of whether you implement standard S&OP or Flexible Planning, we encourage you to read the entire chapter because several features and functionalities are applicable to both planning types and are eventually covered (not necessarily in the same section). For example, we cover the forecasting functionality within Flexible Planning, but it's also available and can be used in standard S&OP for a material or material group. We've also dedicated a separate section on forecasting. This is also true for events and rough-cut planning. An event tends to have an impact on planning figures. A rough-cut planning profile provides better visibility on capacity, material, or PRT situations. Similarly, we cover aggregation/disaggregation in standard S&OP but not in Flexible Planning, although the option is available in both planning types.

In the following sections, we'll cover the important concepts and fundamentals you need to understand S&OP and we'll also provide a comparison of standard S&OP with Flexible Planning, as Section 9.2 will then be on Flexible Planning. Additionally, the concepts that you'll learn in the following sections regarding standard S&OP will also be applicable to Flexible Planning.

Figure 9.1 shows that S&OP can have key figures from one of the following three available options:

- **Sales Information System (SIS)**
  SIS takes information from sales history to propose a sales plan.
- **Profitability Analysis (CO-PA)**
  Information from the CO-PA component is used to help the planner makes a sales plan. The system derives this information from Sales and Profit Planning.
- **Forecasting**
  Historical data is used to come up with a sales plan. Forecasting is covered in Section 9.4.

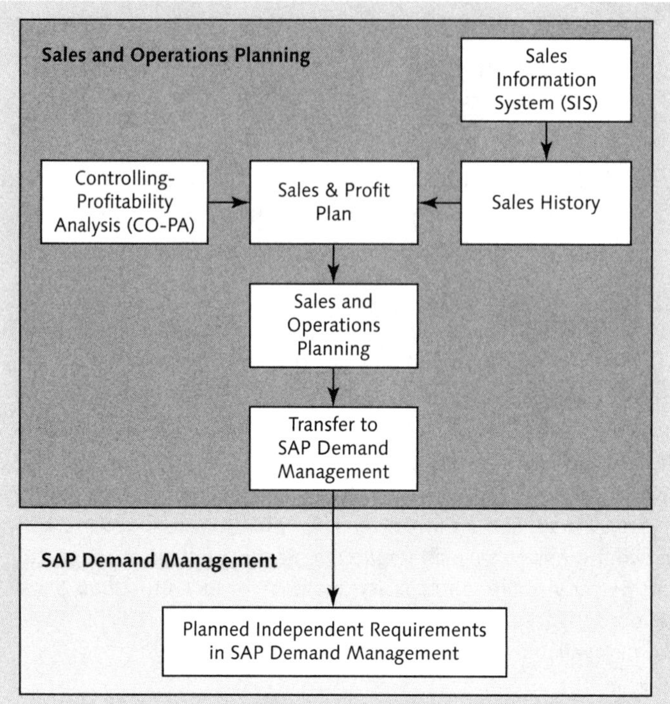

Figure 9.1 Process Overview of Sales and Operations Planning

The figures from S&OP are eventually transferred to SAP Demand Management in the form of PIRs, which form the basis of MRP.

We list the objects used in S&OP (standard S&OP and Flexible Planning) and provide their logical relationship in Figure 9.2. The objects in S&OP form an integral part in the planning process, so it's important to have a comprehensive understanding about them.

- **Information structure (info structure)**
  This is the data structure that stores the important planning parameters. The planning data is stored in key figures for the CVCs. As previously explained, characteristics can be selection criteria based on which the system brings up relevant key figures. Key figures can be quantities or values, for example, number of quality inspection lots, total purchase value of raw materials, operation quantities, or production quantities.

- **Planning method**
  The storage, aggregation, and disaggregation of data with regard to the planning level occur either as consistent planning or as level-by-level planning. In S&OP, the system looks for the planning method that is defined in the info structure.
- **Planning hierarchy**
  The planning hierarchy contains the CVCs for the characteristics of the info structure.
- **Planning table**
  This is where the planner carries out the actual and interactive planning.
- **Planning type**
  The planning type defines the layout or format of the planning table.

See Figure 9.2 to see how each of the objects is linked in S&OP.

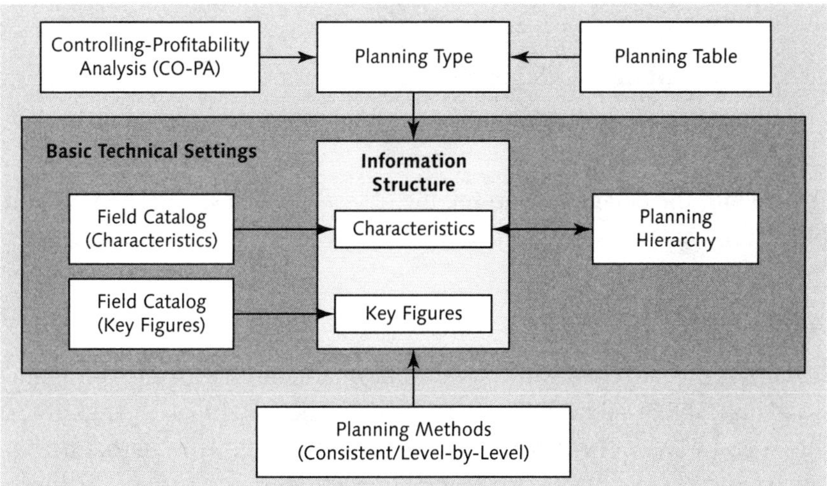

Figure 9.2 Objects of Sales and Operations Planning

Figure 9.3 provides a graphical comparison between the planning methods available for standard S&OP and Flexible Planning.

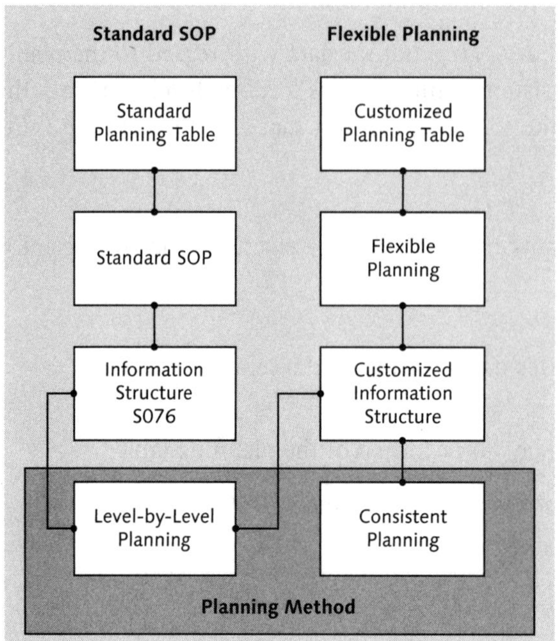

Figure 9.3 Standard S&OP, Flexible Planning, Level-by-Level Planning, and Consistent Planning

Now let's dive into the details concerning the objects used in standard S&OP and in Flexible Planning.

9.1.2 Information Structures

You can create and change the information (info) structures with the configuration (Transaction SPRO) menu path, Logistics General • LIS • Logistics Data Warehouse • Data Basis • Information Structures • Maintain Custom Information Structures. You can also use Transaction MC21 to create info structures and Transaction MC22 to change them.

Figure 9.4 shows Info Struct. S076, which is used in standard S&OP.

The planning result is stored for each CVC among the six key figures listed in the figure. Values for other key figures, such as special production or sales order, for example, can't be stored in this (standard) info structure.

In standard S&OP, you can't change many settings, but we'll review those you can to provide better comprehension of the parameters used in standard S&OP for info structures and key figures.

Display Info Structure: Fields

General features...

Info Struct.	S076	Sales & Operations Planning	Active
Application	01	Sales and Distribution	Saved
Info Struc. Type		Standard	
Plng Possible	☑		

Characteristics	Key figures	Unit	SID	FID
Prod.group/material	Sales	01	☑	☐
Plant	Production	01	☑	☐
Production version	Stock level	01	☐	☐
	Target stock level	01	☐	☐
	Days' supply	00	☐	☐
	Target days' supply	00	☐	☐

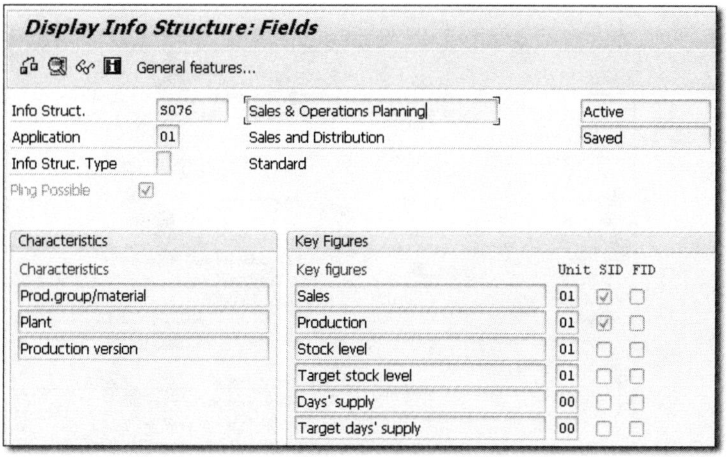

Figure 9.4 Standard SAP Info Structure S076

To view or set parameters for info structures and key figures, follow the configuration (Transaction SPRO) menu path, LOGISTICS • PRODUCTION • SALES & OPERATIONS PLANNING (S&OP) • MASTER DATA • SET PARAMETERS FOR INFO STRUCTURES AND KEY FIGURES, or use Transaction MC7F.

In Figure 9.5 ❶, double-click on the info structure (TABLE) S076, or select the same and choose the DETAILS (magnifying glass) icon, which takes you to the CHANGE VIEW "INFO STRUCTURE PLANNING PARAMETERS": DETAILS screen ❷. This area stipulates that the planning method used in standard S&OP is level-by-level planning (denoted by I in the PLANNING METHOD field).

Choose the KEY FIGURE PLANNING PARAMETERS folder on the left-hand side of the screen, which takes you to the screen shown in Figure 9.6 ❶. The lower half lists the key figures available for standard info structure S076. When you double-click on the ABSAT field, a pop-up screen appears ❷. Here, you control whether the key figure can be used for forecasting and determine the type of aggregation (summation) of the entered data in the planning table.

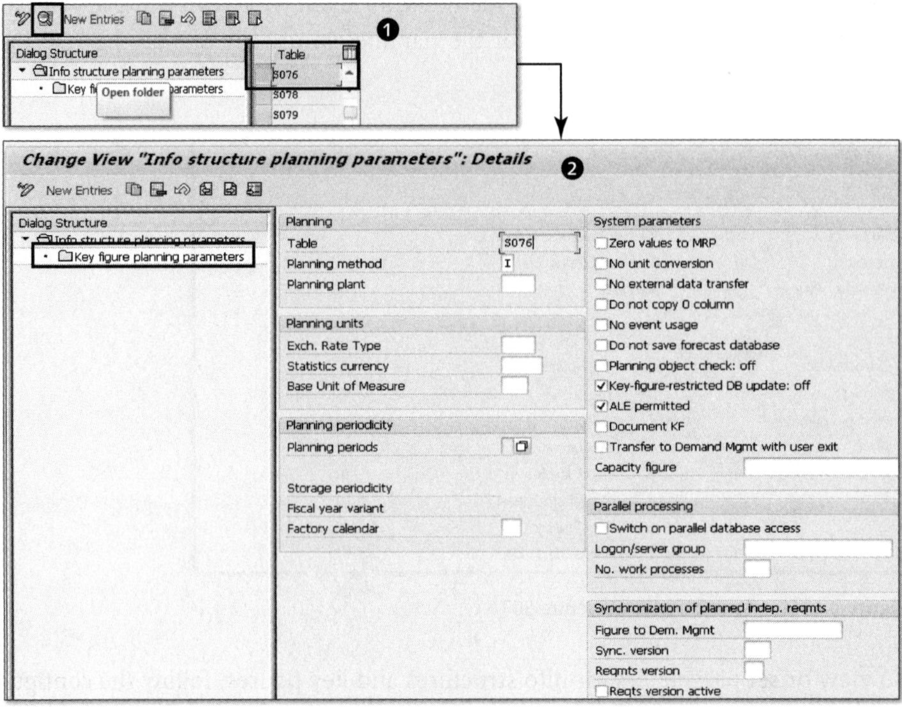

Figure 9.5 Planning Parameters of the Info Structure

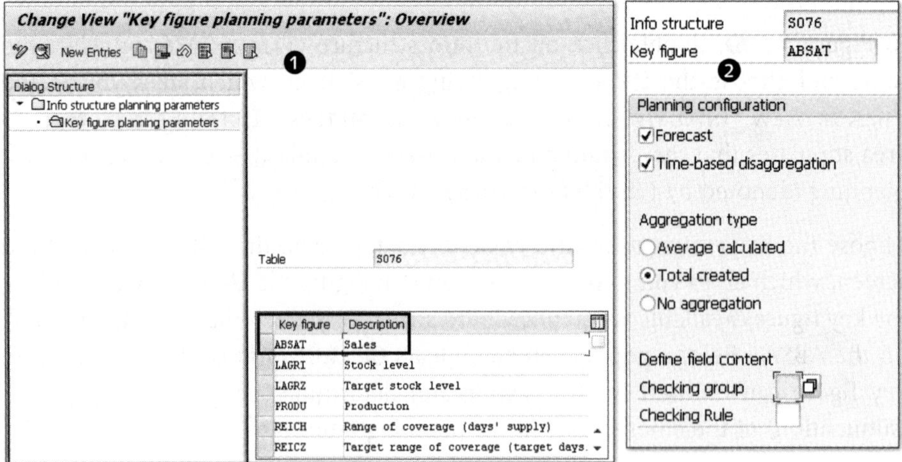

Figure 9.6 Parameters for Key Figures in the Info Structure

9.1.3 Planning Methods

In the beginning of this chapter, we mentioned that with standard S&OP, the configuration and other settings are predefined, with limited options to customize parameters to meet your business requirements. For example, in standard S&OP, the configuration for info structure S076 is preset with characteristics and key figures, as well as the planning table in which you enter the planning figures. The planning is either based on individual materials or group of materials (i.e., product groups) in which you can define proportional factors.

On the other hand, with Flexible Planning, you can set up self-defined info structures with desired characteristics and key figures, and the planning is based on planning hierarchies. You can set up self-defined planning tables as well, and you can perform either level-by-level planning or consistent planning.

Table 9.1 provides a comparison of planning methods available in standard S&OP and in Flexible Planning.

Standard S&OP	Flexible Planning
Preset configuration	Individual configuration based on business processes
Planning based on product groups/ proportional factors	Planning based on planning hierarchies
Standard planning table for key figures entry	Customized planning tables
Level-by-level planning	Consistent planning or level-by-level planning

Table 9.1 Features of Standard S&OP and Flexible Planning

With standard S&OP, you can only plan using level-by-level planning (see Table 9.2), whereas Flexible Planning can use both level-by-level as well as consistent planning. In the level-by-level planning, the planning data is maintained at the specified level only. Because it doesn't automatically aggregate or disaggregate planning data, you have to manually perform these functions.

Table 9.2 provides a comparison of consistent planning and level-by-level planning in detail. To use consistent planning, you have to define planning hierarchies in the system. A planning hierarchy enables you to define proportional factors for each characteristic. For example, if there are three sales organizations, you can define how much (percentage) each sales organization (a characteristic) will contribute in the overall planning hierarchy. Similarly, changes made to any planning level automatically updates and accounts for changes in other planning levels. For example, initially you defined 10 materials each having a 10% proportion to the planning results. Now, if you add a new material, the system automatically updates the proportional factor (100% ÷ 11 materials = 9.09% for each material). The system also automatically aggregates (adds up) and disaggregates (divides up) the planning data to present a consistent planning position in the planning table.

Consistent Planning	Level-by-Level Planning
Planning hierarchy	Product group
Storage at the lowest level; planning data at all levels	Storage at each level; planning data only at maintained level
Automatic aggregation and disaggregation	Aggregation and disaggregation as a planning step

Table 9.2 Features of Consistent Planning and Level-by-Level Planning

9.1.4 Planning Types in Standard Sales and Operations Planning

The *planning type* defines the layout or the format of the planning table and thus represents the link between the planning table, where the actual planning is carried out, and the info structure in which the planning data is stored. You can create several planning types for one info structure. For example, two different planning types may contain planning data for two seasons (autumn and winter). The data is stored in the info structure, so the different planning types depend on each other because they use the same set of data. If you use level-by-level planning, you must create a separate planning type for each planning level.

Standard S&OP has been configured with three planning types, which are used automatically in standard S&OP planning:

- **SOPKAPA**
  Planning type for the planning of individual product groups.
- **SOPKAPAM**
  Planning type for the planning of individual materials.
- **SOPDIS**
  Planning type for the dual-level planning of product group hierarchies.

In the following sections, we'll go through the steps you need to follow to create a product group that you can use in standard S&OP. We also show how you can transfer the planning results of a product group to SAP Demand Management as PIRs.

Create a Product Group

Figure 9.7 shows that in this example, we're making use of info structure S076 in standard S&OP to create a product group called Pump_SOP for plant 3000. This product group has two materials, 1729 and 1731, with a proportional factor of 60% and 40%, respectively. While we explained aggregation and disaggregation earlier, the proportional factor is specific to level-by-level planning only, and you can either maintain the proportional factor manually or allow the system to propose the proportional factor.

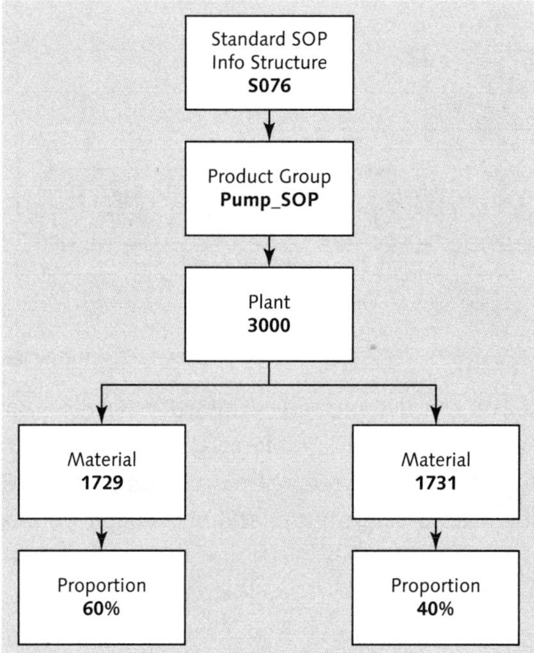

Figure 9.7 Product Group Pump_SOP

For this example, you now need to create the product group in the system. To create a product group, follow the SAP menu path, LOGISTICS • PRODUCTION • SOP •

PRODUCT GROUP • CREATE, or use Transaction MC84. Figure 9.8 shows the initial screen ❶ to define the PRODUCT GROUP PUMP_SOP for PLANT 3000. Enter the materials, "1729" and "1731" and their proportional factors of 60% and 40% ❷.

Choose the PRODUCT GRP. GRAPHIC button to view the product group graphically ❸.

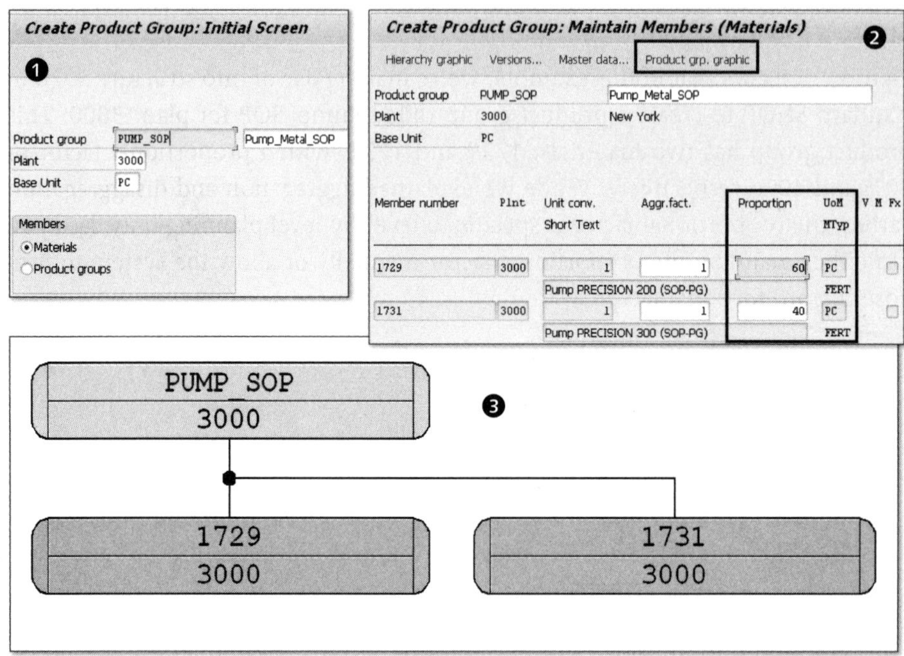

Figure 9.8 Create a Product Group

In addition to the interactive entry, you can also calculate proportional factors on the basis of historical data via the menu path, EDIT • CALCULATE PROPORTIONAL FACTORS. You can use this option to see how well your self-defined (manual) proportional factors compare with what the system proposes and may enable you to prepare a realistic plan. Moreover, you can equally distribute the proportional factors via EDIT • DISTRIBUTE PROPORTIONAL FACTORS. Save the product group.

Create a Plan for the Product Group

To plan the product group in standard S&OP, use the menu path, LOGISTICS • PRODUCTION • SOP • PLANNING • FOR PRODUCT GROUP • CREATE, or use Transaction MC81.

Figure 9.9 shows the initial screen ❶ of the rough-cut plan for the product. Enter the PRODUCT GROUP "PUMP_SOP" and PLANT "3000", and then press Enter to go to the DEFINE VERSION dialog box ❷. In this screen, enter the planning VERSION "001" and the VERSION DESCRIPTION as "Pump_SOP_PG", and again press Enter to go to the CREATE ROUGH-CUT PLAN screen ❸.

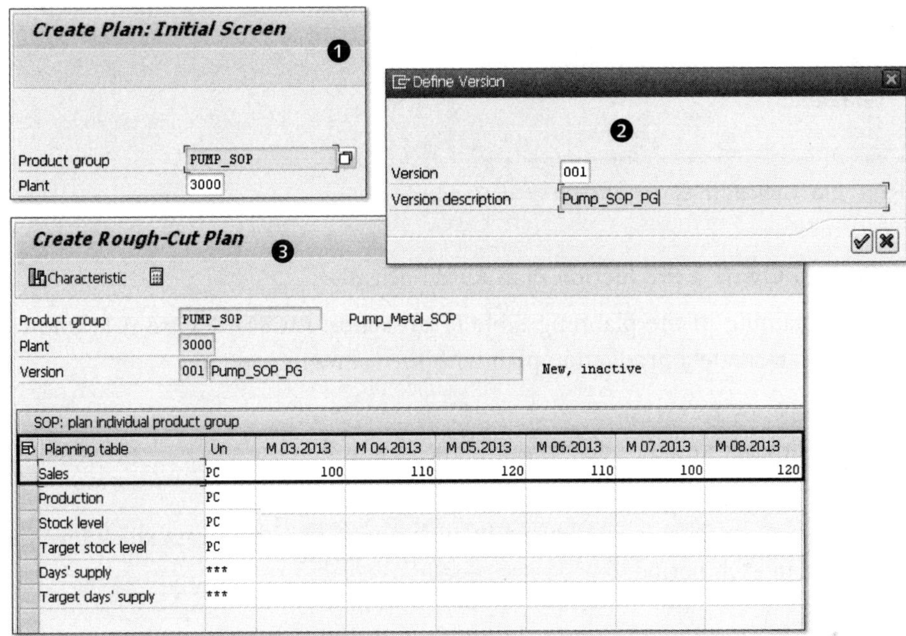

Figure 9.9 Planning in Standard S&OP

If the historical key figure data is already available in the system, you can import it to create a sales plan. Alternatively, you can manually enter the SALES quantities in the PLANNING TABLE, as we've done in this example, by entering sales data for the next six months. Figure 9.10 shows that you have several options to create a sales plan for the product group (or for an individual material, where applicable):

- Transfer plan from SIS
- Transfer CO-PA plan
- Forecast
- Transfer product group proportional from production or from sales

The prerequisite for using any of the options entail that the system has significant historical data to help in effective and reliable planning.

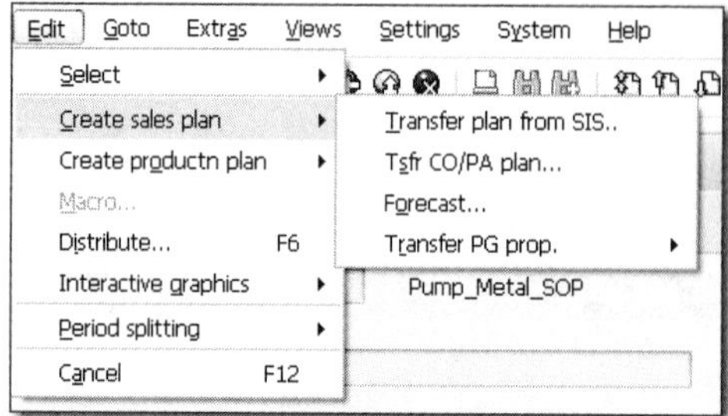

Figure 9.10 Options to Create a Sales Plan in Standard S&OP

Options to Create a Production Plan Automatically

While remaining in the planning table (Transaction MC81), Figure 9.11 shows how you can create a production plan with the following options:

- SYNCHRONOUS TO SALES
  Sales figures are used as operations plans.
- TARGET STOCK LEVEL
  The operations plan is configured in such a way that the target stock level is reached in each period.
- TARGET DAYS' SUPPLY
  The operations plan is configured in such a way that the target stock level is reached in each period.
- STOCK LEVEL = ZERO
  The operations plan is configured in such a way that the entire stock level is consumed.

Depending on your business requirement, you can choose if you want the production plan to be equal to the target stock level or if you eventually want the target stock to be zero at the end of the period. In this example, we want to create the production plan synchronous to the sales plan. This production plan has also taken the target stock levels into account at the end of each period.

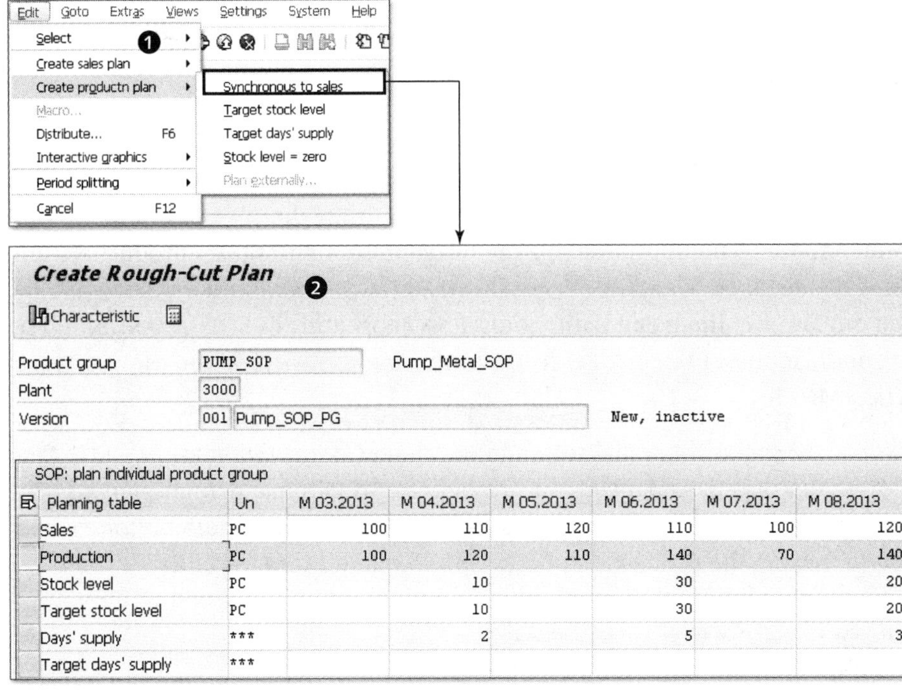

Planning table	Un	M 03.2013	M 04.2013	M 05.2013	M 06.2013	M 07.2013	M 08.2013
Sales	PC	100	110	120	110	100	120
Production	PC	100	120	110	140	70	140
Stock level	PC		10		30		20
Target stock level	PC		10		30		20
Days' supply	***		2		5		3
Target days' supply	***						

Figure 9.11 Options to Create a Production Plan in Standard S&OP

It's important to note that these production figures have no direct implications on the production or procurement. The purpose of entering or incorporating these numbers (key figures, such as sales or production) is to come to a common understanding, where relevant stakeholders, be it sales or production, can transfer the *finalized* key figures to SAP Demand Management in the form of PIRs. For example, several planning versions are created in the system from which input and feedback from sales and production are taken. The sales or the production figures are revised either in the same planning version or by creating a new version. The finalized version eventually transfers to SAP Demand Management and reflects as PIRs with the ACTIVE indictor selected.

The next step is to transfer the production plan to SAP Demand Management, where the requirements of individual members (materials) of the product group are reflected as PIRs. When the transfer of the product group quantities to SAP Demand Management takes place, the system disaggregates at the material level according to the proportional factor defined previously. In this example, we've given a proportional factor of 60% and 40% for the two materials. In the next section, we show how the system disaggregates (divides up) the total planning quantities of the product group in a 60 to 40 ratio.

Transfer to SAP Demand Management

You can directly transfer the planning figures from the planning table to SAP Demand Management by selecting the relevant key figures (e.g., production) and using the menu path, EXTRAS • TRANSFER TO DEMAND MANAGEMENT. Alternatively, you can also use the menu path, LOGISTICS • PRODUCTION • SOP • PLANNING • FOR PRODUCT GROUP • TRANSFER PG TO DEMAND MANAGEMENT, or you can use Transaction MC75.

Figure 9.12 shows the initial screen in which you assign the VERSION "001" ❶ of the product group and plant for transfer to SAP Demand Management. The PROD.PLAN FOR MAT. OR PG MEMBERS AS PROPORTION OF PG radio button enables you to transfer the production plan to SAP Demand Management.

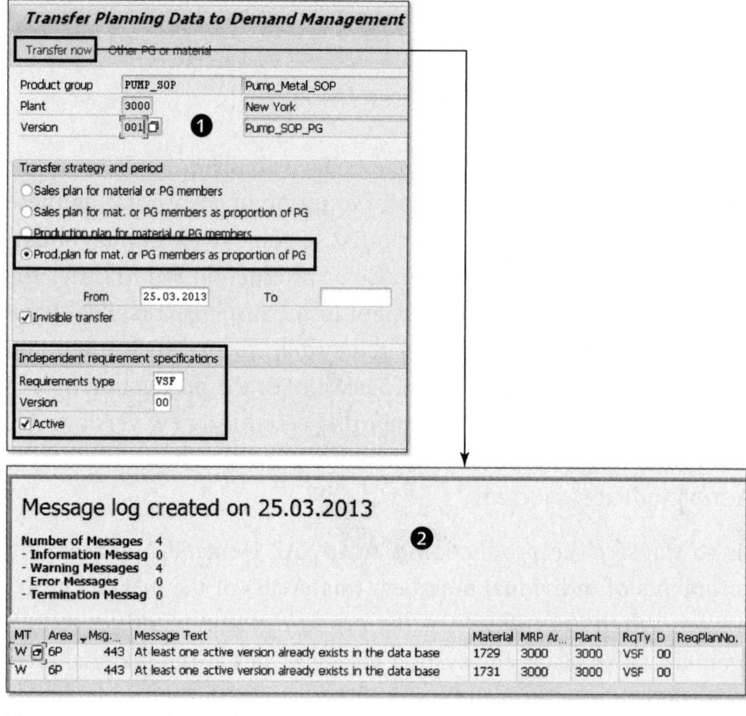

Figure 9.12 Transfer Product Group to SAP Demand Management

You can also specify the requirement type and the version of PIR within which the system transfers the values. In this example, assign REQUIREMENT TYPE as "VSF" and VERSION as "00", and choose TRANSFER NOW. A message log appears after the transfer ❷.

Planned Independent Requirements in SAP Demand Management

To check and confirm whether production figures of the product group were successfully transferred in SAP Demand Management as PIRs, follow the SAP menu path, LOGISTICS • PRODUCTION • DEMAND MANAGEMENT • PLANNED INDEPENDENT REQUIREMENTS • DISPLAY, or use Transaction MD63.

Figure 9.13 shows the initial screen ❶ in which you enter the selection parameters such as PRODUCT GROUP "PUMP_SOP", PLANT "3000", REQUIREMENTS TYPE "VSF", and SELECTED VERSION "00", which were all previously defined in Figure 9.12. Press Enter, and Figure 9.13 ❷ shows how the product group breaks down into individual members (materials) with the transferred quantity divided in a ratio of 60% to 40%.

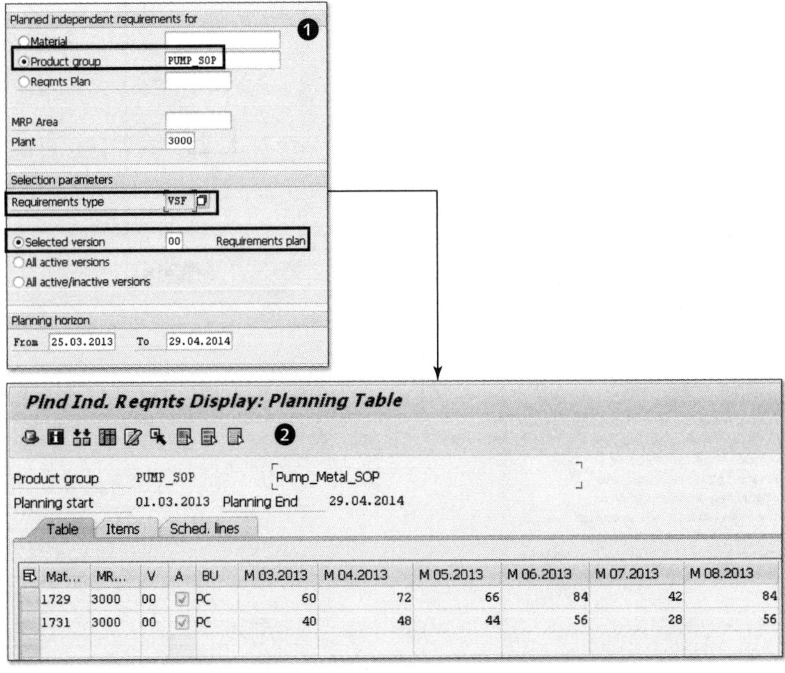

Figure 9.13 Planned Independent Requirements in SAP Demand Management

9.1.5 Distribute Key Figures

During the course of planning the key figures, often the planner needs to make mass changes to several key figures at once. This can be tedious if done one by one or manually. You can use DISTRIBUTE to make mass changes to a key figure in the same row or column. The following are some of the changes that you can make:

- Perform mathematical calculations such as addition, subtraction, multiplication, and so on to all of the values (or a selection) of a key figure.
- Increase or decrease all of the values (or a selection) of a key figure by a defined percentage.
- Distribute a value across the time series and replace the old values.

In Figure 9.14, the option to distribute key figures is available from the menu EDIT • DISTRIBUTE ❶, by pressing F6 or by clicking the CALCULATOR (DISTRIBUTE) icon in the planning table.

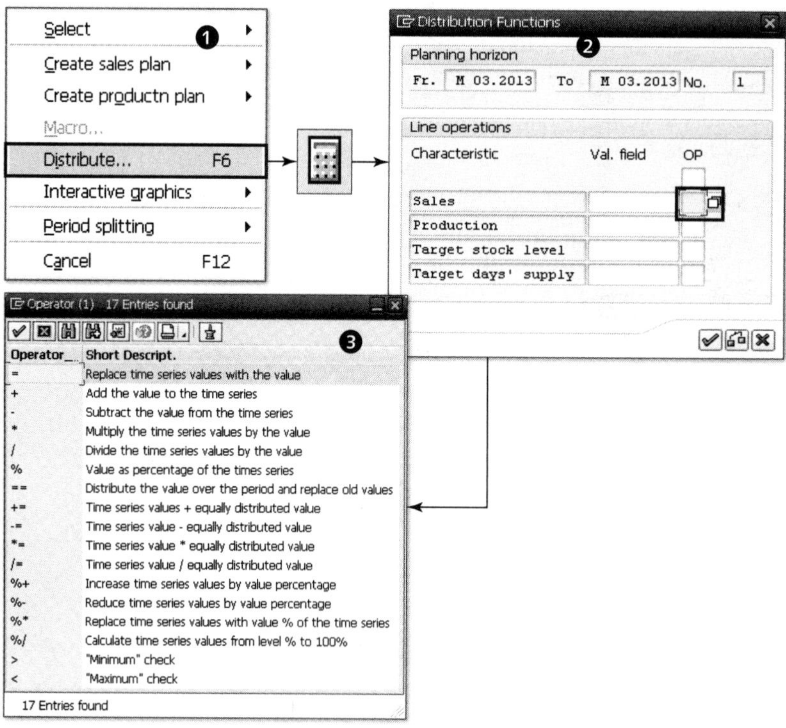

Figure 9.14 Distribution Functionality to Key Figures in S&OP

In the Distribution Functions dialog box ❷ that appears, you can enter a number in the Val. field followed by selecting the right operand (OP). The list of available operands appears ❸.

The following sections cover aggregation and disaggregation as well as tools and features that are available not only standard S&OP but also in Flexible Planning.

Aggregation and Disaggregation

During the course of planning, available tools such as aggregation and disaggregation of key figures help the planner to plan at one organizational (characteristics) level and transfer the values (key figures) to other characteristics levels, based on the defined proportions. Automatic aggregation and disaggregation is a tool available for the planner to help prepare a reliable and realistic sales (or production) plan. For example, the planner wants to consolidate (aggregate) the production quantities (key figures) at a plant level (characteristic). In this scenario, the aggregation tool is helpful to quickly consolidate the desired information. Another example of aggregation occurs when the sales organization is at the highest level, then the distribution channel, then the division, then the material level, and finally the plant level.

Automatic aggregation and disaggregation are consistently available between all of the planning levels in consistent planning, whereas in level-by-level planning, you need to manually trigger them. Because standard S&OP doesn't offer consistent planning, only level-by-level planning is possible.

The system enables you to drill down data from higher levels to lower levels (disaggregation) or progressively sum up the data from the lower levels to the higher levels (aggregation of sum or average). The ultimate objective of planning, either in standard S&OP or in Flexible Planning, is to obtain a plan at the lowest level of the characteristic combinations (e.g., at the product group or material levels) and forward the plan to PP.

The planning hierarchy consists of CVCs and their disaggregation percentages. For example, if you have a CVC of sales organization, distribution channel, division, material, plant, statistics currency, and document currency, then you can view the historical sales data or a forecast plan for given key figures of the characteristics. The system disaggregates if the data is viewed from the lower levels of the hierarchy and aggregates the same, if the data is viewed from the higher levels, using the same set of disaggregation proportions.

Automatic Calculation of Proportional Factors

When you've maintained the planning hierarchy, you can run Transaction MC9B (Using Planning Hierarchy) for consistent planning to allow the system to calculate the proportional factors based on the disaggregation proportions. These proportional factors become the base in consistent planning for the aggregation and disaggregation.

Time-Based Disaggregation

When rough-cut S&OP transfers the data in SAP Demand Management, the system requires a *time-based disaggregation*. In S&OP, the planning is carried out in periods of weeks or months, whereas the PIR needs a current date for its planning activities. If the business process is simple, then the requirement of the period is scheduled for the first working day of the period without any further configuration. However, if you need an in-depth and complex distribution of quantities, then you need to define distribution strategies and period splits. The distribution strategy is assigned to the period split for weekly and monthly distribution. The period split, in turn, is assigned to the plant and the planning group.

Planning Hierarchy in Standard S&OP

The planning hierarchy in standard S&OP defines the proportion of one material with respect to others. For example, a material with a 40% proportional factor enables the system to divide all planning figures of this material to 40% of the total planning figures. You can create planning hierarchy master data in standard S&OP by maintaining product groups. You can also generate the master data automatically using Transaction MC8P for info structure S076. Alternatively, you can also run Program RMCPSOPP (using Transaction SE38) to generate the master data for info structure S076. When you're in the product group as in standard S&OP or planning hierarchy in Flexible Planning, you have to define proportional factors. The system then uses these defined proportional factors either in level-by-level planning or in consistent planning.

Table 9.3 provides an overview of the usage of proportional factors in consistent planning and level-by-level planning.

Planning Mode	Calculation Method	Usage of Proportional Factors	Number of Levels
Level-by-level planning	–	Direct	Single
Consistent planning	Calculation based on the planning hierarchy	Multiplication of proportional factors of multiple levels	Multiple
Consistent planning	Calculation based on actual data	–	Multiple

Table 9.3 Usage of Proportional Factors in Disaggregation

To aggregate or disaggregate planning figures in standard S&OP, follow the menu path, LOGISTICS • PRODUCTION • SOP • DISAGGREGATION • BREAK DOWN PG PLAN, or use Transaction MC76. Having selected the product group, plant, and the version, Figure 9.15 shows the initial screen, in which you can make changes to the proportional factor of each member of the product group and also select the relevant checkboxes to run the standard macros when the planning table screen for standard S&OP appears.

Press [Enter] to be taken to the planning table screen of standard S&OP.

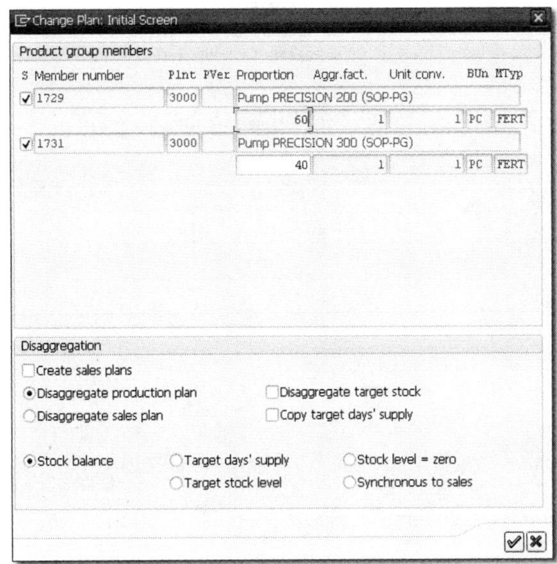

Figure 9.15 Disaggregation in the Product Group

9.1.6 Working with Macros

Macros enable you to automate the linking of key figures in the planning table by using calculation rules. In standard S&OP, the system provides macros for the calculation of the projected stock level and for the creation of an operations plan on the basis of the sales plan. You also can use macros in aggregation and disaggregation during level-by-level planning. You can't define your own macros in standard S&OP.

In Figure 9.16, first select the SALES line item, and then click on the MACRO icon ❶, which brings up the list of standard macros available to select from (see Figure 9.16 ❷). For this example, choose AGGREGATION: SALES, which aggregates the individual sales figures for material 1731 ❸. The individual sales figures of the second material (1729) aren't shown.

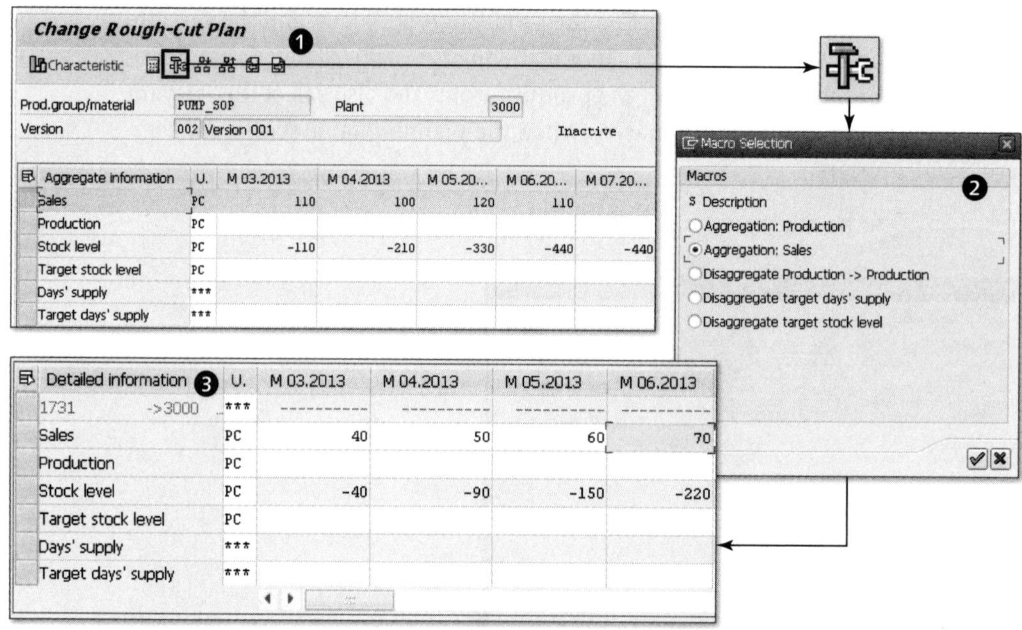

Figure 9.16 Macros in Standard S&OP

[»] **Note**

Self-defined macros in Flexible Planning are covered in greater detail in Section 9.2.6.

9.2 Flexible Planning

It's amply evident that standard S&OP offers limited options in terms of solutions to diverse and complex business scenarios and processes. If your company intends to carry out very lean and production-focused forecasting of sales figures, then you should check whether standard S&OP is able to fulfill your needs.

For more complex planning scenarios, Flexible Planning can help. In Flexible Planning, you can plan and process planning data from any organizational unit, whether it's a sales organization, a distribution channel, or a production plant. Flexible Planning begins with designing an info structure. An info structure defines a structure with organizational units and business planning values, for example, forecast values in different periods, their relationships, and different views of planning values from different organizational units' perspectives, whether it's sales, production, or procurement. Gaining an in-depth understanding of how to define an info structure to fulfill business planning requirements is very important for successful implementation of Flexible Planning.

You can also use info structures for product allocation. For example, in the automobile industry, it's common to assign sales quotas to different regions and to automotive dealers to ensure that a company's automobile is in every region it serves. In an info structure for production allocation, you maintain dealer-wise sales quota. During sales order creation for a specific dealer, the system then checks the information maintained in the info structure for product allocation to see if the dealer is still eligible to place orders.

Similar to standard S&OP where everything is preconfigured, in Flexible Planning, a table is designed using the basic steps of configuring a self-defined info structure, setting the parameters for the info structure and key figures, and configuring the planning type.

In this section, we not only show you how to successfully set up Flexible Planning but also specifically cover the update fields (key figures) involved in Flexible Planning. These update fields enable a planner to better use the Flexible Planning functionality when features such as macros, aggregation or disaggregation, events, and rough-cut planning aren't involved. The update fields apply to standard analysis in Flexible Planning.

S&OP offers the creation of a planning hierarchy and maintenance of disaggregation percentages for the CVCs in a hierarchy. As mentioned previously, a planning hierarchy is available in Flexible Planning only.

The section also covers self-defined macros, aggregation/disaggregation, events, and rough-cut planning profiles because all of these features hold significant value to the planner.

9.2.1 Creating a Self-Defined Info Structure

Because we've already covered the salient features of the info structure for Flexible Planning in Section 9.1.2, we'll now create the self-defined info structure in Flexible Planning.

You can create and change these structures using the configuration (Transaction SPRO) menu path, General Logistics • LIS • Logistics Data Warehouse • Data Basis • Information Structures • Maintain Custom Information Structures. You can also use Transaction MC21 to create info structures and Transaction MC22 to change them.

Figure 9.17 shows the initial screen to create an info structure. For this example, define the Info structure as "S655" and enter the description as SOP: Info Structure. It's important to select the application because the resulting catalog (consisting of characteristics and key figures) will then appear accordingly. For this example, set the Application as "01" for SD. It's also important to choose Plng Possible to enable the user to perform planning functions in Flexible Planning.

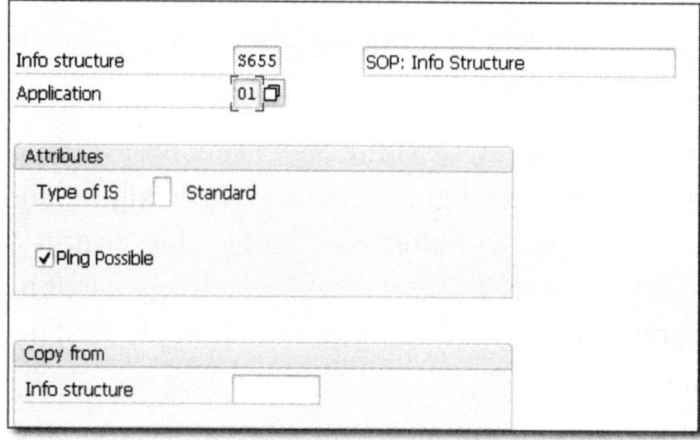

Figure 9.17 Self-Defined Info Structure Creation in Flexible Planning

Press [Enter] to open the screen shown in Figure 9.18. Click on the CHOOSE CHARACT button ❶, which brings up the option to select characteristics from the catalogs. For this example, select the following characteristics from three catalogs:

- SD: SALES ORGANIZATION (ORDER)
  - SALES ORGANIZATION
  - DISTRIBUTION CHANNEL
  - DIVISION
- SD: ARTICLE (ORDER)
  - MATERIAL
  - PLANT
- SD: UNITS (DELIVERY)
  - DOCUMENT CURRENCY
  - STATISTICS CURRENCY

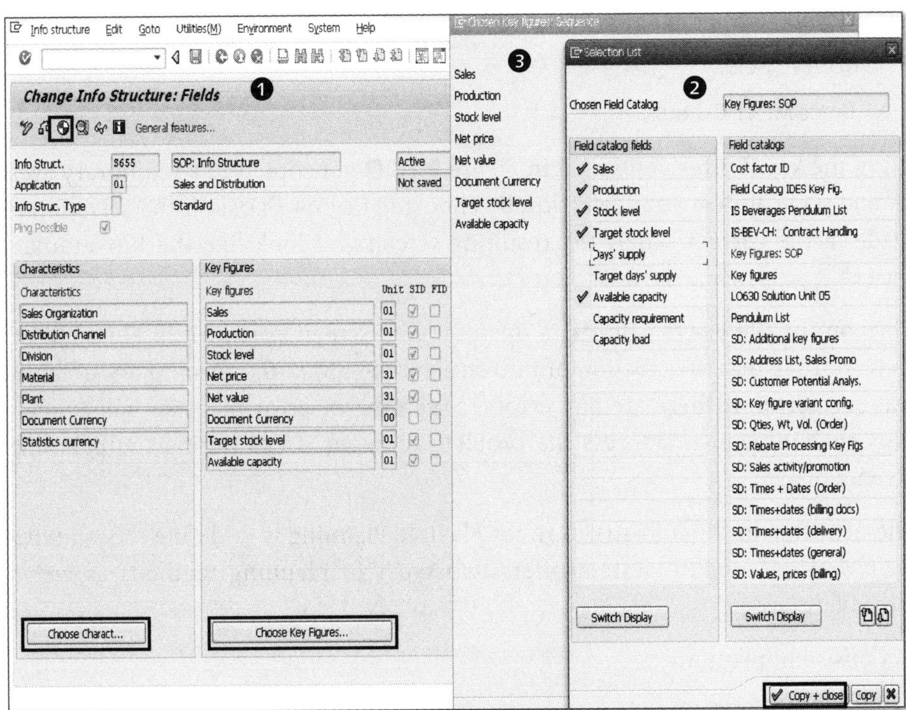

Figure 9.18 Characteristics and Key Figures in Flexible Planning

If the selection is made in the correct order, the resulting screen will look like the CHARACTERISTICS area ❶.

Next, to select the key figures, click on CHOOSE KEY FIGURES, and select the key figures from the following catalog.

For this example, select the following key figures from three different catalogs:

- KEY FIGURES: S&OP
  - SALES
  - PRODUCTION
  - STOCK LEVEL
  - TARGET STOCK LEVEL
  - AVAILABLE CAPACITY
- SD: VALUES, PRICES (ORDER)
  - NET PRICE
  - NET VALUE
- PENDULUM LIST
  - DOCUMENT CURRENCY

All of the key figures you select in Figure 9.18 ❷ get copied to a temporary area ❸, and you confirm your selection by choosing COPY + CLOSE. If the selection is made in the correct order, the resulting screen will look like the KEY FIGURES area ❶.

Click on the GENERATE icon to open the LOG screen as shown in Figure 9.19. Here you see that the creation of info structure S655, in the form of a SAP table, was successful. If there are any errors, a warning or error message will appear, thus enabling you to address the problem to avoid issues during the planning activity phase.

The next step in the info structure for Flexible Planning is to define the *planning method*. The SAP ERP system offers two types of planning methods as we've already discussed:

- Consistent planning
- Level-by-level planning

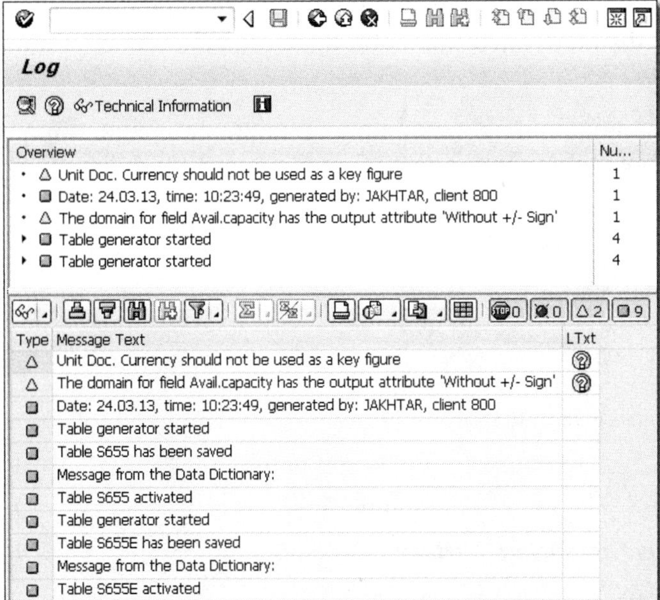

Figure 9.19 Successful Generation of an Info Structure in Flexible Planning

The planning method determines the method of disaggregation of data and the storage of data at various levels of planning. Refer to Section 9.1.5 for details.

To set up a planning method for the self-defined info structure, follow the configuration (Transaction SPRO) menu path, Production • Sales & Operations Planning (SOP) • Master Data • Set Parameters for Info Structures and Key Figures, or use Transaction MC7F.

Figure 9.20 shows an overview screen of the various info structures available. Double-click on the info structure (Table) S655.

The screen shown in Figure 9.21 appears, where you assign the Planning method as "K" (for consistent planning) and the Planning plant as "3000" so that the system automatically uses this plant during transfer of key figures to SAP Demand Management.

Two other important fields are Statistics currency and Base Unit of Measure. For this example, assign "USD" and "PC", respectively, to these two fields. The Statistics currency field enables the system to use it as a base currency, whereas the Base Unit of Measure is the smallest unit of measure in which the inventory figures are shown. If a material has a different base unit of measure than what is being used in planning, then the conversion factor must be defined.

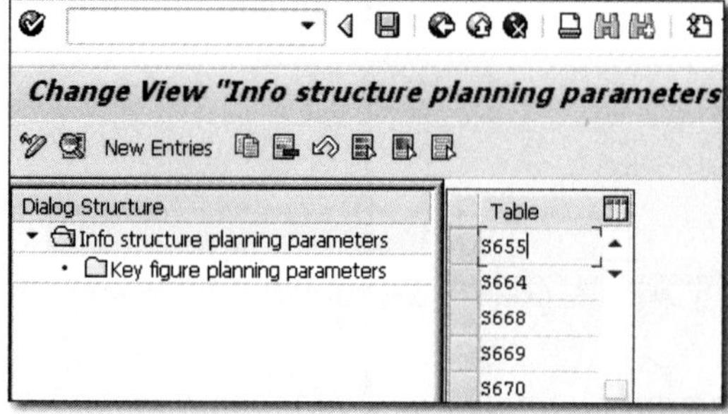

Figure 9.20 Planning Parameters in the Info Structure

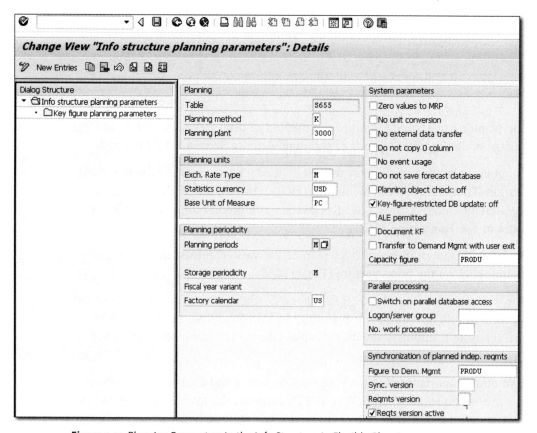

Figure 9.21 Planning Parameters in the Info Structure in Flexible Planning

By using the CAPACITY FIGURE field value "PRODU", you can work with production line values from the planning table while carrying out rough-cut capacity planning. Similarly, by using the FIGURE TO DEM. MGMT value "PRODU", you ensure that the system uses the production line values while transferring the planned values to SAP Demand Management.

Finally, assign "US" in the FACTORY CALENDAR field, which will enable the system to refer to the working days and holidays during rough-cut capacity planning.

Note [«]

You can perform rough-cut capacity planning in a planning table to see if the planned forecast values are within the constraints, for example, available capacity at a critical work center. If they aren't, you can adjust the planned forecast values until the capacity load value is shown within permissible limits. For rough-cut capacity planning, you use the production quantity from the planning table. The system selects the production quantities of different periods from the PP line in the planning table to calculate the capacity requirement and display the capacity load as a percentage value.

The screen shown in Figure 9.22 appears when you choose the KEY FIGURE PLANNING PARAMETERS folder in Figure 9.21.

The lower half of Figure 9.22 lists all the key figures ❶ you selected during the info structure creation. Double-click ABSAT, and in the screen that appears ❷, choose between two options:

- FORECAST
  Choose this checkbox to allow forecasting. If FORECAST isn't selected for the key figures, then forecasting isn't possible for the key figure in the planning table.
- TIME-BASED DISAGGREGATION
  This determines the type of aggregation with three possible options: average value of the total key figures value of a lower level is aggregated to the higher level, the total of key figure values of the lower level is aggregated to the higher level, or no aggregation is to be carried out at all from the lower level to the higher level. It's always recommended to use the total aggregation method.

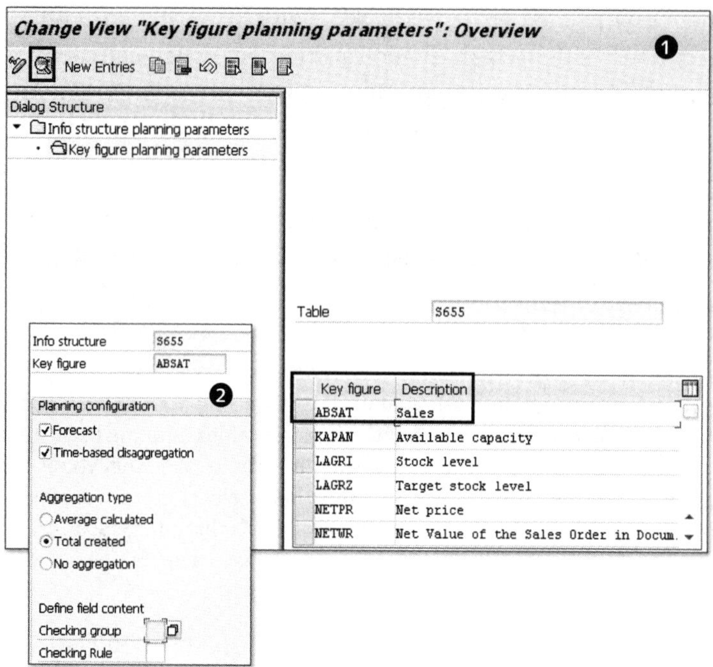

Figure 9.22 Key Figures in the Info Structure

9.2.2 Planning Hierarchy

In standard S&OP, only level-by-level planning is possible. Although the characteristics of the info structure already fine-tune the planning levels, you must still assign characteristic values to them to carry out planning at those levels. For this example, the custom info structure S655 contains the characteristics sales organization, distribution channel, division, material, plant, document currency, and statistics currency.

A planning hierarchy also enables the planner to define the proportional factor of each of the characteristics and in the same sequence of the characteristics in which they are supposed to be planned. You create a planning hierarchy for the following two important reasons:

- To let the SAP ERP system knows how the characteristics values are involved in S&OP.
- To let the SAP ERP system knows the percentage proportional weight of each of the characteristics values in the organizational hierarchy. This is known as the

disaggregation percentage, which is used by the system to disaggregate or aggregate the data across various levels during planning and consolidation. We've already covered aggregation and disaggregation in the Section 9.1.5.

In level-by-level planning, the disaggregation percentage is used directly for disaggregation/aggregation purposes. In consistent planning, the disaggregation percentages are used to calculate the proportional factors used in aggregation and disaggregation.

You need to fulfill the following prerequisites before you can successfully create a planning hierarchy:

- Create an info structure with an active status.
- Configure the planning parameters for an info structure.
- Create the update rules and activate these rules (if applicable).

You can only create one planning hierarchy for a given info structure. Also, note that you can only use one info structure in consistent planning, whereas in level-by-level planning, you can use one or multiple info structures.

To maintain the planning hierarchy for the self-defined info structure S655, use SAP menu path, LOGISTICS • LOGISTICS CONTROLLING • FLEXIBLE PLANNING • PLANNING HIERARCHY • CREATE, or use Transaction MC61.

Enter the INFO STRUCTURE "S655" as shown in Figure 9.23. In the ensuing CHARACTERISTICS pop-up ❶, enter all of the characteristics details. Note that if you don't enter the details after a specific level, the system triggers the planning hierarchy to maintain the proportional factor. For example, if you don't enter any details for DIVISION, then the next three characteristics, MATERIAL, PLANT, and STATISTICS CURRENCY, will enable you to reflect the planning hierarchy in which you can, for example, enter multiple material numbers with a proportional factor for each.

In the CREATE PLANNING HIERARCHY: STATISTICS CURRENCY screen ❷, maintain the statistics CURRENCY characteristic as "USD" with a PROPORTION of "100".

You can navigate from the higher level to the next level by selecting the characteristics values and choosing the NEXT LEVEL icon or by just double-clicking on a characteristic value. You can also check the graphical representation by clicking the HIERARCHY GRAPHIC button in the MAINTAIN PLANNING HIERARCHY screen or directly with Transaction MC67.

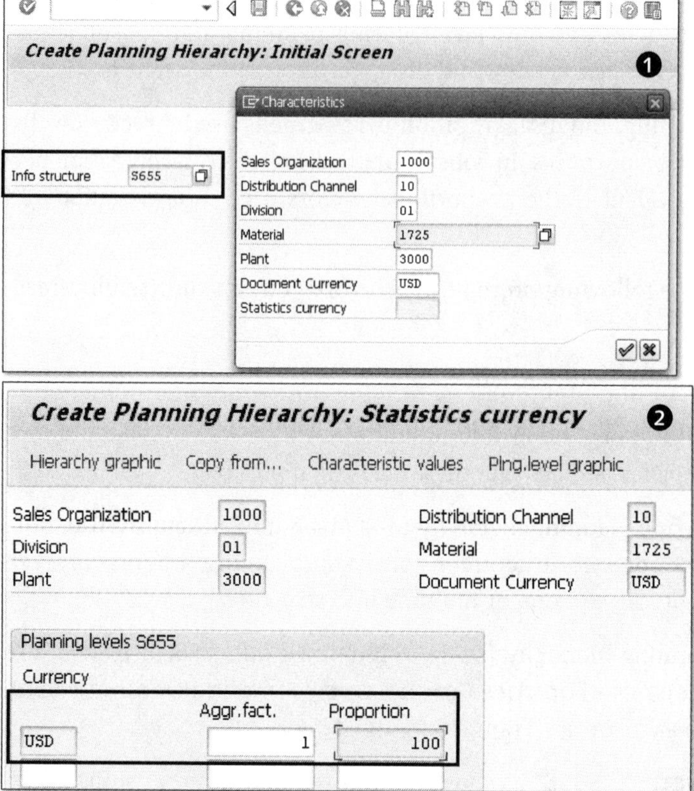

Figure 9.23 Creation of Planning Hierarchy

[+] **Tips & Tricks**

You can also have the system automatically create the planning hierarchy along with the disaggregation percentages based on the historical data available in the info structure and its version by using Transaction MC9A (Generate Master Data). You can also simulate the creation of a planning hierarchy.

9.2.3 Planning Type

With the planning hierarchy already in place, the next step is to maintain the planning type, which then becomes an underlying template or layout in the planning table.

As mentioned previously, in maintaining a planning type for Flexible Planning, you can define your own macros, assign events, allocate rough-cut capacity planning, and deploy planning versions. The planning type then becomes a format, layout, or template of the planning table, containing key figures and arranged in the desired way. A planning type is similar to a Microsoft Excel program, where you can, for example, include the headings on the left-hand side of the worksheet and the data for the headings on the right-hand side of the sheet.

You can also create multiple planning types for a given info structure. Each planning type arranges the key figures of the info structure as desired on a planning table that has its own macros to convert or translate the available data of an info structure. Having multiple planning types enables the planner to plan the same data arranged in different templates, and from different perspectives, to arrive at a feasible operations plan. Alternatively, if you're going to use only one planning type, you can save the data under different planning versions and then finalize on one final planning version, activate it, and transfer the finalized planning data to SAP Demand Management.

Note [«]

In Flexible Planning, it's only possible to transfer the finalized planning to SAP Demand Management if you have the plant characteristic and the material characteristic included in the different characteristics defined in the planning hierarchy.

You create the planning types with reference to an info structure (in this example, it's S655). The name of the planning table must start with a Z or Y only. The name of the planning type can be up to 11 characters.

To create a planning type, use SAP menu path, LOGISTICS • LOGISTICS CONTROLLING • FLEXIBLE PLANNING • TOOLS • PLANNING TYPE/MACRO • CREATE, or use Transaction MC8A. In the PLANNING TYPE: CREATE screen shown in Figure 9.24, enter the PLANNING TYPE "Z655" ❶, select the LINE LAYOUT and MACROS checkboxes, and press [Enter]. When the pop-up appears, enter the INFO STRUCTURE "S655", and press [Enter] again to open the DEFINE PLANNING TYPE dialog box ❷.

Define the planning horizon by entering the PLANNING START date, the FUTURE PERIODS involved in forecasting, and the HISTORICAL PERIODS data that needs to be pulled in from the info structure.

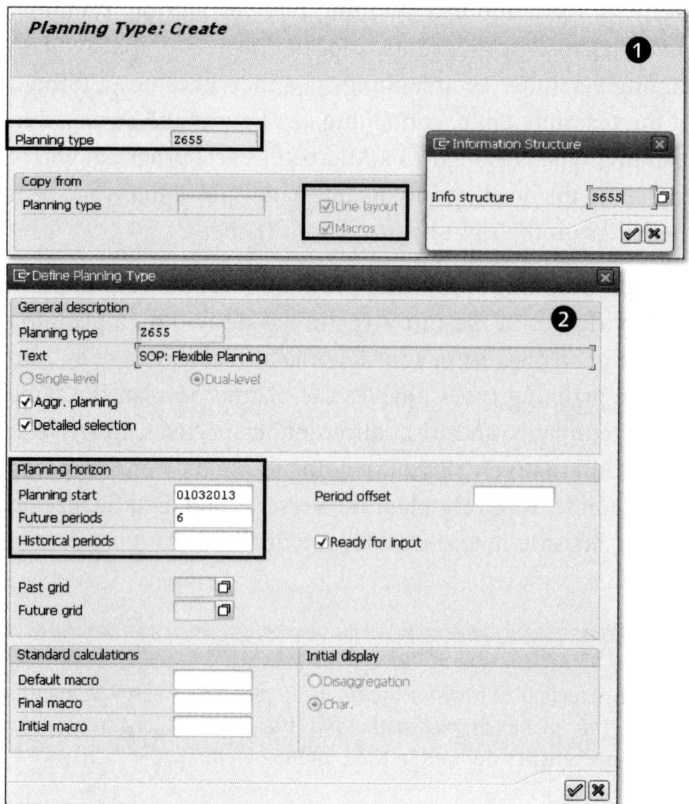

Figure 9.24 Creation of Planning Type

The planning start date is the date or the period from which you want to begin planning. This is the start of the future periods; that is, if you've opted for 6 future periods starting from 01.03.2013 (March 01, 2013), and the period is set as monthly periods in the configuration for info structure, the planning table would show 6 monthly columns starting from 03.2013 to 09.2013. Similarly if the historical data is set for 6 months, the system only pulls in the historical data for the past 6 months, from 09.2012 to 02.2013. This historical data will be used to forecast for the future periods. Press [Enter] and you'll see the screen shown in Figure 9.25.

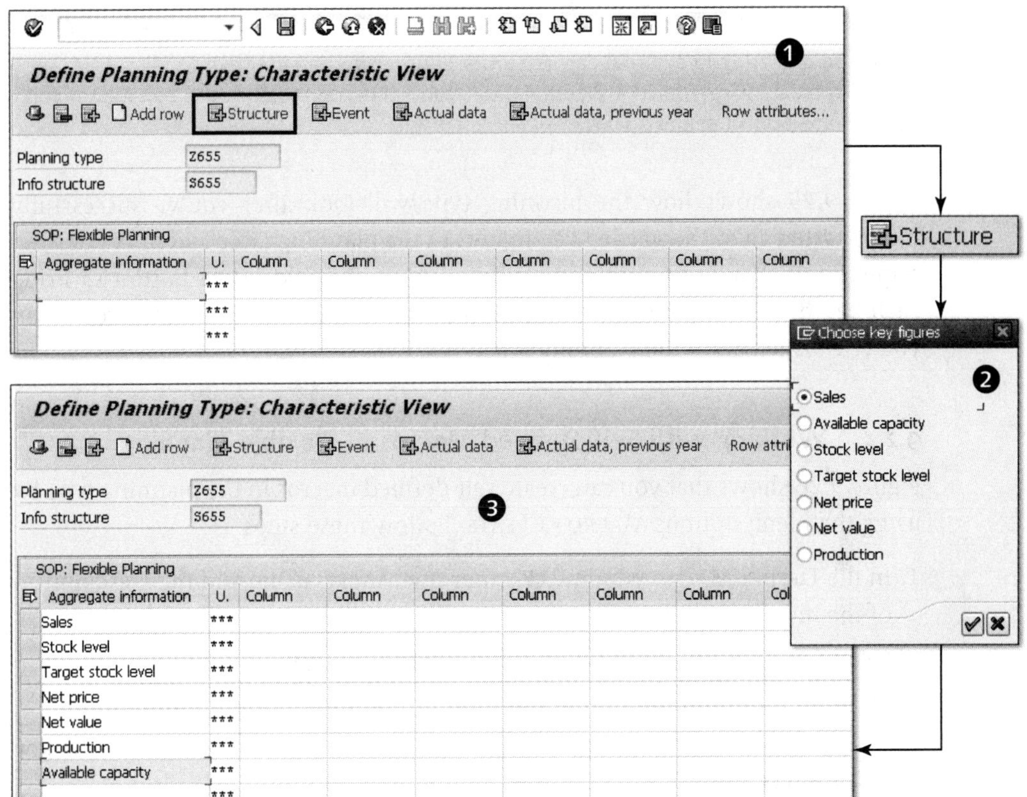

Figure 9.25 Structure Assignment in Planning Type

You can add or remove key figures to the planning type by clicking the STRUCTURE button ❶ to add the key figures, clicking the ADD ROW button to insert a blank row, and clicking the ACTUAL DATA button to add a row that will bring in the actual data in the planning table belonging to version "000" (which contains the actual data) of the info structure. To have the actual data brought in for a key figure, you should first keep your cursor on the key figure for which you want to have the actual data brought in, and then choose the ACTUAL DATA button. For example, if you keep your cursor on the SALES key figure ❷ and then choose ACTUAL DATA, the system adds a row called ACTUAL SALES below the SALES key figure. The actual data row is a noneditable row, although you can use it in formulas (macros) to arrive at the data of other rows. You can use the actual data row to compare the actual data to the planned data.

[+]

Tips & Tricks

You also can bring in external data for a key figure using the user exit EXIT_SAPMMCP6_001. To create a row for the external data, you keep your cursor on a key figure, and then go to EDIT • USER EXIT DATA.

Figure 9.25 shows how the planning type will look after you've successfully incorporated all of the needed key figures in the planning table and in the desired sequence ❸. You'll need to repeatedly click on the STRUCTURE button to bring each key figure into the planning table in the desired order or sequence. When you're finished, save the planning type.

9.2.4 Working with Self-Defined Macros in Flexible Planning

Figure 9.26 shows that you can create self-defined macros in the planning type by using the menu options MACRO • CREATE. Follow these steps:

1. In the DEFINE MACRO window ❶, enter the MACRO name and the DESCRIPTION of the macro, and press [Enter].
2. Select the DISPLAY and CALC. SELECTED AREA checkboxes.
3. Place the cursor on the OPERAND field ❷, and either use drilldown or press [F4] to select a key figure in the ensuing window. For this example, select the key figure SALES, and choose CONTINUE.
4. Use the drilldown for the OP field, and in Figure 9.26 ❸, you can see that all of the normal operands, such as addition, multiplication, and so on, are available to choose from to perform mathematical calculations.
5. Enter the value "0.90" as the second OPERAND, and you'll see PRODUCTION key figure in the RESULT field ❹.

   In other words, whenever you manually enter values or make available the data in the key figure SALES of the planning table and execute the ADJ_PROD. macro, the key figure SALES will be multiplied by 0.9, and the results will be shown in the PRODUCTION key figure.
6. Save the macro.

For this example, define a second macro, TOTAL, which multiplies the key figure SALES with the NET PRICE to result in a NET VALUE key figure as shown in Figure 9.27.

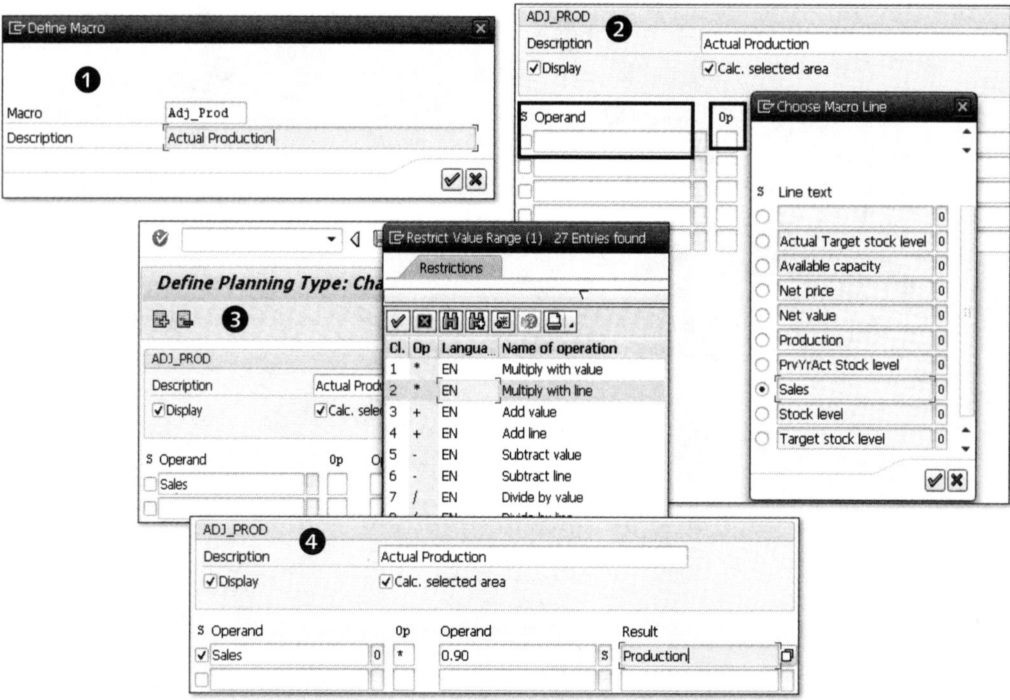

Figure 9.26 First Self-Defined Macros in Flexible Planning

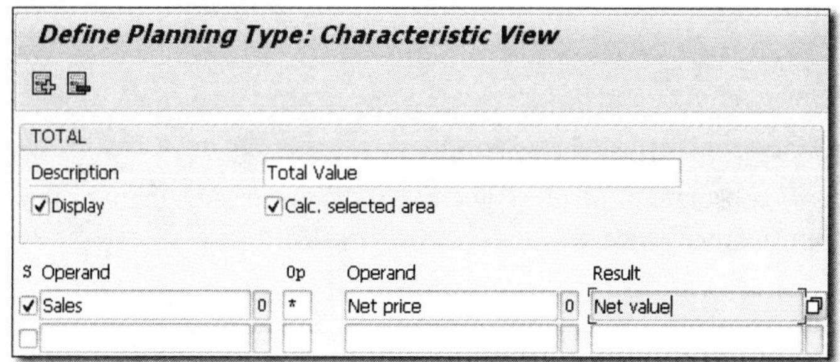

Figure 9.27 Second Self-Defined Macro

After you manually enter the SALES quantity in Figure 9.28 ❶, you can simulate the self-defined macros by choosing MACRO • EXECUTE, which brings up the option to select a macro to execute ❷. Execute both macros one by one, and examine the results ❸.

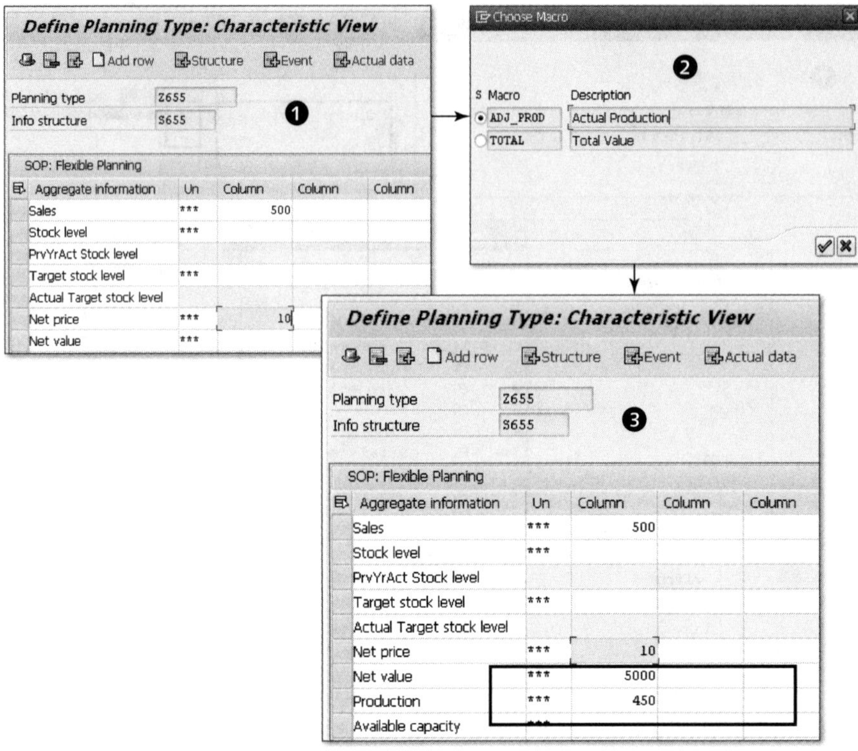

Figure 9.28 Testing Self-Defined Macros

[+]

Tips & Tricks

You can even record the macros by choosing MACRO • RECORD.

If needed, you can delete the obsolete or incorrectly created macros or edit them from the same menu path within the planning type.

9.2.5 Row Attributes in a Planning Type

For any of the key figures, you can double-click on the key figure or select it and choose the ROW ATTRIBUTE icon to set the properties or the attributes of the key figure as in Figure 9.29 ❶. You can mark the key field as a PLANNING LINE (input required), an OUTPUT LINE (display only), a TEXT LINE (information only key figure), or an AUXILIARY LINE, which can be used as a variable in macros. You can also determine the aggregation type of the key figure. In Figure 9.29, we changed the standard description of key figure PRODUCTION to ADJUSTED PRODUCTION ❷.

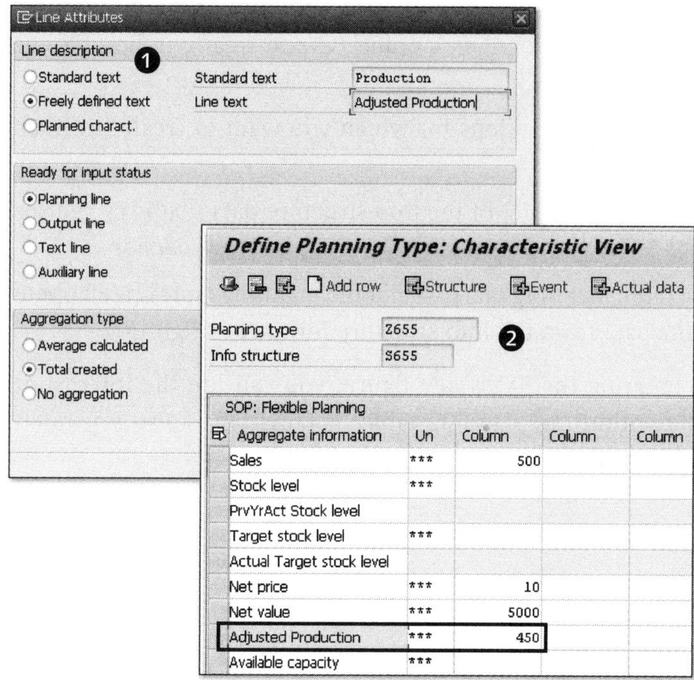

Figure 9.29 Line Attributes in Planning Type

9.2.6 Planning in the Planning Table

Apart from entering the sales, calculating the production quantities, or running macros, you can also perform the following functions (all covered in this chapter, with the exception of transfer to CO-PA):

- Forecasting historical data in the planning table (see the "Execute Forecast" section, next, as well as Section 9.4)
- Transfer to SAP Demand Management (see Section 9.4)
- Capacity planning (rough-cut planning) (see Section 9.5)
- Events (see Section 9.6)
- Transfer to CO-PA

To create a new plan in Flexible Planning for the self-defined info structure, use the menu path, Logistics • Logistics Controlling • Flexible Planning • Planning • Create, or use Transaction MC93. Follow these steps:

1. Enter the planning type (not shown), which for this example is "S655". You can also use, for example, "Z655" for the planning type to clearly indicate a link with info structure Z655.
2. Enter the characteristics combinations for which you want to create the planning table (see Figure 9.30 ❶).
3. Enter the VERSION number, "001", of the info structure data that is to be used in the planning table. The planning version 001 is an inactive version.
4. Press Enter to open the planning table. When the planning table screen opens ❷, you can enter the data from the info structure for the key figures.

Instead of manually entering the SALES key figure, you can use the forecasting functionality available within the planning table. Forecasting is also separately covered in Section 9.4.

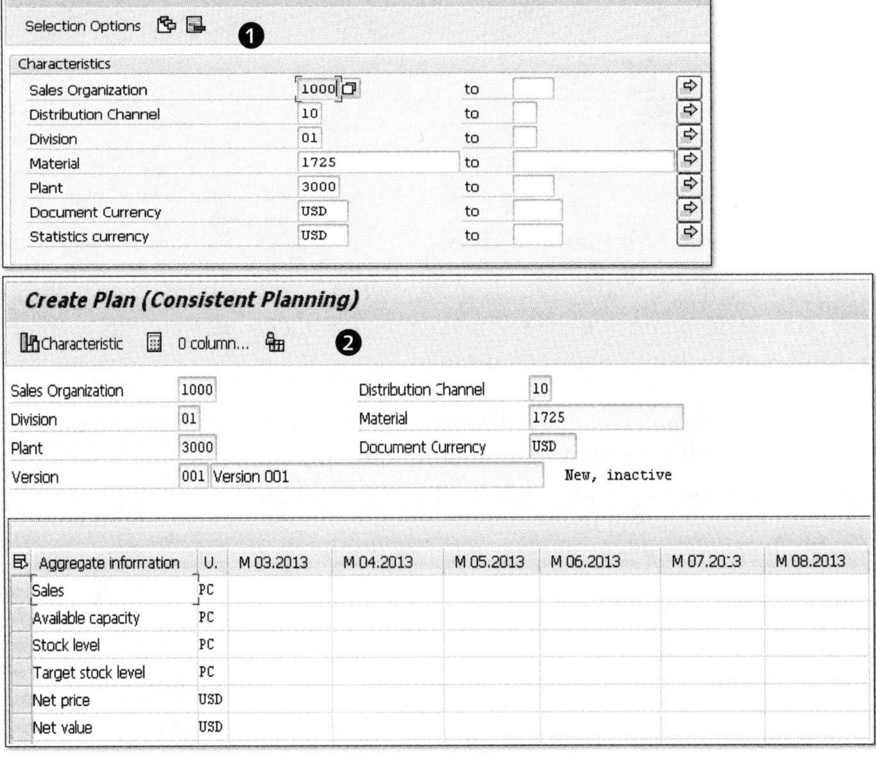

Figure 9.30 Creating a Plan in Flexible Planning

In the following sections, we'll cover how to execute a forecast, check and validate the self-defined macros, and transfer the results of forecast to demand management.

Execute Forecast

Select the SALES key figure row, and proceed with executing the forecast from the menu in the planning table EDIT • FORECAST.

In Figure 9.31 ❶, you can manually enter the PERIOD INTERVALS, with both FORECAST as well as HISTORICAL DATA. For this example, use historical data from the past six months to execute the forecast for the next six months. Alternatively, you can also give the same details by choosing NO. OF PERIODS and assigning NO. OF FORECAST PERIODS and NO. OF HISTORICAL VALUES.

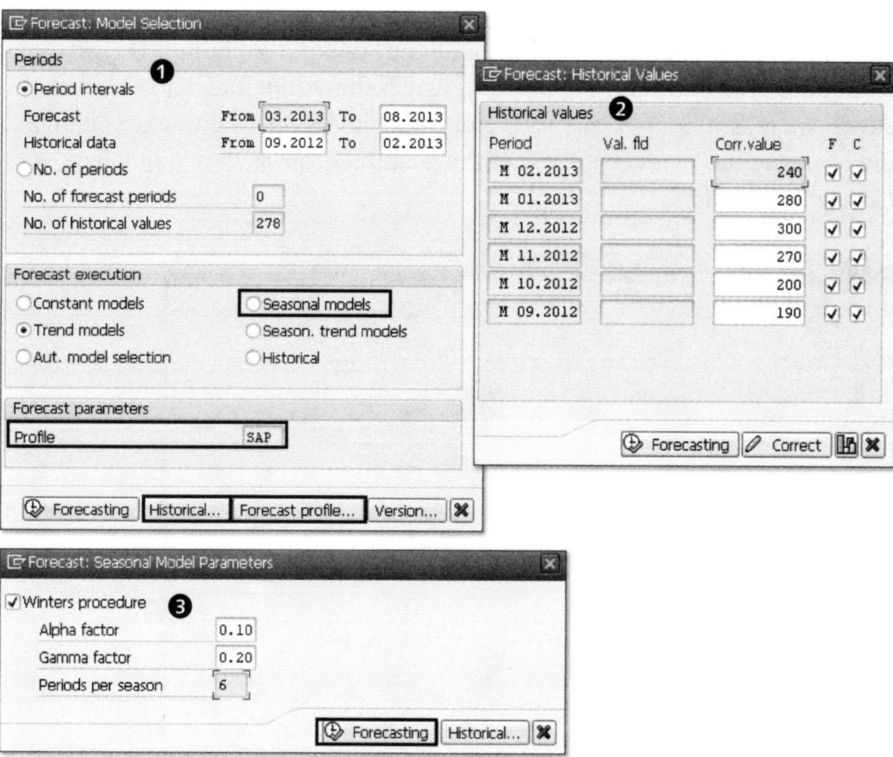

Figure 9.31 Executing a Forecast in Flexible Planning

Next, in the FORECAST EXECUTION area, select the forecast model, which for this example is SEASONAL MODELS.

You can change or update the historical values by choosing the HISTORICAL button. You also can manually enter or correct the historical values, as we've done in this example ❷.

When you choose the FORECASTING button ❷, depending on the forecast model selected, you need to enter further details, such as ALPHA FACTOR and GAMMA FACTOR ❸.

[+] **Tips & Tricks**

To save yourself from entering all of the preceding details, you can use the FORECAST PROFILE button shown in Figure 9.31 ❶. We cover creating forecast profiles in Section 9.4.2.

You can execute the forecast by clicking on the FORECASTING button ❸. This takes you to the screen shown in Figure 9.32, which shows the forecast values for the next six months. Choose CONTINUE, and all of the forecast figures are automatically copied to the SALES key figure of the planning table as shown in Figure 9.33 ❶ in the next section.

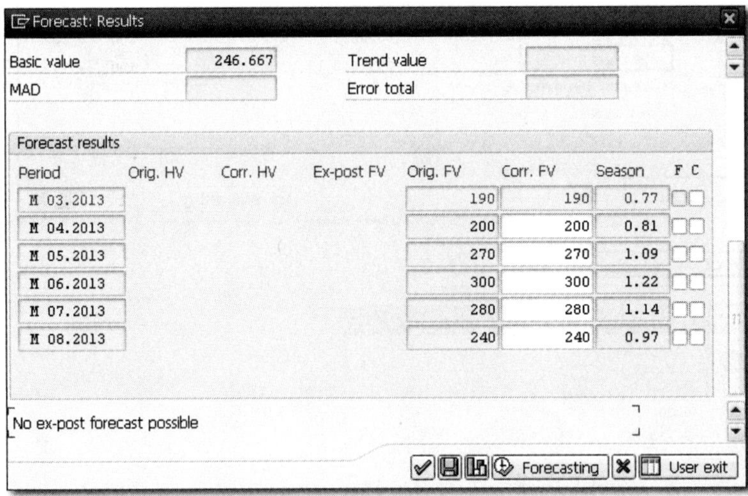

Figure 9.32 Results of the Forecast

[«]

Note

Because we've separately covered forecasting in sufficient detail in this chapter, we've intentionally limited the discussion here to only the basic steps involved in forecasting in the planning table.

Validate Self-Defined Macros

As shown in Figure 9.33, after the SALES figures are in place from the forecasting activity and you've manually entered the NET PRICE, you execute both the macros, and the system automatically fills in the NET VALUE and ADJUSTED PRODUCTION key figures. Figure 9.33 shows the three steps (❶, ❷, and ❸ in order) involved in running or executing macros.

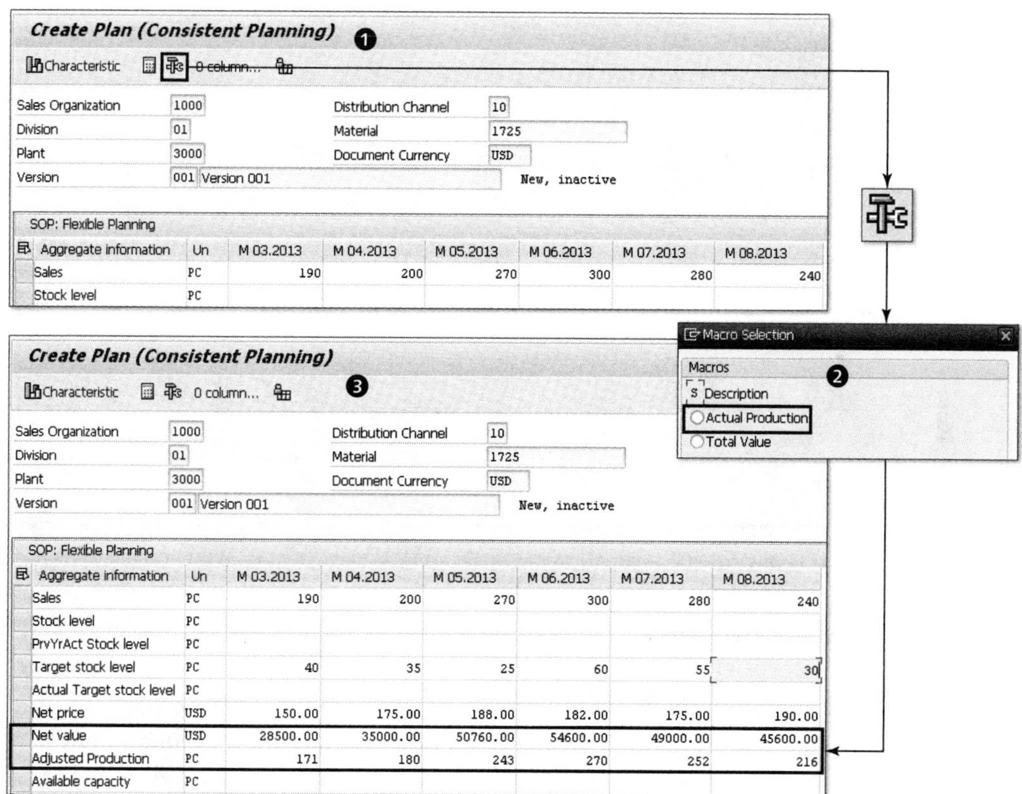

Figure 9.33 Validate Self-Defined Macros in Flexible Planning

With production figures also now in place within the planning table, you transfer the same to SAP Demand Management, in which the system will reflect the quantities of the production key figures as PIRs.

Transfer Key Figures to SAP Demand Management

While remaining within the planning table, select the PRODUCTION row, and choose EXTRAS • TRANSFER TO DEMAND MANAGEMENT (see Figure 9.34). (Alternatively, you can use the menu path, LOGISTICS • LOGISTICS CONTROLLING • FLEXIBLE PLANNING • ENVIRONMENT • TRANSFER MATERIAL TO DEMAND MANAGEMENT, or use Transaction MC90.)

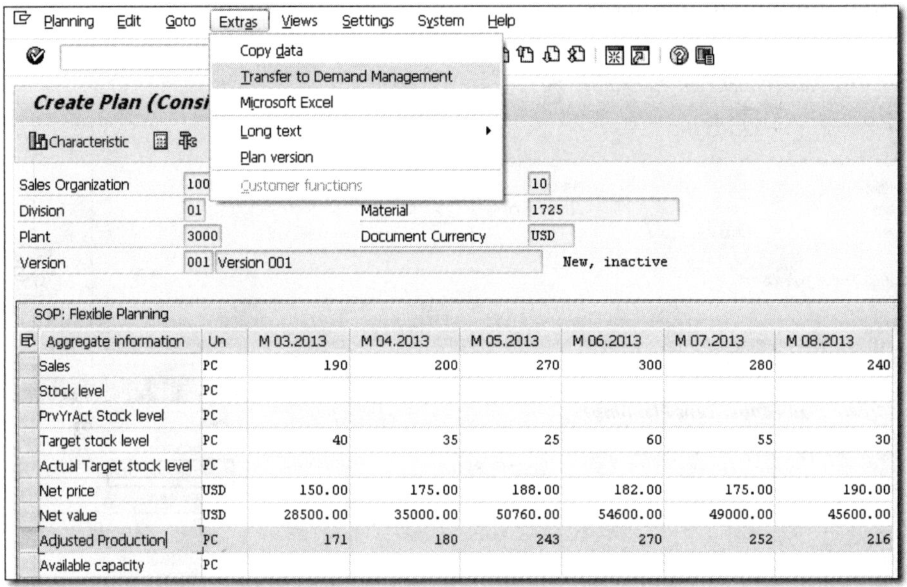

Figure 9.34 Transferring Planning Data to SAP Demand Management from the Planning Table

The screen shown in Figure 9.35 appears, in which you assign the VERSION "001" of INFO STRUCTURE "S655" for transfer to SAP Demand Management. The key figure for this example is PRODU (production), and the transfer horizon begins from 24.03.2013 (March 24, 2013). Further, you can also specify the requirement type and the version of PIR within which the system transfers the values. For this example, assign REQUIREMENT TYPE "VSF" and VERSION "00", and then choose the TRANSFER NOW button.

Transfer Planning Data to Demand Management

Transfer now

Material	1725	Pump PRECISION 100 (Flex Plan)
Plant	3000	New York
Info structure	S655	
Version	001	Version 001

Transfer horizon and key figure

Key figure	PRODU		
From	24.03.2013	To	

☑ Invisible transfer

Independent requirement specifications

Requirements type	VSF
Version	00

☑ Active

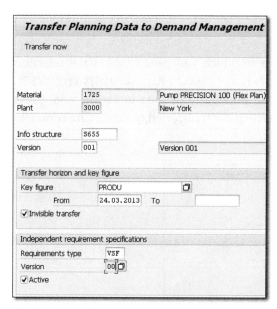

Figure 9.35 Initial Screen for Transfer of Planning Data to SAP Demand Management

The message log shown in Figure 9.36 appears after the transfer of key figures to SAP Demand Management.

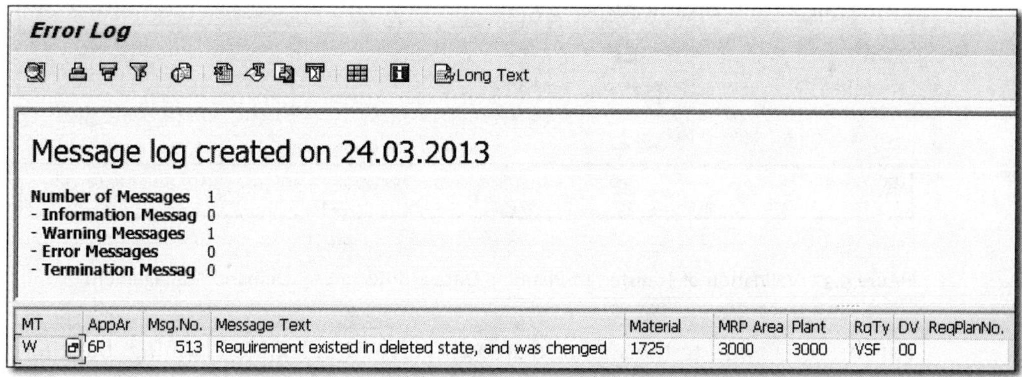

Figure 9.36 Message Log after Transfer of Planning Data to SAP Demand Management

Transfer Results Validation in SAP Demand Management

To check and confirm whether production figures were successfully transferred to SAP Demand Management as PIRs, follow the SAP menu path, LOGISTICS •

PRODUCTION • DEMAND MANAGEMENT • PLANNED INDEPENDENT REQUIREMENTS • DISPLAY, or use Transaction MD63.

Figure 9.37 shows the initial screen in which you need to enter the selection parameters such as MATERIAL "1725", PLANT "3000", REQUIREMENTS TYPE "VSF", and VERSION "00" ❶, which were all previously defined in Figure 9.35. Press [Enter], and in the next screen, notice that the transferred quantities are complete and correct ❷.

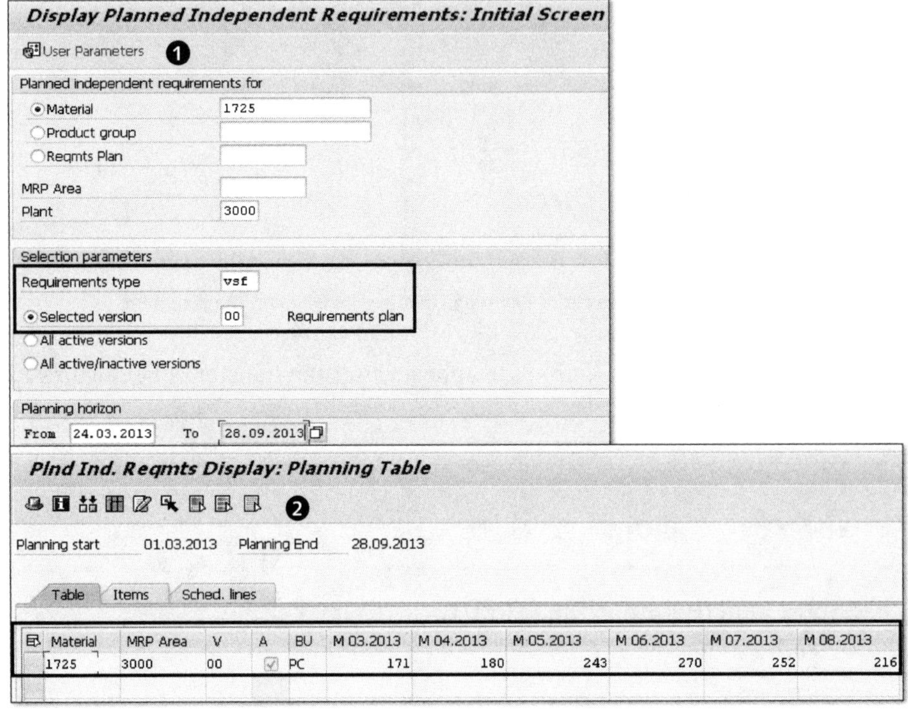

Figure 9.37 Validation of Transferred Planning Data as PIRs in SAP Demand Management

This brings an end to the standard steps involved in Flexible Planning, but there are a few salient features of the planning table, especially those related to planning hierarchy, that we need to discuss next.

9.2.7 Additional Features of Planning Tables

Following are some additional features related to planning tables:

- **Checking header information**
  To check the header information, which shows the planning hierarchy levels, choose SETTINGS • CHOOSE HEADER INFO in the planning table.
- **Firming/unfirming key figure data**
  During consistent planning and to ensure that aggregation and disaggregation won't change the key figures data, you can firm or unfirm the key figure data. Choose SETTINGS • FIRMING • SWITCH ON. You also need to follow the following sequence in the firming process:
  - Activate fixing.
  - Change the value.
  - Deactivate fixing.

  You can deactivate fixing by choosing SETTINGS • FIRMING • UNFIRM PERIOD.
- **Determining decimal places and rounding**
  You can determine the number of decimal places and also allow rounding of key figures by pressing F8 or choosing SETTINGS • PLACES AFTER DECIMAL POINT.
- **Copying key figure data from another info structure**
  You can copy data for a key figure from another info structure by choosing EXTRAS • COPY DATA. The copy of key figure data takes place from the source info structure for a given period to the target info structure.
- **Showing row totals**
  You can show the row totals by choosing SETTINGS • SHOW ROW TOTALS.
- **Transferring all planning table data**
  You can transfer the entire data of the planning table to Excel by choosing EXTRA • MICROSOFT EXCEL.
- **Moving up or down the planning hierarchy**
  You can use the following icons to move up or down the planning hierarchy within the planning table:
  - MEMBER icon
  - OWNER icon
  - NEXT MEMBER icon
  - PREVIOUS MEMBER icon
  - ALL MEMBERS icon

9.2.8 Info Structure Entries in SAP Database Tables

Because info structures are SAP tables with self-defined characteristics and key figures, you can use Transaction SE16N to view the entries of the self-defined info structure.

Figure 9.38 shows the GENERAL TABLE DISPLAY initial screen (ABAP table), in which you can enter the self-defined info structure TABLE "S655" ❶. Press F8 or choose the EXECUTE icon to see the next screen ❷, which contains the six entries for the planning VERSION "001" (first column to the left), the NET PRICE, and the NET VALUE, among other fields.

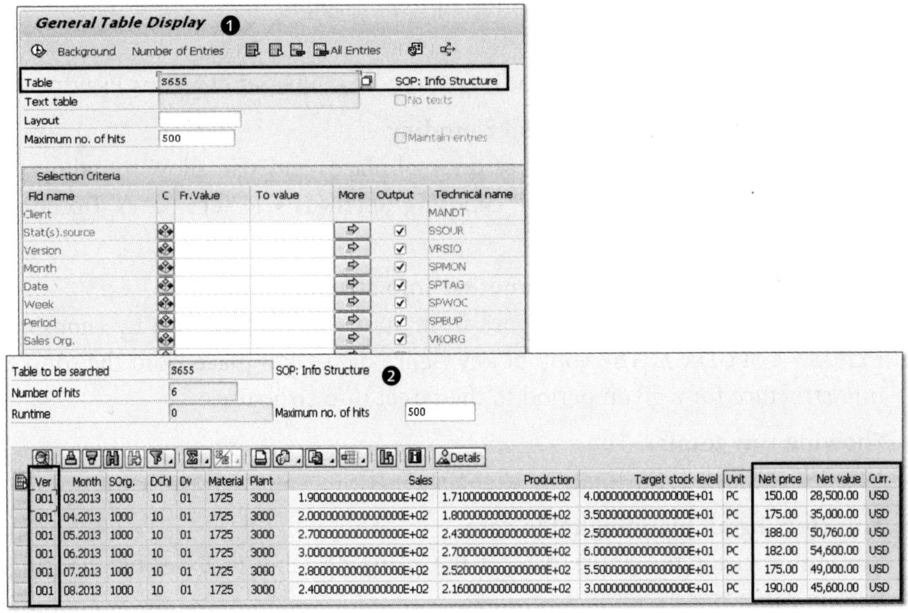

Ver	Month	SOrg.	DChl	Dv	Material	Plant	Sales	Production	Target stock level	Unit	Net price	Net value	Curr.
001	03.2013	1000	10	01	1725	3000	1.9000000000000000E+02	1.7100000000000000E+02	4.0000000000000000E+01	PC	150.00	28,500.00	USD
001	04.2013	1000	10	01	1725	3000	2.0000000000000000E+02	1.8000000000000000E+02	3.5000000000000000E+01	PC	175.00	35,000.00	USD
001	05.2013	1000	10	01	1725	3000	2.7000000000000000E+02	2.4300000000000000E+02	2.5000000000000000E+01	PC	188.00	50,760.00	USD
001	06.2013	1000	10	01	1725	3000	3.0000000000000000E+02	2.7000000000000000E+02	6.0000000000000000E+01	PC	182.00	54,600.00	USD
001	07.2013	1000	10	01	1725	3000	2.8000000000000000E+02	2.5200000000000000E+02	5.5000000000000000E+01	PC	175.00	49,000.00	USD
001	08.2013	1000	10	01	1725	3000	2.4000000000000000E+02	2.1600000000000000E+02	3.0000000000000000E+01	PC	190.00	45,600.00	USD

Figure 9.38 ABAP Tables for Custom Info Structure S655

9.3 Maintaining Version Management

In the info structure, the planning data is retained in versions. This enables the planner to carry out simulations. You can assign any number to a version, and these remain inactive with the exception of two specific versions having defined meanings and functions: version 000 always contains actual data, and version A01 is the active version.

Every version that is created and for which data is stored in the planning table, all of the values for each of the key figures, and characteristics combinations entered either manually or evaluated by macros, events, or forecasts are directly updated in the info structure against the respective version number.

You can copy one version's data to another version as well as delete the obsolete or unwanted versions, as you'll see in the following sections.

9.3.1 Copy a Version

Depending on whether you're using standard S&OP or Flexible Planning, you can use the following menu path or transactions to copy one version to another:

- **Standard S&OP**
  Choose Logistics • Production • SOP • Planning • Version Management • Copy, or use Transaction MC78.
- **Flexible Planning**
  Choose Logistics • Logistics Controlling • Flexible Planning • Planning • Version Management • Copy, or use Transaction MC8V.

To copy the PIR details of the source version to the target version of an info structure, enter the Source version (in this example, it's version "001") to the Target version (in this example, it's "A01"), and choose Copy at the top of the screen (see Figure 9.39). You can also give the date range (Period) that you want to use to copy the PIR details of the source version to the target version.

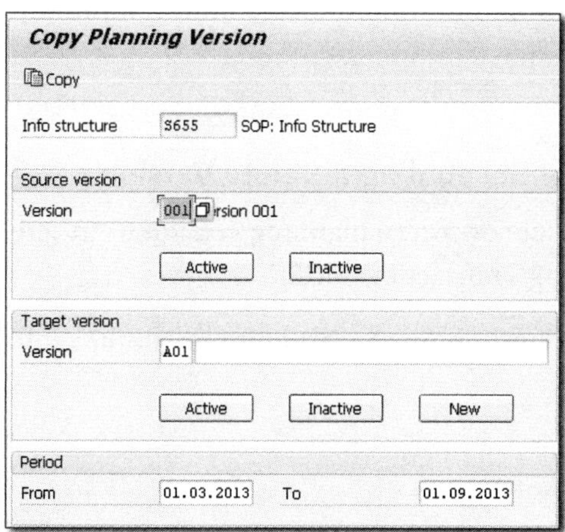

Figure 9.39 Copy Planning Version

9.3.2 Delete a Version

Depending on whether you're using standard S&OP or Flexible Planning, you can use the following menu path, or transactions to delete planning versions:

- **Standard S&OP**
  Choose Standard SOP: Logistics • Production • SOP • Planning • Version Management • Delete, or use Transaction MC78.
- **Flexible Planning**
  Choose Logistics • Logistics Controlling • Flexible Planning • Planning • Version Management • Delete, or use Transaction MC8W.

To delete one or several versions of an info structure, select the relevant checkboxes next to the version numbers, and choose Delete at the top of the screen (see Figure 9.40).

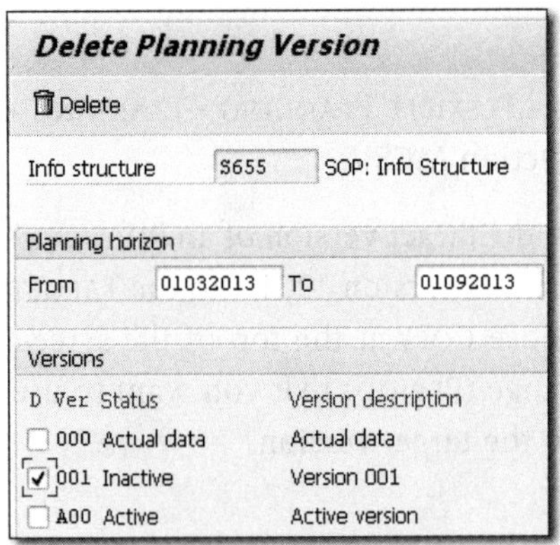

Figure 9.40 Delete Planning Version

9.3.3 Scheduling a Copy Version or Scheduling a Delete Version

You can schedule a background job to delete or copy a planning version or to activate an existing planning version by using Transaction MC8K.

To schedule a copy from one target version of one info structure to another info structure, use Transaction MC8Q.

9.4 Forecasting

A *forecast* is the prediction of future requirements, based on the past consumption trend in a given period of time. These consumptions can be, for example, material withdrawal against the production order, process order, or even cost center. Whenever you create a new material in the system, the system automatically incorporates the CONSUMPTION tab in the ADDITIONAL DATA area of the material master. The consumption for finished goods can be against the sales order. Hence, forecasting offers enormous benefits to plan any material type, whether it's a raw material, a consumable, a packing material, a semifinished good, or even a finished good, based on historical consumption figures. While each withdrawal or consumption updates the information in the system, you can also manually enter the past consumption figures of a material to help drive and facilitate the forecasting.

A detailed analysis of the consumption figures provide a pattern or other valuable information on the consumption pattern, that is, a constant consumption, a moving average, or a seasonal trend model. Therefore, the SAP ERP system offers several standard forecasting models to choose from. If you're not sure which forecast model to select, then the SAP ERP system allows the planner to select AUTOMATIC MODEL SELECTION to facilitate forecasting. However, using this option has its disadvantage.

Warning! [!]

SAP doesn't recommend using AUTOMATIC MODEL SELECTION for forecasting because it eventually ends up incorporating one of the several forecast models (such as constant, moving average, or seasonal) available in the material master. The SAP-recommended forecasting model may not be in line with your business scenario.

Tips & Tricks [+]

When using the forecasting tool, take a couple of materials as test cases and run the entire forecasting cycle. This forecasting cycle includes maintaining historical values, selecting the right forecasting model, and maintaining additional forecasting parameters. Finally, execute the forecast and see if the forecast results closely match with what you expect the system to do. If it doesn't, you need to consider evaluating other available forecasting models or refining the forecasting parameters. After you've gained the necessary experience with a limited number of materials, you can replicate the same steps for other similar materials. Be sure to periodically review and update the forecasting parameters to ensure reliable material planning results.

This section covers the forecasting options available in S&OP (which are standard and Flexible Planning) as well as the options available in the material master in case you want to run forecasting on a consumable material or a packing material, which isn't part of S&OP.

9.4.1 Forecasting View in Material Master

You can enter the forecast parameters in the forecasting view of the material master with Transaction MM02. Figure 9.41 shows the FORECASTING view of material 1725 for plant 3000. Details of all of the important fields on the screen are given in the subsequent sections of this chapter.

When you run the forecast for a material–plant combination (Transaction MP30) or for the entire plant (Transaction MP38), the system uses information from the material master and offers no option to make changes in the forecast parameters.

[»] **Note**

You can navigate to the FORECAST menu in SAP ERP system via the menu path, LOGISTICS • PRODUCTION • PRODUCTION PLANNING • MATERIALS FORECAST.

You can execute forecasting in the following:

- S&OP for materials or for product groups
- Flexible Planning

When executing a forecast in either of these areas, the system provides you with the option to select the forecast profile as well as make changes to the default data defined in the forecast profile. You can also carry out forecasting for consumption-based materials, as essentially the concepts and the fundamentals remain the same.

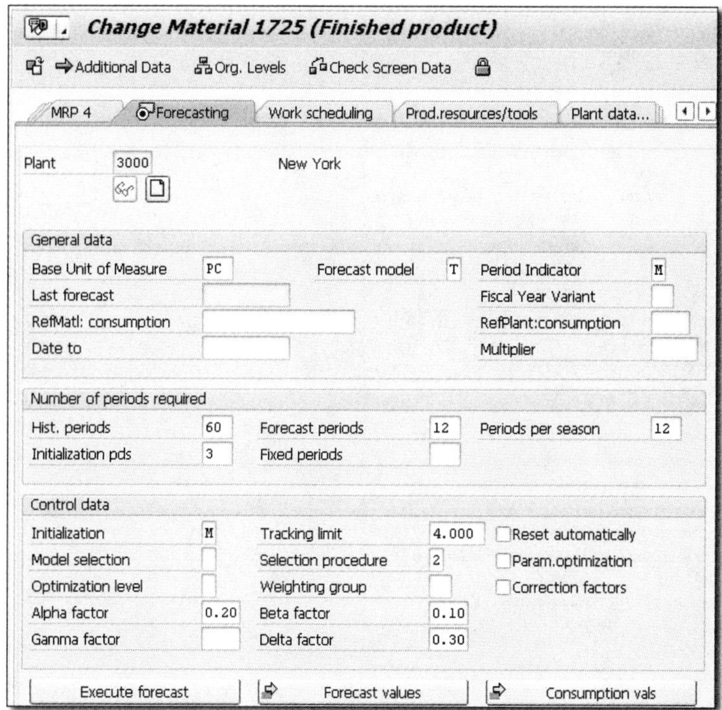

Figure 9.41 Forecasting View in the Material Master

9.4.2 Forecast Profile

A forecast profile consists of the user-defined values (parameters), which then become available during the execution of forecasting. You can, however, still make changes to these values (parameters) in the forecast profile before executing the forecast.

To create a forecast profile, use configuration (Transaction SPRO) menu path, Production • Sales & Operations Planning (SOP) • Functions • Forecasting • Maintain Forecasting Profiles, or use Transaction MC96.

Figure 9.42 ❶ and ❷ shows basically the same initial forecasting screens that are now available to create a forecast profile. For this example, create a new profile "ZSAP".

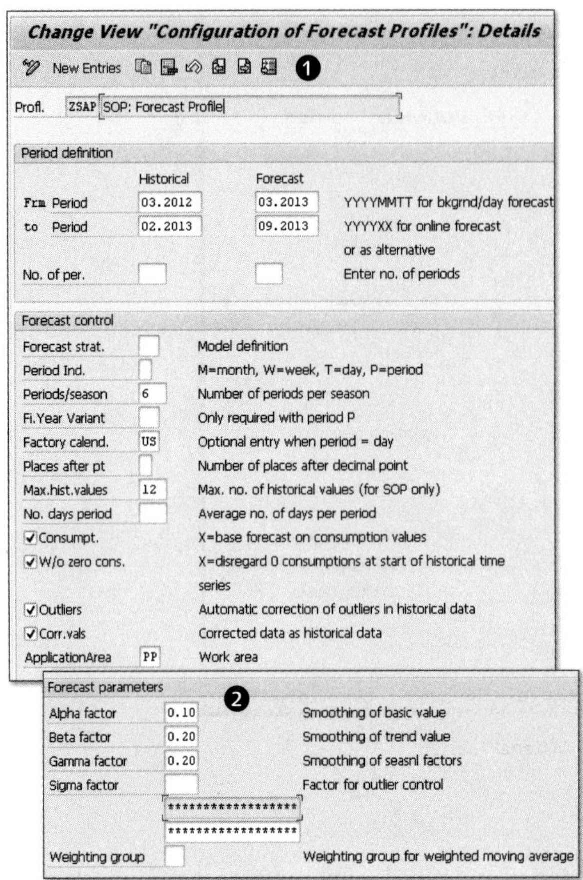

Figure 9.42 Configuration of the Forecast Profile

The details of the important fields in the forecast profile are given in the following list.

- Frm Period/to Period
  The period or intervals for the historical data and for forecast data, which you'll use during forecasting. For this example, use 12 months for the historical data (from 03/12 to 02/13) and use 7 months for the forecast interval (from 03/13 to 09/13).
- Period Ind.
  The period or interval of the required future periods for which the forecasting will trigger.

- FORECAST STRAT.
  The forecasting strategies/models that the system will use to carry out the forecast. Forecasting strategies are covered in Section 9.4.3.
- FACTORY CALEND.
  The factory calendars specify the holidays or the fiscal year variant. For this example, assign factory calendar US.
- PERIODS/SEASON
  The number of periods in a season if there is a seasonal trend, which for this example is "6".
- MAX.HIST.VALUES
  Maximum number of historical periods; for this example, enter "12". The system will consider the past 12 months' historical values to suggest forecast values.
- CONSUMPT.
  This indicator is used to base the forecast of historical consumption as in standard S&OP's level-by-level planning forecast based on the material consumption history.
- FORECAST PARAMETERS
  Corrects historical values for the following (see Figure 9.42 ❷):
  - ALPHA FACTOR: Used to smooth the basic value.
  - BETA FACTOR: Used to smooth the trend value.
  - GAMMA FACTOR: Used to smooth the seasonal index.
  - SIGMA FACTOR: Used to correct the historical figures that lie outside the predefined upper and lower tolerances. Lower the sigma factor, greater the control on forecast figures. SAP recommends using a value between 0.6 to 2.0.

9.4.3 Forecast Strategy

Table 9.4 lists the relationship of various forecast models and the corresponding forecast strategies. For each forecast strategy (or model) selection, you have to give details of the parameters, such as ALPHA FACTOR, BETA FACTOR, or GAMMA FACTOR, which is highlighted in the Hints column of the table. For example, if the demand of a product is generally constant throughout the year, then you can select the constant model. Similarly, if you want to maintain an average quantity of material in the warehouse, then you can use the moving average model. In moving average, it adds up the total quantity of the material from historical figures and then divides this total quantity with the period under consideration to arrive at an average number. So, you need to evaluate your consumption figures for a given historical period to which the forecasting model fits best to meet your business requirement.

Time Series Development	Forecast Model	Forecast Strategy	Hints
Constant progression	Constant model	10	The end results are constant figures.
Constant progression	Constant model with first-order exponential smoothing	11	You give an alpha factor to smooth the basic values.
Constant progression	Constant model with automatic alpha smoothing factor adoption	12	First-order exponential smoothing is used.
Constant progression	Moving average model	13	It averages out the values, based on the number of historical values.
Constant progression	Weighted moving average model	14	You need to maintain a weighting group.
Trend-like pattern	Linear regression	20	It evaluates the trend in values to come up with plausible trend values.
Trend-like pattern	First-order exponential smoothing	21	You need to maintain alpha and beta factors to smooth out the basic and trend values.
Trend-like pattern	Trend model of second-order exponential smoothing	22	You need to maintain alpha and beta factors to smooth out the basic and trend values.
Trend-like pattern	Automatic alpha smoothing factor adoption	23	You need to maintain second-order exponential smoothing with or without model parameter optimization.
Season-like pattern	Seasonal model (winters procedure)	31	You need to maintain alpha factors, gamma factors, and periods per season for basic and seasonal smoothing.

Table 9.4 Forecast Models and Forecast Strategies

Time Series Development	Forecast Model	Forecast Strategy	Hints
Season-like pattern	Seasonal model of first-order exponential smoothing	40	You need to maintain alpha, beta, and gamma factors and periods per season.
Automatic model selection		50	The system takes its own calculation and factors into account and determines/suggests the appropriate model.
Automatic model selection		52	This is more time- and resource-consuming, but it provides more reflective results.
Historical data way		60	The system will use the past or historical values as is, without automatically making any corrections or adjustments.

Table 9.4 Forecast Models and Forecast Strategies (Cont.)

Figure 9.43 shows the dropdown list of forecast strategies available for you to choose from and assign the same in the forecast profile.

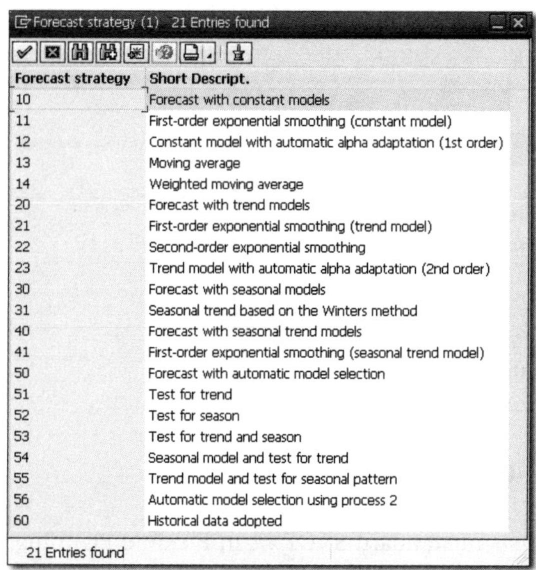

Figure 9.43 Forecast Strategy

9.4.4 Using the Forecast Profile

When you choose the FORECAST PROFILE icon (refer to Figure 9.31 ❶) while executing the forecast, the FORECAST: FORECAST PROFILE pop-up appears as shown in Figure 9.44, in which you can select the relevant forecast profile. The system considers all of the forecast parameters during forecasting.

Figure 9.44 Forecast Profile Selection

If you still want to make any last-minute changes to the values defined in the forecast profile, then choose the CHANGE icon in Figure 9.44, and the screen shown in Figure 9.45 appears in which you can make the desired changes and execute the forecast. This screen enables the planner to make any last-minute changes to the forecast profile and proceed with executing the forecast.

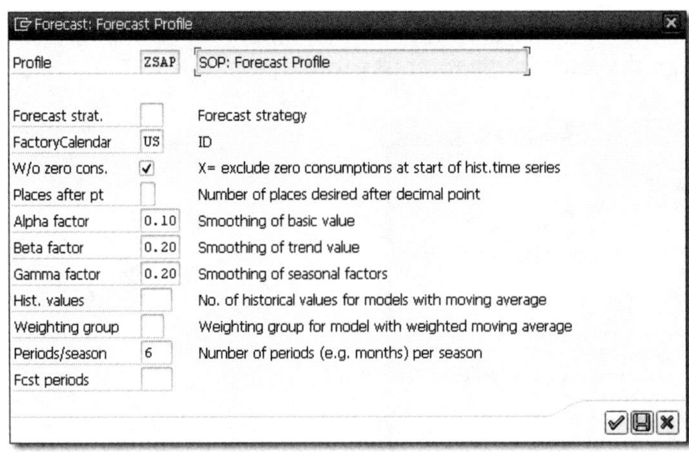

Figure 9.45 Options to Make Changes in the Forecast Profile

9.5 Rough-Cut Planning Profile

For you to carry out capacity analysis in standard S&OP or in Flexible Planning, you need master data for the capacity requirements for production of a base quantity of the material or product group. This master data is known as a *task list*, and it can be a routing, rate routing, master recipe, or rough-cut planning profile. Normally, a rough-cut planning profile is created for the material–plant combination because it provides a rough estimation of the capacity requirements.

[!]

Warning!

An important point to consider is that not every info structure created or used in S&OP has plant as a characteristic. This makes the capacity planning very difficult. It's for this reason that a planning plant should be entered in the SET PARAMETERS FOR INFO STRUCTURE AND KEY FIGURES configuration screen (Transaction MC7F). Additionally, without reference to the plant and material, transfer of key figures to PIRs isn't possible.

Because the capacity evaluation is performed at the plant level, it's advisable to include it in the info structure and percentage splits given to them for disaggregation and aggregation purposes.

Also in the SET PARAMETERS FOR INFO STRUCTURE AND KEY FIGURE screen (Transaction MC7F), you should enter a key figure in the CAPACITY FIGURE field, whose planned figures in the planning table will be considered in capacity planning. For this example, we used the PRODU value in the capacity figure (refer to Figure 9.21).

Based on the details given in the rough-cut planning profile, the resource leveling calculates the resource load for the operations plan and thus enables you to quickly estimate feasibility in terms of capacity.

Next, we'll cover the rough-cut planning profile and how the system reflects pegged requirements.

9.5.1 Create a Profile

To create a rough-cut planning profile, use SAP menu path, LOGISTICS • LOGISTICS CONTROLLING • FLEXIBLE PLANNING • TOOLS • ROUGH CUT PLANNING PROFILE • CREATE, or use Transaction MC35.

On the initial screen of a rough-cut planning profile, enter info structure "S655", and click on the Execute icon, which opens the pop-up shown in Figure 9.46. Enter all of the characteristics details, and press Enter. The GENERAL DATA screen shown in Figure 9.47 appears.

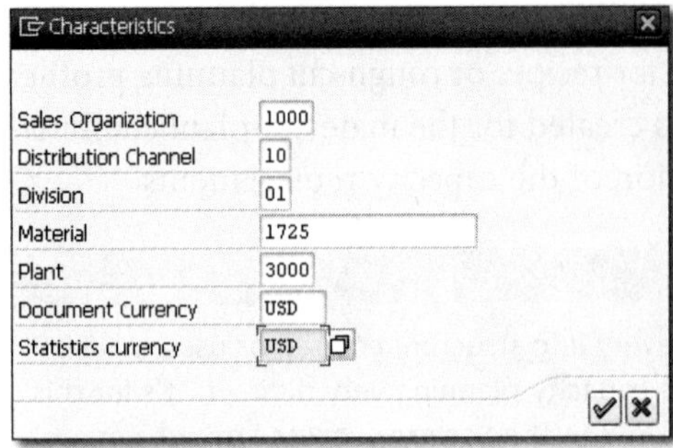

Figure 9.46 Initial Screen for Characteristics Assignment

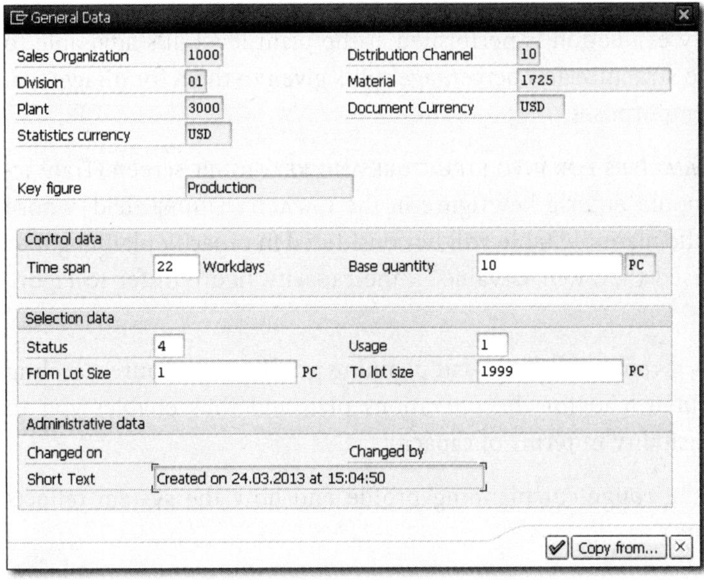

Figure 9.47 General Data in a Rough-Cut Planning Profile

In this screen, enter "22" in the TIME SPAN field, and enter "10" PC in the BASE QUANTITY field. Set the STATUS to "4" (for RELEASED), set the USAGE to "1" (for production), and press [Enter].

[+]

Tips & Tricks

You can change the parameters in the GENERAL DATA screen when needed. For example, you can change the number of working days to account for any national or unannounced holiday in a month.

In the screen that appears as shown in Figure 9.48, double-click the first line of the RESOURCE TABLE column ❶. In the RESOURCE TYPE dialog box, choose WORK CENTER. As a result, the system displays the RESOURCE pop-up in which you can enter the WORK CENTER as "1310", PLANT as "3000", and UNIT as "H" (for hours). Press Enter to go to the next screen ❷.

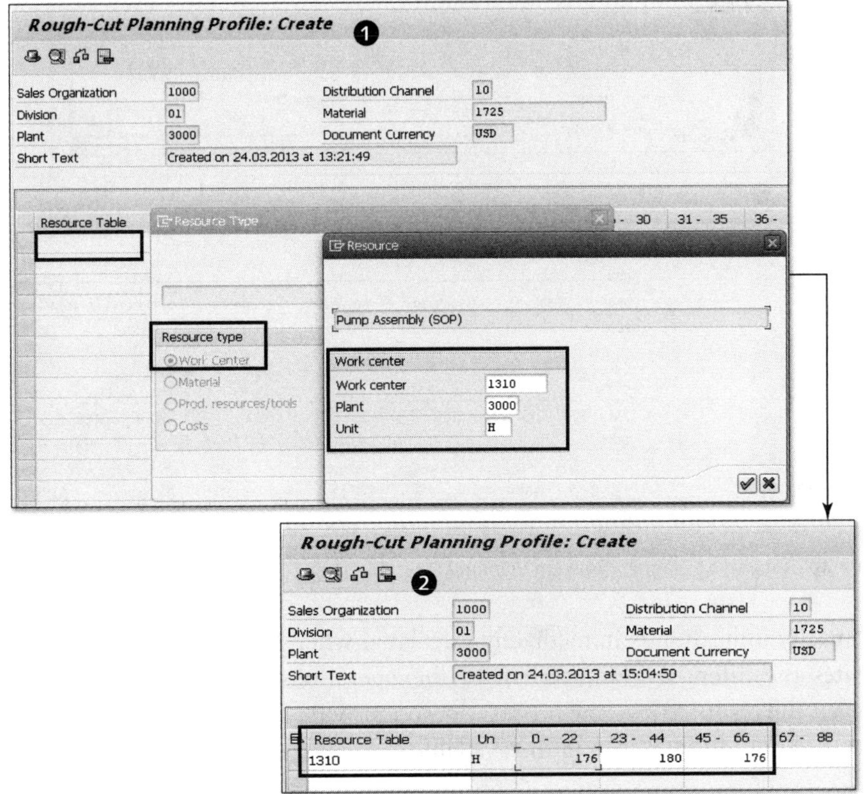

Figure 9.48 Creation of a Rough-Cut Planning Profile

Note

Although work center is used as a resource in this example, you can also use other resources such as PRT or materials.

The planning table allows you to display the available capacity and the capacity requirement via menu path, VIEWS • CAPACITY SITUATION • ROUGH-CUT PLANNING • SHOW (see Figure 9.49 ❶).

The PLANNING TABLE area in this screen ❷ shows that based on the monthly production figures, the system shows the capacity requirements against the available capacity at the work center 1310 for plant 3000.

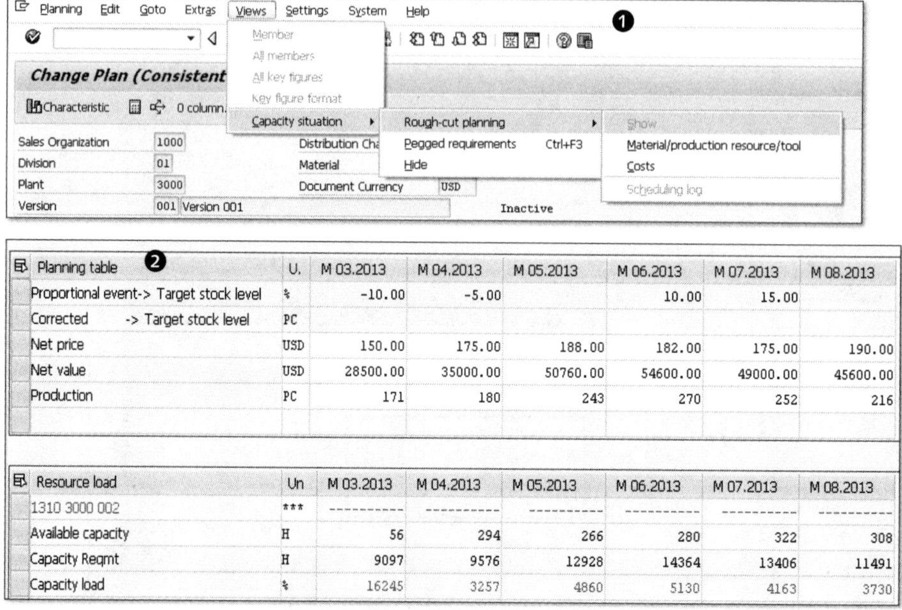

Planning table	U.	M 03.2013	M 04.2013	M 05.2013	M 06.2013	M 07.2013	M 08.2013
Proportional event-> Target stock level	%	-10.00	-5.00		10.00	15.00	
Corrected -> Target stock level	PC						
Net price	USD	150.00	175.00	188.00	182.00	175.00	190.00
Net value	USD	28500.00	35000.00	50760.00	54600.00	49000.00	45600.00
Production	PC	171	180	243	270	252	216

Resource load	Un	M 03.2013	M 04.2013	M 05.2013	M 06.2013	M 07.2013	M 08.2013
1310 3000 002	***	----------	----------	----------	----------	----------	----------
Available capacity	H	56	294	266	280	322	308
Capacity Reqmt	H	9097	9576	12928	14364	13406	11491
Capacity load	%	16245	3257	4860	5130	4163	3730

Figure 9.49 Rough-Cut Planning View in Planning

The capacity utilization is immediately calculated when the plan changes, which facilitates the interactive adjustment of the operations plan to the capacities. However, the resource-leveling process doesn't provide any standard function for adapting the production plan to the available capacity.

If a capacity is overloaded, a corresponding message is displayed in the footer. For this example, this is the case for March 2013 (see Figure 9.50).

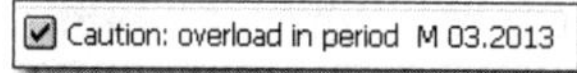

Figure 9.50 System Message in Case of Capacity Overload

9.5.2 Pegged Requirements

Within the planning table, pegged requirements offer a quick way to gain a better understanding of a workload on a resource, whether it's a work center, a material, or a PRT, to evaluate pegging. The term "pegged" is synonymous with "association" or "attachment," which means one planning element is attached to another. For this example, the production quantity of 180 PC of material 1725 at work center 1310 will have a capacity requirement of 9576.0 hours for the month of April 2013. To view the pegged requirements within the planning table, choose VIEWS • PEGGED REQUIREMENTS to open the SOURCE OF REQUIREMENTS screen (see Figure 9.51).

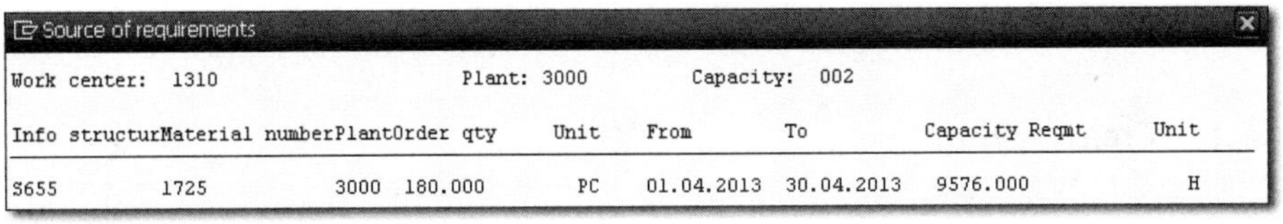

Source of requirements

Work center: 1310 Plant: 3000 Capacity: 002

Info structur	Material number	Plant	Order qty	Unit	From	To	Capacity Reqmt	Unit
S655	1725	3000	180.000	PC	01.04.2013	30.04.2013	9576.000	H

Figure 9.51 Pegged Requirements

9.6 Events

Major events, national holidays, games, planned promotions, and festivals can affect sales quantities (by increasing or decreasing), so understanding events in advance will help your planning efforts. You can increase or decrease your sales quantities using a functionality called *events*.

The events functionality is applied to the sales key figures in all possible cases or to any other key figures that give you the sales quantities or sales values. The system provides the option to assign events across all of the characteristics levels or for a specific characteristics level for which the event should be applied.

For example, let's say initially you used the forecasting tool and seasonal model to come up with the planned sales figures. Later, your company decides to participate in an important and relevant trade show that should have a positive impact on your company's sales. To attend to this anticipated surge in product sales (or demand), you can incorporate this event (the trade show) into your planned sales figures to enable the system to account for the increase. The events are assigned to the planning types and thus come into play in the planning table.

[«]

Note

Creating an event isn't a mandatory activity in the planning process, so you can opt out of the event option in the planning situation, if needed.

The SAP ERP system provides two ways of modeling an event: *cumulative modeling* and *percentage modeling*. For cumulative events, the sales quantity affected by the event is maintained as a difference quantity in absolute figures. The overall sales quantity is represented as the total of normal sales and includes the event. For percentage events, the sales quantity affected by the event refers to the quantity planned in the SALES key figure.

In the following sections, we'll show you how to create events and assign them to planning. We'll also show how an event's assignment impacts materials planning.

9.6.1 Create Events

You can create events using the menu path, LOGISTICS • LOGISTICS CONTROLLING • FLEXIBLE PLANNING • TOOLS • EVENTS • CREATE, or by using Transaction MC64.

The initial screen to create an event is shown in Figure 9.52. For this example, enter the EVENT NUMBER as "S655: SOP EVENT" ❶, and then press Enter.

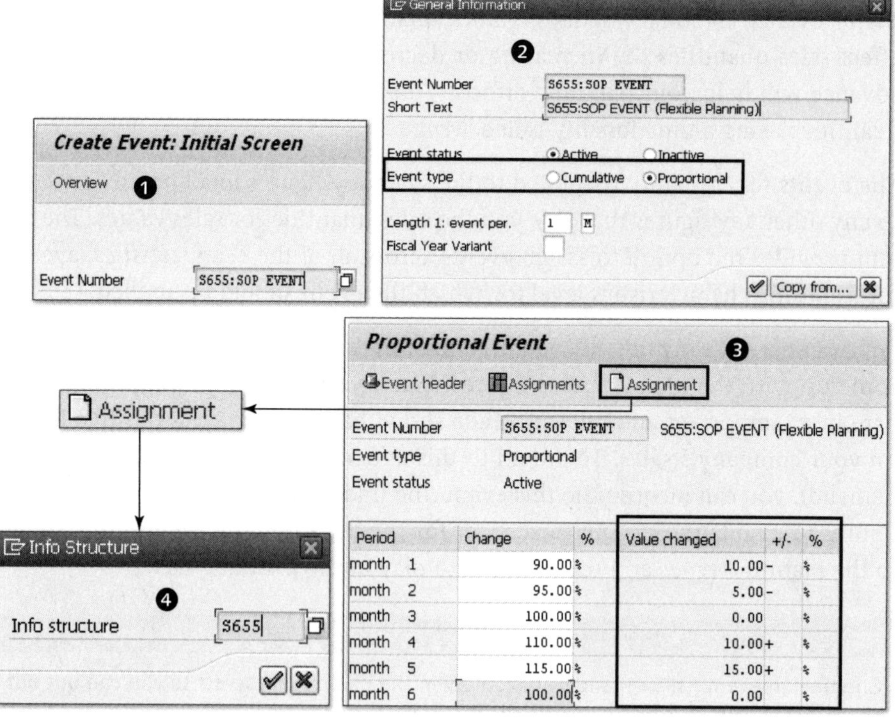

Figure 9.52 Creation of an Event

In the General Information pop-up ❷, select the Proportional radio button, and press Enter again. In the Proportional Event screen ❸ that appears, enter the monthly change in percentage, and choose the Assignment icon, which opens the Info Structure pop-up. In this pop-up, assign the Info structure as "S655" ❹.

Now that you've created the event, you move straight into assigning that event.

9.6.2 Assignment of Events

The screen shown in Figure 9.53 appears after you've finished creating the event in the previous section. In this screen, the characteristics of the info structure automatically appear and are filled in, such as Sales Organization and Division.

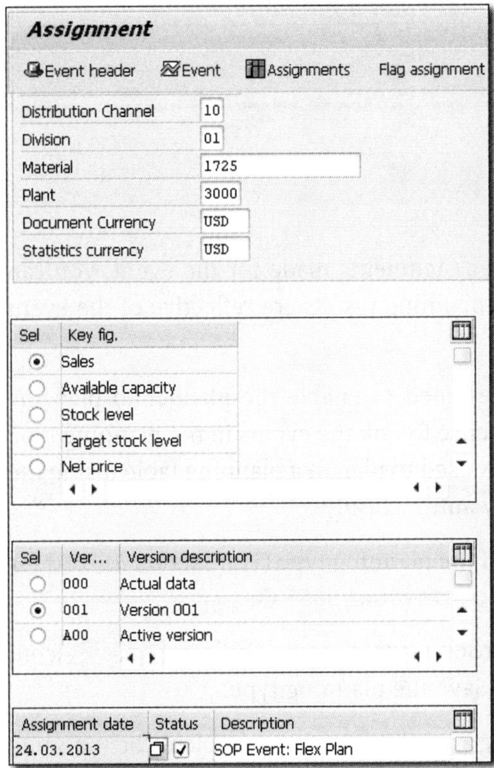

Figure 9.53 Assignment of the Event

For this example, select the SALES key figure radio button, and then select the 001 radio button for the inactive version. Finally, the system proposes the current data for the event, which you can change. You also need to choose the STATUS checkbox to indicate that the event status is active.

Finally, choose the ASSIGNMENTS icon, which takes you to Figure 9.54. This screen provides the list of event assignments, which you can change or even delete.

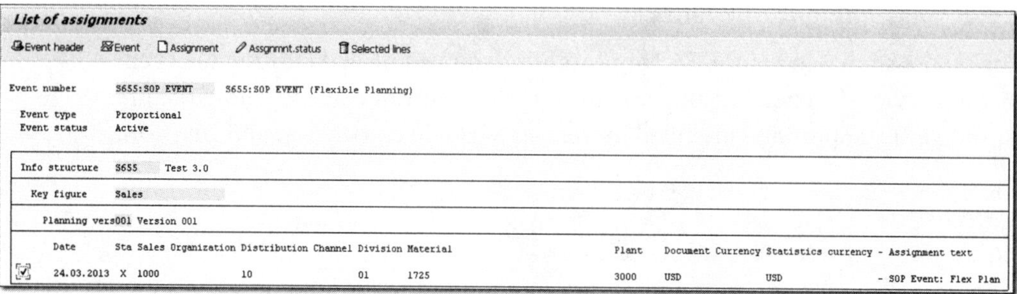

Figure 9.54 List of Event Assignments

9.6.3 Events in Planning

With all of the necessary settings and assignments made for the event, you can now test the same to see whether the planning results are reflective of the event in the planning type.

The planning type table is a frame designed to enable the planner to plan the planning figures. Therefore, it makes sense to link the events in the planning type itself, so that when the planning type is used to plan in a planning table (using the key figures), the events come into play automatically.

Figure 9.55 shows the change screen of the planning type (Transaction MC8A) for the same PLANNING TYPE Z655 and INFO STRUCTURE S655 ❶.

For this example, select the TARGET STOCK LEVEL and click on the Event icon, which shows you the event details ❷. Save the planning type.

The next step is to change the existing plan (or create a new plan) for the same planning type Z655 and info structure S655 and see how the entered stock figures reflect the impact of an event.

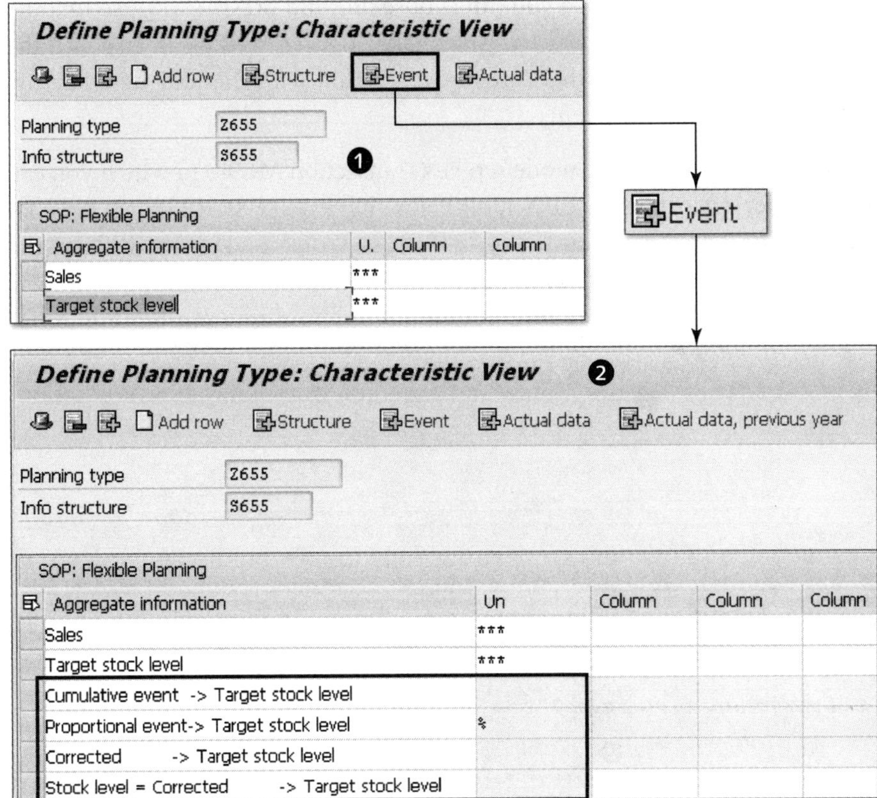

Figure 9.55 Events in the Planning Type

In this step, you need to use a key figure from the existing set of key figures to represent "events" in the planning type. Normally, such a figure is an unused or unwanted key figure from the list. If you already know that you're going to use events in planning, we suggest having one or two unwanted key figures that aren't related to any planning activities to be exclusively used for purposes such as an event. If you've already used all of the key figures in the planning type, you may need to delete the key figure, which you're going to use for adding the event, so that when you click on the Event icon, the key figure will be shown as available for addition.

In the planning type, keep your cursor on a key figure (location where you want the event-related key figures to appear—preferably below the Actual Sales or Sales Quantities), and then choose Event. This action shows you a list of unused key figures, from which you can choose one and press Enter, so that the system draws in four other new event-related key figures in its place. In short, the system requires a key figure that can be overwritten with the four new key figures required for planning events.

Figure 9.56 shows the change mode screen (Transaction MC94) in which you can manually enter quantities in each month for the TARGET STOCK LEVEL field ❶. Notice that the PROPORTIONAL EVENT -> TARGET STOCK LEVEL field reflects the details of the event (Event Number: S655:SOP Event).

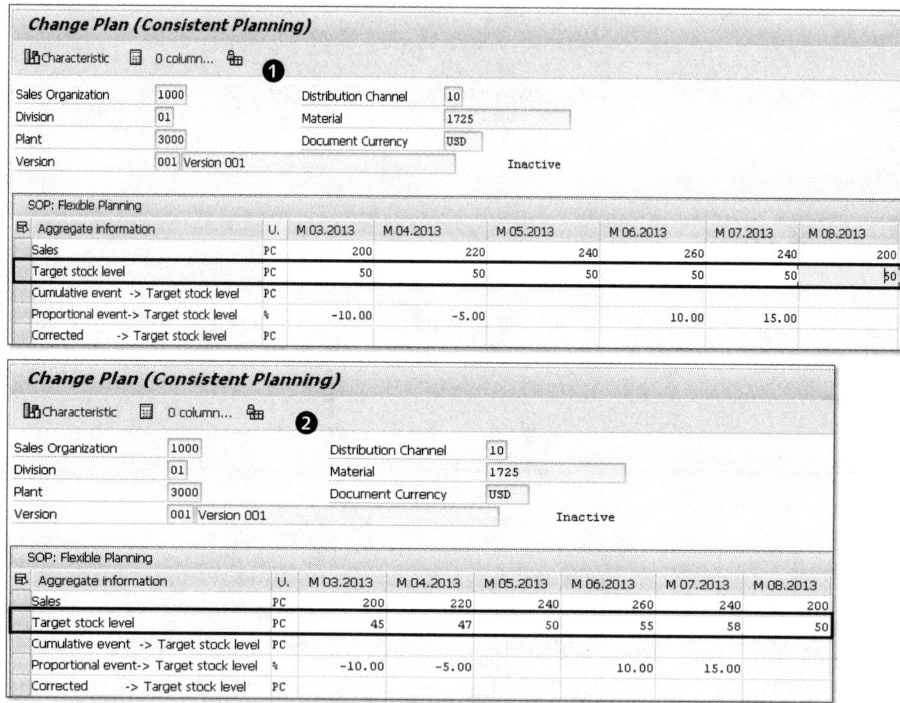

Figure 9.56 Event Working in Flexible Planning

[»] **Note**

It's very important to note that the impact of an event won't be visible in the planning table unless you make changes to quantities in the TARGET STOCK LEVEL field and save the planning table. Then, the next time you enter the planning table, you should be able to see the planning table values populated in the event-based key figures, which for this example is the TARGET STOCK LEVEL field ❷.

9.7 Mass Processing in Sales and Operations Planning

As a planner attending to a large and diverse range of materials and products, it becomes imperative to maintain several of them during the planning phase. In many planning steps, it's crucial that the planner performs the requisite functions online and interactively; however, planners can automate a few planning steps in the mass processing functionality of S&OP. Some of the mass processing steps that you can perform are listed here:

- Forecasting
- Determining the opening stock level
- Running a macro
- Transferring planning data to SAP Demand Management
- Copying key figure information from one info structure to another
- Transferring S&OP key figure data to CO-PA

Because these steps eventually become routine work for a planner, it certainly makes sense to use the mass processing function.

In mass processing, you must define activities that describe the individual steps that are involved and carried out, as well as their assignment to mass processing jobs. Before you create a mass processing job in S&OP, you need to create a planning activity.

9.7.1 Planning Activity

A *planning activity* defines the function that you'll execute through a given mass processing batch job. The planning activity consists of function profiles, which define the functions in detail. You need to create each of the function profiles separately before the creation of the planning activity in SAP ERP.

To create a planning activity, use the configuration (Transaction SPRO) menu path, PRODUCTION • SALES & OPERATIONS PLANNING (SOP) • FUNCTIONS • MASS PROCESSING • DEFINE ACTIVITIES, or use Transaction MC8T.

In Figure 9.57, enter the name of the planning activity in the KEY F.PLNG.ACTIVITY field (for this example, "SOP"), the PLANNING TYPE ("Z655"), the planning activity type, and the corresponding activity profile, which require execution through this planning activity. For a given function that requires mass processing execution, you need to choose the relevant activity type. After you select the relevant ACTIVITY, the corresponding activity profiles, such as MACRO, TRANSFER PROFILE, or COPY PROFILE, also become available for you to select from.

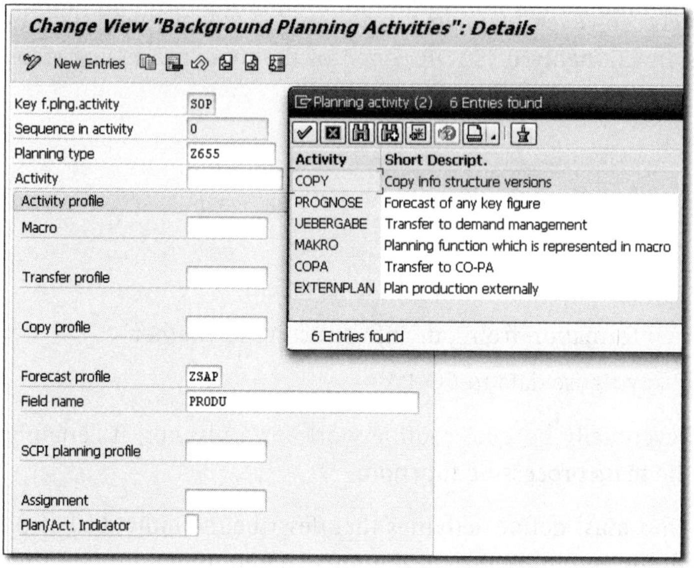

Figure 9.57 Planning Activity Configuration

9.7.2 Setting Up a Mass Processing Job

To set up the mass processing job, you need to perform the following steps in the given sequence:

1. Assign a name to the mass processing job.
2. Assign the info structure and version.
3. Assign the planning type.
4. Assign a name to the variant.
5. Select the selection screens.
6. Define the characteristic values, activity, and aggregation level.
7. Save the variant.

To create the mass processing job, follow the SAP menu path, LOGISTICS • LOGISTICS CONTROLLING • FLEXIBLE PLANNING • PLANNING • MASS PROCESSING • CREATE, or use Transaction MC8D. (You can use Transaction MC8E to change it and Transaction MC8F to delete it.)

In Figure 9.58, the following steps are involved:

1. Create a planning job. For this example, enter the Job number as "SOP_SAP" and a short description of the Job name as "SOP_SAP_Flex_Plan" ❶.
2. Choose Execute, and maintain the Info structure "S655" and Version "001" ❷, on which to apply the mass processing job.
3. Choose Execute, and maintain the Plng.type "Z655" ❸, which is the specific planning type within the info structure on which the system will perform the mass processing function.
4. Select a planning type, and then maintain the Variant Name for the selection screen details.
5. Enter the variant name in the Create variant field ❹, which in this example is "SOP_Flex_Plan", and choose Create to create the variant and enter the selection screen details.
6. Select the Created checkbox ❺, and choose Continue to proceed with maintaining the selection screen details.

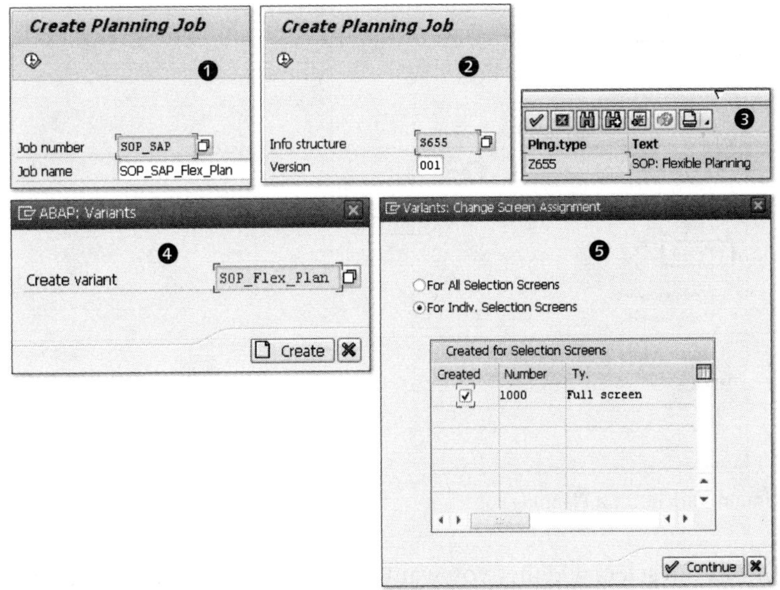

Figure 9.58 Steps in Creating a Planning Job

After you're finished creating the planning job, the screen shown in two parts in Figure 9.59 ❶ and ❷ appears, in which you perform the following functions:

1. Select the characteristics levels on which the system will run the mass processing job, the AGGREGATION LEVEL and the PLANNING ACTIVITY. You configured the planning activity in Section 9.7.1. The planning activity consists of functions that the system performs through mass processing. It contains the profiles for the respective functions that are configured.
2. Click on the ATTRIBUTES button to go to the VARIANT ATTRIBUTES pop-up shown in Figure 9.60.

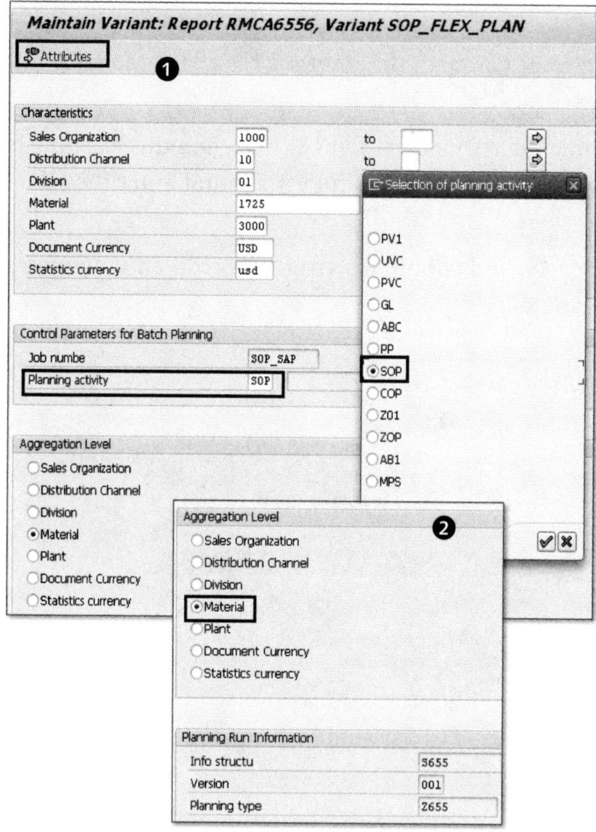

Figure 9.59 Maintain Variant for a Planning Job

In Figure 9.60, you can select whether you want to run the mass processing in the background only by selecting the relevant checkbox. Save the variant attributes.

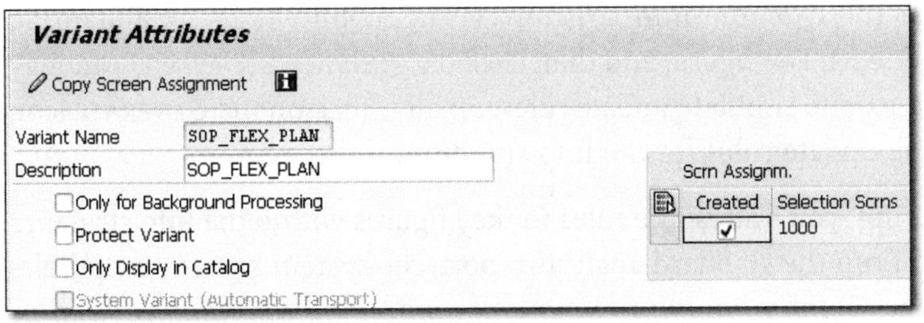

Variant Attributes

Copy Screen Assignment

Variant Name	SOP_FLEX_PLAN
Description	SOP_FLEX_PLAN

☐ Only for Background Processing
☐ Protect Variant
☐ Only Display in Catalog
☐ System Variant (Automatic Transport)

Scrn Assignm.

Created	Selection Scrns
☑	1000

Figure 9.60 Variant Attributes in Background Job Planning

9.7.3 Scheduling the Mass Processing Job

You can schedule the mass processing job using Transaction MC8G. You need to specify the variant name and use the standard batch scheduling function.

You can also monitor the mass processing jobs using Transaction SM37. You can use Transaction MC8I to obtain an overview of a specific job. To delete the mass processing job, you can use Transaction MC8F.

9.8 Standard Analysis in Flexible Planning

Now that you've learned all options and features available in Flexible Planning, we'll cover a lesser-known functionality of standard analysis in Flexible Planning. In Chapter 19, we show how you can take advantage of the large number of standard analysis reports available using predefined and preconfigured info structures in all the logistics components of the SAP ERP system. However, this section covers a user-defined info structure (not a predefined info structure) that you can still use in standard analysis. User-defined standard analysis helps when you want to evaluate all the desired key figures in one report (analysis report) by using one info structure. This eliminates the need to refer to two or more predefined info structures to obtain the desired information.

The following summarizes the steps involved in creating a standard analysis in Flexible Planning:

1. Maintain a self-defined info structure (Transaction MC21).
2. Set parameters for the info structure and key figures (Transaction MC7F).

3. Create update rules for the info structure (Transaction MC24). Maintain an update rule for each key figure, and then click on Activate updating. You may use the Check icon to simulate update rules. Finally, click on the Generate icon to generate the update rules for the info structure.

In Flexible Planning, you update the rules for key figures within the info structure so that when you run the standard analysis report, the system brings up the relevant information.

The step to define the update rule is ensured right after creating and activating the info structure. Using Transaction MC24, you can create the update rules for the characteristics and key figures of the info structure.

Update rules enables the system to do the following:

- Know which characteristics and key figures the system must update when the user performs the business transactions.
- Know the SAP tables from where the system needs to fetch the data for given key figures and characteristics. It also includes the information for the dates, which the system should use for records or data that are populated in the structures.

The *update group* is equally important because it defines the business areas and the tables that you'll use for updating the info structure. You can maintain multiple update rules for a given key figure by copying an existing rule.

The SAP ERP system offers application areas such as MM, SD, PP, QM, Project Systems (PS), and PM, among others.

[!]

Warning!

You should only include the key figures for which you're updating the info structure from SAP ERP system tables of the application area to ensure that the system isn't overburdened.

[»]

Note

To update groups in configuration (Transaction SPRO), follow menu path, Logistics Information System (LIS) • Logistics Data Warehouse • Updating • Updating Definition • Specific Definition Using Update Rules • Maintain Update Rules.

Before the activation is done, you should also check the consistency of the set update rules by choosing the GENERATE icon to check for any errors or warning messages and eliminate them. An info structure with errors won't be generated, and no data will be populated in the info structure.

After attending to the update rules, you need to configure the parameters to update data in the info structure. The period split must match with the period for planning, that is, SET PLANNING PARAMETERS FOR THE INFO STRUCTURE (see Section 9.2.2).

1. Perform the business transactions such as creation of sales order or production order.
2. You can create the necessary transaction data in the SAP ERP system to test the update of the info structure in the SAP tables. For example, create a sales order with order quantity and order value (if these two key figures are used in the self-defined info structure), and see if the data is automatically updated in the info structure (table) by using Transactions SE16N or SE11 and giving the details of the self-defined info structure.
3. Run standard analysis for the self-defined info structure (Transaction MC9C). On the initial screen, enter the self-defined info structure, and enter VERSION "000" (actual data). Press Enter, and on the next screen, your defined characteristics will be available to choose from.

Tips & Tricks [+]

You can check the update log by using Transaction MC30. Enter the user ID, and choose EXECUTE. Keep the cursor on the info structure number, and choose the DETAILS icon to check the data updated in detail. Choose the ANALYSIS icon to check the programs used in updating the info structure.

Note [«]

All of the master data and transactions for Flexible Planning are listed in one place, under the following SAP menu: LOGISTICS • LOGISTICS CONTROLLING • FLEXIBLE PLANNING.

9.9 Summary

Sales and operations planning (S&OP) offers significant optimization to the business processes by helping the planner use standard S&OP or Flexible Planning. For standard S&OP, the product group is used to define the percentage or the proportional factor of each material within the group or hierarchy, whereas in Flexible Planning, the planning hierarchies are used.

You can make extensive use of standard or self-defined macros and also create large numbers of planning versions. The final operations plan gets transferred to SAP Demand Management as PIRs. The forecasting tool is also available in standard S&OP and in Flexible Planning to facilitate proposing realistic planning figures. Events help you take any specific spike or dip in planning figures into account. The rough-cut planning profile provides better visibility on capacities and materials commitments. You can perform a large amount of routine planning work by using the mass processing functionality. You can also use Flexible Planning to perform standard analyses for the self-defined info structure.

The next chapter covers SAP Demand Management.

You need to establish balance between anticipating the demand for the products and ensuring that the right production, procurement, and distribution processes are in place to fulfill them for eventual consumption. You can take advantage of the several planning tools available to come up with realistic planning figures in SAP Demand Management.

10 SAP Demand Management

Demand planners in companies are always striving to maintain a balance between the anticipated demands of a product from customers and ensuring its timely availability. SAP Demand Management today is much more than using the forecasting tool to predict what to produce and when. In fact, it has evolved into taking a much closer and critical look at each step of the entire sales, production, procurement, and distribution element for ensuring an optimal balance. This balance is necessary so that the optimum inventory of raw material is maintained in the company's warehouses, and the machines' capacities aren't booked on those products that won't sell well or suddenly have several unexpected and unanticipated customer-specific orders, resulting in undue long lead time and delayed deliveries.

We begin this chapter by discussing a few important and frequently used planning strategies, as well as where these strategies are assigned in the material master. We then cover the requirements types and requirements classes of both planned independent requirements (PIRs) and customer independent requirements and how these are interrelated to the planning strategy. Next, we cover the creation of PIRs, customer independent requirements, and dependent requirements, and run the material requirements planning (MRP) to see the planning results. Finally, we evaluate the PIR of a material, its allocation to customer independent requirements, and how the PIR is reduced. We also show how you can reorganize the PIRs to remove old and redundant data from the planning process.

10.1 Planning Strategy

The SAP Demand Management process is comprised of assessing and evaluating each and every product that the company produces and then assigning the right planning strategy. A *planning strategy* is a unique identifier to denote whether the material follows make-to-stock (MTS) production, follows make-to-order (MTO) production, uses planning at the assembly level, or uses planning parameters of another material.

Table 10.1 lists a selection of the most commonly used planning strategies in SAP Demand Management.

Planning Strategy (Key)	Planning Strategy Description
10	MTS production
20	MTO production
40	Planning with final assembly
50	Planning without final assembly (MTO)
52	Planning without final assembly (MTS)
60	Planning with planning material (MTO)
63	Planning with planning material (MTS)
70	Subassembly planning

Table 10.1 Selected Planning Strategies

A strategy group is a set of planning strategies. You assign the strategy group to a material in the STRATEGY GROUP field in the MRP 3 view of the material master (Transaction MM02). In the following sections, we'll cover an example of each production process—MTS and MTO—by using a strategy group. We'll then discuss some of the main factors and elements of planning strategies.

10.1.1 Planning with Final Assembly

Figure 10.1 shows the MRP 3 view of the MATERIAL 100-300 and PLANT 3000 combination with the planning STRATEGY GROUP 40 (PLANNING WITH FINAL ASSEMBLY). Usually, you assign this strategy 40 to a finished good. The use of planning strategy 40 is more prevalent on those products with generally stable production and demand in MTS production. However, it does account for unexpected or unforeseen high customer requirements and also covers for it. For this reason, the system consumes the PIR as soon as it books a sales order.

When the system runs the MRP, it primarily absorbs or consumes the PIR. However, PIR reduction or deletion doesn't happen until goods are issued against the sales order.

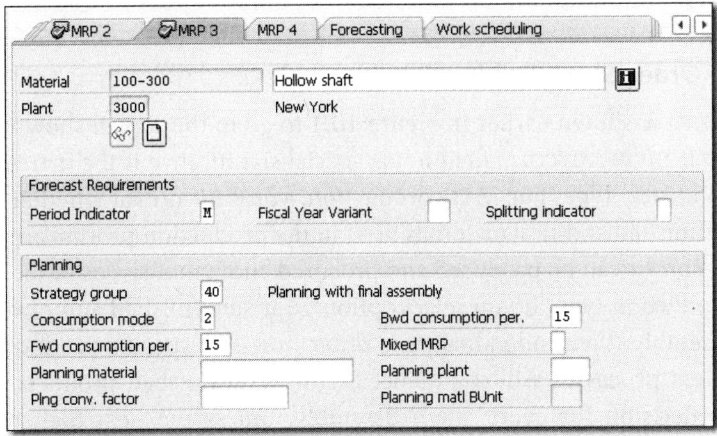

Figure 10.1 Strategy Group in the MRP 3 View of the Material Master

The CONSUMPTION MODE field offers the following four options (available as dropdown options when you place the cursor on the field and press F4):

- 1 – BACKWARD CONSUMPTION ONLY
- 2 – BACKWARD/FORWARD CONSUMPTION
- 3 – FORWARD CONSUMPTION ONLY
- 4 – FORWARD/BACKWARD CONSUMPTION

The assigned consumption mode helps to determine how the system should consume the PIRs; that is, the sales orders should consume (reduce) the PIRs in the backward direction in time starting at the availability of the sales order (option 1), forward only (option 3), or first look to consume in the backward direction and then look for forward PIRs (future dates) if the PIRs (past dates) is unable to fulfill the demand (option 2). Depending on the entry made in the CONSUMPTION MODE field, you need to enter the number of days in the BWD CONSUMPTION PER. or FWD CONSUMPTION PER. field.

[»]

Note

Although this isn't always applicable, we suggest that the number of days that you enter for each period must realistically account for consumption to avoid overproduction to meet the customers' requirements. On the other hand, overproduction is sometimes encouraged to ensure maximum use of resources. The consumption intervals rather relate to the time granularity of PIRs.

10.1.2 Make-to-Order/Stock Production

Click on the MRP 4 view shown earlier in Figure 10.1 to go to the screen shown in Figure 10.2. The INDIVIDUAL/COLL. field holds special significance if the material is an MTO production type. For MTO production, you can control whether the associated semifinished and raw materials used in the production process are also sales order-specific or can be produced and procured anonymously, as in the case of the MTS production type. If you select option 1 for semifinished and raw materials of the assembly, then individual sales order line item-specific production and procurement processes will commence. With MTO finished products, you can make the decision for every single assembly, unless the next higher assemblies in the structure hand down collective requirements, in which case, all of the dependent components are also collective.

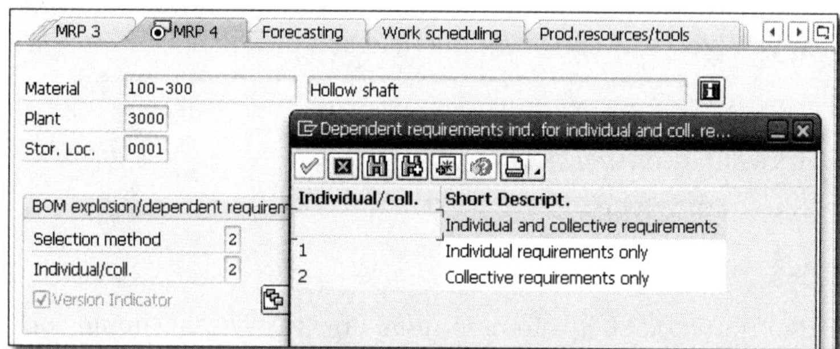

Figure 10.2 Individual or Collective Options in the MRP 4 View of the Material Master

Figure 10.3 again shows the MRP 3 view of the material master of a different finished material, in which you assign a planning strategy (for this example, it's STRATEGY GROUP 60), which uses the planning data of another material. The planning material is exclusively used for planning and is never actually manufactured. For example, if a company manufactures finished products in MTO with almost the same set of components but with different packaging specifications/materials for different customers, then all except the packaging specifications/materials can be planned beforehand, and the packaging material is procured as soon as the customer requirement comes in.

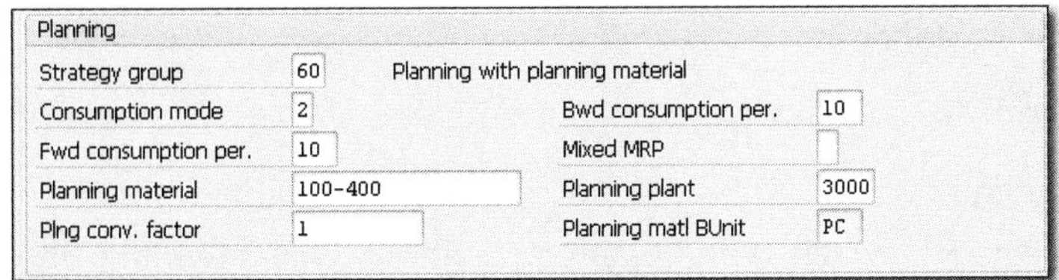

Planning			
Strategy group	60	Planning with planning material	
Consumption mode	2	Bwd consumption per.	10
Fwd consumption per.	10	Mixed MRP	
Planning material	100-400	Planning plant	3000
Plng conv. factor	1	Planning matl BUnit	PC

Figure 10.3 Planning Material and Planning Plant

[«]

Note

The importance of planning data in SAP Demand Management can be gauged from the fact that during the MRP run, the system looks for the planning figures in SAP Demand Management, and the results of the MRP run are the order proposals (planned orders), which the demand planner can convert into production orders (or process orders) for in-house production or purchase requisitions for external procurement.

The planning strategy must be entered both in the planning material and in the end product. In the material master of the end product, you must also enter the planning material, the planning plant, and the conversion factor between the planning material and the end product. This planning material also needs to have a valid bill of materials (BOM). For this example, enter "100-400" in the PLANNING MATERIAL field, enter "3000" in the PLANNING PLANT field, and enter "1" in the PLNG CONV. FACTOR field for a STRATEGY GROUP 60 (PLANNING WITH PLANNING MATERIAL).

Following are two other frequently-used MTO planning strategies:

- **Planning strategy 20**
  Planning strategy 20, which is MTO, is a straightforward case in which the customer's sales order forms the basis for the production process to begin. When the customer places the sales order along with the MRP run, the system creates planned orders not just for the finished goods but also for the semifinished goods, which are then converted into production orders. All procurement elements are sales order-specific. When a finished good is produced against the

sales order, it's shown as sales order stock in the system. Goods issuance against a sales order is also made with reference to the same sales order stock.

[»] **Note**

See Chapter 20 in which we show you how to put strategy 20 to practical use in an MTO scenario. It covers the integration aspect of SAP ERP Production Planning (PP) with SAP ERP Sales and Distribution (SD).

- **Planning strategy 50**

  In planning strategy 50, which is for planning without final assembly, the actual value addition to the product is at the finished good level. You maintain the strategy group 50 in the material master and at the same time, you set the INDIVIDUAL/COLL. field to "2" in the MRP 4 view of the material master to reflect the collective requirements of the components. These settings ensure that all procurement and production follows the MTS process up to the components level, but for the finished goods, the strategy changes to MTO.

  You can enter quantities of the finished good in the PIR and run MRP. The system creates planned orders for the finished good as well as the assemblies and raw materials. The planned orders of assemblies are convertible to production orders and purchase requisitions, but the planned orders for finished goods aren't convertible to production orders. This means that the system carries out production and procurement one level below the finished good.

  Meanwhile, for the finished good, each incoming sales order consumes the PIR and at the same times reduces the quantities of the unconvertible PIR of the finished good. It creates a new convertible planned order (of the sales order quantity) into the production order when you again run MRP. If all sales orders are unable to consume the PIR, the remaining unconvertible planned orders remain in the system. If there are more sales orders than PIRs, the system automatically increases the unconvertible planned orders into convertible ones. This process is known as *netting*.

10.1.3 Requirements Class and Requirements Type

In this section, we'll discuss the factors and elements that make up a planning strategy that is eventually assigned to the material master.

In Figure 10.4, the requirements class of the PIRs and its corresponding requirements type is assigned to the planning strategy to control the consumption behavior of the PIRs. At the same time, the planning strategy also contains the customer requirements class and its associated requirements types. The planning strategy forms part of the planning strategy group and is eventually assigned to the material master. The requirements class for customer requirements is maintained using Transaction OVZG. The only common parameter it contains is the consumption indicator, which is described there as an allocation indicator. Otherwise, the requirements class for customer requirements contains sales-relevant parameters customizing (Transaction OVZG) and is beyond the scope of this book. However, we cover the requirements class for PIRs in Section 10.1.5. Later in this chapter, we cover each of these elements in greater detail.

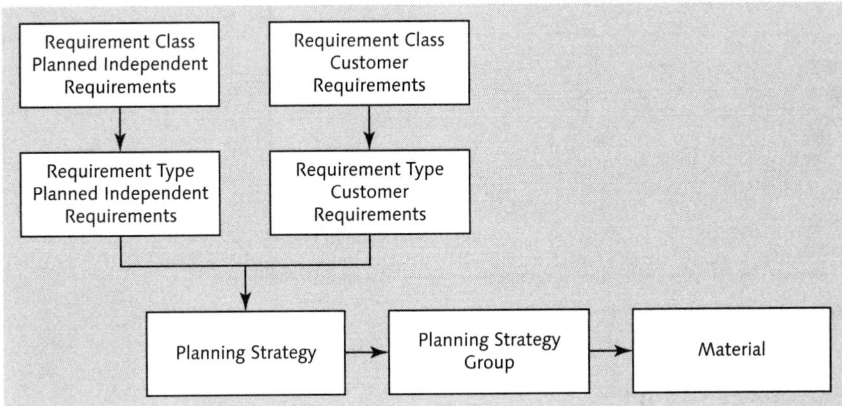

Figure 10.4 Relationships among Requirements Type, Requirements Class, and Planning Strategy

To maintain the planning strategy, follow the configuration (Transaction SPRO) menu path, PRODUCTION • PRODUCTION PLANNING • DEMAND MANAGEMENT • PLANNED INDEPENDENT REQUIREMENT • PLANNING STRATEGY. Here you can see the detailed view of planning STRATEGY 40 (PLANNING WITH FINAL ASSEMBLY) in which the REQMTS TYPE FOR INDEP.REQMTS is VSF, its corresponding REQMTS CLASS is 101 and is directly assigned to the REQMT TYPE OF CUSTOMER REQMT as KSV, and the corresponding REQUIREMENTS CLASS is 050, which is for consumption for the warehouse (stock) (see Figure 10.5).

Because several standard strategy types are delivered with relevant settings, there isn't much you need to change or modify in the strategy type.

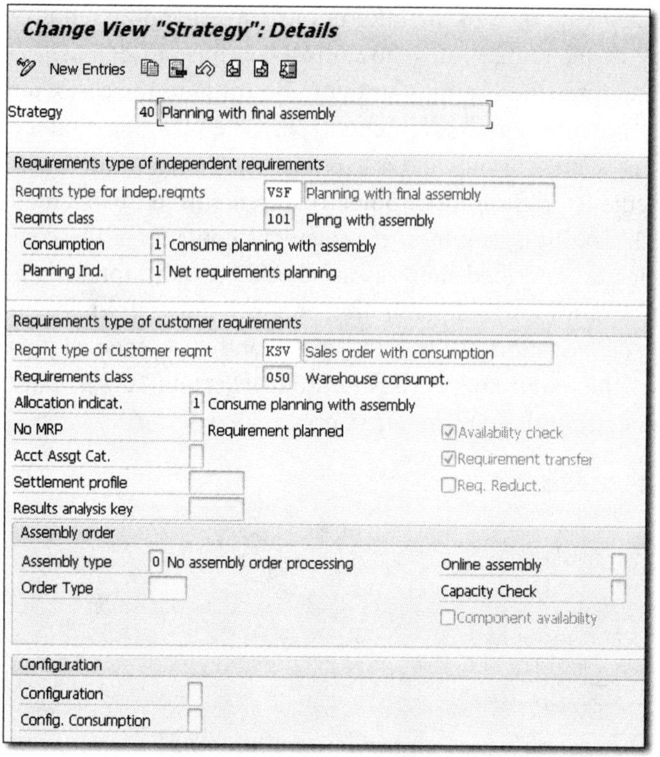

Figure 10.5 Configuration Settings of Strategy 40

10.1.4 Strategy Groups

The screen shown in Figure 10.6 contains STRATEGY GROUP 50 (PLANNING WITHOUT FINAL ASSEMBLY) and has MAIN STRATEGY as 50 along with alternate strategies of 40 and 20. During the planning process, the system first plans the material according to the main strategy (which in this case is 50). The system uses the main strategy for proposing the requirements types whenever a new requirement is created. You can, however, manually overwrite these defaults (e.g., in Transactions MD81/MD61) with the requirements types of the alternative strategies of the strategy group. This option is helpful if, for example, you normally work with MTS, but for one special sales order, you want to trigger MTO. Then you have to manually select the requirements type of the alternative strategy. The system never chooses these alternatives automatically.

To maintain the strategy group, follow the configuration (Transaction SPRO) menu path, PRODUCTION • PRODUCTION PLANNING • DEMAND MANAGEMENT • PLANNED INDEPENDENT REQUIREMENT • PLANNING STRATEGY • DEFINE STRATEGY GROUP.

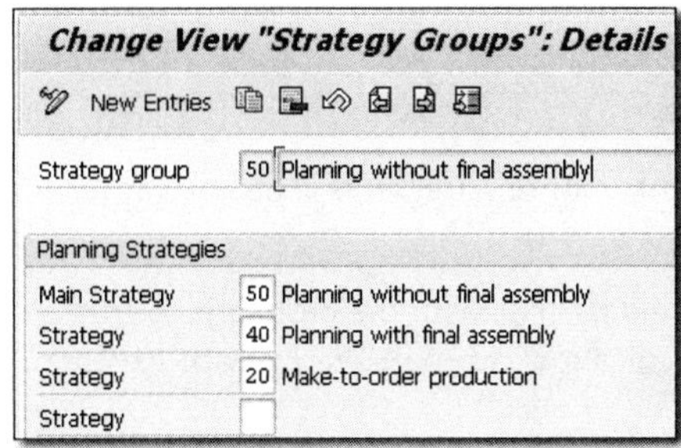

Figure 10.6 Strategy Groups

You can assign a strategy group to an MRP group as shown in Figure 10.7. The advantage of assigning a strategy group to an MRP group is that by using the MRP group, you can easily maintain many materials with an identical planning strategy. Moreover, you can maintain exceptions in the MRP 3 view directly because this entry has priority over the MRP group setting for the strategy group.

To assign the strategy group, follow the configuration (Transaction SPRO) menu path, PRODUCTION • PRODUCTION PLANNING • DEMAND MANAGEMENT • PLANNED INDEPENDENT REQUIREMENT • PLANNING STRATEGY • ASSIGN MRP GROUP TO STRATEGY GROUP.

Change View "Strategy Group for MRP Group": Overview

Plnt	MRP Group	MRP group description
3000	0000	No planning / no requirements transfer
3000	0010	Make-to-stock production
3000	0011	Make-to-stock prod./gross reqmts plnning
3000	0020	Make-to-order production
3000	0021	Make-to-order prod./ project settlement
3000	0025	Make-to-order for configurable material

Figure 10.7 Assignment of a Strategy Group to an MRP Group and Plant

10.1.5 Maintain Requirements Class for Planned Independent Requirements

As covered previously in this chapter, the requirements class for PIRs contains a large number of parameters on how the system should account for the consumption.

[!]

Warning!

SAP recommends the following when maintaining your requirements classes:

- Never change existing requirements classes.
- If necessary, try to choose an existing requirements class with the desired properties. First check to see if your requirements specification can be covered with existing strategies/requirements classes.
- If there is no requirements class with the desired properties, copy an existing one and change the copy.
- In this case, create a new requirements type and strategy as well. Don't change existing strategies and requirements types.

To maintain the requirements class, use configuration (Transaction SPRO) menu path, LOGISTICS • PRODUCTION • PRODUCTION PLANNING • DEMAND MANAGEMENT • PLANNED INDEPENDENT REQUIREMENTS • MAINTAIN REQUIREMENTS CLASSES, or use Transaction OMPO. Figure 10.8 shows the options to maintain the requirements class for the PIRs.

Each requirement class contains the following control functions:

- PLANNING INDICATOR (PI)
  This indicator controls whether the MRP is performed as net requirements planning, as gross requirements planning, or by an MTO segment. We cover these in detail in Chapter 11.
- CONSUMPTION INDICATOR (CONI)
  This indicator controls if and how the planning is consumed by sales orders.
- CONFIGURATION INDICATOR (C)
  This indicator determines how the consumption for a configurable material takes place.
- INDICATOR FOR REQUIREMENTS REDUCTION (RRD)
  This indicator controls if the PIRs for anonymous MTS production are reduced.

- Requirement Category (Cat)

  The requirement types category controls whether the requirements class is relevant for PIRs or for customer requirements.

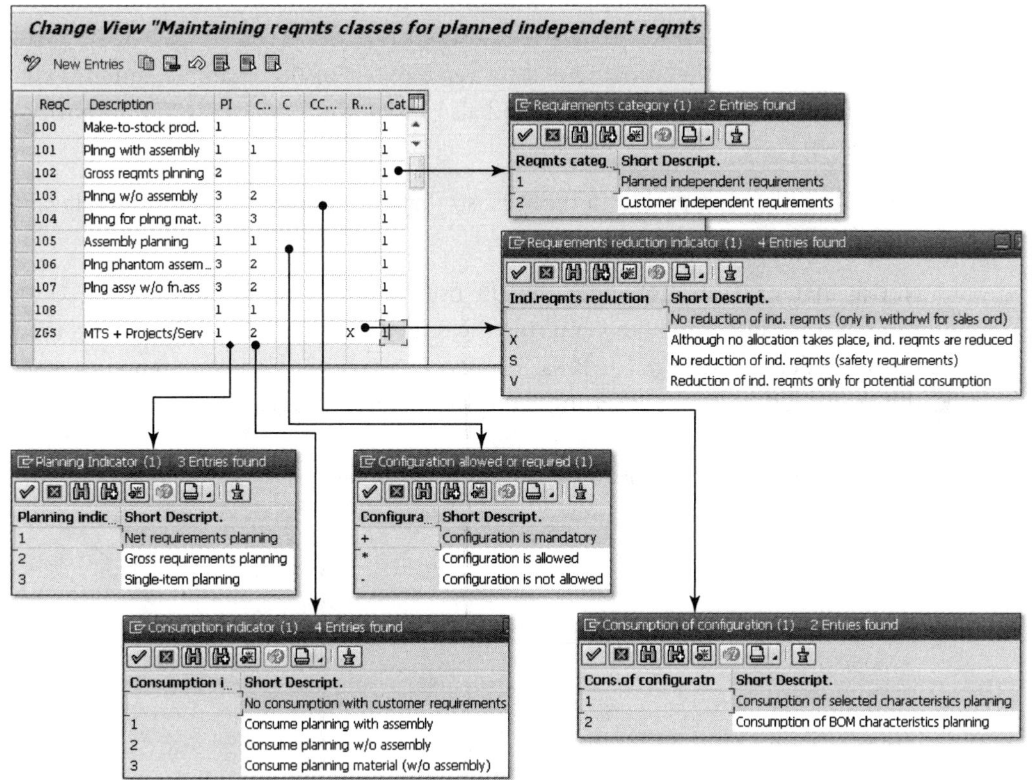

Figure 10.8 Requirements Class for Planned Independent Requirements

10.2 Planned Independent Requirements

PIRs are the direct outcome of sales and operations planning (S&OP) and form an integral part of SAP Demand Management. During the MRP run, the system refers to PIRs with an active version to come up with procurement proposals.

[«]

Note

Refer to Chapter 9 for information about S&OP and how the data and information flows into SAP Demand Management.

You can also create the PIR independently. In fact, SAP Demand Management offers a far greater number of options to create PIRs than S&OP does. To create PIRs, follow the SAP menu path, LOGISTICS • PRODUCTION • PRODUCTION PLANNING • DEMAND MANAGEMENT • PLANNED INDEPENDENT REQUIREMENTS • CREATE, or use Transaction MD61.

Figure 10.9 shows the initial screen to create PIRs. For this example, enter MATERIAL "100-300" and PLANT "3000", and also assign the VERSION as "GP" (PRODUCTION PLAN).

Define the planning period in months, so that the system brings up the relevant field for data input.

Notice that you can also enter default user parameters by choosing the USER PARAMETERS button and then entering the relevant details. These parameters then form the default settings for PIRs. However, if you preset a requirements type, this overrules the strategy settings and is therefore not recommended.

Figure 10.9 Initial Screen of Planned Independent Requirements

Press [Enter] and the PIR screen appears in which you manually enter monthly production figures for the VERSION GP (Figure 10.10). Notice the checkbox for column A, which indicates that the PIR for material 100-300, plant 3000, and version GP is active. An *active* version indicates that the system will consider it during the operational planning run (MRP). This compares with Long-Term Planning (LTP) MRP in which the system can also consider an inactive version to come up with procurement proposals (simulated).

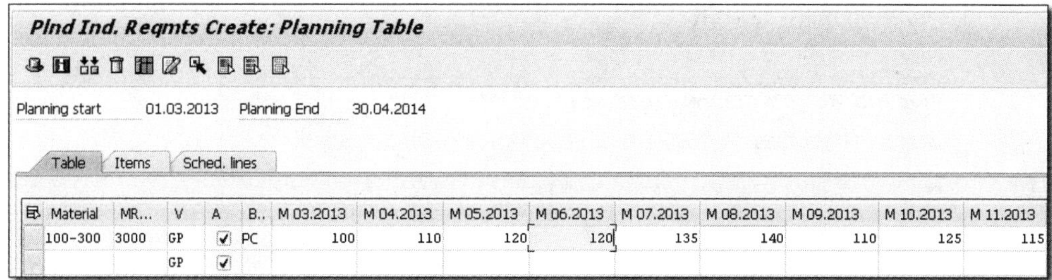

Figure 10.10 Planned Independent Requirements (PIRs)

Figure 10.11 shows the ITEMS tab of the PIR for MATERIAL 100-300, wherein the system assigns the requirement types of PIRs based on the strategy type defined in the material master.

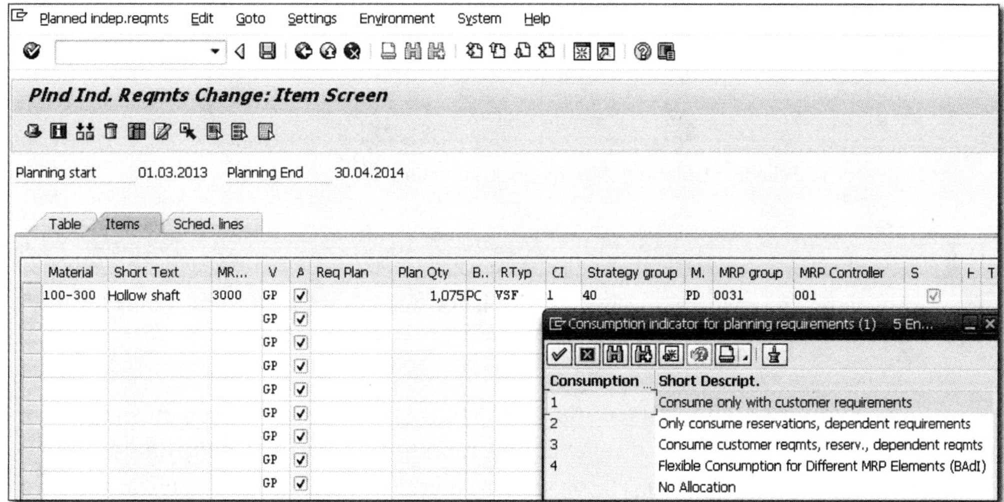

Figure 10.11 Consumption Type in PIRs

For this example, the system assigns requirement type VSF, which is for planning strategy 40. Pressing F4 while keeping your cursor in the CI (consumption indicator) field reveals that CONSUMPTION 1 to CONSUME ONLY WITH CUSTOMER REQUIREMENTS option applies.

Figure 10.12 shows the detailed SCHED. LINES view of the PIRs. Because you chose the planning period to be in months on the initial PIRs screen, the same can be changed, if needed. The system also shows the planning (financial) values of the quantities. However, there are other options available to change the time granularity to daily or weekly.

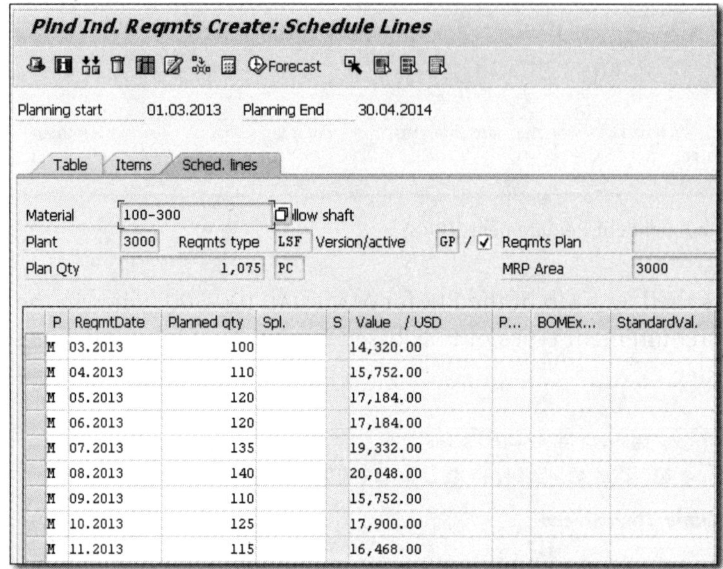

P	ReqmtDate	Planned qty	Spl.	S	Value / USD	P...	BOMEx...	StandardVal.
M	03.2013	100			14,320.00			
M	04.2013	110			15,752.00			
M	05.2013	120			17,184.00			
M	06.2013	120			17,184.00			
M	07.2013	135			19,332.00			
M	08.2013	140			20,048.00			
M	09.2013	110			15,752.00			
M	10.2013	125			17,900.00			
M	11.2013	115			16,468.00			

Figure 10.12 Schedule Lines in PIRs with Financial Values

[»]

Note

The system enables you to leverage the objects and layouts for display in the PIRs. We encourage you to explore these options.

While you manually entered the production quantities in the screen shown earlier in Figure 10.10, you also have the option to automatically copy PIRs from several other planning tools. This is shown in Figure 10.13, where you can copy the results from the forecast of the material (the forecast on the material must already be executed and results saved), from a sales plan, or from a production plan. You can even copy results from simulative planning (LTP).

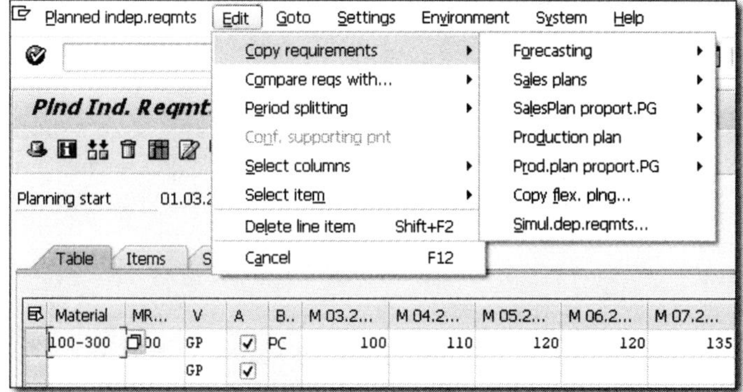

Figure 10.13 Options Available to Create PIRs

Note

LTP is covered in Chapter 12.

[«]

Another highly intuitive feature of PIR creation is that you can also compare the PIR quantities with consumption, with the forecasting results, and with the sales or production plans. To do this, click EDIT • COMPARE REQS WITH....

When you enter the quantities in the PIR or transfer them for S&OP, you have to assign a version number, which denotes different demand plans. There may be several inactive versions and even some simulation versions available in the system. The inactive or simulation versions are used in a simulation tool such as LTP, whereas an active version becomes part of operative planning (MRP). The recommendation for the SAP ERP system is to have version 00 as the one and only active version, whereas all other versions must remain either inactive or used in simulations (LTP).

While the standard version numbers available are generally sufficient for the business processes of the company, you can also create new version numbers if needed, or you can amend the existing ones to make them more reflective of the business needs. To create a new version, use configuration (Transaction SPRO) menu path, LOGISTICS • PRODUCTION • PRODUCTION PLANNING • DEMAND MANAGEMENT • PLANNED INDEPENDENT REQUIREMENTS • DEFINE VERSION NUMBERS (see Figure 10.14).

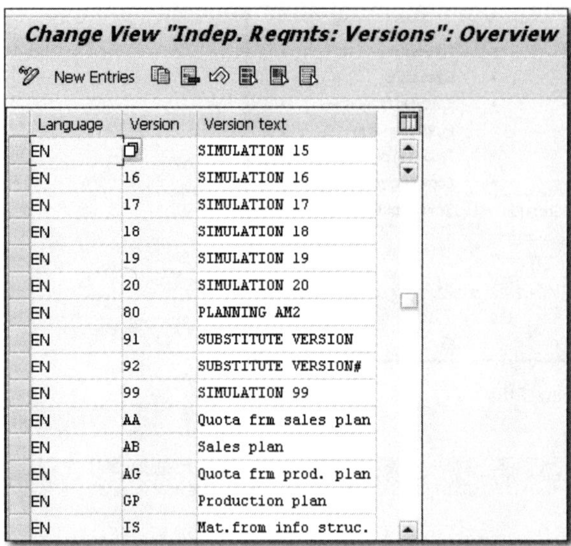

Language	Version	Version text
EN		SIMULATION 15
EN	16	SIMULATION 16
EN	17	SIMULATION 17
EN	18	SIMULATION 18
EN	19	SIMULATION 19
EN	20	SIMULATION 20
EN	80	PLANNING AM2
EN	91	SUBSTITUTE VERSION
EN	92	SUBSTITUTE VERSION#
EN	99	SIMULATION 99
EN	AA	Quota frm sales plan
EN	AB	Sales plan
EN	AG	Quota frm prod. plan
EN	GP	Production plan
EN	IS	Mat.from info struc.

Figure 10.14 Versions for PIRs

10.3 Customer Independent Requirements

When the user enters the sales order in the system, the system immediately adjusts it with the PIR as consumption. While sales orders and associated activities such as outbound delivery and post goods issue form the basis of consumption and PIR adjustments, it's also possible to independently create customer independent requirements in SAP Demand Management without reference to a sales order. This is usually done when testing the PP component in a project phase without having the SD component available.

In the following sections, we'll cover the various requirements, such as customer independent requirements and planned independent requirements, and their evaluations.

10.3.1 Creating Customer Independent Requirements

To create customer independent requirements, follow the menu path, LOGISTICS • PRODUCTION • PRODUCTION PLANNING • DEMAND MANAGEMENT • CUSTOMER

REQUIREMENTS • CREATE, or use Transaction MD81. Figure 10.15 shows the initial screen to create the independent requirement. When you drill down the REQUIREMENTS TYPE field (by pressing F4 while keeping the cursor on the field), a large number of standard requirement types specific to the customer will be available to choose from. If you've assigned a strategy group in the MRP 3 view of the material master, the system automatically assigns a requirement class to the customer independent requirements.

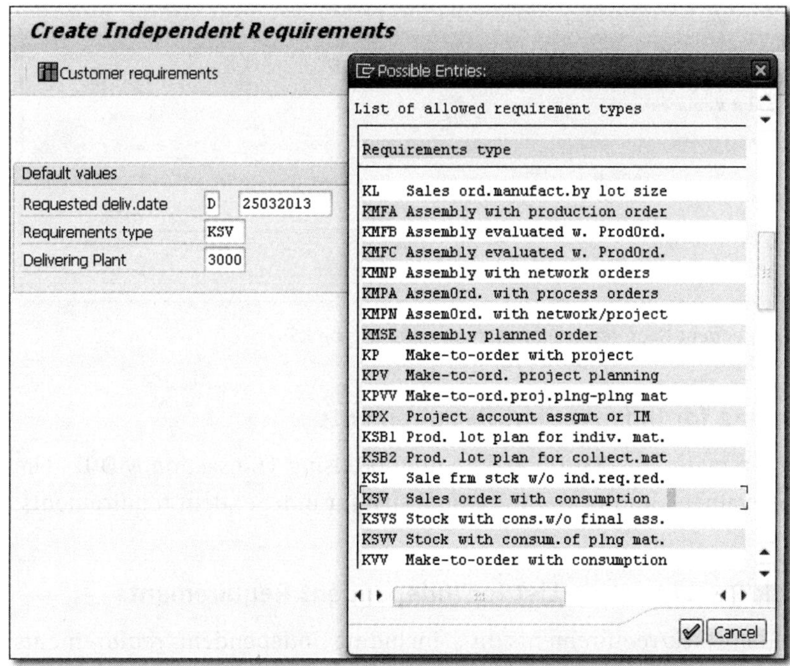

Figure 10.15 Creation of Independent Requirements

Note [«]

While the system automatically fills in the REQUIREMENTS TYPE field whenever you've assigned a strategy group to the material master, you can overwrite it if you have alternative strategies defined in the strategy group.

Press [Enter], and the screen shown in Figure 10.16 appears where you can enter MATERIAL number "100-300" and an ORDER QUANTITY of "50" "PC" for PLNT "3000". Save the customer independent requirements, and the system generates an internal number for the customer requirement, which for this example is 50000033.

Next, you'll run the MRP to see how the system accounts for the customer requirements in the planning run.

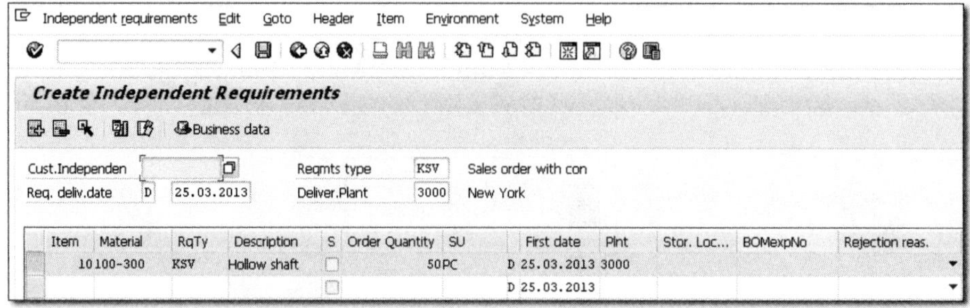

Figure 10.16 Independent Requirements with Requirement Type KSV

10.3.2 Planning for Independent Requirements

Run the MRP on material 100-300 and plant 3000 using Transaction MD02. The next step is to evaluate the stock/requirements list for independent requirements.

10.3.3 Stock/Requirements List for Independent Requirements

To evaluate the stock/requirements list, including independent requirements after the MRP run, use the SAP menu path, LOGISTICS • PRODUCTION • PRODUCTION PLANNING • DEMAND MANAGEMENT • ENVIRONMENT • STOCK/REQUIREMENTS LIST, or use Transaction MD04. As shown in Figure 10.17, the system creates an MRP ELEMNT planned order 37088 for quantity 50 against the customer requirement 50000033. The customer requirement is shown as CUSREQ in the MRP ELEMNT column.

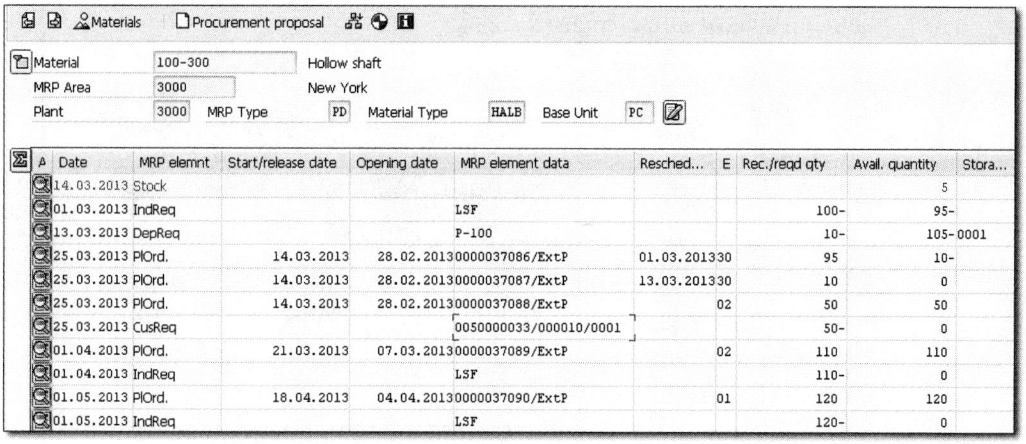

Materials | Procurement proposal

Material	100-300	Hollow shaft					
MRP Area	3000	New York					
Plant	3000	MRP Type	PD	Material Type	HALB	Base Unit	PC

A	Date	MRP elemnt	Start/release date	Opening date	MRP element data	Resched...	E	Rec./reqd qty	Avail. quantity	Stora...
	14.03.2013	Stock							5	
	01.03.2013	IndReq			LSF			100-	95-	
	13.03.2013	DepReq			P-100			10-	105-	0001
	25.03.2013	PlOrd.	14.03.2013	28.02.2013	0000037086/ExtP	01.03.2013	30	95	10-	
	25.03.2013	PlOrd.	14.03.2013	28.02.2013	0000037087/ExtP	13.03.2013	30	10	0	
	25.03.2013	PlOrd.	14.03.2013	28.02.2013	0000037088/ExtP		02	50	50	
	25.03.2013	CusReq			0050000033/000010/0001			50-	0	
	01.04.2013	PlOrd.	21.03.2013	07.03.2013	0000037089/ExtP		02	110	110	
	01.04.2013	IndReq			LSF			110-	0	
	01.05.2013	PlOrd.	18.04.2013	04.04.2013	0000037090/ExtP		01	120	120	
	01.05.2013	IndReq			LSF			120-	0	

Figure 10.17 Results of MRP with Customer Requirements

10.3.4 Total Independent Requirements: Evaluation

At any given time, you can evaluate the total PIRs of a material and its consumption. The system displays the PLANNED QTY. as available in the system, whereas the TOTAL ASSIGNMENT column shows the consumption figures. Use SAP menu path, LOGISTICS • PRODUCTION • PRODUCTION PLANNING • DEMAND MANAGEMENT • EVALUATIONS • DISPLAY TOTAL REQMTS, or use Transaction MD73. Here you see the initial screen (see Figure 10.18), wherein you enter the MATERIAL "100-300", enter the PLANT "3000", and select the OUTPUT LIST 1 option from the dropdown list.

Press [Enter], and the DISPLAY TOTAL REQUIREMENTS screen shown in Figure 10.19 appears. Here you see the monthly planned quantity you previously entered in Figure 10.10 in the upper half of the screen, and you see the customer requirement of 50 PC in the lower half.

When you have only the dynamic consumption, then this is shown in the TOTAL ASSGMNT column. If the entire process is finished—that is, the sales order is delivered—the system shows the quantity in the WITHDRAWAL QTY column.

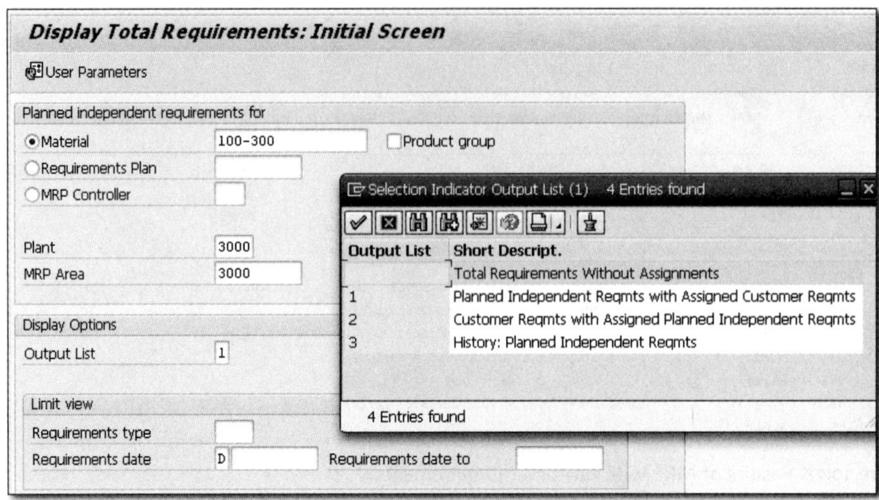

Figure 10.18 Initial Screen for Total Requirements of Material 100-300

Display Total Requirements

Material	Short Text	MRP Area	RqTy	DV	ReqPlanNo.	Total planned qty	BUn	Ac	Txt
P Reqmts dt.	Planned quantity	Withdrawal qty Txt							
100-300	Hollow shaft	3000	LSF	GP		1,075	PC	☑	☐
M 03.2013	100	☐							
M 04.2013	110	☐							
M 05.2013	120	☐							
M 06.2013	120	☐							
M 07.2013	135	☐							
M 08.2013	140	☐							
M 09.2013	110	☐							
M 10.2013	125	☐							
M 11.2013	115	☐							
100-300	Hollow shaft		KSV		0050000033	50	PC	☐	☐
D 25.03.2013	50.000	☐							

Figure 10.19 Total PIR and Requirements for Material 100-300

10.3.5 Total Independent Requirements: Reorganization

Although the PIRs are automatically reduced, you must regularly and periodically evaluate the PIRs in the system to see if the same needs to be eliminated (or adjusted). This helps ensure that old and nonrelevant planning data no longer becomes part of the planning process during the MRP run.

To evaluate the PIR, use the SAP menu path, LOGISTICS • PRODUCTION • PRODUCTION PLANNING • DEMAND MANAGEMENT • ENVIRONMENT • INDEPENDENT REQUIREMENTS: REORGANIZATION • ADJUST REQUIREMENTS, or use Transaction MD74. In the initial screen, you can give the initial parameters, so that when the system is executed, it only brings up the relevant results for consideration and action. Notice that you also have the flexibility to evaluate the results of the organization in the TEST MODE (NO DATABASE CHANGES) (see Figure 10.20). This helps in evaluating the results in simulative mode before actual reorganization.

The KEY DATE field refers to the date *until* which the system should bring up the PIR data for reorganization. For this example, it's 14.06.2013 (June 14, 2013), and the system should consider all of the PIRs *until* the month of June, 2013. It's important to highlight that such PIR reorganization is ensured for past/old PIRs only, which the company no longer wants to plan for or account for in PP.

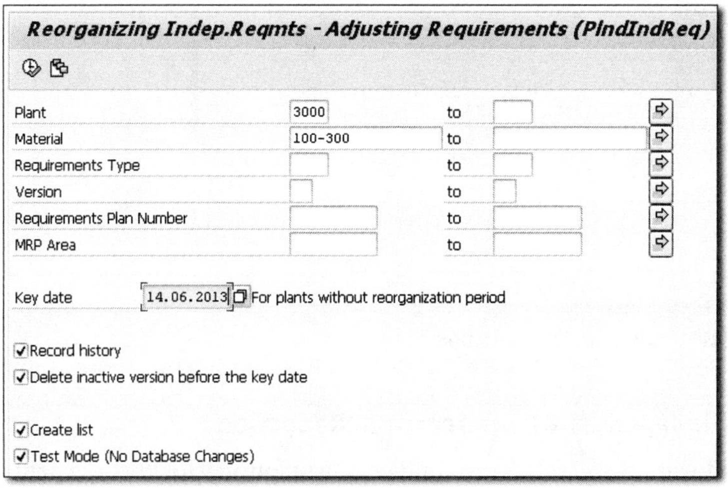

Reorganizing Indep.Reqmts - Adjusting Requirements (PIndIndReq)

Plant	3000	to	
Material	100-300	to	
Requirements Type		to	
Version		to	
Requirements Plan Number		to	
MRP Area		to	

Key date 14.06.2013 For plants without reorganization period

☑ Record history
☑ Delete inactive version before the key date

☑ Create list
☑ Test Mode (No Database Changes)

Figure 10.20 Initial Screen for Reorganizing PIRs

[«]

Note

You can define the plant-level PIR reorganization using Transaction OMP8, where you can define the number of days. The system subtracts these days from today's date to come up with the reorganization date. The system reorganizes all PIR data older than the reorganization date.

After you enter the initial parameters, press F8 or choose EXECUTE to open the screen shown in Figure 10.21. This screen brings up the data for the four months, until June 2013, to which the system was able to find PIRs for the material 100-300 and plant 3000 containing quantities. After evaluation, you can organize (delete) this, as described in the next section.

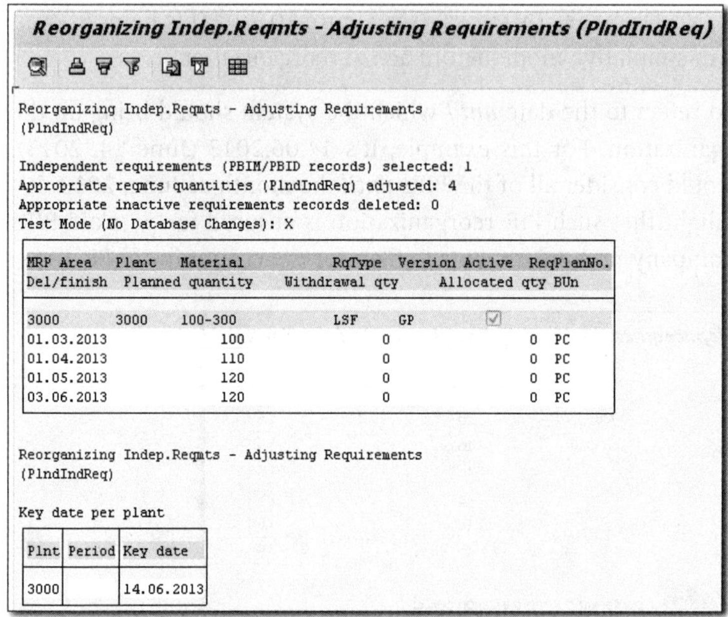
Reorganizing Indep.Reqmts - Adjusting Requirements (PlndIndReq)

Reorganizing Indep.Reqmts - Adjusting Requirements (PlndIndReq)

Independent requirements (PBIM/PBID records) selected: 1
Appropriate reqmts quantities (PlndIndReq) adjusted: 4
Appropriate inactive requirements records deleted: 0
Test Mode (No Database Changes): X

MRP Area	Plant	Material	RqType	Version	Active	ReqPlanNo.
Del/finish	Planned quantity	Withdrawal qty	Allocated qty	BUn		
3000	3000	100-300	LSF	GP	☑	
01.03.2013	100	0	0	PC		
01.04.2013	110	0	0	PC		
01.05.2013	120	0	0	PC		
03.06.2013	120	0	0	PC		

Reorganizing Indep.Reqmts - Adjusting Requirements (PlndIndReq)

Key date per plant

Plnt	Period	Key date
3000		14.06.2013

Figure 10.21 Simulative Results of PIR Organization

10.3.6 Planned Independent Requirements: Reduction

PIRs are reduced during the goods issue. Unlike consumption, the PIRs are actually reduced and not just consumed. During MTS production using strategy 10, the oldest PIR is reduced first. However, the reduction can also affect PIRs in the future if relevant strategies are used with consumption such as reduction in consumption strategies 40, 50, 60, and so on.

Here, reduction simply takes place according to the current consumption situation when posting the goods issue. For this strategy to work, you need to set the configuration data accordingly. The configuration is performed using Transaction OMJJ. For nonconsumption strategy 10, you have this setting in the requirements class. This transaction is in the domain of SAP ERP Materials Management (MM), wherein you make relevant settings per movement type. For nonconsumption strategy 10, these settings are available in the requirements class.

After executing Transaction OMJJ, you can provide the movement type; for example, "601" (goods issue against delivery). In the UPDATING CONTROL area shown in Figure 10.22, you can select the IND. RQMTS REDUCTION checkbox to enable the system to reduce the PIRs at the time of goods issuance (e.g., movement type 601) against a sales order.

During assembly planning, the PIR is reduced when goods are withdrawn for the production order.

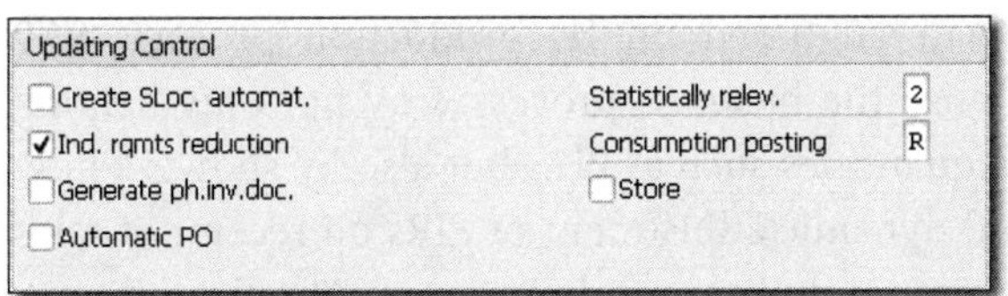

Figure 10.22 Configuration Settings of PIR Reduction

Table 10.2 provides an overview of common (selected) planning strategy types to which the system reduces the PIRs, based on specific transactions such as goods issue against a sales order or production order.

Planning Strategy	Planning Strategy Description	PIRs Reduction
10	MTS production	Goods issue (sales order, MTS)
20	MTO production	Goods issue sales order
30	Production by lot size	Goods issue (sales order, MTS)
40	Planning with final assembly	Goods issue sales order
50	Planning without final assembly (MTO)	Goods issue sales order
52	Planning without final assembly (MTS)	Goods issue production order
60	Planning with planning material (MTO)	Goods issue sales order
63	Planning with planning material (MTS)	Goods issue production order

Table 10.2 Relationship of Selected Strategy Types and PIR Reduction

Planning Strategy	Planning Strategy Description	PIRs Reduction
70	Subassembly planning	Goods issue production order
82	Assembly planning with production order	Goods issue sales order

Table 10.2 Relationship of Selected Strategy Types and PIR Reduction (Cont.)

10.4 Summary

SAP Demand Management acts as a central point to account for both the planning for production processes and the consumption to sales processes. With the right strategy type used for each material based on its production process, SAP Demand Management can optimize the business processes by not only helping the user enter realistic planning quantities such as PIRs but also by showing how these relate to the automatic and dynamic adjustment of PIRs on receipt of sales orders and their further processing, including goods issuance. The flexibility to assign different planning and consumption strategies broadly mitigates the insecurities and other risks involved. At the same time, the system keeps a realistic check on the quantities entered in PIRs compared to the actual demand situation by making appropriate adjustments.

The next chapter covers material requirements planning (MRP).

Material requirements planning is an integral part of SAP ERP Production Planning and SAP ERP Materials Management. In this chapter, you'll learn how to use this tool to optimize your logistics and supply chain planning processes.

11 Material Requirements Planning

Material requirements planning (MRP) is a highly versatile and intuitive planning tool that generates procurement proposals to fulfill in-house production and external procurement demands to help the production and procurement planners optimize their business processes. Furthermore, it generates these procurement proposals for all levels of production-based materials and also takes lead times, scrap quantities, external procurements, special procurements, planning cycles, planning calendars, and lot sizes into account. MRP isn't limited to quantities calculations—it includes scrap and stock considerations, and also provides in-depth scheduling solutions and capacity requirements.

To effectively use the results of MRP, you need to ensure that you provide complete, correct, and comprehensive master data as often as possible. To facilitate this initiative, the system also offers several helpful tools: planned delivery time calculation, for external procurement; statistical analyses, for in-house production times or external procurement times; forecasting tools; calculation of safety stock; and reorder point planning. With the passage of time and with more experience and reliable data, you can continue to update the planning data (the MRP data in four different MRP views of a material) and continue the refining and optimization process.

This chapter begins with a process overview of MRP, including the important role of MRP in all of supply chain planning. It covers both types of MRP—that is, MRP and consumption-based planning—and their variants. We'll cover important influencing factors involved in ensuring calculation of reliable planning results, such as lot sizes, MRP types, scrap, and stocks. Wherever applicable, we also cover the necessary configuration to get MRP up and running in your system. We also cover consumption-based and forecast-based planning in detail.

Next, we cover the types of planning available in MRP following by scheduling, net requirements calculation logic, and procurement proposals (both interactive creation and manual). We move on to cover the steps involved in running the MRP and then analyze the planning results in detail. The chapter concludes with a discussion of two important areas of planning: the planning calendar and MRP areas.

[»] **Note**

Although this chapter primarily helps to build your concepts and understanding of MRP, see Chapter 12 on Long-Term Planning (LTP) to see how to put MRP to actual use by using a real-life example from the fertilizer industry. We also suggest that you see Chapter 20, which covers the integration of SAP ERP Production Planning (PP) with SAP ERP Sales and Distribution (SD) and SAP ERP Project Systems (PS), with regards to make-to-order (MTO) and engineer-to-order (ETO) production methods.

11.1 Process Overview

Figure 11.1 provides an overview of the role of MRP in the planning and production areas of the SAP ERP system.

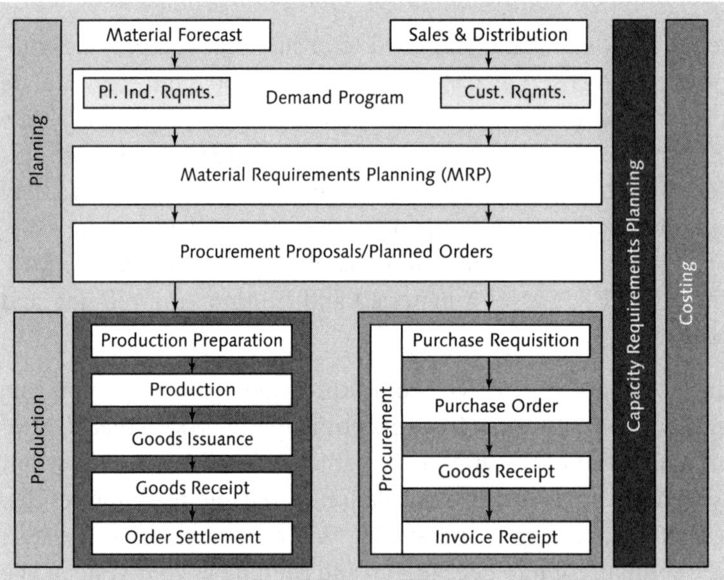

Figure 11.1 MRP in Planning

The planning process starts with requirements coming from SD or from a material forecast and becomes a part of SAP Demand Management. The system also takes customer requirements and planned independent requirements (PIRs) into account. The PIR with an active version consists of quantities of the material on which you want to run MRP. The system runs the MRP based on the details given in the demand program, and the outcome of the MRP run is the creation of procurement proposals. These procurement proposals can be planned orders (order proposals), purchase requisitions for external procurement, or delivery schedules for scheduling agreements with vendors. The procurement proposals distinctively divide up into two separate yet interlinked areas: procurement for externally procured materials and production for in-house produced materials.

Figure 11.2 reflects the fact that you can categorize MRP into two major subareas:

- **Material requirements planning**
  For in-house produced material with high value (type A), you use MRP. In this case, the system explodes the bill of materials (BOM) and routing (or master recipe) and plans materials at all assembly and subassembly levels. It calculates basic dates or even lead time scheduling, as well as capacity requirements.
- **Consumption-based planning**
  For B- and C-type materials, you categorize the materials as consumption-based and plan them accordingly. Reorder point planning can be either automatic or manual and finds more prevalent use in planning for B- and C-type materials.

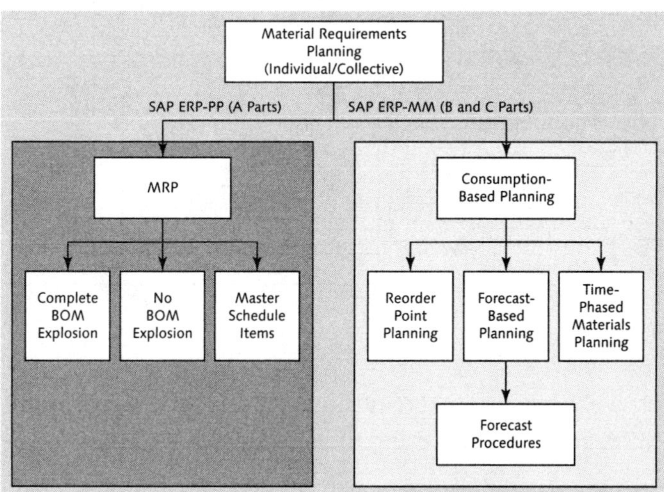

Figure 11.2 MRP and Consumption-Based Planning

MRP consists of the following steps as shown in Figure 11.3:

❶ Check the planning file to avoid planning materials that aren't needed.

❷ Determine the material shortage quantity (or surplus quantity) by carrying out a net requirements calculation.

❸ Calculate the procurement quantity to include lot sizes for planned orders or purchase requisitions.

❹ Schedule basic dates or lead time (for in-house production only).

❺ Select the source of supply by determining the procurement proposal.

❻ Determine the requirement of subordinate parts on the basis of the BOM explosion by carrying out a dependent requirements determination.

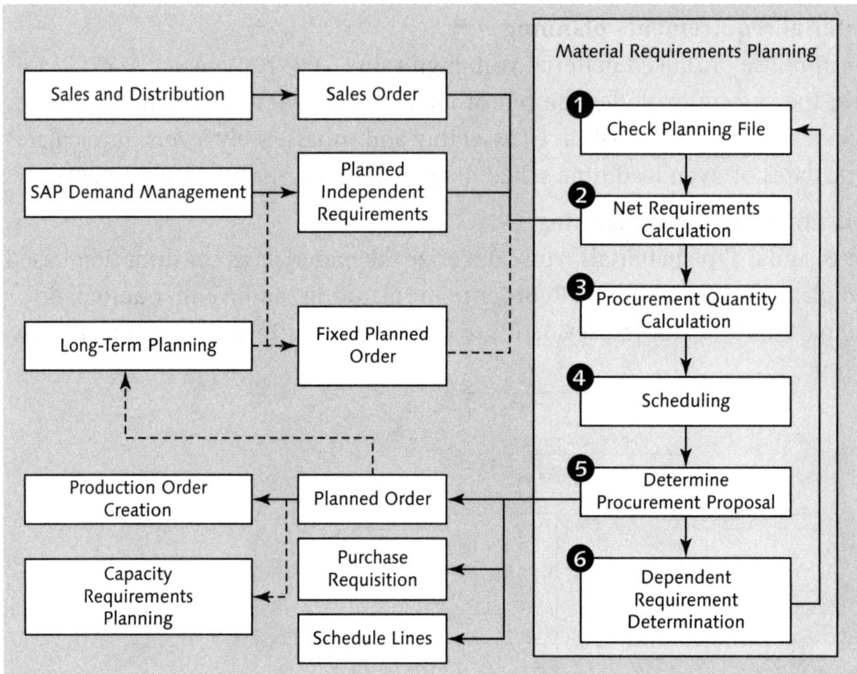

Figure 11.3 Process Steps in MRP

In the following sections, we'll cover the prerequisites and some of the important influencing factors in MRP, while also covering the necessary configuration settings you need to make.

11.1.1 Prerequisites

During an SAP ERP system implementation project, or even later on when you want to start taking advantage of planning tools such as MRP, you need to ensure that you follow a step-by-step approach in identifying materials that you want to plan with MRP. You can plan just about any kind of material with MRP. Broadly, you should consider the following factors and/or take the following steps:

1. To begin, prepare a comprehensive list of all materials (e.g., in a Microsoft Excel file). This Excel file may be a template that your SAP ERP system's consultant will make available for you. Materials can be finished goods, components, assemblies, raw materials, packing materials, consumables, or maintenance spares. You can also plan materials with special procurement types, such as subcontracting materials, phantom assemblies, pipeline materials, procurement from another plant, production in another plant, stock transport orders (interplant transfer), and so on. (See Chapter 13, in which we cover special procurement types and show how you can take advantage of MRP with the same.)
2. Segregate the materials that you want to plan from those that aren't worth the time, effort, or cost involved in planning, and then eliminate them from the Excel file.
3. For materials that you want to plan, segregate them further by classifying them as A, B, or C types. Type A materials are high values and generally cover finished goods, semifinished/assemblies, raw materials, packaging materials, and so on. Type B and C are medium- to low-value materials such as consumables, routine maintenance spares, and so on.

For all A-type materials, individually evaluate each and every important factor involved in MRP to see how these fit in the planning and give you the desired results. For example, ask yourself questions such as the following:

- What should the safety stock of this material be?
- How many days of planning window do I need so that the system doesn't make any automatic changes to my production (or procurement) plan?
- When there is a requirement for a material, how much should I plan (exact quantity, minimum quantity, maximum quantity, etc.)?
- How should the incoming sales orders consume my planning figures?
- Is this material make-to-stock or make-to-order?
- Do I need to produce this material for a specific project?

- Is this assembly produced in-house or produced at another plant of the company?
- How much scrap (percentage) does the production of a material generate for a product or subassembly, so I can incorporate that in planning?
- Should I also first ensure to forecast material requirements by using a forecasting tool and adjust forecast values before proceeding with material planning?

You need to be asking these (and many more) questions for each material and correspondingly enter or update the information in the Excel file. In this chapter, you'll find some factors that MRP will consider to come up with the planning results.

For raw materials or packaging materials, and because these are externally procured, you need to know answers to questions such as the following:

- What is the planned delivery time of this material?
- What is the goods receipt processing time?
- How much underdelivery and overdelivery of a material is allowed?
- Is there a scheduling agreement or a contract (quantity or value) with a vendor for a material?
- Is there a quota arrangement in place, in case of multiple vendors for the same material?

We suggest that you involve your procurement personnel (SAP ERP Materials Management [MM]) to help you answer these (and many more) questions because they have firsthand knowledge and experience of the procurement.

[»] **Note**

See Chapter 20, in which we show the integration of PP with MM, including how you can leverage MRP to automatically create procurement proposals of a material when a scheduling agreement with a vendor is in place.

For B- and C-type materials, consider using consumption-based planning. For consumption-based planning, you can choose whether you want to use reorder-based planning that can manually or automatically suggest the reorder point of each material based on historical consumption. For manual reorder point planning, you have to enter the reorder point quantity, which may be based on your past experience or from consumption figures available in another planning program (not the SAP ERP system). Further, you can also take advantage of the forecasting tool to help you plan your future consumption. Forecasting also uses historical consumption data available in the system.

Because planning is a continuously evolving and ongoing process, you can continue to fine-tune the planning parameters of materials to help the system come up with planning results that you can use in your production and procurement processes almost immediately (and without need of significant adjustments). For example, when you first entered the planned delivery time of a material, it was five days, but as the system built up a significant database of this information (planned delivery time) in the past year, the average delivery time turned out to be seven days. You can update this information in the MRP view of the material master so that future planning results of MRP are more reflective of this reality. The system also makes available a tool (Transaction WPDTC) to provide you with a comparison between the information that you've entered in the material master and the actual planned delivery time that the system recorded for each delivery. Several such analysis reports and tools are available. Chapter 19 covers standard analysis for the PP component.

11.1.2 Influencing Factors in Material Requirements Planning

Whether it's MRP or consumption-based planning, you'll define a large number of factors in the four different MRP views of the material master to influence the planning results of MRP. In fact, the system refers to this detailed information in the MRP views to come up with procurement proposals. Some of these factors may be how the system performs the net requirements calculations, what percentage of safety stock to consider in planning, the lot-sizing procedures to use, if forecasting is a necessary perquisite to planning, and how to manage special procurement types. In the following sections, we cover the important factors and parameters that you need to set up to achieve the desired planning results from MRP and that you can subsequently use in your production and procurement activities.

11.1.3 Lot Sizes

When the system determines a material shortage after performing a net requirements calculation, it looks up to the lot-sizing information that is maintained in the MRP 1 view of the material master. The lot sizing specifies the quantity that the system should propose when it comes across a quantity shortage of a material. For the system to create the correct lot size as a procurement proposal, it also looks for other information, such as safety stock, maximum or minimum lot size, dynamic safety stock, scrap percentage, and rounding value.

There are three different lot-sizing procedures in the SAP ERP system, which we'll discuss in the following sections:

- Static lot-sizing procedures
- Periodic lot-sizing procedures
- Optimum lot-sizing procedures

Static Lot-Sizing Procedures

The static lot-sizing procedures can further be categorized as exact lot sizing, fixed lot sizing, fixed lot sizing with splitting, and replenish to maximum stock level:

- **Exact lot sizing (EX)**
  The system creates planned orders or procurement proposals to cover the exact shortage requirement only. For example, if a new requirement quantity of a material is 110 units, the system creates the procurement proposal for 110 units only.
- **Fixed lot sizing (FX)**
  The system proposes the total requirement in multiples of fixed lots. For example, if the lot size is fixed as 50 units, and the shortage quantity is 250, then the system creates 5 planned orders to cover the shortages.
- **Fixed lot sizing with splitting (FS)**
  The system proposes the total requirement in multiples of fixed lots with overlap. This lot sizing is generally preferred if you can't produce or procure a large quantity at once and hence must split the quantity with overlap. With this lot-sizing procedure, you have to assign a fixed lot quantity together with a rounding value. The rounding value must be a multiple of the fixed lot quantity. For example, if the lot size is fixed as 100 units, and the rounding value is 20 units, then the system creates 5 planned orders to cover the shortages.
- **Replenish to maximum stock level (HB)**
  The system creates planned orders to fill it to the maximum stock level. You have to define the maximum stock level in the relevant field of the MRP 1 view of the material master. For example, if the maximum stock level is 300 units,

and the demand quantity is 50 units, the system creates a planned order of 350. It first ensures that that the 300 units of stock fill up, and then the demand quantity is added to it.

Periodic Lot-Sizing Procedures

The periodic lot-sizing procedures are further categorized into daily lot sizing, weekly lot sizing, and monthly lot sizing. You can configure the period of the lot size in Customizing. For example, if the lot-sizing procedure is daily, then the system considers demand with all of the receipts for that day and creates a daily shortage proposal in case a shortage exists. If it's a weekly lot size, then the system combines all of the requirements for a week and comes up with one procurement proposal. You can even choose if the system should plan a material's availability at the beginning, middle, or end of the week.

The requirements that lie in a given period are combined together. The system makes the material available on the start date of the first requirement in the procurement proposals. However, you have the option to change the requirement date to the period start date or period end date. You can use Transaction OMI4 to set the availability date as either the start or the end of the period.

Lot Size with Splitting

You can split the static or period-based lot sizes into rounding values. It's also possible to shift the orders by a certain *takt* time, which means a control parameter for scheduling based on takts (i.e., physical areas of a production line). This procedure avoids the simultaneous scheduling of numerous orders and thus ensures that the actually required lead time is observed. If you use splitting in combination with a fixed lot size, the rounding value must be a multiple integer of the fixed lot size. The takt time indicates the offset between order dates in terms of workdays. The different orders overlap in time and are shifted against each other by the takt time.

Figure 11.4 displays the entries required for this procedure in the MRP 1 view of the material master (Transaction MM02). For this example, assign LOT SIZE as "FS" (fixed lot with splitting) and give a FIXED LOT SIZE of "100". Next, set the ROUNDING VALUE as "25" (it should be a multiple of the fixed lot quantity). Further, give a TAKT TIME of "3" days (workdays). When the system runs the MRP, assuming that there is no existing stock of the material in the plant, it will create four planned orders of 25 each for the total quantity of 100 and will also delay the start date of each planned order by 3 workdays (takt time).

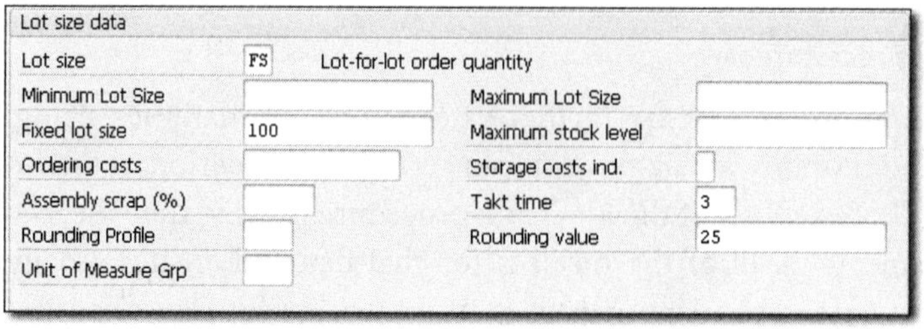

Lot size data

Lot size	FS	Lot-for-lot order quantity	
Minimum Lot Size		Maximum Lot Size	
Fixed lot size	100	Maximum stock level	
Ordering costs		Storage costs ind.	
Assembly scrap (%)		Takt time	3
Rounding Profile		Rounding value	25
Unit of Measure Grp			

Figure 11.4 Lot Size with Splitting

Optimum Lot-Sizing Procedures

The optimum lot-sizing procedures attend to economical lot-sizing requirements of the company and include the Groff lot-sizing method and part-period balancing. In these kinds of procedures, the system takes into account the costs of the procurement as a well as storage costs.

In the optimum lot-sizing procedure, the production or purchase of materials entails fixed lot size costs such as setup costs for machines or purchase order costs and variable costs, which are referred to as capital tie-up due to stockholding. To minimize the variable costs, you need to ensure that you procure quantities as small as possible. To minimize the fixed costs, you should procure quantities as large as possible. Optimum lot-sizing procedures consider variable and fixed costs and determine optimal lot sizes according to different procedures based on these costs. To determine the variable costs, you need to use the storage costs percentage specified in the STORAGE COSTS IND.

You maintain both the MRP lot size and the STORAGE COSTS IND. by following the configuration Transaction OMI4 (see Figure 11.5).

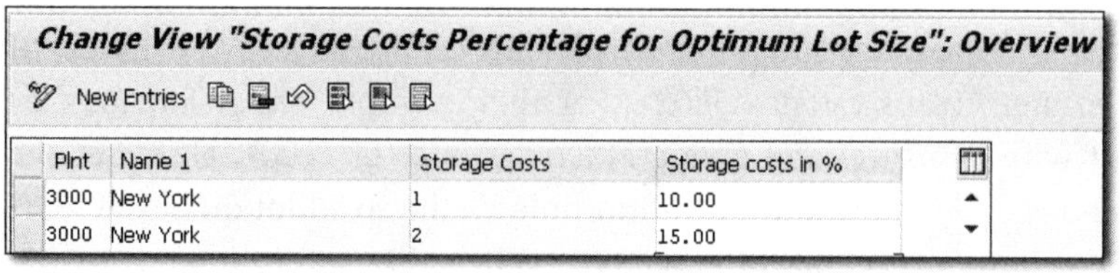

Change View "Storage Costs Percentage for Optimum Lot Size": Overview

New Entries

Plnt	Name 1	Storage Costs	Storage costs in %
3000	New York	1	10.00
3000	New York	2	15.00

Figure 11.5 Storage Costs for Optimum Lot Size

You enter the storage costs indicator along with the ordering costs in the MRP 1 view of the material master. The system calculates the variable costs according to the following formula:

Variable costs = Requirement × Price × Storage costs [%] × Storage duration ÷ 365

Figure 11.6 shows the MRP 1 view of the material master in which you enter the STORAGE COSTS IND. as well as ORDERING COSTS.

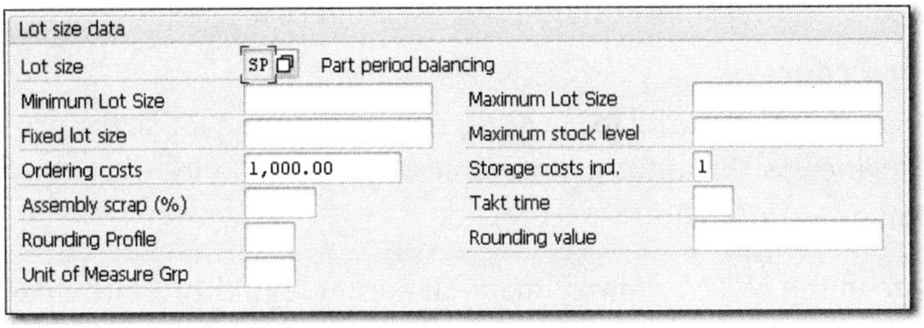

Lot size data			
Lot size	SP	Part period balancing	
Minimum Lot Size		Maximum Lot Size	
Fixed lot size		Maximum stock level	
Ordering costs	1,000.00	Storage costs ind.	1
Assembly scrap (%)		Takt time	
Rounding Profile		Rounding value	
Unit of Measure Grp			

Figure 11.6 Part-Period Balancing-Based Lot Sizing

The price is maintained in the ACCOUNTING 1 view of the material master (see Figure 11.7). Because for this example, the PRICE CONTROL is standard price (S), the system considers the standard price of $370.29 in its calculation.

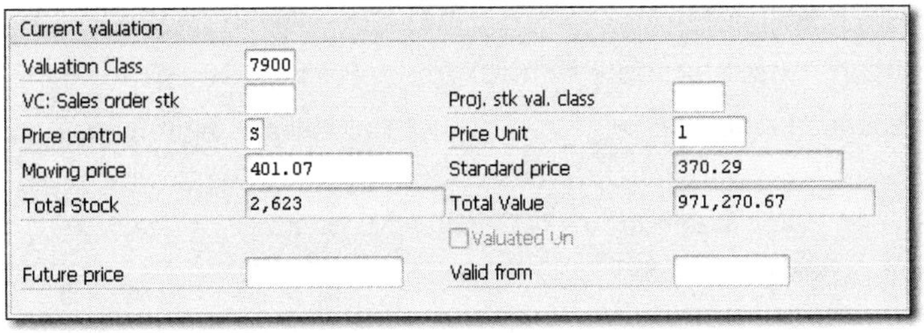

Current valuation			
Valuation Class	7900		
VC: Sales order stk		Proj. stk val. class	
Price control	S	Price Unit	1
Moving price	401.07	Standard price	370.29
Total Stock	2,623	Total Value	971,270.67
		Valuated Un	
Future price		Valid from	

Figure 11.7 Accounting Parameters for Part-Period Balancing

The lot-sizing procedures of part-period balancing, least-unit cost procedure, and Groff dynamic lot size creation use the fixed and variable costs to determine the lot size in different ways:

- **Part-period balancing**
  Part-period balancing (MRP lot size SP) determines the lot size in which the amount of variable costs is just under that of the fixed costs. For example, the lot size for the first lot was 250 with variable costs of US $400 and is below

fixed costs of US $500. The variable costs of the next lot size of 300 units (US $800) would exceed the amount of fixed costs.

- **Least-unit cost procedure**
  The least-unit cost procedure (MRP lot size WI) is based on the minimum amount of unit costs.
- **Dynamic lot size creation**
  The dynamic lot size creation (MRP lot size DY) combines different requirements until the additional variable costs exceed the fixed costs.
- **Groff lot-sizing procedure**
  The Groff lot-sizing procedure (MRP lot size GR) is an optimizing lot-sizing procedure in which the increase in additional variable costs corresponds to the gradient of decreasing ordering costs.

You assign the lot size in the MRP 1 view of material master. Table 11.1 provides a selection of standard MRP lot sizes with lot-sizing procedures.

MRP Lot Size	Lot-Sizing Procedure	Lot Size Indicator
EX: Exact lot size calculation	Static lot size	E: Exact lot size
FK: Fixed lot size for customer MTO production	Static lot size	F: Fixed lot size
FS: Fixed/splitting	Static lot size	S: Fixed with splitting
FX: Fixed order quantity	Static lot size	F: Fixed lot size
HB: Replenish to maximum stock level	Static lot size	H: Replenishment to maximum stock level
MB: Monthly lot size	Lot size according to flexible period length	M: Monthly lot size
PK: Lot size according to flexible period length and planning calendar	Lot size according to flexible period	K: Period based on planning calendar
TB: Daily lot size	Lot size according to flexible period length	T: Daily lot size
WB: Weekly lot size	Lot size according to flexible period length	W: Weekly lot size
WI: Least unit cost procedure	Optimization lot size	W: Least unit cost procedure

Table 11.1 Selection of MRP Lot Sizes and Lot-Sizing Procedures

MRP Lot Size	Lot-Sizing Procedure	Lot Size Indicator
DY: Dynamic lot size creation	Optimization lot size	D: Dynamic planning
GR: Groff lot-sizing procedure	Optimization lot size	G: Procedure according to Groff
SP: Part-period balancing	Optimization lot size	S: Part-period balancing

Table 11.1 Selection of MRP Lot Sizes and Lot-Sizing Procedures (Cont.)

When you select a certain lot size, some of them require additional information and must be maintained in the material master (see Table 11.2).

MRP Lot Size	Fields
HB: Replenish to maximum stock level	MAXIMUM STOCK LEVEL
FS: Fixed/splitting	FIXED LOT SIZE, ROUNDING VALUE, TAKT TIME
FX: Fixed order quantity	FIXED LOT SIZE
GR: Groff lot-sizing procedure	ORDERING COSTS, STORAGE COSTS IND.
DY: Dynamic lot size creation	ORDERING COSTS, STORAGE COSTS IND.
SP: Part-period balancing	ORDERING COSTS, STORAGE COSTS IND.
WI: Least unit cost procedure	ORDERING COSTS, STORAGE COSTS IND.

Table 11.2 Required Data in Fields Based on MRP Lot Size Selection

11.1.4 Configuring Material Requirements Planning Lot Size

Although standard lot sizes available in the SAP ERP system are generally sufficient to fulfill most business needs, you can configure the MRP lot size to attend to any specific business process.

Example [Ex]

Phosphoric acid is the main component used in the production of specialized fertilizer, Di-Ammonium Phosphate (DAP). The company wants to replenish to maximum stock of phosphoric acid at the time of placing new orders. However, there are supplier requirements regarding minimum and maximum order limits for shipment. To cater to this business need, you create a new MRP lot size (which may also be a copy of MRP lot size HB: Replenish to maximum stock level), and also select the CHECK MIN. (MINIMUM) LOT SIZE and CHECK MAX. (MAXIMUM) LOT SIZE checkboxes. You assign this self-defined MRP lot size to the MRP 1 view of the material master. The system now prompts you to enter the minimum and maximum lot sizes of the material. Not only will the system now make sure to replenish the material to the maximum stock level, but it will also take the minimum and maximum lot sizes into account while creating procurement proposals.

To illustrate this example further, the storage tank capacity of phosphoric acid is 40,000 MT, the minimum lot size is 20,000 MT, and the maximum is 25,000 MT. Currently, the stock in the storage tank has dropped to 10,000 MT. The system creates one planned order for 25,000 MT, which ensures that this is the maximum lot size it can go up to and still remain within the maximum storage tank capacity of 40,000 MT. If you hadn't defined the minimum or maximum lot sizes, the system would have created a planned order of 30,000 MT to replenish the existing stock of 10,000 MT to the maximum of 40,000 MT.

You can configure the MRP lot size in Transaction SPRO using menu path, LOGISTICS • PRODUCTION • MATERIAL REQUIREMENTS PLANNING • PLANNING • LOT-SIZE CALCULATION • CHECK LOT-SIZING PROCEDURES, or use Transaction OMI4.

Figure 11.8 displays the definition of MRP LOT SIZE EX for an exact lot size calculation.

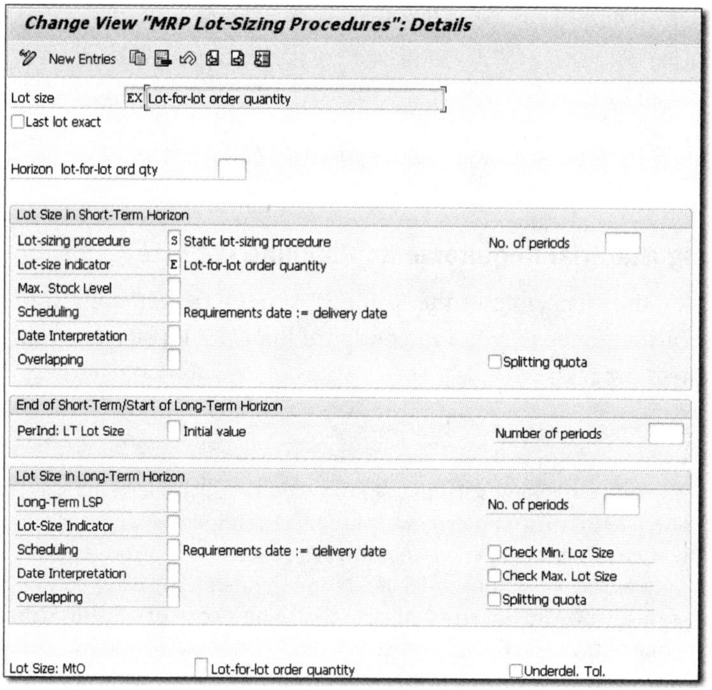

Figure 11.8 Lot Size Configuration

You can also set different or applicable lot-sizing parameters for short-term and long-term horizons. Procurement/production always takes place with the lot size of the short-term horizon. The long-term horizon enables the forecast function (mostly aggregated, e.g., via monthly periodic lot size). This is often useful in reducing the number of requirement coverage elements, especially if fixed lot sizes are used that are smaller than the average requirement quantity. Typically, a period-based lot size is used for the long-term horizon.

Tips & Tricks [+]

If fixed lot sizes are used for financial or practical reasons, and without any technical need, you can avoid surplus procurements by setting the LAST LOT EXACT checkbox in the configuration of the MRP lot size. This is particularly applicable for higher value goods.

11.1.5 Rounding

In some cases, it's useful to round the quantities of the procurement proposals. In the simplest business scenario, you can use a rounding value. In this case, the order quantity of the requirement coverage element always represents a multiple of the rounding value. In more complex scenarios where the rounding value depends on the requirement quantity, you can use a rounding profile. The rounding profile enables you, for example, to model different pack sizes.

The system offers static and dynamic rounding profiles. The *static* rounding profiles round up the value after a threshold value has been reached. If the rounding value exceeds the threshold, the system continuously rounds the value until it reaches the next threshold. You can create rounding profiles by using Transaction OWD1.

Example [Ex]

If the rounding value is 30 units, the procurement quantity is 75 units, and the minimum lot size is 15 units, then the system always rounds up the planned order to a multiple of 30 units, specifying the rounding to a quantity of 30 units. In this example, the system rounds off the procurement quantity of 75 units to 90. You can use the ROUNDING VALUE option in the MRP 1 view of the material master in almost all lot-sizing procedures.

11.1.6 Static Rounding Profile

The static rounding value method may not fulfill all of the business scenarios in which a given rounding value is used and the procurement proposals are rounding to that value. There may also be business processes in which the rounding value may change based on the size of the procurement proposal. In such instances, you can use a static rounding profile.

For example, you can configure the system to have a rounded value of 10 units if the procurement quantity is at least 1 unit. For the procurement quantity of 31 units or more, the system should round this off to 50 units.

Using Transaction OWD1, this example in Figure 11.9 ❶ shows that a ROUNDING PROFILE PP01 has been created for PLANT 3000. Click on the STATIC button, and the next screen ❷ pops up. Enter the threshold values and the rounding values. Click on the SIMULATION button to bring up the SIMULATION SELECTION dialog box ❸. Enter the QUANTITY SIMULATION of "1" unit TO QTY "150", and choose CONTINUE. The system then displays the simulation results ❹.

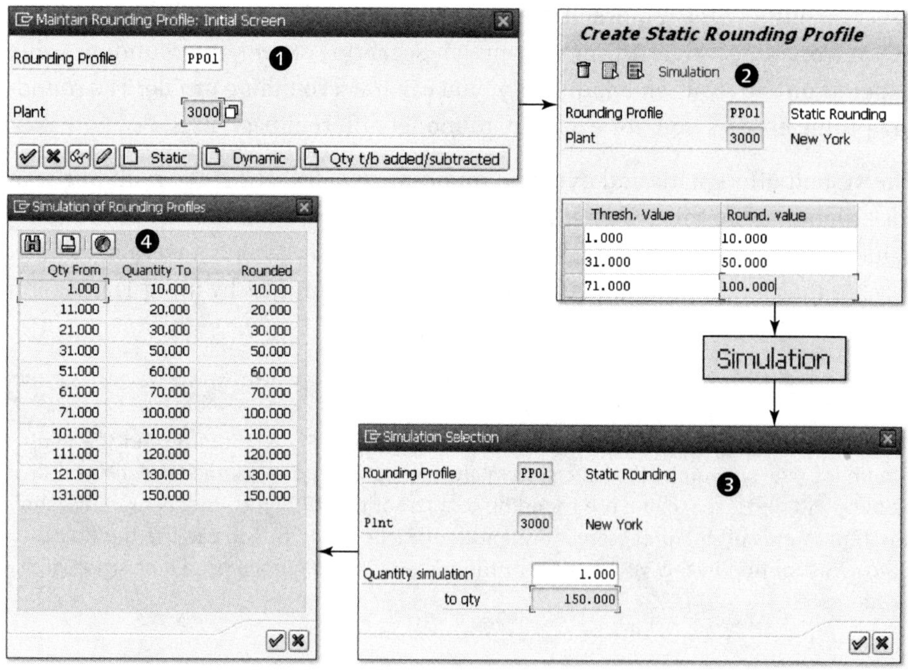

Figure 11.9 Rounding Profile (Static)

If you want to specify how the system acts when it reaches a certainly threshold value by either adding up or reducing the quantity to the defined percentage, go back to Figure 11.9 ❶. This time, click on the QTY T/B ADDED/SUBTRACTED button, and the screen shown in Figure 11.10 appears. You can define the threshold values and then enter the percentage that the system should add or subtract to the procurement quantity. So, if the procurement quantity is less than or equal to 100 PC, the system will add up to 10% to it. When the procurement quantity is greater than 100 but less than or equal to 200, it will reduce the quantity by 15%.

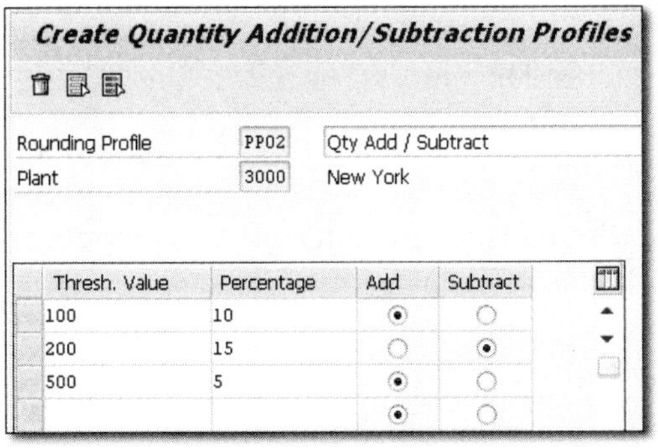

Create Quantity Addition/Subtraction Profiles

Rounding Profile PP02 Qty Add / Subtract
Plant 3000 New York

Thresh. Value	Percentage	Add	Subtract
100	10	◉	○
200	15	○	◉
500	5	◉	○
		◉	○

Figure 11.10 Quantity Addition/Subtraction Profiles

In a dynamic rounding profile, you need to define the rounding off method as well as the rounding rule. The rounding off method may be the multiple of order unit or sales order. You can also use an available customer exit to define any unique rounding method. For a rounding rule, you can either round up or down the threshold value by the defined percentage.

11.2 Scrap

Companies are often faced with problems relating to materials they can't use, known as scrap, whether from defective finished goods produced, excess consumption of raw material due to wastage in the production process, or operational inefficiencies. The SAP ERP system offers several options to enter scrap or waste details at every level—the assembly or the component (raw material) level—so that production and procurement planning are closely synchronized with actual and practical situations.

Based on historical data or practical experience, you can maintain assembly and component scrap as master data, which subsequently facilitates production and procurement processes. Scrap is treated differently in the SAP ERP system. Scrap in real life can be available or even sold but isn't treated as inventory in the SAP ERP system.

The two types of scrap are described here:

- **Assembly scrap**
  This is completely unusable finished or semifinished product for which the raw material was issued in accordance with a BOM. Assembly scrap is different from co-products or by-products, which are inventory managed in the SAP ERP system.
- **Component scrap**
  This allows the system to increase the issuance quantity of the component against a production order by the defined percentage to account for scrap or wastage during production.

Successfully assigning assembly and component scrap in master data enables an SAP ERP system to consider scrap not only during production planning—either by increasing production quantity to account for assembly scrap or increasing the issuance quantity to account for component scrap—but also during procurement planning. During the MRP run, the system takes the scrap percentage into account and reflects it in the planning proposals generated.

To enter the details of assembly or component scrap in the MRP view of the material master, follow the menu path, Logistics • Materials Management • Material Master • Material • Change • Immediately, or use Transaction MM02.

In the following sections, we'll provide additional details on the different types of scrap and also explain how to configure different uses.

11.2.1 Assembly Scrap

You can enter assembly scrap in the MRP 1 view of the material master. Here, enter the assembly scrap in the form of a percentage in the Assembly scrap (%) field. If this material is then used as an assembly, the system considers it during scrap calculation by correspondingly increasing the production order quantity. For example, say you've defined an assembly scrap of 10% for the material. When you create a production order, say of quantity 100 PC, the system automatically increases the production order quantity to 110 PC to account for 10% scrap. At the same time, it also issues an information message. Because the production order quantity increased to 110 PC, the system accordingly increases the components' quantities also, as it reads this information from the material's BOM (and multiplies the production order quantity, which in this case is now 110 PC with the components' quantities).

11.2.2 Component Scrap

You can enter component scrap in the MRP 4 view of the material master in the form of a percentage. If this material is then used as a component in the BOM, the system considers it during scrap calculation. For example, entering 10% scrap for 100 kg of component material increases the component issuance quantity in the production order to 110 kg (see Figure 11.11).

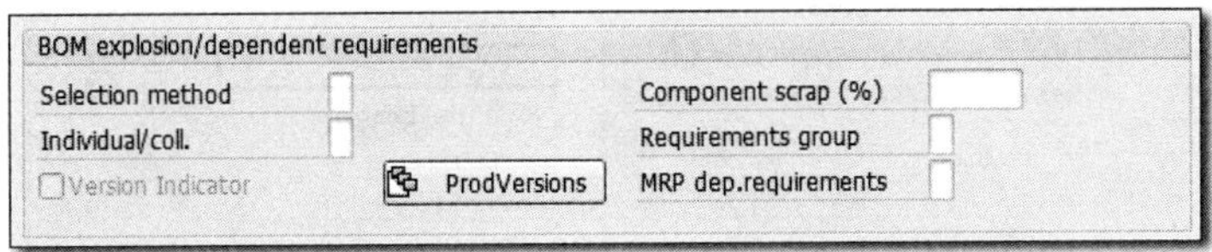

Figure 11.11 Component Scrap

11.2.3 Operations and Component Scraps in Bill of Materials

In Transaction CS02, to change a BOM for a material, click on the ITEM OVERVIEW column (Figure 11.12 ❶). In the screen that appears ❷, you can enter operation scrap as well as component scrap in percentages. When you enter operation scrap in a percentage, you also select the NET ID checkbox.

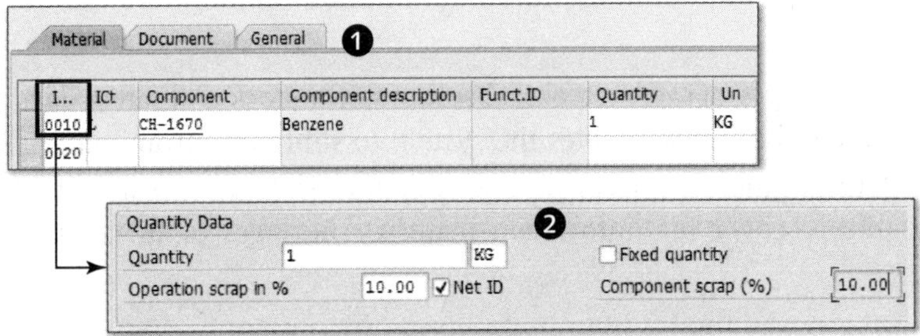

Figure 11.12 Operation and Component Scrap in BOM

If you enter both operation scrap and component scrap in the BOM, the system takes *both* of them into account. For example, for the component quantity of 1.00 kg, the system calculates the final component issuance quantity as 1.21 kg, after first taking operation scrap and then component scrap into account. That is, it used the calculations 1.00 kg × 1.10 = 1.10 kg and then again 1.10 kg × 1.10 = 1.21 kg.

11.2.4 Scrap in Routing

You use Transaction CA02 to change a routing for a material. In Figure 11.13 ❶, click on the ITEM OVERVIEW column to go to the GENERAL DATA area ❷. You can enter operation scrap as a percentage. When you create a production order, the system increases the component issuance quantity by the defined percentage to account for operation scrap in the routing.

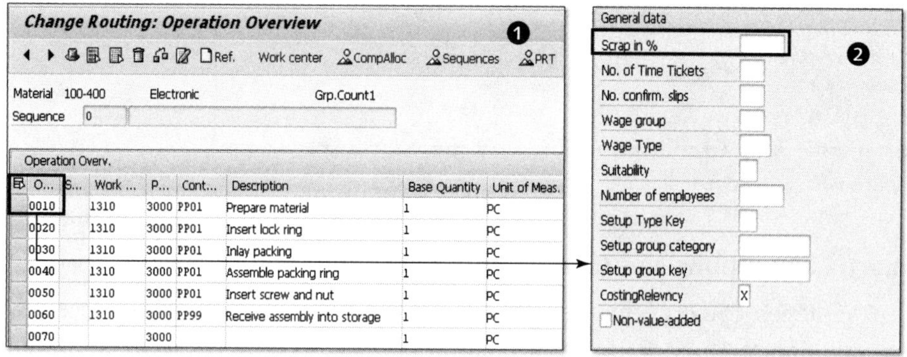

Figure 11.13 Scrap in Routing

11.3 Safety Stock

The system offers two types of safety stocks: absolute safety stock and safety days' supply. The absolute safety stock enables the system to subtract it from material availability calculations (net requirements calculation). The safety stock must always be available to cover for unforeseen material shortages or unexpected high demand.

For days' supply/safety time, the system plans the goods receipt in advance by the period specified as safety time. Thus, planned days' supply of the stock, in fact, corresponds to the number of days specified as safety time. The system shifts backs the date of the receipts by the number of working days and also takes the factory calendar into account.

You make the relevant entries in the MRP 2 view of the material master. The system refers to the planning of a safety days' supply as safety time. To consider the safety time, you need to set the SAFETY TIME IND. to "1" (for independent requirements) or "2" (for all requirements), and then also give the SAFETY TIME/ACT.COV. in days. Further, you can define deviations from the constant safety time using a safety time period profile (STIME PERIOD PROFILE field) (Figure 11.14).

Net requirements calculation

Safety Stock	50	Service level (%)	
Min safety stock		Coverage profile	
Safety time ind.	2	Safety time/act.cov.	3 days
STime period profile			

Figure 11.14 Safety Stock and Safety Days' Supply

In the following sections, we'll cover how the safety stock availability can be used in net requirements calculation and also the selection method that the system uses for receipts.

11.3.1 Safety Stock Availability

To avoid the unnecessary creation of planned orders for a smaller quantity, which can otherwise be covered by the safety stock, you can define the percentage (share) of safety stock that the system can use in such business scenarios. For example, during the MRP run, the system assesses a shortage of 2 PC of a material. You have, however, maintained a safety stock of 40 PC for this material, which the system doesn't consider in the MRP run. If you define that a 10% safety stock can be used to account for smaller shortages, which for this example means that 4 PC come from safety stock, then the system can cater to this small requirement or shortage of 2 PC from the 4 PC available from safety stock and no longer creates a procurement proposal.

To define the safety stock parameters, use configuration (Transaction SPRO) menu path, LOGISTICS • PRODUCTION • MATERIAL REQUIREMENTS PLANNING • PLANNING • MRP CALCULATION • STOCKS • DEFINE SAFETY STOCK AVAILABILITY. For this example, the system can consider 10% of the safety stock for PLNT 3000 and MRP

GROUP 0010 for net requirements calculation (see Figure 11.15). You eventually assign the MRP group in the MRP 1 view of the material master.

Change View "Safety Stock Availability": Overview

Plnt	Plant Name	MRP group	Share SStk
3000	New York	0000	
3000	New York	0010	10
3000	New York	0011	
3000	New York	0020	
3000	New York	0021	

Figure 11.15 Share of Safety Stock

11.3.2 Master Data Selection

For material that's produced in-house, you also create a BOM and routing (or master recipe). There might be several alternatives available for a BOM or a routing, so you can control the selection of alternatives. This ensures that the system is able to assign the relevant BOM and routing to the created planned orders, and it enables you to perform functions such as availability checks, and so on.

With the right control in place, the system automatically assigns the correct selection or alternate in the planned orders created from the planning run. This is accessed and managed in the MRP 4 view of the material master (Transaction MM02) as shown in Figure 11.16.

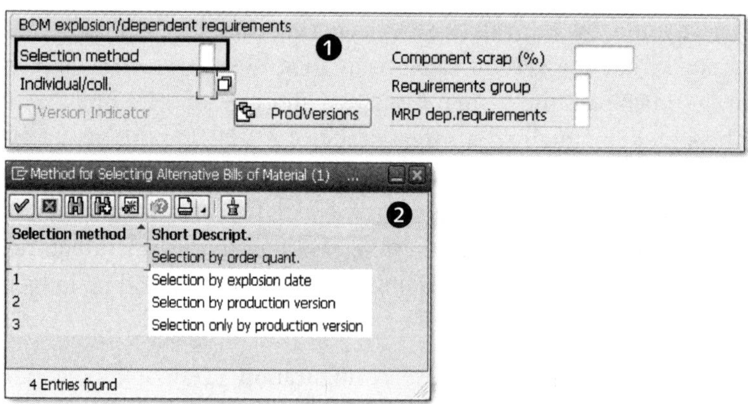

Figure 11.16 BOM Explosion Selection Method

11.4 Material Requirements Planning Procedures

When you decide to implement MRP in your company and on a specific set of materials (whether it's finished goods, raw materials, assemblies, consumables, packing materials, or spare parts), you have to determine the MRP procedures that you're going to follow. For example, you don't want to plan at all for a few materials because these materials aren't worth the time, effort, and resources involved in planning them. Similarly, for some very critical components of the supply chain, you want to accord them much higher priority than the normally planned materials. Further, to bring sanity and stability to your production and procurement plans, you want a time horizon in which the system won't make any changes to the procurement proposals between the various MRP runs.

This section begins with covering the important MRP types that cater to a large number of business scenarios and also covers the configuration settings you need to meet your business needs.

11.4.1 Material Requirements Planning Types

When you decide to plan a material using MRP, one of the very first parameters that you define in the material master is the MRP type. An MRP type is a control function used to control several subsequent steps. For example, if you define the MRP type as "ND" for no planning, the system won't plan the material at all. If you use MRP types that starts with P* (e.g., PD, P1, P2, etc.), you can control how the system takes the net requirements calculations into account and how it firms the procurement proposals for MRP during the MRP runs. The net requirements calculation takes the plant stock and all of the receipts into account and then subtracts it with requirement quantities. If the net requirements calculation comes up with a shortage or deficit quantities, the system creates procurement proposals during the MRP run. If it evaluates that there is no shortage, then the system doesn't create procurement proposals.

The same planning logic applies for MRP types that begin with M* (M0, M1, M2, etc.), but this is specific to Master Production Scheduling (MPS) materials only. MPS enables the company to optimally plan cost-intensive resources and to avoid production bottlenecks by efficiently planning the master schedule items. The best practices in planning entail that the companies first run MPS to plan critical components and then proceed with the MRP run.

If you use reorder point planning types (automatic or manual), then you have to enter the reorder point quantity of that material.

The following sections detail some of the commonly used MRP types.

Material Requirements Planning Type PD

This type of planning is carried out for the quantities planned in PIRs or quantities planned in incoming sales orders (MTO or MTS), dependent requirements, and stock transfer requirements—that is, any requirement that is MRP-relevant. In this type of net requirements calculation and its planning, the following statements apply:

- The system doesn't consider nor ask for any forecast information or historical consumption values. Material planning is ensured only on the basis of absolute requirements or demands in hand.
- The system determines the available stock in the planning run, which is required to meet the demands, with the following logic:

  Plant stock (+) scheduled receipts from production and purchase (–) all of the demands, for example, from sales order, material reservations, and PIRs

If the available stock is unable to fulfill the demand, then depending on the settings of the planning run, it creates planned orders for in-house production as well as planned orders, delivery schedules/scheduling lines, or purchase requisitions for external procurement.

Material Requirements Planning Types with the Planning Time Fence and Firming Logic

To ensure greater stability to the production and procurement plans, you can take advantage of firming logic and the planning time fence. The *firming logic* defines how the system should handle the existing or new procurement proposals by automatically firming the planned orders, allowing an option for manual firming, or allowing neither of those. The *planning time fence* allows for the definition of a range within which MRP doesn't create, change, move, or delete any procurement proposals. This can be a great reliability factor for the production department if a planning time fence such as the following is used: for the next five days starting from today, the production plan shall not be changed, not even if there are new sales orders.

You can maintain the planning time fence in the MRP 1 view of the material master or in Customizing. The planning time fence is the number of workdays (and not calendar days) within which you can protect your master plan. The system firms the procurement proposals to protect them from any changes. However, as an MRP planner, you still have the flexibility to make any last-minute manual changes, if needed. Firming of the procurement proposals (planned orders, purchase requisitions, delivery schedules) falling within the planning time fence prevents the proposals from being adapted in the next planning run. In other words, in the planning time fence, the system doesn't create or delete any procurement proposals nor does it change the existing proposals.

The system calculates the planning time fence from the current date and adds up the number of working days incorporated as the planning time fence in the material master (MRP 1 view). Furthermore, you can also define the planning time fence in the configuration settings of MRP GROUP because it's easier for you to just assign this MRP group to the material master. When the system moves forward in the time scale, it's logical that any procurement proposal, which was previously outside this planning time fence, will start to move within the planning time fence one by one. When they move in to this time fence, they are automatically firmed for protection from changes. That means the horizon keeps rolling with time.

There two firming options for procurement proposals are automatic firming and manual firming.

There are nine MRP types with firming available in the system (M0, M1, M2, M3, M4, P1, P2, P3, and P4), and these differ with regard to the automatic firming of elements within the planning time fence and also with regard to the creation of new elements.

It's important to note that firming type M0 behaves just like MRP type PD; that is, the planning time fence is inactive for this MPS time.

Note [«]

The firming type M refers to MPS, whereas firming type P refers to MRP. When working with the planning time fence, you need to set the firming methods and the scheduling methods of the procurement proposals within the planning time fence.

Table 11.3 shows the similarities of firming types for both MRP (firming type P) and MPS (firming type M). Let's consider an example of firming type P1 to explain how firming works. In firming type P1, the system automatically firms the existing procurement proposals as long as these are within the defined planning time fence. Additionally, it creates new order proposals when it evaluates that there is going to be a shortage of material within the planning time fence. However, the system pushes the new procurement proposals outside the planning time fence to ensure that the production plan isn't disturbed. When the newly created procurement proposals begin to fall within the planning time fence, the system automatically firms them and makes them a part of the production plan.

Firming Types	0	1	2	3	4
MRP Types	PD M0	P1 M1	P2 M2	P3 M3	P4 M4
Firming methods of existing procurement proposals (within the planning time fence)					
No firming of procurement proposals within the planning time fence.	✓				
Automatic firming of procurement proposals within the planning time fence.		✓	✓		
Manual firming of procurement proposals within the planning time fence (no automatic firming of procurement proposals).				✓	
Firming methods of new order proposals (within the planning time fence)					
Automatic creation of new procurement proposals to cover the shortages within the planning time fence. However, the system pushes the dates of these new procurement proposals outside the planning time fence.		✓		✓	
No new procurement proposals created to cover the material shortages within the planning time fence.			✓		✓

Table 11.3 Firming Types in Planning

You can also set how the system should account for roll-forward. A roll-forward defines the number of workdays starting from today (both forward and backward) in which the system can proceed to delete the firmed planned orders if you didn't convert them to either production orders or purchase requisitions. For example, you want to delete all of the firmed planned orders older than five days during the next MRP run. To do this, you select the MRP type with firming (e.g., P1) and select the DELETE FIRM PLANNED ORDERS option from the ROLL FORWARD field. Next, you define "–5" in the ROLL FORWARD field in MRP group parameters for the specific plant (Transaction OPPR). The negative value (number of days from today) is for deletion of the past and firmed planned order, and a positive value is to manage deletion of the firmed planned orders in the future. In the same MRP group settings (Transaction OPPR), you can also assign the number of workdays (from today) that you want the system to firm the planned orders for the relevant firming type.

Figure 11.17 shows the MRP 1 view of the material master and shows the MRP TYPE P3 with a PLANNING TIME FENCE of 3 days.

MRP procedure			
MRP Type	P3	MRP, fixing type -3-	
Reorder Point		Planning time fence	3
Planning cycle		MRP Controller	001

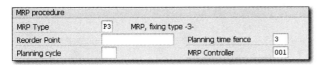

Figure 11.17 Planning Time Fence

11.4.2 Configuring Material Requirements Planning Types

You can configure the MRP type in Transaction SPRO by following the menu path, PRODUCTION • MATERIAL REQUIREMENTS PLANNING • MASTER DATA • CHECK MRP TYPES. Here you see the detailed view of MRP TYPE PD (see Figure 11.18).

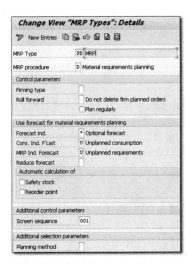

Change View "MRP Types": Details

New Entries

MRP Type	PD	MRP
MRP procedure	D	Material requirements planning
Control parameters		
Firming type		
Roll forward		Do not delete firm planned orders
	☐	Plan regularly
Use forecast for material requirements planning		
Forecast ind.	*	Optional forecast
Cons. Ind. F'cast	U	Unplanned consumption
MRP Ind. Forecast	U	Unplanned requirements
Reduce forecast		
Automatic calculation of		
☐ Safety stock		
☐ Reorder point		
Additional control parameters		
Screen sequence	001	
Additional selection parameters		
Planning method		

Figure 11.18 Configuration (Transaction SPRO) of MRP Type PD

You can control the MRP procedure that the system uses in the MRP type. Place the cursor on the MRP PROCEDURE field, and click on the dropdown option or press F4. Figure 11.19 shows the resulting available list of MRP procedures.

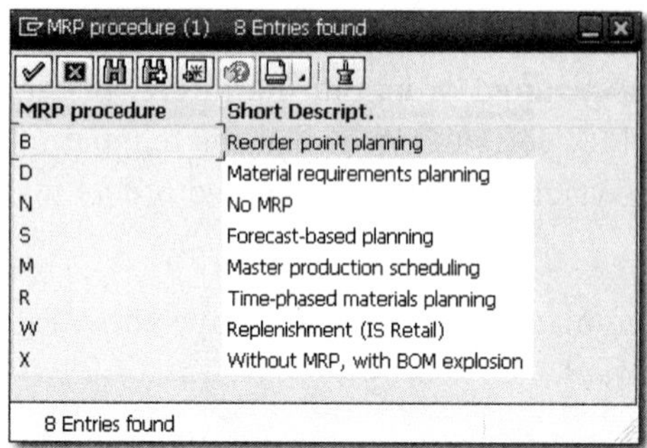
MRP procedure (1) 8 Entries found

MRP procedure	Short Descript.
B	Reorder point planning
D	Material requirements planning
N	No MRP
S	Forecast-based planning
M	Master production scheduling
R	Time-phased materials planning
W	Replenishment (IS Retail)
X	Without MRP, with BOM explosion

8 Entries found

Figure 11.19 MRP Procedures

11.5 Consumption-Based Planning

In consumption-based planning, the replenishment is triggered based on the consumption of stock. The system uses past consumption history, and if you take advantage of forecasting functionality, then the system can predict and suggest both the reorder level and safety stock level. A good example of consumption-based planning is reorder point planning, in which the planning of a material is triggered when the stock falls below a reorder point. On the other hand, you can also have forecast-based planning where planning is carried out based on forecasted figures for the material (based on historical data).

[»] **Note**

For consumption-based planning, you need to ensure that the company has a well-functioning and integrated Inventory Management functionality of the MM component in the SAP ERP system. Further, consumption-based planning requires constant replenishment lead time (RLT) and more or less constant (stable) consumption of the material under consideration.

The MRP procedures of consumption-based planning are primarily used in areas without in-house production or for planning B and C parts, as well as operating supplies in manufacturing organizations. The type of procurement proposal that is automatically generated during MRP depends on the procurement type of the material. In-house production always requires the creation of a planned order. For external procurement, the MRP controller can choose between creating a planned order and a purchase requisition. If the MRP controller wants to use a planned order, the planned order must be converted into a purchase requisition and then into a purchase order. For example, say you've set a manual reorder point planning VB for a material and maintained a reorder point level at 50. During the MRP run, the system finds that the stock of this material has fallen below 50 and creates a procurement proposal.

Following are the available planning types in consumption-based planning:

- Reorder point planning
- Forecast-based consumption planning (Section 11.6)
- Time-phased materials planning (Section 11.6.3)

In reorder point planning, the system compares the available stock at the plant level with the reorder point. The reorder point is made up of the safety stock and the average material required during the RLT. The reorder point stock ensures that there is sufficient stock available to cover the demand during the period it takes to replenish the stock. Accordingly, when defining the reorder point, you need to take into account the safety stock, the historical consumption, or the future requirements and RLT. The safety stock needs to cover the excess material consumption during the RLT as well as the additional requirements if delivery is delayed. Manual reorder point planning requires the users to define the reorder point and safety stock by themselves and store these parameters in the corresponding material master record. In automatic reorder point planning, the reorder point and safety stock are determined by the integrated forecast program.

Using historical material consumption values, the program determines the future requirements. Depending on the service level defined by the MRP controller and on the RLT of the material, the reorder level and the safety stock are then calculated and transferred to the respective material master record. Because the forecast program is run at regular intervals, the reorder point and safety stock are adapted to the current consumption and delivery situation and thus help to reduce the stock level.

Note [«]

While we only discuss reorder point planning here, you can find discussions of the other types of planning in Section 11.6.2 and Section 11.6.3.

In the following sections, we'll cover planning types that are frequently used in consumption-based planning.

11.5.1 Type VB: Manual Reorder Point Planning

In manual reorder point planning (MRP type VB), when the plant stock and firmed receipts of a material fall below the reorder point, the system triggers planning for the material. Figure 11.20 shows the principle of reorder point planning. You need to give due consideration to the historical consumption, as well as the future requirement and the timeliness of the supplier's or production department's service provision when defining the safety stock.

In reorder point planning, you normally maintain a safety stock, which you enter in the MRP 1 view of the material master, although it isn't mandatory. Safety stock accounts for any additional unplanned consumption during the RLT. You can also make the system calculate the safety stock automatically. To do so, select the relevant checkbox in the configuration of the MRP type. It greatly helps the planner if the RLT is maintained as accurately as possible.

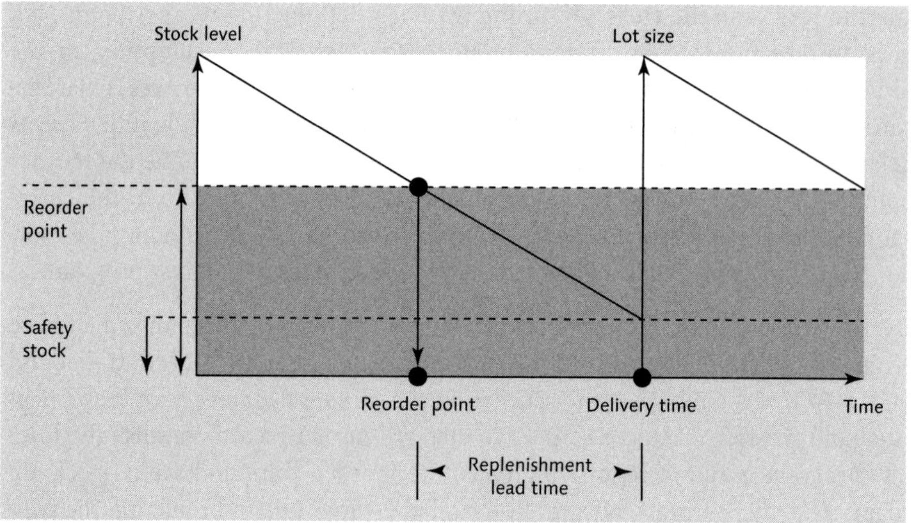

Figure 11.20 Reorder Point Planning

For reorder point planning, the system doesn't take incoming demand into account, unless you select the reorder point planning with external requirements. In fact, the demand plays no role in planning or creating procurement proposals.

The system creates the procurement proposal for a quantity equal to the reorder point or equal to a fixed lot size, if it's maintained in the material master. If procurement proposals already exist for the material with quantities greater than the required quantity, the system doesn't create a new procurement proposal.

[+]

Tips & Tricks

Use the fixed lot-sizing procedure for reorder point based planning so that the system procures the fixed lot size, instead of only creating a proposal for the deficit quantity.

You need to enter the reorder level manually in the MRP 1 view of the material master for manual reorder point planning. For automatic reorder point planning, the system automatically calculates the reorder point.

You can configure the MRP type in Transaction SPRO by following the menu path, Production • Material Requirements Planning • Master Data • Check MRP Types.

Figure 11.21 shows the configuration screen of manual reorder point planning with external requirements (MRP Type V1). Notice that you can take advantage of the system's functionality of automatically calculating safety stock as well as reorder point. For configuration of the MRP type, refer to Section 11.4.2.

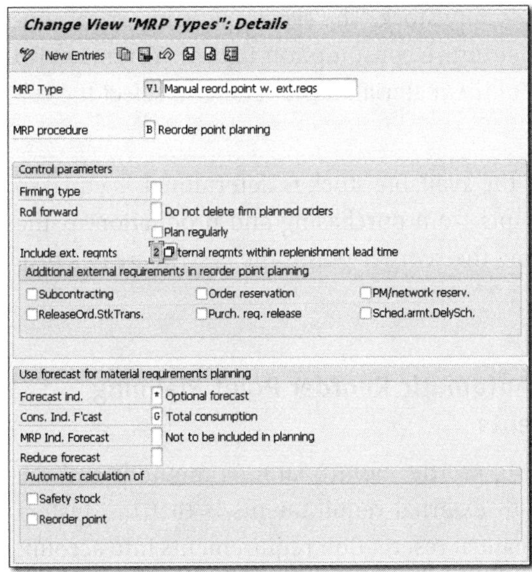

Figure 11.21 Configuration Settings for Reorder Point Planning (Manual)

Reorder point planning with external requirements represents a special case. This MRP procedure uses sales orders and manual reservations to calculate the available stock. The corresponding MRP types are V1 for manual and V2 for automatic reorder point planning. Choosing the MRP type allows you to define whether you want to consider the external requirements only within the RLT or in the entire horizon.

The dropdown options become available when you place the curser in the INCLUDE EXT. REQMTS field as shown in Figure 11.22.

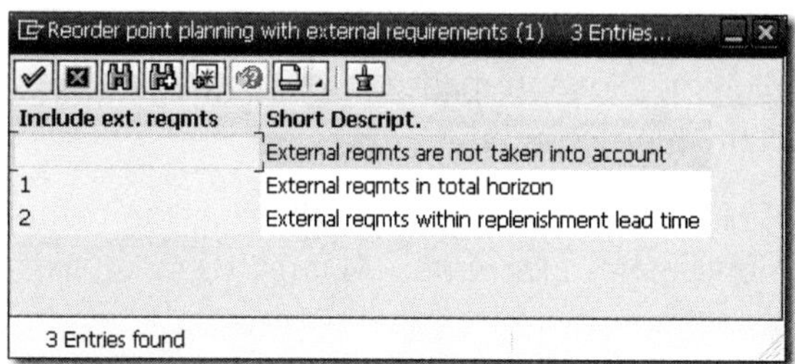

Figure 11.22 External Requirements in Reorder Point Planning

11.5.2 Type VM: Automatic Reorder Point Planning

In automatic reorder point planning (MRP type VM), the system automatically calculates the recorder point and the safety stock. The system uses historical consumption data to predict and arrive at future consumption figures. You may also need to extend the forecasting view of the material master and also select the relevant forecasting model.

In the net requirement calculations, the available stock is determined as equal to the plant stock and the firmed receipts from purchasing and production. If the existing stock of a material falls below the reorder point, the system creates procurement proposals.

11.5.3 Type V1/V2: Manual or Automatic Reorder Point Planning with External Requirements

The difference between the preceding reorder point planning methods and the reorder point planning methods with external requirements is that the system takes sales order requirements and manual reservation requirements into account when calculating the net requirement.

The system considers external requirements for the period within the RLT or within the total planning horizon. You can configure specific settings in the Customizing of the MRP type.

11.6 Forecast-Based Consumption Planning

Some of the consumption-based MRP procedures require that you execute and save the forecast before the system carries out MRP. In such cases, the FORECASTING indicator in the relevant MRP type is marked as mandatory/obligatory, rather than optional. When you execute the forecast on a material, the system looks for the forecasting parameters that you've maintained in the forecasting view of the material master, such as forecasting model and historical or forecast (future) periods. Further, depending on the model that you've selected, it also considers smoothing factors such as alpha, beta, or gamma factors. In forecasting, the system considers the planned and unplanned consumption quantities of the material for the periods under evaluation and comes up with forecast (future) figures. When the forecast results are saved, the MRP for forecast-based consumption planning uses these forecast figures to create procurement proposals after the MRP run. This type of planning is generally applicable to those materials where the consumption is unpredictable and varies greatly, and the company has to rely on past consumption figures to arrive at realistic future consumption figures.

The following MRP procedures require executing a forecast as a prerequisite to running MRP:

- Forecast-based planning
- Time-phased materials planning

The system updates the consumption data in the material master as total consumption (planned consumption and unplanned consumption) or unplanned consumption. To view the consumption details of the material, go to BASIC DATA view of a material master (Transaction MM02), and click on the ADDITIONAL DATA icon. Of the several tabs, the last tab holds consumption quantities. Per the MRP reservations, the system updates the consumption data of the planned consumption. If consumption isn't planned per MRP reservations, or if the material issuance quantity is greater than the reservation quantities, then the system updates the unplanned consumption.

In MRP type configuration, you can choose the type of historical consumption of a material for planning whether the system uses total consumption or unplanned consumption. You can also choose the reduction method for forecast requirements. The default system setting is reduction by consumption.

In the following sections, we'll show you how to set up forecasting and then use it in materials planning.

11.6.1 Basics of Forecasting

We've already covered forecasting in detail in Chapter 9. In this section, we'll only briefly cover details relevant to forecast-based consumption planning. The consumption forecast for components is based on material withdrawals. You can maintain the forecast parameters in the FORECASTING view of the material master, as shown in Figure 11.23.

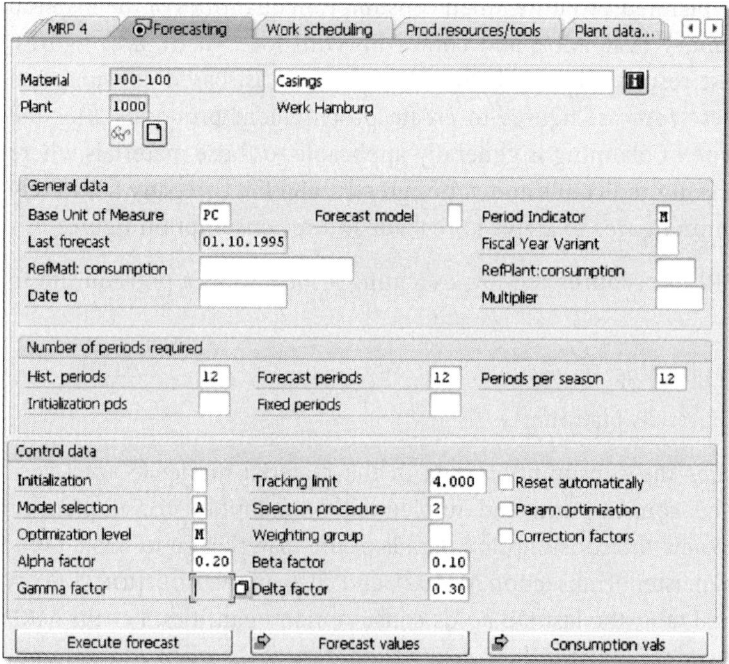

Figure 11.23 Forecasting View of the Material Master

The parameters include the granularity of the forecast, that is, whether it's carried out on the basis of monthly, weekly, or daily values. The granularity can be defined using the PERIOD INDICATOR. Other entries specify the number of historical periods based on which the forecast is carried out, as well as the number of forecast periods. You can either specify the forecast model directly or have it determined automatically. Possible forecast models include the following:

- Constant model (D)
- Trend model (T)
- Seasonal model (S)
- Seasonal trend model (X)
- Moving average (G)

The EXECUTE FORECAST button enables you to carry out a forecast in the material master, while the other two buttons allow you to edit the historical values and the forecast values.

You can run an individual forecast for a material or as an overall forecast for several materials. For an individual material forecast, use Transaction MP30. After executing the forecast, you can change the forecast results and also correct forecast values interactively. You can change the forecast using Transaction MP31 and display the forecast using Transaction MP32.

For the overall forecast, use Transaction MP38 to restrict the planning scope by plants, materials, and ABC indicators. The ABC indicator can be maintained in the MRP 1 view of the material master.

In the following sections, we cover forecast-based and time-phased materials planning.

11.6.2 Type VV: Forecast-Based Planning

In forecast-based planning, you have to execute material forecasts first. Based on the results of the forecast, the system calculates and plans future consumption quantities. You can select the forecasting periods to be in days, weeks, months, or for the entire year.

You can specify the number of historical periods that the system uses in forecasting and also the future forecast horizon (periods) for each material. After the planning run, the system makes material available in the beginning of the period. You have the option to further divide or split these material requirements into finer detail by using the Splitting indicator in the MRP 3 view of the material master.

The system can consider safety stock in the net requirement calculations. The system takes receipts into account to ensure the demand from the forecast is met. If the demand isn't met, then the system creates procurement proposals.

The period pattern for the forecast (day, week, month, or posting period) and the number of prediction periods can be defined separately for each material. However, the period pattern of the forecast may not be detailed enough for MRP because it may not provide the required granularity. For this reason, you can use the Splitting indicator to define for each material how the forecast requirement values may further be distributed to a more detailed period pattern for MRP purposes.

You can maintain the Splitting indicator in configuration (Transaction SPRO) by following the menu path, Production • Requirements Planning • Forecast • Define splitting of Forecast Requirements for MRP. The Splitting indicator contains the number of periods to be considered as well as the number of days and/or weeks for which the splitting is to be carried out (see Figure 11.24).

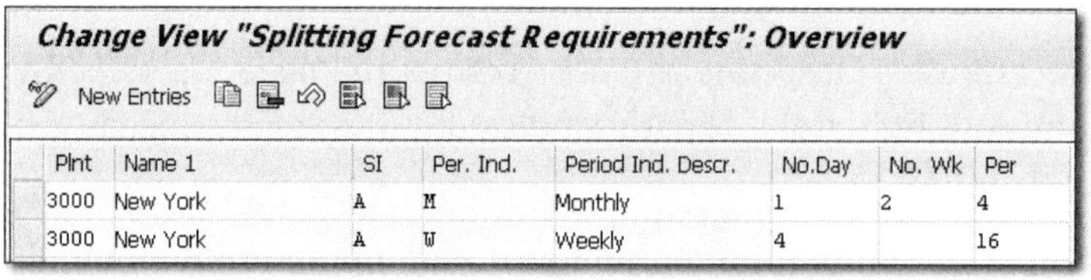

Change View "Splitting Forecast Requirements": Overview

Plnt	Name 1	SI	Per. Ind.	Period Ind. Descr.	No.Day	No. Wk	Per
3000	New York	A	M	Monthly	1	2	4
3000	New York	A	W	Weekly	4		16

Figure 11.24 Distributing the Forecast Values to Different Periods

In the MRP 3 view of the material master, you can assign the Splitting indicator in the relevant field (Figure 11.25).

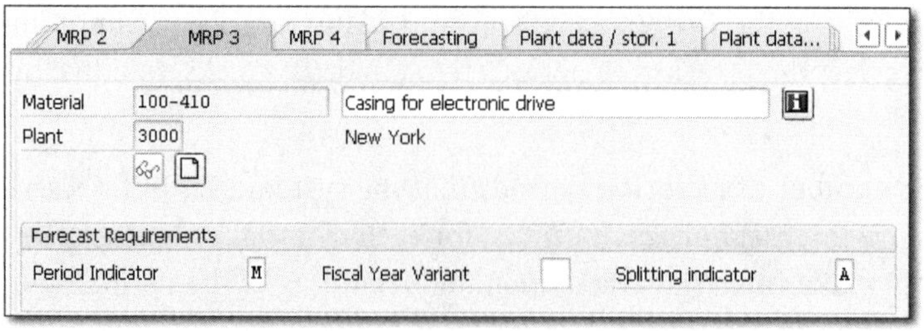

Figure 11.25 Splitting Indicator in the Material Master (MRP 3 View)

11.6.3 Type R1: Time-Phased Planning

If a vendor always delivers a material on a certain day of the week due to its own internal route and delivery planning, then it makes sense to carry out your company's planning run that follows the same cycle and is displaced by the delivery time. Similarly, if the company procures or produces certain material during specific times of the year only, then time-phased planning is used. For example, chocolate manufacturers in a few hot and humid climatic countries cease production during the summer months. In such cases, all production and procurement planning are confined to the defined planning cycle only.

This can be done using the MRP procedure of time-phased materials planning. To plan a material based on time phases, you must use Transaction MM02 to enter the MRP type for time-phased planning and the planning cycle in the material master (see Figure 11.26). A forecast is also used in this planning concept, and the general procedure is very similar to forecast-based planning with the difference that the procurement intervals are predefined.

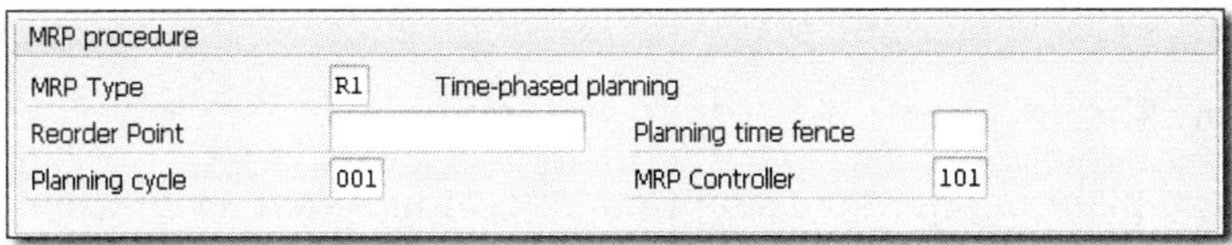

Figure 11.26 Time-Phased Materials Planning

For configuration of a planning cycle, refer to Section 11.13 in this chapter. The configuration basics of setting up a planning cycle is same as setting up a planning calendar.

11.7 Types of Planning Runs

Depending on your business process, you can use one of the several types of planning runs available in the SAP ERP system. A planning run denotes the level and mode at which you want to run MRP. For example, you may want to run MRP on an individual plant or group of plants, individual material or group of materials (product group), or in online mode or in background mode. We also cover how you can run MRP on MTO or ETO production methods.

[»] **Note**

To learn how to perform MRP for a single-item, multi-level planning run in an actual business case in the fertilizer industry, see Chapter 12 on LTP. LTP makes extensive use of MRP functionalities and at the same time offers greater similarities with MRP, including navigation.

Some of the planning runs include the following:

- Single-item, single-level
- Single-item, multi-level
- Total planning online
- Total planning background
- Single-item planning, sales order
- Single-item planning, project

We discuss each of the planning runs in the following sections.

11.7.1 Single-Item, Single-Level

You can use this type of planning run for a single material at the plant level, when you only want to plan one single material without disturbing the planning situation of all other dependent materials (no BOM explosion). You use Transaction MD03 for a single-item, single-level planning run and its BOM explosion. For example, due to sudden changes in the production plan of a single material only, you want to run MRP to attend to the changed planning situation and enable the system to create new procurement proposals. Because the components for this material (the material BOM) are already available in stock, you don't want to disturb their planning, so you plan only the single item at the single level.

11.7.2 Single-Item, Multi-Level

This type of planning run is used when you want to plan a single material, including its BOM. Because the assemblies in the BOM may also contain additional assemblies and components, these are also planned.

For single-item, multi-level types of planning runs, you use Transaction MD02. Here the system plans the single material (at a plant) that you provided on the initial parameters selection screen.

When using this planning run option, you can use the DISPLAY PLANNING RESULTS BEFORE THEY ARE SAVED option to enable you to check the planning results before saving. During the planning run, the system stops at each breaking point, which allows you to save the planning results created so far and move to the next breaking point. Alternatively, you can select the option to enable the system to continue planning without stopping at each breaking point.

You can also select the DISPLAY MATERIAL LIST option, and the system displays the following information at the conclusion of the planning run:

- Number of materials that have been planned
- Planning parameters that have been used in the selection screen
- Number of materials that have been planned with exceptions
- Number of planned orders created
- List of materials that were planned
- Start time of the planning run
- End time of the planning run

11.7.3 Total Planning Online

In this type of planning run, the system carries out a multi-level planning run, that is, for all of the materials in a plant, for all of the materials in multiple plants, or for MRP, as defined through SCOPE OF PLANNING, which is available in the selection screen. Total planning online is carried out using Transaction MD01.

Warning! [!]

Due to the fact that online usage of scope of planning as well as total planning online consumes enormous system resources and can have performance issues, running it in the foreground isn't recommended. Rather, it should be run in the background and also during the slow working hours (e.g., in the evening or overnight).

If your business process requires you to consider and use other parameters in the MRP run, you can either define or use a user exit. For example, if you want to carry out the MRP run for a set of MRP controllers only or for a specific procurement type (E: In-house managed, F: Externally procured, X: Both), then you use a user exit. The user exit enhancement is `M61X0001`, and you need to engage an ABAP consultant for this.

11.7.4 Total Planning Background

With the total planning background method, you can carry out the planning in the background, as opposed to the foreground or online. Using Transaction MDBT, you can create a variant of the Transaction MD01 screen and schedule it in the background on a periodic basis. The program name for Transaction MDBT is RMMRP000. Use Transaction SE38, give the program name, and define the variant to execute the MRP in the background.

[+]

Tips & Tricks

Running MRP is a time- and system-intensive activity; it's normally scheduled during slow working hours (in the evening or overnight). Therefore, total planning as a background job is the standard case for MRP.

11.7.5 Single-Item Planning, Sales Order

For MTO production, you can use single-item planning for a sales order. You need to give the sales order number together with the line item of the sales order and use Transaction MD50 to execute the planning. The system not only creates a procurement element or receipt with the sales order reference for the finished good but also plans all of the components defined in the BOM of the finished good.

[»]

Note

To see how the single-item planning, sales order planning run works with an actual business case (based on a sales order and its corresponding line item), see Chapter 20, in which we cover integration aspects of the PP component with the SD component.

11.7.6 Single-Item Planning, Project

You can also execute separate planning for ETO materials. If the material—finished good or a spare part of a raw material—is assigned to a work breakdown structure (WBS) element, then you can use Transaction MD51 to execute planning.

> **Note** [«]
>
> To see how the single-item planning, project planning run works with an actual business case (based on a WBS element), see Chapter 20 where we cover integration aspects of the PP component with the PS component.

11.8 Scheduling

For the system to automatically schedule procurement proposals during the MRP run, you need to ensure that you enter the requisite information in the MRP views of the material master. The requisite master data is listed here:

- In-house production time (applicable for in-house production)
- Planned delivery time (applicable for external procurement)
- Schedule margin keys for float times (optional)
- Interoperation times (optional)
- Planned goods receipt times (optional)
- Setup times and teardown times (optional for in-house production)

We'll explain each of these master data steps in the following sections.

11.8.1 Scheduling In-House Production

You can schedule the in-house production in the planning run, and the system creates planned orders in two different ways:

- **Basic date determination**
  The system schedules the dates based on the production time defined in the material master.
- **Lead time scheduling**
  The system schedules the dates (and time) based on the information in the routing (or master recipe).

When you run the MRP, you can select the option for the system to schedule based on basic date determination or lead time scheduling.

11.8.2 Basic Date Determination

Basic date determination is a type of scheduling that determines the basic dates within which production is to take place. For this reason, the in-house production time defined by the basic dates should be bigger than or equal to the in-house production time that is defined by the routing explosion. In MRP, the system first calculates the basic order dates from the data in the material master. Then, in detailed planning, it uses the data in the routing. In the context of basic date scheduling, no operation dates are defined, and no capacity requirements are generated. The determination of the in-house production time can be carried out as quantity dependent or independent. In both cases, the system reverts to the entries in the material master and takes the factory calendars into account.

[+] **Tips & Tricks**

You can use Transactions CA96 and CA97 to update the IN-HOUSE PRODUCTION parameter in terms of DAYS in the MRP 2 view of the material master (see Figure 11.27). At the same time, Transaction CA97 helps you to update the ASSEMBLY SCRAP field in the material master and is taken from the total operation scrap in the routing of material.

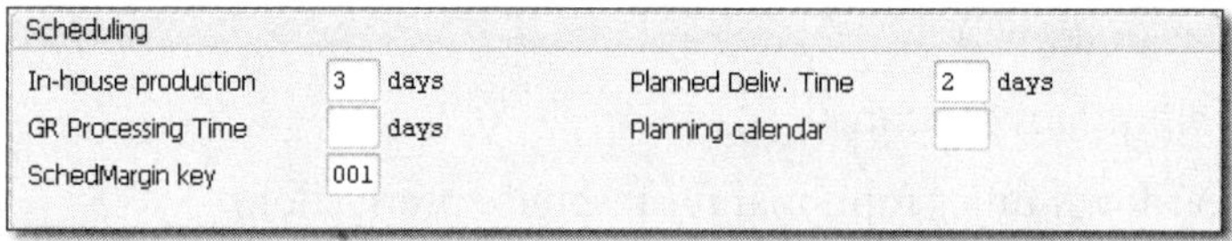

Figure 11.27 In-House Production Time and Scheduling Margin Key

[»] **Note**

The scheduling margin key (SMK) is the float consisting of the number of days to release the production order and the number of days before and after production when the material finally becomes available. SMKs are buffer times for scheduling.

The next two sections cover in-house production times that are either dependent or independent of quantity produced.

Quantity-Independent In-House Production Time

In this case, the in-house production time doesn't depend on the order quantity; that is, small quantities require the same amount of time as large quantities. This assumption is valid for planning if the order quantity is limited by the lot-sizing procedure, if experience has shown that order quantities usually don't differ very much, or if the daily granularity can cover the fluctuations of the order quantity.

Quantity-Dependent In-House Production Time

If you can no longer ignore the influence of the order quantity of the in-house production time, you can also calculate the in-house production time as quantity-dependent. For this purpose, you need to carry out a detailed evaluation of the in-house production times and maintain the Setup time, Interoperation, and Processing time fields as well as the Base quantity in the Work Scheduling view of the material master (see Figure 11.28). The interoperation time represents a summary of move times, wait times, and safety times. Only the processing time is quantity-dependent because it depends on the base quantity. The setup and interoperation times are quantity-independent and are added to the processing time.

The following formula describes the quantity-dependent in-house production time:

> In-house production time = Setup + (Order quantity ÷ Base quantity) × Processing time + Interoperation time

You can maintain parameters either for the quantity-independent or for the quantity-dependent in-house production time (see Figure 11.28).

In-house production time in days

Lot size dependent				Lot size independent	
Setup time	1	Interoperation		InhseProdTime	3
Processing time	1	Base quantity	10		

Figure 11.28 In-House Production Time in Days

11.8.3 Planned Order Dates

In backward scheduling, you can derive the dates for the planned order from the requirement date. To do this, you must deduct the various periods, including opening period, in-house production time, and goods receipt processing time, from the requirement date.

In forward scheduling, the system shifts the order start date to the current date, and the opening period is omitted. Forward scheduling occurs in MRP only if the requirement date is too close to the current date.

Figure 11.29 shows the DATES area for the planned order for the material, in which it takes the various dates into account to arrive at the start and finish dates of the order. Because this is backward scheduling, enter the ORD.FINISH (order finish) date as "4/30/2013". The system uses SMK 001 to come up with the basic date calculations. From the finish date 4/30/3013, it first deducts a float before production of 2 days and a release period of 5 days to come up with the START date of 04/22/2013. The system also deducts 10 more days to account for the PLND OPEN. (planned opening) period. All dates are in workdays (Monday through Friday) per the factory calendar (US).

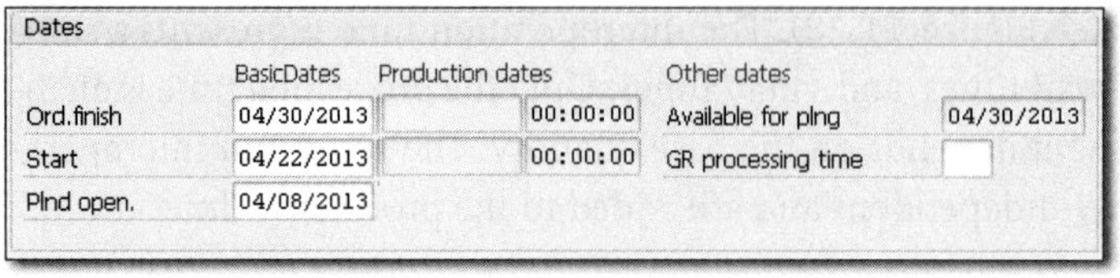

Dates

	BasicDates	Production dates		Other dates	
Ord.finish	04/30/2013		00:00:00	Available for plng	04/30/2013
Start	04/22/2013		00:00:00	GR processing time	
Plnd open.	04/08/2013				

Figure 11.29 Basic Dates Determination in a Planned Order

While remaining in the CHANGE PLANNED ORDER screen, you can also perform scheduling by clicking on the relevant icon. In scheduling, the system accurately calculates the time it takes to perform each activity at every operation of the work center.

11.8.4 Scheduling External Procurement

When planned orders for externally procured materials or purchase requisitions are created, they are often not immediately assigned to a supplier. For this reason, the PLANNED DELIV. TIME parameter in the material master is used for scheduling (see Figure 11.30).

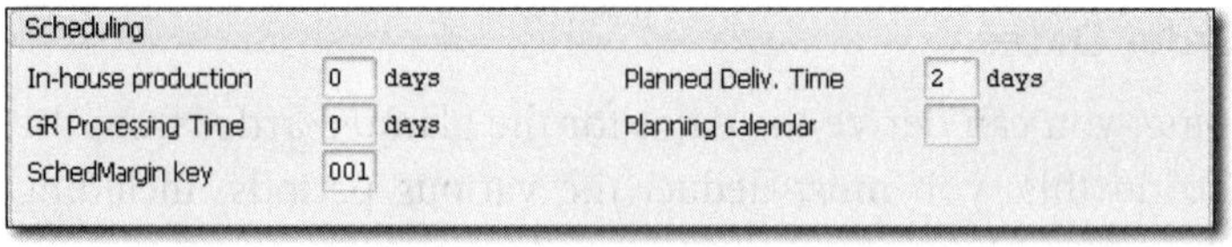

Scheduling

In-house production	0 days	Planned Deliv. Time	2 days
GR Processing Time	0 days	Planning calendar	
SchedMargin key	001		

Figure 11.30 Scheduling for Planned Delivery Time

The planned delivery time refers to workdays that appear in the factory calendar, provided you haven't specified a planning calendar. As an alternative to determining the planned delivery time from the material master, you can determine the planned delivery time from the purchasing info record. If you want to do this, this scheduling type must be allowed in the plant parameters. Moreover, the source list for the material must contain a relevant entry.

Tips & Tricks [+]

You can use Transaction WPDTC to calculate and compare the planned delivery time of a material (or group of materials) in the material master, PIR, and vendor.

The scheduling process for external procurement is carried out similarly to that of in-house production except that it uses four periods (for backward scheduling):

- Opening period
- Purchasing department processing time
- Planned delivery time
- Goods receipt processing time

These periods are deducted from the requirement date.

The planned delivery time corresponds to the in-house production time in in-house production, and the opening period and goods receipt processing time are similarly maintained and used. The purchasing department processing time is maintained in the PLANT PARAMETERS area (see Figure 11.31). These options are described in Section 11.11.3.

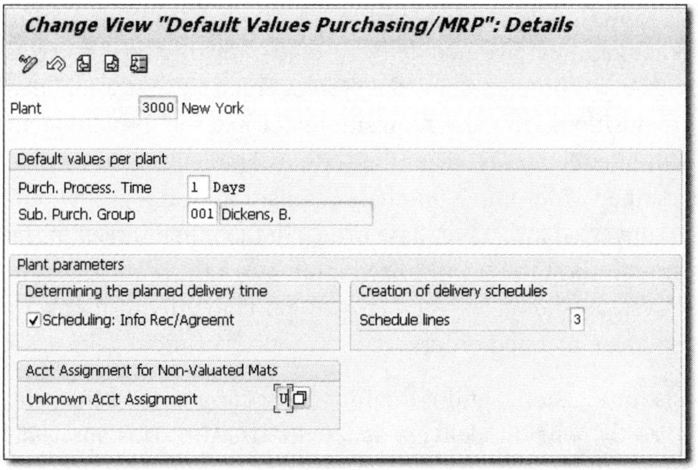

Figure 11.31 Purchase Processing Time in Configuration

11.8.5 Forward and Backward Scheduling

You can make a distinction between the scheduling types regarding whether you want to use backward scheduling or forward scheduling, and you can also choose whether production takes place in-house or externally.

Similarly, the following interdependencies also exist for different MRP procedures:

- Materials planned in forecast-based planning
- Material requirements planning

For the two MRP procedures, the start dates are known, and the planning should ensure that materials are available on these dates. The determination of basic dates for in-house production and of the release date for purchasing therefore occurs as *backward scheduling* in these MRP procedures. Only if the basic date determined during scheduling is in the past does the system automatically switch to *forward scheduling* from today's date (when the MRP run took place).

In reorder point planning, the dates are defined by means of forward scheduling. After the reorder point is fallen short of during the planning run, you need to initiate the procurement activities. The availability date of the materials is determined based on the date of the material shortage.

For external procurement, the order finish date corresponds to the delivery date, while the order start date corresponds to the release date. Here, the processing time in purchasing is taken into account as well as the goods receipt processing time in terms of workdays and the planned delivery time of the material in terms of calendar days. For external procurement, the system creates either a planned order or a purchase requisition. Purchase requisitions for external procurement plan the external procurement quantity. After the system completes the planning, you can convert the planned order into a purchase requisition, and then the purchase requisition gets converted into a purchase order. In the initial screen of the planning run, you can use the CREATION indicator for purchase requisitions to define whether the system should directly create purchase requisitions or whether it should first create planned orders.

If a delivery plan exists for a material and if the source list contains an MRP-relevant entry, you can directly generate delivery schedules in MRP. This applies if you have a scheduling agreement of materials with the vendor. You can do this by specifying the CREATION indicator for delivery schedules in the initial screen of the planning run.

11.9 Procurement Proposals

When you run MRP in the system, the system looks for procurement type information to decide if the ensuing procurement proposals entail in-house production, external procurement, or both. The PROCUREMENT TYPE in the MRP 2 view of the material master defines which of the alternatives applies to the material in the plant (see Figure 11.32). If both in-house production and external procurement are allowed, MRP assumes in-house production by default. For example, the raw material is generally not sold or produced in-house, so it makes sense to maintain the procurement type as external. Similarly, the finished product isn't procured but instead is produced in-house, so it makes sense assign it an in-house procurement type. Finally, the semifinished good may be produced in-house as well as procured externally, so you may assign it the procurement type as both.

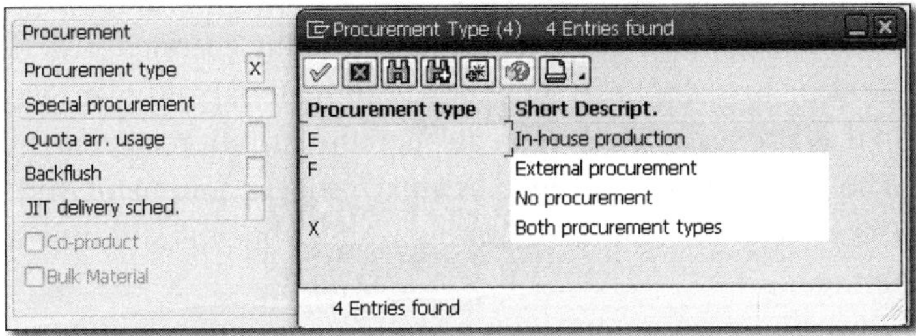

Figure 11.32 Procurement Type in the Material Master

In the planning run, the system creates planned orders. These planned orders are converted into a production order (or process order) for in-house production. For external procurement, the planned order can be converted into a purchase requisition, which in turn is converted into a purchase order or purchase order item.

In the following sections, we'll discuss how to interactively create planned orders from a stock/requirements list. We'll also show you how to independently create a planned order for a material and the planned order profile it uses. We'll also show you how to check for components' availability, including available-to-promise (ATP) logic.

11.9.1 Planned Orders

Planned orders contain an order quantity, a start date, and a finish date. The start and finish dates are the result of the scheduling process and are based on the factory calendar. The availability date can occur after the finish date if a goods receipt processing time has been maintained.

In addition, the planned order contains the dependent requirements from the BOM explosion as well as organizational data.

There are a couple of ways in which you can manually create planned orders:

- While remaining in the stock/requirements list (Transaction MD04)
- By using Transaction MD11, which is the planned order creation transaction

For this example, we show you how to create a planned order in the stock/requirements list. Later, we'll also show how to manually create a planned order using Transaction MD11.

Planned Order Creation with Interactive Planning

You can carry out interactive planning by specifically creating planned orders using Transaction MD11 or by using interactive MRP for material and plant via Transaction MD43. In the latter case, the system displays the current requirement and stock situation. The PLANNING button enables you to trigger automatic MRP (see Figure 11.33). The system displays the planning result, which is still only a simulation at this point.

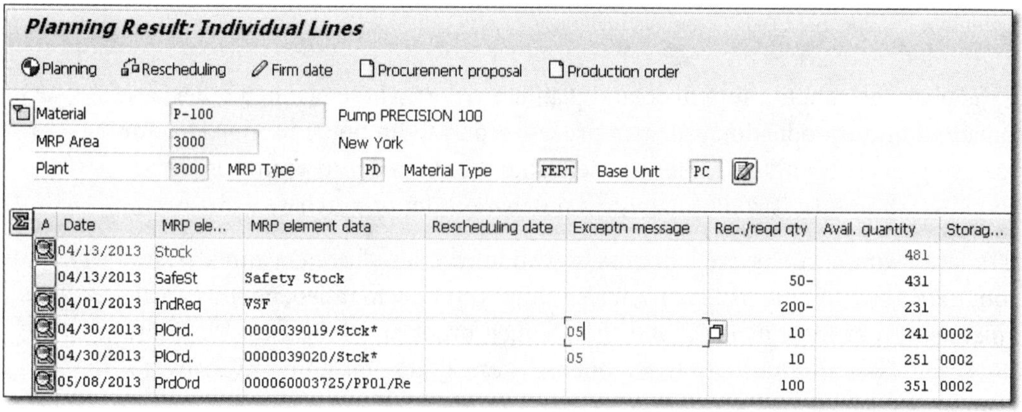

Planning Result: Individual Lines

Planning | Rescheduling | Firm date | Procurement proposal | Production order

Material P-100 Pump PRECISION 100
MRP Area 3000 New York
Plant 3000 MRP Type PD Material Type FERT Base Unit PC

A	Date	MRP ele...	MRP element data	Rescheduling date	Exceptn message	Rec./reqd qty	Avail. quantity	Storag...
	04/13/2013	Stock					481	
	04/13/2013	SafeSt	Safety Stock			50-	431	
	04/01/2013	IndReq	VSF			200-	231	
	04/30/2013	PlOrd.	0000039019/Stck*		05	10	241	0002
	04/30/2013	PlOrd.	0000039020/Stck*		05	10	251	0002
	05/08/2013	PrdOrd	000060003725/PP01/Re			100	351	0002

Figure 11.33 Planning Results

The system doesn't create the planned order until you save the planning results. Moreover, Transaction MD43 also enables you to create and reschedule specific procurement proposals manually, when you click on the PROCUREMENT PROPOSAL, PRODUCTION ORDER, or RESCHEDULING buttons.

Figure 11.33 shows the planning results of the MATERIAL P-100 and PLANT 3000, after you click on the EXECUTE MRP RUN icon.

While remaining in interactive planning, choose the PROCUREMENT PROPOSAL button (see Figure 11.34). In the SELECT PROCUREMENT PROPOSAL pop-up ❶, select the procurement proposal, which for this example, is a planned order with the profile LA. Press Enter to open the PLANNED ORDER creation pop-up ❷. Enter the details such as order quantity, start date and so on, and choose CONTINUE. The system incorporates a new planned order for the material ❸. For this example, it creates the planned order number 39021.

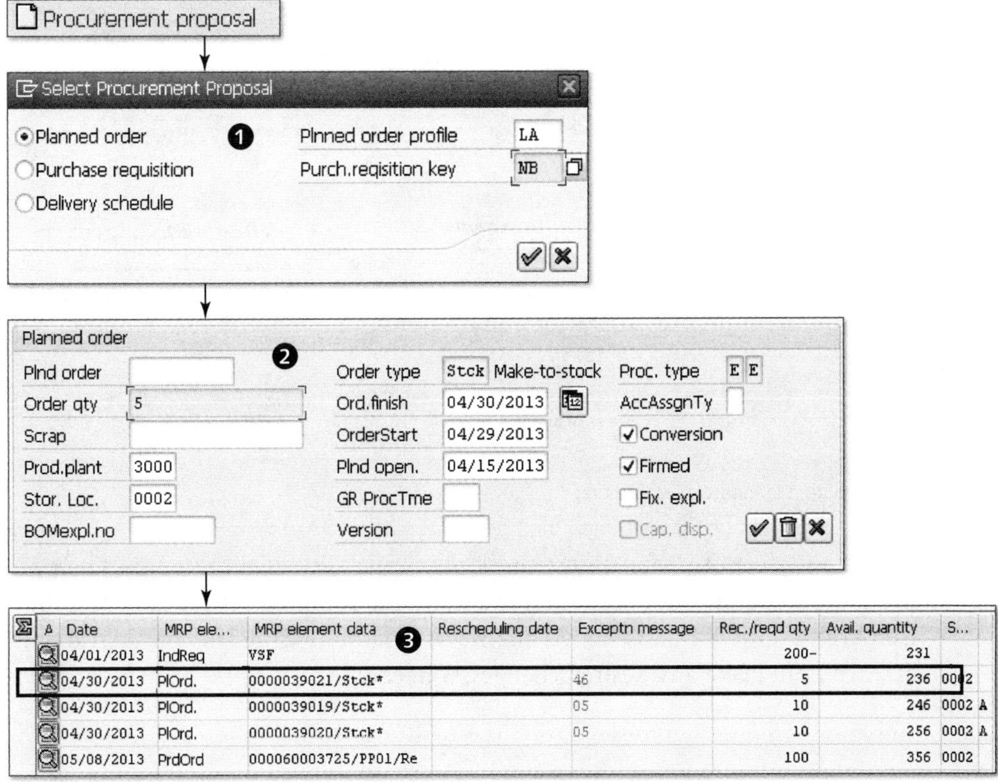

Figure 11.34 Planning Results after Planned Order Creation (Manual)

Manual Creation of Planned Order

You can use Transaction MD11 to manually create a planned order. Figure 11.35 ❶ shows the initial screen in which you enter the PLNNED ORDER PROFILE. For this example, use profile "LA", which supports MTS production. Press [Enter], and provide the material number, the plant, the order quantity, and the basic dates in the screen that appears ❷. A manually created planned order is always firmed, and the system denotes it with an asterisk "*" in the stock/requirements list (Transaction MD04). The CONVERSION INDICATOR ensures that the planned order can eventually be converted either into a production order (or a process order) for in-house production or into a purchase requisition for external procurement.

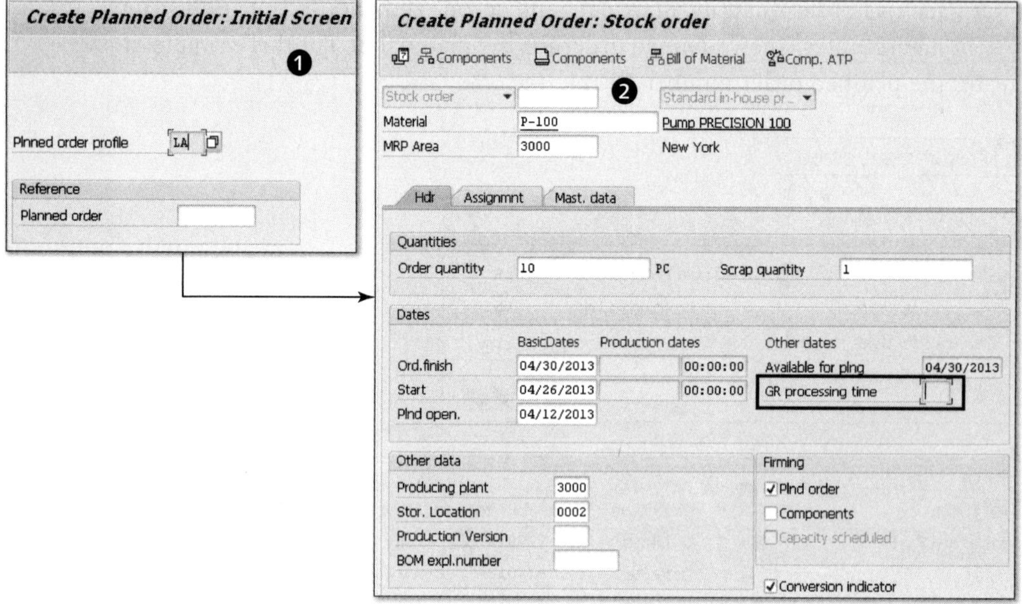

Figure 11.35 Planned Order Creation

Next we perform two functions: check the availability of the material's components to ensure smooth production and check the components available-to-promise (ATP). Figure 11.36 ❶ shows the BOM (component) details of MATERIAL P-100. You can make any desired changes, if needed.

Components

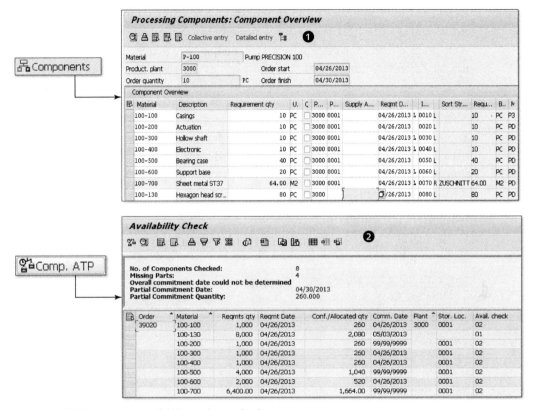

Processing Components: Component Overview ❶

Collective entry Detailed entry

Material	P-100	Pump PRECISION 100	
Product. plant	3000	Order start	04/26/2013
Order quantity	10 PC	Order finish	04/30/2013

Component Overview

Material	Description	Requirement qty	U.	C	P...	P...	Supply A...	Reqmt D...	I...	Sort Str...	Requ...	B..	M
100-100	Casings	10	PC		3000	0001		04/26/2013 L	0010 L		10	PC	P3
100-200	Actuation	10	PC		3000	0001		04/26/2013	0020 L		10	PC	PD
100-300	Hollow shaft	10	PC		3000	0001		04/26/2013 L	0030 L		10	PC	PD
100-400	Electronic	10	PC		3000	0001		04/26/2013 L	0040 L		10	PC	PD
100-500	Bearing case	40	PC		3000	0001		04/26/2013	0050 L		40	PC	PD
100-600	Support base	20	PC		3000	0001		04/26/2013 L	0060 L		20	PC	PD
100-700	Sheet metal ST37	64.00	M2		3000	0001		04/26/2013 L	0070 R	ZUSCHNITT	64.00	M2	PD
100-130	Hexagon head scr...	80	PC		3000			/26/2013	0080 L		80	PC	PD

Comp. ATP

Availability Check ❷

No. of Components Checked: 8
Missing Parts: 4
Overall commitment date could not be determined
Partial Commitment Date: 04/30/2013
Partial Commitment Quantity: 260.000

Order	Material	Reqmts qty	Reqmt Date	Conf./Allocated qty	Comm. Date	Plant	Stor. Loc.	Avail. check
39020	100-100	1,000	04/26/2013	260	04/26/2013	3000	0001	02
	100-130	8,000	04/26/2013	2,080	05/03/2013			01
	100-200	1,000	04/26/2013	260	99/99/9999		0001	02
	100-300	1,000	04/26/2013	260	04/26/2013		0001	02
	100-400	1,000	04/26/2013	260	04/26/2013		0001	02
	100-500	4,000	04/26/2013	1,040	99/99/9999		0001	02
	100-600	2,000	04/26/2013	520	04/26/2013		0001	02
	100-700	6,400.00	04/26/2013	1,664.00	99/99/9999		0001	02

Figure 11.36 Components Availability and ATP Check

Increase the planned order quantity of MATERIAL P-100 to "1000" units, and then choose the COMP. ATP icon to open the AVAILABILITY CHECK screen ❷. Notice that the available component quantities are shown with green lines, while the system reflects the shortages with yellow lines. When you choose the BACK icon in Figure 11.36 ❷, it leads to the screen shown in Figure 11.37.

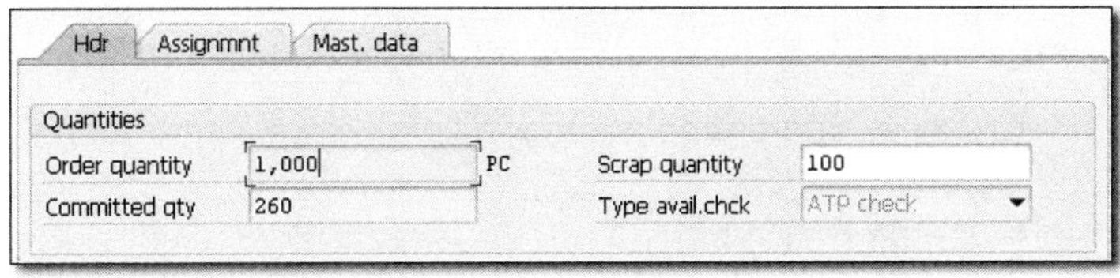

Figure 11.37 Committed Quantity in Planned Order after ATP Check

In Figure 11.37, notice that the system is only able to commit a material's quantity of 260 against the planned order quantity of 1,000 because it uses ATP CHECK. ATP CHECK enables the system to advise the quantity that it can commit or promise, based on the current available stock and other reservations/issuances.

11.9.2 Planned Order Profile

You've already seen during the planned order creation that you have to enter a planned order profile. You can define the order profile in configuration (Transaction SPRO) by following the path, PRODUCTION • MATERIAL REQUIREMENTS PLANNING • PROCUREMENT PROPOSALS • PLANNED ORDERS • DEFINE ORDER PROFILE. The order profile is used for both planned orders (OBJECT TYPE 1) and purchase requisitions (OBJECT TYPE 2), and it contains information on the permitted procurement types and special procurement types (see Figure 11.38).

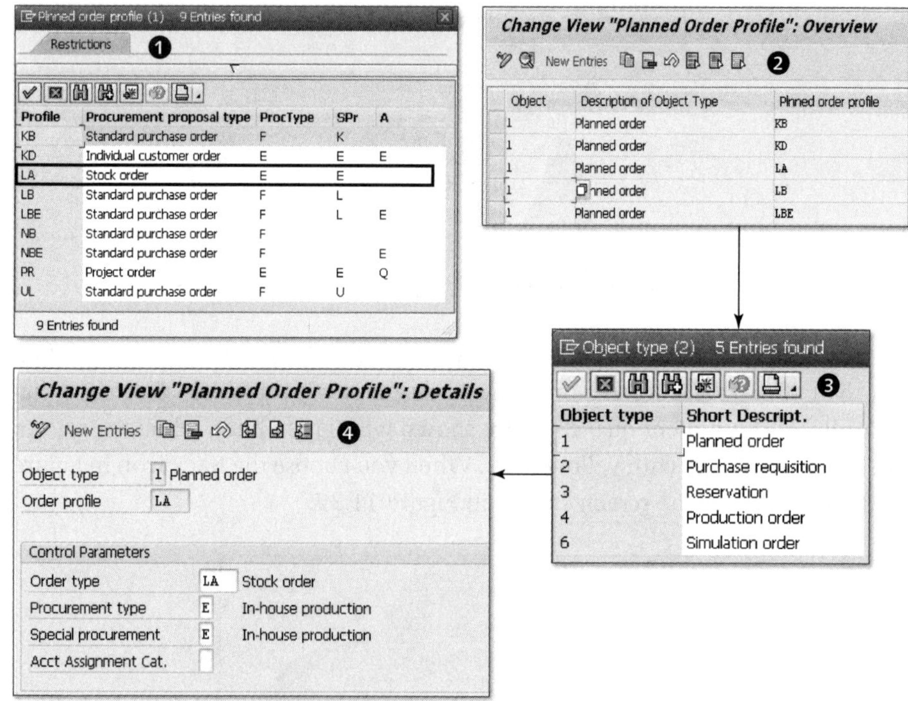

Figure 11.38 Planned Order Profile

Other available object types include the following:

- Reservation
- Production order
- Simulation order

Figure 11.38 ❶ shows the available planned order profile. To create a new planned order profile, choose New Entries ❷. Enter a name for the planned order profile, and select the object type ❸. Finally, in the detailed screen ❹, enter the procurement type or even special procurement control parameters.

11.9.3 Purchase Requisitions

You can also directly create a purchase requisition with Transaction ME51N and change it with Transaction ME52N. To create a purchase order with reference to a purchase requisition, you use Transaction ME21N, and then use Transaction ME22N to change it. This is in the domain of the MM component and isn't covered further in this book.

11.10 Executing Material Requirements Planning

Now that we've covered how you can create procurement proposals in MRP, we'll move on to some of the more technical steps and explanations you'll need when actually executing MRP in your SAP ERP system.

The following sections begin by ensuring that you create the planning file entry to enable the system to consider planning-relevant materials during the next MRP run. We also cover the net requirements logic that the system uses to calculate and come up with requirements. Finally, we cover how you can also use special stocks, such as stock in transfer, to enable the system to include it in plant stock calculation.

11.10.1 Planning File Entry and the Selection of Materials for Planning

The system automatically adds a material to the planning file list after the necessary MRP configuration settings are in place and you've assigned the relevant MRP types to the material masters.

The system considers three planning options for an MRP run:

- **Regenerative planning (NEUPL)**
  Regenerative planning (NEUPL) takes care of planning for all planning-relevant materials for a plant and includes all materials in the planning file. However, use of this option should be very limited because it tends to unnecessarily drain a system's resources and performance.
- **Net change planning (NETCH)**
  Net change planning (NETCH) is a preferred and frequently used planning method. Whenever there is a change of a material with regard to the MRP-relevant situation, the system automatically creates a planning file entry. This enables the system to consider this material during the planning run. To avoid overloading the precious system resources, this type of planning (NETCH) only takes delta to the planning situation before. Not only does the system consider new requirements/receipts, but it also considers changes to the MRP views in the material masters, changes to the BOM, and so on.
- **Net change planning in the planning horizon (NETPL)**
  Net change planning in the planning horizon (NETPL) takes into account only changes that occur during the predefined planning horizon. This option is quite efficient and effective, especially when you use NEUPL or NETCH from time to time as well.

MRP-relevant changes generate an entry in the planning file. You can display the planning file using Transaction MD21. In addition to the planning entries, the file also contains information on the low-level code (000 being the highest low-level code) as well as an indication of whether the material is a master schedule item. This low-level code is an internal number that tells the MRP algorithm how to sort the materials for effective planning to get a short runtime (see Figure 11.39). We discuss the lower-level code later in this section.

You can also manually create a planning file entry for a material. To do so, follow the SAP menu path, Logistics • Production • MRP • Planning • Planning File Entry • Create, or use Transaction MD20 (see Figure 11.39 ❶).

For this example, enter the Material as "P-100" and Plant as "3000", and select the Plng file entry NETPL checkbox. The MRP run principle takes into account the NETPL checkbox and therefore NETCH-run resets NETPL and NETCH, whereas NETPL-run resets only NETPL.

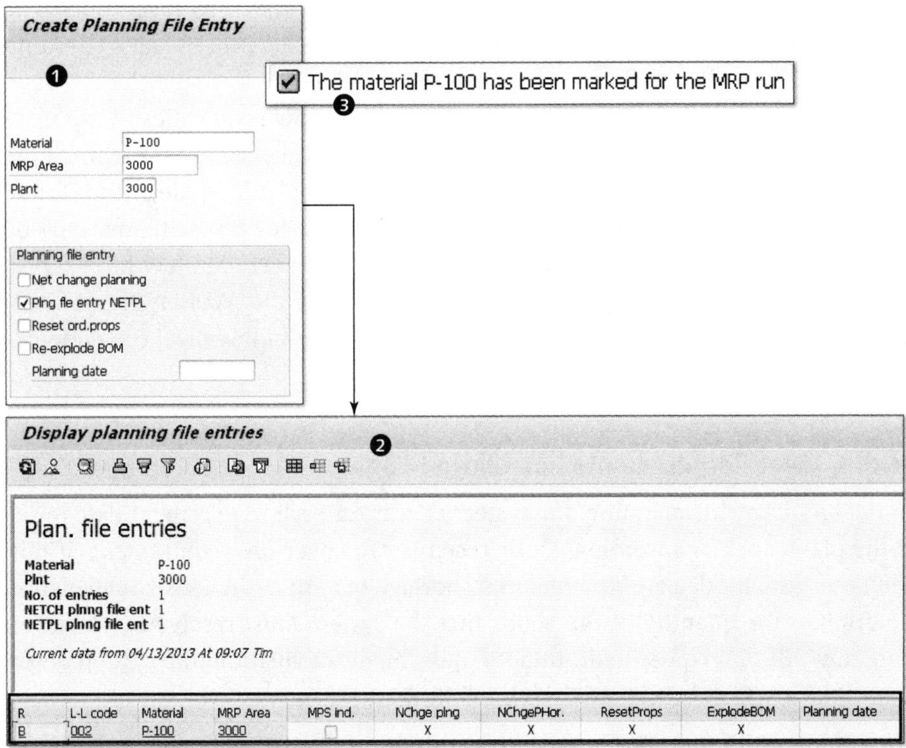

Figure 11.39 Creating the Planning File for a Material

Press Enter to open the DISPLAY PLANNING FILE ENTRIES screen ❷, wherein it displays the low-level code. When you save, the system displays a message ❸.

While working with a *BOM structure*, you must distinguish between the *production structure* and the *MRP structure* in the product structure description. The production structure stipulates the quantity of a material or assembly, which must be available at the given point in time during the production process (assignment of the material or assembly to a production level). The procedure of *low-level coding (LLC)* takes into account that a material may appear in several products and at more than one production level of one or different products. The low-level code is the lowest level that a material appears in any product structure. The system waits until it knows about all of the requirements that might occur for a specific material. This is the case at the lowest level (which is the highest number of LLC), and the material can be planned only once to cover all of the different requirements.

The materials are planned according to the sequence of their low-level codes. The highest low-level code, that is, the saleable product, is low-level code. The components that are directly used for the saleable product are low-level code 1, low-level code 2, and so on. The system stores all of the low-level codes in the planning file. The planning file enables the MRP controller to view the structure of the low-level codes at any time. Thus, it's possible to view planning file entries for all materials of a plant, to display the planning file entry for a specific material by entering the material number, to restrict the selection to certain low-level codes, or to view only the master schedule items by setting the MASTER PRODUCTION SCHEDULING flag. You use Transaction MD21 to view the low-level codes details in the planning file.

11.10.2 Net Requirements Calculation Logic

In net requirements planning, the system checks for each requirement date to see if the plant stock or incoming goods receipts can cover the requirement. If not, then the system calculates the material shortage quantity. The lot size calculation determines the quantity of the goods that the system must receive by the given date to fulfill the requirement. In net requirements calculation, the system determines the available stock based on the following formula:

> Plant stock – safety stock + open order quantity (purchase orders, production orders, fixed procurement proposals) – requirement quantity (PIRs, customer requirements, material reservations, forecast requirements for unplanned additional requirements) = Available stock

In gross requirements planning, a BOM explosion, including lot size calculation and scheduling, is carried out *without* considering warehouse and plant stocks. The planning run merely includes the expected receipts (planned orders and purchase requisitions). For materials for which you want to carry out gross requirements planning, you must specify the 2 GROSS REQUIREMENTS PLANNING indicator in the MIXED MRP column in the material master (see Figure 11.40).

[»] **Note**

The planning strategy 11 supports gross requirements planning. Refer to Chapter 10 on SAP Demand Management, in which we cover planning strategies.

Planning			
Strategy group			
Consumption mode		Bwd consumption per.	
Fwd consumption per.		Mixed MRP	
Planning material		Planning plant	
Plng conv. factor		Planning matl BUnit	

Mixed MRP indicator (1) 3 Entries found

Mixed MRP	Short Descript.
1	Subassembly planning with final assembly
2	Gross requirements planning
3	Subassembly planning without final assembly

Figure 11.40 Mixed MRP in Planning

Depending on the company's business processes, you can define whether or not you want to include the stock in transfer and blocked stock in the plant stock. For example, during the transportation of perishable products such as ice cream, it makes sense to include it in plant stock, as long as the material is in transit (stock in transfer). The supplying plant had already performed quality tests on the ice cream shipment before dispatching, so as soon as it's received in the receiving plant, it can be dispatched on to distributors and dealers. The receiving plant can consider this stock in transfer, as if it's already in its plant stock, to avoid creation of unnecessary procurement elements.

Use the following configuration (Transaction SPRO) path, Production • Material Requirements Planning • Planning • MRP Calculation • Stocks • Define Availability of Stock in Transfer/Blocked Stock/Restricted Stock (see Figure 11.41).

Change View "Available Stocks": Overview

Plnt	Name 1	Stock in transfer	Blocked Stk	Restricted-use stock
3000	New York	☑	☐	☐

Figure 11.41 Stock Availability for MRP Calculation

11.10.3 Planning Control Parameters in Materials Requirements Planning

You can run MRP for a single material, for an entire plant, or for multiple plants, if you've configured scope of planning (see Section 11.11.2 for scope of planning configuration). The planning scope in the SAP ERP system stipulates making use of parallel processing. Hence, you need to ensure that this mandatory setting is in place. To set up parallel processing, follow the configuration (Transaction SPRO) path, PRODUCTION • MATERIAL REQUIREMENTS PLANNING • DEFINE PARALLEL PROCESSING IN MRP.

In the ensuing screen, enter the destination of the parallel processing and number of parallel sessions. You may need to engage your SAP NetWeaver (Basis) resource to set up a destination for parallel processing (if not done already).

While running the MRP, you need to maintain a standard set of parameters on the initial selection screen or in the variant for the background job. The outcome of this activity is that the system then only plans and presents results based on the input parameters.

Figure 11.42 shows the initial screen for single-item, multi-level MRP (Transaction MD02).

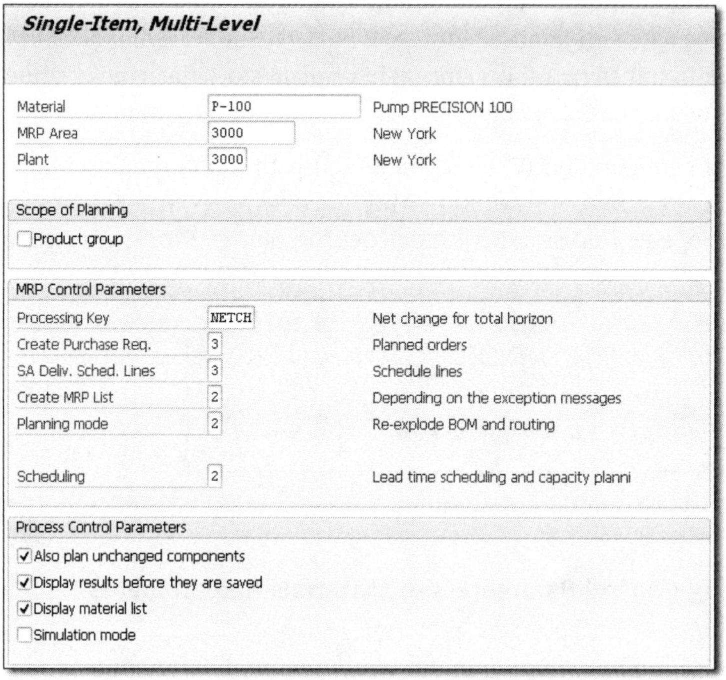

Single-Item, Multi-Level

Material	P-100	Pump PRECISION 100
MRP Area	3000	New York
Plant	3000	New York

Scope of Planning

☐ Product group

MRP Control Parameters

Processing Key	NETCH	Net change for total horizon
Create Purchase Req.	3	Planned orders
SA Deliv. Sched. Lines	3	Schedule lines
Create MRP List	2	Depending on the exception messages
Planning mode	2	Re-explode BOM and routing
Scheduling	2	Lead time scheduling and capacity planni

Process Control Parameters

☑ Also plan unchanged components
☑ Display results before they are saved
☑ Display material list
☐ Simulation mode

Figure 11.42 Single-Item, Multi-Level MRP (Initial Screen)

We cover some of the important selection parameters of MRP in the following sections.

Processing Key

The processing key enables the system to decide the type of planning run it needs to carry out for the material. The options for the planning run are NETCH, NETPL, and NEUPL, as discussed in Section 11.12.1.

Purchase Requisition Key

If external procurement is implied, the system considers this setting during the planning run. If necessary settings are in place, then the system can directly create purchase requisitions for all externally procured materials. Alternatively, the system can create planned orders that you can convert into purchase requisitions.

Delivery Schedules

If you've maintained a source list for the material and have incorporated scheduling agreements with vendors in the source list, the system can create schedule lines. You can also control whether you want the system to create schedule lines only in the opening period, in the planning horizon, or in planned orders instead.

Creation of Material Requirements Planning List

The system includes the results of the planning run in the MRP list after the planning run. An MRP list is a static or frozen list of stock/requirements situations with the date and time immediately after the planning run. This indicator controls how the system should create the MRP list. The available options are for the system to create an MRP list for all planned orders, create an MRP list for planned orders with exceptions, or don't create any MRP list at all. Exceptions are alerts in the system that the planner needs to keep an eye on. We cover this later in Section 11.12.2. You can also control which exception messages the system should create for MRP lists in the configuration step for MRP: DEFINE AND GROUP EXCEPTION MESSAGES.

Planning Mode

During the planning run, the system plans various procurement elements, such as the unfirmed planned orders, purchase requisitions, and scheduling agreements, once again in the next planning run. You can control how the system should proceed with the planning data that are already there (from the latest planning run or due to interactive activities of the MRP planner). Three options are available:

- ADOPT PLANNING DATA
  If there is any change in requirement quantities, dates, or lot-sizing procedures for the unfirmed planned orders (or other procurement elements), the system reexplodes the BOM for the new quantities in the MRP run.
- RE-EXPLODE BOM AND ROUTING
  The system reexplodes the BOM and routing if there is a change in the BOM master data, routing, production versions, or BOM explosion numbers. This helps to ensure that the system reexplodes the BOM for the existing unfirmed planned orders. The procurement proposals created in the previous planning aren't deleted but are adjusted to take the changes into account.
- DELETE AND RECREATE PLANNING DATA
  The system deletes the existing procurement proposals (i.e., the entire planning data of the previous planning run), unless they are firmed and reexplodes the BOM and routing, thereby creating altogether new procurement proposals.

Scheduling Types

For in-house produced materials, the system offers two types of scheduling options:

- Basic date scheduling
- Lead time scheduling

After you enter the parameters on the initial screen of Transaction MD02, and press Enter twice (first to confirm the MRP parameters causing the entered parameters to turn red and then to execute the MRP), the system takes you to the screen shown in Figure 11.43, which is the planning result of a large number of materials. In this result, not only is the finished good (P-100) planned but also all of the components of the BOM. You can, however, view the planning results of a material by double-clicking it.

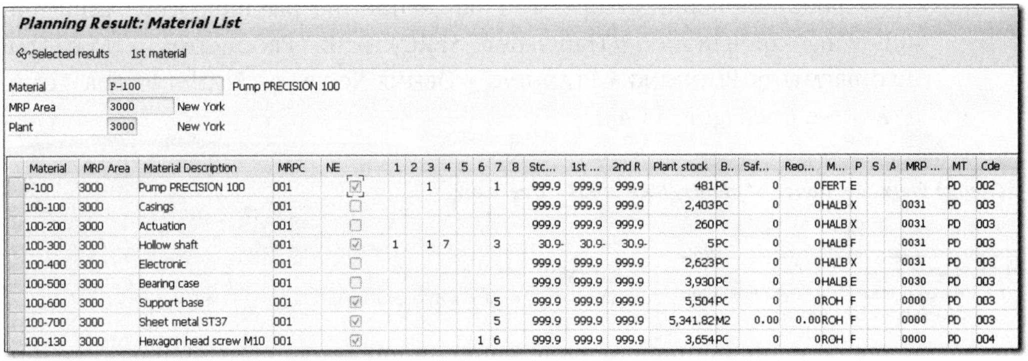

Planning Result: Material List

Selected results 1st material

Material P-100 Pump PRECISION 100
MRP Area 3000 New York
Plant 3000 New York

Material	MRP Area	Material Description	MRPC	NE	1	2	3	4	5	6	7	8	Stc...	1st ...	2nd R	Plant stock	B..	Saf...	Reo...	M...	P	S	A	MRP ...	MT	Cde
P-100	3000	Pump PRECISION 100	001	☑			1				1		999.9	999.9	999.9	481	PC	0	0	FERT	E				PD	002
100-100	3000	Casings	001	☐									999.9	999.9	999.9	2,403	PC	0	0	HALB	X			0031	PD	003
100-200	3000	Actuation	001	☐									999.9	999.9	999.9	260	PC	0	0	HALB	X			0031	PD	003
100-300	3000	Hollow shaft	001	☑	1		1	7			3		30.9-	30.9-	30.9-	5	PC	0	0	HALB	F			0031	PD	003
100-400	3000	Electronic	001	☐									999.9	999.9	999.9	2,623	PC	0	0	HALB	X			0031	PD	003
100-500	3000	Bearing case	001	☐									999.9	999.9	999.9	3,930	PC	0	0	HALB	E			0030	PD	003
100-600	3000	Support base	001	☑							5		999.9	999.9	999.9	5,504	PC	0	0	ROH	F			0000	PD	003
100-700	3000	Sheet metal ST37	001	☑							5		999.9	999.9	999.9	5,341.82	M2	0.00	0.00	ROH	F			0000	PD	003
100-130	3000	Hexagon head screw M10	001	☑						1	6		999.9	999.9	999.9	3,654	PC	0	0	ROH	F			0000	PD	004

Figure 11.43 Results of the MRP Run

11.11 Configuration Settings for Material Requirements Planning

In the following sections, we'll show you the configuration settings that you need to make to ensure that MRP runs at the plant level:

- MRP activation
- Scope of planning configuration
- Plant parameters configuration
- MRP group configuration

11.11.1 Material Requirements Planning Activation

This configuration step allows you to choose the plant for which you want to activate MRP. To do so, follow the configuration (Transaction SPRO) menu path, LOGISTICS • PRODUCTION • MRP • PLANNING FILE ENTRY • ACTIVATE MRP AND SET UP PLANNING FILE, or use Transaction OMDU.

11.11.2 Scope of Planning Configuration

When you run plant-level MRP with Transaction MD01, the scope of MRP generally refers to a specific plant. You can also refer to several plants if you define a corresponding planning scope. You can define the planning scope using the following configuration (Transaction SPRO) path, PRODUCTION • MATERIAL REQUIREMENTS PLANNING • PLANNING • DEFINE SCOPE OF PLANNING FOR TOTAL PLANNING (see Figure 11.44).

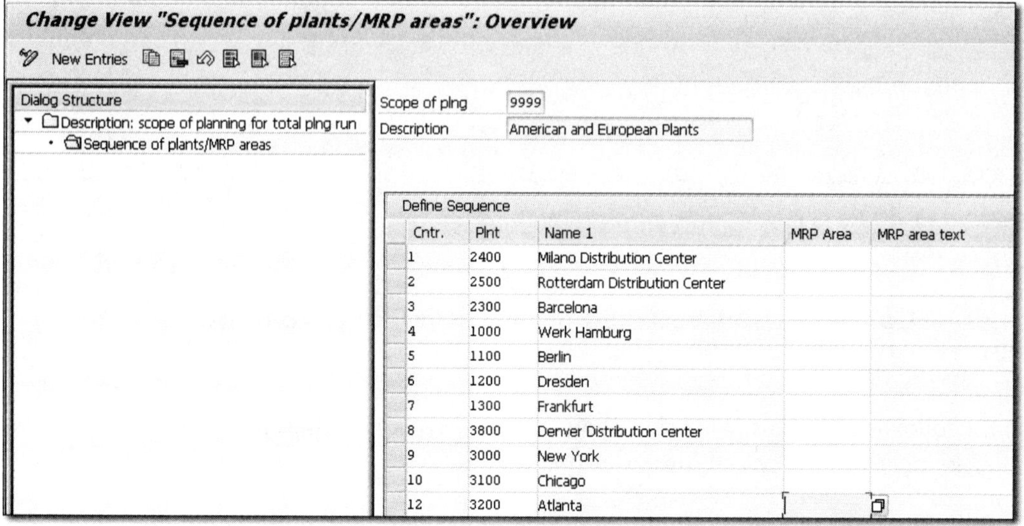

Figure 11.44 Scope of Planning at the Plant Level

11.11.3 Plant Parameters Configuration

You can configure MRP up to the following three levels:

- Plant parameters
- MRP groups
- Material master (MRP views)

Material master settings have higher priority than MRP group parameters, which in turn have higher priority than plant parameters. No setting can be found on all three levels, but they can be on up to two of them. Therefore, plant parameters are usually used to set defaults, which can be overruled if necessary.

The plant parameters of MRP contain a view of numerous configuration settings that are relevant to MRP. The plant parameters screen shows whether certain configuration settings have been maintained or whether they are still in their initial status. Using this screen enables you to navigate directly to the relevant configuration area.

In the plant parameter configuration step, following are some of the configurations:

- Reference Plant
- Number ranges
- Direct Procurement
- Available Stock
- Rescheduling
- External Procurement
- Conversion of planned orders
- Dependent Requirements availability
- Floats (Schedule Margin Keys)
- BOM Explosion
- BOM/Routing Selection

To configure plant parameters, follow the configuration (Transaction SPRO) menu path, Logistics • Production • Material Requirements Planning • Plant Parameters, or use Transaction OPPQ. Click on the Maintain button. A pop-up appears in which you enter the Plant as "3000", and then click on the Maintain icon. You'll then see a detailed screen where you can maintain the individual plant parameters.

11.11.4 Material Requirements Planning Group Configuration

You can group materials using MRP groups and then assign the MRP group in the MRP 1 view of the material master. Each of the MRP groups can have its own set of MRP parameters. When you work with an MRP group, you have the following options, which are specific to a material or group of materials:

- Strategy group
- Settlement type and horizon

- Rescheduling horizon and planning horizon
- Production storage location selection
- Conversion order types of planned orders
- Planning horizon
- Planning time fence and roll-forward periods
- BOM and task list selection IDs
- Direct procurement parameters
- Planned order scheduling parameters
- Start number of days allowed in the past
- Availability checking groups
- Period split for distributing the independent requirement
- Maximum MRP intervals
- Availability of safety stock for MRP
- BOM explosion
- Direct procurement of nonstock items
- Creation indicators for purchase requisitions
- Scheduling of external procurement according to the information record
- Checking rule for dependent demand

To configure MRP groups, follow the configuration (Transaction SPRO) menu path, Logistics • Production • Material Requirements Planning • MRP Groups • Carry out Overall Maintenance of MRP Groups, or use Transaction OPPR. Figure 11.45 ❶ shows the Material Requirements Planning (MRP) Group initial screen. Enter the Plant as "3000", and click on the Maintain button. In the pop-up that appears ❷, enter the MRP group "0040", and click on the Maintain button. The detailed screen appears for maintaining individual parameters of MRP group ❸.

When you've configured the MRP group, save the group and assign it in the MRP 1 view of the material master (Figure 11.46).

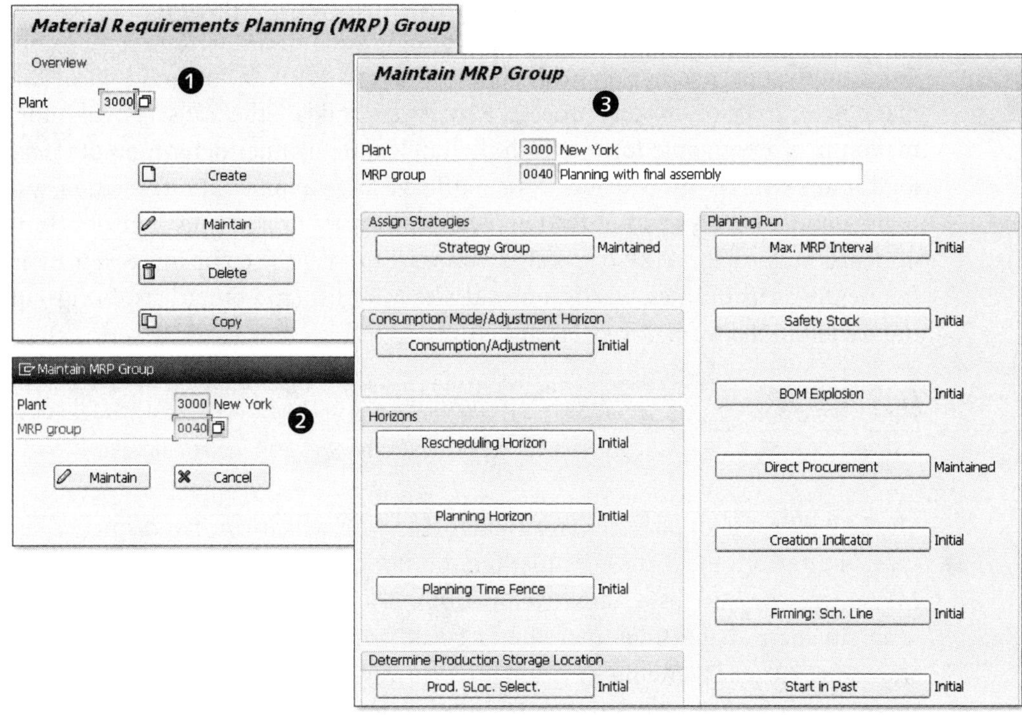

Figure 11.45 MRP Group Settings

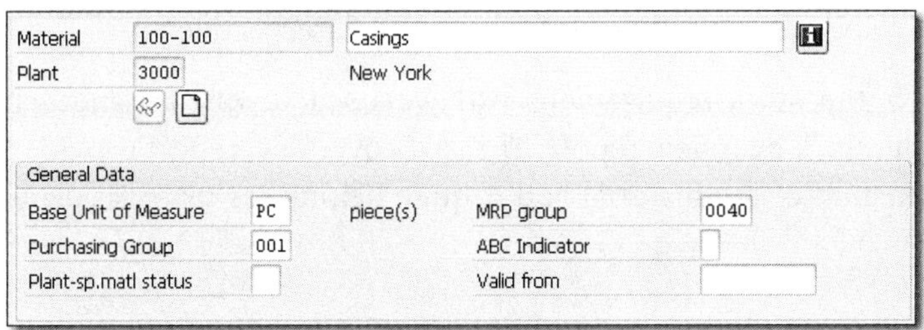

Figure 11.46 MRP Group in the Material Master

Note [«]

While this example covers the maintenance of an MRP group, you can also create a new MRP group, delete an existing MRP group, or copy an MRP group from an existing one. You can also assign the MRP group to a material type by following the configuration (Transaction SPRO) menu path, Logistics • Production • Material Requirements Planning • MRP Groups • Define MRP Group for Each Material Type.

11.12 Material Requirements Planning Run Analysis

For a production planner or MRP planner, reports form an integral part of the planning or decision-making process. Reports consolidate the transactional information in a presentable form, which then helps the planner determine the next logical step to take. The SAP ERP system offers a large number of information systems, standard analyses, and flexible analyses reports in all of the logistics components. Because the MRP run is also a transaction that the company performs as frequently as its business needs demand, the MRP list (and other lists) brings up the output in the form of reports or lists.

[»] **Note**

Refer to Chapter 19 in which we cover reporting in the SAP ERP system in detail.

The SAP ERP system provides two similar lists with which you can display, evaluate, and rework the planning situation: the stock/requirements list (Transaction MD04) and the MRP list (Transaction MD05). While the MRP list shows the planning situation at the time after the last MRP run, the stock/requirements list shows the current planning situation. The current situation can deviate from the situation of the last planning run because changes might have occurred in the meantime; for example, due to goods receipts and issues, confirmations, or sales orders. You can compare the two lists via the menu path, GOTO • MRP LIST COMPARISON or GOTO • STOCK/REQUIREMENTS LIST COMPARISON, respectively.

But before we cover the stock/requirement list, we need to discuss another important report—the stock overview report. The stock overview report provides the current stock situation of the material and is quite helpful for material planners.

11.12.1 Stock Overview

You can use Transaction MMBE to access the stock overview report (see Figure 11.47). In the initial parameter screen, enter the MATERIAL number, together with the PLANT, and select other parameters, such as STORAGE LOCATION or SPECIAL STOCK INDICATOR. With the stock overview report, you can only view the stock details of one material at a time. Choose EXECUTE or press F8. This leads to the screen shown in Figure 11.48.

Note [«]

If you want to view the warehouse stock of a large number of materials all at the same time, you can use Transaction MB52.

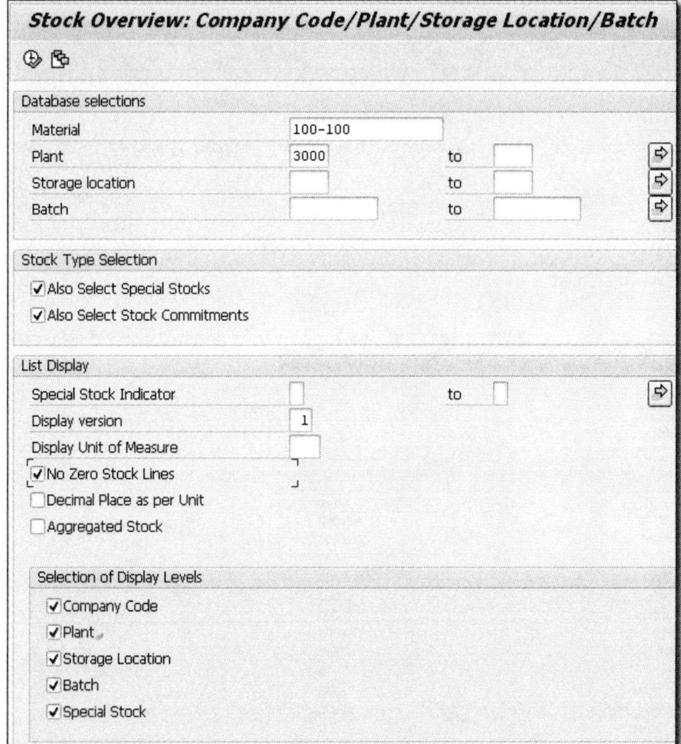

Figure 11.47 Initial Screen of the Stock Overview Report

Figure 11.48 shows the overview of the stock situation for MATERIAL 100-100 in plant 3000, along with the various stock types such as unrestricted use stock. Click on the DETAILED DISPLAY button.

Figure 11.49 provides comprehensive details on the various stock types and available quantities in each (if any).

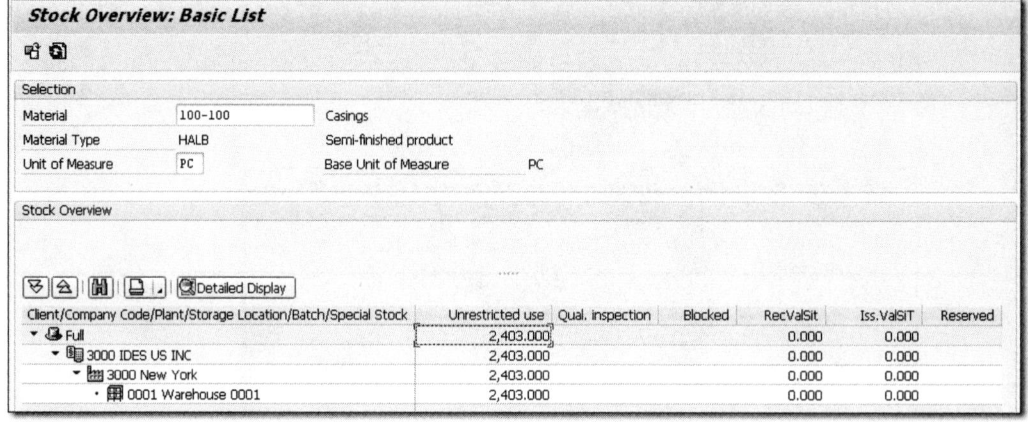

Figure 11.48 Stock Overview: Basic List

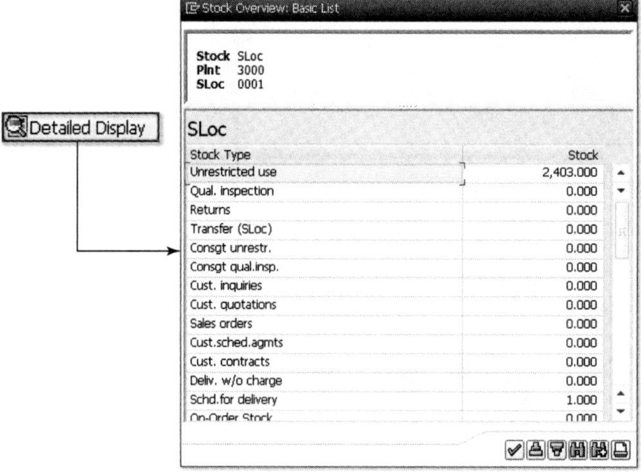

Figure 11.49 Detailed Display of Material Stock

11.12.2 Stock/Requirements List

Figure 11.50 illustrates some of the options provided by the stock/requirements list (Transaction MD04):

- Display of exception messages
- Jump into the detail view
- Jump into the order report

- Display goods receipt date or availability date
- Show the RLT
- Change the MRP element
- Interactive planning

The stock/requirements list describes all MRP-relevant elements for a material within a plant. These include the stock, customer requirements and PIRs, dependent demands, planned orders, and production orders. In addition, it shows parameters that are relevant to MRP such as the safety stock or the planning time fence. For all elements, details can be displayed and changed, if this is permitted. However, you can't create procurement proposals interactively. Usually, the system displays the availability date of a goods receipt. But you can also display the goods receipt date instead, for example, to plan actions in case of delays.

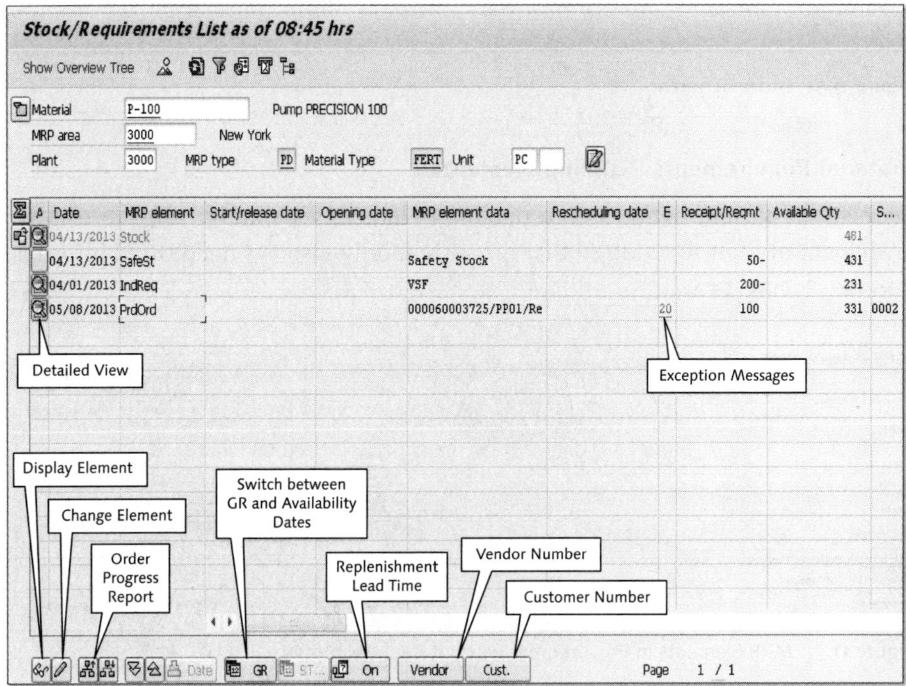

Figure 11.50 Select Functions in the Stock Requirements List

We now briefly explain a few of the available options and functions of the stock/requirements list.

Order Report

Figure 11.51 shows the order report of each element of the stock/requirements list. In this example, it shows that production order 60003725 has component 100-130, which doesn't provide the requirement coverage. The green icon represents a satisfactory stock situation, whereas the red icon represents a signal for the planner to take action.

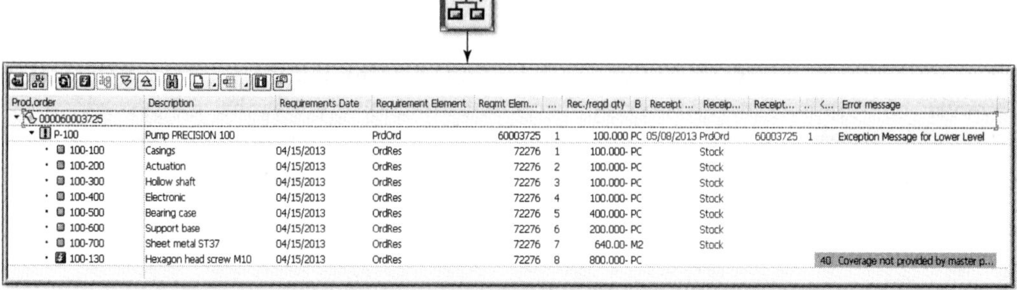

Figure 11.51 Order Report

Material Requirements Planning Elements

Because the system denotes each planning element separately, you can use the dropdown or filter function so that the system only displays the desired information (see Figure 11.52).

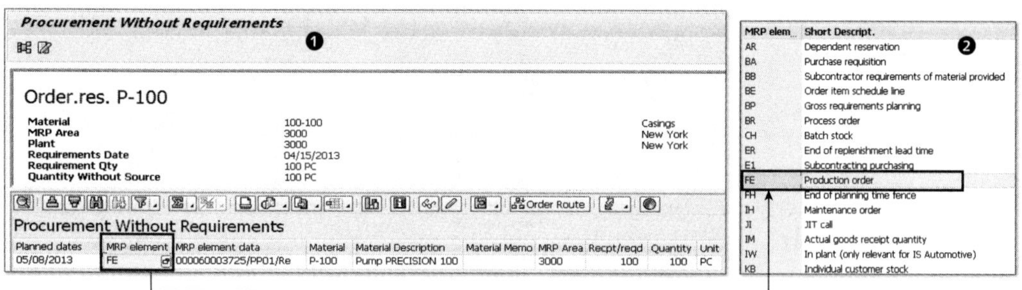

Figure 11.52 MRP Elements in Procurement without Requirements

Header Details of Stock/Requirements List

Figure 11.53 shows that when you're already on the stock/requirements list, you can click on the HEADER DATA icon to find all of the parameters used by the system in planning the material. In other words, you no longer have to separately view the MRP parameters in the material master but can do so while remaining in the stock/requirements list.

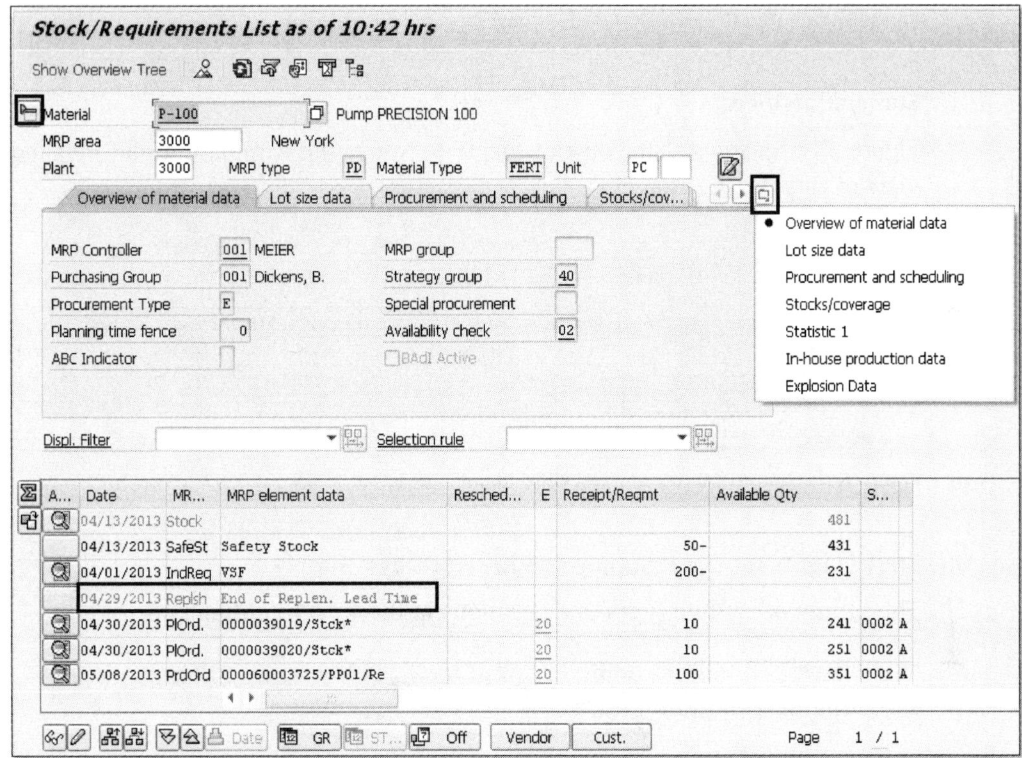

Figure 11.53 Replenishment Lead Time

Interactive Conversion of Procurement Elements

Figure 11.54 shows that when you click on the ADDITIONAL DATA FOR MRP ELEMENT icon of any procurement element, for example, a planned order, the system opens a pop-up wherein you can perform the interactive conversion of a planned order either into a purchase requisition for external procurement or a production/process order for in-house production.

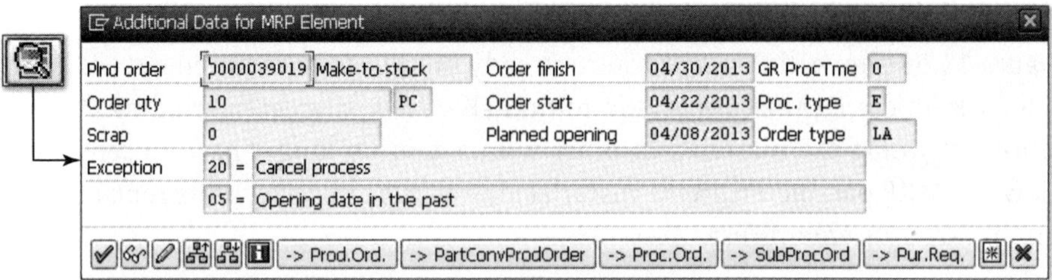

Figure 11.54 Interactive Detail View for Converting a Planned Order

Summation View

Figure 11.55 shows that you can toggle between the summation and detailed view of all planning elements (receipts, issuances/requirements, etc.).

Days | Weeks

A	Period/s...	Plnd ind....	Require...	Receipts	Avail. quantity	ATP quantity	Actual cover...
	Stock				431	481	999.9
	04/01/13	200-	0	0	231	0	999.9
	04/30/13	0	0	20	251	20	999.9
	05/08/13	0	0	100	351	100	999.9

Figure 11.55 Summation View of the Stock Requirement List

Traffic Lights for Days' Supply and Receipt Days' Supply

The system offers two types of evaluations in the stock/requirements list:

- Stock days' supply
- Receipt days' supply

In the stock/requirement list or in the MRP list, the system calculates the number of days on the time axis for which the existing stock of material can fulfill the requirements.

Receipt days' supply indicates how long the current stock and the expected receipts can cover the requirements. The system offers two different types of receipt days' supply:

- Receipt days' supply 1
- Receipt days' supply 2

You can select the receipt elements for both types of receipt days' supplies by following the configuration (Transaction SPRO) menu path, PRODUCTION • MATERIAL REQUIREMENTS PLANNING • EVALUATION • DEFINE RECEIPT ELEMENTS FOR RECEIPT DAYS' SUPPLY. In the configuration, you can select which procurement elements the system should consider or exclude in its calculation for days' supply.

Figure 11.56 ❶ shows the traffic light for MATERIAL P-100 in the stock/requirements list. You can save traffic light settings not only for days' supply (stock days' supply and receipt days' supply) but also for exception messages. We cover exception messages in the next section.

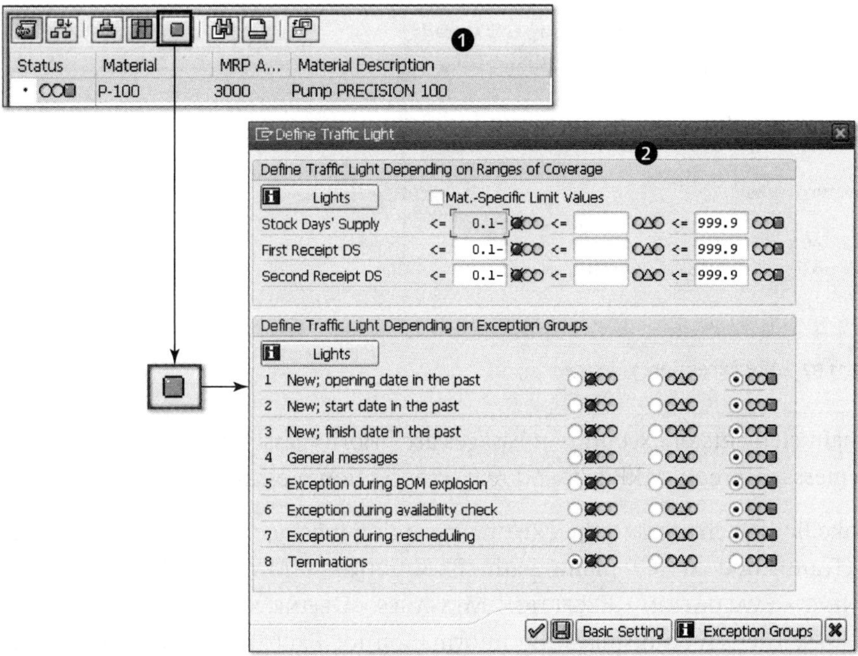

Figure 11.56 Traffic Lights

Exception Messages

The system generates exception messages in MRP when it encounters alerts during the planning run. The system displays exception messages for each relevant procurement element in the stock/requirements list. You can also view the exception elements by choosing EDIT • FIND IN THE LIST (see Figure 11.57 ❶ and ❷).

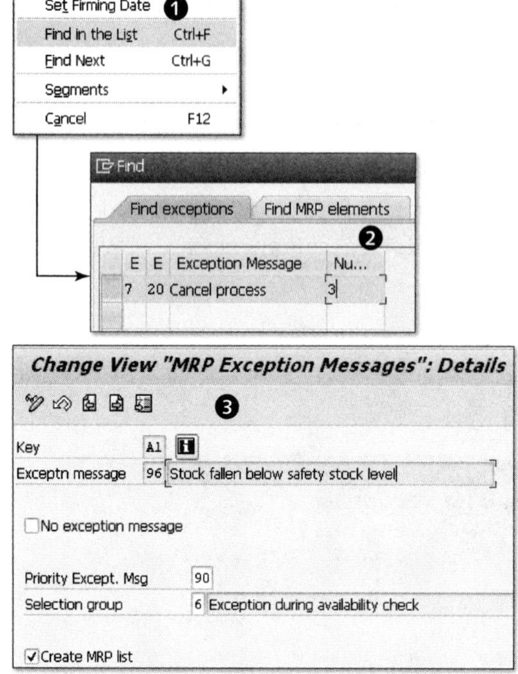

Figure 11.57 MRP Exception Messages

The main elements of exception messages are priority, assignment to group, suppress message, create MRP list, and text of exception message.

To make limited changes to the existing exception message, follow the configuration (Transaction SPRO) menu path, Production • Material Requirements Planning • Evaluation • Exception Messages • Define and Group Exception Messages. The screen displays the information for Exceptn message 96: Stock fallen below safety stock level ❸.

[+] **Tips & Tricks**

You can change the text of exception messages to denote the importance of an exception. For example, you can change text to "Warning: stock fallen below safety stock level" to help you quickly segregate exception messages with a warning from exception messages with information.

Table 11.4 provides a selection of available exception messages in the system.

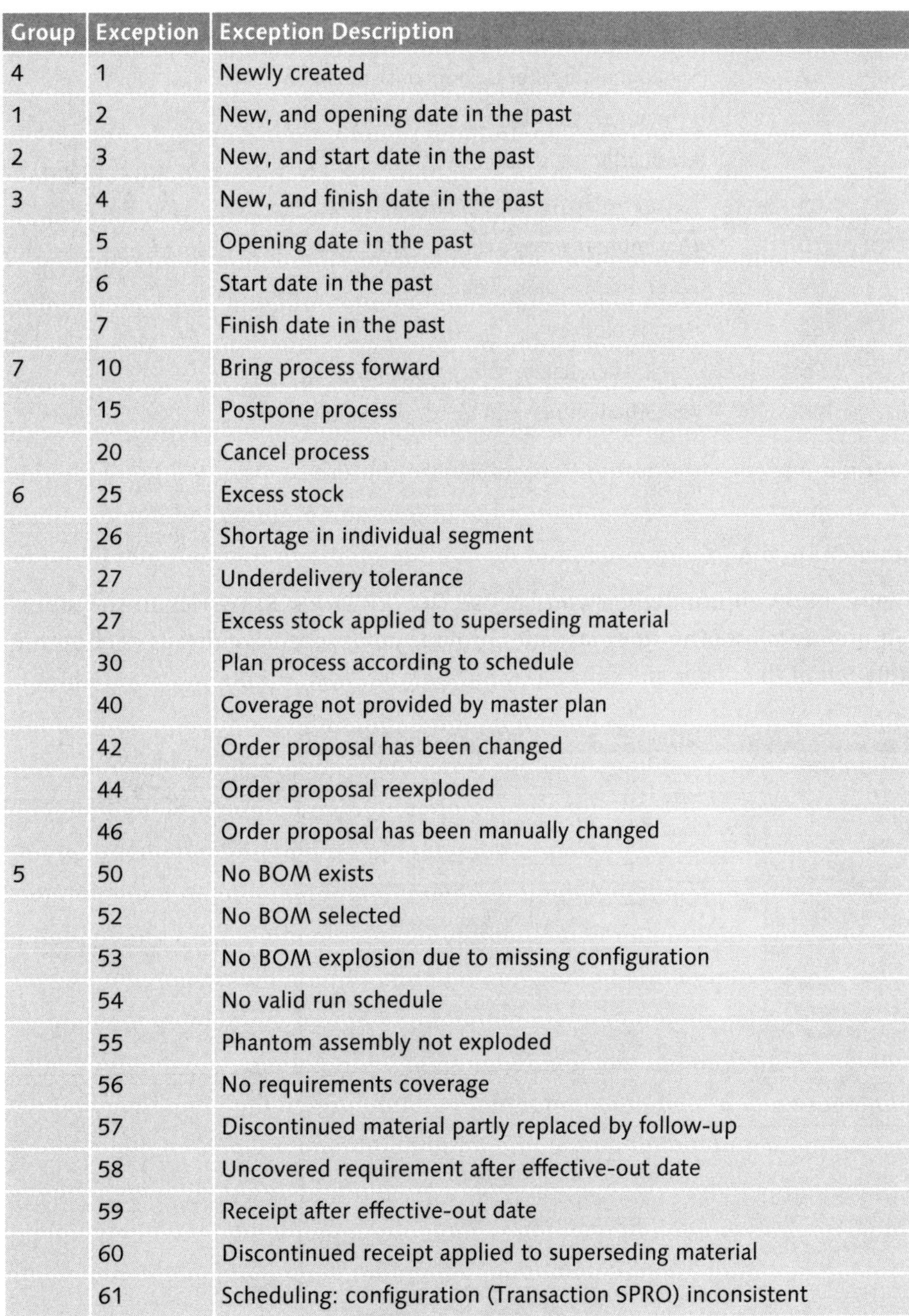

Group	Exception	Exception Description
4	1	Newly created
1	2	New, and opening date in the past
2	3	New, and start date in the past
3	4	New, and finish date in the past
	5	Opening date in the past
	6	Start date in the past
	7	Finish date in the past
7	10	Bring process forward
	15	Postpone process
	20	Cancel process
6	25	Excess stock
	26	Shortage in individual segment
	27	Underdelivery tolerance
	27	Excess stock applied to superseding material
	30	Plan process according to schedule
	40	Coverage not provided by master plan
	42	Order proposal has been changed
	44	Order proposal reexploded
	46	Order proposal has been manually changed
5	50	No BOM exists
	52	No BOM selected
	53	No BOM explosion due to missing configuration
	54	No valid run schedule
	55	Phantom assembly not exploded
	56	No requirements coverage
	57	Discontinued material partly replaced by follow-up
	58	Uncovered requirement after effective-out date
	59	Receipt after effective-out date
	60	Discontinued receipt applied to superseding material
	61	Scheduling: configuration (Transaction SPRO) inconsistent

Table 11.4 A Selection of Exception Messages after the MRP Run

Group	Exception	Exception Description
	62	Scheduling: master data inconsistent
	63	Production start before order start
	64	Production finish after order finish
	69	Recursive BOM components possible
	70	Maximum release quantity – quota exceeded
	80	Reference to retail promotion
	82	Item is blocked
	96	Stock fallen below safety stock level
8	98	Abnormal end of requirements planning

Table 11.4 A Selection of Exception Messages after the MRP Run (Cont.)

Stock Statistics

Figure 11.58 appears when you choose GOTO • STOCK STATISTICS in the stock/requirements list. The stock statistics provide comprehensive details of the stock situation of the material.

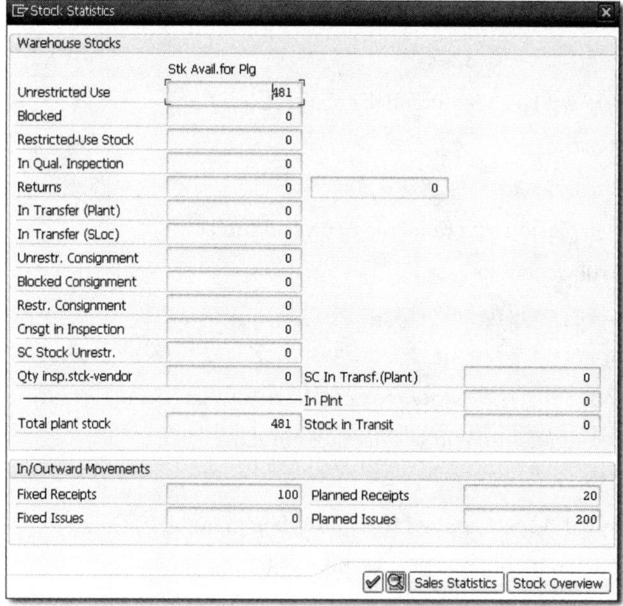

Figure 11.58 Stock Statistics

User Settings

You can save individual display settings in the stock/requirements list to suit your preferences or business needs. Figure 11.59 appears when you choose SETTINGS • SETTINGS while in the stock/requirements list.

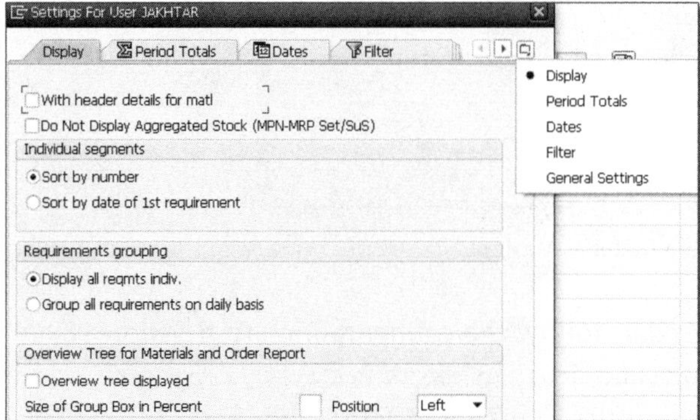

Figure 11.59 Settings in the Stock Requirements List

[«]

> **Note**
>
> We encourage you to explore the several other features and functionalities of stock/requirements lists by clicking on different icons and choosing from the menu options.

11.13 Planning Calendar

There are several instances in which either the company receives goods on specific days of the week or the vendor supplies goods only during specific days of the week. In other words, the planning of material differs from the factory calendar of the company. For example, several agricultural products used in the production process of edible products are produced only during specific periods of the year. The company must procure them and also store them to ensure they are available for consumption during the production process. For example, red chilies, an agricultural product, is only produced from January to April. The company uses the planning calendar to ensure the system creates procurement proposals of this raw material (red chilies) within the four-month time period.

It's also possible to specify a planning date in the planning run. This option is useful because it enables you to bring the planning run forward to an earlier date. If the planning run is set for Monday, for example, you can perform this planning run on Friday.

You can assign the planning calendar in the MRP 2 view of the material master (Transaction MM02). Figure 11.60 shows the PLANNING CALENDAR field, in which you enter the planning calendar.

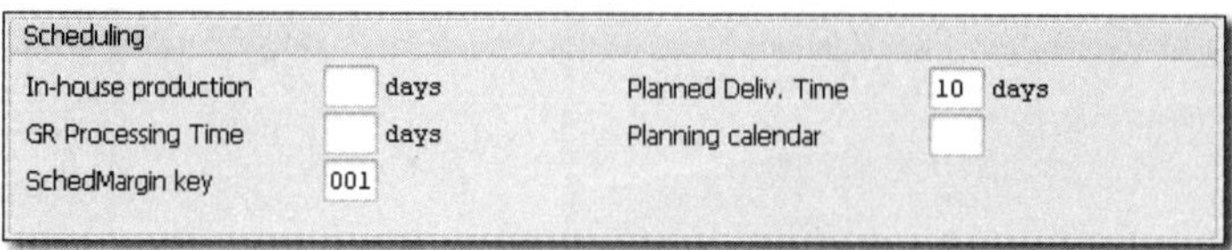

Figure 11.60 Planning Calendar in the Material Master

To create a new planning calendar, follow the configuration (Transaction SPRO) menu path, PRODUCTION • MATERIAL REQUIREMENTS PLANNING • MASTER DATA • MAINTAIN PLANNING CALENDAR, or use Transaction MD25.

In Figure 11.61 ❶, after you define PLANNING CALENDAR as "0001" for PLANT "3000", select the FOLLOWING WORKING DAY radio button if the period start date isn't a working day.

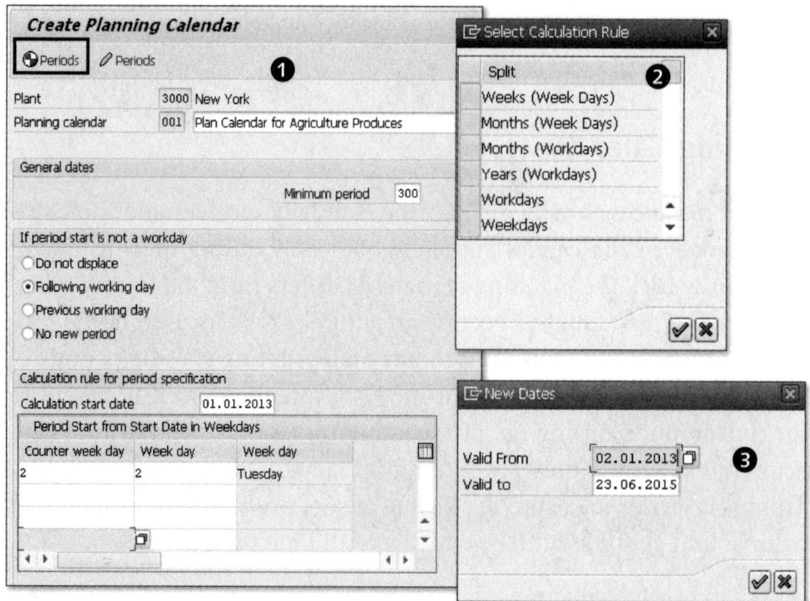

Figure 11.61 Creation of the Planning Calendar

You can also define the minimum number of periods that the system must generate in the planning calendar. When you press [Enter], a pop-up appears ❷, in which you can select weekdays, workweek, workdays, work months, and so on. Click on the GENERATE PERIODS icon, and the NEW DATES pop-up appears ❸, in which you enter the start and end date of the planning calendar. Choose CONTINUE.

In Figure 11.62, notice that all period (planning) starts from Tuesday, and this is the same as defined in the lower half of the first screen shown in Figure 11.61 ❶.

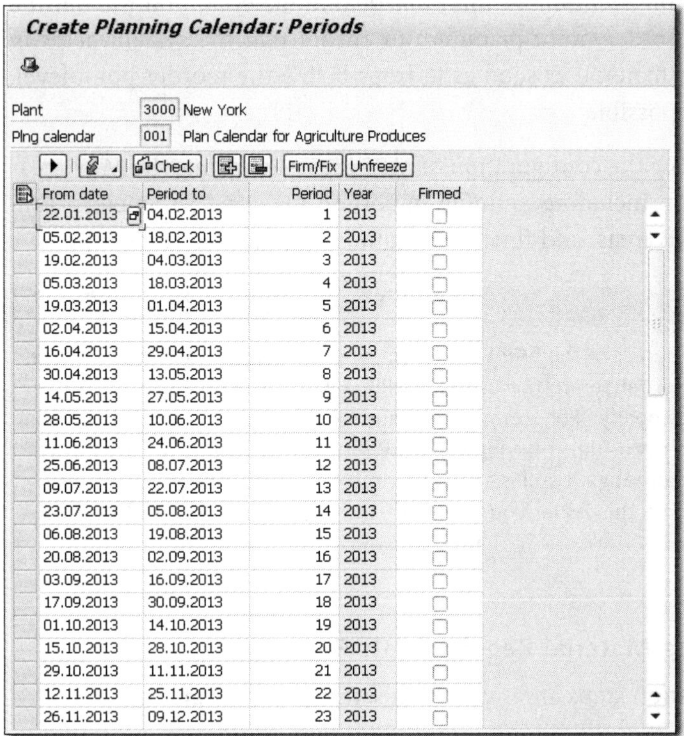

Create Planning Calendar: Periods

Plant 3000 New York
Plng calendar 001 Plan Calendar for Agriculture Produces

Check | Firm/Fix | Unfreeze

From date	Period to	Period	Year	Firmed
22.01.2013	04.02.2013	1	2013	☐
05.02.2013	18.02.2013	2	2013	☐
19.02.2013	04.03.2013	3	2013	☐
05.03.2013	18.03.2013	4	2013	☐
19.03.2013	01.04.2013	5	2013	☐
02.04.2013	15.04.2013	6	2013	☐
16.04.2013	29.04.2013	7	2013	☐
30.04.2013	13.05.2013	8	2013	☐
14.05.2013	27.05.2013	9	2013	☐
28.05.2013	10.06.2013	10	2013	☐
11.06.2013	24.06.2013	11	2013	☐
25.06.2013	08.07.2013	12	2013	☐
09.07.2013	22.07.2013	13	2013	☐
23.07.2013	05.08.2013	14	2013	☐
06.08.2013	19.08.2013	15	2013	☐
20.08.2013	02.09.2013	16	2013	☐
03.09.2013	16.09.2013	17	2013	☐
17.09.2013	30.09.2013	18	2013	☐
01.10.2013	14.10.2013	19	2013	☐
15.10.2013	28.10.2013	20	2013	☐
29.10.2013	11.11.2013	21	2013	☐
12.11.2013	25.11.2013	22	2013	☐
26.11.2013	09.12.2013	23	2013	☐

Figure 11.62 Planning Calendar Created

11.14 Material Requirements Planning Areas

Apart from running MRP at a single plant level, at a material level, or at a product group level, you can activate MRP areas if you want to run MRP on the following:

- Multiple plants
- Single storage location or a group of storage locations
- Subcontractor

The advantage of setting up MRP areas is that you can plan material at the storage location level, which allows you to plan material differently from the MRP planning of the material at the plant level. For example, if you have a consumable material that is stored at several storage locations in a plant, then you can set the MRP type of this material as "PD" (planning). However, for two specific storage locations with high consumption of this consumable material, you can set the MRP type as "VB" (reorder point planning) to ensure that the system helps in replenishment of this material as soon as it drops below the reorder point level. MRP areas make this possible.

In this section, we show the configuration of MRP areas; activating the MRP areas in the material master, including defining important planning parameters; running the MRP for MRP areas; and finally evaluating the planning run's results.

[!]

Warning!

After you activate MRP areas in the SAP ERP system, you can't deactivate them. Hence, you need to extensively deliberate the use of MRP areas before proceeding to implement them in your company. For example, because subcontracting stock is special stock, the system won't consider it during the MRP run, neither will it show up as the company's own stock (although it still is and is only moved to the vendor's premises for processing). To overcome this issue, you may have to use Business Add-In (BAdI) MD_SUBCONT_LOGIC.

11.14.1 Configuring Material Requirements Planning Areas

Three main configuration steps are involved in setting up MRP areas:

1. Set the MRP Area Active checkbox.
2. Define the MRP areas.
3. Convert a planning file for use on the MRP areas level once.

For this example, we'll show you how to set up MRP areas for a storage location. To activate MRP areas, use SAP configuration (Transaction SPRO) menu path,

Production • Material Requirements Planning • Master Data • MRP Areas • Activate MRP for MRP Areas. Here, select the MRP area active checkbox and save (see Figure 11.63).

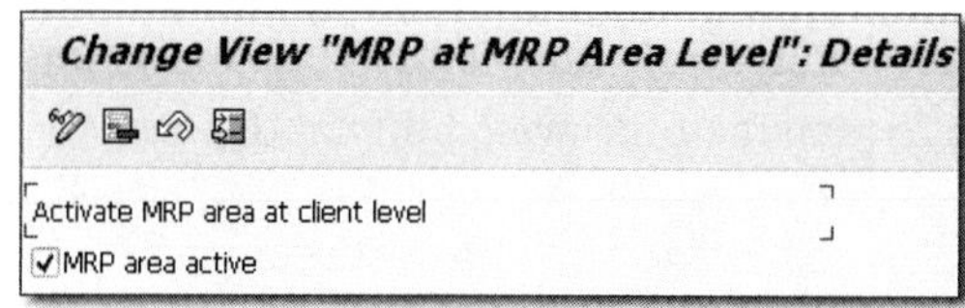

Figure 11.63 Activate MRP Areas at the Client Level

To define MRP areas, use SAP configuration (Transaction SPRO) menu path, Production • Material Requirements Planning • Master Data • MRP Areas • Define MRP Areas. Choose New Entries, set MRP Area as "3000-0001", and provide a description. For MRP area type, select 02 from the dropdown menu, which is for storage location MRP. Assign the Plant as "3000", and press Enter. Next, on the right-hand side of the screen, double-click on Assign storage locations, and this leads to the screen shown in Figure 11.64.

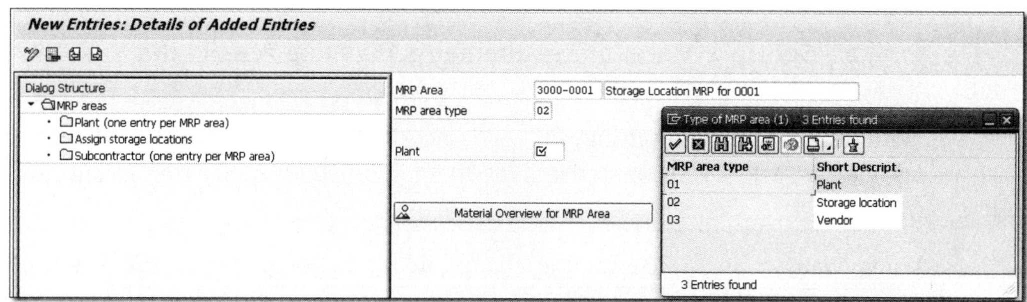

Figure 11.64 Configuration Settings for MRP Areas for Storage Location

[«]

Note

The MRP area type 02 is the standard case to define additional MRP areas on the storage location level. MRP area type 01 is for the plant MRP areas and is created automatically by converting the planning file for MRP areas. The MRP area type 03 is for subcontracting, in which you maintain your company's stock at vendors' premises and ensure materials' planning.

In Figure 11.65, assign the STOR. LOC. (storage location) as "0001" for PLANT "3000", and save the entry. For this example, we used a single storage location for the plant, but you can also group several storage locations under one MRP area. This storage location (or locations) will be planned separately and independently from planning at the plant level. Additionally, you can define subcontracting MRP areas within a plant to enable you to plan material availability at the subcontractor (vendor) level.

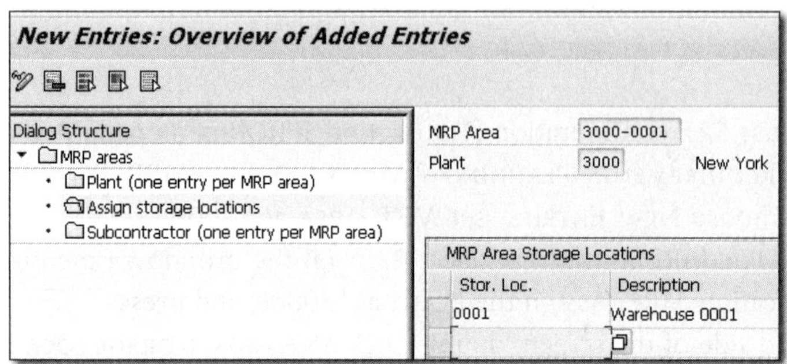

Figure 11.65 Assigning Storage Location MRP

11.14.2 Set Up a Material Requirements Planning Area in the Material Master

With the configuration settings for MRP areas in place, you can activate the storage location MRP for the material 100-130 for plant 3000 and for storage location 0001.

Using Transaction MM02 for the change in the material master and entering MATERIAL "100-130" for PLANT "3000" and for STOR. LOC. "0001", select the MRP 1 view. Figure 11.66 reflects that you can activate MRP areas. Also note that for this example, we've entered MRP TYPE ND for this material because we only want to plan this material using MRP areas (and not at the plant level). MRP type at the plant level is completely independent of the MRP areas level. For this example, the system won't plan the material at the plant level because we assigned its MRP TYPE as ND (No planning). Rather, the system will now look for the planning information at the MRP areas level to come up with planning results after the MRP run.

Figure 11.66 MRP Areas in Material Master

[«]

Note

You should carefully evaluate the materials that you want to plan at the plant level and at the MRP areas level.

Click on the MRP AREAS button to open the screen shown in Figure 11.67. Select the configured MRP AREA 3000-0001, and also set the MRP PROFILE as "VB01". An MRP profile consists of default MRP settings that the system considers during planning. The MRP profile eventually reduces the data entry efforts in individually maintaining the fields in MRP views.

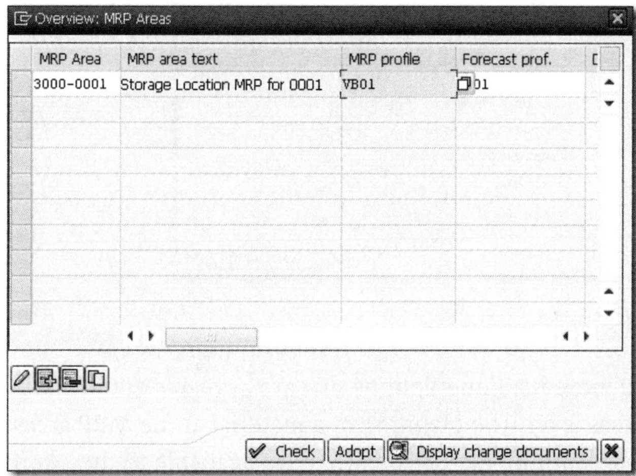

Figure 11.67 MRP Area Settings for the Storage Location in the Material Master

When you click on the CHECK icon, it leads to the screen shown in Figure 11.68. The screen shows the MRP 1 view of MATERIAL 100-130, PLANT 3000, and MRP AREA 3000-0001. Because we assigned the MRP type as VB for manual reorder point planning at the MRP area level, set REORDER POINT as "100". You can assign other parameters such as MRP CONTROLLER and LOT SIZE as shown in Figure 11.68.

It's important to note that with MRP areas, you now have the option to treat a material separately in the storage location MRP areas and in the plant MRP areas.

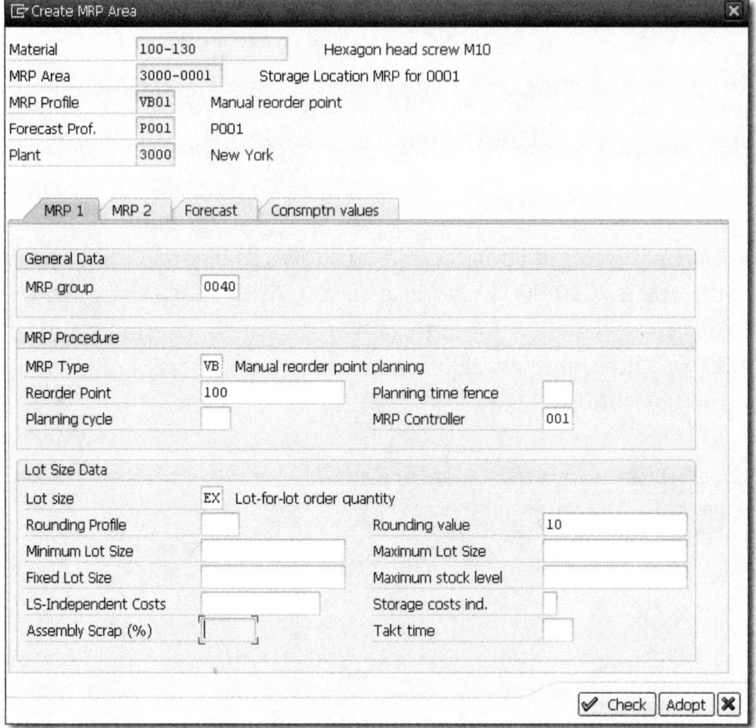

Figure 11.68 MRP 1 View of MRP Areas

Click on MRP 2 to see some more planning options that are available at the material level and at the MRP areas level. The planning of a material at the MRP areas level doesn't offer the same depth or the options that are available at the plant level. Enter a planned delivery time of "10" days, and click on the FORECAST tab.

The FORECAST tab shown in Figure 11.69 enables you to enter forecast parameters and also interactively execute a forecast of the material. Finally, click on the CONSMPTN VALUES tab.

In Figure 11.70, you can enter the planned and unplanned consumption values. Click on the ADOPT icon in MRP areas, and save the settings of the material master.

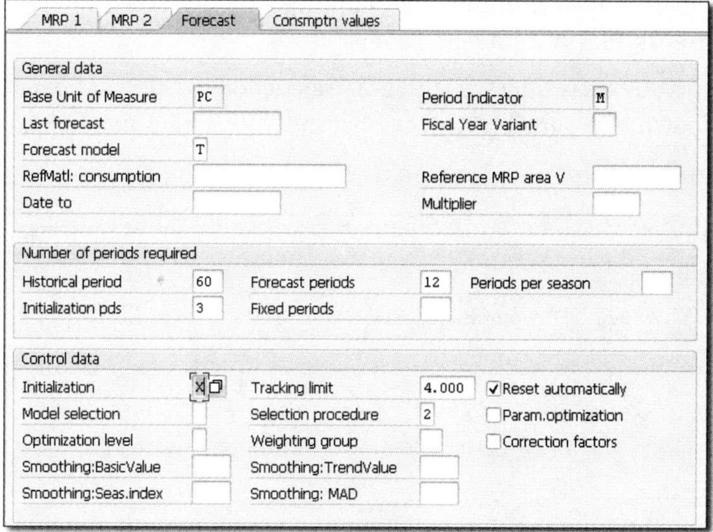

Figure 11.69 Forecast View of MRP Areas

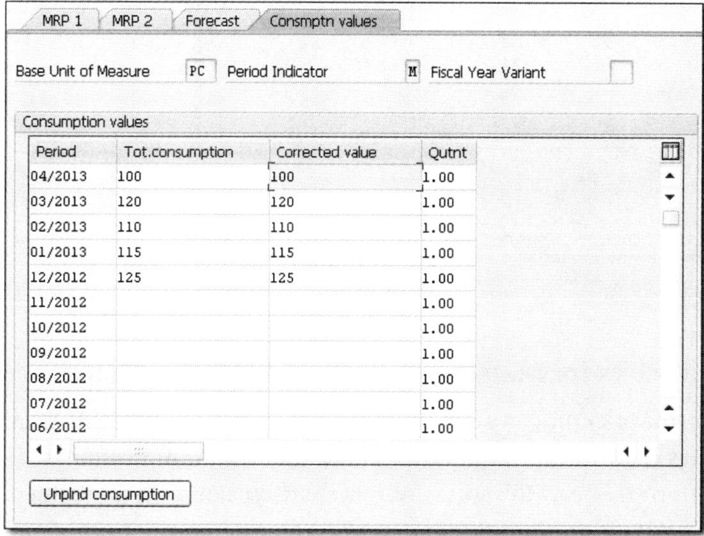

Figure 11.70 Consumption View of MRP Areas

11.14.3 Running Material Requirements Planning at the Material Requirements Planning Areas Level

To run the MRP for MRP areas interactively, use Transaction MD02, and enter the parameters as shown in Figure 11.71. Note that the MRP area is maintained accordingly. Press Enter twice.

Single-Item, Multi-Level

Material	100-130	Hexagon head screw M10
MRP Area	3000-0001	Storage Location MRP for 0001
Plant	3000	New York

Scope of planning

☐ Product group

MRP control parameters

Processing key	NETCH	Net change for total horizon
Create purchase req.	3	Planned orders
Delivery schedules	3	Schedule lines
Create MRP list	2	Depending on the exception messages
Planning mode	2	Re-explode BOM and routing
Scheduling	2	Lead time scheduling and capacity planni

Process control parameters

☑ Also plan unchanged components
☑ Display results before they are saved
☑ Display material list
☐ Simulation mode

Figure 11.71 Run MRP on MRP Areas

11.14.4 Planning Results for Material Requirements Planning Areas

Figure 11.72 shows the planning results of Material 100-130 in MRP Area 3000-0001 and shows that because there is an existing stock of 30 in the warehouse (storage location of the MRP areas), the system creates a new planned order 39022 for a quantity of 70 to ensure the stock level reaches the target quantity of 100. To test the functionality and check the results without saving it, we used the planning results option. Use Transaction MD04 or Transaction MD05 to evaluate the planning results of MRP.

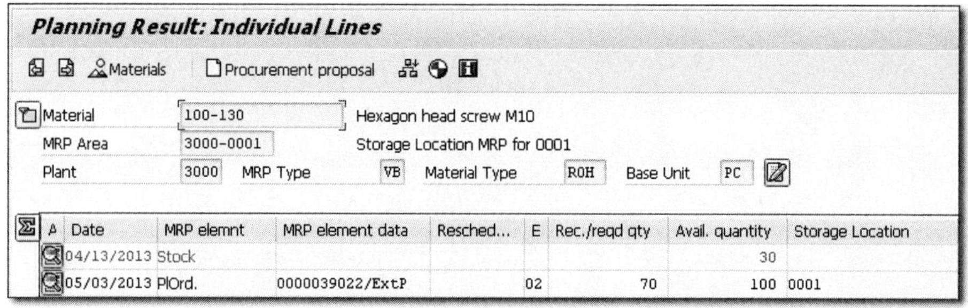

Figure 11.72 Results of MRP Areas Planning

11.15 Summary

You can take enormous advantage of MRP functionality in the SAP ERP system to help improve and optimize the business processes. You can plan all types of materials, including high-, medium-, and low-value, as well as materials with special procurement type status. The planning calendar helps you plan materials that are specific to particular calendars only. Further, you can give individual attention to a material if it has any specific or unique business need. You can also use MRP area level planning to independently plan materials, which are specific to a storage location (or group of storage locations) or at the subcontractor level.

The next chapter covers Long-Term Planning (LTP) in the SAP ERP system.

Long-Term Planning allows you to test various business models to make the best choices without affecting the standard planning database. The beneficiaries of Long-Term Planning are various stakeholders in the company—the production planner, the procurement in-charge, the inventory controller, and the capacity planner.

12 Long-Term Planning

Long-Term Planning (referenced as PP-MP-LTP in the SAP ERP system, but referred to as LTP in this book) is used to simulate various business scenarios to help in production and procurement planning using existing master data and other information. Companies can use this tool to test various hypothetical assumptions affecting business decisions. When compared with material requirements planning (MRP), LTP doesn't affect the database of results that was created while running normal (operative) MRP. However, you can use simulation mode in operational MRP to get similar results as in LTP.

LTP is a separate planning area where you can undertake all simulation-related planning. However, if the results from simulative planning of LTP are satisfactory, you can also transfer them to operative planning in standard MRP. While it's easy to set up LTP, it's equally easy to delete the planning scenarios when you no longer need them. For example, you quickly set up a planning scenario to test procurement quantities of materials that you need for the next six months, including the associated financial values (capital tie-up). When you no longer need this information, you can delete the planning scenario from the system.

LTP results in the following benefits for specific business roles:

- Capacity planners are better able to plan their machine resources and manpower resources.
- The purchasing department uses the information on the future requirements quantities to estimate and plan future procurement orders.
- The inventory controller gains greater comprehension of warehousing requirements.

- Vendors are able to get a preview of a company's future procurement needs, which enables them to foresee and take action to meet impending demands of the company.
- The product costing team can get a preview of the associated costs of producing products.

In this chapter, we show the steps you need to take to prepare the master data and planning data for LTP using a real-life business case from the fertilizer industry. We also cover the Logistics Information System (LIS) with particular reference to LTP, focusing on the Purchasing and Inventory Controlling subsets of LIS. Finally, we also discuss further planning options offered by LTP to cater to various business scenarios.

[»]

Note

LTP isn't confined to just the long-term planning needs of the company but can be equally beneficial for short-term to medium-term planning.

For short-term simulation, you have the option to include sales orders, firm receipts from production, or purchase orders, whereas for long-term simulation, you can switch off scrap calculations and other checks to gain a broader view of the planning.

12.1 Long-Term Planning Master Data and Planning Data

LTP derives its detail from existing information in the SAP ERP system, including the planning data in the four different MRP views of the material master. Similarly, when you perform multi-level planning of the material in LTP, the system makes use of bills of material (BOMs) and routing. To run an LTP simulation, you need to ensure that the planning data is set up in the SAP ERP system on which you intend to run the simulation. For example, you need to enter month-wise planning quantities of the finished good as planned independent requirements (PIRs). You define all of these quantities in a specific planning version number. The same material can have different month-wise planning quantities separately defined in different version numbers. Using inactive versions in LTP, you can execute various simulative LTP scenarios and evaluate the results separately. You need the following data in your SAP system before you can run LTP:

- **Master data**
  Material for LTP (i.e., material to be planned using LTP), BOM, and routing of the material. The planning scenarios are also part of master data that you need to set up.
- **Planning data**
  The planning quantities of the material that you'll enter as PIRs.
- **Planning scenario**
  The planning scenario that forms an integral part of master data setup for LTP.

We now cover master and planning data in detail. In the example, you'll perform LTP on unpacked area bulk, which is the main material, and its associated components (ammonia hydrous and UF-85 for urea plant) in the BOM, as shown in Table 12.1. All of these materials and their planning data in the MRP views must already be maintained in the system, including the material BOM and routing.

Number	Material number	Material description
1	1120000003	Unpacked bulk urea
2	1115000002	Ammonia hydrous
3	1110000001	UF-85 for urea plant

Table 12.1 Unpacked Urea Bulk and Its Components in a BOM

12.1.1 Master Data: Bill of Materials

Table 12.2 lists the details of the BOM for unpacked urea bulk. As previously explained, LTP makes use of the same planning data that you had defined in the MRP views of materials, so you don't need to define any additional data.

Number	Material Number	Material Description	Quantity	Unit of Measure	LTP Planning (Y/N)
1	1120000003	Unpacked urea bulk	1	Metric ton	Yes
2	1115000002	Ammonia hydrous	0.57	Metric ton	No
3	1110000001	UF-85 for urea plant	0.009	Metric ton	Yes

Table 12.2 BOM Details of Unpacked Urea Bulk and Its Two Components

[Ex]

Example

For component ammonia hydrous, the MRP TYPE defined in the MRP 1 view of the material master is ND, which means NO PLANNING. Therefore, we'll eliminate it from all planning activities in LTP. It won't be part of the operational MRP run either. Only in exceptional circumstances would you not plan material in MRP or LTP; otherwise, we suggest that you ensure your data modeling for MRP or LTP to include all components of a material. For example, in the caustic soda production, one of the raw materials is raw salt. The raw salt is not only very cheap but also readily available in countries with huge salt mines. Further, there's no specific warehousing requirement for raw salt so it can be placed in open space. With no significant capital tie-up and no Inventory Controlling requirement, it makes sense not to plan this material (raw salt).

Other planning parameters that you've defined in the MRP views are applicable in LTP. For example, if you've defined a fixed lot size or minimum lot size for a material, then the system will consider the lot-sizing procedures during LTP. This also applies if, for example, you've defined a planning calendar for external procurement.

The BOM for unpacked urea bulk was also previously defined, and once again LTP makes use of this information. The two materials for LTP are therefore unpacked urea bulk and UF-85.

[+]

Tips & Tricks

We recommend that you consider creating a new BOM exclusively for LTP simulation. This provides greater visibility from a planning perspective. A dedicated BOM for LTP not only eliminates the chances of error while creating a normal production BOM of a material but also ensures that it's used for LTP purposes only. It then becomes easier for the person managing the production planning master data to segregate the normal production BOM from the LTP BOM. BOM usage is maintained in configuration. The selection ID enables you to use separate BOMs for LTP purposes. To do this, you must create a separate BOM usage and assign it to a selection ID by using the order of priority for BOM usages (Transaction OS31).

In Section 12.1.4, we address the significance of defining a separate BOM usage for LTP. Refer to Chapter 3, in which we explain how to define and set up BOMs and routing selection.

Figure 12.1 shows the initial screen for BOM display. It includes a BOM USAGE field, which indicates the application of the BOM, whether it's a production BOM, a costing BOM, an engineering/design BOM, or another usage. For this example, enter BOM USAGE "1", which indicates production use. To display the BOM, follow the menu path, LOGISTICS • PRODUCTION • MASTER DATA • BILL OF MATERIAL • BILL OF MATERIAL • DISPLAY, or use Transaction CS03.

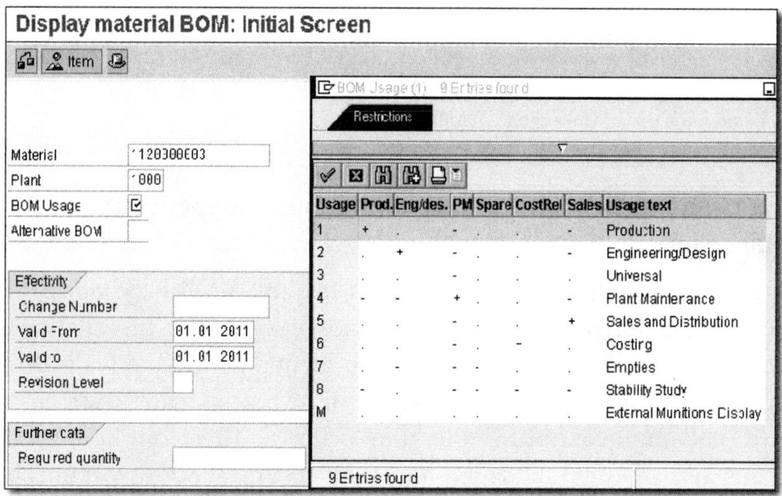

Figure 12.1 Production BOM with Usage 1

12.1.2 Planning Data: Planning Quantity

Table 12.3 presents the monthly planning quantities for the main material (unpacked urea bulk). You'll enter these quantities in the PIRs for this example. LTP will simulate the production and procurement requirements based on this information (PIRs).

Number	Month/Year	Quantity (Metric Tons)
1	02.2011	30,000.00
2	03.2011	45,000.00
3	04.2011	50,000.00
4	05.2011	52,000.00
5	06.2011	54,000.00
6	07.2011	49,000.00
7	08.2011	48,000.00

Table 12.3 Planning Quantities of Unpacked Urea Bulk

Number	Month/Year	Quantity (Metric Tons)
8	09.2011	55,000.00
9	10.2011	35,000.00
10	11.2011	40,000.00
11	12.2011	25,000.00

Table 12.3 Planning Quantities of Unpacked Urea Bulk (Cont.)

12.1.3 Planning Data: Version Number of Planned Independent Requirements

The purpose of having multiple versions of PIRs in the SAP ERP system is to account for various planning situations and scenarios, each of which is identified by its version number. A simulation version is the identified demand plan. One demand plan can be a sales plan while another can be a production plan, each having its own planning quantities of the same material. There can be several inactive versions available for simulation and comparison, but we recommend only having one active (operational) version.

For this example, use VERSION "02" (simulation 2) for LTP (see Figure 12.2). To enter planning quantities in the PIR, follow the menu path, LOGISTICS • PRODUCTION • PRODUCTION PLANNING • DEMAND MANAGEMENT • PLANNED INDEPENDENT REQUIREMENTS • CREATE, or use Transaction MD61.

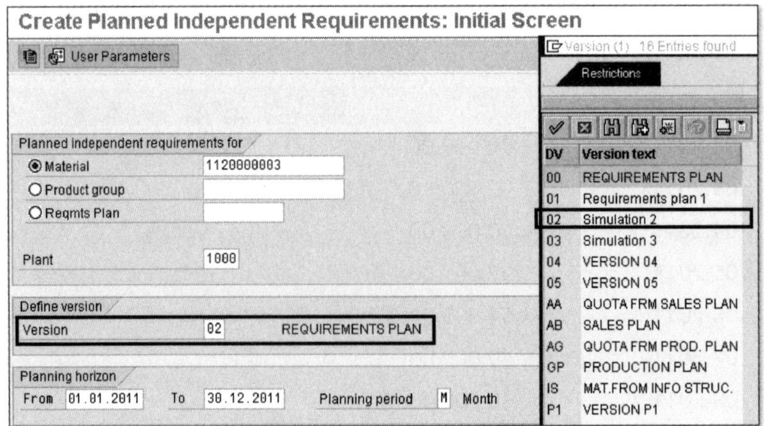

Figure 12.2 Standard Versions Available in the SAP ERP System

[«]

Note

Although the standard SAP ERP system provides several PIR versions, you can create more versions if you need them to attend to specific business processes in configuration Transaction OMP2.

12.1.4 Create a Planning Scenario

To create a planning scenario, follow the menu path, LOGISTICS • PRODUCTION • PRODUCTION PLANNING • LONG-TERM PLANNING • PLANNING SCENARIO • CREATE, or use Transaction MS31. These are the important steps in creating a planning scenario:

1. On this first screen, define the planning scenario. For this example, define the PLANNING SCENARIO as "001", and also give a short description (Figure 12.3).
2. Choose the LONG-TERM PLANNING radio button to denote that it's LTP, and press [Enter].

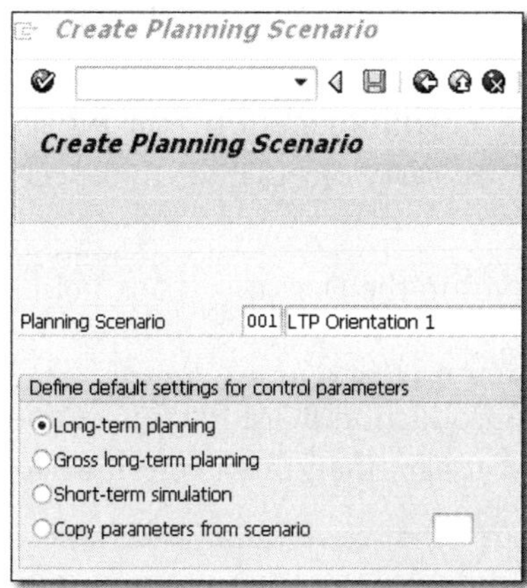

Figure 12.3 Initial Screen to Create a Planning Scenario

12.2 Long-Term Planning: Business Process

Having covered the master data as well as the planning data that you need to set up for LTP, we now cover the business processes involved in LTP. After you create the planning scenario, you need to perform the following series of activities in sequential order:

1. Enter the PIR in a simulative version.
2. Run LTP (simulative MRP), and save the results.
3. Evaluate the LTP stock/requirements list.

[»] **Note**

When you're working with LTP in your system, it may take several rounds to achieve a satisfactory production plan. It's helpful, then, to know that the basics and the underlying principles involved in LTP are very similar to MRP—adjusting, planning, evaluating, and comparing until the results are satisfactory.

The LTP example that we present in this chapter is a straightforward one, and the objective is to help you understand the sequence of steps involved in running LTP. For example, you can set up the planning quantities in the PIR much earlier than setting up LTP. We present it here to show that creating a PIR is one of the several steps that you need to undertake in the initial stages of master data setup for LTP.

12.2.1 Create the Planning Scenario

We'll now explain the step-by-step procedure for running the LTP process, which you'll also find in more detail in upcoming sections:

1. Enter "01.01.2011" and "31.01.2011" in the PLANNING PERIOD FOR INDEP. REQUIREMENTS area of the screen. The system will carry out LTP for the specified period only.
2. Enter "2" (PLANT STOCK AT THE TIME OF PLANNING) in the OPENING STOCK field.
3. Select other parameters as deemed appropriate.
4. Select the BOM SELECTION ID 01, which is the production BOM for this example. If you've configured an LTP-specific BOM usage, then this is the time to assign it here.
5. Click on the PLANNED INDEPENDENT REQUIREMENTS button located on the top section of the CREATE PLANNING SCENARIO screen (see Figure 12.4 ❶). Assign the simulative version (use 02 for this example).
6. Click on the + icon to add the version entry, add the FROM and TO dates, and then choose CONFIRM.
7. Select the plants for which the LTP will be applicable. The PLANTS button ❷ is located on the top section of the CREATE PLANNING SCENARIO screen. Click on the + icon to add plant entry, and then choose CONFIRM. For this example, we use plant 1000.

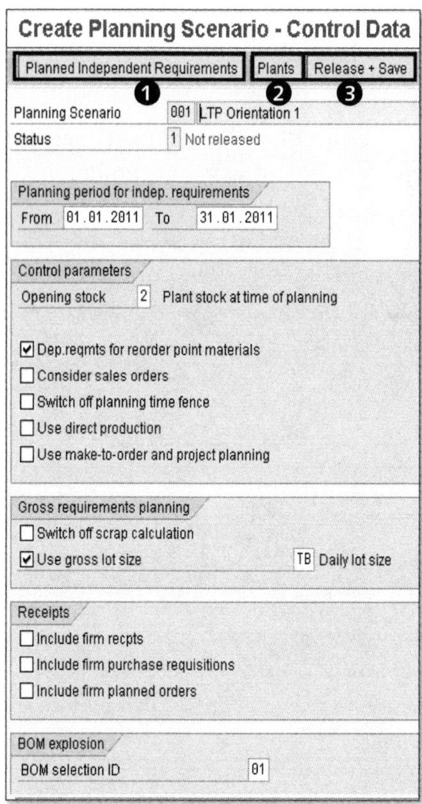

Figure 12.4 Planning Scenario Creation Screen to Define Parameters

8. Choose the RELEASE + SAVE button ❸ to save the planning scenario. After a planning scenario has been released, you can no longer change its parameters. However, you can undo the release within the planning scenario.

   The system automatically creates planning file entries in due course. When you release the planning scenario, the system creates entries in the planning files. However, when you cancel the release of the planning scenario, the system correspondingly deletes the entries from the planning file.

12.2.2 Enter Planned Independent Requirements for the Simulative Version

After you create the planning scenario, which is 001 for this example, the next step is to incorporate all of the planning quantities that we defined in Table

12.3 as PIRs. Follow the menu path, LOGISTICS • PRODUCTION • PRODUCTION PLANNING • LONG-TERM PLANNING • PLANNED INDEPENDENT REQUIREMENTS • CREATE, or use Transaction MD61.

Figure 12.5 shows the initial screen ❶ where you enter the relevant material number (for this example, unpacked urea bulk), plant, and version number.

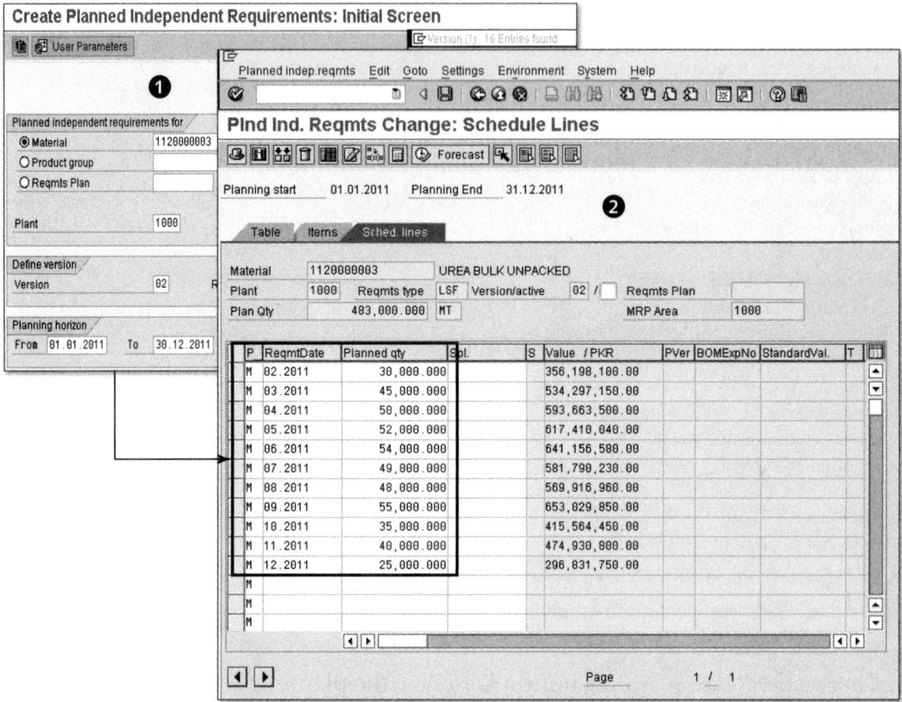

Figure 12.5 PIR Details of Urea Unpacked Bulk in Version 02

Area ❷ shows the tabular view after incorporating all of the planning quantities for the unpacked urea bulk, as initially defined in Table 12.3. After incorporating all of the planning quantities, save the results by choosing SAVE.

12.2.3 Run Long-Term Planning (Simulative Material Requirements Planning)

To run the LTP interactively (simulative MRP), follow the menu path, LOGISTICS • PRODUCTION • PRODUCTION PLANNING • LONG-TERM PLANNING • LONG-TERM

PLANNING • SINGLE ITEM, MULTI-LEVEL, or use Transaction MS02. Here, you define the initial parameters to run the LTP. Enter the PLANNING SCENARIO as "001", as well as the MATERIAL code of the unpacked urea bulk and the PLANT (see Figure 12.6).

Just like the MRP, you can define the parameters, which eventually influence the results of LTP. For this example, define that during LTP, the system should reexplode the BOM and routing to read the latest data and perform lead time scheduling to account for capacity planning (see Figure 12.6).

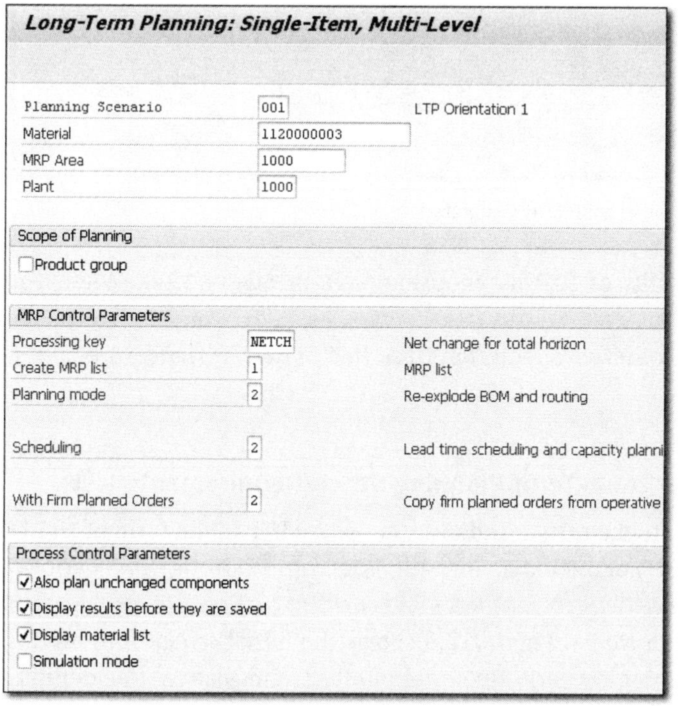

Figure 12.6 Initial Screen to Run LTP

Press Enter, and the system issues a warning message to check all of the planning parameters. Press Enter again to confirm, and the system runs the LTP.

Figure 12.7 shows the planning results for the materials selected for planning in the form of a list. The material list is the output after LTP is run, and the system displays it because the DISPLAY MATERIAL LIST checkbox was selected. The results are also saved. In other words, we defined in Table 12.2 that planning is only applicable to the materials unpacked urea bulk and UF-85, and now Figure 12.7 shows the same in the form of a material list. Notice that because ammonia anhydrous wasn't marked for planning, it's neither planned nor available in the material list.

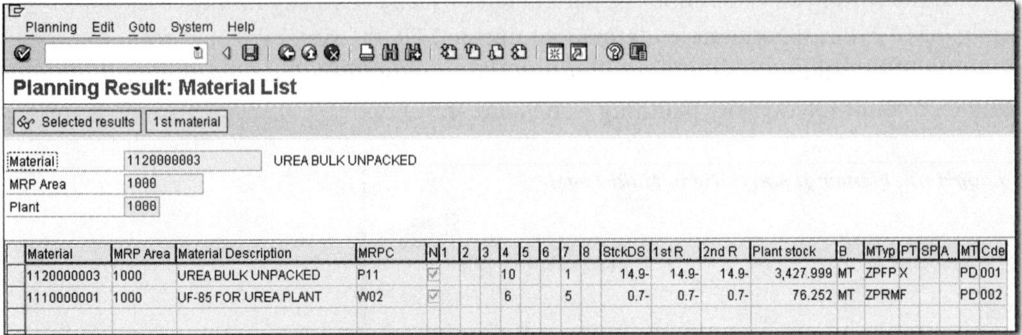

Material	MRP Area	Material Description	MRPC	N	1	2	3	4	5	6	7	8	StckDS	1st R	2nd R	Plant stock	B	MTyp	PT	SP	A	MT	Cde
1120000003	1000	UREA BULK UNPACKED	P11	✓				10			1		14.9-	14.9-	14.9-	3,427.999	MT	ZPFP	X			PD	001
1110000001	1000	UF-85 FOR UREA PLANT	W02	✓				6			5		0.7-	0.7-	0.7-	76.252	MT	ZPRMF				PD	002

Figure 12.7 LTP List (Simulative MRP List) Generated

Save the planning results of LTP by choosing SAVE in Figure 12.6. When you select the DISPLAY RESULTS BEFORE THEY ARE SAVED checkbox, you have to specifically save the results; otherwise, you don't use this option in standard business processes.

12.2.4 Evaluate the Long-Term Planning Stock/Requirements List

It's now time to evaluate the results and outcome of the LTP in the form of an LTP stock/requirements list. Follow the menu path, LOGISTICS • PRODUCTION • PRODUCTION PLANNING • LONG-TERM PLANNING • EVALUATIONS • STOCK/REQUIREMENTS LIST, or use Transaction MS04. Figure 12.8 shows the initial screen for entering parameters with planning scenario, material number, and plant. After defining the initial parameters ❶, press Enter or choose CONTINUE.

[»] **Note**

In the standard display layout of a stock/requirements list, the system shows individual planning elements, including independent requirements, planned orders, and so on. However, we've changed the layout to show how the monthly quantities entered in the PIR are shown in the stock/requirements list. Notice that the first column is the monthly PERIOD/SEGMENT, whereas the second column is the PIR (PLND IND.REQMTS).

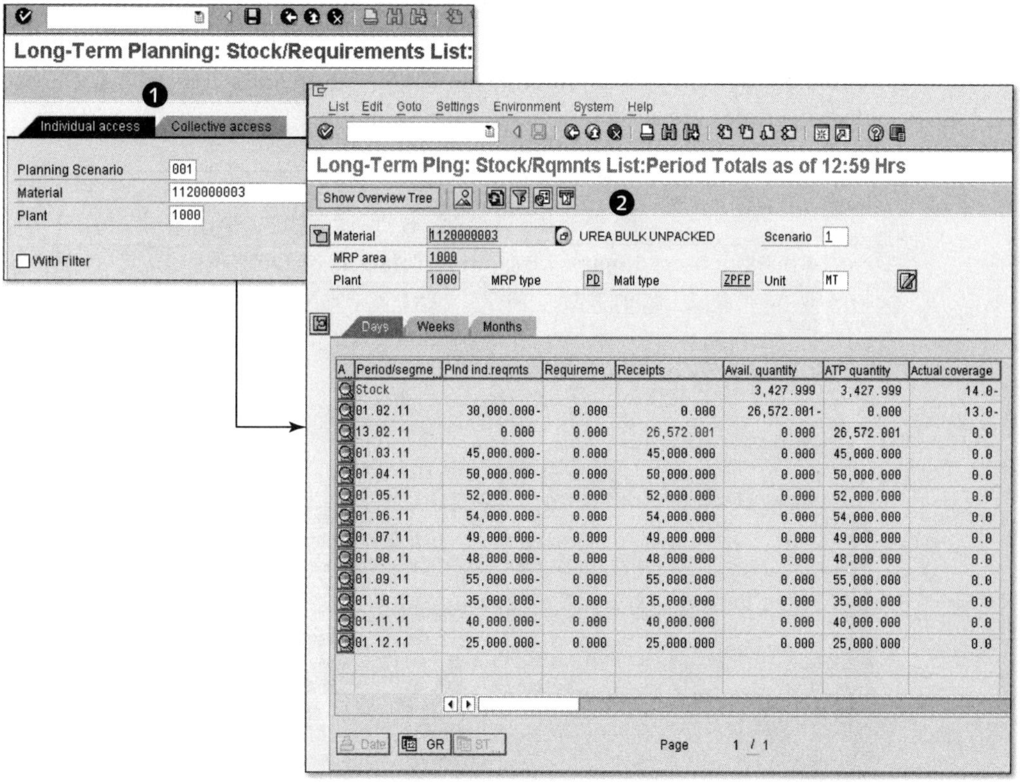

Figure 12.8 Stock/Requirement List of Urea Bulk Unpacked

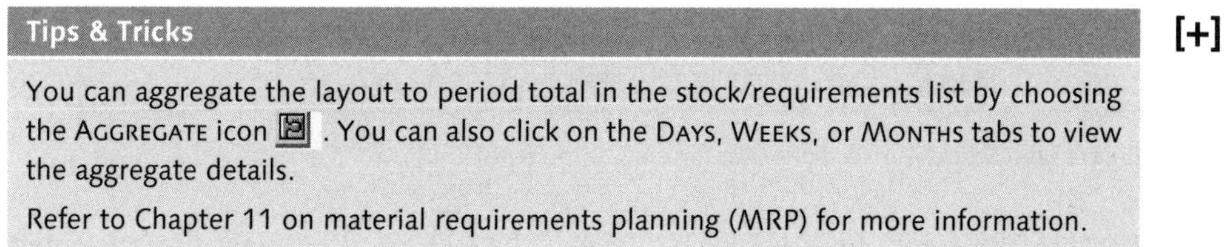

Tips & Tricks [+]

You can aggregate the layout to period total in the stock/requirements list by choosing the AGGREGATE icon . You can also click on the DAYS, WEEKS, or MONTHS tabs to view the aggregate details.

Refer to Chapter 11 on material requirements planning (MRP) for more information.

Figure 12.8 has an opening stock of 3,428 metric tons ❷. This is because we selected the existing stock as the opening stock balance while creating planning scenario 001. All figures are available as shown in Table 12.3, which shows the planning quantities for unpacked urea bulk.

Figure 12.9 shows the results of component UF-85 of the finished product (unpacked bulk urea). The dependent requirements of UF-85 calculated by LTP are based on the PIRs and the BOM explosion details given in Table 12.3. The calculated quantities are shown in the RECEIPTS column.

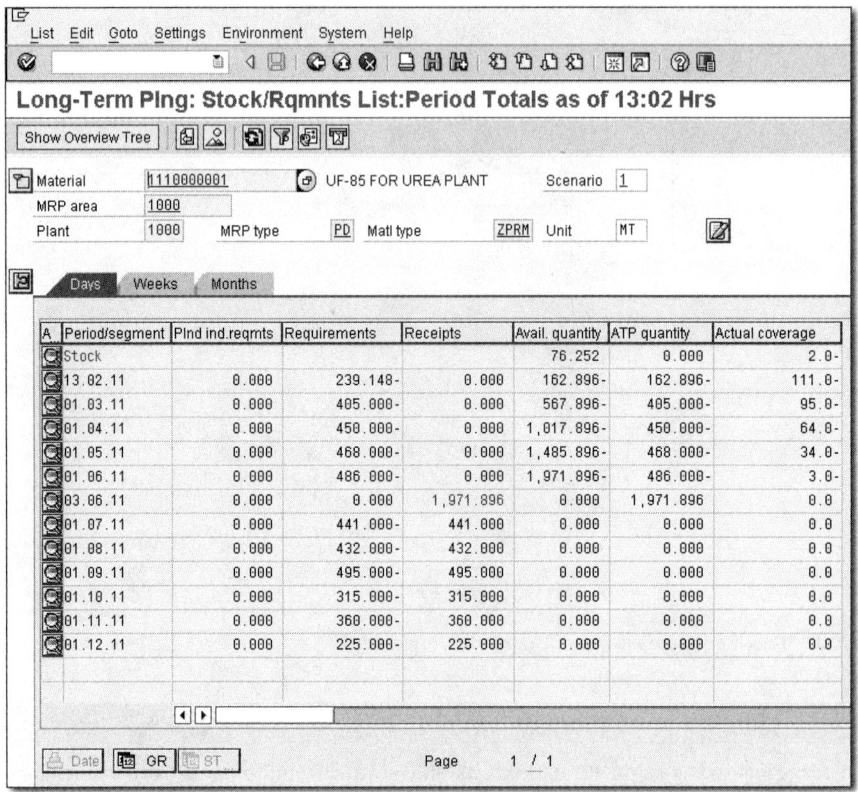

A	Period/segment	Plnd ind.reqmts	Requirements	Receipts	Avail. quantity	ATP quantity	Actual coverage
	Stock				76.252	0.000	2.0-
	13.02.11	0.000	239.148-	0.000	162.896-	162.896-	111.0-
	01.03.11	0.000	405.000-	0.000	567.896-	405.000-	95.0-
	01.04.11	0.000	450.000-	0.000	1,017.896-	450.000-	64.0-
	01.05.11	0.000	468.000-	0.000	1,485.896-	468.000-	34.0-
	01.06.11	0.000	486.000-	0.000	1,971.896-	486.000-	3.0-
	03.06.11	0.000	0.000	1,971.896	0.000	1,971.896	0.0
	01.07.11	0.000	441.000-	441.000	0.000	0.000	0.0
	01.08.11	0.000	432.000-	432.000	0.000	0.000	0.0
	01.09.11	0.000	495.000-	495.000	0.000	0.000	0.0
	01.10.11	0.000	315.000-	315.000	0.000	0.000	0.0
	01.11.11	0.000	360.000-	360.000	0.000	0.000	0.0
	01.12.11	0.000	225.000-	225.000	0.000	0.000	0.0

Figure 12.9 Stock/Requirements List for the Component UF-85

Because the component UF-85 is planned, its LTP results reflect the calculated results based on the BOM for unpacked urea bulk in Table 12.2 (0.009). Table 12.4 represents the calculated value of the component UF-85 in LTP based on the BOM of unpacked urea bulk.

Month/Year	Unpacked Urea Bulk (Quantity in Metric Tons)	UF-85 (Quantity in Metric Tons)
02.2011	30,000.00	270 (30,000 × 0.009 = 270)
03.2011	45,000.00	405
04.2011	50,000.00	450
05.2011	52,000.00	468
06.2011	54,000.00	486
07.2011	49,000.00	441
08.2011	48,000.00	432
09.2011	55,000.00	495
10.2011	35,000.00	315
11.2011	40,000.00	360
12.2011	25,000.00	225

Table 12.4 Results of LTP for Component UF-85 of Unpacked Urea Bulk

12.3 Further Options in Long-Term Planning

You now know how to successfully run LTP. The following are further options that are available in LTP:

- Manually create a simulative planned order
- Firm the simulative planned order using a firming date
- Calculate average plant stock
- Copy PIRs or firmed planned orders from LTP to operative planning

Some of these options aren't commonly known; we'll explain these in the following sections.

12.3.1 Manually Create a Simulative Planned Order

While LTP provides complete automation in planning and proposal generation, including the creation of planned orders, it's often necessary to manually create a simulative planned order to meet any additional requirements during the simulation phase. For example, while evaluating results from LTP, you realize that you need to account for additional demand for a specific component. Instead of making changes to the PIR of a finished product and then reexplode the BOM to come up with the component's quantity, you can quickly create a simulative planned order for that specific component. This simulative planned order then becomes available in all evaluations (such as the stock/requirements list) and information systems (such as Purchasing and Inventory Controlling) and is also firmed. The system won't make changes during the next LTP run to this manually created simulative planned order.

To create a simulative planned order, follow the menu path, LOGISTICS • PRODUCTION • PRODUCTION PLANNING • LONG-TERM PLANNING • SIMULATIVE PLANNED ORDER • CREATE, or use Transaction MS11. Figure 12.10 shows the screen for the creation of a simulative planned order. For this example, create a simulative planned order for component UF-85 for a quantity of 60 metric tons and also associate it with planning scenario 1. Also, set the FIRMING indicator so that during the next LTP run, the details of this simulative planned order aren't changed.

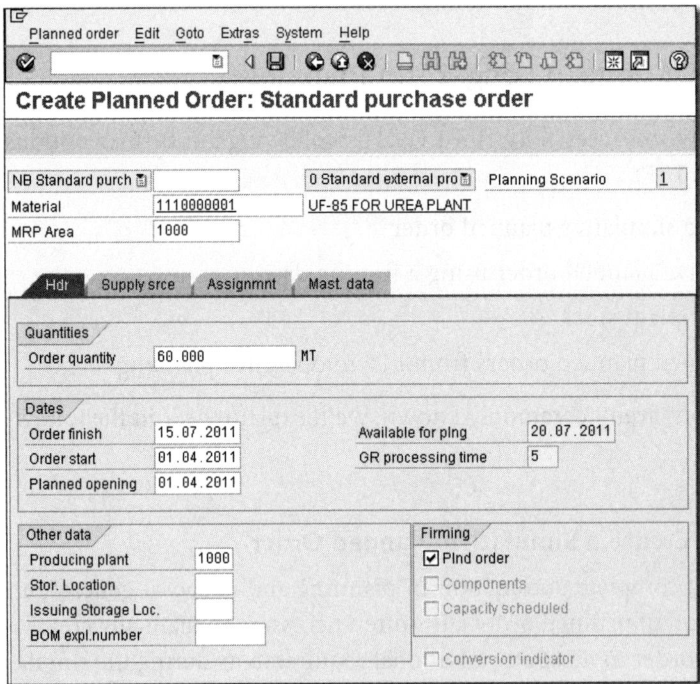

Figure 12.10 Manual Creation of a Simulative Planned Order

12.3.2 Firm the Simulative Planned Order Using a Firming Date

The process of manually firming planned orders ensures that all future LTP runs don't impact or overwrite the existing simulative planning data. In firming, a date is specified, and all of the simulative planned orders generated from LTP are then firmed until that date. During the next LTP run, these manually firmed simulative planned orders aren't changed or deleted. While remaining in the stock/requirement list (Transaction MS04), choose Edit • Manual Firming Date (Figure 12.11).

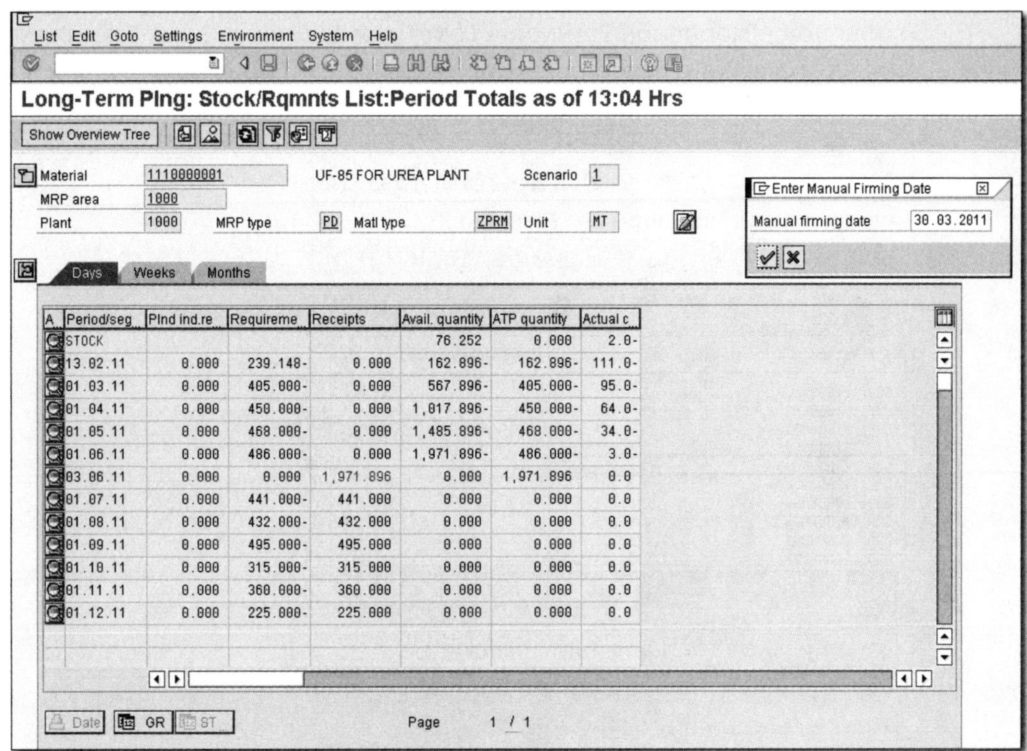

A	Period/seg	Plnd ind.re	Requireme	Receipts	Avail. quantity	ATP quantity	Actual c
	STOCK				76.252	0.000	2.0-
	13.02.11	0.000	239.148-	0.000	162.896-	162.896-	111.0-
	01.03.11	0.000	405.000-	0.000	567.896-	405.000-	95.0-
	01.04.11	0.000	450.000-	0.000	1,017.896-	450.000-	64.0-
	01.05.11	0.000	468.000-	0.000	1,485.896-	468.000-	34.0-
	01.06.11	0.000	486.000-	0.000	1,971.896-	486.000-	3.0-
	03.06.11	0.000	0.000	1,971.896	0.000	1,971.896	0.0
	01.07.11	0.000	441.000-	441.000	0.000	0.000	0.0
	01.08.11	0.000	432.000-	432.000	0.000	0.000	0.0
	01.09.11	0.000	495.000-	495.000	0.000	0.000	0.0
	01.10.11	0.000	315.000-	315.000	0.000	0.000	0.0
	01.11.11	0.000	360.000-	360.000	0.000	0.000	0.0
	01.12.11	0.000	225.000-	225.000	0.000	0.000	0.0

Figure 12.11 Manual Firming Date Set as 30.03.2011

[«]

Note

In the stock/requirements list for operative MRP (Transaction MD04), the system displays the manual firming date when it's active.

12.3.3 Calculate Average Plant Stock

Based on the period under evaluation, LTP offers the option to calculate the average plant stock of all the materials that went through the LTP process. The average plant stock is calculated by adding up the total stock for the period under evaluation divided by the number of periods (months) under evaluation. The benefit of providing the average plant stock to Purchasing and Inventory Management departments is that they can better coordinate with vendors for timely deliveries, while warehouses have an advanced preview of space to be made available for this incoming stock. To calculate average plant stock, follow menu path, LOGISTICS • PRODUCTION • PRODUCTION PLANNING • LONG-TERM PLANNING • LONG-TERM PLANNING • AVERAGE PLANT STOCK, or use Transaction MS29 (Figure 12.12).

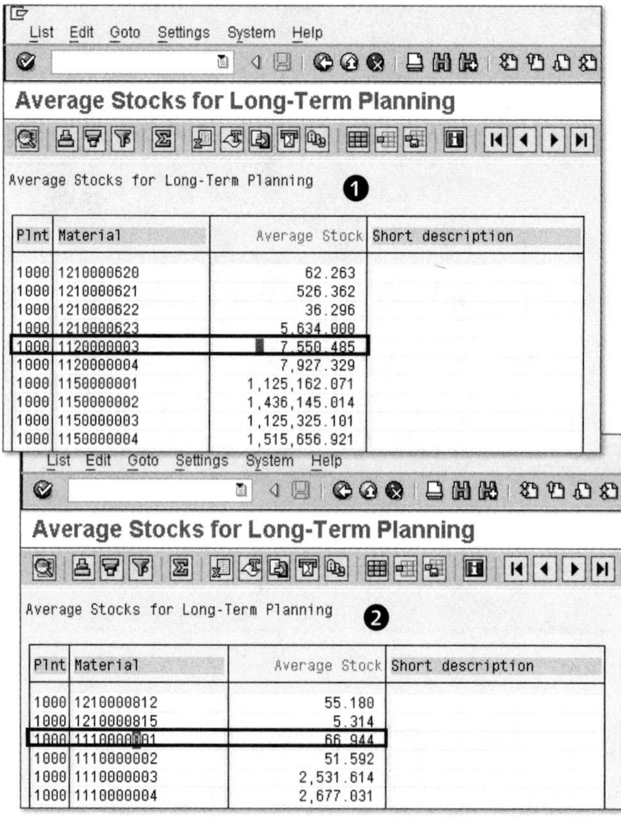

Figure 12.12 Average Plant Stock of Unpacked Urea Bulk and UF-85

The average plant stock for unpacked urea bulk was calculated by the SAP ERP system to be 7,550.485 metric tons for the period under evaluation ❶. The average plant stock for UF-85 is 66.944 metric tons for the period under evaluation ❷.

12.3.4 Copy Long-Term Planning Results to Operative Planning

After all of the necessary simulations (LTP) have been completed, the business often needs to move the results of one of the simulations of LTP to operative planning, on which the actual MRP run can be executed. For this example, proceed with the understanding or assumption that you're satisfied with the results of LTP and want to copy the PIR to operative planning. In an actual business scenario, there may be several rounds of PIR updating and planning simulations within LTP before you copy the agreed-upon PIR into operative planning. The agreed-upon PIR for transfer to operative planning will be the one with which you're satisfied with the LTP simulation results. While this example focuses on copying PIRs, you can also copy firmed (simulative) planned orders to operative planning.

The option to copy a PIR of simulative planning to operative planning saves an enormous amount of work already done during the LTP process and eliminates redundancy in data entry. Further, during the copy function, you have the flexibility to make changes in operative planning with respect to LTP. After the PIR is in operative planning, assign the Active status to it, and run the MRP.

To copy the LTP's PIR to operative planning, follow the menu path, Logistics • Production • Production Planning • Long-Term Planning • Planned Independent Requirements • Copy Version, or use Transaction MS64. Figure 12.13 shows the initial screen to define parameters to transfer the results of the simulative planning of version 02 to the operative planning in version 03.

Tips & Tricks [+]

Although we're deliberately taking the longer route in this example to show the steps involved in the copy function, you can use Transaction MS32 (Change Planning Scenario) and use the Active Independent Requirements icon to activate PIRs.

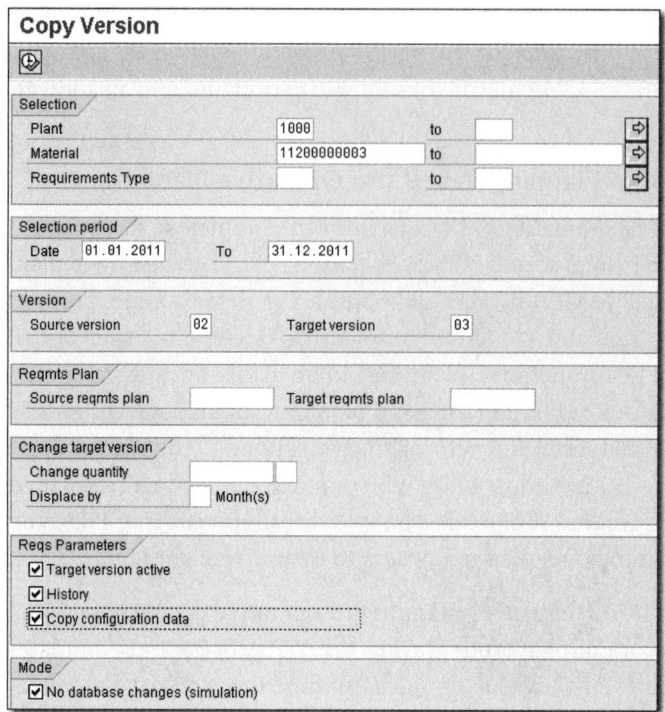

Figure 12.13 Parameters Screen to Copy Source Version 02 to Target Version 03

You specified that source version 02 should be copied to target version 03 for the material unpacked urea bulk. Other parameters to define include dates of transfer from version 02 to 03. You can also make changes in quantities between the source and target versions as well as the number of months by which you want to displace the planning data to operative data during transfer. The displacement is the moving of quantities forward or backward in time from version 02 to version 03 during copying of simulative data to operative data. Also select the No Database Changes (Simulation) checkbox to enable the copy function in the simulation mode and to check for any errors or other deviations before performing the actual transfer.

After execution (by pressing F8), Figure 12.14 shows the simulated copied results to target version 03 with all of the urea bulk quantity successfully transferred to version 03. If you go back to the previous screen, deselect the No Database Changes (Simulation), and press F8 again, this time the copied results are available in the PIR under version 03.

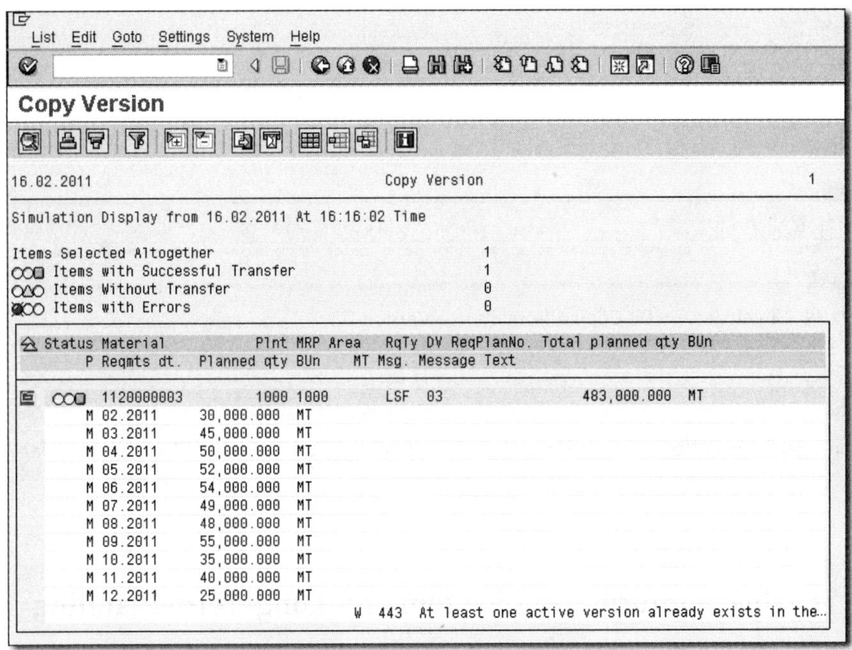

Figure 12.14 Results after Using the Copy Version Functionality

Note [«]

While this example shows the transfer of a PIR from version 02 to version 03, we suggest that you use version 00, which is the active version in the system to transfer LTP PIRs to operative planning (MRP).

After copying the LTP planning results of source version 02 to operative planning in version 03, check to see if the outcome of the copy function is that the LTP PIR data of version 02 are now available in version 03. To check the PIR, follow the menu path, LOGISTIC • PRODUCTION • PRODUCTION PLANNING • LONG-TERM PLANNING • PLANNED INDEPENDENT REQUIREMENTS • CHANGE, or use Transaction MD62. Looking at the screen shown in Figure 12.15, you can confirm that the system successfully executes the copy function, and the simulative figures of version 02 are now transferred in 03. Refer to the third column (with the heading V) to note that it's version 03 and consider the activated ACTIVE field (the checkbox) right next to it.

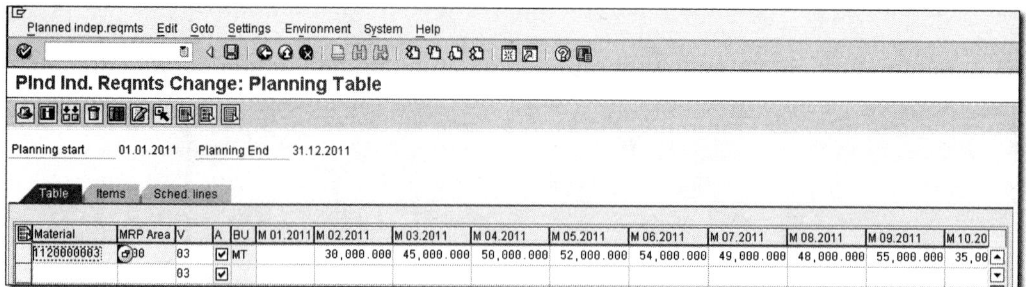

Figure 12.15 Target Version 03 Copied from Source Version 02

[+] **Tips & Tricks**

You can use Transaction MD74 to delete or reorganize old or inactive PIRs on a regular basis.

12.4 Evaluate Information Systems for Long-Term Planning

The Logistics Information System (LIS) is available for all core logistics components in the SAP ERP system. The LIS for each component provides a multitude of standard analysis reports. It derives the data and information from all of the transactions performed in the system, thus ensuring the availability of comprehensive information for evaluation and decision-making purposes. An information system is a reporting option available in the system to display the desired information based on the user-defined criteria. Therefore, there is also a comprehensive information system for LTP, catering to both the Purchasing as well as Inventory Controlling functions.

In the following sections, we'll explain how to set up and evaluate the results of the information systems for Purchasing and Inventory Controlling in LTP. You have to first set up the data for these two information systems before the system displays the planning results of a planning scenario. You don't have to set up any data for capacity requirements for work centers for a given planning scenario.

[»] **Note**

To learn how to use features available in LIS, refer to Chapter 19 in which we cover reporting in the SAP ERP system.

12.4.1 Setting Up a Purchasing Information System for Long-Term Planning

You have to set up data before the purchasing information system for LTP is ready for use and evaluation. To get to the initial screen for the data setup (Figure 12.16), follow the menu path, LOGISTICS • PRODUCTION • PRODUCTION PLANNING • LONG-TERM PLANNING • EVALUATIONS • PURCHASING INFORMATION SYSTEM • SET UP DATA, or use Transaction MS70.

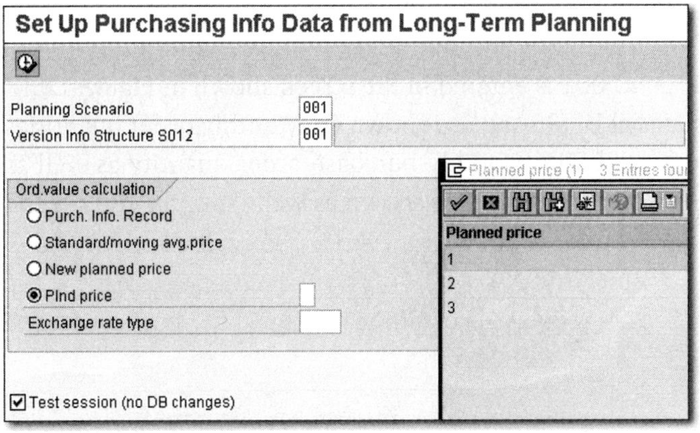

Set Up Purchasing Info Data from Long-Term Planning

Planning Scenario	001
Version Info Structure S012	001

Ord.value calculation
- ○ Purch. Info. Record
- ○ Standard/moving avg.price
- ○ New planned price
- ◉ Plnd price
- Exchange rate type

☑ Test session (no DB changes)

Planned price (1) 3 Entries fou

Planned price
1
2
3

Figure 12.16 Set Up Screen for the Purchasing Information System for LTP

To set up the data, select the relevant parameters (e.g., PLANNING SCENARIO). For this example, select PLANNING SCENARIO 001. For order value calculation, choose STANDARD/MOVING AVG.PRICE, which the system will read from the material master. If you want the planned prices of the material to form the basis for purchasing value calculation, then you can choose PLND (planned) PRICE and enter from one of the three planned prices available. Press [F8] to execute.

Warning! [!]

Make sure you deselect the TEST SESSION (NO DB CHANGES) checkbox shown in Figure 12.16 to execute in the system.

12.4.2 Evaluating with the Purchasing Information System for Long-Term Planning

The purchasing information system for LTP is now ready for use for evaluation purposes. Follow the menu path, LOGISTICS • PRODUCTION • PRODUCTION PLANNING • LONG-TERM PLANNING • EVALUATION • PURCHASING INFORMATION SYSTEM • MATERIAL, or use Transaction MCEC. Enter the relevant parameters for component UF-85 as shown in Figure 12.17 ❶. These parameters are PLANNING SCENARIO, STANDARD/MOVING AVG.PRICE (or other value calculation), MATERIAL number, PLANT, and the PERIOD TO ANALYZE.

Note that PLANNING SCENARIO 001 is entered in the screen shown in Figure 12.17 ❶, and the results for material UF-85 are also shown ❷. Monthly order quantities along with values are reflected. The monthly purchase order quantity as well as the monthly total value in local currency ❷ is shown as well. You can also see the per unit price in local currency of UF-85.

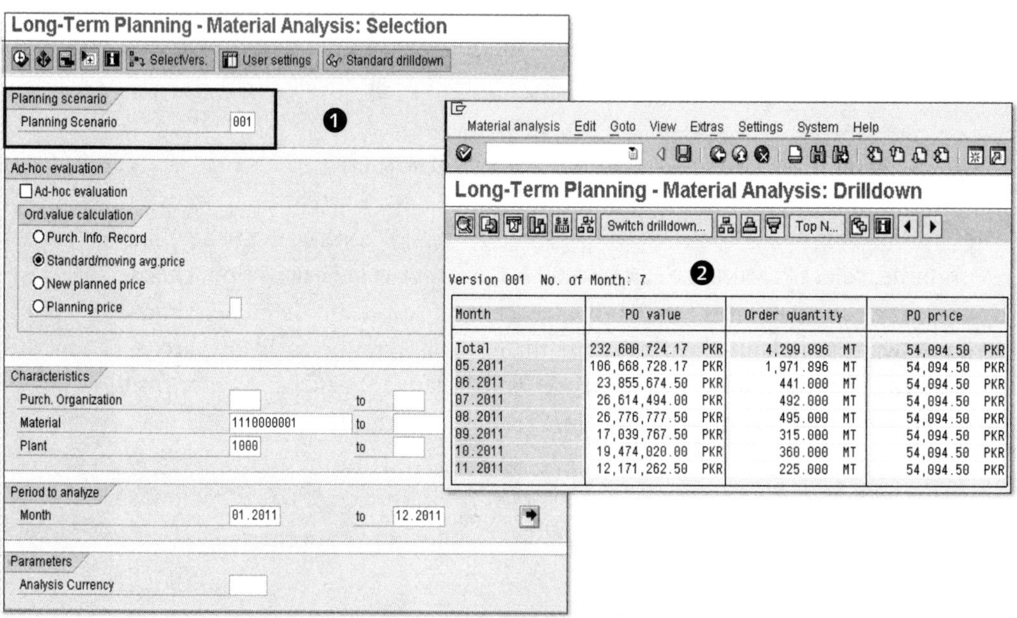

Figure 12.17 Evaluation of Component UF-85 in the Purchasing Information System for Materials

You can use the results of your simulative planning data of the planning scenario with operative planning for comparison purposes. You use Transaction MS44 to bring up the comparison, whether it's material specific or plant specific. You can choose whether you want to run a comparison with the current operative planning situation (corresponding to the stock/requirements list), with the last operative planning run (corresponding to the MRP list), or with another planning scenario. You can define the presentation of the comparison by means of a layout; in this case, we chose the standard layout, SAPSOP. This layout displays the issues, receipts, and available quantity for LTP and operative planning (indicated by comparative data in this case).

12.4.3 Setting Up an Inventory Controlling Information System for Long-Term Planning

Similar to setting up a purchasing information system for LTP, you have to set up data before the Inventory Controlling information system for LTP is ready for use. To set up the data, follow the menu path, LOGISTICS • PRODUCTION • PRODUCTION PLANNING • LONG-TERM PLANNING • EVALUATIONS • INVENTORY CONTROLLING • SETUP DATA, or use Transaction MCB&.

In Figure 12.18, the selection parameters are MATERIAL, PLANT, evaluation PERIOD, and PLANNING SCENARIO (which is 001 for this example). After the parameters are defined, choose the EXECUTE icon or press F8. This is all that you have to do to set up data for the Inventory Controlling information system for LTP.

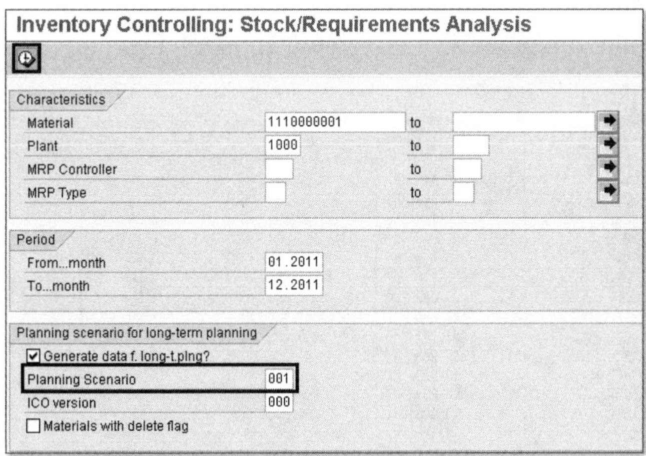

Figure 12.18 Setup of Inventory Controlling Information System for LTP

12.4.4 Evaluating the Inventory Controlling Information System for Long-Term Planning

The Inventory Controlling information system for LTP is now ready for use for evaluation purposes. To get to the INVENTORY CONTROLLING screen, follow the menu path, LOGISTICS • PRODUCTION • PRODUCTION PLANNING • LONG-TERM PLANNING • EVALUATIONS • INVENTORY CONTROLLING • EVALUATION, or use Transaction MCB). Figure 12.19 ❶ reflects the inventory situation of component UF-85 for individual months. The second screen ❷ shows the monthly inventory of material UF-85. The total monthly requirement of U-85 is shown first, followed by the stock situation, and finally by how much quantity must be received per month against procurement proposals generated by MRP/LTP (as shown in the GDSRECEIPT (MRP) column).

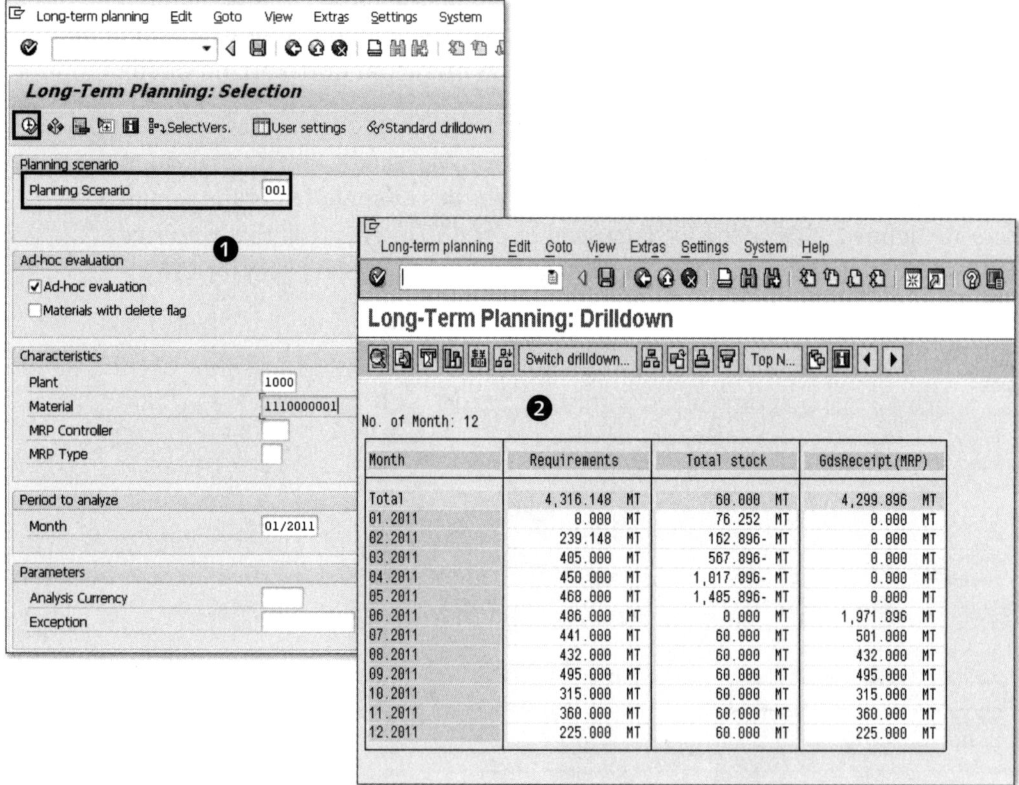

Month	Requirements	Total stock	GdsReceipt(MRP)
Total	4,316.148 MT	60.000 MT	4,299.896 MT
01.2011	0.000 MT	76.252 MT	0.000 MT
02.2011	239.148 MT	162.896- MT	0.000 MT
03.2011	405.000 MT	567.896- MT	0.000 MT
04.2011	450.000 MT	1,017.896- MT	0.000 MT
05.2011	468.000 MT	1,485.896- MT	0.000 MT
06.2011	486.000 MT	0.000 MT	1,971.896 MT
07.2011	441.000 MT	60.000 MT	501.000 MT
08.2011	432.000 MT	60.000 MT	432.000 MT
09.2011	495.000 MT	60.000 MT	495.000 MT
10.2011	315.000 MT	60.000 MT	315.000 MT
11.2011	360.000 MT	60.000 MT	360.000 MT
12.2011	225.000 MT	60.000 MT	225.000 MT

Figure 12.19 Inventory Controlling of Component UF-85

12.4.5 Capacity Planning

You can gain greater visibility of the capacity situation for various planning scenarios, and you don't have to set up any data for it. Further, the simulated capacity planning functionality helps you plan for future capacity expansions in case the existing capacity setup is unable to meet the impending capacity requirements.

To access the CAPACITY PLANNING screen, follow the menu path, LOGISTICS • PRODUCTION • PRODUCTION PLANNING • LONG-TERM PLANNING • EVALUATIONS • CAPACITY REQUIREMENTS • WORK CENTERS, or use Transaction MC38. On the initial screen, enter the planning scenario for which you want to evaluate the planning situation. Press [Enter], and the system brings up comprehensive capacity details for you to evaluate.

Note [«]

Refer to Chapter 14 on capacity requirements planning for more information.

Because PP completely integrates with Product Costing, you can also leverage LTP to evaluate and transfer planned activity requirements to production. Similarly, you can transfer scheduled activities to PP business processes.

To access the ACTIVITY REQUIREMENT submenu, follow the menu path, LOGISTICS • PRODUCTION • PRODUCTION PLANNING • LONG-TERM PLANNING • ENVIRONMENT • ACTIVITY REQUIREMENT.

12.5 Summary

LTP is able to attend to several of the business scenarios in simulation mode, as well as to attend to questions such as the average plant stock, the production quantities, the procurement quantities and values, and the capacities over a given period of time. LTP uses the same concepts and fundamentals as MRP. Even the system navigation and functions of LTP and MRP offer great similarities.

The next chapter begins Part V and covers special procurement types.

PART V

Optimizing Production Planning

The SAP ERP system provides special procurement types that you can use to attend to unique business scenarios. This might be where the production of assembly and procurement of components are nontraditional in nature and involve complex and diverse logistics processes.

13 Special Procurement Types

A traditional production process involves procuring components from suppliers and vendors, producing them in-house, and eventually selling them to customers. However, in a truly globalized economy, both small companies and companies with giant production setups across many countries and locations must deal with diverse, challenging, and complex logistics and supply chain processes. The same processes also need to be mapped in the SAP ERP system for effective planning of procurement and production processes.

Consider the following actual and real-time business processes and the complexities involved:

- You have a vendor who keeps its material's stock in your warehouse, but you only pay the vendor when your company actually consumes the material.
- You have a product in which few of the components become part of the assembly, yet they are part of the overall product offering. During the packing process, you want all of these components available at the same time and place.
- You have a product in which some of the production steps are performed in-house, while the others are performed by external vendors/service providers.

These business scenarios and more are catered to with *special procurement types* in the SAP ERP system. These business processes vary from handling phantom assembly during production, subcontracting consignment to material production at another plant, to direct production or procurement.

When any special procurement is involved, you must ensure that you assign the relevant special procurement type key, either in the MRP 2 view of the material master or in the detailed view of the component in the bill of materials (BOM). In this chapter, we'll first give you an overview of special procurement types in the SAP ERP system, and then we'll discuss each type of special procurement.

13.1 Overview

A *special procurement type key* is the control function that the system looks for during the planning of the material to bring forth the relevant results (after planning) for immediate execution. The special procurement type key is plant-specific, and you can assign this key at two levels, depending on the business processes:

- Material master (in the MRP 2 view)
- BOM in the detailed view of the component

[»]

Note

While we cover the maximum details of each of the special procurement type business processes in this chapter, we suggest that you engage an SAP ERP Materials Management (MM) resource/consultant to have end-to-end comprehension of the processes involved.

Figure 13.1 shows the MRP 2 view of MATERIAL P-100 and PLANT 3000. Assign the special procurement type key in the SPECIAL PROCUREMENT field by placing your cursor on the field and pressing F4 or clicking on the dropdown menu.

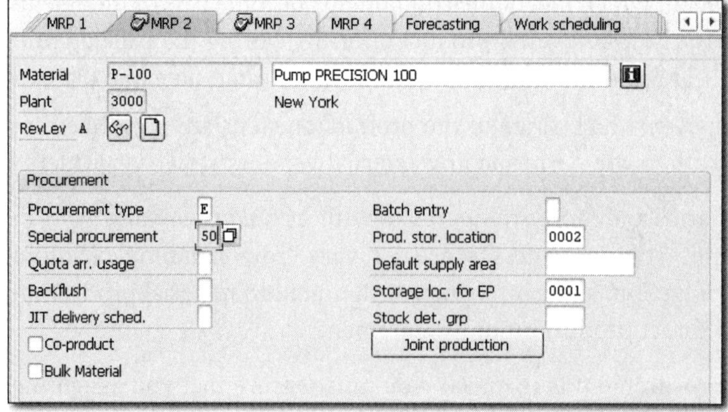

Figure 13.1 Special Procurement Type Field in the Material Master

This leads to the pop-up that contains the list of several standard procurement types delivered by the SAP ERP system in its standard offering, as well as additional special procurement types created to fulfill the specific business needs.

[«]

Note

You define the special procurement key using the configuration (Transaction SPRO) path, PRODUCTION • MATERIAL REQUIREMENTS PLANNING • MASTER DATA • DEFINE SPECIAL PROCUREMENT TYPE (see Figure 13.2).

Change View "Special Procurement": Overview

New Entries

Plnt	Name 1	Sp.Pr.Type	Special procurement type description
3000	New York	10	Consignment
3000	New York	30	Subcontracting
3000	New York	40	Stock transfer (proc.from alter.plant)
3000	New York	42	Stock transfer (proc.from plant 3200
3000	New York	45	Stock transfer (proc.from plant 1000)
3000	New York	50	Phantom assembly
3000	New York	52	Direct production/Collective order
3000	New York	60	Planned independent requirements
3000	New York	70	Reservation from alternate plant
3000	New York	72	Reservation from alternate plant 3200
3000	New York	80	Production in alternate plant
3000	New York	82	Production in alternate plant (3200)

Figure 13.2 Configuration of the Special Procurement Type Key for Plant 3000

[+]

Tips & Tricks

In a nonproduction SAP ERP system, whenever you assign a special procurement type key in the MRP 2 view of the material master or to a component in the BOM and then perform the necessary business transaction such as creating a production order or purchase requisition, you can always run material requirements planning (MRP) on that material/component (Transaction MD02) to test how the system reflects the planning results of that specific special procurement type key. The same testing logic applies when you create a new special procurement key to cater to a business requirement.

You should now have a general understanding of the special procurement type processes. In the actual business processes of the company, preference should be given to making better and effective use of MRP results, so that the predecessor–successor relationship of the entire chain of events is available. Now let's consider each of the special procurement types in detail.

13.2 Phantom Assembly

The special procurement type key for phantom assembly is 50. A *phantom assembly* is the logical grouping of one component or many different components, which forms an integral part of a final or superior product's offering. Examples of phantom assembly are the accompanying speakers, connecting wires, and so on when you buy a stereo system that you can install as and when needed. All of the components of phantom assembly are mandatory for the production process. Due to similarity to the production processes, it makes sense to group the components for availability. Hence, these logical grouping are purely organizational in nature to better manage the production processes. Also, note that the components in phantom assembly are never combined with each other but are made available at the same time for an efficient production.

You don't have to maintain routing for a phantom assembly because it's not produced, but you do have to define the BOM, which is then eventually entered as a component in the material's BOM. Phantom assembly doesn't have stock of its own but that of components which make up the phantom assembly. Also, because no routing is available for phantom assembly, you can't record the machine or the labor duration to reflect the same in Cost Accounting (Product Cost Controlling [CO-PC]). The superior product's routing should account for the machine or labor hours involved in phantom assembly.

Figure 13.3 shows the configuration screen for phantom assembly for PLANT 3000 and special procurement key (SP.PR. TYPE) 50. The PHANTOM ITEM checkbox is also selected here.

With the special procurement type key 50 assigned to the material master, Figure 13.4 shows the COMPONENT OVERVIEW screen of the production order for MATERIAL 1300-120, which contains the phantom assembly 1300-100. Phantom assembly 1300-100 has a grayed out line item, and PHANTOM ITEM is checked. The phantom assembly explodes, and individual components are listed directly below it. Any changes made to the quantity of phantom assembly 1300-100 automatically enable the system to calculate the components' quantities accordingly (as defined in the BOM of the phantom assembly).

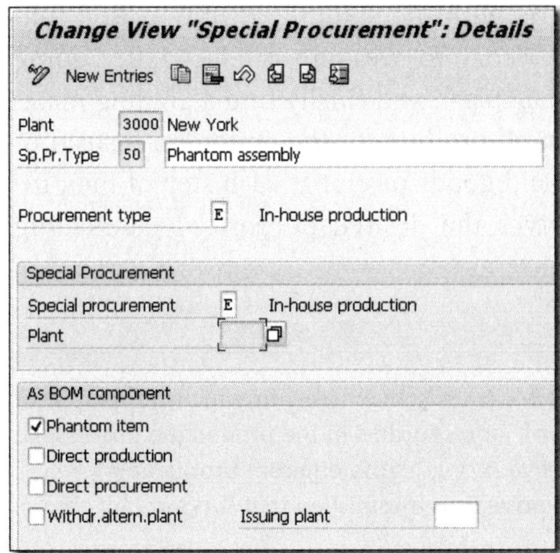

Figure 13.3 Special Procurement Type Key 50 for Phantom Assembly

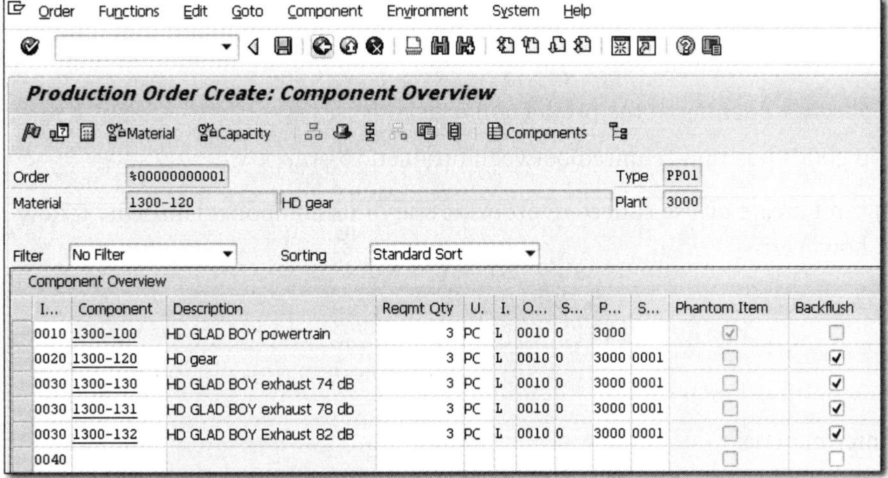

Figure 13.4 Phantom Assembly of Component 1300-100

13.3 Direct Production

The special procurement type key for direct production is 52. *Direct production* means that there is no stock posting among the various stages of the production processes. An example of direct production is that during the textile make-up of a garment, the production process starts with the spinning of raw cotton into weft material, which in turn is converted into weaving product (after going through several intermediate production steps), and finally into a grading material for onward production into a garment product for the customer. Instead of repeatedly performing goods issuance and goods receipt at each step of the production process, direct production serves the desired purpose of eliminating these steps.

[»] **Note**

Direct production is alternatively referred to as *collective order*, in which the parent–child or superior–subordinate relationship of various orders in the production processes exists. The network of orders in the collective order, across different production levels, such as finished product, assembly, or component, is established that supports synchronized actions in the network of orders.

Some of the other functions available in direct production are listed here:

- Quantity changes in the leading order applied to the entire collective order
- Collective scheduling (optional)
- Collective opening of the production orders
- No goods postings required between production orders

You can't create or use collective orders if one of its components has the following, however:

- Co-product
- By-product
- Discontinued material
- Inter-material

[»] **Note**

See Chapter 16 on handling co-products and by-products for more information.

The highest material of the direct production doesn't contain the special procurement type key 52, whereas all of the subordinate materials do (the components defined in the material BOM of the finished/highest material). All subordinate materials of the collective order have their independent BOMs and routings. With a collective order, you get to see an integrated view of the entire production process. Each order within the collective order offers its own comprehensive visibility, including the assignment of a separate order number. Further, it saves time and effort because you don't have to remove and place produced components during various production processes. The confirmation process at each individual order level is enough to move the produced component to the next (higher) order level. Finally, if you make changes to the collective order, for example, in quantity, the system automatically makes the necessary quantity adjustments in all of the subordinate orders. In a collective order, you just have to perform goods receipt of the topmost order and not for all of the subordinate orders.

Tips & Tricks [+]

To view a collective order, use Transaction CO02, and choose COLLECTIVE ORDER.

Figure 13.5 shows the configuration screen for direct production/collective order for PLANT 3000 with a SP.PR. TYPE of 52. The DIRECT PRODUCTION checkbox has also been selected here.

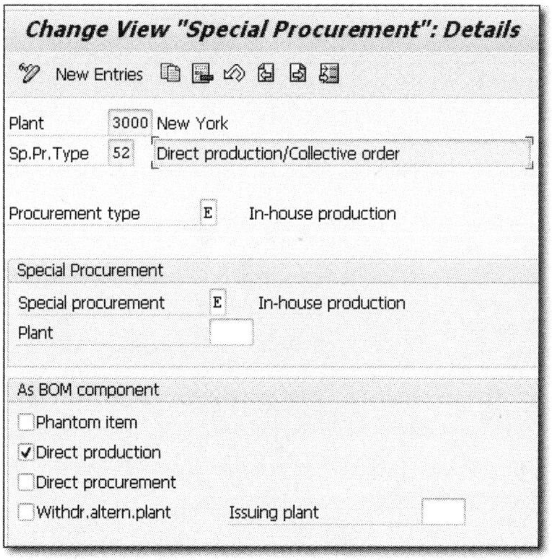

Change View "Special Procurement": Details

New Entries

Plant 3000 New York
Sp.Pr.Type 52 Direct production/Collective order

Procurement type E In-house production

Special Procurement
Special procurement E In-house production
Plant

As BOM component
☐ Phantom item
☑ Direct production
☐ Direct procurement
☐ Withdr.altern.plant   Issuing plant

Figure 13.5 Special Procurement Type Key 52 for Direct Production

[»]

Note

You also need to ensure that in Transaction OPJH, the COLL. (COLLECTIVE) ORDER WITH GOODS MOVEMENT checkbox is checked for the relevant order type.

With special procurement type key 52 assigned to the material masters undergoing direct production, Figure 13.6 shows the header screen of the production order for MATERIAL 400-100. Notice the DATES IN COLLECTIVE ORDER area in the GENERAL tab to denote that it's a collective order (direct production). Choosing the COMPONENT OVERVIEW icon opens the PRODUCTION ORDER CREATE: COMPONENT OVERVIEW screen shown in Figure 13.7.

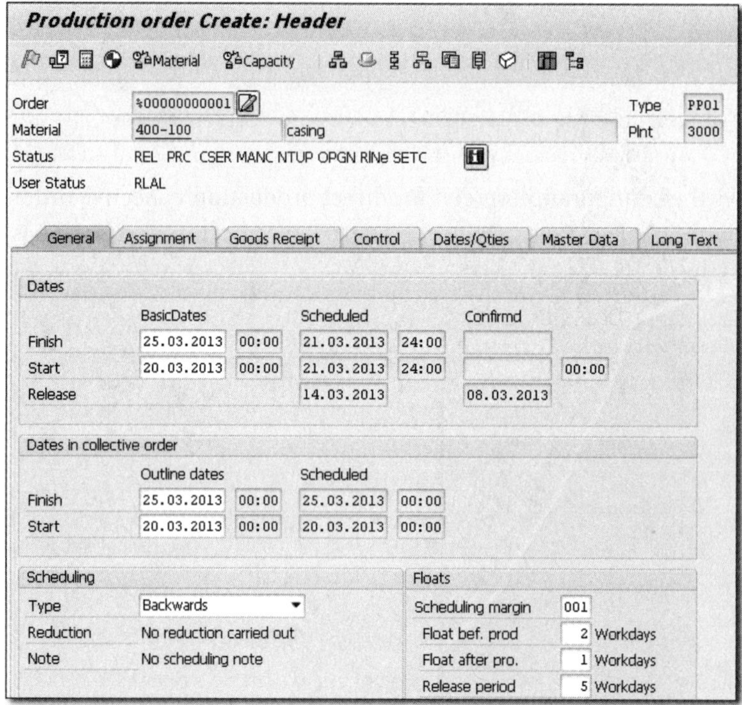

Figure 13.6 Collective Order Dates (Scheduling) in a Production Order

The last two components shown in Figure 13.7, 400-140 and 400-150, are grayed out, and the DIR. PROCUREMENT column reflects 2, denoting direct production.

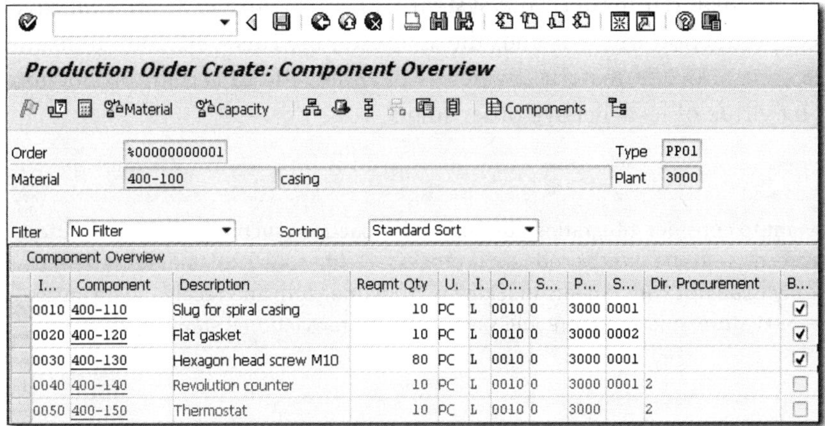

Figure 13.7 Components of Direct Production

Save the production order, and it will generate a production order number. In the change mode of the production order (Transaction CO02) shown in Figure 13.8, the system shows the collective ORDER for the main MATERIAL 400-100 as 60003529, whereas individual production orders were created for each of the direct production materials: production ORDER 60003527 for MATERIAL 400-140, and production ORDER 60003528 for MATERIAL 400-150, respectively.

Production Order Change: Collective Order Overview

Material	Order	Sys.Status	Target qty	SchedStart	Sched.Fin.
400-100	60003529	REL PRC MANC OPGN RlNe SETC	10	21.03.2013	21.03.2013
400-140	60003527	REL PRC MANC OPGN SETC	11	21.03.2013	21.03.2013
400-150	60003528	REL PRC CSER MANC OPGN SETC	12	21.03.2013	21.03.2013

Figure 13.8 Direct Production of Components 400-140 and 400-150

Alternatively, if a component is generally not a part of a collective order, you have the option to assign the special procurement key for direct production directly in the BOM item and not in the material master. For example, in one production process, the component is part of the collective order, whereas in another production process, it's not. If you assign the special procurement type key 52 in the material master of the component, the system will make it a part of all of the collective orders in which this component is used. However, if you assign the special procurement type key to the component's detailed view of the material BOM (and not in the material master), the system will only consider it for collective order/direct production where it finds the assigned key. This way, you can maintain better control of the material, which is only specific to certain production processes by virtue of its collective order status.

[+] **Tips & Tricks**

You may want to consider integrating automated transactions such as backflush or auto-GR (goods receipt) in the direct production process. Doing so will enable the system to automatically issue out components (backflush) from one store and receive them (auto-GR) in the next store, and therefore will save you two transactional steps.

13.4 Direct Procurement

The special procurement type key for direct procurement is 51. In *direct procurement*, the system automatically creates a purchase requisition or initiates the procurement process the moment you create a manual production order. If you use the MRP run's results, the system correctly reflects the material for direct procurement, as long as it has the relevant special procurement type key 51 assigned to the component.

Figure 13.9 shows the configuration screen for direct procurement for PLANT 3000, along with the selected DIRECT PROCUREMENT checkbox.

Figure 13.9 Special Procurement Type Key 51 for Direct Procurement

[«]

Note

You can use the default values for direct procurement to control whether requirement coverage elements are generated through dependent requirements of the planned orders (in the corresponding MRP run) or by the requirements that result from production orders. Figure 13.10 shows such default values; you can maintain them via the Customizing path, PRODUCTION • MATERIAL REQUIREMENTS PLANNING • PLANNING • DIRECT PROCUREMENT • SETTINGS FOR DIRECT PROCUREMENT, or via Transaction OPPB.

On the initial screen to maintain settings for direct procurement (see Figure 13.10), choose the PLANT icon to open the screen showing the three options ❶ available for plant 3000. Selecting option 3 enables the system to automatically trigger direct procurement planning during the MRP run. Save the settings, and click on the BACK icon at the top of the screen, which again brings you back to the initial screen. This time, click on the MRP GROUP icon to see the group information ❷. In this screen, you can make the relevant settings specific to the MRP group. This option helps when different MRP controllers want the system to generate direct procurement proposals per their business needs. Save the settings.

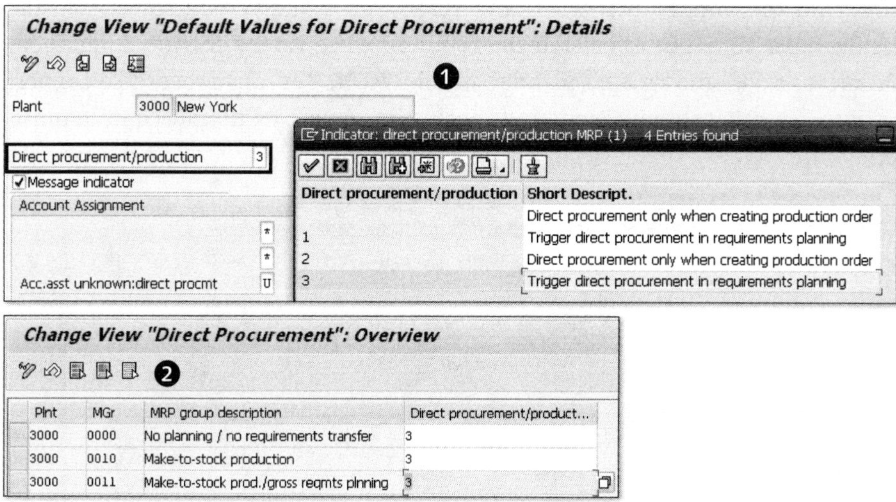

Figure 13.10 Configuration Settings for Direct Procurement

With the special procurement type key 51 assigned to the material masters undergoing direct procurement, Figure 13.11 shows the COMPONENT OVERVIEW screen of the production order for MATERIAL P-100. Here you can see that COMPONENT

100-300 is the direct procurement material. This is denoted by the assignment of 1 in the DIR. PROCUREMENT column, which corresponds to EXTERNAL PROCUREMENT, in the dropdown list shown in the pop-up. Save the production order.

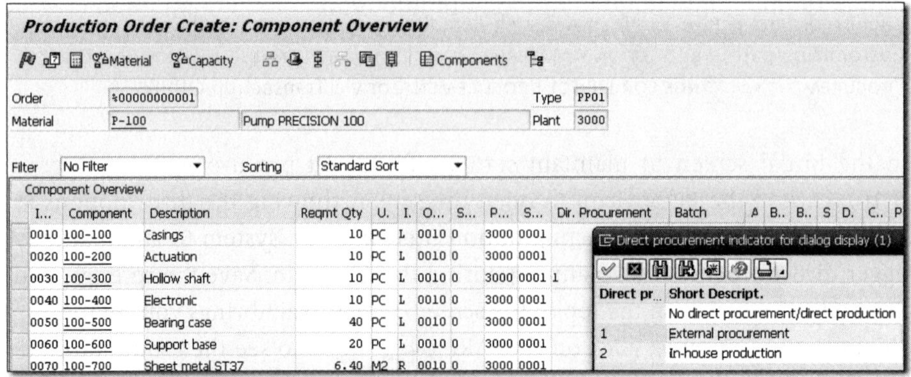

Figure 13.11 Component 100-300 with Direct Procurement (External)

Figure 13.12 shows the planning results when the user runs MRP (Transaction MD02) for the MATERIAL 100-300 and PLANT 3000. The system considered the existing stock of 5 PC, calculated the requirement quantity for the component 100-300 as 5 PC, and created planned order 37074. You can proceed to convert this into a purchase requisition, and the system generates the purchase requisition number.

Planning Result: Individual Lines

Materials 1st material Procurement proposal

Material 100-300 Hollow shaft
MRP Area 3000 New York
Plant 3000 MRP Type PD Material Type HALB Base Unit PC

A	Date	MRP ele...	MRP element data	Rescheduli...	E...	Rec./reqd qty	Avail. quantity	S...
	09.03.2013	Stock					5	
	13.03.2013	DepReq	P-100			10-	5-	0001
	22.03.2013	PlOrd.	0000037074/ExtP	13.03.2013	30	5	0	

Figure 13.12 MRP Results of Material 100-300 for Direct Procurement

Back to the production order containing the direct procurement's component 100-300, Figure 13.13 shows the detailed component view of the MATERIAL 100-300 and the corresponding purchase requisition number 10013837 ❶.

Choose the PURCHASE REQUISITION icon ❶ to open the DISPLAY PURCHASE REQ. 10013837 screen ❷. Here, all procurement details are automatically copied into the purchase requisition. Also note that the ACCOUNT ASSIGNMENT CATEGORY F (column A) denotes that it's an order (production order).

The remaining procurement process belongs to the MM component and is beyond the scope of this chapter.

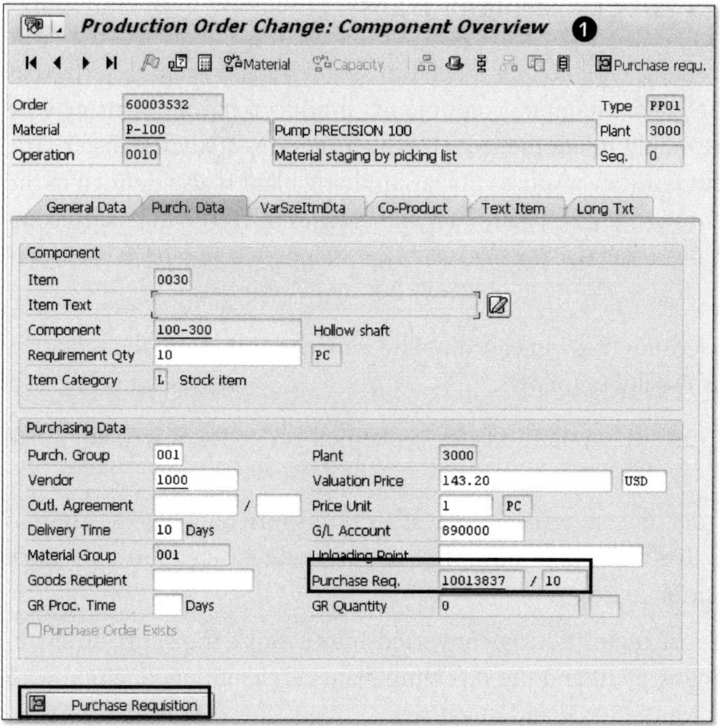

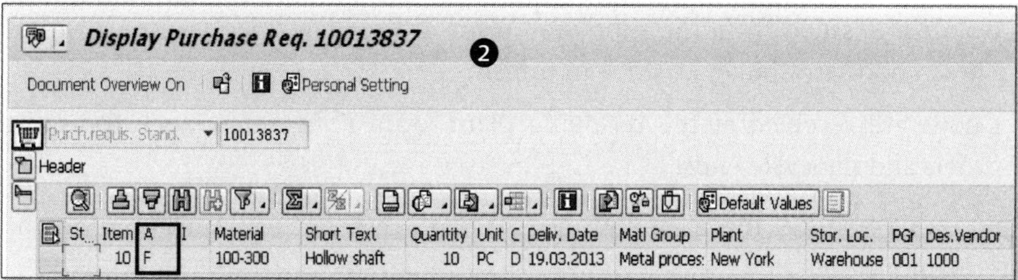

Figure 13.13 Purchase Requisition with Direct Procurement

If a component is generally not a part of direct procurement, you have the option to assign the special procurement key for direct procurement directly in the BOM item and not in the MRP 2 view of the material master. In this way, you can maintain better control on the material, which is only specific to certain production processes by virtue of its direct procurement status.

13.5 Stock Transfer (Interplant Transfer)

The special procurement type key for stock transfer is 40 and is characterized as an *interplant transfer*. For example, components produced in one plant may be required for the assembly in another plant, which requires a stock transfer. The special procurement type key requires that a supplying plant is also defined in the special procurement type key and plant (which becomes the receiving plant) combination. Additionally, SPECIAL PROCUREMENT U is assigned to the special procurement type key.

Depending on the settings made in coordination with the MM team, the sequence of steps involved is usually as follows:

1. After the MRP run, the system creates a planned order for interplant stock transfer.
2. The planned order is converted into a stock transport requisition. Alternatively, you can also make the relevant settings for direct creation of a stock transport requisition.
3. The stock transport requisition is converted into a stock transport order, in which the supplying plant and the receiving plants are mentioned, along with other details such as material and quantities.
4. Goods are issued from the supplying plant against the stock transport order. These goods are shown as stock in transit.
5. Goods are received at the receiving plant against the same stock transport order, and the cycle ends.

Figure 13.14 shows the minimum business process steps in stock transfer or interplant transfer.

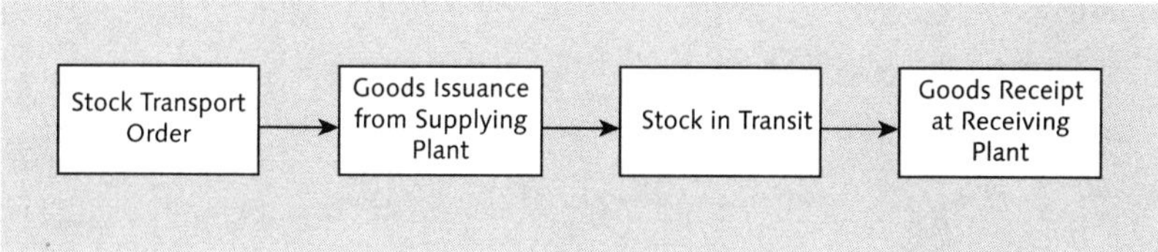

Figure 13.14 Business Processes in the Stock Transfer or Stock Transport Order

Figure 13.15 displays the configuration screen for stock transfer (procurement from an alternate plant) for the receiving PLANT 3000 and SP.PR. TYPE key 40. U is assigned in the SPECIAL PROCUREMENT field, and the supplying PLANT is 3100.

Figure 13.15 Special Procurement Type Key 40 for Stock Transfer (Interplant Transfer)

After this screen is filled in, access Transaction ME21N (see Figure 13.16) for the initial screen for the stock transport order (for interplant transfer). Select STOCK TRANSP. ORDER from the dropdown list on the top-left side of the screen. Assign the SUPPLYING PLANT as "1100", and then enter the Item category as "U" (denoted in the column I), the MATERIAL as "400-300", the PO QUANTITY as "2", and the receiving plant (PLNT) as "1400" (STUTTGART). Save the stock transport order. The next step is to issue the material from the supplying plant against the same stock transport order number.

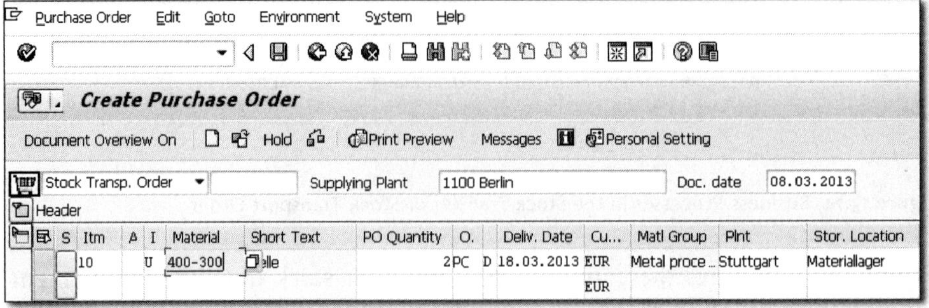

Figure 13.16 Stock Transport Order

Access the screen shown in Figure 13.17 via Transaction MIGO, and select GOODS ISSUE and then PURCHASE ORDER from the dropdown lists at the top. Then enter the purchase order (stock transport order) number. The movement type for issuance against a stock transport order is 351 and is shown on the top-right corner of the screen. Save the entries, and the stock are shown as in transit.

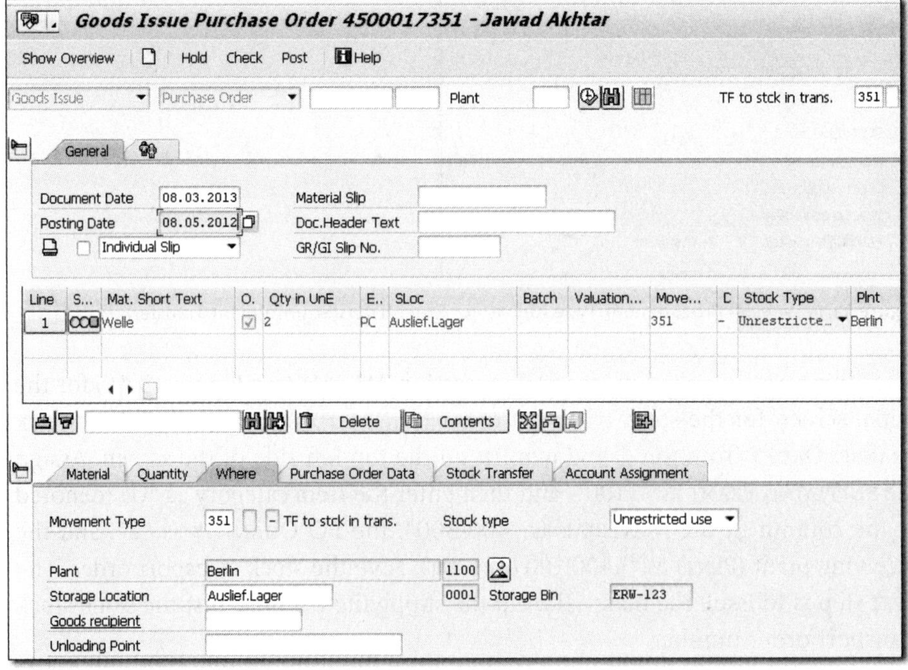

Figure 13.17 Goods Issue against Stock Transport Order

Transaction MB5T takes you to the screen shown in Figure 13.18 after you've entered the initial parameters such as SUPPLYING PLANT, ISSUING PLANT, and so on.

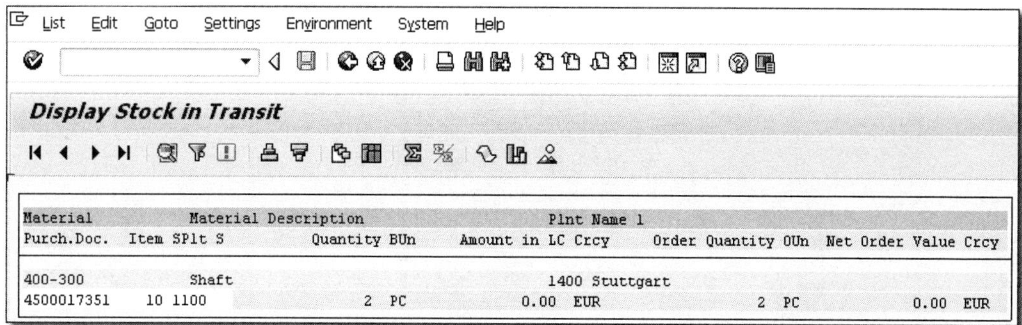

Figure 13.18 Stock in Transit

The stock remains in transit until the receiving plants receive goods against the same stock transport order. The relevant movement type for this is 101.

13.6 Withdrawal from Alternate Plant

The special procurement type key for withdrawal from an alternative plant is 70. This special procurement type key is assigned to the specific component of the BOM, which is withdrawn from the alternative plant. This special procurement type key works well when some components of the BOM of assembly are procured or withdrawn from another plant, which is in close physical and geographical proximity. While with stock transfer (interplant transfer), you have the option to enter additional transportation and other costs, this isn't possible with special procurement type 70. You also use this special procurement type when there is significantly less time involved in the transportation of goods. When configuring the special procurement type key for a business process, you have to select the WITHDR.ALTERN.PLANT checkbox and also assign the issuing plant.

During production order creation of an assembly for one plant, the system automatically suggests alternative plants for issuance of the component.

Figure 13.19 shows the configuration screen for WITHDRAWAL FROM ALTERNATE PLANT for PLANT 3000 with the SP.PR. TYPE of 70. The WITHDR.ALTERN.PLANT checkbox is selected, and the ISSUING PLANT is set as 3100.

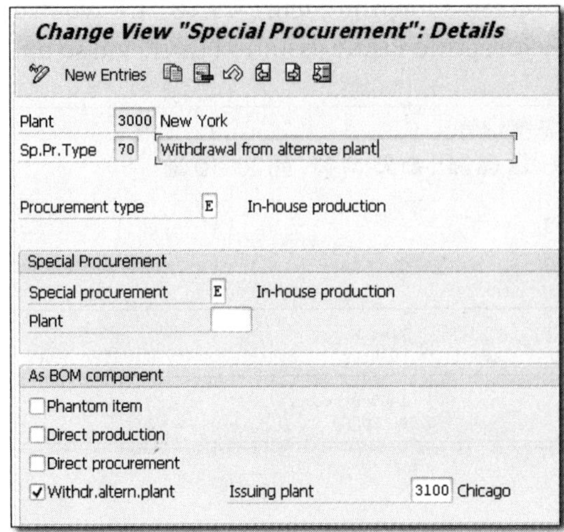

Figure 13.19 Special Procurement Type Key 70 for Withdrawal from Plant 3100 for Plant 3000

With special procurement type key 70 assigned to the component, Figure 13.20 shows the COMPONENT OVERVIEW screen of the production order, in which COMPONENT 400-500 shows the issuance PLANT (withdrawal plant) as 3100, compared to PLANT 3000 for all of the remaining components.

Order Functions Edit Goto Component Environment System Help

Production Order Create: Component Overview

Material Capacity Components

Order %00000000001 Type PP01

Material P-400 Pump (Collective Order) Plant 3000

Filter No Filter Sorting Standard Sort

Component Overview

I...	Component	Description	Reqmt Qty	U.	I.	O...	S...	Plant	Stor. Loc.
0020	400-200	Fly wheel CI	10	PC	L	0010	0	3000	0001
0030	400-300	Shaft	10	PC	L	0010	0	3000	0001
0050	400-500	Bearing case	10	PC	L	0010	0	3100	0001
0060	400-600	Support base	10	PC	L	0010	0	3000	0001
0070	400-700	Sheet metal ST37	6.40	M2	R	0010	0	3000	0001
0080	400-800	Color blue	24	KG	L	0010	0	3000	0001
0010	400-100	casing	10	PC	L	0010	0	3000	
0040	400-400	Electronic	10	PC	L	0010	0	3000	

Figure 13.20 Component 400-500 for Withdrawal from Plant 3100

13.7 Production in Alternate Plant

The special procurement type key for production in an alternate plant is 80. In this special procurement type, the entire planning of producing a material is done in one plant, whereas the actual and the physical production take place at another plant. The two plants are differentiated as "planning plant" and "production plant" in all of the MRP-relevant elements, whether they are planned orders, production orders, or MRP lists.

In the special procurement type key for production in an alternate plant, you specify the plant in which the production will take place, along with the SPECIAL PROCUREMENT field as P.

Figure 13.21 shows the configuration screen for PRODUCTION IN ALTERNATE PLANT for PLANT 3000 and SP.PR. TYPE 80. The SPECIAL PROCUREMENT key is set as P for PROD. OTHER PLANT, and the production PLANT is 3100.

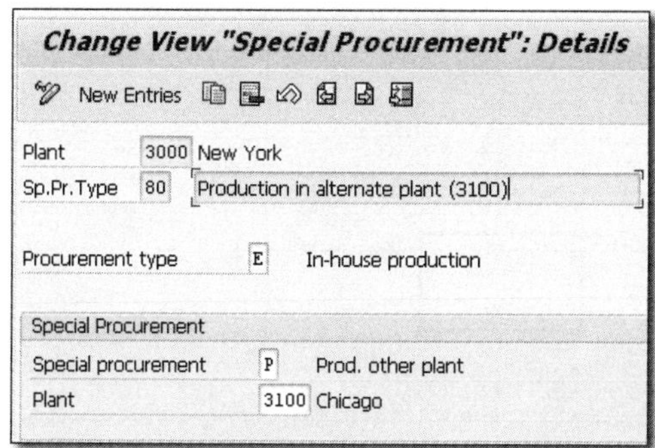

Figure 13.21 Special Procurement Type Key 80 for Production in Alternate Plant 3100 for Planning Plant 3000

With the special procurement type key 80 assigned to the relevant component, Figure 13.22 shows the COMPONENT OVERVIEW of the production order for MATERIAL T-20000 and COMPONENT P-100.

Nowhere does it reflects that this material is for production at an alternate plant, but when you run the MRP on the material P-100 and plant 3300 (Transaction MD02), the results of the MRP are shown in Figure 13.23. The results of MRP on MATERIAL P-100 for PLANT 3300 are given ❶. Notice that after the MRP run, the planned order 37057 shows the supplying PLANT as 3000.

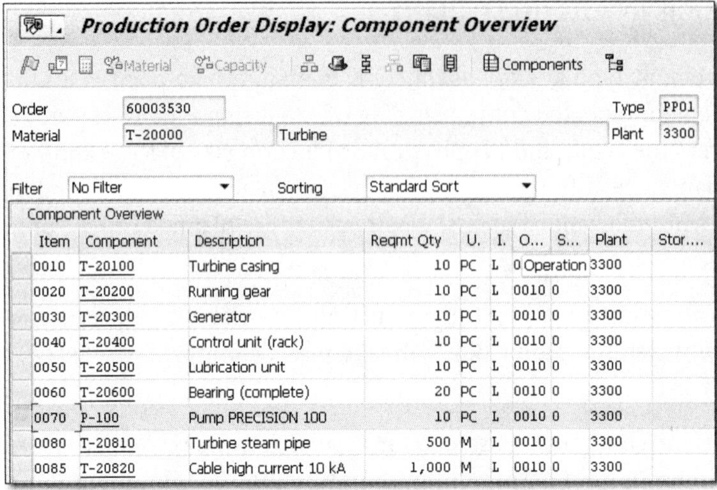

Figure 13.22 Components of the Production Order

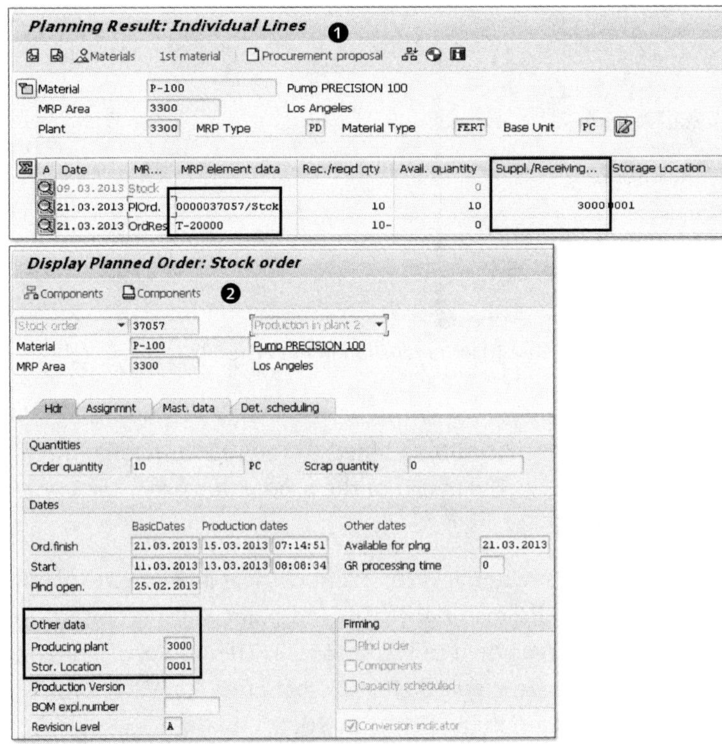

Figure 13.23 MRP Results for Component P-100 for Plant 3300 with Production Plant as 3000

The detailed view of planned order 37057 (Transaction MD13) ❷ for MATERIAL P-100 and PLANT 3300 shows the PRODUCING PLANT as 3000.

13.8 Subcontracting

The special procurement type key for subcontracting is 30. The *subcontracting* process begins when the company hands over (issues) the components to a subcontractor, who is usually a vendor, against the subcontracting purchase order, either to assemble or to add value to the product. The company takes the components' inventory out of its books, and the same is reflected in the vendor's inventory. When the vendor has done what it's contracted to do, the company receives the goods against the same subcontracting purchase order from the vendor. The vendor invoices the company for the services rendered and receives payment.

For the subcontracting process to work effectively, the following prerequisites in the system must be met:

1. The material master of the assembly must have the special procurement type key assigned as 30.
2. A subcontracting BOM must exist in the system. You don't need a routing.
3. The purchasing view of the material master must be maintained, including choosing the SOURCE LIST checkbox. You'll also have to maintain a source list for this material.
4. A purchasing information record (i.e., info record) is needed between the material and vendor to reflect the source of supply. A *purchasing information record* is the purchasing relationship between a material and a vendor. When creating the info record, be sure to choose the SUBCONTRACTING option on the initial screen.

Figure 13.24 is the end-to-end business processes of subcontracting.

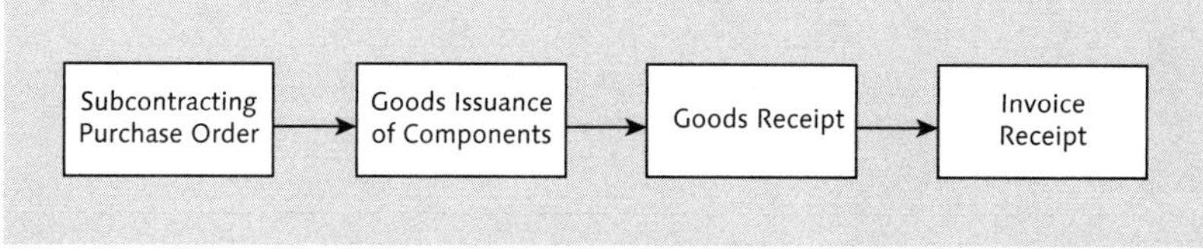

Figure 13.24 Business Processes in Subcontracting

Figure 13.25 shows the configuration screen for subcontracting for PLANT 3000 and SP.PR. TYPE 30. The SPECIAL PROCUREMENT field is assigned as L SUBCONTRACTING.

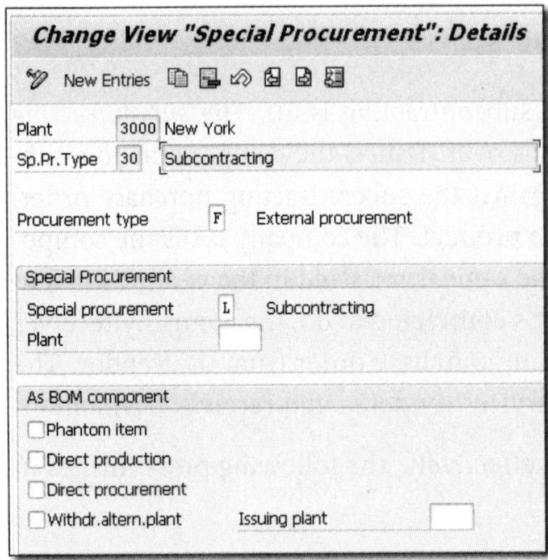

Figure 13.25 Special Procurement Type Key 30 for Subcontracting

Figure 13.26 shows the CREATE PURCHASE ORDER screen (Transaction ME21N), where you assign the VENDOR as "1000", and assign the item category (denoted by column I, in the center of the figure) as L to denote that it's a subcontracting order. Also, enter the assembly MATERIAL code as "103-100", the PO QUANTITY as 1 PC, the NET PRICE as "100.00" USD, and the PLNT as "3000" (for NEW YORK).

At the bottom of the screen, choose the COMPONENTS icon to open the COMPONENT OVERVIEW screen.

In the COMPONENT OVERVIEW of the subcontracting purchase order shown in Figure 13.27, issue the component MATERIAL 100-120 in quantity of 1 PC to the vendor, so that the vendor brings back the assembled item 103-100. Save the subcontracting purchase order. The next step is to issue the component 100-120.

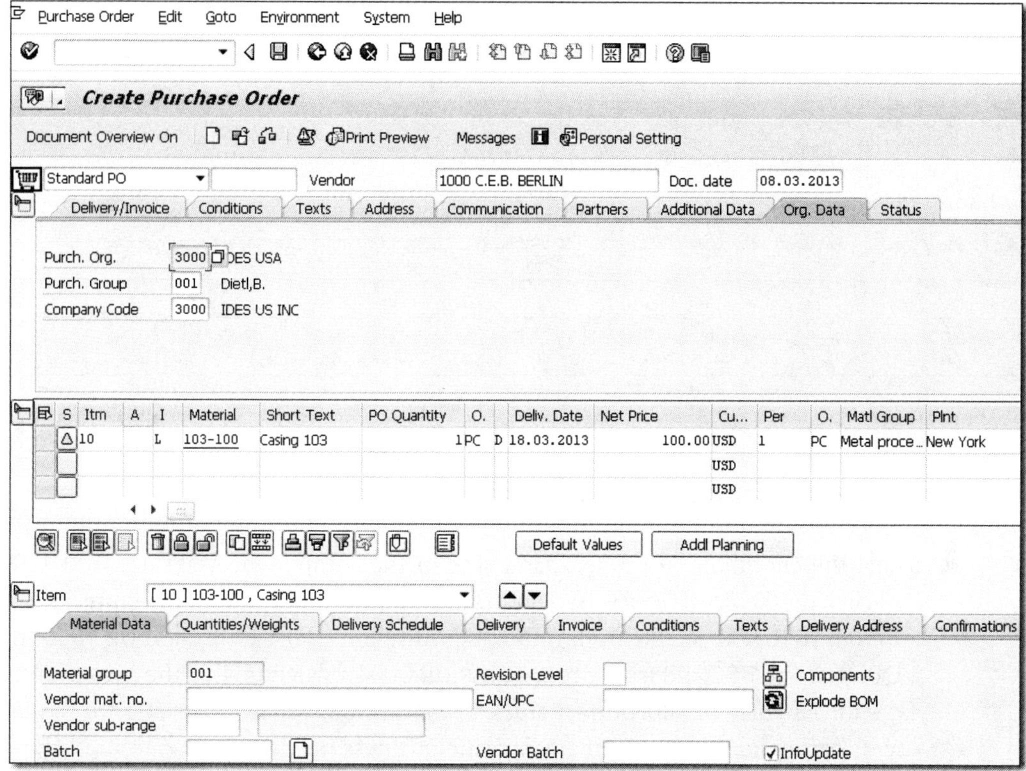

Figure 13.26 Subcontracting Purchase Order

Processing Components: Component Overview

Collective entry Detailed entry

Material	103-100	Casing 103	
Plant	3000	Release Date	07.03.2013
Quantity	1.000	Delivery date	18.03.2013

Component Overview

Material	Description	Requirement ...	U.	C	Plant	Prod.stor.loc.	Supply A...	Reqmt D...	I...
100-120	Flat gasket	1	PC	☐	3000	0001		07.03.2013	0020 L

Figure 13.27 Component Overview of the Subcontracting Purchase Order

After you enter the VENDOR as "1000" on the initial parameters selection screen for the subcontracting order for vendor (Transaction ME2O), the screen shown in Figure 13.28 appears. It shows the component you need to issue to the vendor against the subcontracting order, which in this example is MATERIAL 100-120 with a quantity of 1 PC.

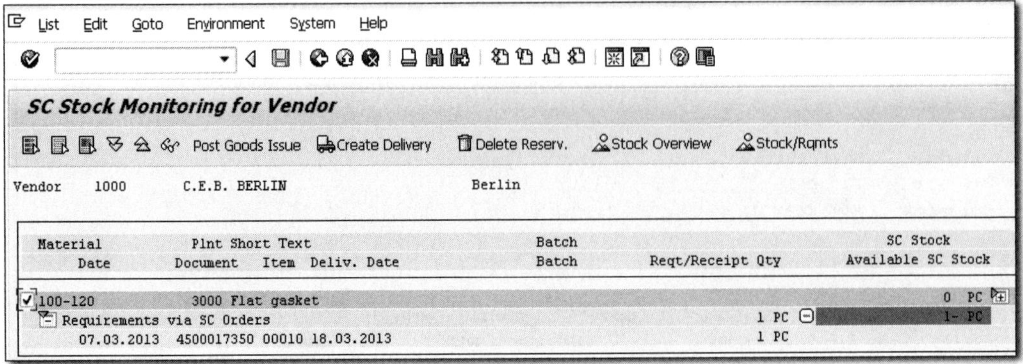

Figure 13.28 Stock Monitoring for the Subcontracting Purchase Order

Select the checkbox on the left-hand side of the component MATERIAL 100-120 and click on the POST GOODS ISSUE button, and the pop-up shown in Figure 13.29 appears. Here you see the option to issue MATERIAL 100-120 to VENDOR 1000 in a QUANTITY of 1 PC and from STOR. LOCATION 0001. Also note that the MOVEMENT TYPE for issuance to subcontract stock is 541. You also have the option to issue excess components, in case of any specific business needs.

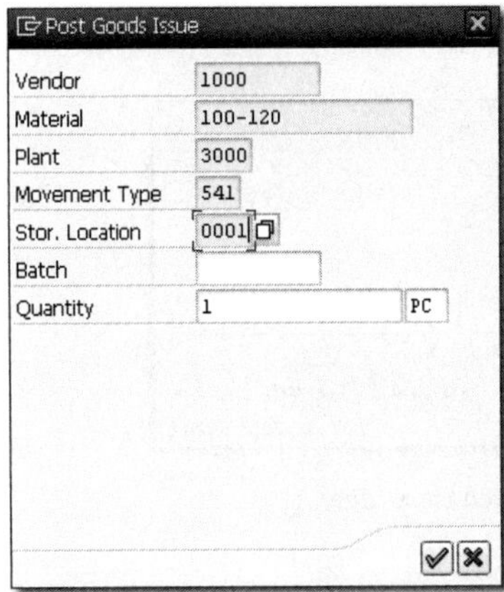

Figure 13.29 Stock Transfer for a Subcontracting Purchase Order

Tips & Tricks [+]

You can also use Transaction MB1B for transfer posting the in-house stock to the vendor's subcontracting stock with reference to the purchase order and using the same movement type 541.

Figure 13.30 shows the stock overview of MATERIAL 100-120 (Transaction MMBE), in which the system shows the STOCK PROVIDED TO VENDOR as a separate line item. In this example, this is 923 PCS.

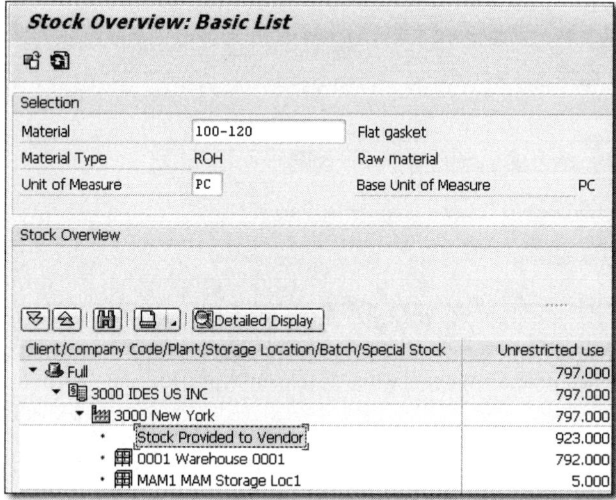

Figure 13.30 Stock Overview with Details of Stock Provided to (Subcontractor)

The last step is to receive the goods against the same subcontracting order. Figure 13.31 shows the initial screen to receive goods against the subcontracting purchase order (Transaction MIGO). Notice that you not only can receive the assembly material with movement type 101, but you can also record the actual consumption of components by vendor with movement type 543 O for goods issuance against the subcontracted stock.

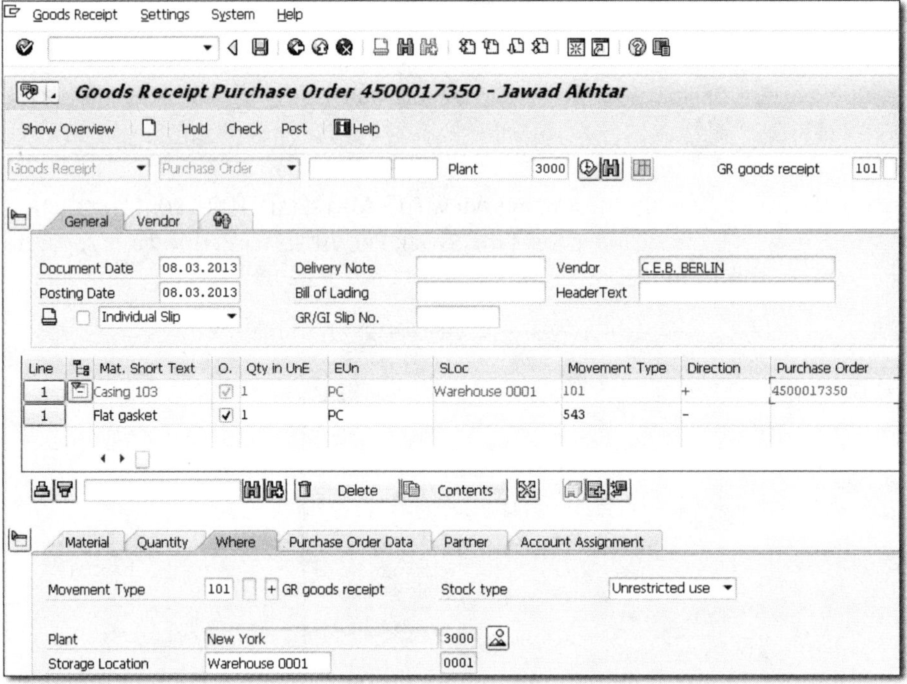

Figure 13.31 Goods Receipt for Subcontracted Material with Movement Types 101 and 543 O

13.9 Consignment

The special procurement type key for consignment is 10. In *consignment*, the vendor provides the customer with material that isn't billed until the customer removes it from the consignment store. For example, a vendor places its goods on your premises for you to consume. When you record consumption of this consignment material against a production order or even a cost center, you become liable to pay the vendor. For the consignment process to work effectively, the following prerequisites in the system must be met:

1. The material master must have the special procurement type key assigned as 10.
2. The purchasing view of the material master must be maintained, including choosing the SOURCE LIST checkbox. You also need to maintain the source list for the material.

3. The vendor must already exist in the system.
4. A purchasing information record is needed between the material and vendor to reflect the source of supply. When creating the info record, be sure to choose the CONSIGNMENT option on the initial screen.
5. The source list of the material and plant combination must be maintained.

Figure 13.32 shows the configuration screen for consignment for PLANT 3000 and SP.PR. TYPE 10, along with the SPECIAL PROCUREMENT assigned as K.

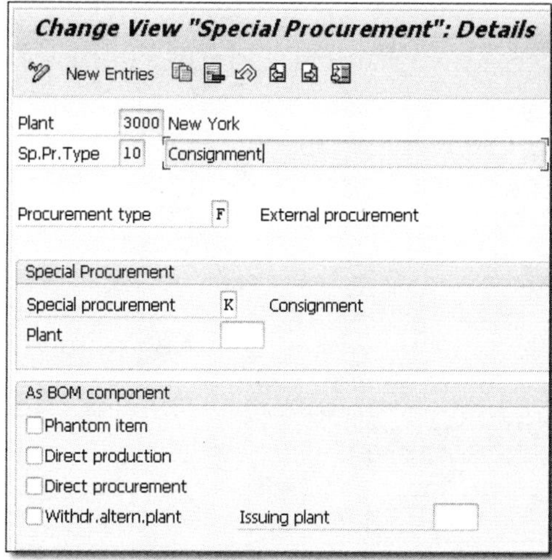

Figure 13.32 Special Procurement Type Key 10 for Consignment

After the special procurement key 10 is assigned to the material master, Figure 13.33 shows the COMPONENT OVERVIEW screen of the production order for MATERIAL AS-100 ❶. The component AS-400 is a consignment material, which is denoted by the value K (MATERIAL PROVIDED BY CUSTOMER) in the MAT. PROV. IND. field shown further down on the screen ❷.

Save the production order, and run MRP on component AS-400 and plant 3000 with Transaction MD02.

After the MRP is run and results are saved, you can view the stock/requirements list (Transaction MD04) for material AS-400 and plant 3000. Figure 13.34 shows the purchase requisition for the same MATERIAL AS-400 and PLANT 3000, with the relevant item category K (denoted by the column I).

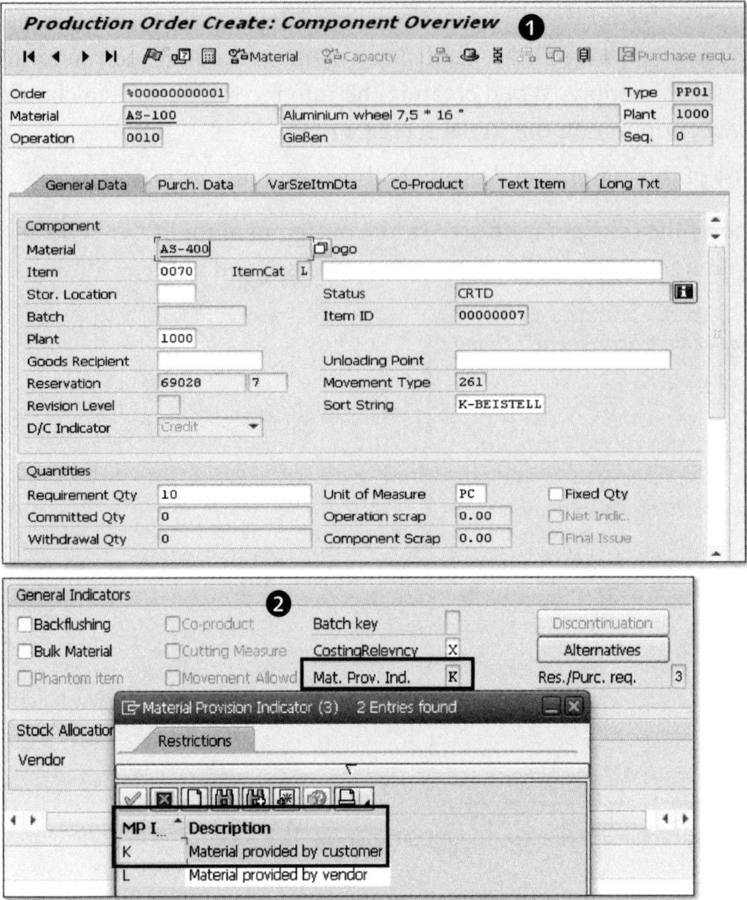

Figure 13.33 Consignment Component AS-400 Provided by the Customer

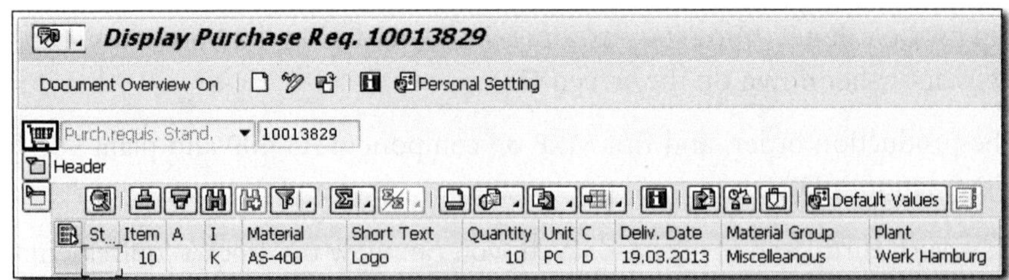

St...	Item	A	I	Material	Short Text	Quantity	Unit	C	Deliv. Date	Material Group	Plant
	10		K	AS-400	Logo	10	PC	D	19.03.2013	Miscelleanous	Werk Hamburg

Figure 13.34 Purchase Requisition for Consignment

[«]

Note

The remaining steps of the procurement process are beyond the scope of this chapter.

13.10 Pipeline Material

Pipeline material doesn't require any special procurement type key. *Pipeline material* denotes that the material is in a pipeline and is always available. Customers only pay for the pipeline material upon consumption. Examples of pipeline material are natural gas or heavy water, supplied by various vendors.

You need to set up the following data in the system for pipeline material:

1. The material type of the material master is pipeline material and has purchasing views activated, in addition to other views such as BASIC DATA and ACCOUNTING/COSTING views.
2. The PURCHASING view of the material master must be maintained, including choosing the SOURCE LIST checkbox.
3. The vendor must already exist in the system.
4. A purchasing information record is needed between the material and vendor to reflect the source of supply. When creating the InfoRecord, choose the PIPELINE option on the initial screen.
5. The source list of the pipeline material and plant combination must be maintained.

13.11 Summary

Apart from the traditional and routine production and procurement processes, the SAP ERP system offers a large number of options to manage complex and diverse businesses processes involved in the entire supply chain and logistics functions. Some of these are subcontracting, stock transfer, phantom assembly, production in alternate plant, procurement from alternate plant, and consignment materials. If correctly configured and implemented in coordination with the MM team, these can bring about great improvements in optimizing the planning, procurement, and production processes.

The next chapter covers capacity requirements planning (CRP).

Apart from materials availability, capacity availability plays a central role in the entire production process. Capacity requirements planning helps companies ensure that the relevant resources become available at the required time for optimum and uninterrupted production.

14 Capacity Requirements Planning

Building on the concepts and understanding that you developed in Chapter 3 and Chapter 6 on capacity requirements planning (CRP), this chapter focuses on the *short-term* capacity planning after the creation or release of production orders. We've covered the medium- to long-term capacity planning, vis-à-vis sales and operations planning (S&OP) (Chapter 9) and Long-Term Planning (LTP) (Chapter 12). The CRP details for discrete manufacturing that we cover in this chapter are equally relevant and applicable to the process and, to a certain extent, in repetitive manufacturing (REM) types. For example, when deciding to assign relevant profiles for a capacity evaluation or capacity leveling, the system offers profiles with "REM" in their naming conventions for repetitive manufacturing and with "PI" in their name for process manufacturing. For discrete manufacturing, the profiles have a naming convention with "SFC" (for shop floor control). Because the system uses the same principles that we cover in this chapter in SAP ERP Plant Maintenance (PM) and SAP ERP Project System (PS), the available profiles have naming conventions as "PM" or "PS", respectively.

We'll start this chapter with an overview of the processes involved in capacity planning and then cover the capacity requirements and capacity evaluations. We cover both the standard as well as variable capacity evaluations. Next, we cover how you can cumulate capacity requirements at different work centers and also perform real-time checks on capacity availability, either during order creation or order release.

The chapter then moves to cover dispatching and the associated profiles, the dispatching sequence, sequence-dependent setup, and finally the midpoint scheduling. We also cover how you can use the mass processing function to expedite the business processes, such as dispatching or deallocation. Finally, we cover some of the more important features available in both the graphical and tabular versions of capacity planning tables.

14.1 Process Overview

The objective of CRP is to schedule operations within the availability capacity and to check their feasibility when creating logical sequences of operations.

CRP can be implemented for production orders and also for planned orders from material requirements planning (MRP) and LTP. You can make the relevant configuration settings to enable the system to also consider planned orders in capacity planning. This option works best if your company's business processes are mature enough, and you have significantly high confidence in the planning results that MRP creates. Because there are often frequent and at times high numbers of last-minute changes between the planning and actual production and execution, it may not be wise to book capacities based only on planning results. In doing so, you're compromising the capacity availability, which in turn may have an impact on the actual production schedule. At the same time, there may be a genuine business need that entails converting planned orders earlier on into production orders to book the capacities based on production orders. Still others prepare a comprehensive and feasible production plan based only on planned orders.

Apart from scheduling operations, CRP involves two important processes:

- **Capacity evaluation**
  The *capacity evaluation* provides a comparison between available capacities with capacity requirements, with reference to the work center or specific production order.
- **Capacity availability check**
  The *capacity availability check* uses the same information to check whether the available capacity at the requisite work center is sufficient to manage the operation of the production order. Depending on the configuration settings, the system can perform the capacity availability check during production order creation or during its release. If there is insufficient capacity, three options are available:

- The system won't create or release the production order.
- The system will create or release the production order.
- The system prompts you to decide if you still want to create or release the production order, despite the capacity shortage.

[«]

Note

Refer to Chapter 6, Section 6.3.8, in which we covered available capacity.

If there is insufficient capacity or to account for capacity restrictions, the system performs *finite scheduling* to look for periods when sufficient capacity is available. Insufficient capacity at the work center leads to rescheduling the operation, but you still have to perform the *dispatch* function to give a formal go-ahead to proceed with production on the given work center. You can perform the dispatch function either from the capacity planning table or from the tabular capacity planning table. If you have an incorrect or inadvertent dispatch, you can interactively *deallocate* the operation.

On dispatch, the system performs the following four functions:

- Determines the dispatch dates
- Creates the dispatching sequence
- Performs finite scheduling of the operation to be dispatched
- Performs midpoint scheduling of nondispatched operations of the production order

On dispatch of an operation, the system assigns it the status DISP (dispatched) and no longer schedules the operation automatically.

Now that you have a basic understanding of how CRP works, let's delve into the details you'll need to know to understand how it works in your system, alongside your business requirements.

[«]

Note

Implementing CRP makes greater sense in companies that have a diverse range of products and also want to ensure that all of its products are produced in sufficient quantities for availability in the market. If a company produces products that use up all of its available capacities or where there are single products such as fertilizer, cement, caustic soda, or iron bars, then implementing CRP doesn't add much value to business processes.

During SAP ERP system implementation, it's best to discuss and evaluate how implementing CRP will improve business processes and then decide on whether to implement CRP.

14.2 Capacity Requirements and Capacity Evaluation

The system reflects the day-to-day business processes in place, including creation and release of production orders, their confirmations, the delays in productions, and the extra time it takes to set up or process the product in capacity requirements. The capacity evaluation enables you to view the capacity load or overload at various work centers in a large number of standard and variable evaluation options. These two business processes, that is, capacity requirements and capacity evaluation, are essential before getting started with the actual CRP process, and we'll discuss them in more detail in the following sections.

In the following sections, we'll cover how to understand the capacity requirements that the system creates, their evaluations, and the steps you can take for capacity leveling.

14.2.1 Capacity Requirements

As already mentioned, the system uses the settings that you've already configured in Chapter 3 for CRP and also the information that you entered in the SAP ERP Production Planning (PP) master data, such as work center and routing from Chapter 6. Here, we briefly cover these details.

The system determines the capacity requirements during lead time scheduling if you've set the Det. Cap. Reqmnts option in the control key. Furthermore, the system takes formulas that you entered in the work center, the default values in the routing, and the order quantity into consideration. It's important to emphasize that the formulas for scheduling aren't used in CRP. For each component of an operation, such as setup, processing, and wait times, there is an earliest and latest date. It's set to one of the operation dates. In the standard settings, this is the latest setup start date or processing start date, if the operation doesn't provide for any setup time. An operation can span several periods, and because the system bases the capacity evaluation on this information, the distribution key determines how the system distributes capacity to different periods.

The *distribution key* itself consists of the distribution strategy and the distribution function. The distribution key defines the operation dates and on what basis it carries out the distribution, for example, on factory calendar, the Gregorian calendar, or on operation time. You can maintain the distribution key, distribution strategy, and distribution function through the configuration (Transaction SPRO) menu path, PRODUCTION • CAPACITY REQUIREMENTS PLANNING • OPERATIONS • CAPACITY REQUIREMENT • DISTRIBUTION.

The system determines the capacity requirement during lead time scheduling and has to go through the capacity availability check and finally the dispatching process. If the system confirms the operation during the capacity availability check, it assigns the capacity requirement with the system status BSTKZ. It does the same when you dispatch the operation. In this way, the system converts the capacity requirement into a component called a *basic load*.

The basic load has an important role to play in the capacity availability check because available capacity represents the difference between the available capacity and the basic load. The system reduces the capacity requirements when you perform any of the following:

- Confirm an operation.
- Set the status of a production order as technically complete.
- Lock the production order.
- Set the deletion flag in the production order.

The system also updates the capacity requirements when you partially confirm an operation, and it correspondingly reduces part of the capacity requirements for the operation.

14.2.2 Standard Evaluation of Capacity Utilization

The SAP ERP system offers several methods of *capacity evaluation*. Each standard evaluation has its own transaction code, whereas for variable evaluation, you select the relevant profile to enable the system to display the requisite information.

In standard evaluations, the system displays the capacity requirements in direct comparison to the available capacities. Following are some of the standard capacity evaluations:

- **Standard overview (Transaction CM01)**
  The system aggregates the load on a weekly basis.
- **Detailed capacity list (Transaction CM02)**
  The system shows the load per order and week.
- **Pool of orders/operations (Transaction CM03)**
  The system presents only the load of released operations.
- **Overload (Transaction CM05)**
  The system only shows work centers with overload.
- **Extended selection, work center view (Transaction CM50)**
  The system displays an aggregated view of the daily load.
- **Extended selection, order view (Transaction CM52)**
  The system displays an aggregated view of the weekly load for all work centers that the selected orders occupy.

To access the screen where you can work with the capacity planning evaluation, follow the SAP menu path, Logistics • Production • Capacity Planning • Evaluation • Work Center View • Load, or use Transaction CM01. Figure 14.1 shows the standard overview (load) of the Work center BAF and Plant 3000 combination. It shows the capacity requirements for each calendar week that result from the creation or release of production orders. The screen also shows the available capacity (AvailCap.) of the work center and the resulting percentage of load (CapLoad), as well as the remaining available capacity (RemAvailCap).

For example, in calendar week 27, a capacity requirement of 69.15 hours (H) leads to a utilization (or an overload) of 197% based on an available capacity of 35.20 H. The system displays the periods with an overload in red. The evaluation doesn't determine whether or not the system checked the capacity requirement or whether it has been dispatched.

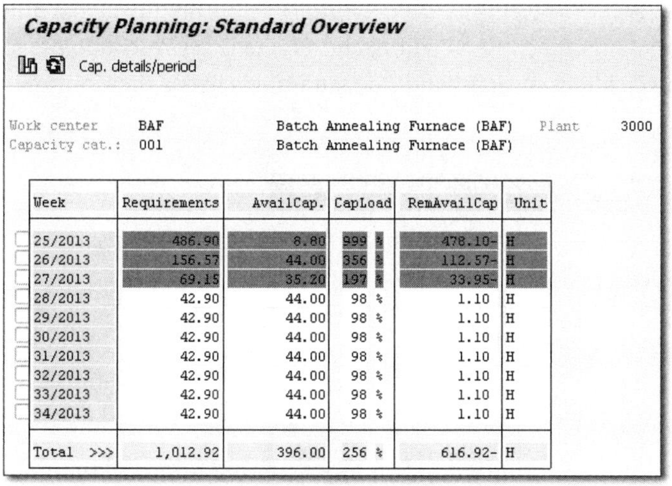

Capacity Planning: Standard Overview

Cap. details/period

Work center BAF Batch Annealing Furnace (BAF) Plant 3000
Capacity cat.: 001 Batch Annealing Furnace (BAF)

Week	Requirements	AvailCap.	CapLoad	RemAvailCap	Unit
25/2013	486.90	8.80	999 %	478.10-	H
26/2013	156.57	44.00	356 %	112.57-	H
27/2013	69.15	35.20	197 %	33.95-	H
28/2013	42.90	44.00	98 %	1.10	H
29/2013	42.90	44.00	98 %	1.10	H
30/2013	42.90	44.00	98 %	1.10	H
31/2013	42.90	44.00	98 %	1.10	H
32/2013	42.90	44.00	98 %	1.10	H
33/2013	42.90	44.00	98 %	1.10	H
34/2013	42.90	44.00	98 %	1.10	H
Total >>>	1,012.92	396.00	256 %	616.92-	H

Figure 14.1 Capacity Evaluation of the Work Center

Select the checkbox for 26/2013 (week 26 of the year 2013), and choose the CAP. DETAILS/PERIOD button (or use Transaction CM02). The screen shown in Figure 14.2 appears, showing the evaluation list of the capacity requirement of each operation, including the production order number, the header material, and the order quantity. The CHOOSE FIELDS button enables you to display additional fields of the operation, such as its start time. You can also directly jump into the production order from here.

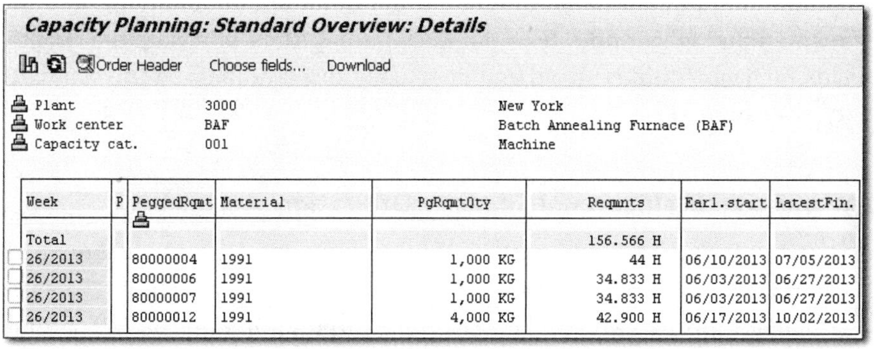

Capacity Planning: Standard Overview: Details

Order Header Choose fields... Download

Plant 3000 New York
Work center BAF Batch Annealing Furnace (BAF)
Capacity cat. 001 Machine

Week	P	PeggedRqmt	Material	PgRqmtQty	Reqmnts	Earl.start	LatestFin.
Total					156.566 H		
26/2013		80000004	1991	1,000 KG	44 H	06/10/2013	07/05/2013
26/2013		80000006	1991	1,000 KG	34.833 H	06/03/2013	06/27/2013
26/2013		80000007	1991	1,000 KG	34.833 H	06/03/2013	06/27/2013
26/2013		80000012	1991	4,000 KG	42.900 H	06/17/2013	10/02/2013

Figure 14.2 Detailed Capacity Evaluation for a Week

In the SELECT FIELDS screen shown in Figure 14.3, the system lists already-displayed fields on the left-hand hand side of the screen, whereas you can choose from a large number of fields available on the right-hand side. Simply select the fields that you want to show in the standard capacity evaluation, choose the DISPLAY icon to move them from the right-hand side to the left-hand side, and then choose the CONTINUE icon.

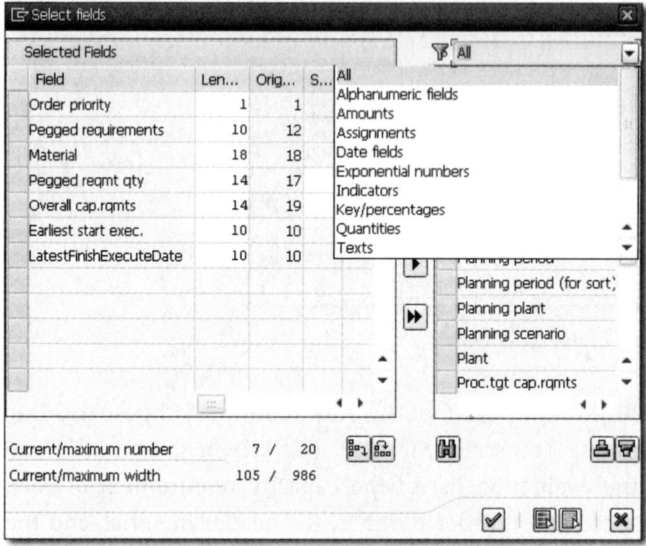

Figure 14.3 Fields for Selection in Capacity Evaluation

Because you can select many fields, you can use the filter option to narrow down the available choices. For example, if you use the QUANTITIES filter available in the dropdown on the right-hand side of the screen, the system brings up the quantities fields for you to choose from and eventually displays them in the standard capacity evaluation.

14.2.3 Variable Evaluation of Capacity Utilization

In *variable evaluations*, you can base the capacity evaluation on individual requirements. In this process, the system requires an overall profile that defines the evaluation. Standard profiles are available for this purpose, or you can define your own profiles.

For a variable evaluation of capacity utilization, follow the SAP menu path, LOGISTICS • PRODUCTION • CAPACITY PLANNING • EVALUATION • VARIABLE, or use

Transaction CM07. The PLANNING screen in Figure 14.4 shows the variable evaluation using overall profile SAPSFC010. In contrast to the standard overview (Transaction CM01 or overall profile SAPX912), this profile displays only the requirements of the basic load.

Here, select OVERALL PROF SAPSFC010, and in the ensuing screen, enter the WORK CENTER "BAF" and PLANT "3000", and then press [Enter] or choose the VARIABLE OVERVIEW icon.

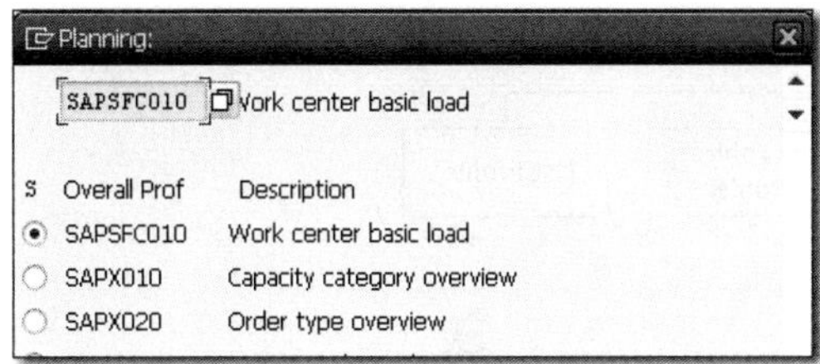

Figure 14.4 Selection of the Variable Evaluation Overall Profile SAPSFC010

The VARIABLE OVERVIEW screen that appears (see Figure 14.5) is similar to the standard evaluation, but it displays only those requirements that affect the basic load. The system shows the capacity requirements that have been dispatched.

Capacity Planning: Variable Overview

Version Overview of order categories
Unit H

Week	Wk.orders	Pl.orders	Sum	Avail.cap.	Load in %
25/2013	486.90	0.00	486.90	8.80	5,532.96
26/2013	156.57	0.00	156.57	44.00	355.83
27/2013	69.15	0.00	69.15	35.20	196.46
28/2013	42.90	0.00	42.90	44.00	97.50
29/2013	42.90	0.00	42.90	44.00	97.50
30/2013	42.90	0.00	42.90	44.00	97.50
31/2013	42.90	0.00	42.90	44.00	97.50
32/2013	42.90	0.00	42.90	44.00	97.50
33/2013	42.90	0.00	42.90	44.00	97.50
34/2013	42.90	0.00	42.90	44.00	97.50

Figure 14.5 Variable Evaluation Using Overall Profile SAPSFC010

You can define the variable evaluation using profiles, which we'll discuss next.

Overall Profile

The *overall profile* consists of four profiles: the *selection profile, option profile, list profile*, and *graphic profile*. These profiles, in turn, contain additional profiles. Figure 14.6 provides an overview of the profile structure. We'll discuss the first two profiles, while the remaining two profiles (list profile and graphic profile) contain additional settings to display a standard evaluation.

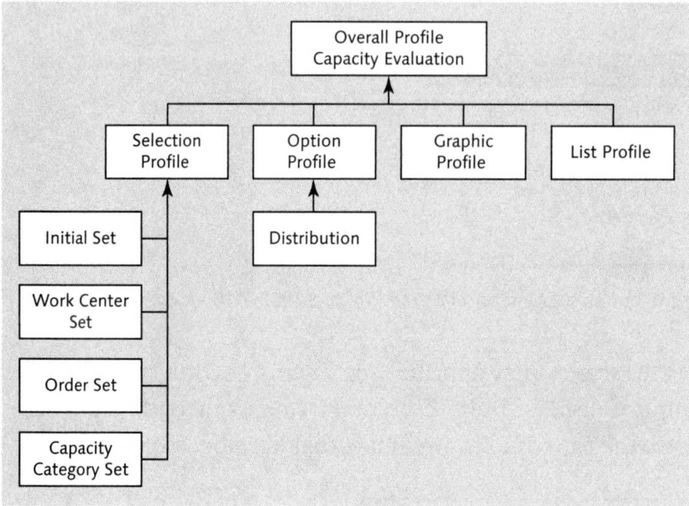

Figure 14.6 Profiles for Variable Evaluation

To create or make changes to an overall profile, follow the configuration (Transaction SPRO) menu path, PRODUCTION • CAPACITY REQUIREMENTS PLANNING • EVALUATION • PROFILES • DEFINE OVERALL PROFILES, or use Transaction OPA6. Figure 14.7 shows the assignment for OVERALL PROF. SAPSFC010.

Change View "Overall Profile": Details

New Entries

Overall prof. SAPSFC010 Work center basic load

Subprofiles

Sel. profile	SAPSFCA010	Work center/basic load
Option profile	SAPB020	Weeks (8)
List profile	SAPC010	Standard overview
Graph. profile	SAPD030	Overview with columns, Gantt latest dat.

Figure 14.7 Overall Profile for Variable Evaluation

Selection Profile

The *selection profile* determines which capacity requirements the system should read. To create or make changes to a selection profile, follow the configuration (Transaction SPRO) menu path, PRODUCTION • CAPACITY REQUIREMENTS PLANNING • EVALUATION • PROFILES • DEFINE SELECTION PROFILES, or use Transaction OPA2. Figure 14.8 shows SEL. PROFILE SAPSFCA010, which it then assigns to an overall profile SAPSFC010.

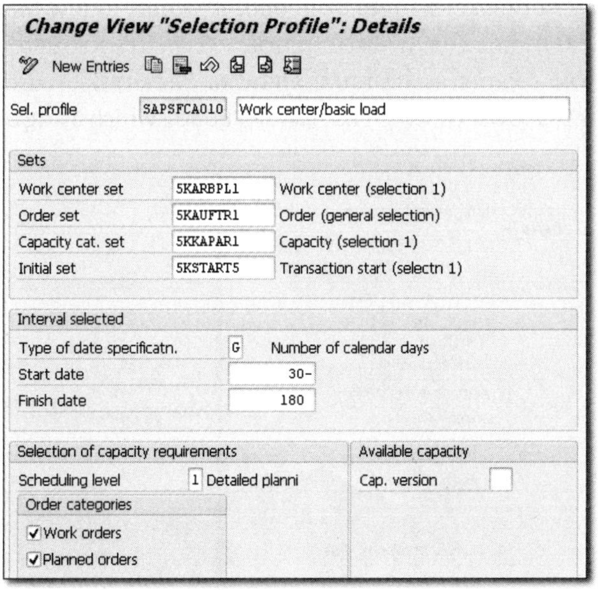

Figure 14.8 Selection Profile for Variable Evaluation

The objects for which the system reads the capacity requirements are determined on the basis of the set combinations for the work center, capacity category, and the order. You can also determine them on the basis of the *order categories*, that is, whether it's a production order or a planned order. The initial set determines which restrictions the system makes during the call. This information is output in the SELECTION PROFILE screen (see Figure 14.8).

You can create or change the set combinations using either the configuration (Transaction SPRO) menu path, PRODUCTION • CAPACITY REQUIREMENTS PLANNING • EVALUATION • SELECTION SET • DEFINE SETS, or Transactions CMS1 or CMS2, respectively. The sets often use input variables that you can create or change using configuration (Transaction SPRO) menu path, PRODUCTION • CAPACITY REQUIREMENTS PLANNING • EVALUATION • SELECTION SET • DEFINE VARIABLES, or Transaction CMV1 or CMV2, respectively.

Option Profile

The *option profile* contains items to include in backlogs, to cumulate the capacity requirements and the available capacity (see Section 14.2.4), and to distribute the capacity requirements. To create or make changes to an option profile, follow the configuration (Transaction SPRO) menu path, PRODUCTION • CAPACITY REQUIREMENTS PLANNING • EVALUATION • PROFILES • DEFINE OPTIONS PROFILES, or use Transaction OPA3. Figure 14.9 shows OPTION PROFILE SAPB020, which is also used for standard evaluation.

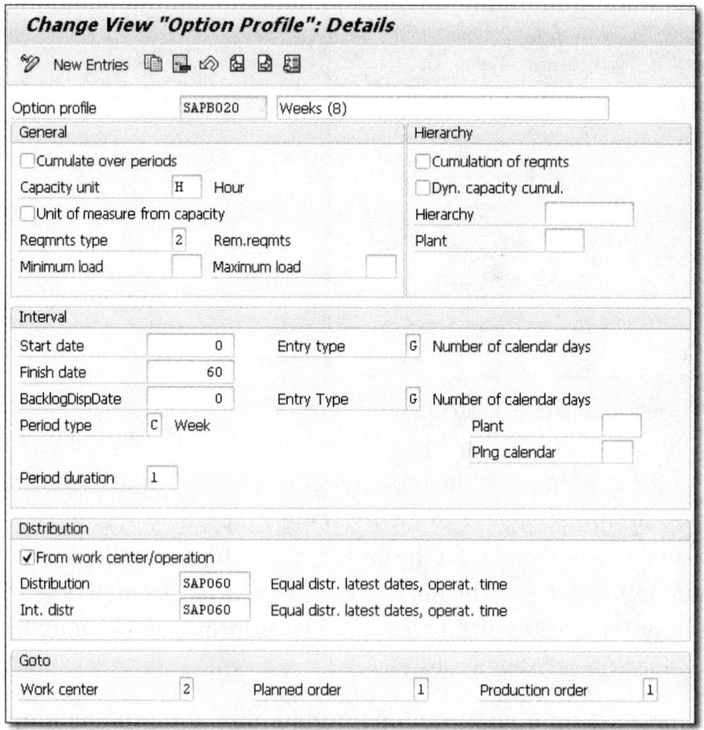

Figure 14.9 Option Profile for Variable Evaluation

The *list profile*, which can be set up by using Transaction OPA5, and the *graphic profile*, which can be set up by following the configuration (Transaction SPRO) menu path or using Transaction OPA5, both contain additional settings to display the standard evaluation.

14.2.4 Cumulating the Capacity Requirements

For evaluation purposes, especially when large numbers of orders at various work centers are taken into consideration, it's often helpful to cumulate the capacity requirements. This provides a broader overview of the capacity situation at each work center. You can cumulate capacities either statically or dynamically from work center hierarchies. When you're statically cumulating the capacity, the system only refers to the work center and its available capacity. To carry out capacity evaluation in cumulated form, that is, to cumulate both available capacity and capacity requirements, you must cumulate capacity requirements planning dynamically.

In the HIERARCHY area shown previously in Figure 14.9, the options to cumulate capacities dynamically as well as the HIERARCHY work center for the plant are available.

14.2.5 Checking Capacity Availability

When you create a production order, you can check for capacity availability in the order before dispatching it. This step helps you to decide whether it's possible to dispatch the order or not. The system considers period-based information by checking it to find out whether the available capacity in the period of operation is bigger than or equal to the capacity requirement of the operation. The length of the period is set in the period profile, which we'll cover later in this chapter. The shortest possible period is one day. When the system performs a period-based check, it only informs that the capacity is available (or unavailable) within the given period and not the planned operation date.

The available capacity represents the difference between the permitted capacity overload defined in the work center and the basic load. The basic load doesn't contain all of the capacity requirements. By default, only dispatched operations and those for which the capacity has been confirmed affect the basic load. You can display the basic load using the variable evaluation (Transaction CM07) and profile SAPSFC010.

[»] **Note**

If no capacity availability check has been carried out over a long period of time, or if it's being used for the first time, you can create the basic load on the basis of the existing (unchecked) capacity requirements using Transaction CM99. In this case, all capacity requirements will be confirmed.

Often due to delays in actual production, the system creates a backlog (e.g., non-confirmed operations from the past). You can use the OVERALL CAPACITY LOAD checkbox in the strategy profile and the backlog dispatching date (BACKLOGDISP-DATE 0; refer to Figure 14.9) to define whether you want to include them in the determination of the available capacity. We cover strategy profiles a bit later in this chapter. If you do include them, the capacity requirements of the past are temporarily scheduled forward into the current period and, if necessary, into subsequent periods and added to the basic load.

An operation can last for several periods. In this case, the distribution key determines the way in which the system distributes the capacity requirement to different periods. The capacity check uses the distribution key that is assigned to the *evaluation profile* (see the next section). The capacity availability check is carried out for each individual order, and all operations are checked one after the other, according to their numerical order. A restriction on this capability is that only those operations are checked that exist in work centers relevant to finite scheduling. Therefore, it makes sense to define critical work centers (work centers with bottleneck capacities) as finite in order to ensure there are enough degrees of freedom in the capacity planning.

You can carry out the capacity availability check interactively at the production order creation or release. Alternatively, you can perform an interactive capacity availability check from the production order, as shown in Figure 14.10. The CAPACITY AVAILABILITY pop-up appears when you choose the CAPACITY button in the production order header. The system checks for capacity during production order creation of the steel sheet rerolling process.

In this example in Figure 14.10, there is at least one operation where the work center doesn't have sufficient capacity. More detailed information is available as well as the option to use finite scheduling to shift the operations into periods with sufficient capacity. Choose the DETAILED INFO button to see the detailed information for this capacity availability check provided in the CAPACITIES WITH OVERLOAD screen that appears (see Figure 14.11).

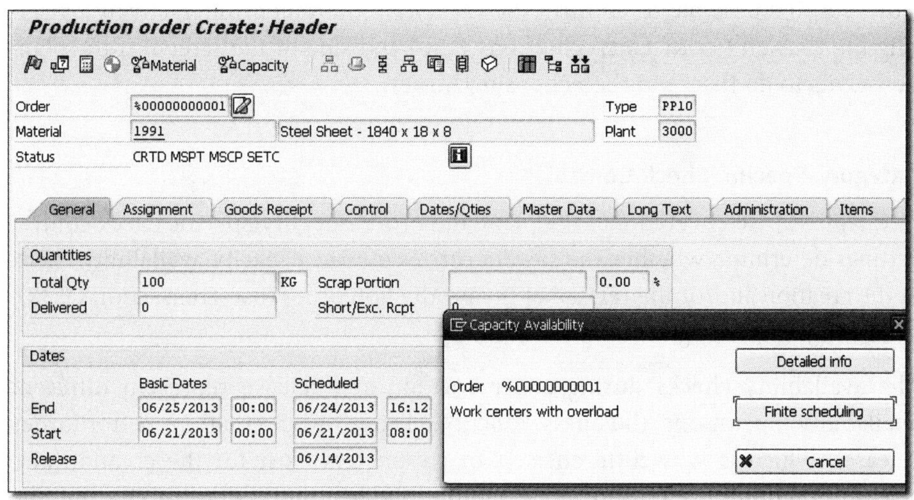

Figure 14.10 Interactive Capacity Availability Check

Here, the capacity problem exists in the first Operation at Work center BAF, and in the Period 25/2013 (week 25 of the year 2013) because the requirement of 182.90 hours far exceeds the available capacity of 44.00 hours. Because this situation leads to a load of 415.7%, the system can't confirm the operation.

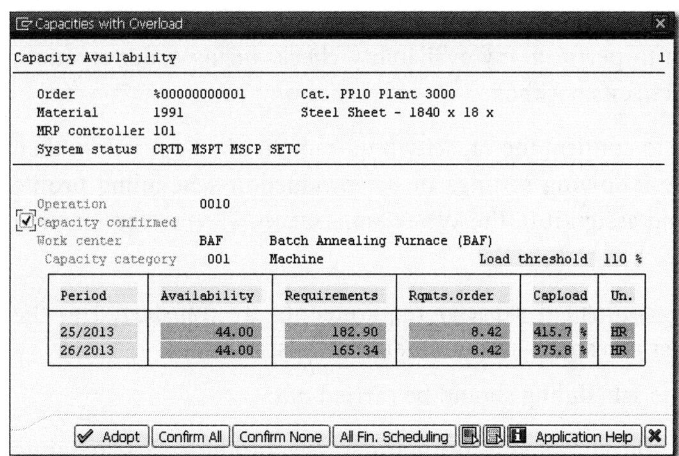

Figure 14.11 Detailed Information on Available Capacities, Including Overload

The system determines the load threshold on the basis of the permitted overload specified in the capacity master data and can exceed 100%.

However, it's possible to confirm the capacity requirement interactively by setting the CAPACITY CONFIRMED checkbox (left-hand side of the screen) if the planner wants to do that, and then choosing ADOPT.

Category-Specific Check Control

In Chapter 3, we covered that the plant and order category-specific check control settings determine whether the system carries out the capacity availability check at the creation and/or the release of the production order (use Transaction OPJK). We briefly cover it again here.

The availability checks during order creation and release represent different availability operations (business function 1 for order creation, 2 for order release), which is why different sets of parameters exist for the creation and release functions. In addition to deciding whether the system carries out the capacity availability check, you also need to assign the overall profile for the capacity availability check. If you don't make any overall profile at this stage, the system uses the standard profile SAPSFCG013. Further, you also need to define whether the production order creation or release is possible with unconfirmed capacity requirements.

The screen shown in Figure 14.12 displays the check control at order creation (use Transaction OPJK) in the CAPACITY AVAILABILITY area of the screen. If you don't want the system to perform any availability check during order creation, you can select the NO CHECK checkbox.

If the system is unable to confirm the capacity requirement of an operation, then you need to make the following settings in the production scheduling profile (Transaction OPKP), and assign it in the WORK SCHEDULING view to the material master:

- Define whether the confirmed capacity requirements are transferred to the basic load or whether all capacity requirements are discarded.
- Define whether finite scheduling should be carried out.

Figure 14.13 displays the corresponding fields in the production scheduling profile (PROD. SCHED. PROFILE) 000001 for PLANT 3000. These settings are only relevant to collective processing. Here you set the OVERALL PROFILE for capacity leveling and define the parameters for AVAILABILITY CHECK.

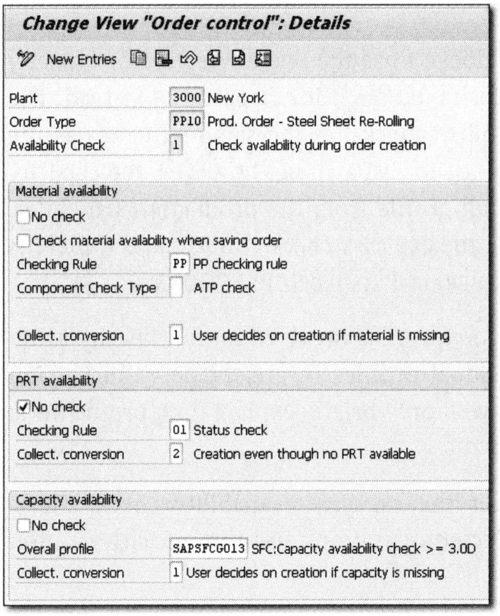

Change View "Order control": Details

New Entries

Plant 3000 New York
Order Type PP10 Prod. Order - Steel Sheet Re-Rolling
Availability Check 1 Check availability during order creation

Material availability
[ ] No check
[ ] Check material availability when saving order
Checking Rule PP PP checking rule
Component Check Type ATP check
Collect. conversion 1 User decides on creation if material is missing

PRT availability
[x] No check
Checking Rule 01 Status check
Collect. conversion 2 Creation even though no PRT available

Capacity availability
[ ] No check
Overall profile SAPSFCG013 SFC:Capacity availability check >= 3.0D
Collect. conversion 1 User decides on creation if capacity is missing

Figure 14.12 Check Control for the Capacity Availability Check

Change View "Production Scheduling Profile": Details

New Entries

Plant 3000 New York
Prod. Sched. Profile 000001 CREATE AND RELEASE ORDER

Automatic Actions
On Creation
[ ] Release
[ ] Document Links - Material
[ ] Document Links - BOM
On Release
[ ] Execute Printing
[x] Schedule Order
[x] Document Links - Material
[x] Document Links - BOM
[ ] Create Control Instructions

Material Availability Check
[ ] Confirm Available Partial Qty
Goods Receipt
[ ] Automatic Goods Receipt

Capacity Planning
Leveling
Overall Profile SAPSFCG011 SFC: View work center/capacity (3 graf.)
Availability Check
[x] Confirm Capacity
[x] Finite Scheduling

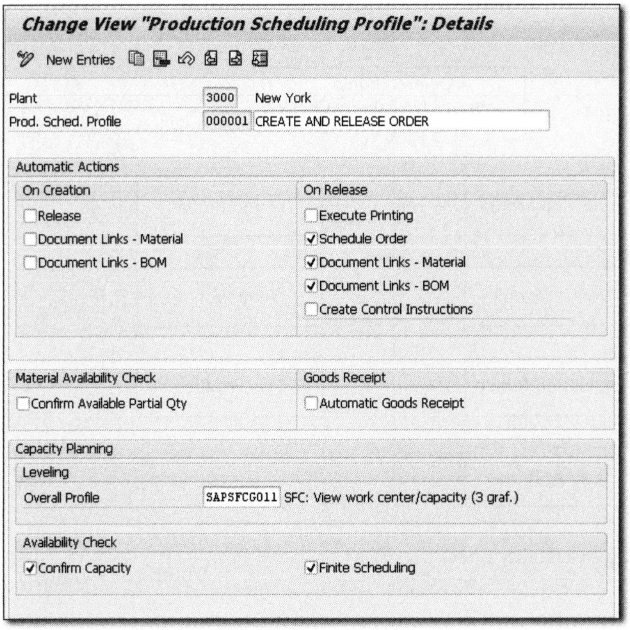

Figure 14.13 Production Scheduling Profile

Profiles

It's important to note that the system doesn't use the overall profile for capacity leveling for the capacity availability check but uses it for dispatching instead. The overall profile for the capacity availability check has the same structure as the overall profile for dispatching. In contrast to dispatching, however, the capacity availability check doesn't use the overall profile from the production scheduler. Instead, it uses the overall profile from the capacity check control, and if there is no profile available there, it uses the OVERALL PROFILE SAPSFCG013.

Figure 14.14 provides an overview of the profiles involved in the capacity availability check. Because we cover individual profiles in greater detail in Section 14.4.2, which deals with dispatching, we'll only briefly explain each profile here:

- **Selection profile**
  Defines the selection of objects for the capacity availability check. These include the work centers of the selected operations and the capacity requirements of the basic load.

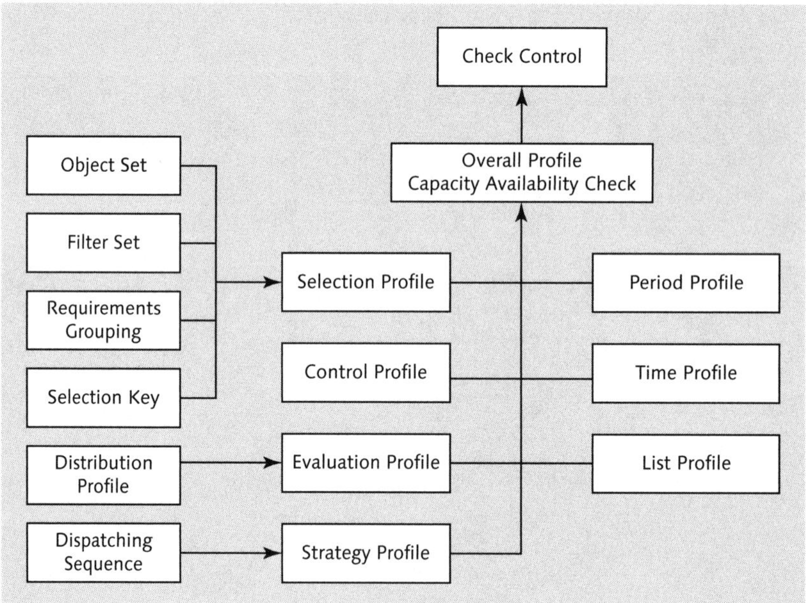

Figure 14.14 Profiles for Capacity Availability Check

- **Control profile**
  Defines the preparation of the data.
- **Evaluation profile**
  Assigns the distribution key for distributing the capacity requirements, among other parameters.
- **Strategy profile**
  Contains the sequence and search direction for finite scheduling, among other parameters.
- **Period profile**
  Defines the length of the period for the check.
- **Time profile**
  Defines the time fences for the capacity requirements in question.
- **List profile**
  Contains settings for the result output.

The result of the capacity availability check is the confirmation of the capacity requirements, if the available capacity is sufficient. If the available capacity within the period is insufficient, you can trigger the finite scheduling process to find a period with a sufficient capacity and to confirm the capacity requirement in that available new period. The system doesn't automatically dispatch the operation either by the capacity availability check or by finite scheduling. You have to manually dispatch the operation, either individually or in the mass processing step. We'll cover both later in Section 14.4.1 and Section 14.4.6, respectively.

14.3 Finite Scheduling

If the available capacity is insufficient to meet the capacity requirement of an operation on the requested date, you can use the *finite scheduling* option to find the next possible date on which the available capacity is sufficient for the operation. You can perform finite scheduling either in the capacity availability check or during dispatching.

While performing finite scheduling, the system refers to the strategy profile to search for the direction in which to perform the scheduling (backward or forward scheduling). A search in the direction of today's date is based on backward scheduling and represents the standard case, whereas a search into the future is based on forward scheduling. However, much of this also depends on the company's business processes and needs, where only forward scheduling is considered to look for free or available capacity in the future. If insufficient capacity is found in one direction, then you can also reverse the search direction.

The screen shown in Figure 14.15 appears during the capacity availability check, and then in the next capacity overload pop-up, you click on the Finite Scheduling icon (not shown). The system proposes the new latest date (NewLtStart) and time (NewLtTim), after it performs finite scheduling. At this stage, you can confirm the capacity by selecting the Cap. confirm. checkbox and then choosing Adopt.

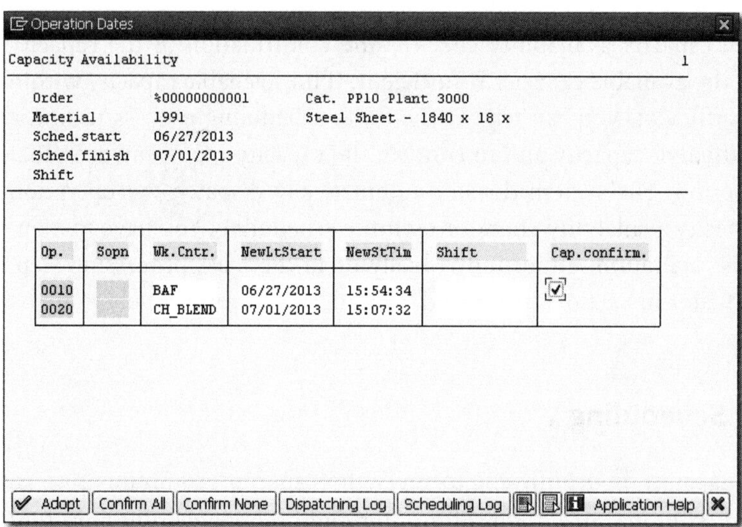

Figure 14.15 Finite Scheduling Results (After the Capacity Availability Check)

When you perform the dispatch function for an operation (see Figure 14.16 ❶), the system automatically performs finite scheduling and assigns the operation immediately next to the already assigned operation ❷.

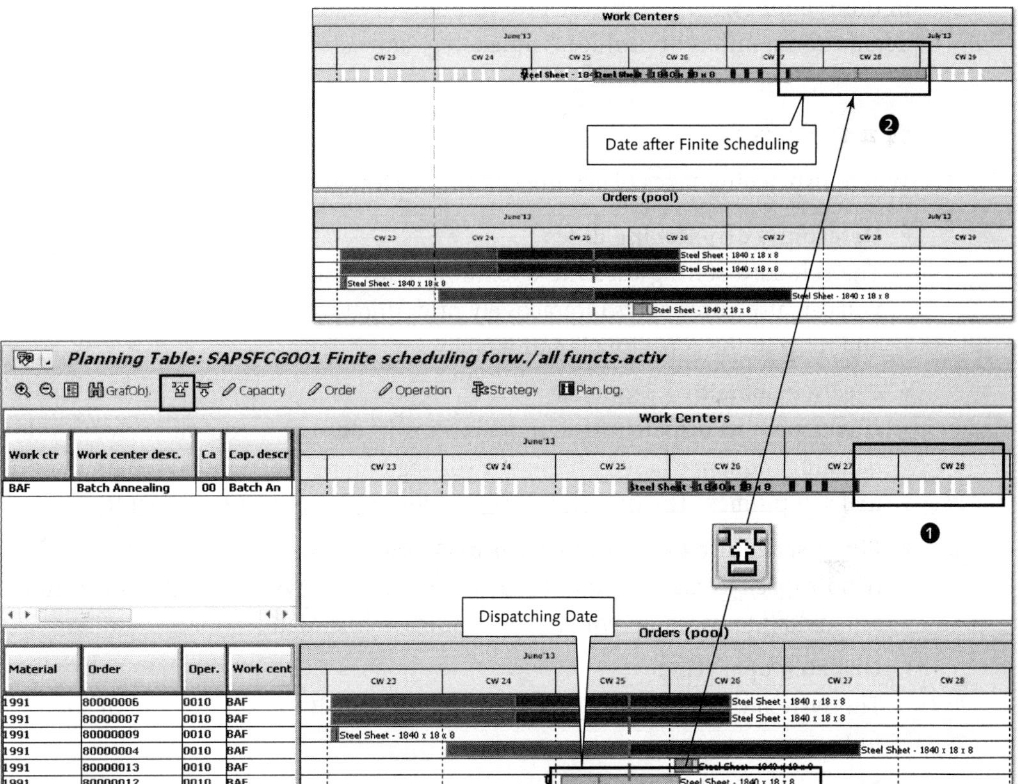

Figure 14.16 Finite Scheduling Results (Triggered by Dispatching)

14.4 Dispatching

In the *dispatching* process, the system creates a workable or feasible production plan in which the capacity is available. To achieve this objective, there should be a sequence of dates set up for the individual operations in the work centers. The nondispatched operations form the pool of orders/operations for the capacity planner/production scheduler to work on. To dispatch an operation, the capacity planner selects the operation, which is ready for dispatch from the capacity planning table or from the tabular capacity planning table, and initiates the dispatching function.

In this section, we'll go over the process steps of dispatching and then get into the details of the different profiles, sequencing, and setup.

14.4.1 Process Steps

When dispatching takes place, the system carries out the following steps:

- **Determine dispatching dates**
  The standard setting uses the operation dates as dispatching dates. However, you can also enter dates interactively or define the dispatching date using drag and drop in the capacity planning table.
- **Create dispatching sequence**
  If you want to dispatch several operations in one step, you must create a dispatching sequence based on specific sorting criteria (Section 14.4.3) or on optimal setup times. The default settings stipulate dispatch at the latest date.
- **Finite scheduling if no capacity is available**
  If no capacity exists on the dispatching date, the system needs to run finite scheduling to come up with a feasible dispatching date.
- **Dispatch operation**
  The system dispatches the operation and assigns it the status DISP (if successful).
- **Midpoint scheduling**
  You need to adjust the dates of the nondispatched operations in such a way that the system carries out the midpoint scheduling process for them. In midpoint scheduling, the system schedules the nondispatched operations so that earlier ones are scheduled backward and later ones forward. We cover midpoint scheduling later in Section 14.4.5.

[»]

Note

It may be impractical from your particular practice or business point of view, but you can also base the dispatching processing on an infinite capacity assumption. To do this, you need to deselect the finite scheduling option in the strategy profile. However, with bottleneck capacities at critical work centers hardly permit this option.

During dispatching, you may realize that it's not possible to meet the basic dates of the order. Because the system carries out dispatching for each operation, it's also possible to compromise on the sequence of operations for the dispatch. You can use the strategy profile to define the extent to which the system allows inconsistencies within the operations of an order. However, the capacity planner needs to monitor the external dependencies such as the date-based dependency between orders and customer requirements to ensure timely production.

Whenever you perform the dispatch or the deallocate function, the system updates the planning and the scheduling logs. In fact, it issues a message to inform the capacity planner to check the planning and/or scheduling logs. You can call the planning log from the respective application, for instance, from the capacity planning table or from finite scheduling in the interactive capacity availability check. Figure 14.17 displays a sample planning log that contains two information messages and no warnings.

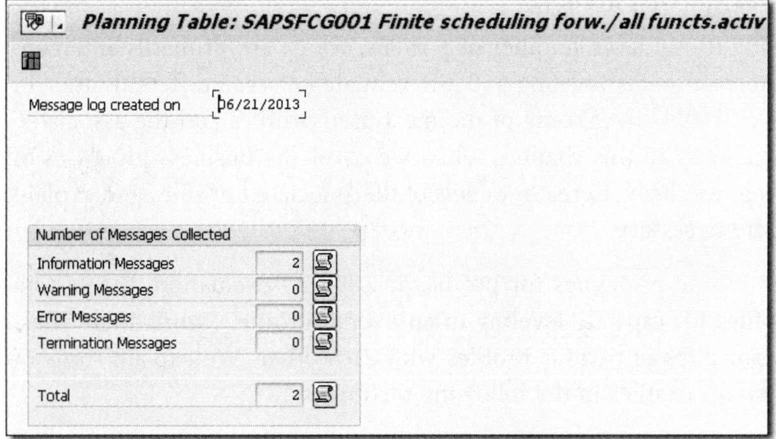

Figure 14.17 Planning Log Summary after Finite Scheduling

Double-click on the information or warning messages, and you'll see a log that displays information messages, for instance, that the system successfully dispatched operation 0010 and provides the dispatching date and time as well (in the second information message).

Figure 14.18 shows the scheduling log that appears when you use the menu path, Extras • Log • Scheduling, in the capacity planning table. Here it issues two warning messages: scheduling conflict and nonmaintenance of capacity category in the scheduling view of the work center.

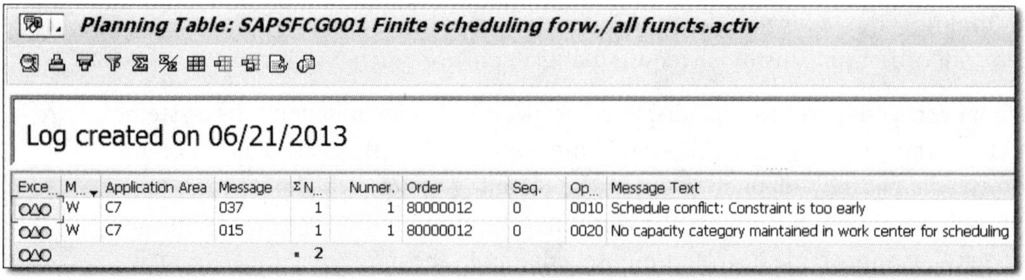

Planning Table: SAPSFCG001 Finite scheduling forw./all functs.activ

Log created on 06/21/2013

Exce...	M...	Application Area	Message	ΣN...	Numer.	Order	Seq.	Op...	Message Text
	W	C7	037	1	1	80000012	0	0010	Schedule conflict: Constraint is too early
	W	C7	015	1	1	80000012	0	0020	No capacity category maintained in work center for scheduling
				• 2					

Figure 14.18 Detailed Scheduling Message Log after Finite Scheduling

14.4.2 Profiles for Dispatching

Capacity leveling uses a large number of profiles, which are primarily interconnected, to perform capacity leveling and differentiate between different tasks. In this section, we briefly cover some of the most-used profiles and the associated settings in each. Later in this chapter, when we cover the business processes of capacity leveling, we'll cover greater details of the associated profiles and explain their roles where necessary.

Building on the same principles for profiles in capacity evaluation, the system cumulates profiles for capacity leveling in an overall profile. Figure 14.19 illustrates the relationships of various profiles with each other. We explain some of the more important profiles in the following sections.

Overall Profile

To create or make changes to an overall profile in capacity leveling, follow the configuration (Transaction SPRO) menu path, Production • Capacity Requirements Planning • Capacity Leveling and Extended Evaluation • Define Overall Profile, or use Transaction OPDO.

The screen shown in Figure 14.20 displays the standard Overall profile SAPSFCG011 as well as different subprofiles.

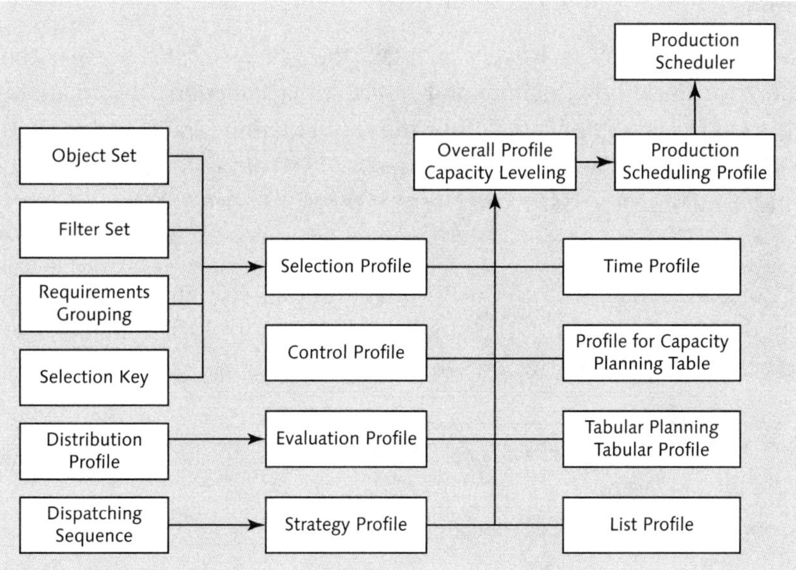

Figure 14.19 Profiles for Capacity Leveling

Change View "Capacity Leveling - Overall Profiles": Details

New Entries

Overall profile: SAPSFCG011 — SFC: View work center/capacity (3 graf.) — Check

Subprofiles

Subprofile	Profile	Description
Selection profile	SAPSFCS001	SFC: View work center -> cap.(3 graf.)
Control profile	SAP___C002	Gen: Planning table
Time profile	SAP___Z002	Gen: Medium-term (3 months)
Evaluation profile	SAP___A001	Gen: Standard evaluation profile
Strategy prof.	SAP___T001	Finite scheduling forw./all functs.activ
Period profile		
Planning tab.profile	SAPSFCL010	SFC: only main cap. (3 graf.)
Prof. pln.tab. (tab)		
List profile	SAPSFCI001	SFC: Production planning

Figure 14.20 Overall Profile for Dispatching

You don't have to maintain the PERIOD PROFILE in the overall profile because it's relevant for a capacity availability check only. The following sections describe some of the individual profiles shown in this screen.

Strategy Profile

It's very important to comprehensively maintain the *strategy profile* because the system refers to it during dispatching and rescheduling functions. To create or make changes to the strategy profile, follow the configuration (Transaction SPRO) menu path, PRODUCTION • CAPACITY REQUIREMENTS PLANNING • CAPACITY LEVELING AND EXTENDED EVALUATION • STRATEGY • DEFINE STRATEGY PROFILE, or use Transaction OPDB.

Figure 14.21 displays strategy profile SAP_T001, including its numerous parameters. The important parameters of the strategy profile control the functions for dispatching, scheduling, and creating a sequence of operations.

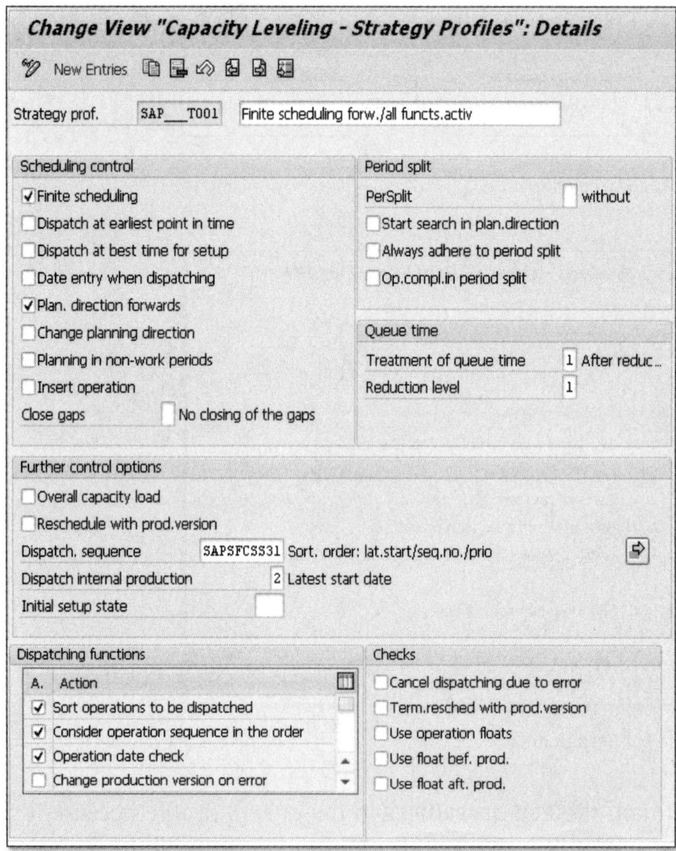

Figure 14.21 Strategy Profile

The SCHEDULING CONTROL parameters shown at the top of the screen allow you to make the following settings, for example:

- You can set whether you want to carry out finite scheduling during dispatching. If the FINITE SCHEDULING parameter isn't set, the dispatch process assumes that unlimited capacity is available. If there is insufficient available capacity, the dispatching attempt becomes unsuccessful.
- When you use the DISPATCH AT EARLIEST POINT IN TIME parameter, the system doesn't consider interdependencies with other operations and dispatches the operation as early as possible.
- The parameter DISPATCH AT BEST TIME FOR SETUP chooses the dispatch date on the basis of the existing setup sequence in such a way that the increase in the overall setup time is kept as small as possible. This entails that the system should have the provision to shift the already-dispatched operations. To attend to this need, you also need to set the parameter INSERT OPERATION. The system supports this function in the capacity planning table and makes sense only if setup matrices for sequence-based setup are in place. We explain the setup matrices later in Section 14.4.4.
- The DATE ENTRY WHEN DISPATCHING parameter ensures that instead of using the operation date during dispatching, a pop-up window appears and prompts the user for an interactive date entry.
- The default setting for the planning direction is backward of the dispatch date. However, you can change this setting when you activate the parameter PLAN. DIRECTION FORWARDS. When the system is unable to find available capacity within the planning horizon, and you've also set the CHANGE PLANNING DIRECTION parameter, the system continues to search for available capacity in the opposite direction.
- The PLANNING IN NON-WORK PERIODS parameter overrides the nonwork periods in the work center and suggests 24-hour capacity availability with a full rate of capacity utilization to the dispatch process.
- The CLOSE GAPS parameter enables you to automatically close gaps that are caused by deallocations or rescheduling. This again entails rescheduling of already-dispatched operations and is only possible in the capacity planning table.

The parameters in the DISPATCHING FUNCTIONS area at the bottom of the screen define which options are available for dispatching:

- SORT OPERATIONS TO BE DISPATCHED determines how to create the right sequence of operations, if you want the system to dispatch several operations simultaneously.
- The CONSIDER OPERATION SEQUENCE IN THE ORDER parameter enables the system to adhere to the sequence of operations in the production order. When you use this parameter, then in the case of forward scheduling, the system doesn't dispatch the operation prior to its predecessor. Similarly, with backward scheduling, the system makes sure that it doesn't dispatch the operation after its successor.
- If you use OPERATION DATE CHECK, you can check whether the new operation dates lie within the floats that can be selected using the following parameters:
  - USE OPERATION FLOATS
  - USE FLOAT BEF. PROD. (use before production)
  - USE FLOAT AFT. PROD. (use after production)
- If you want the system to immediately perform the midpoint scheduling process, then you need to select the MIDPOINT SCHEDULING parameter. When you select this parameter and depending on the operations, the system immediately carries out the midpoint planning function; otherwise, it carries it out when you save the planning results.
- SETUP TIME OPTIMIZATION creates the dispatching sequence according to the setup time minimization criterion.
- If you set the CANCEL DISPATCHING DUE TO ERROR parameter, the system doesn't allow dispatch of operations outside the floats.

Selection Profile

The *selection profile* enables you to define which capacity requirements the system should consider for dispatch. To create or make changes to the selection profile, follow the configuration (Transaction SPRO) menu path, PRODUCTION • CAPACITY

REQUIREMENTS PLANNING • CAPACITY LEVELING AND EXTENDED EVALUATION • DEFINE SELECTION PROFILE, or use Transaction OPD1 (see Figure 14.22).

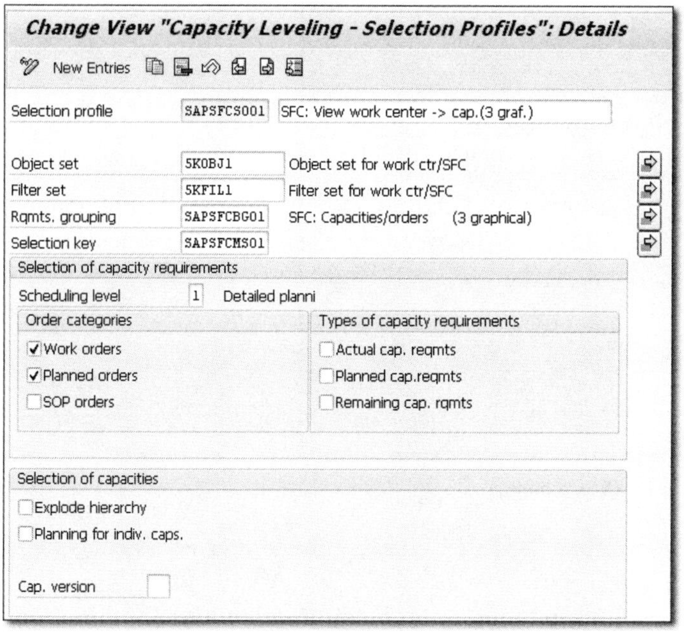

Figure 14.22 Selection Profile for Dispatching

Control Profile

The *control profile* defines how the system presents the data. It can be presented continuously over time for the capacity planning table, or it can be based on periods for the tabular capacity planning table. An important parameter of the control profile defines whether you can modify the operations and at which stage the system should set the locks to avoid any further changes to the operations.

To create or make changes to the control profile, follow the configuration (Transaction SPRO) menu path, PRODUCTION • CAPACITY REQUIREMENTS PLANNING • CAPACITY LEVELING AND EXTENDED EVALUATION • DEFINE CONTROL PROFILE, or use Transaction OPDE (see Figure 14.23).

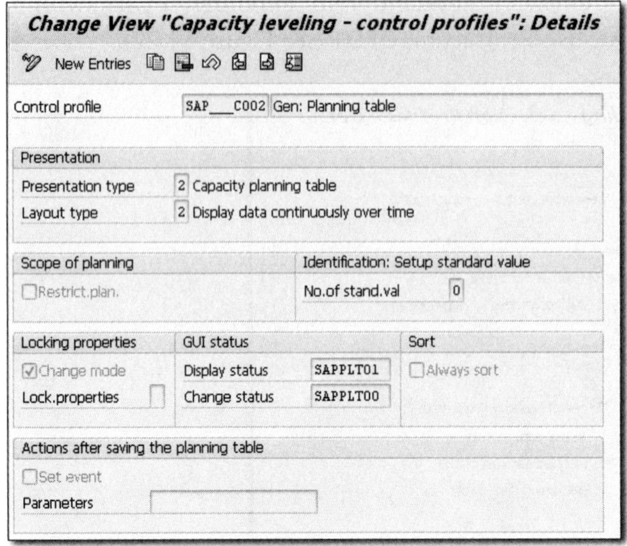

Figure 14.23 Control Profile for Dispatching

Evaluation Profile

The *evaluation profile* enables you to maintain the distribution keys for the capacity requirements, the units of measure, and the details to cumulate the capacity requirements. To create or make changes to the evaluation profile, follow the configuration (Transaction SPRO) menu path, PRODUCTION • CAPACITY REQUIREMENTS PLANNING • CAPACITY LEVELING AND EXTENDED EVALUATION • STRATEGY • DEFINE EVALUATION PROFILE, or use Transaction OPD3.

Time Profile

The *time profile* contains the time horizons for importing the data records as well as for evaluation and planning. The planning horizon must be smaller than or equal to the evaluation period, which in turn must be smaller than or equal to the database read period. Another entry concerns the date for dispatching the backlog on which the system dispatches all backlog capacity requirements—provided it considers the backlog.

To create or make changes to the time profile, follow the configuration (Transaction SPRO) menu path, PRODUCTION • CAPACITY REQUIREMENTS PLANNING •

Capacity Leveling and Extended Evaluation • Define Time Profile, or use Transaction OPD2. Figure 14.24 displays Time profile SAPPM_Z002.

Change View "Capacity Leveling - Time Profile": Details

New Entries

Time profile SAPPM_Z002 PM: 2 months + 5 days in the past

Database read period			
Entry type	J Number of calendar months	Start date	-3
		Finish date	3

Evaluation period			
Entry type	G Number of calendar days	Start date	-5
		Finish date	60

Planning period			
Entry type	G Number of calendar days	Start date	-1
		Finish date	60

Dispatching of backlog			
Entry Type	G Number of calendar days	BacklogDispDate	0

Figure 14.24 Time Profile for Dispatching

14.4.3 Dispatching Sequence

Generally, you'll carry out the dispatching function manually, individually, and interactively. But when you have a large and a drastic change in the production plan, which also greatly impacts the operations involved, it makes sense to dispatch several operations according to specific rules. The easiest and most straightforward rule that most companies follow in dispatching is according to the start dates and times of operations. You can implement similar rules by using the sorting function, whereas the layout key enables you to implement sorting from a large number of available fields.

You can create the layout key using Transaction CY39 in Customizing (see Figure 14.25). On this screen, you can select the fields that you want to sort and then eventually make available in dispatching. After you select the fields, choose the Sequence/Heading button, and the screen shown in Figure 14.26 appears.

The field details in Figure 14.26 allow you to select sorting in ascending or descending order. You then assign the layout key to the strategy profile.

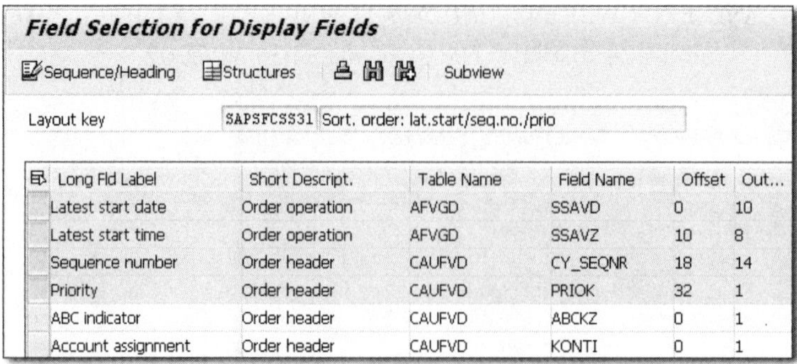
Field Selection for Display Fields

Sequence/Heading Structures Subview

Layout key SAPSFCSS31 Sort. order: lat.start/seq.no./prio

Long Fld Label	Short Descript.	Table Name	Field Name	Offset	Out...
Latest start date	Order operation	AFVGD	SSAVD	0	10
Latest start time	Order operation	AFVGD	SSAVZ	10	8
Sequence number	Order header	CAUFVD	CY_SEQNR	18	14
Priority	Order header	CAUFVD	PRIOK	32	1
ABC indicator	Order header	CAUFVD	ABCKZ	0	1
Account assignment	Order header	CAUFVD	KONTI	0	1

Figure 14.25 Layout Key for Sequence Creation during Dispatching

Field Selection for Display Fields

Select Subview

Layout key SAPSFCSS31 Sort. order: lat.start/seq.no./prio

Long Fld Label	Desc. sort.	Output length	Structure	Short Descript.	Offset	Table Na...	Field Name
Latest start date		10	AFVG	Order operation	0	AFVGD	SSAVD
Latest start time		8	AFVG	Order operation	10	AFVGD	SSAVZ
Sequence number		14	AFKO	Order header	18	CAUFVD	CY_SEQNR
Priority		1	AFKO	Order header	32	CAUFVD	PRIOK

Figure 14.26 Sort Order for Fields during Dispatching

[+]

Tips & Tricks

In addition to the layout key, you can also use user exit CYPP0001 to sort the capacity requirements. You'll need to engage an ABAP resource to define the parameters in the user exit.

The sequence in which the operations are dispatched may not necessarily match the sequence in the work center, as certain conditions of the dispatching process can trigger the system to dispatch some operations differently. For example, if a long-duration operation doesn't fit into any gap that has been created on the basis of operations already dispatched, the system dispatches that long-duration operation further in the planning horizon.

14.4.4 Sequence-Dependent Setup

In some of the production processes, the setup time depends on the sequence in which the product is produced or the physical state of the machine left behind after an operation. For example, if you have a paint shop and paint yellow first and then a dark color, this might not be that critical because the setup time for dark might be of standard length. But if you first paint the dark color and yellow afterwards, the setup time for yellow might be much higher because the paint shop machines need to be cleaned very thoroughly from the dark color.

You can configure the dependency of setup work on a processing sequence by assigning the setup groups in respective operation of the routing. The assignment takes place at two levels: the setup group category and the setup group key. A setup group category can contain several setup group keys; for example, a setup group category can be color, and the setup group keys can be black, white, and yellow. The fact that the assignment occurs at two different levels has the advantage of enabling both a detailed maintenance between setup group keys and a less detailed maintenance between setup group categories.

Setup Group Categories

You can configure group categories and set up group keys using Transaction OP43. Figure 14.27 shows the list of setup groups available for Plant 3000.

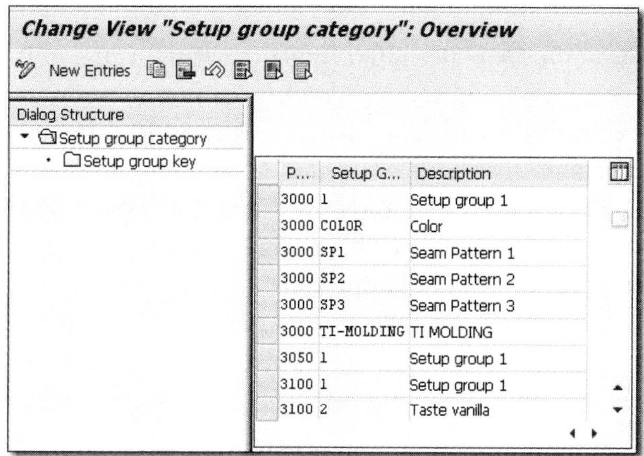

Figure 14.27 Setup Group Category

Select the SETUP GROUP CATEGORY folder, and double-click on the SETUP GROUP KEY folder on the left-hand side of the figure.

Finally, in the operations details of the routing (Transaction CA02), you assign the setup group categories and keys (see Figure 14.28).

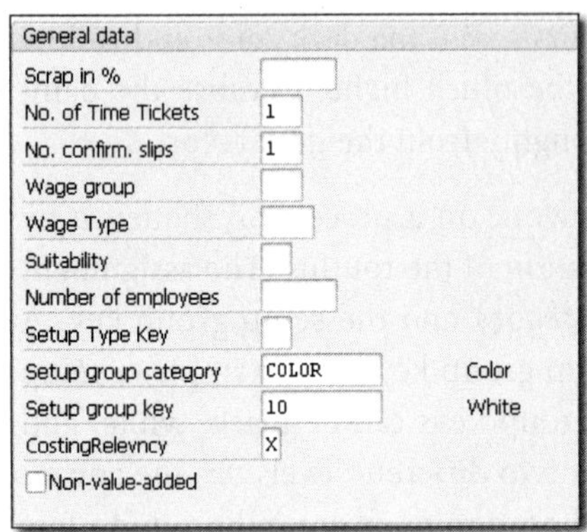

Figure 14.28 Assignment of Setup Group Category and Setup Group Key in Routing

You can maintain the setup time between the setup group categories and the setup group key in a setup matrix through the configuration (Transaction SPRO) menu path, PRODUCTION • CAPACITY REQUIREMENTS PLANNING • MASTER DATA • ROUTING DATA • DEFINE SETUP MATRIX, or by using Transaction OPDA. The setup matrix is unique for each plant.

Figure 14.29 illustrates an example in which the setup time from a previous operation with setup group category 1 and setup group key 10 to a subsequent operation with setup group category 1 and setup group key 20 is 10 minutes. However, the setup time for the opposite direction; that is, from setup group category 1 and setup group key 20 to setup group category 1 and setup group key 10 amounts to 25 minutes. Maintaining a setup matrix enables you to assign them to individual operations in the routing, saving time and effort in master data maintenance. The last column (not shown) shows the standard value (activity type) that the system can use for setup time. Examples of activities in the standard value key (SVK) are setup time, machine time, and labor time. Note that we covered SVK in Chapter 3.

Change View "Setup Matrix": Overview

New Entries

Plnt	Pred.group	PreSubgrp	Succ.group	SuccSub...	Standard ...	Unit	S
3000	*		SP3		10	MIN	1
3000	1	10	1	20	10	MIN	1
3000	1	20	1	10	25	MIN	1

Figure 14.29 Setup Matrix

Setup Time

When you're in the capacity planning table, the system offers the following planning functions for planning with sequence-dependent setup times:

- Interactive change of default values
- Automatic adjustment of default values
- Dispatch at best time for setup
- Setup time optimization

In the capacity planning table, you can also make automatic or manual changes to the setup time by following the menu path, FUNCTIONS • ADJUST SETUP TIME • MANUALLY. The screen shown in Figure 14.30 appears, in which you can make changes to the setup time in the SETUP field, which in this example is 10 minutes.

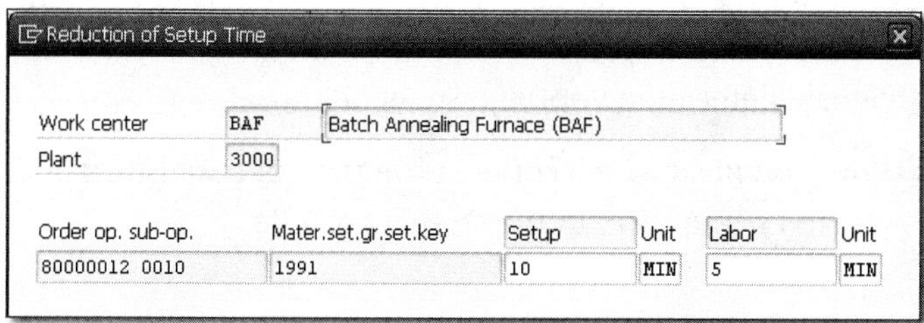

Figure 14.30 Manual Adjustment in Setup Time

You can increase or decrease the setup time. When you make this setup time adjustment, the system carries out a new scheduling process. If an operation was already dispatched, the system deallocates it when the setup adjustment takes place. With automatic setup time adjustment, the system calculates the setup times of operations on the basis of their sequence and reschedules them accordingly. To carry out the automatic setup function, you must first set the PLAN.

DIRECTION FORWARDS and INSERT OPERATION parameters in the strategy profile. To close gaps that occur when the system reduces the setup times, you also need to set the CLOSE GAPS parameter in the strategy profile.

When you use the DISPATCH AT BEST TIME FOR SETUP function, the system doesn't try to dispatch the operation at the predefined dispatch date. It will do this at a point when the overall setup time for all operations increases is minimal. To be able to use this function, you must set the DISPATCH AT BEST TIME FOR SETUP and INSERT OPERATION parameters in the strategy profile and make sure that the parameter for the SETUP TIME OPTIMIZATION dispatch function is *not* set.

To enable the system to propose an optimum time for setup, it compares the dispatch times of operations with each other. If several optimal setup times exist, the system uses the latest of the optimal setup times. During dispatch, the system doesn't adjust the setup time of operations according to the sequence of operations. Instead, it requires an automatic setup time adjustment in the next step. To dispatch several operations at the same time, the system uses the dispatch sequence defined in the strategy profile.

The setup time optimization establishes an optimal setup sequence for a group of operations. The system dispatches the sequence in the work center without gaps and also automatically adjusts the setup times. The setup time optimization function is useful to redesign the planning because it ensures that the sequence of operations is optimal even after the dispatch. But if the dispatch period contains operations that have already been dispatched, the system deviates the new sequences from the previously determined optimal sequence.

For setup time optimization, you need to select the SETUP TIME OPTIMIZATION and the PLAN. DIRECTION FORWARDS parameters in the strategy profile.

14.4.5 Midpoint Scheduling

Midpoint scheduling behaves in the same way as lead time scheduling, except that the basis of scheduling isn't on order start or finish dates. Instead, it starts with the date of the dispatched operation of the order and carries out a backward or forward scheduling process from the operation, according to the planning direction that you've defined in the strategy profile. The system considers the other dispatched operations of the order as fixed. The dispatch occurs for each individual operation. During the dispatch of an operation, the dates of the operation can change. In such cases, you can use *midpoint scheduling* to adjust the dates of nondispatched operations to the changed dates of a dispatched operation.

For example, out of three operations, if the dispatch of the second operation took place, then midpoint scheduling can adjust the date of the nondispatched operations, which are the first and third ones. See Figure 14.31 for midpoint scheduling of three operations of an order.

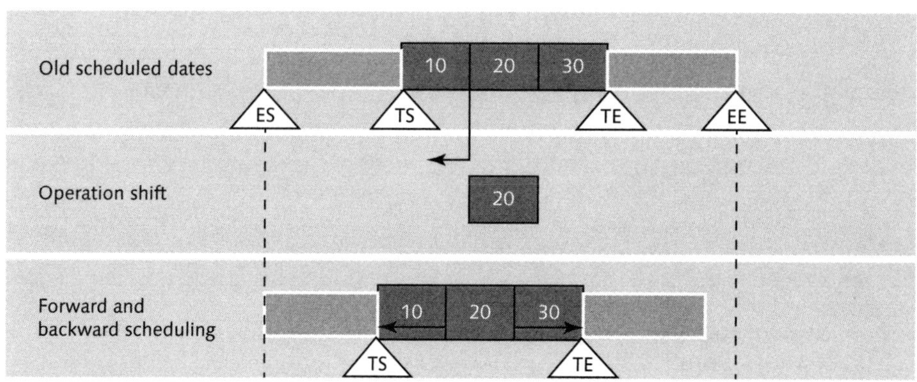

Figure 14.31 Midpoint Scheduling

14.4.6 Mass Processing

Often where there are several operations or orders for which you want to perform a mass function, such as dispatch or deallocation, you can use the mass processing functionality available in capacity planning. To maintain the mass processing settings, follow the SAP menu path, Logistics • Production • Capacity Planning • Leveling • Variable • Background • Execute, or use Transaction CM40.

Figure 14.32 shows the initial screen in which you assign the Overall profile SAPSFCG011 for capacity leveling ❶. Enter the Job name as "PP10", and choose Save parameters in the Call up function area of the screen. Next, click on the Save icon, or press Ctrl+S ❷. On saving parameters for the first time, the system prompts you to enter work center parameters, in which you can enter text in the Work center, Plant, Capacity category, or Capacity planner grp fields ❸.

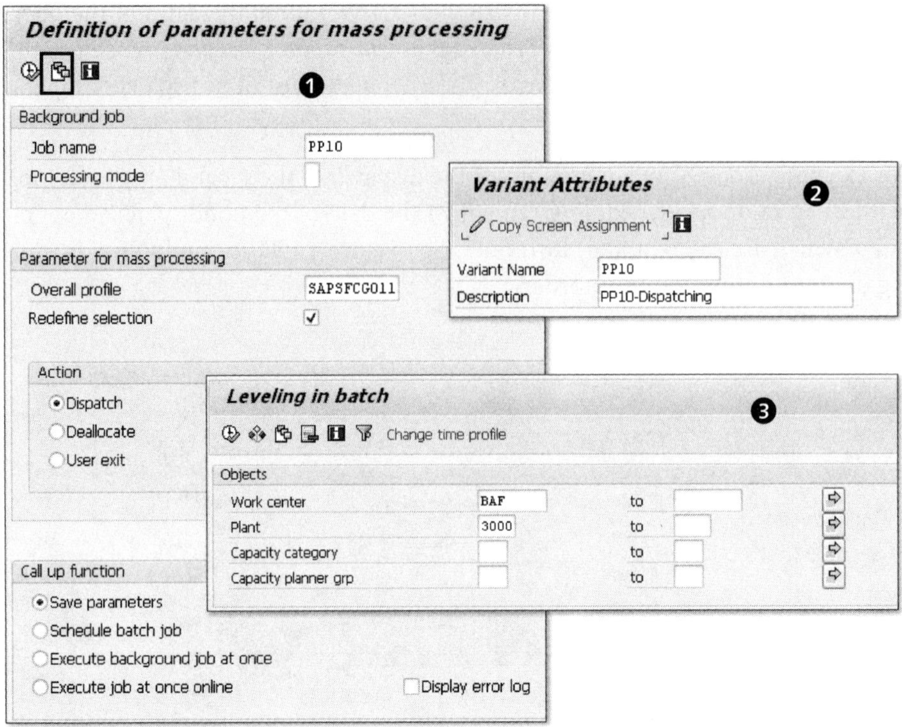

Figure 14.32 Mass Processing

Next time, when you use Transaction CM40, you can use one of the three available actions, such as DISPATCH, DEALLOCATE, or USER EXIT (i.e., user exit CY190001) in the ACTION area of the screen. In the CALL UP FUNCTION area of the screen, you can define whether you want to dispatch the job at a later stage (SAVE PARAMETERS), dispatch it directly, or execute it directly.

Transaction CM41 enables you to display the logs of the dispatch jobs.

14.5 Capacity Planning Table

Two types of capacity planning tables are available in the system: graphical and tabular. Our focus will primarily remain on the graphical capacity planning table, but all of the concepts and functions available in the graphical capacity planning table are also available in the tabular capacity planning table.

The highly interactive and intuitive nature of the graphical capacity planning table enables a capacity planner to perform several activities and functions:

- Dispatch operations
- Deallocate operations
- Make changes to an order
- Make changes to an operation
- Make changes to capacity

While we discuss the first two functions in the preceding list in the following sections, there are relevant icons available in the graphical capacity planning table for the remaining functions. For example, if you want to make changes to a specific order, choose the CHANGE ORDER icon, and the system navigates to the production order change screen (Transaction CO02), in which you can then make the desired changes. The options to change operations and capacity details are also available.

You can also make changes in the strategy profile in the capacity planning table to perform the desired functions and achieve the requisite planning results.

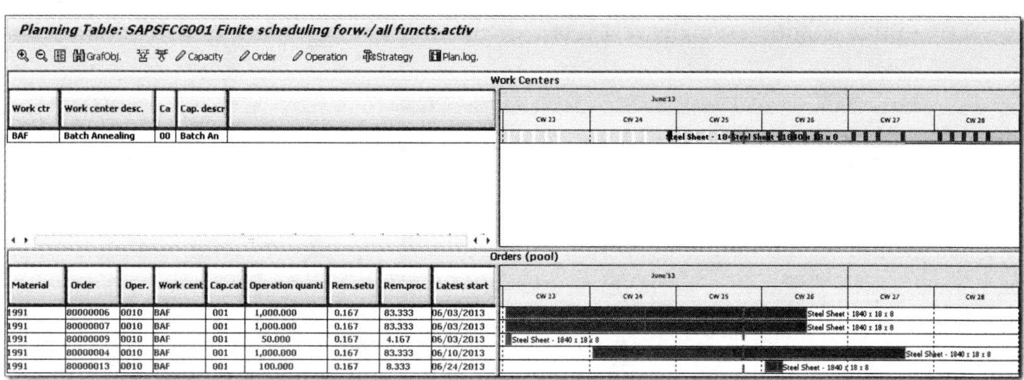

Figure 14.33 Capacity Planning Table

To access the capacity planning table, follow the SAP menu path, LOGISTICS • PRODUCTION • CAPACITY PLANNING • LEVELING • WORK CENTER VIEW • PLANNING TABLE (GRAPHICAL), or use Transaction CM21. Figure 14.33 shows the capacity planning table that appears after you enter the initial parameters, such as the WORK CENTER and PLANT, and choose EXECUTE. If the system is able to find capacity requirements in the given period for work centers or orders, it displays the capacity information.The upper-left pane displays the work centers, whereas the upper-right pane displays the dispatched operations. The lower-left pane displays the pool of operations and the corresponding production orders. The lower-right pane displays nondispatched operations.

You can define the sort profile of the pane to enable the system to sort from the pool of orders/operations. The system uses the order start date in the standard setting for sorting. However, the sort sequence gets disturbed when you deallocate the operations because the system adds new objects to the end of the list. You can choose EDIT • SORT NEW to reestablish the previous sort sequence.

You can modify the *time-based scaling* of the planning table by choosing SETTINGS • SCALE or by pressing the [Ctrl] key and simultaneously dragging a section of the screen.

The search function is helpful in navigating through the planning table, especially in finding relevant information where there is a large amount of data or information available. You can call the search window via EDIT • SEARCH or by pressing the key combination [Ctrl]+[F]. Then you must enter the production order number in this window (see Figure 14.34).

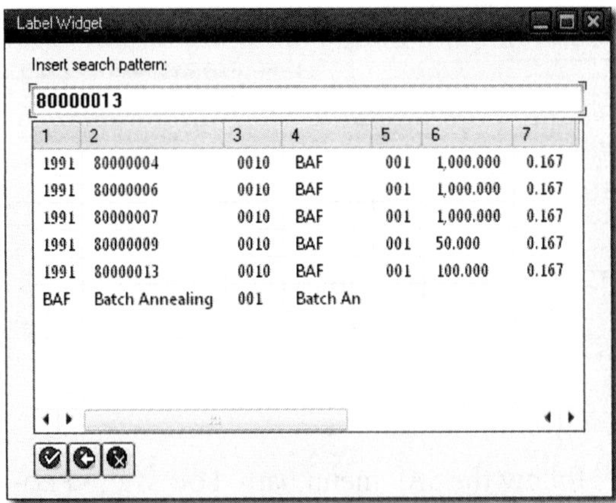

Figure 14.34 Search in Capacity Planning Table

The system facilitates your navigation in the capacity planning table by highlighting objects that belong together. Let's say you want the system to highlight the orders that belong to the same material (which in this example is 1991). To do this, choose EDIT • HIGHLIGHT • OBJECTS THAT BELONG TOGETHER (see Figure 14.35). The lower-right pane now shows all of the objects belonging together have turned dark-green.

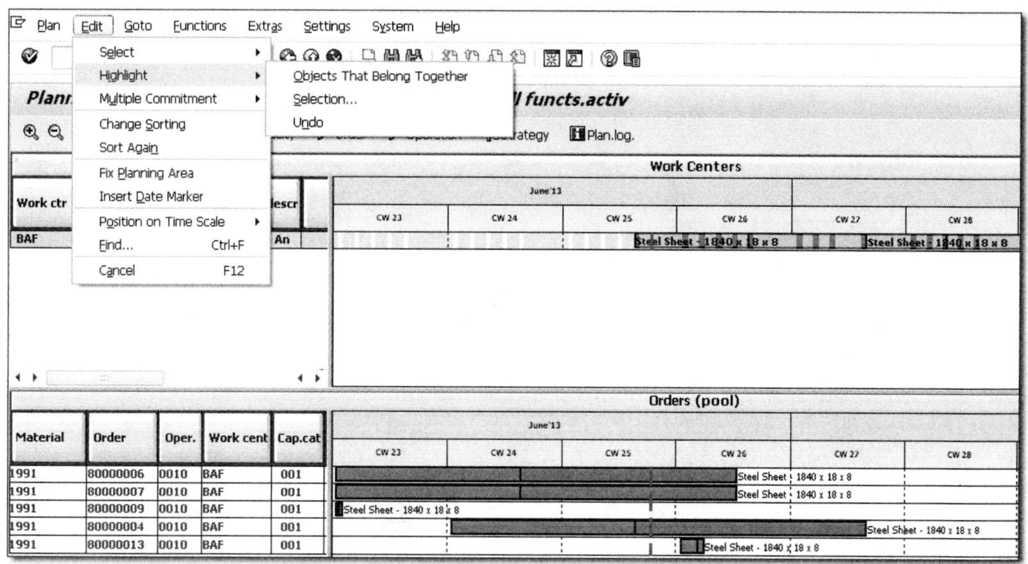

Figure 14.35 Highlighting All Objects Belonging Together

You can also highlight the objects that belong together with other criteria, such as order start date, target quantity, or planner group. To do this, choose EDIT • HIGHLIGHT • SELECTION. In the resulting screen, you can check the selection field you want to group and enter the field value that matches your purpose. Choose the CONTINUE icon, and you'll see the groupings, again in dark green.

14.5.1 Dispatch Operations

In the capacity planning table, there are two ways to dispatch operations. You can call the dispatch function by clicking on the relevant icon from the menu bar, or you can drag and drop the operation using the mouse. In both cases, you need to first select the operations that you want to dispatch from the pool of orders/operations. The screen shown in Figure 14.36 displays the dispatch using the dispatch function in the menu bar ❶. From the lower-right pane, select the operation that you want to dispatch, and then click on the DISPATCH icon, and the system moves it to the upper-right pane ❷.

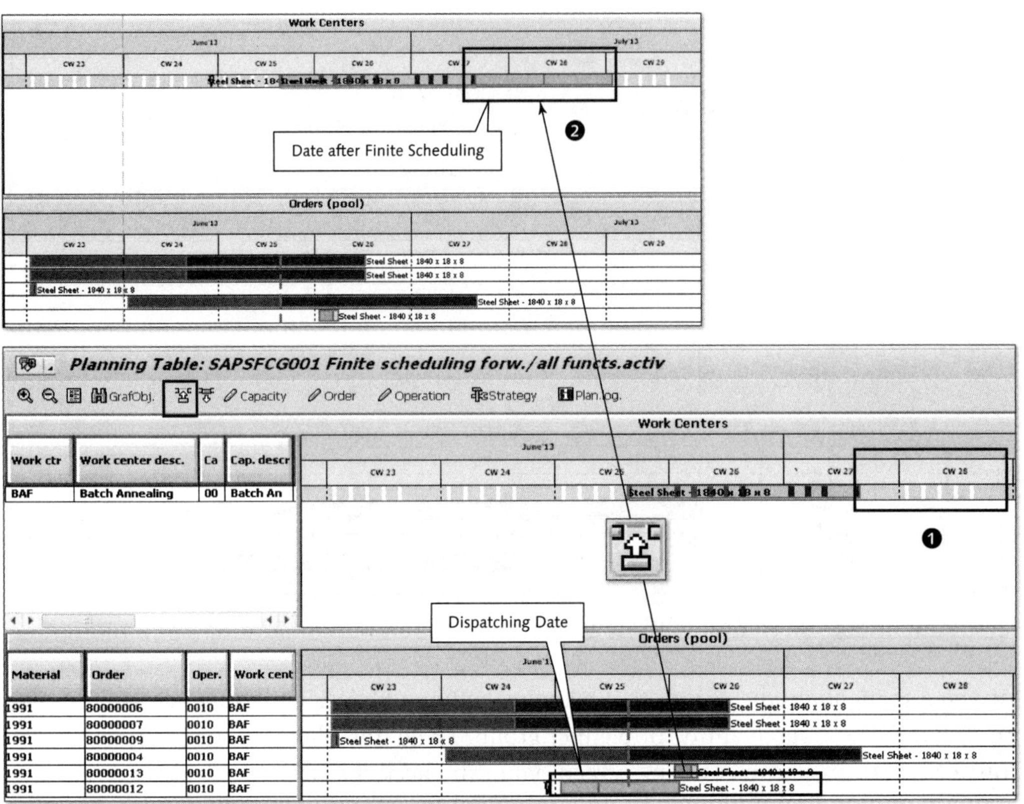

Figure 14.36 Dispatching in the Capacity Planning Table

The dispatch date is the scheduled operation date, which by default is the latest start date. If capacity is available on that date, the system dispatches the operation on the dispatch date. If capacity is unavailable on the dispatch date, finite scheduling determines the next possible date on which capacity is available, and the system then accordingly dispatches the operation on that date. If you use the drag-and-drop method for dispatching the operation, the "drop" date represents the dispatch date.

14.5.2 Deallocate

The *deallocation* process is similar to the dispatching process. You can deallocate either through the DEALLOCATION icon in the menu bar or by using drag and drop in the pool of orders/operations pane.

14.5.3 Options in the Graphical Planning Table

We've already covered in the sequence-dependent setup that the system requires specific settings in the strategy profile to enable it to plan effectively in the capacity planning table. The capacity planning board allows you to change the strategy profile when you click on the STRATEGY icon. You also can make temporary changes to the selected strategy profile. These changes then apply to the dispatches that you carry out immediately after that, but the system won't store these settings when you leave the capacity planning table.

If the available capacity is insufficient, you can increase it from within the capacity planning table, for example, by adding another shift or an interval. You can adjust the available capacity from the capacity planning table by choosing GOTO • CAPACITY • CHANGE. Similarly, you can also make changes to the production order or to the operations from within the capacity planning table by clicking on the relevant icons available in the menu bar.

You can display the planned increase in material stock in the capacity planning table. To display the MATERIAL STOCK pane for selected materials, choose EXTRAS • MATERIAL STOCK • SHOW. As shown in Figure 14.37, this pane appears toward the bottom of the capacity planning table.

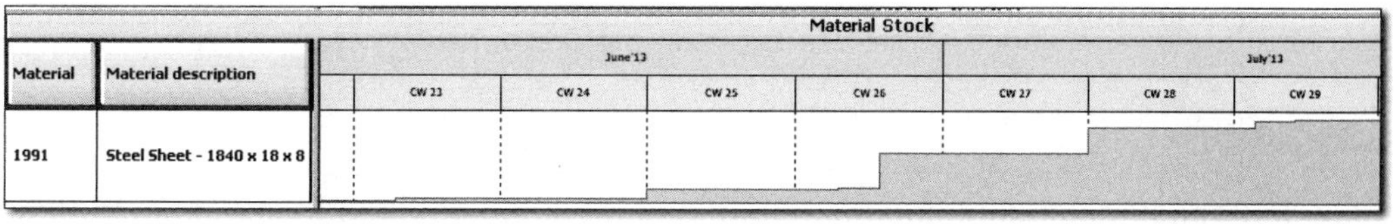

Figure 14.37 Material Stock Pane

Depending on how you've set up the capacity master data, several operations can occupy a work center at the same time. In such cases, a green line below the operations indicates a multiple commitment. You can choose EDIT • MULTIPLE COMMITMENT • SHOW to display the multiple commitments of the selected work centers.

[»]

Note

The system enables you to leverage the objects and layouts for display in the capacity planning table. We encourage you to explore these options, especially from the SETTINGS menu. Here you'll find options for SCALE, INCREMENT, SHIFT CALENDAR, and TIME LINE, among others.

In addition to the graphical capacity planning table, the system also provides a tabular capacity planning table that contains nearly all functions of the capacity planning table. The only exceptions are the functions for sequence-dependent setup and the associated settings for the planning strategy. Figure 14.38 displays the initial screen of the tabular capacity planning table.

To access the tabular capacity planning table, use the SAP menu path, LOGISTICS • PRODUCTION • CAPACITY PLANNING • LEVELING • WORK CENTER VIEW • PLANNING TABLE (TABULAR), or use Transaction CM22.

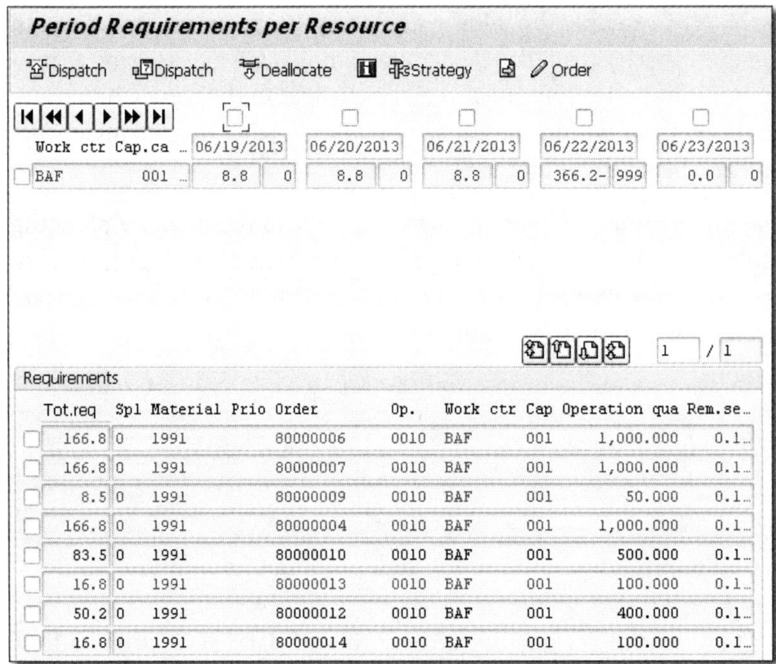

Tot.req	Spl	Material	Prio	Order	Op.	Work ctr	Cap	Operation qua	Rem.se...
166.8	0	1991		80000006	0010	BAF	001	1,000.000	0.1...
166.8	0	1991		80000007	0010	BAF	001	1,000.000	0.1...
8.5	0	1991		80000009	0010	BAF	001	50.000	0.1...
166.8	0	1991		80000004	0010	BAF	001	1,000.000	0.1...
83.5	0	1991		80000010	0010	BAF	001	500.000	0.1...
16.8	0	1991		80000013	0010	BAF	001	100.000	0.1...
50.2	0	1991		80000012	0010	BAF	001	400.000	0.1...
16.8	0	1991		80000014	0010	BAF	001	100.000	0.1...

Figure 14.38 Tabular Capacity Planning Table

Due to the extensive similarities between the capacity planning tables in graphical form and tabular form and now that you've learned about all of the properties and functions, you can easily conduct your planning process in the tabular capacity planning table, if necessary.

14.6 Summary

CRP helps the planner ensure that the system helps to make important business decisions, from dispatching the right operation at the right work center to evaluating the possibilities of increasing a shift or an interval. You can also ensure that the system performs finite scheduling if there is a capacity shortage or you can perform midpoint scheduling to account for on-ground production and capacities realities.

The next chapter discusses the classification system.

The classification system helps to search for information in the system based on user-defined parameters and criteria. The standard features and options available are sufficient to set up the classification system in almost no time with hardly any configuration. The classification system is also used across SAP's logistics components.

15 Classification

Classification finds extensive usage and applications in all of the logistics components of the SAP ERP system. It offers various options to categorize and organize the master data for quick and easy search, whether it's a material master, equipment, Document Management System (DMS) document, batch, or inspection method in the SAP ERP Quality Management (QM) component.

You can independently implement and integrate the classification system with your existing business processes at any time deemed necessary. For example, let's say your company already runs the SAP ERP system and has massive master data that it wants to better categorize and organize for easy and quick search. Another example might be that you have materials for which there's no provision in the material master to enter the material's attribute/characteristic details. For example, in the steel sheets rerolling industry, there are six specific quantity and qualitative attributes based on which the company produces several variations of materials. There's no provision to enter these specific attributes in the material master, but with the classification system, this is possible.

You can implement the classification system as a standalone and an independent project. Integrating the classification system with the existing business processes consists of three main activities:

- Creating characteristics
- Creating classes and assigning characteristics
- Assigning classes to the relevant master data

[+] **Tips & Tricks**

The standard SAP ERP system offers the provision to use two separate batch input programs to upload large numbers of characteristics, create classes, and assign characteristics to them.

For an efficient upload activity, we suggest that you first use the batch input to upload characteristics. Use the second batch input program to not only create new classes but also assign the previously created characteristics in the same step.

In this chapter, we show you how to create a characteristic, create a class, assign already-created characteristics to the class, and finally assign the class to the material master. We also show you how the assigned value in the material master is eventually used in the search function. The second example consists of equipment to which the relevant class is assigned.

We also briefly cover how the classification system plays a central role in the Process Management subcomponent of SAP ERP Production Planning for Process Industries (PP-PI).

15.1 Classification System

The classification system is primarily comprised of classes and characteristics. A *characteristic* refers to the master data used to capture the property or attribute of an object for easy identification in searches. For example, a material master can be identified by its physical properties such as density, viscosity or thickness, or chemical properties, including flammability or corrosiveness.

The following steps are involved in integrating the classification system in the material master:

1. Create a characteristic for each property.
2. Create a class with class type 001, which is specifically used for the material master.
3. Link the characteristics and the class.
4. Assign the class to the material master.

[«]

Note

Exactly the same steps are involved if you want to use the classification system for other objects, such as equipment or functional locations.

15.1.1 Characteristics

Characteristics are defined as attributes and properties of an object, both qualitative and quantitative. For example, a qualitative characteristic color may have three values: red, green, and yellow. Similarly, a quantitative characteristic may be the pH of an acidic solution in the range of 1 – 6, while the pressure in the vessel is denoted in the range of 100 – 125 bar (a unit of measure defined for the characteristic pressure). For a quantitative characteristic, you define the format as numeric and specify the number of decimal places as well as the unit of measure, if applicable. Following is an example of how to create a quantitative characteristic.

Follow the menu path, CROSS-APPLICATIONS COMPONENT • CLASSIFICATION SYSTEM • MASTER DATA • CHARACTERISTICS, or use Transaction CT04. Enter the name of the characteristic you want to create, which in this example is "C002", and then choose the CREATE icon.

In the next screen, information related to the characteristics is defined such as description, data type, number of characteristics, decimal places (if any), and unit of measure. Table 15.1 lists and describes the important fields that you'll see.

Field Name	Field Description and Value
DESCRIPTION	Description of the characteristic, for example, pump capacity.
CHARS GROUP	Groups together similar characteristics in one group. For PP-PI, you have to assign the characteristic into a characteristic group, as either process instruction characteristics or process message characteristics. If not done, the system will issue a warning message during Process Management settings and configuration.
STATUS	Status of the characteristics, for example, RELEASED, to indicate it's ready for use.

Table 15.1 Details of Individual Fields of a Characteristic

Field Name	Field Description and Value
DATA TYPE	Indicates the type of data this characteristic has, for example, date, time format, or numeric format.
NUMBER OF CHARS	Indicates the number of characters a value of this characteristic can contain.
DECIMAL PLACES	Indicates the number of decimal places if the format is numeric.
UNIT OF MEASURE	Unit of measure of the characteristic (quantitative/numeric), if any.
SINGLE-VALUE/ MULTIPLE VALUES (radio button)	Option in case a single value or multiple values are allowed for a characteristic. An example of multiple values usage is when you can select a surface finish as coarse, pitted, and uneven. For multiple values, the system offers a checkbox for each value to select from the dropdown list.
NEGATIVE VALS ALLOWED (checkbox)	Option if negative values are allowed. You can't enter a negative value for a characteristic if you don't select the checkbox.
ENTRY REQUIRED (checkbox)	Option if the user must enter a value for this characteristic. If it's not selected, then entering a value is optional.

Table 15.1 Details of Individual Fields of a Characteristic (Cont.)

You'll then switch over to the VALUES tab and enter "10 m3", "20 m3", and "30 m3" under the CHAR. VALUE column as shown in Figure 15.1. These refer to the pump's capacity values, with m^3 being the unit of measure of capacity, which in this case is a cubic meter. The characteristic provides an option to record the pump value, which can then be searched in the material master along with any other characteristic. Select the ADDITIONAL VALUES checkbox to ensure that values other than 10, 20, or 30 are also allowed for entry as an input value.

Now the characteristic C002 has three values available to choose from: 10 M3, 20 M3, and 30 M3. You can select the DEFAULT VALUES checkbox (the D column) on any specific value, which will then automatically appear when the characteristic is used and therefore reduce the data entry efforts. However, you can overwrite the default value if necessary.

Click on the ADDNL DATA tab to see the options shown in Figure 15.2.

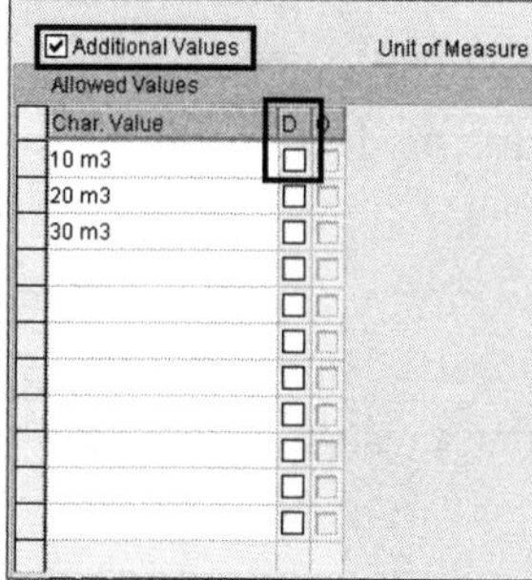

Figure 15.1 Creation of Characteristics

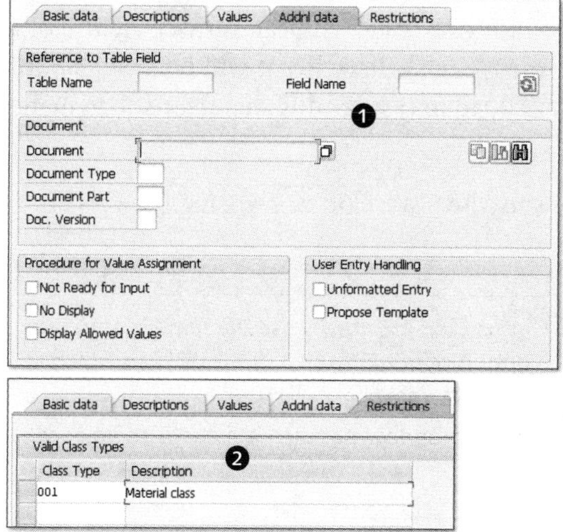

Figure 15.2 Additional Data and Restrictions of Characteristics

If applicable, you can enter the relevant DMS documents number, document type, document part, and document version ❶. Following are the details of the various checkboxes and their functions:

- Not Ready for Input
  You can't enter a value in this characteristic. In other words, a preassigned default or calculated value automatically fills the characteristic field, and the user has no option to make an entry.
- No Display
  The system won't display a characteristic.

- Display Allowed Values
  If several values are already available from the Values tab of the characteristic, the system displays those values to facilitate the user during data entry.
- Unformatted Entry
  If this checkbox is selected, the system makes the provision to set the characteristic's field length to a maximum of 30 so the user can make multiple entries, separated by colons. If not selected, then the system displays the exact length of the characteristic for data entry.
- Propose Template
  If selected, the system proposes or displays the template to facilitate the user during data entry. Click back to the Basic data tab to view the template details.

Finally, click on the Restrictions tab in Figure 15.2. You can restrict the use of a characteristic to a specific class type (or types); otherwise, it will be available for use in all class types. Enter Class Type "001" to restrict this characteristic to material class only ❷.

Save the characteristic C002 by choosing the Save icon or pressing [Ctrl]+[S].

[»]

Note

If you create a characteristic with a date format and eventually use the same in any class, the system will *not* provide the standard option of bringing up a calendar from which to choose a date. The user has to manually enter the date.

If you want to have the calendar option in the dropdown of the date characteristic, you need to engage the assistance of an ABAP consultant.

15.1.2 Create a Class and Assign Characteristics

A class is also master data of the classification system. You assign characteristics in classes. A class is mainly identified by its class type and acts as a controlling function. Different class types are available in the SAP system to achieve different purposes. Some of the available class types are listed here:

- Class type 001 for material master
- Class type 002 for equipment master
- Class type 022 for Batch Management (BM)

- Class type 300 for variant classification
- Class type 017 for DMS
- Class type 019 for work center

To create a class, follow the menu path, CENTRAL FUNCTION • CLASSIFICATION • MASTER DATA • CLASSES, or use Transaction CL02 (Figure 15.3 ❶).

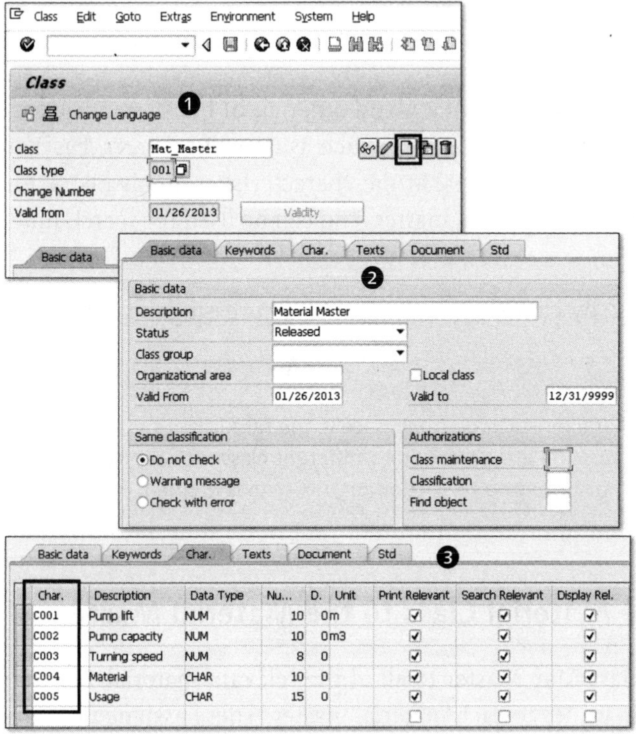

Figure 15.3 Creation of Class and Characteristics Assignment

For this example, enter the CLASS as "Mat_Master" and the CLASS TYPE as "001" ❶ to indicate that it's a material master class (class type 001). Click the CREATE icon in the icon bar to the right to create the class and go to the BASIC DATA tab ❷. Table 15.2 defines some of the fields.

Field Name	Field Description and Value
DESCRIPTION	Short description of the class
STATUS	Available statuses to choose from, for example, RELEASED status makes the class available
VALID FROM	The date when the class becomes available for use
VALID TO	The date until which you can use the class

Table 15.2 Field Names and Descriptions for the Basic Data Tab

Next, click the CHAR. (for characteristics) tab, and enter all of the previously created characteristics. Select all of the checkboxes such as PRINT RELEVANT, DISPLAY REL., and SEARCH RELEVANT ❸ to ensure that the characteristics are available for printing, can be displayed in the material master, and can be used for search functions, respectively.

Save the class MAT_MASTER by choosing the SAVE icon or pressing [Ctrl]+[S].

[+]

Tips & Tricks

You can't assign the same characteristic more than once in the same class and class type. You can, however, assign the same characteristic to different classes of the same class type or different class types, provided no restriction on its usage is defined.

15.2 Assigning the Material Class to the Material Master

You can now assign the class Mat_Master to all of the relevant materials, as well as assign the characteristic values to each material master. These assigned characteristic values in material masters will then be used in search functions.

To assign a class to the material master, follow the menu path, LOGISTICS • MATERIALS MANAGEMENT • MATERIAL MASTER • MATERIAL • CREATE (GENERAL) • IMMEDIATELY, or use Transaction MM01. Enter the MATERIAL number "CH-3000" (Figure 15.4 ❶), and press [Enter] to bring up the SELECT VIEWS of the material master ❷. Select the CLASSIFICATION view, and press [Enter] again, which brings up the CLASS TYPE pop-up window ❸. Select MATERIAL CLASS (class type 001), and press [Enter].

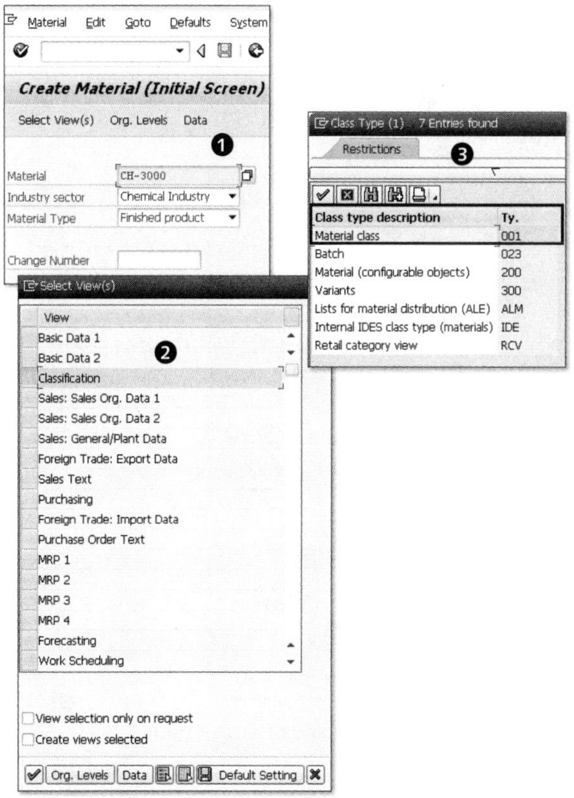

Figure 15.4 Assignment of a Material Class to the Material Master CH-3000

[«]

Note

If the CLASSIFICATION view of the material master already exists in the system, use Transaction MM02 to proceed further with assigning a class to the material master.

The CLASSIFICATION view of material master CH-3000 appears as shown in Figure 15.5. In the CLASS column, enter "mat_master" ❶, and press Enter to see the five characteristics previously assigned to the class ❷. You can now enter the relevant values for each of the characteristics as shown in Figure 15.6.

In Figure 15.6, you enter all of the five characteristic values of the pump, which you can then use in the search functions. Save the material master by clicking on the SAVE icon or pressing Ctrl+S.

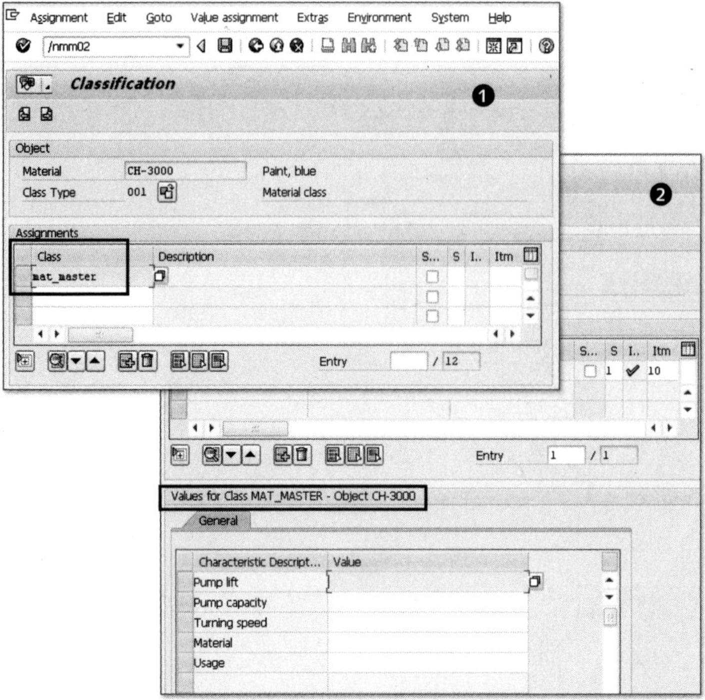

Figure 15.5 Material Class Mat_Master in the Material Master and Characteristics

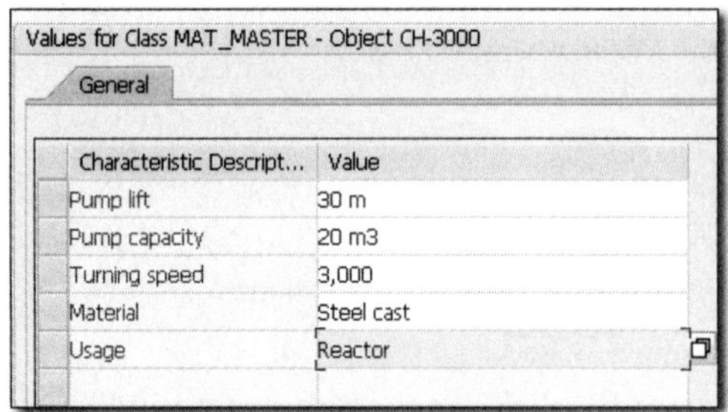

Figure 15.6 Characteristic Values Entered for Material Master CH-3000

[»]

Note

You also use the classification system when you want to integrate Variant Configuration in your business processes. You can refer to the book *Variant Configuration with SAP* (2nd ed., SAP PRESS, 2012) at *www.sap-press.com/2889*.

Advance functions in the classification system such as precondition, action, and selection condition are known as *object dependencies*. Object dependencies enable the user to maintain logical dependencies and grouping of various characteristics with each other. For example, when you place an order for a new Ford Mustang in black, you can only choose gray or black seat covers. With a red Ford Mustang, the seat cover options offered are different.

15.3 Finding Objects in Classes

Now that the characteristics values are already assigned to respective material masters, let's proceed with taking one characteristic, pump capacity, and search for all of the materials that contain a specific characteristic value of 20m3.

To find objects in classes, follow the menu path, CROSS-APPLICATION COMPONENTS • CLASSIFICATION SYSTEM • FIND • FIND OBJECTS IN CLASSES, or use Transaction CL30N (Figure 15.7).

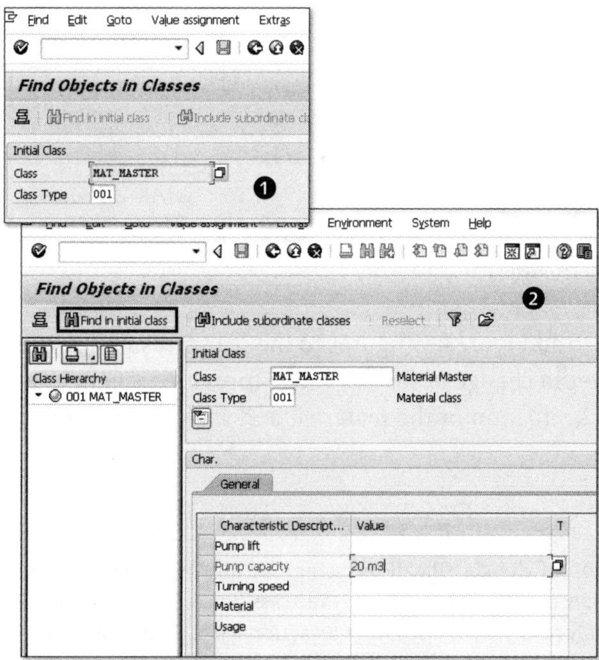

Figure 15.7 Characteristic Search for Pump Capacity

Enter the CLASS "MAT_MASTER" and CLASS TYPE as "001" ❶, and press Enter to bring up the detailed view of all of the available characteristics to use in the search function ❷. Enter "20 m3" as the characteristic VALUE for the characteristic PUMP CAPACITY, and click on the FIND IN INITIAL CLASS button.

The bottom half of Figure 15.8 shows an example of the three results that match the search criteria. Each of the three materials contains the characteristic value of 20 M3. The pump capacity is shown in the third column from the left with the column title PUMP CAPACITY.

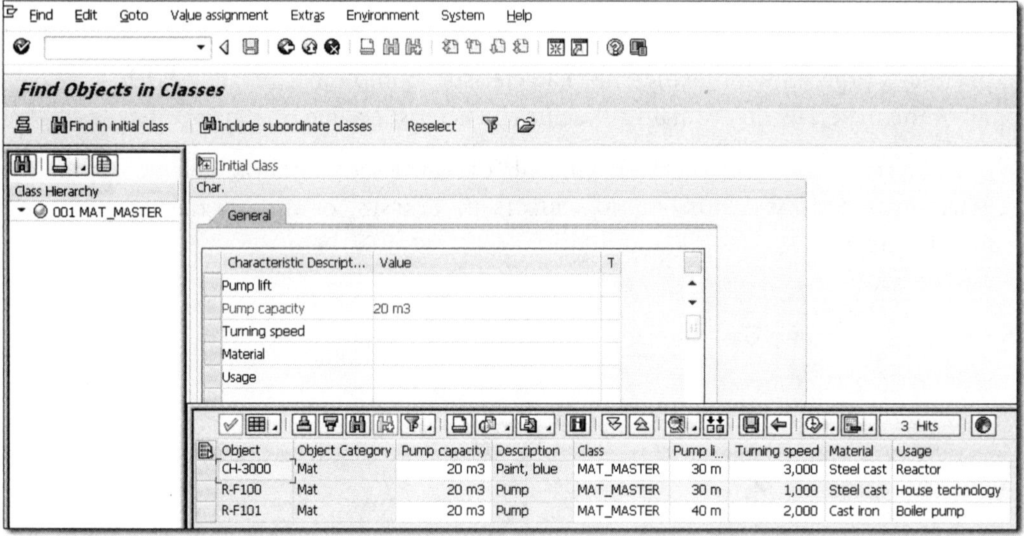

Figure 15.8 Search Results for Characteristic Pump Capacity

Several options are available to add more key figures to the search results, directly navigate to the material master in change or display mode (provided you have the authorization), view the stock situation of the material, and save the current layout as the default settings.

[+]

Tips & Tricks

You can use the search function of the classification system for any of the classes and class types using the same Transaction CL30N. With the classification system implemented, you can search the material by class also. In Transaction MM03 (Display Material), place the cursor on the MATERIAL field, press F4 (or use the dropdown), and then select the tab or option to search MATERIALS FOR CLASS. This will take you back to the screen shown earlier in Figure 15.7 in which you can perform the same activities in that screen and the screen shown in Figure 15.8.

Figure 15.9 displays the option available to search MATERIAL FOR CLASS, for example, when you want to search a material master (Transaction MM02 for change).

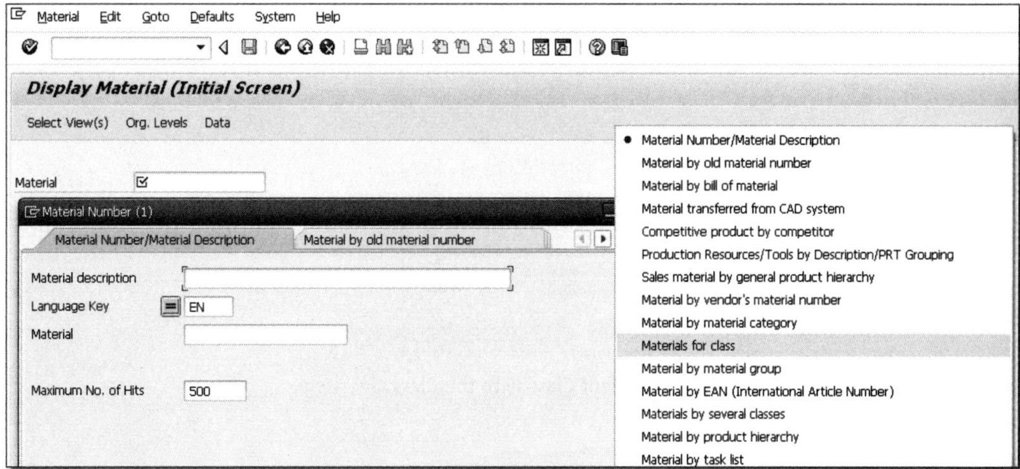

Figure 15.9 Option to Search for Material for Class

15.4 Assigning an Equipment Class to Equipment

In the second example, by using Transaction CL02, you can create another class named "Equip" with class type 002 and assign all of the previously created five characteristics. Class type 002 is specific to equipment.

Next, you create the equipment class by using Transaction IE01 and assigning the previously created class EQUIP. In Figure 15.10 ❷, click on the CLASS OVERVIEW button. Then, enter the EQUIPMENT class "Equip" ❶, and press Enter.

Figure 15.11 ❷ shows all of the characteristics assigned to the class EQUIP and CLASS TYPE 002 ❶. Enter the values as shown in Figure 15.11 ❷, and save the equipment.

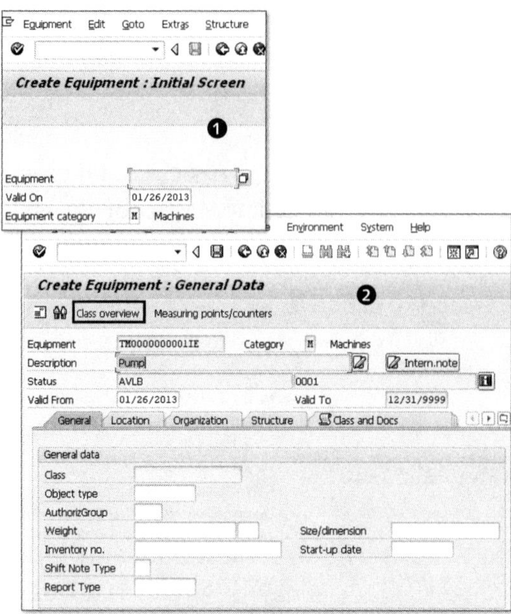

Figure 15.10 Creating an Equipment Class with the Class Overview

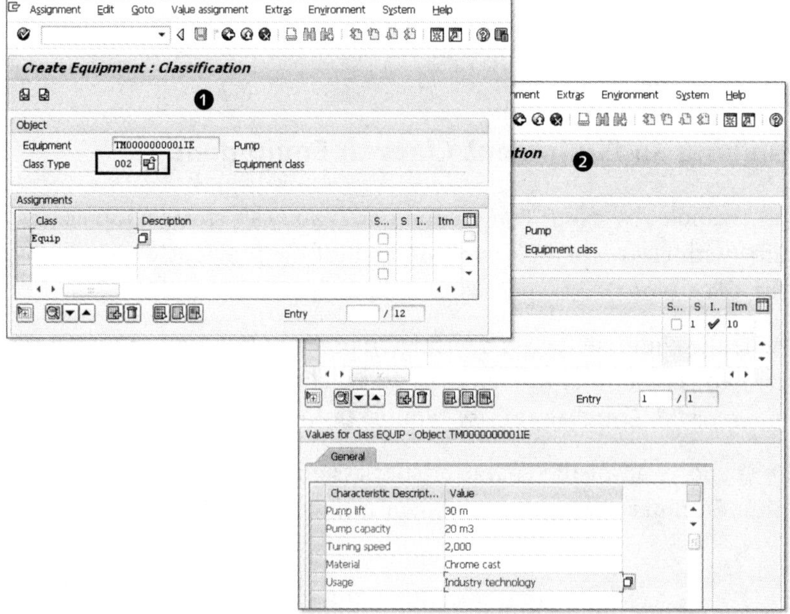

Figure 15.11 Class Equip of Class Type 002 Assigned to Equipment

Using Transaction IE03 (Display Equipment), place the cursor on the EQUIPMENT field and press F4, or click on the dropdown option to see that one of the options in which you can search for equipment by classification (see Figure 15.12).

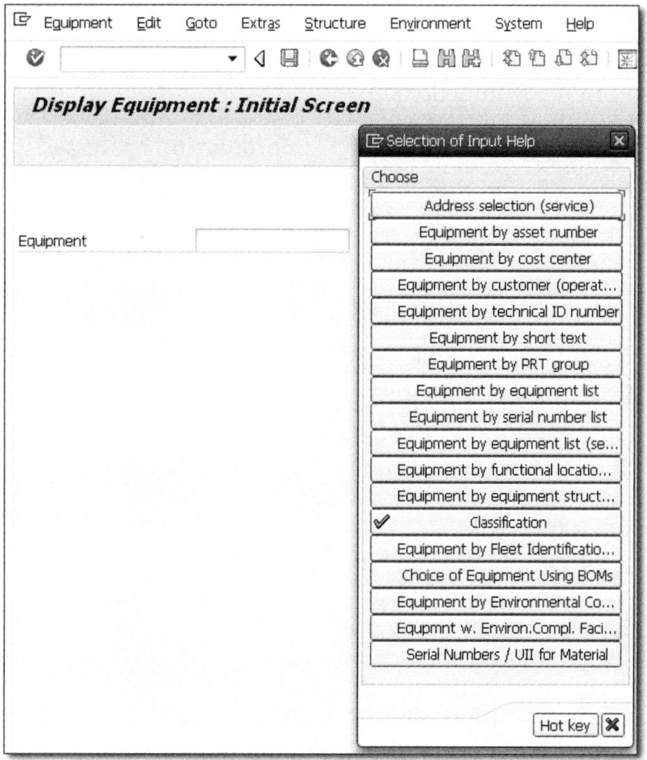

Figure 15.12 Option to Search for Equipment by Classification

Note [«]

You can browse the CLASSIFICATION menu in the SAP ERP system to evaluate which other search and reporting functions can benefit your business processes. For example, you can use the reporting option WHERE-USED for characteristics to find all of the classes in which a specific characteristic is in use. This option helps when, for example, you want to replace an old characteristic with a new one and must know the classes in which it's presently in use.

The menu path is CROSS-APPLICATIONS COMPONENT • CLASSIFICATION SYSTEM • FIND CROSS-APPLICATIONS COMPONENT • CLASSIFICATION SYSTEM • REPORTING.

15.5 Summary

This chapter shows how quick and easy it is to set up the classification system in any business processes of a company. Implementing the classification system serves two important purposes. It provides better categorization and cataloging of objects and, at the same time, provides the user with the options to use the same in search functions. The classification system isn't just restricted to the PP component and finds usage even in the BM functionality of the MM component.

The next chapter covers co-products and by-products in production processes.

Companies often produce auxiliary products in addition to the main material that's made during the production process of semifinished or finished goods. This chapter explains co-products and by-products side by side.

16 Co-Products and By-Products in Production Processes

Some companies produce additional products in addition to their main product range, some of which hold significant financial value. These additional products are either classified as a co-product or a by-product. While co-products are desirable due to the fact they can be used in other production processes or even sold as standalone offerings, by-products have significantly lower financial value. Generally, the co-product option offers better controls. Therefore, if the material is financially significant, declaring the material a co-product offers more checks and balances as compared to the by-product option.

Let's get a better understanding of what these different products mean with some examples. In the petrochemical industry, when a company produces goods from its main input, which is crude oil, it can also generate co-products, including petrol, diesel, kerosene, and naphthalene. A co-product entails joint production, and the cost of production can be apportioned between the main product and co-products.

Examples of by-products are the metal strips produced during the rerolling mill process in the steel industry or ethanol as a by-product of the sugar industry, both of which have insignificant commercial value.

Table 16.1 shows some of the key differentiators between a co-product and a by-product to help you decide how to treat these additional products in your SAP ERP system. It shows the steps involved in declaring a material as co-product or by-product and the associated controls available. When you perform goods receipt of a co-product or a by-product, the system reflects this inventory movement (goods receipt) with the relevant movement type.

Serial Number	Key Differentiator	Co-Product	By-Product
01	Significant financial value to the company	X	
02	Can be planned in production processes	X	
03	Setup of costing as joint production	X	
04	Ease of master data setup in the SAP ERP system		X
05	MATERIAL MASTER checkbox in the SAP ERP system	X	
06	BILL OF MATERIALS (BOM) checkbox	X	
07	Joint production setup/appropriation structure	X	
08	Negative quantity in BOM	X	X
09	Movement type to record goods receipt against order	101	531

Table 16.1 Key Differentiators of a Co-Product with a By-Product

[»] **Note**

Usually a company's Product Cost Controlling (CO-PC) team declares a material as either a co-product or a by-product. The primary responsibility of implementing it ends up with SAP ERP Production Planning (PP) and SAP ERP Materials Management (MM) teams in coordination with the CO-PC team. Co-products and by-products are each handled differently in an SAP ERP system.

In this chapter, we present side-by-side descriptions of a co-product and a by-product to show their similarities and differences so that you can see which option best suits your business requirements. The descriptions include details such as the required material master and BOM master data. Sample process orders show the impact and behavior, including costing when there is a co-product or by-product. Subsequent production transactions include goods issuance against the process order, confirmation, and finally goods receipt for a process order. We also cover some of the standard material document information available in a process order along with planned versus actual costs in analyses and itemized forms.

[«]

Note

Although this chapter covers the production cycle of process industries, the same concepts and functionalities apply to discrete manufacturing.

We now cover the master data that you need to set up co-products and by-products in your SAP ERP system.

16.1 Check in Material Master

To declare a material as a co-product, you set the Co-product flag in the material master of all materials that you want to designate as co-products, as well as in the BOM. All co-products include the main co-product (BOM header) and the co-products in the BOM items. There's no such requirement for a by-product.

16.1.1 Co-Product

To activate a co-product in the material master and to change the material master, follow the menu path, Logistics • Materials Management • Material Master • Material • Change • Immediately, or use Transaction MM02. Then follow these steps:

1. Enter the Material number, and click the Select view(s) icon.
2. In the pop-up window, select the Costing views and choose Continue.
3. Select the Organizational levels icon. Select the Organizational level, that is, plant. For this example, enter Plant "Z001".

[+]

Tips & Tricks

You can also activate co-products in the MRP 2 view or the Costing view. Whichever view is activated first causes it to be shown in the other view in the material master.

4. Press [Enter] or click the ENTER icon. This opens the COSTING 1 view of the material master as shown in Figure 16.1. Select the CO-PRODUCT checkbox to declare that the material is a co-product.

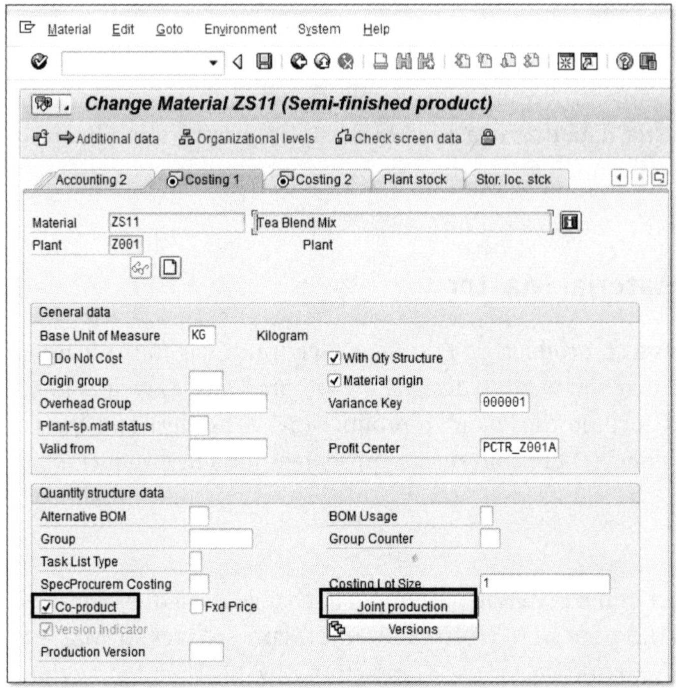

Figure 16.1 Costing View of the Material Master with the Co-Product Checkbox Flagged

5. Click the JOINT PRODUCTION button to open the screen shown in to Figure 16.2, which is the cost apportionment structure for co-products. When a process order or a production order is created in an SAP ERP system, the system generates an order item for each co-product. The actual costs for goods issues and confirmed internal activities are collected on the order header. Goods receipts for the co-products are entered with reference to the order item. At the end of the period, the actual costs incurred for the order are distributed to the co-products as equivalence numbers or as percentages.

6. Enter "40" percent as the equivalence number in the NO column, followed by a short DESCRIPTION ❶.

7. Select the apportion structure (STRUC.) as PI, which is for a process order. This means that 40% of the total cost of production of material ZF01 will be allocated to the material ZS11.

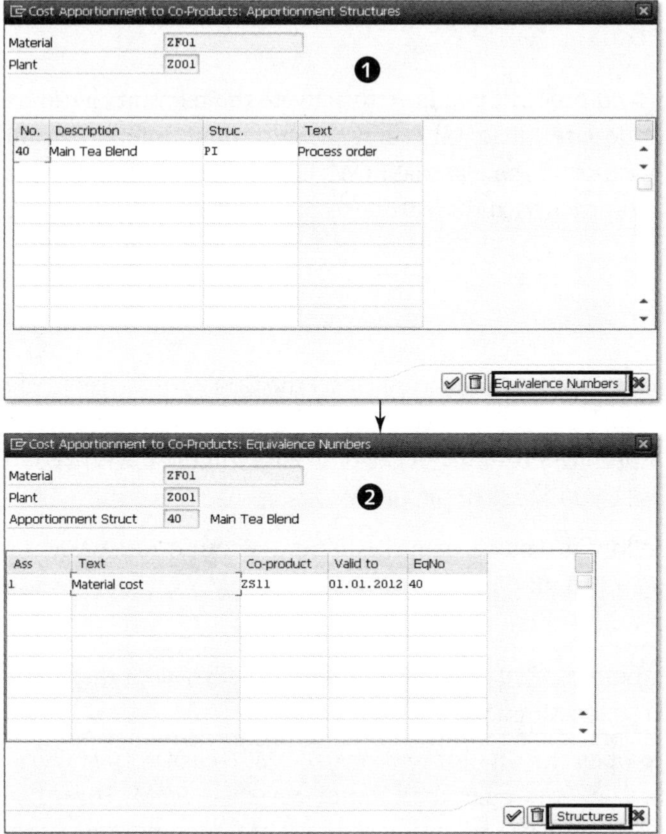

Figure 16.2 Cost Apportionment Structure to Co-Products and Equivalence Number

8. Keep the cursor in the first row of the table in the screen ❶, and click the EQUIVALENCE NUMBERS button. The screen that appears ❷ requires details of the assignment of the apportion structure, that is, the cost of the finished good (ZF01) that will be allocated to the co-product (ZS11). In this example, it's 40%.
9. Click CONTINUE, and save the material master.

16.1.2 By-Product

No such check or settings in the material master are needed for a by-product.

16.2 Bill of Materials

To declare a material as a co-product, you have to activate the relevant checkbox in the material BOM. To declare a material as a by-product, there is no provision or need to select any checkbox in the material BOM. The quantity defined in the material BOM for both co-products and by-products is negative, for example, –2 units.

16.2.1 Co-Product

If the material is marked as a co-product, you must set the Co-product flag in the BOM of the product in which the co-product exists. Both the main product in the BOM header and the co-products with the negative quantity in the BOM need to have the Co-product flag set in the material master.

To get to the screen, follow the menu path, Logistics • Production • Master Data • Bill of Material • Bill of Material • Material BOM • Change, or use Transaction CS02. Then follow these steps:

1. Enter the Material number, "ZF01" in this example, together with the Plant "Z001" and Alternative BOM usage "1" for production.
2. Click the Item icon to open the Change material BOM: General Item Overview screen, or press [Enter] to open the Change material BOM: Item: All data screen. You'll see the item overview of the BOM; enter the negative Quantity for Material ZS11, which is a co-product. Other components such as ZP01 and ZP01 are defined in positive quantities and are consumed during production.
3. Select the line item 0010, and click the Item icon or press [F7]. This leads to the screen shown in Figure 16.3. Select the Co-product checkbox to declare material ZS11 as a co-product. Save the BOM by pressing [Ctrl]+[S] or by clicking the Save icon.

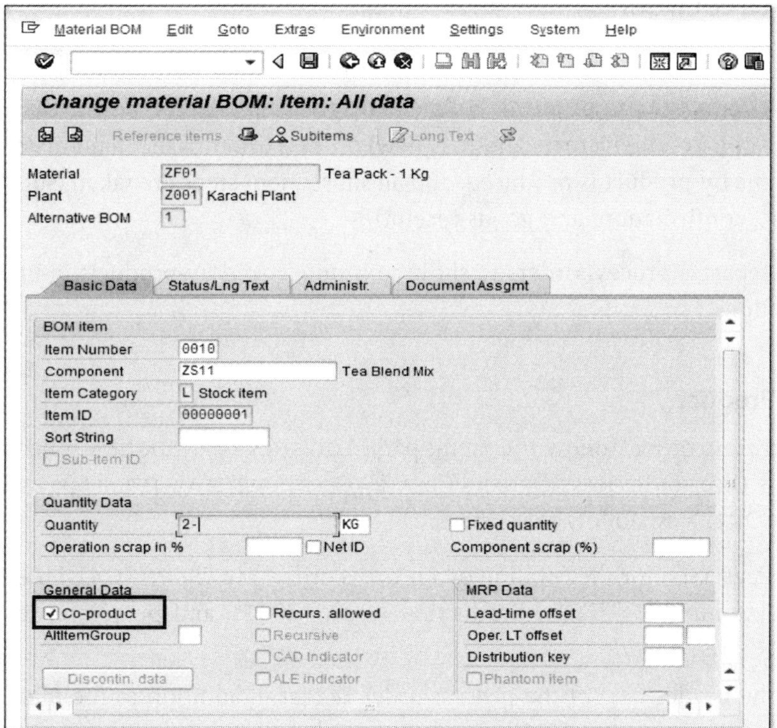

Figure 16.3 BOM for Material ZF01 with Co-Product ZS11

16.2.2 By-Product

For a by-product, just a quantity in the negative is sufficient in a BOM, as shown previously in Figure 16.3. The CO-PRODUCT checkbox in the BOM should not be checked.

This is all that is needed to set up the relevant master data for both co-products and by-products. The recipe is identical to recipes without co-products. However, to reflect how co-products and by-products are handled in transactional data, we show the entire process in the following section.

[«]

Note

For discrete manufacturing, the master recipe is replaced by a routing.

16.3 Process Order

For transactional data, you first create a process order of the main material in which the co-product is also produced followed by goods issuance, confirmation, and goods receipt. You then create another (new) process order of the main material in which the by-product is produced, and all subsequent steps are taken (such as goods issue, confirmation, and goods receipt).

You'll create separate process orders to show co-products and by-products in the following sections.

16.3.1 Co-Product

To create a process order, follow the menu path, LOGISTICS • PRODUCTION – PROCESS • PROCESS ORDER • PROCESS ORDER • CREATE • CREATE WITH MATERIAL, or use Transaction COR1. Now follow these steps:

1. On the initial screen for the creation of a process order (see Figure 16.4), enter the MATERIAL number "ZF01", production PLANT "Z001", and process order TYPE "TI01". Press [Enter].

   Notice in the header data of the process order ❶ that in addition to the main product ZF01, a co-product ZS01 is also being produced. Therefore, the system automatically selects the MULTIPLE ITEMS checkbox.
2. Click the MATERIALS button to see the screen showing the material list view of a process order ❷. Here, the co-products are automatically reflected by the relevant check as well as by the fact that the quantity of both the product (main product = item 0000) and co-product are shown in the negative. In a process order without a co-product, the main material isn't shown in the material list. The positive quantities are the consumption quantities of materials in the process order.
3. Click the RELEASE icon ❷, and save by pressing [Ctrl]+[S] or clicking the SAVE icon. The system generates a process order number.

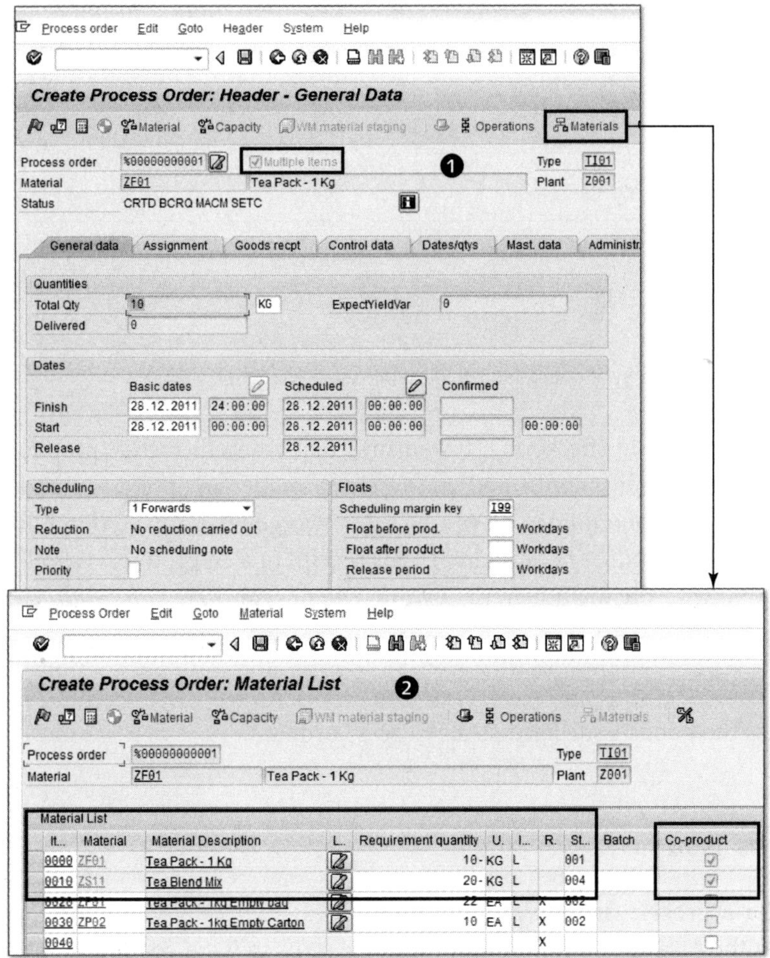

Figure 16.4 Process Order Creation with Co-Product

16.3.2 By-Product

In the case of a by-product, follow the same process to get to the screen shown in Figure 16.4. Then follow these steps:

1. Click the MATERIALS button in Figure 16.4 ❶, which takes you to the screen shown in Figure 16.5. For a by-product, the system shows just a negative quantity.

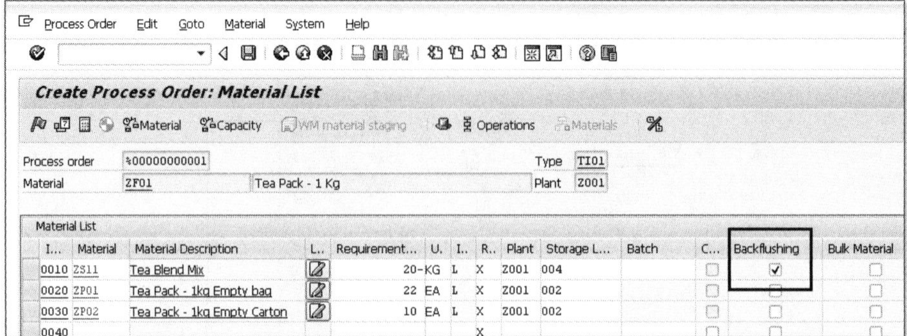

Figure 16.5 Material List with By-Product and Backflushing Selected

2. Select the BACKFLUSHING checkbox. This ensures that when the actual production of the process order is confirmed, the system provides an option to simultaneously record the amount of a by-product that was generated. You don't do the same for the co-product because the goods receipt of a co-product is simultaneously done with the main material by movement type 101.
3. Release the process order by clicking the RELEASE icon, and save the process order by pressing Ctrl+S.

The next step is the goods issue against the process order.

16.4 Goods Issue

Goods issue for a process order is the standard process whereby you issue the components of the process order. We'll go over the steps to issue goods for both a co-product and by-product in the following sections.

16.4.1 Co-Product

To issue goods against a process order for a co-product, use the menu path, LOGISTICS • MATERIALS MANAGEMENT • INVENTORY MANAGEMENT • GOODS MOVEMENT • GOODS MOVEMENT, or use Transaction MIGO. Then follow these steps:

1. In Figure 16.6, select GI FOR ORDER (goods issue for order), and enter the process order number.

2. Press [Enter], which brings up the components to issue. Check the items as OK, and save by pressing [Ctrl]+[S] or clicking the SAVE icon. The system then generates a material document. (A material document is an SAP ERP system document that captures all of the details of the goods movement.) The movement type for goods issuance against order is 261 as shown on the far right side of the screen.

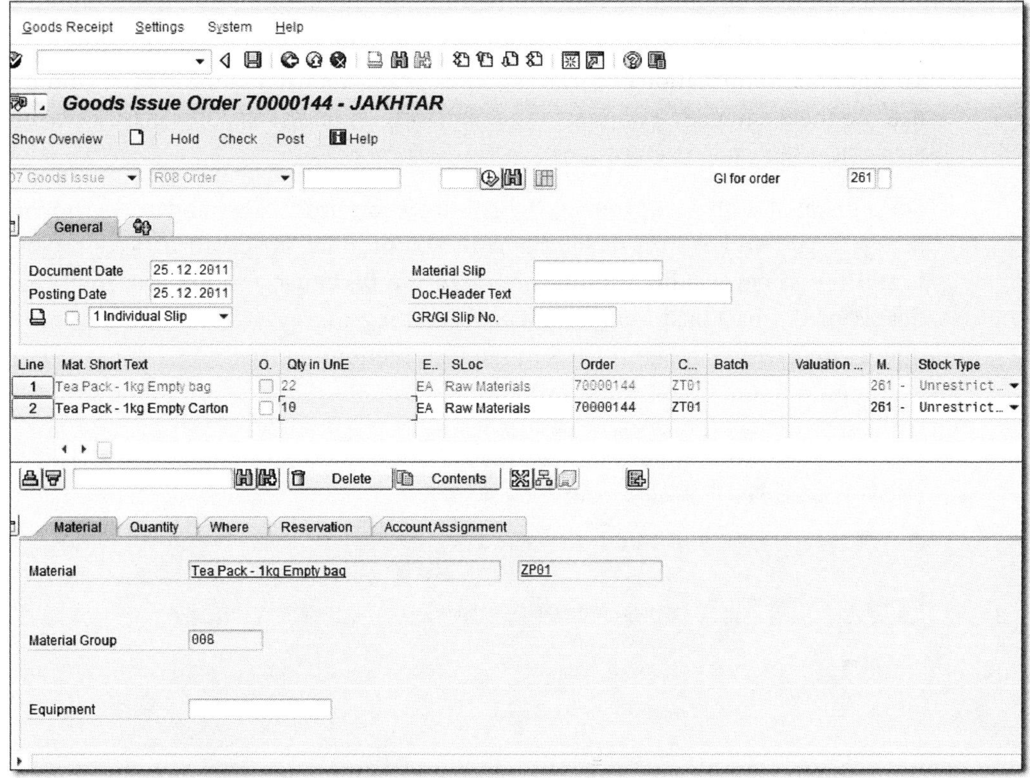

Figure 16.6 Goods Issue of a Process Order

16.4.2 By-Product

For a process order with a by-product, the process of goods issue is the same as shown in Figure 16.6.

16.5 Confirmation

Confirmation is the process used to declare the quantity against the process order that has been produced. No special process is needed for confirmation of a process order with a co-product, but a special process is needed for confirmation of a by-product.

16.5.1 Co-Product

To enter the confirmation of a process order of a main material that has a co-product, use the menu path, LOGISTICS • PRODUCTION – PROCESS • PROCESS ORDER • CONFIRMATION • ENTER FOR ORDER, or use Transaction CORK.

For a material with a co-product, the confirmation process is standard as shown in Figure 16.7 ❶. For a main material with a by-product, the confirmation process is still the same, but because we activated the backflush for a by-product, the details of the backflush, which entails recording the by-product quantity produced, are covered in Figure 16.8 in the next section.

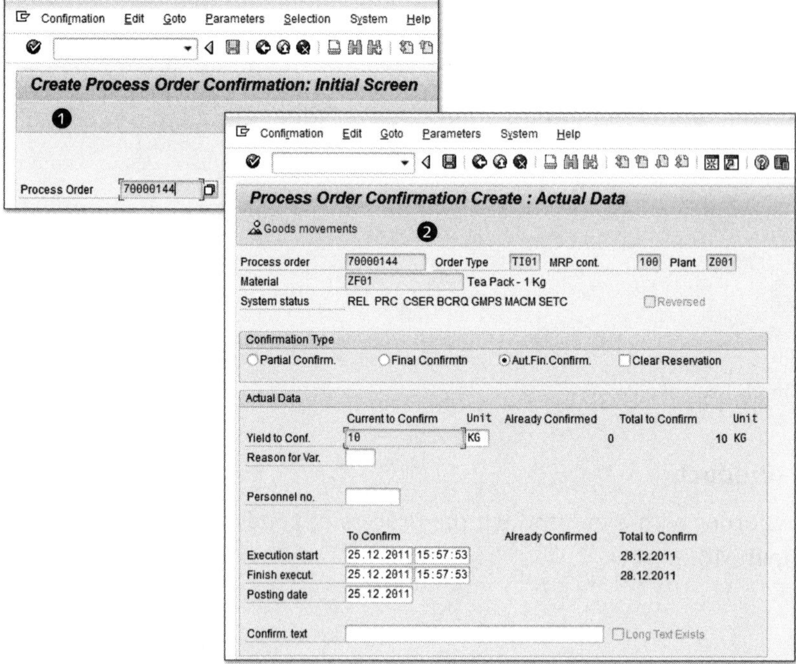

Figure 16.7 Confirmation of Process Order

In the initial screen for the process order confirmation ❶, enter the PROCESS ORDER number and press Enter.

Figure 16.7 shows the process order confirmation screen in which you enter the quantity of goods produced ❷. For this example, enter a quantity of "10" to show that the yield or production quantity of the process order was 10 units. Save the confirmation by pressing Ctrl+S or clicking the SAVE icon.

16.5.2 By-Product

For a process order with a by-product, the process of confirmation is standard. Click the GOODS MOVEMENTS Figure 16.8 button ❶ to go to the GOODS MOVEMENTS screen of a process order ❷, in which you can enter the quantity of the by-product generated. Also, notice that the by-product is denoted by MOVEMENT TYPE 531. Save the confirmation by pressing Ctrl+S or clicking the SAVE icon.

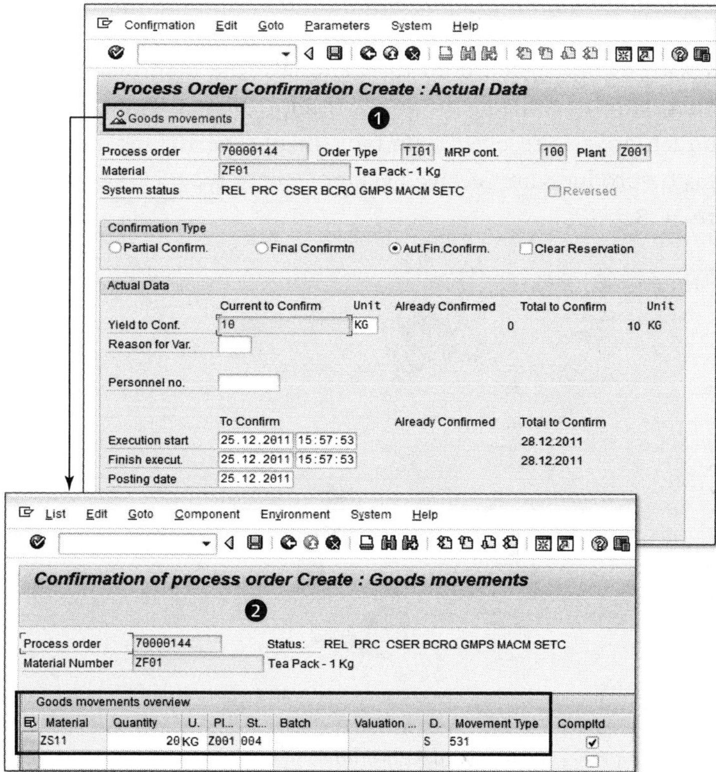

Figure 16.8 Goods Movement of a Process Order with the Backflush Functionality

The goods movement (goods issue at this stage) is made possible by the backflush functionality. When a process order is being confirmed, goods issue (through backflushing) is also performed simultaneously. Because this is a by-product with movement type 531, during the actual confirmation process of main material ZF01, we're declaring that 20 kg of material ZS11 was produced as a by-product.

16.6 Goods Receipt

After the goods are produced for a process order, you can perform the goods receipt by following the menu path, Logistics • Materials Management • Inventory Management • Goods Movement • Goods Movement, or using Transaction MIGO (Figure 16.9). After this, the process differs between the co-product and the by-product, which we'll discuss respectively in the following sections.

16.6.1 Co-Product

For process orders with a co-product, both the materials (the main material as well as the co-product) will be available for goods receipt. Just as with the main product, the co-products are also received with movement type 101, as shown in Figure 16.9, whereas by-products are received via movement type 531, as shown previously in Figure 16.8.

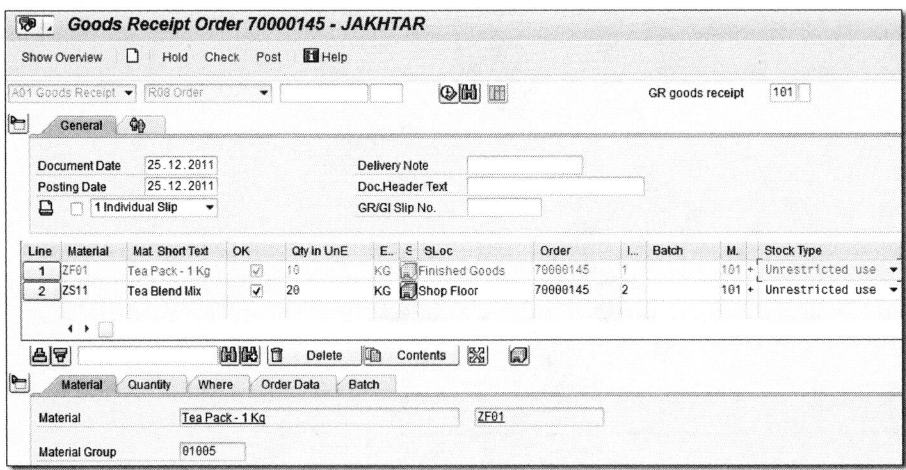

Figure 16.9 Goods Receipt for a Process Order with a Co-Product

Figure 16.9 shows the screen in which the goods receipt for the process order is undertaken. Enter the process order number, and save by pressing Ctrl+S or clicking the SAVE icon.

16.6.2 By-Product

Earlier, we showed how you can use the backflush functionality to record a by-product during the process order confirmation process. However, there is also an alternative available to record the by-product for a process order. You can use this option if you don't want to use the backflush functionality, or you realize there is some additional by-product quantity that you want to record for a process order.

You'll come to the same screen shown in Figure 16.9. In Figure 16.10, select GOODS RECEIPT and OTHER, and then assign a movement type of "531". Now you have to give the component numbers along with quantities, followed by the process order number, which are declared as by-products. This is also a way to receive by-products for a process order (if the backflush functionality in the process order confirmation isn't activated). We mention it here in detail, as an alternative option to declare a by-product as produced.

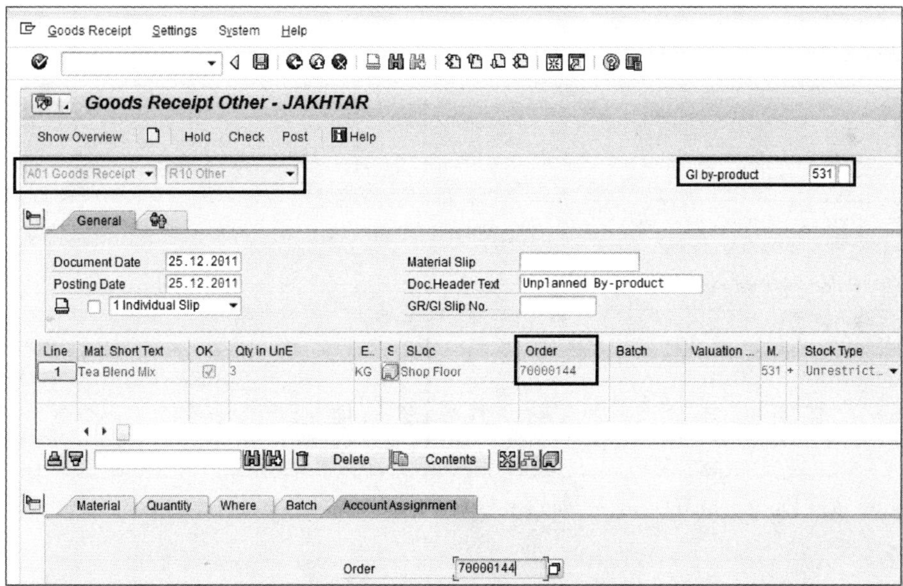

Figure 16.10 Goods Receipt for a By-Product for a Process Order

Save by pressing [Ctrl]+[S] or by clicking the SAVE icon to generate a material document. A material document captures all of the details of the goods movement.

16.7 Documented Goods Movement

A documented goods movement against a process order is the list of all the goods movements that have taken place. This includes goods issuance and goods receipt along with any goods receipt of a co-product or a by-product.

Each goods movement is denoted by the material document number along with the quantity and value. To view the documented goods movement of a process order, follow the menu path, LOGISTICS • PRODUCTION – PROCESS • PROCESS ORDER • PROCESS ORDER • DISPLAY, or use Transaction COR3. Enter the process order number and press [Enter]. On the DISPLAY PROCESS ORDER screen, follow the menu path, GOTO • LISTS • DOCUMENTED GOODS MOVEMENT (see Figure 16.11 ❶).

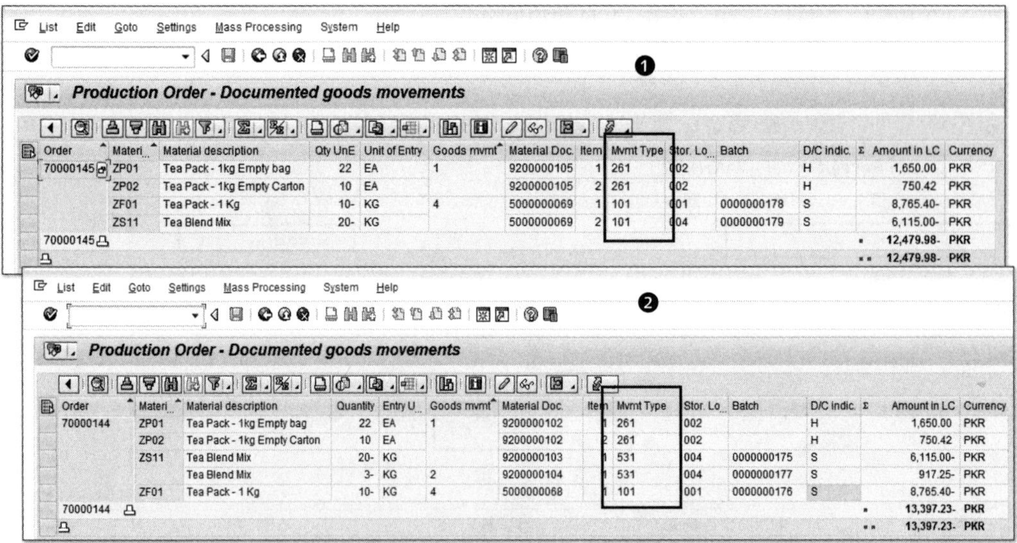

Order	Materi...	Material description	Qty UnE	Unit of Entry	Goods mvmt	Material Doc.	Item	Mvmt Type	Stor. Lo...	Batch	D/C indic.	Σ	Amount in LC	Currency
70000145	ZP01	Tea Pack - 1kg Empty bag	22	EA	1	9200000105	1	261	002		H		1,650.00	PKR
	ZP02	Tea Pack - 1kg Empty Carton	10	EA		9200000105	2	261	002		H		750.42	PKR
	ZF01	Tea Pack - 1 Kg	10-	KG	4	5000000069	1	101	001	0000000178	S		8,765.40-	PKR
	ZS11	Tea Blend Mix	20-	KG		5000000069	2	101	004	0000000179	S		6,115.00-	PKR
70000145												•	12,479.98-	PKR
												••	12,479.98-	PKR

Order	Materi...	Material description	Quantity	Entry U...	Goods mvmt	Material Doc.	Item	Mvmt Type	Stor. Lo...	Batch	D/C indic.	Σ	Amount in LC	Currency
70000144	ZP01	Tea Pack - 1kg Empty bag	22	EA	1	9200000102	1	261	002		H		1,650.00	PKR
	ZP02	Tea Pack - 1kg Empty Carton	10	EA		9200000102	2	261	002		H		750.42	PKR
	ZS11	Tea Blend Mix	20-	KG		9200000103	1	531	004	0000000175	S		6,115.00-	PKR
		Tea Blend Mix	3-	KG	2	9200000104	1	531	004	0000000177	S		917.25-	PKR
	ZF01	Tea Pack - 1 Kg	10-	KG	4	5000000068	1	101	001	0000000176	S		8,765.40-	PKR
70000144												•	13,397.23-	PKR
												••	13,397.23-	PKR

Figure 16.11 Documented Goods Movement for a Process Order

16.7.1 Co-Product

Figure 16.11 in the previous section shows the list of quantities as well as the amount of goods receipt and goods issuance against a process order. The relevant movement types are also shown as well as whether it's a debit entry (S) or a credit entry (H).

16.7.2 By-Product

Figure 16.11 ❷ shows a list of documented goods movements in which a by-product was generated. The by-product received is denoted by movement type 531. The goods receipt of the finished good is denoted by movement type 101. The goods issuance of raw materials against a process order is denoted by movement type 261.

16.8 Cost Analysis

Now we explain how to perform an overall cost analysis versus an itemized view of planned versus actual cost of a process order having either a co-product or a by-product.

Because PP completely integrates with CO-PC, all planned costs are calculated at the time of the creation and release of a process order, including the cost of a main material having a co-product or by-product. These planned costs are then updated with actual costs in the process order when necessary transactional data such as goods issuance, confirmation, and goods receipts are undertaken. All planned versus actual costs are subsequently available for comparison and analysis purposes.

16.8.1 Co-Product

To view the costing details of the process order, follow the menu path, LOGISTICS • PRODUCTION – PROCESS • PROCESS ORDER • PROCESS ORDER • DISPLAY, or use Transaction COR3. Enter the process order number and press Enter.

On the display process order screen, follow the menu path, GOTO • COSTS • ITEMIZATION. Figure 16.12 ❶ shows the quantities of the different elements of the process orders. The activities such as machine and labor are denoted by E, while the goods receipt of a finished good as well as a co-product is denoted by A. The components that will be issued against the process order and their quantities are denoted by M. Finally, simulated goods receipt quantities and values are denoted by H. This is because SAP ERP includes an option to provide a financial value of simulated goods receipts (before actual goods receipt has taken place) by H. Then when the actual goods receipt (A) is undertaken, the values are automatically updated to reflect the difference between the simulated value and actual value.

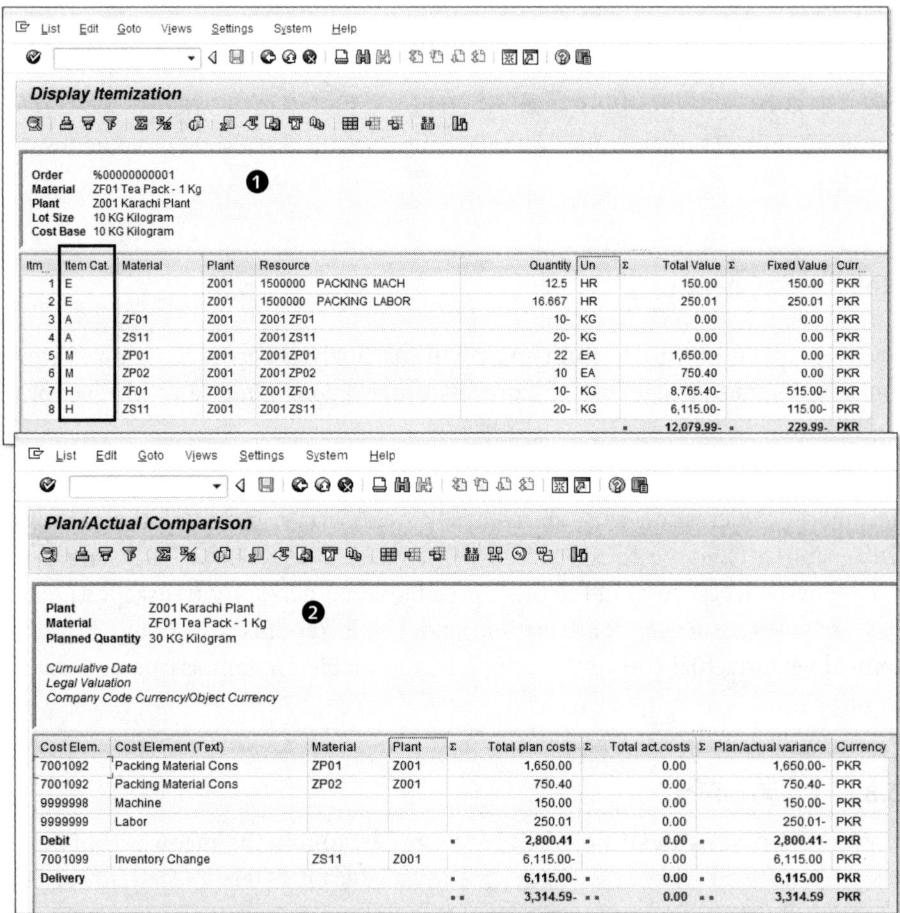

Figure 16.12 Planned Cost of a Process Order: Itemized View

Figure 16.12 ❷ shows the plan versus actual comparison of financial values against a process order. To view this, choose Goto • Costs • Analysis on the Display Process Order screen.

16.8.2 By-Product

All details of planned versus actual values and quantities are automatically updated when you access a process order. You'll be able to see the target versus actual comparison of values as well as quantities against a process order. On the display process order screen, follow menu path, Goto • Costs • Analysis. The resulting screen shows that a semifinished material ZS11, which is also the by-product, was generated in 23 units against the planned quantity of 20 units.

16.9 Summary

The flexibility and options available to record co-products or by-products greatly help in bringing better visibility to the entire production and Inventory Management process. It also ends up reflecting favorably in CO-PC.

The next chapter covers digital signatures.

Using a digital signature can save a business both time and money. This chapter teaches you how to use this important functionality in several important business processes across the SAP ERP system landscape.

17 Digital Signature

The digital signature functionality reduces the number of printed documents that need to be signed within an organization. You can define and then assign digital signatures in an SAP ERP system during various transaction processes, such as changing the status of a document in the Document Management System (DMS), completing a process instruction sheet in SAP ERP Production Planning (PP), or recording results and making usage decisions in SAP ERP Quality Management (QM). Digital signatures can be internal, such as an SAP ERP system user's password, or even external devices such as fingerprint readers or card slides.

Digital signatures can confirm important decisions in QM, validate the correctness of shift reports in SAP ERP Plant Maintenance (PM), or simply confirm the values of process parameters in SAP ERP Production Planning for Process Industries (PP-PI). They do all of this electronically, saving significant time and effort and eventually bringing about improvement in business processes as well as saving money.

Note [«]

If you further integrate the digital signature functionality with SAP ERP system workflow, then the person responsible for digitally signing a particular SAP ERP system document automatically receives the relevant document information via SAP Inbox.

This chapter begins by covering the important steps needed to configure digital signatures. Next, we incorporate signature strategy, which is the collection and sequence of individual signatures in DMS. Then, we run the entire end-to-end business process in DMS to show digital signatures in action. Finally, we show the comprehensive log created by digital signatures.

[»]

Note

The concepts and flows of digital signature in DMS covered in this chapter are equally applicable to all of the other SAP ERP components in which it has its application.

17.1 Configuration Steps to Set Up a Digital Signature

The following three important steps are used in configuring a digital signature:

- Defining authorization groups
- Defining individual signatures
- Defining a signature strategy

[+]

Tips & Tricks

Although this chapter focuses on covering the digital signature in the DMS configuration, the general (component-independent) menu path for digital signatures is SPRO • REFERENCE IMG (or press [F5]) • CROSS-APPLICATION COMPONENTS • GENERAL APPLICATION FUNCTIONS • DIGITAL SIGNATURE.

We'll discuss each of the steps for configuration in the following sections, before moving on to assigning a signature strategy.

17.1.1 Define Authorization Groups

An *authorization group* is used to limit the authorization of a document or document status to a specific user or group of users performing the same role. Once configured, this authorization group can then be controlled through authorization object C_SIGN_BGR. You need to coordinate with your Basis team members to get the right authorization group assigned to the right user.

To create an authorization group, follow SAP ERP menu path, SPRO • REFERENCE IMG (or press [F5]) • CROSS-APPLICATION COMPONENTS • DOCUMENT MANAGEMENT • APPROVAL • DEFINE AUTHORIZATION GROUP. Here you can see the initial screen for the creation of an authorization group for digital signatures (see Figure 17.1).

Choose New Entries to create a new authorization group, DMS-01, and give it a description (DMS initiators' Group, in this example). Save the authorization group by pressing [Ctrl]+[S] or by choosing the Save icon.

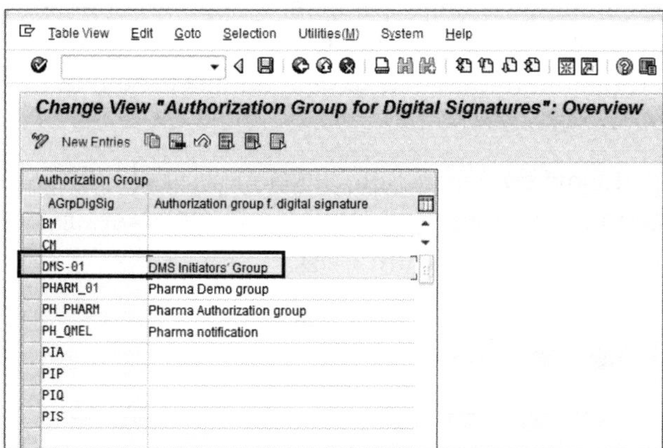

Figure 17.1 Creating a New Authorization Group

17.1.2 Define Individual Signatures

An *individual signature* is a unique identification of the SAP ERP system digital signature to which the user is assigned. To create a new individual signature, follow the menu path, SPRO • Reference IMG (or press [F5]) • Cross-Application Components • Document Management • Approval • Define Individual Signature. In the screen that appears (see Figure 17.2), choose New Entries, and define two individual signatures—DMS-01 and DMS-02—with descriptions. Both individual signatures are assigned to the same authorization group: DMS-01. Save the entries by pressing [Ctrl]+[S] or choosing Save.

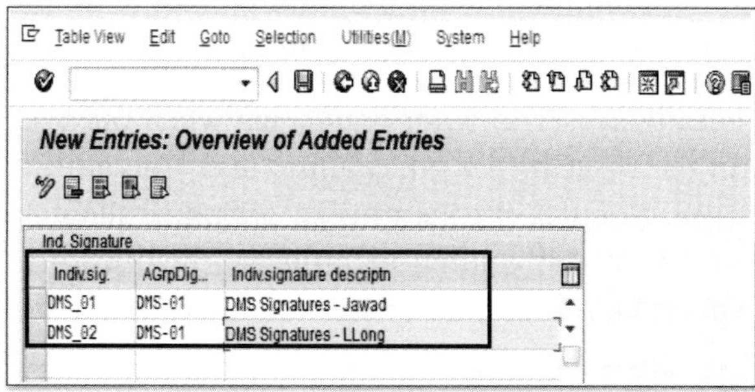

Figure 17.2 Creating New Individual Signatures

17.1.3 Define a Signature Strategy

A *signature strategy* is the combination of different individual signatures that are assigned to an authorization group. A signature strategy also defines the sequence and flow of individual signatures and maintains the predecessor–successor relationships. The signature strategy is used to control the level to which the document can be considered as released. A RELEASED status denotes that the digital signature process is complete in every respect.

To create a signature strategy, follow the menu path, SPRO • REFERENCE IMG (or press [F5]) • CROSS-APPLICATION COMPONENTS • DOCUMENT MANAGEMENT • APPROVAL • DEFINE SIGNATURE STRATEGY (see Figure 17.3).

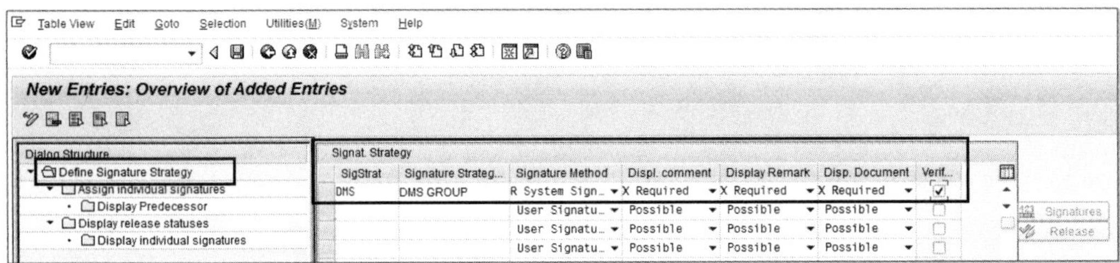

Figure 17.3 Initial Screen to Define a Signature Strategy

Before we show you how to create a new signature strategy, we need to explain some of the columns on the right side of the screen and their functions. The third column, SIGNATURE METHOD, stipulates how SAP ERP verifies a signature. The three possible options are listed here:

- System signature with authorization, using an SAP ERP user name and password
- User signature with external security product with verification
- User signature with external security product without verification

The DISPL. COMMENT column enables a user to enter comments when the user executes the digital signature. In the DISPLAY REMARK column, users can enter a short text before signing. The DISP. DOCUMENT column is an option to allow or disallow viewing of a document. The VERIFY column with the checkbox verifies the sequence of the digital signature, known as the signature strategy, before the system marks the document as complete.

To define a new signature strategy, choose NEW ENTRIES at the top of the overview screen shown in Figure 17.3. In the second screen of the same figure, you define the SIGNATURE STRATEGY with the description "DMS GROUP". Then select the SIGNATURE METHOD option to be SYSTEM SIGNATURE AND PASSWORD, and mark the DISPL. COMMENT box, DISPLAY REMARK box, and DISP. DOCUMENT as REQUIRED entries. Finally, select the VERIFY checkbox to require verification of the signature strategy.

Next, select the signature strategy. Enter "DMS" in the SIGSTRAT field, and double-click the ASSIGN INDIVIDUAL SIGNATURES folder.

Choose NEW ENTRIES (Figure 17.4), and add the previously created individual signatures DMS_01 and DMS_02 to the signature strategy DMS. Save by pressing Ctrl+S or choosing SAVE. Click the BACK icon to return to the overview screen.

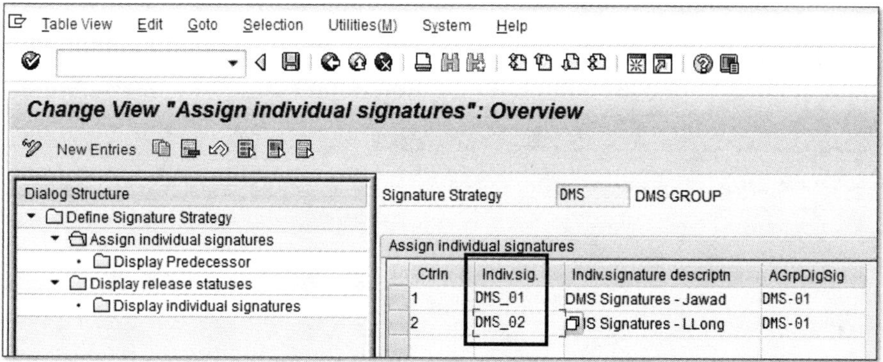

Figure 17.4 Individual Signatures for the Signature Strategy

Figure 17.5 displays the signature strategy overview screen originally displayed in Figure 17.3 but now with a new entry, DMS. Select the newly created signature strategy DMS, and click the SIGNATURES button on the right-hand side.

Here you can also see the SIGNATURE SEQUENCE screen in which individual signature DMS_01 must be signed before individual signature DMS_02. In other words, DMS_01 is the predecessor (denoted as the predecessor in the DMS_01 column), and DMS_02 is the successor. Click CONTINUE to go to the screen shown in Figure 17.6.

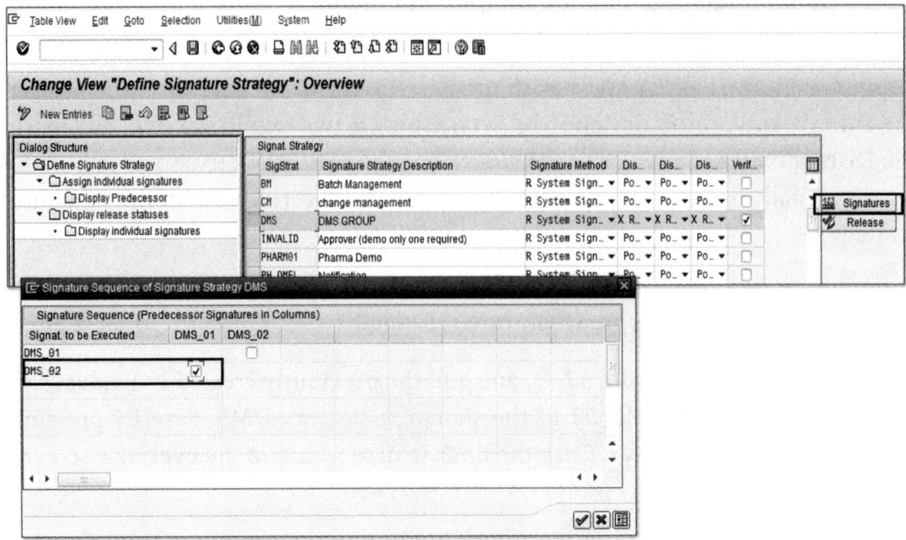

Figure 17.5 Signature Sequence of the Signature Strategy

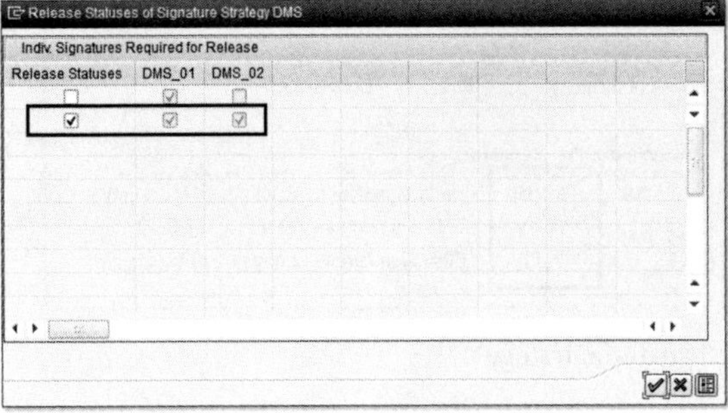

Figure 17.6 Release Statuses of Signature Strategy

Select the signature strategy DMS, and click the RELEASE icon. From the pop-up that appears, select the checkbox that denotes the release status of the document as well as the completion of the signature strategy. A completion strategy means that the required sequence of digital signatures is performed.

For this example, select the checkbox in the second row in which both individual signatures, DMS_01 and DMS_02, must be marked for the signature strategy to declare it as complete.

Click the CONTINUE icon and save. The configuration needed to enable the SAP ERP system digital signature is complete. However, for the digital signature to work in the desired business process, you need to assign the newly created signature strategy, DMS, to the relevant DMS document type DRW, which we cover in the next section.

17.1.4 Assign a Signature Strategy to a Document Management System Document Type

You want to assign the digital signature strategy DMS to DMS document type DRW. To assign the newly created digital strategy to a specific DMS document status, follow the menu path, SPRO • REFERENCE IMG (or press F5) • CROSS-APPLICATION COMPONENTS • DOCUMENT MANAGEMENT • CONTROL DATA • DEFINE DOCUMENT TYPE. Figure 17.7 shows the DMS document type overview screen.

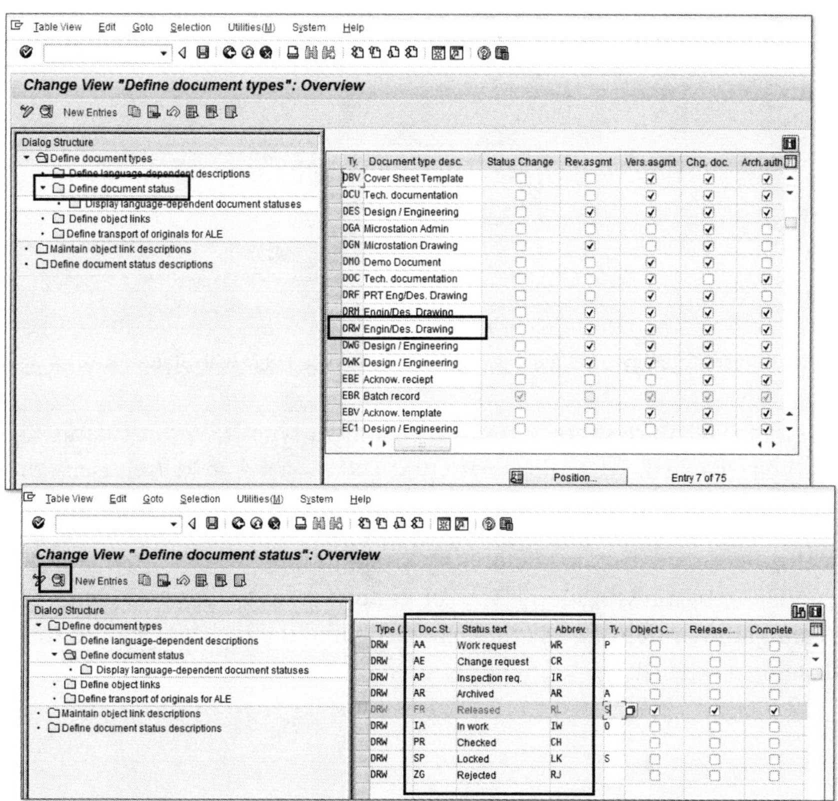

Figure 17.7 DMS Document Type DRW with Different Statuses

Select document type DRW, and click the DEFINE DOCUMENT STATUS folder shown in the lower half of Figure 17.7. This screen contains a large number of user-defined statuses for DMS document type DRW. Notice the different document statuses such as IA for IN WORK status and its abbreviation of IW. Similarly, the document status AA is a WORK REQUEST with abbreviation WR.

Select the document status FR, which is for STATUS TEXT RELEASED, with the abbreviation RL. Click the DETAIL icon to go to the screen shown in Figure 17.8.

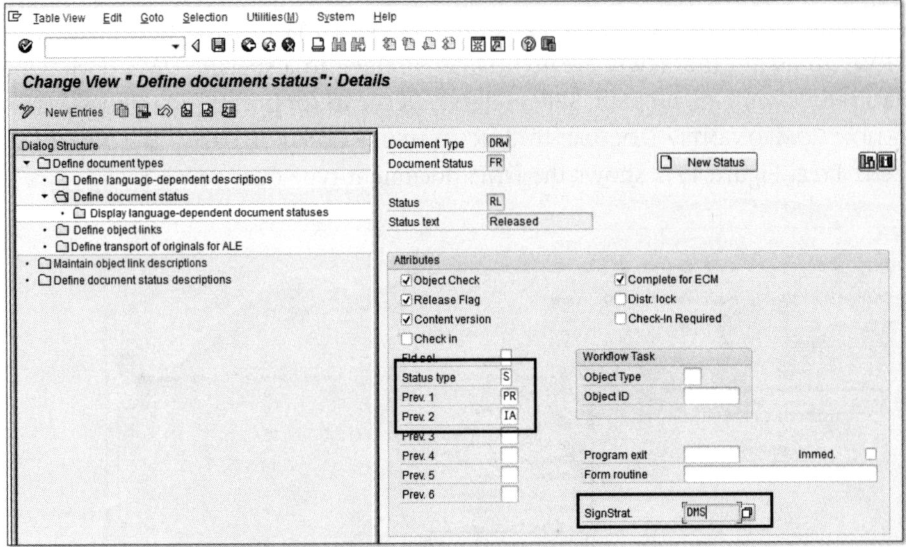

Figure 17.8 Document Strategy DMS Assigned to Document Type DRW with Status FR

Here, assign the signature strategy DMS to document type DRW with status RL. As soon as you change the DMS document type DRW's status to RL (released), the system activates the signature strategy DMS.

Additionally, the previous status before the RL status must be IA. Save the document type DRW by pressing Ctrl+S or choosing SAVE. This completes the needed configuration steps for defining and assigning the digital signature strategy to document type DRW.

17.2 Digital Signature in Action

You first create a new DMS document of document type DRW and then set the initial status to WR. Then, change to the next status IW as stipulated in the status network shown earlier in Figure 17.8. A *status network* is the sequence of status changes that must be followed for any document type. Finally, change the status from IW to RL, to which the signature strategy DMS has been assigned. The DMS document type is also alternatively referred to as a *Document InfoRecord (DIR)*.

To create a new DMS document, follow the menu path, LOGISTICS • CENTRAL FUNCTIONS • DOCUMENT MANAGEMENT SYSTEM • DOCUMENT • CREATE, or use Transaction CV01N. Enter the DOCUMENT TYPE "DRW", and press Enter (see Figure 17.9).

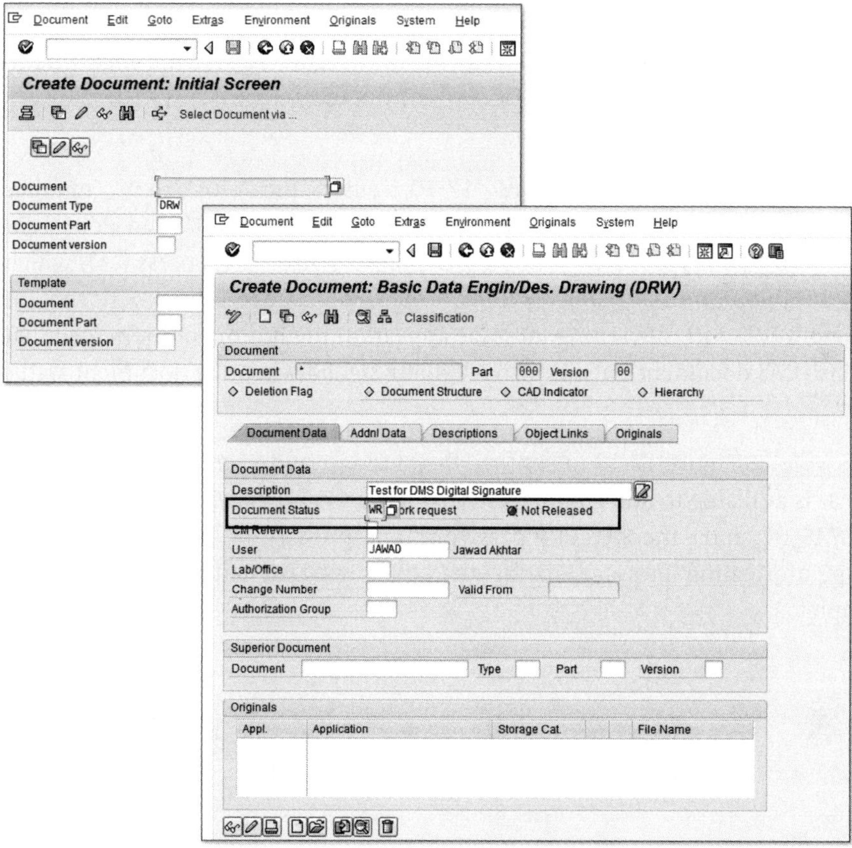

Figure 17.9 Initial Screen of a DMS Document with WR Status

The second screen in Figure 17.9 contains the initial DMS status WR for work request. Enter a short description in the DESCRIPTION field, and save the DMS document. In this example, the system generates document number 10000000258, document type DRW, document part 000, and document version 00.

Next, change the status of the already created DMS document from WR to IW (IN PROCESS status) because it should be the status before the DMS document's status can be changed to RL for RELEASE.

To change the DMS, follow the menu path, LOGISTICS • CENTRAL FUNCTIONS • DOCUMENT MANAGEMENT SYSTEM • DOCUMENT • CHANGE, or use Transaction CV02N. Enter the document number ("10000000258" in this example), document TYPE "DRW", document PART "000", and document VERSION "00". Press [Enter]. Change the document status to IW, and save by clicking the SAVE icon or pressing [Ctrl]+[S].

To again change the status from IW to RL in the DIR, follow the menu path, LOGISTICS • CENTRAL FUNCTIONS • DOCUMENT MANAGEMENT SYSTEM • DOCUMENT • CHANGE, or use Transaction CV02N.

In the screen shown in Figure 17.10, enter the DOCUMENT number "10000000258", document TYPE "DRW", document PART "000", and document VERSION "00", and press [Enter]. In the DOCUMENT STATUS field, press [F4], which brings up a pop-up of available options. When you change the DMS document status from IW to RL, a warning message appears at the bottom of the screen that says this DMS document must now be digitally signed when the document status is RL.

Press [Enter] in Figure 17.10 to confirm the message, and a screen appears with a box that is available to add comments (see Figure 17.11). Select individual signature DMS_01, enter the SAP ERP password, and press [Enter]. You'll then see a pop-up information message stating that the digital signature process is still incomplete.

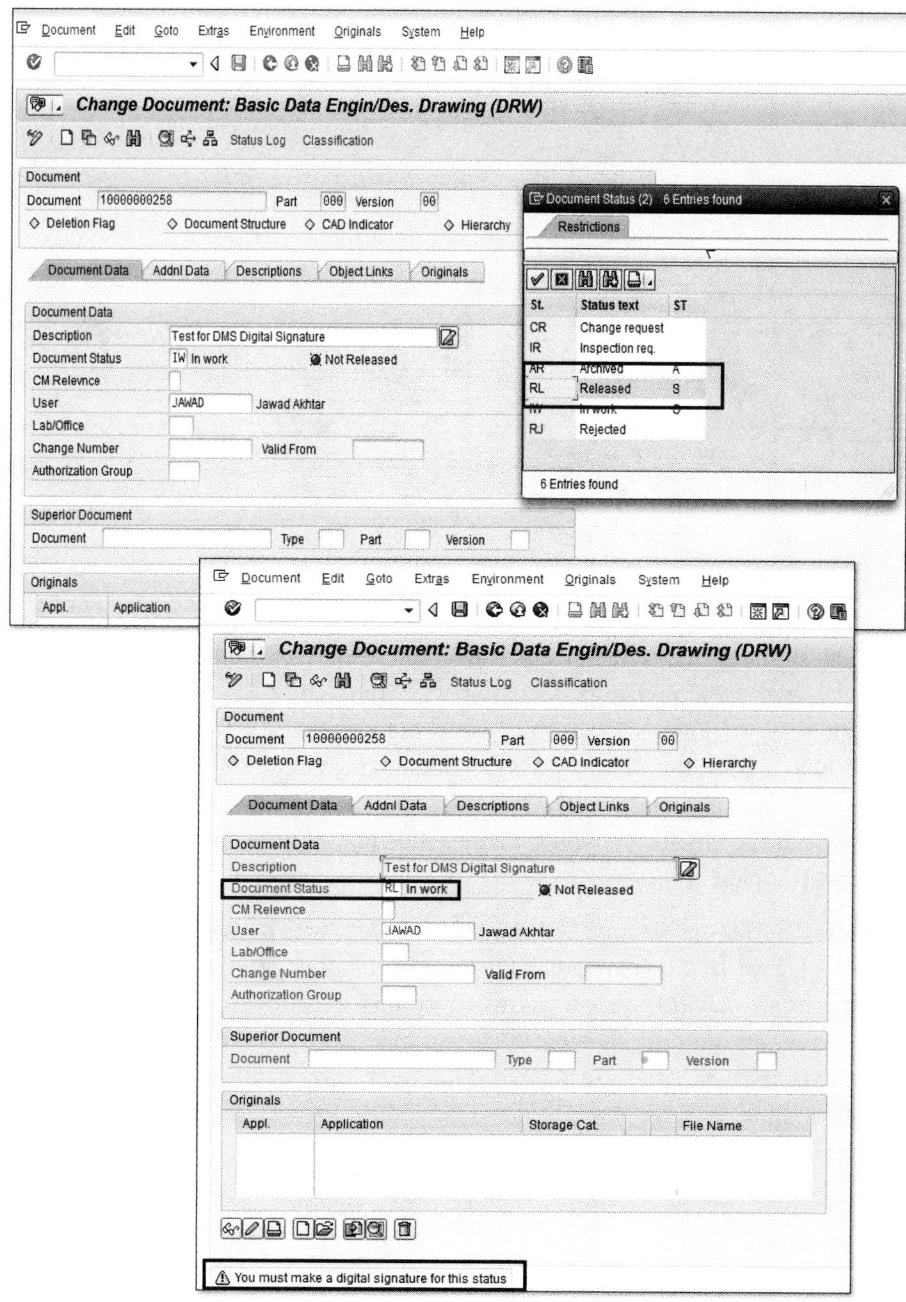

Figure 17.10 Changing the Status of the DMS Document to RL

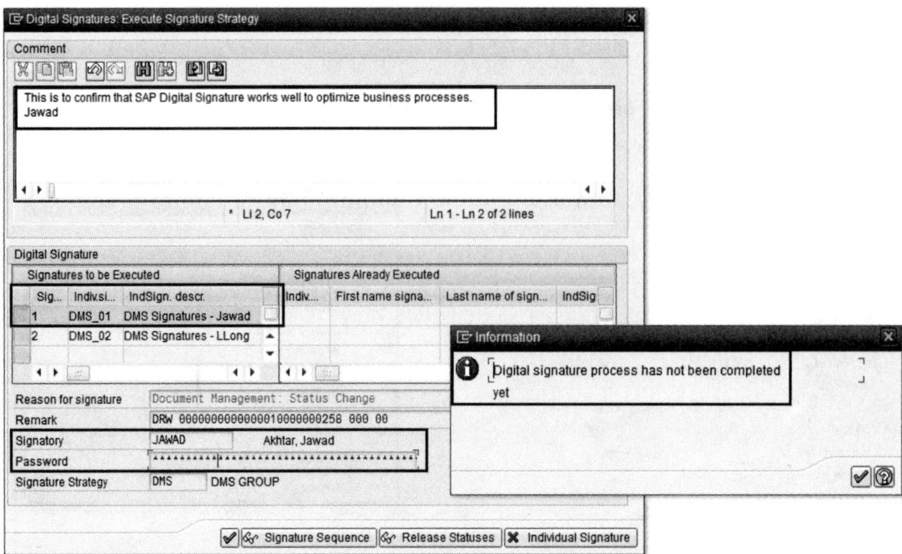

Figure 17.11 First (DMS_01) of the Two Individual Digital Signatures to Be Executed

Now log in to the SAP ERP system with a different user name ("LLong" in this example), and use the change document via Transaction CV02N. Alternatively, you can refer to Figure 17.10 and enter the same DMS Document number ("10000000258"), Type ("DRW"), Part ("000"), and Version ("00"). After you've entered this information, press Enter to go to the screen shown in Figure 17.12. At the bottom of this screen, you get a message that says the digital signature process isn't complete.

Click the Digital Signature button, which takes you to the pop-up window shown in Figure 17.13 where you have to enter the digital signature "DMS_02" in the form of an SAP ERP system password. In this screen, you have the option to enter comments, and the signature that was already executed (DMS_01) appears on the bottom right. After the password has been successfully entered, it leads to the information message that the digital signature process has been successfully completed.

Figure 17.13 shows the second of the two (DMS_02) individual digital signatures to be executed.

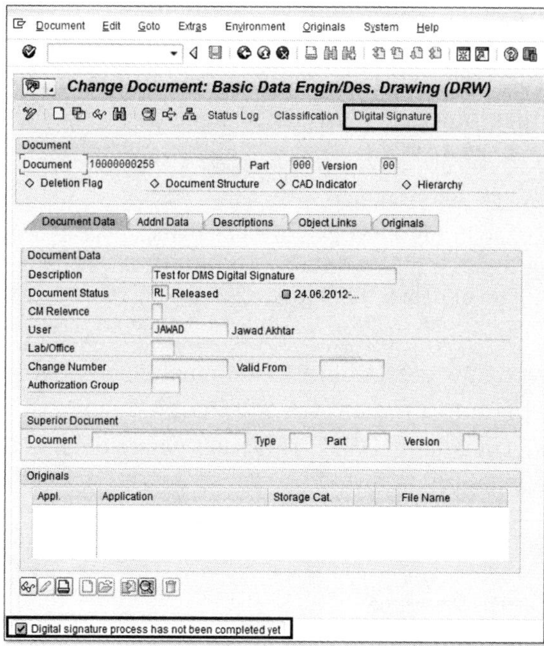

Figure 17.12 Information Message Stating That the Digital Signature Process Is Incomplete

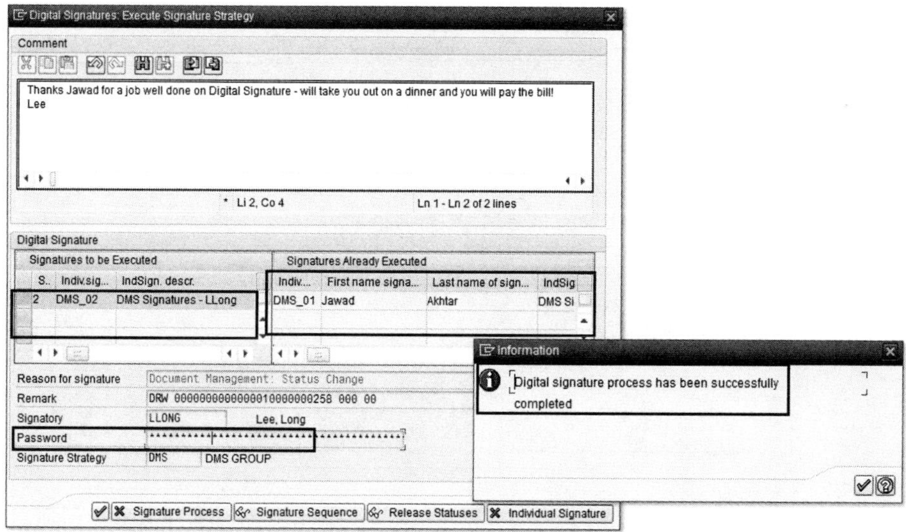

Figure 17.13 Information Message Stating That the Digital Signature Process Is Complete

17.3 Digital Signature Logs

The SAP ERP system provides a comprehensive log that you can view to ensure complete visibility of all the activities undertaken for digital signatures.

[»]

> **Note**
>
> This digital signature log is available not only for DMS but also for all of the applications and components in which digital signatures are used, including PM, QM, Engineering Change Management (ECM), and PP-PI. For details, refer to Section 17.4.

To access the digital signature logs, follow the SAP ERP system menu path, CROSS APPLICATION COMPONENTS • ENGINEERING CHANGE MANAGEMENT • REPORTING • LOG FOR DIGITAL SIGNATURE, or use Transaction DSAL. This takes you to the initial screen to enter relevant parameters (Figure 17.14).

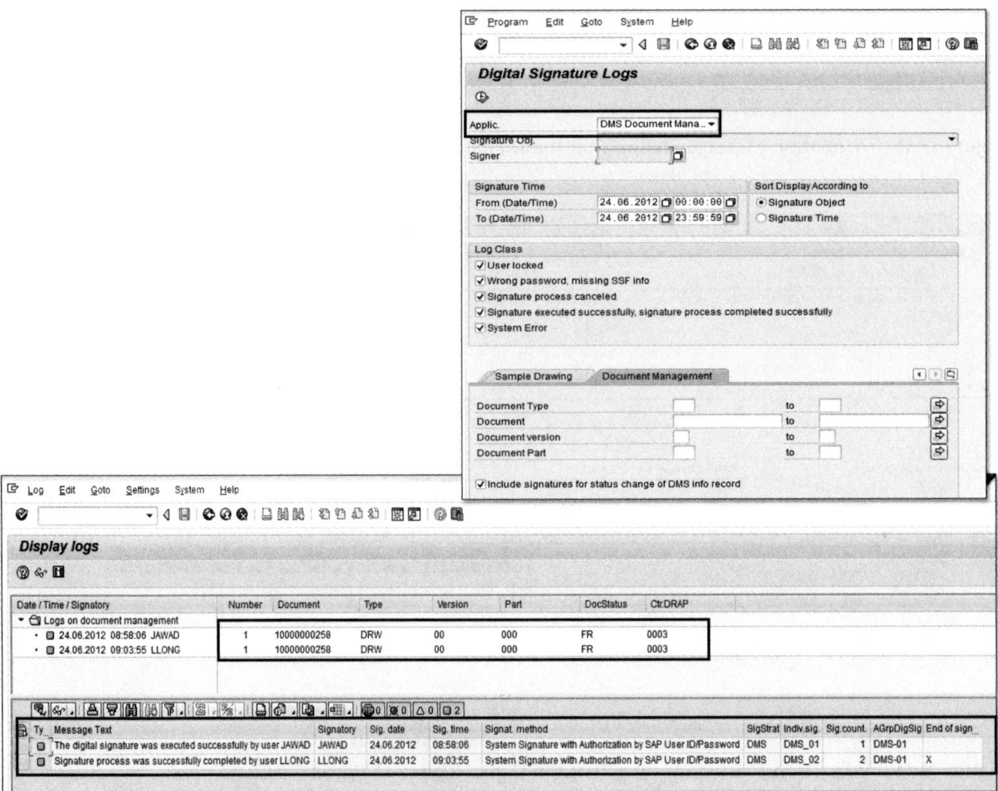

Figure 17.14 Detailed Log for Digital Signatures

In the APPLIC. (application) field, select DMS and don't provide any other selection criteria. Press F8 or choose EXECUTE. The screen provides a comprehensive log of digital signatures for DMS document 10000000258. The bottom half of the screen provides details about who executed the digital signatures and when the signatures were executed. It also provides details of the individual signatures (DMS_01 and DMS_02) as well the signature strategy DMS.

17.4 Application of Digital Signature in SAP ERP Components

The digital signature functionality is used in several components of the SAP ERP system. As explained in the beginning of this chapter, you can integrate the digital signature functionality in several logistics components that we list in the following sections. The steps that we cover in configuring digital signatures apply to SAP ERP system components, which provide the options for the digital signature functionality. Integrating the digital signature process is a step toward optimizing your business processes and reducing (if not eliminating) the need to physically sign the documents. Further, if you integrate digital signatures with SAP Business Workflow, it further brings efficiencies to the business processes by ensuring the relevant SAP ERP object is delivered to the right user for necessary action (digitally signing the SAP ERP object).

The first step is to configure the digital signature as outlined in this chapter, and then assign it to the relevant SAP ERP object. For example, if you want to integrate digital signatures in shift reports, the option to do so is available in the configuration (Transaction SPRO) menu of shift reports.

17.4.1 Production Planning for Process Industries

You can implement digital signatures in process instruction sheets, which is part of the Process Management subcomponent of PP-PI.

You can assign digital signatures in the control recipe destination of a process instruction sheet to one or more of the following functions:

- Activating or deactivating the process instruction sheet
- Locking or unlocking the process instruction sheet

- Marking the process instruction sheet as complete
- Value entry in the process instruction sheet
- Deviations in entered values in the process instruction sheet
- Process steps in the process instruction sheet

You can also integrate digital signatures in shift reports for PP of process manufacturing as well as discrete manufacturing.

17.4.2 Quality Management

You can integrate digital signatures in QM in four different business processes:

- Results recording
- Usage decision
- Physical sample drawing
- Header or item level of QM notification

17.4.3 Plant Maintenance

In PM, you can integrate digital signatures to the following business processes:

- Shift report
- Order type-based digital signature
- Header or item level of PM notification

17.4.4 Document Management System

This chapter already covers the actual example of implementing digital signatures in DMS. You can assign a digital signature to any specific status of a document type in DMS.

17.4.5 Engineering Change Management

ECM consists of two important subareas: engineering change request (ECR) and engineering change order (ECO). You can assign digital signatures in either or both subareas. Following are some of the functions in which you can integrate digital signatures in the ECR/ECO master record:

- Check ECR
- ECR checked
- Approve ECR
- Withdraw approval
- Reject ECR
- Convert ECR
- Order complete
- Order incomplete
- Close order
- Release ECO
- Lock ECR/ECO
- Unlock ECR/ECO

When you set one or more of the following specific statuses within ECR/ECO, you have the option to integrate digital signatures:

- Change possible
- Change unnecessary
- Change impossible
- Complete change
- Incorrect Change
- Error corrected
- Release change

Other areas of SAP ERP in which you can integrate digital signatures include Audit Management and electronic batch record.

17.5 Summary

This chapter covered the necessary configuration needed to implement the use of a digital signature as well as the business processes in which the user can effectively integrate them, including multiple signatures. Using a digital signature eliminates the need to manually and physically sign important documents, thus helping in optimizing the business process.

The next chapter begins Part VI and covers the Early Warning System (EWS) in detail.

PART VI
Monitoring and Evaluation

The Early Warning System informs you whenever any deviation or exception to an important business process occurs, allowing you to take timely action to optimize and monitor your business processes.

18 Early Warning System

The Early Warning System (EWS) is built on the same SAP ERP system logistics information structures as other standard analysis tools and reports, with added flexibility that allows you to monitor and receive alerts for only the specific exceptions that are important to your business processes. It provides real-time updates and can be tailored to individual needs without the involvement of any custom ABAP development.

EWS is delivered as part of the logistics component and can provide alerts to individual business process owners in all of the logistics functions, such as the Logistics Information System (LIS), shop floor information system, purchasing information system, and plant maintenance information system. The alerts help business process owners quickly make important decisions or take actions as deemed necessary.

In this chapter, we'll explain the benefits of EWS, and then provide the steps you need to get it up and running in your system.

18.1 Overview

There are many instances in SAP ERP Production Planning (PP) where EWS can come in handy, depending on what is important to the business. In this chapter, you'll learn how easy and intuitive it is to set up EWS by following the steps in logical and sequential order. If the business requirements change at any time, you can make changes or even delete the EWS that you had set up.

Some examples of where you can use EWS are listed here:

- When you record production scrap at greater than 5% of total production
- When sales figures show a decreasing trend for a high-running, high-profit item for the past three months
- When quality specifications of an important raw material are out by more than 3%
- When a vendor continues to make late deliveries of critical components
- When inventory of precious metal exceeds a certain value

In any of these situations, you can configure EWS to the individual exception so you'll receive alerts and can react immediately. While you can set up EWS in any of the logistics areas, we use an example in PP. The relevant information structures available for PP are listed in Table 18.1.

Information Structure	Description
S021	Production order
S022	Operation
S023	Material
S024	Work center
S025	Run schedule header
S026	Material usage
S027	Product costs
S028	Reporting point statistics
S029	Kanban
S225	Goods receipts: Repetitive manufacturing
S226	Material usage: Repetitive manufacturing
S227	Product costs: Repetitive manufacturing

Table 18.1 Standard Information Structures Available for Production Planning

[»] **Note**

When EWS is set up in logistics, the relevant information structures are automatically made available by the SAP ERP system. For example, if EWS is set up in SAP ERP Materials Management (MM), all MM-related information structures become available.

Setting up EWS entails the creation of an exception by selecting the desired information structure from Table 18.1, followed by the characteristics and key figures. The requirements of each key figure need to be defined next, and, finally, you configure the follow-up function. In summary, the required steps are as follows:

1. Create an exception.
2. Group the exceptions.
3. Create a periodic analysis.

Note [«]

If there are several exceptions to the same information structure, they can be combined to create a *group exception*, which is a list of individual exceptions. Like the first step, the follow-up function for this step involves determining whether an email alert to an individual suffices or if it should be communicated to the entire distribution list.

The third and final step is periodic analysis, in which you define the frequency with which the system monitors the exceptions and issues alerts. Let's move on to discuss the first step in the following section: creating an exception.

18.2 Exceptions

An *exception* is important data that you want to monitor from a business point of view. The sequence of steps in creating an exception is the following:

1. Choose the characteristics.
2. Define the key figures.
3. Define the characteristics values.
4. Choose the requirements.
5. Define the requirements.
6. Define the follow-up processing.

For this example, we've created an exception in PP (shop floor), Z010, using the information structure S021 (production order). In exception Z010, we specified that the system should issue an alert whenever the material CH-3000 has the following characteristics:

- Shows a negative trend of planned versus actual quantity deviation
- Exceeds actual scrap quantity of the material by 10 kg
- Quantity of goods received against a production order is less than 80 kg

If the system records an exception in the SAP ERP system, we want it to highlight with the color red and send an email alert to our SAP Inbox.

In the following sections, we'll cover the steps you need to perform to set up EWS in your SAP ERP system.

18.2.1 Set Up Exceptions

Use the menu path, LOGISTICS • LOGISTICS CONTROLLING • SHOP FLOOR INFORMATION SYSTEM • EARLY WARNING SYSTEM • EXCEPTION • CREATE, or use Transaction MC=1 to set up the exception (see Figure 18.1). Follow the step-by-step process for creating exception Z010:

1. For this example, enter the EXCEPTION "Z010", a short description, and information structure "S021". Click the CHARACTERISTICS icon. (Info structure S021 corresponds to the production order.)

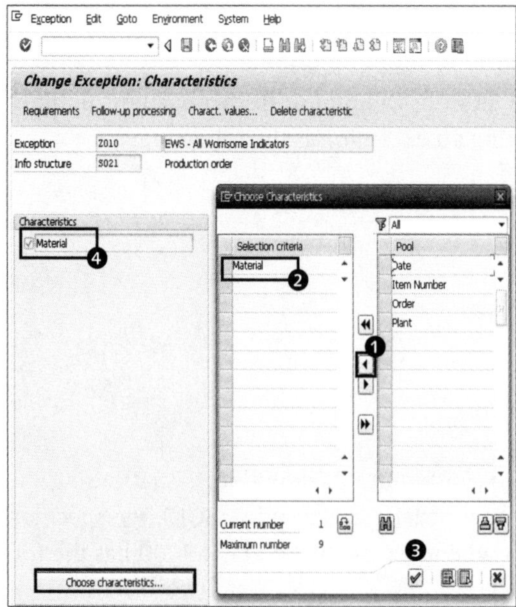

Figure 18.1 Characteristics Setup in EWS

2. In the selection screen, click the CHOOSE CHARACTERISTICS icon to select the characteristics on which you want to activate EWS.
3. In the pop-up screen that appears in Figure 18.1, select MATERIAL ❷, and move it to the left column by clicking the CHOOSE icon ◀ ❶. Choose EXECUTE ❸.
4. For each characteristic in Figure 18.2, provide the corresponding value or a range from which to choose. Click the characteristic MATERIAL ❶, and click the CHARACT. VALUES ❷ button at the top.
5. In the pop-up window that appears, enter the desired value or range for which the system should monitor EWS (Figure 18.2). For this example, enter "CH-3000" ❸ for material. Choose COPY.

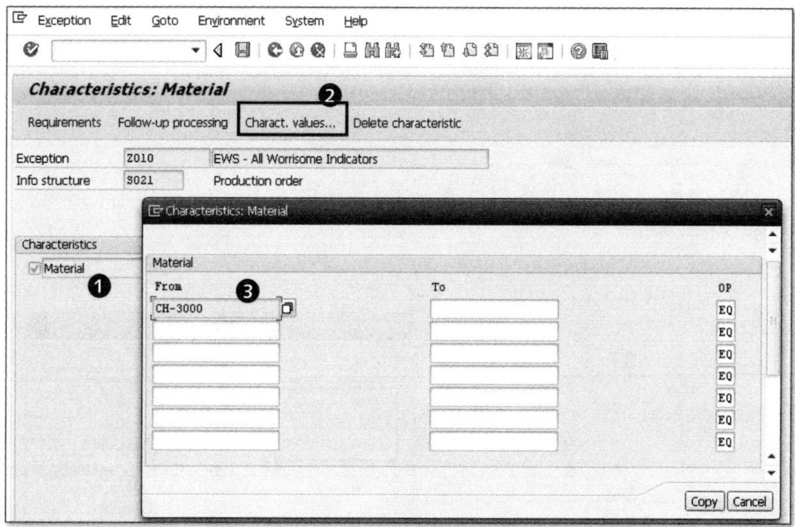

Figure 18.2 Characteristics Value for Material CH-3000

Note [«]

You can give a range of materials for EWS to monitor by filling out both the FROM and TO fields in Figure 18.2. Alternatively, leaving these fields blank means EWS considers all materials in the alert.

18.2.2 Define Requirements

Now that the value for the characteristic material is defined, the next step is to define the requirements—that is, the information that you need EWS to monitor in day-to-day business processes. Select the REQUIREMENTS button shown previously in Figure 18.2 to define requirements for the material.

For this example, choose three key figures for the material (Figure 18.3) by following these steps:

1. On the REQUIREMENTS screen, click the CHOOSE KEY FIGURES button, and from the subsequent pop-up screen, select the key figures that will form the basis of your EWS alerts. For this example, select the three key figures ❶ by moving them from the right-hand side to the left-hand side with the CHOOSE icon ◀.
2. Choose CONTINUE ❷.
3. Define the PERIOD TO ANALYZE fields ❸, which stipulate the past and future periods on which the EWS must evaluate deviations. By defining PREVIOUS PERIODS as "12", you stipulate that all key figures for the past 12 months that meet the conditions must be considered during evaluation.

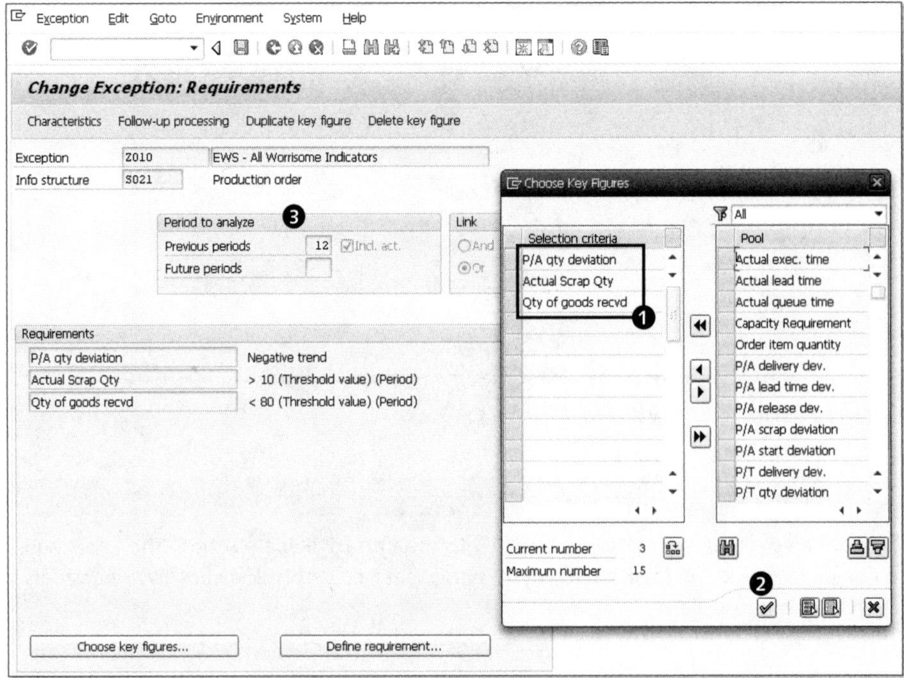

Figure 18.3 Key Figures for EWS

4. After the key figures are selected, click the DEFINE REQUIREMENT button to specify the limits or other details that form the basis for EWS alerts (Figure 18.4). The requirement has to be defined for each key figure. The options available are THRESHOLD VAL. ANAL., which requires a value to be specified, or TREND ANALYSIS, which is used to monitor positive or negative trends and requires you to specify the percentage of deviation in the trend.

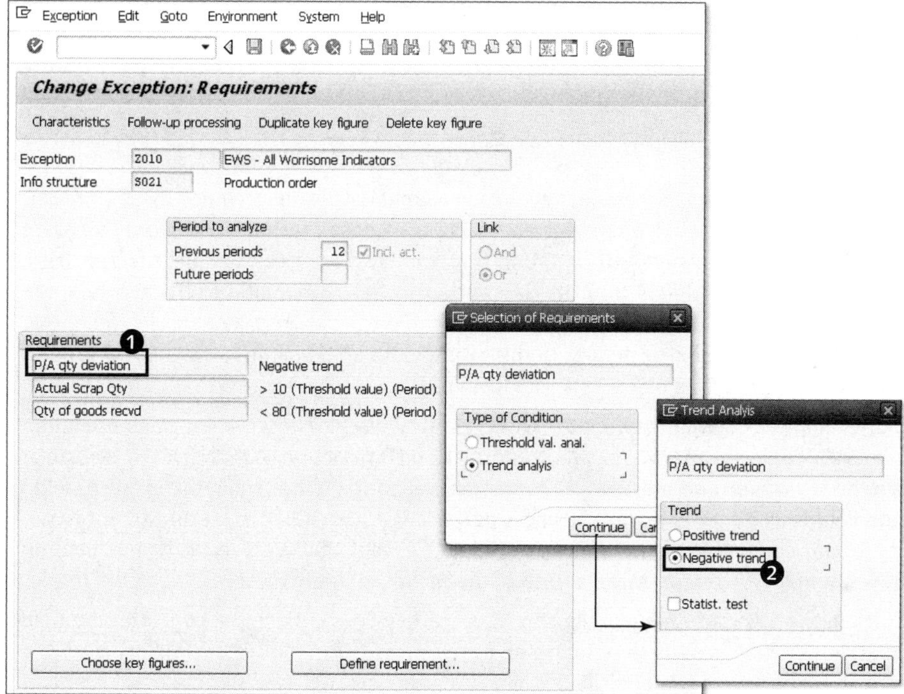

Figure 18.4 Define Requirements for Key P/A Qty Deviation

5. For the first key figure, P/A QTY DEVIATION ❶ in Figure 18.4, place the cursor on the key figure, and double-click it. Alternatively, you can also click on the DEFINE REQUIREMENT button, and a pop-up will appear.

6. In the pop-up window, select the TREND ANALYSIS radio button, and choose the CONTINUE button. In the next pop-up window that appears, click the NEGATIVE TREND radio button ❷, and then choose the CONTINUE button.

7. For the second key figure, follow the same steps as for the first key figure. When the pop-up window appears, enter ">10" for Actual Scrap Qty, and then choose the Continue button.
8. Repeat the preceding step for the third key figure, Qty of goods recvd, with a threshold value of "<80".

[»]

Note

All other standard operands, such as less than, greater than, equal to, and not equal to, are available for use when setting threshold values.

[+]

Tips & Tricks

Given that there are so many key figures available with similar names or descriptions while setting up EWS, it often gets confusing to choose the right key figure to ensure that an alert will be issued for deviation through EWS. For this example, the key figure Actual Scrap Qty. differs from its description in the confirmation screen, where it's mentioned only as Scrap. In such cases, you can ensure correct key figure selection in EWS through its technical names. In this example, the field name (technical) for scrap is XMNGA, and the field element is MC_XMNGA.

For example, you want to monitor scrap quantity during confirmation of an operation via EWS. To do so, and while in the confirmation (Transaction CO11N), place the cursor on the Scrap field and press F1. The system will bring up the screen that explains what the field is all about and what business purpose it serves. Click on Technical Information, and you'll find the field name and data element that are specific to this field (in this example, it's scrap). Make a note of the technical information.

Back in the EWS setup and while choosing the key figure, place the cursor on the field that closely matches the description and then press F1. The system will bring up the screen that explains what the field is all about and what it's used for. Click on Technical Information, and there you'll find the field name and data element that are specific to this field. If this technical information matches with the details that you've previously noted down, then you've chosen the correct key figure. If not, continue the search for other key figures with similar names and check their technical information until you find the right key figure.

18.2.3 Follow-Up Processing

The next step in defining an exception is to define the follow-up processing—that is, what and how the EWS informs you that an exception has occurred. For this example, select the Follow-up processing button to bring up the screen shown in Figure 18.5. Here we've specified that whenever the exception Z010 occurs, the details are to be marked in red and sent as a table to the entered mail recipient. Further, the same needs to be communicated by email to the corresponding SAP Inbox.

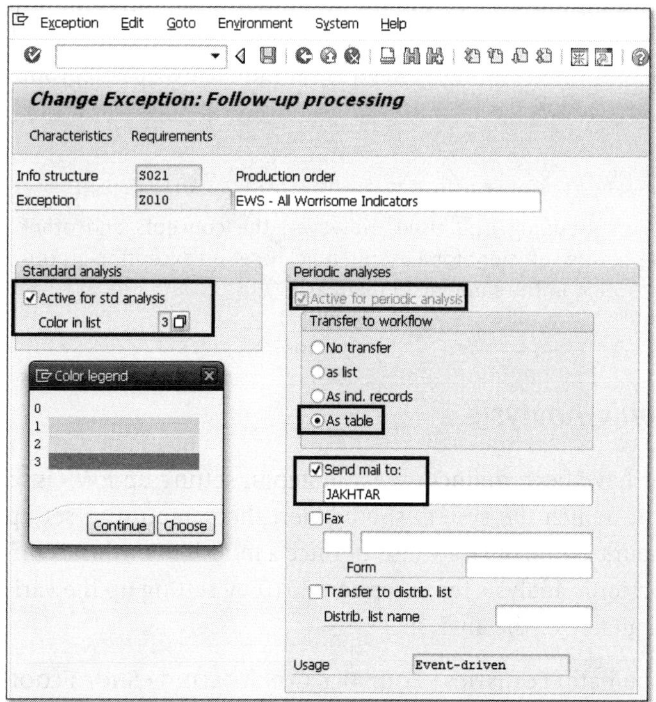

Figure 18.5 Follow-Up Processing to Define the Requisite Parameters

Tips & Tricks [+]

Select both the Active for std (standard) analysis and Active for periodic analysis checkboxes in Figure 18.5, as these will be subsequently checked whenever you run analysis reports in EWS.

18.2.4 Group Exceptions

If you want to monitor more than one exception from the *same* information structure in EWS, it's convenient and practical to group them together in exception groups for ease of reporting. For example, if the same material's scrap is monitored for its upper tolerance limit with the color red and its lower tolerance limit with the color yellow, then grouping these two different exceptions provides details in one list or report in EWS. Let's clarify this further. You define that whenever the material scrap is exceeded by greater than 10% (upper tolerance limit), the system highlights it in red in EWS. When the material scrap is less than 10% (lower tolerance limit), the system highlights it in yellow in EWS.

If no such requirement exists, then you don't need to define the exception grouping.

[»] **Note**

For this example, exception grouping isn't used. However, the concepts and other details covered in this chapter are sufficient for a user to quickly set up exception grouping. Exception grouping is found in the same menu path as EWS.

18.3 Set Up Periodic Analysis

Now that the exceptions have been defined, the last step in setting up EWS is to define the frequency with which the system should alert the person who set up EWS (e.g., every eight hours, daily, once a week, or once a month). For this example, we need to set up periodic analysis for exception Z010 by setting up the variant and defining the frequency of the analysis.

To do so, follow the menu path, Logistics • Logistics Controlling • Shop Floor Information System • Early Warning System • Periodic Analysis • Area to Analyze • Create, or use Transaction MC=7 to get to the Periodic Analysis screen. Then follow these steps:

1. Enter the Exception for periodic analysis, which is "Z010", and click the Create button. A pop-up window appears asking for the name of the variant to create.
2. Define "Z010" as the variant. Click the Create button in the pop-up window to open the Variants: Change Screen Assignment screen (Figure 18.6 ❶).
3. Define this variant as applicable for the individual screens (For Indiv. Selection Screens), which means that the parameters of the variant Z010, material CH-3000, are automatically available for selection on the EWS screen for info structure S021, Number 1000. If you choose the For All Selection Screens radio button, then these parameters are available for all of the different screens of EWS by default.

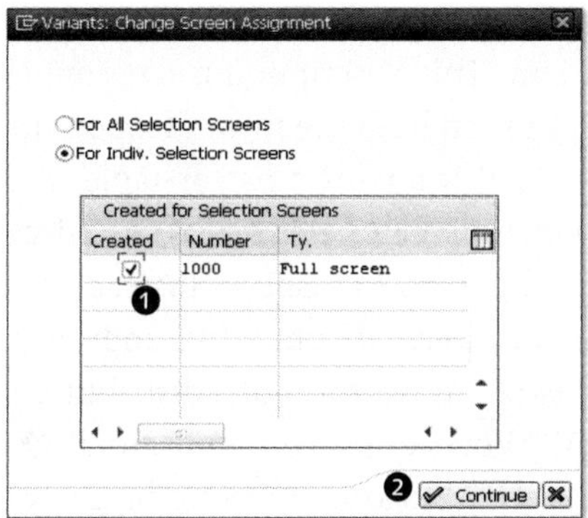

Figure 18.6 Setting Up Periodic Analysis

4. Click the CONTINUE button ❷ to open the MAINTAIN VARIANT screen in Figure 18.7. You can define the initial screen parameters for your variant here. For this example, define that the system should automatically select MATERIAL CH-3000 and make it available as the default.

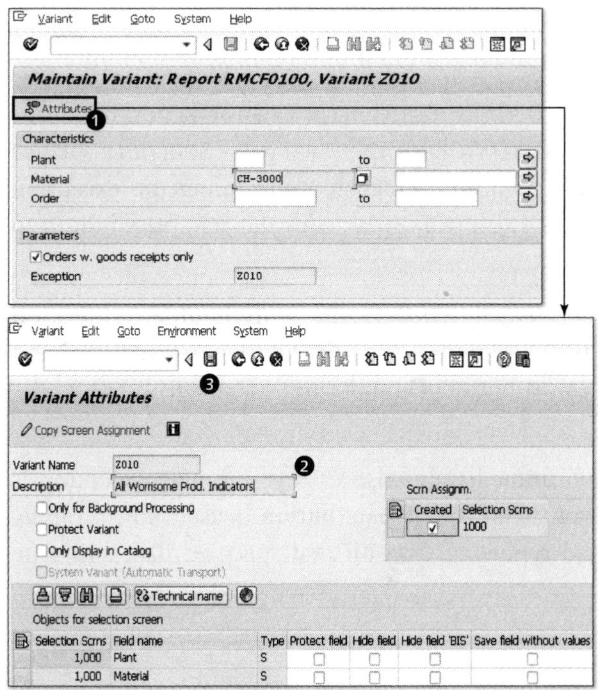

Figure 18.7 Maintain Variant in EWS

5. After defining the parameters, click the ATTRIBUTES button ❶, where you can provide a meaningful description of the variant. This description is communicated whenever EWS needs to alert an exception, in both the form of a pop-up window and also when it's delivered to an SAP Inbox as a list or table. An example of a meaningful description is "Production material is out of specification," which immediately alerts the concerned person to respond instead of a more generic and vague descriptions such as "PP Specs. are out." This approach also helps if the same alert is sent to many people in the form of a distribution list. For this example, use the description "All Worrisome Prod. Indicators" ❷. Finally, click on the SAVE icon ❸ or press Ctrl+S to save the periodic analysis.

   A message appears to confirm that variant Z010 is saved with all of the values on the selection screen.

18.4 Schedule an Early Warning System

The next step is to define the scheduling information on exception Z010 to determine how frequently the system should analyze it for potential exceptions.

Use the menu path, LOGISTICS • LOGISTICS CONTROLLING • SHOP FLOOR INFORMATION SYSTEM • EARLY WARNING SYSTEM • PERIODIC ANALYSIS • SCHEDULE, or use Transaction MC=B to bring up the initial screen for scheduling the exception Z010 (Figure 18.8). After defining EXCEPTION Z010, click the SCHEDULE button to schedule your analysis frequency.

Figure 18.8 shows two options: either schedule the analysis of the exception group immediately ❶, or define it as a periodic job ❷ and then define the period values. Click on the OTHER PERIOD button ❸, and enter "1" (MINUTE) ❹ in the ensuing pop-up.

The option to schedule the job immediately ensures that any new exception is immediately reported. However, if the DATE/TIME button is used, for example, then exception monitoring and reporting start on that specified date and time only, as a scheduled job.

The second option of scheduling a periodic job ❺ helps ensure that system resources aren't unduly overburdened.

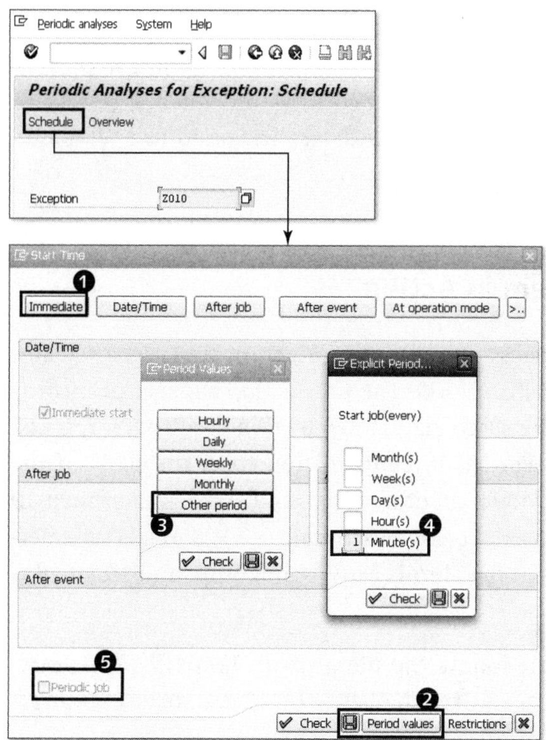

Figure 18.8 Define Schedule Details for EWS

Tips & Tricks [+]

The scheduling frequency should be practical as well as within a reasonable gap, for example, daily. Under the DAILY scheduling option, all exceptions for the entire day are consolidated into one issued alert. This allows the person who set up the alert to attend to all exceptions in one sitting.

Click the SAVE icon to save your scheduling details. After saving, a message appears to confirm that the background job for exception Z010 has been successfully planned.

This ends our discussion on how to set up EWS. The next section shows how EWS works in the system, when the user performs day-to-day business functions, and how the system alerts the user who sets up EWS when an exception occurs. The business users can continue to perform and report their daily activities in the SAP ERP system, such as performing goods receipts against purchase orders, reporting daily production with scrap percentage, and recording the quality results of inspection lots. Actual monitoring occurs and alerts are triggered in EWS based on this day-to-day information.

18.5 Early Warning System in Action

For this example, the business process owner (the production supervisor) creates a production order for the material CH-3000. The production supervisor records the actual production as well as the scrap via the production confirmation transaction in SAP ERP. The details recorded in confirmation form the basis of the EWS. Any negative trend in the production of material, any scrap greater than 10 kg or quantity of the produced material below 80 kg causes EWS to alert the person who set it up. For this example, the alert is sent as a pop-up, and an email is sent to our SAP Inbox.

To create a new production order, follow the menu path, Logistics • Production • Shop Floor Control • Order • Create • With Material, or use Transaction CO01 (see Figure 18.9).

Enter the Material "CH-3000", Production Plant "1100", and Order Type "PP01". Press Enter. On the Production order screen, enter a Total Qty (planned) as "200 KG", and release the production order by clicking on the Release (green flag) icon. Save the production order, and, for this example, the system then generates production order number 60003711.

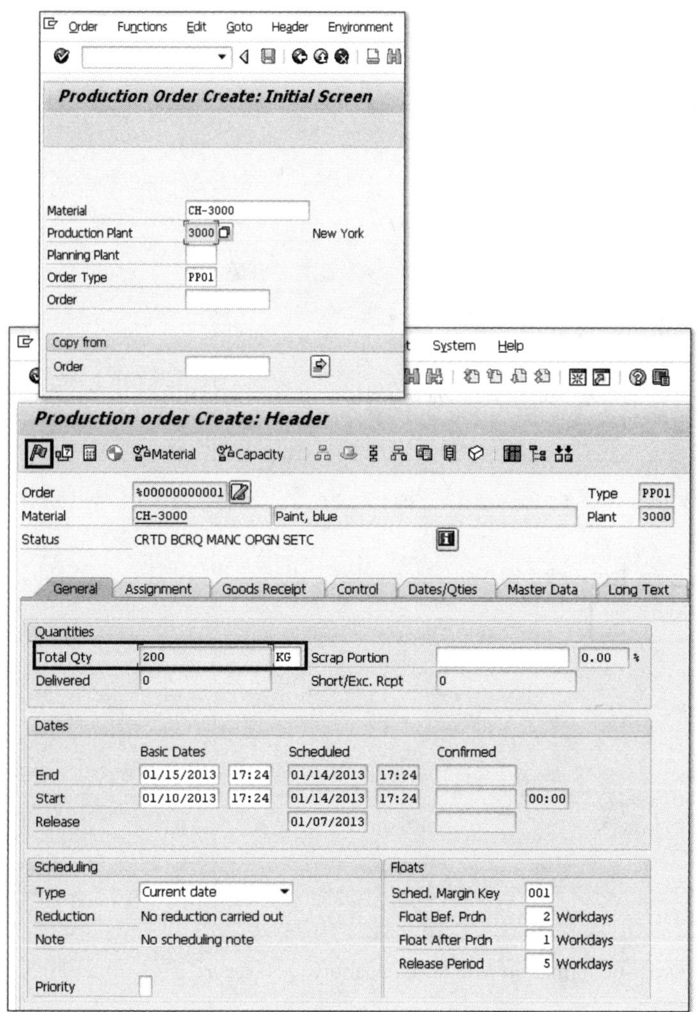

Figure 18.9 Production Order Creation

The next step is to record the production quantities as well as the scrap in confirmation of the production order. To record confirmation of the production order, follow the menu path, LOGISTICS • PRODUCTION • SHOP FLOOR CONTROL • CONFIRMATION • ENTER • FOR ORDER, or use Transaction CO15 (Figure 18.10).

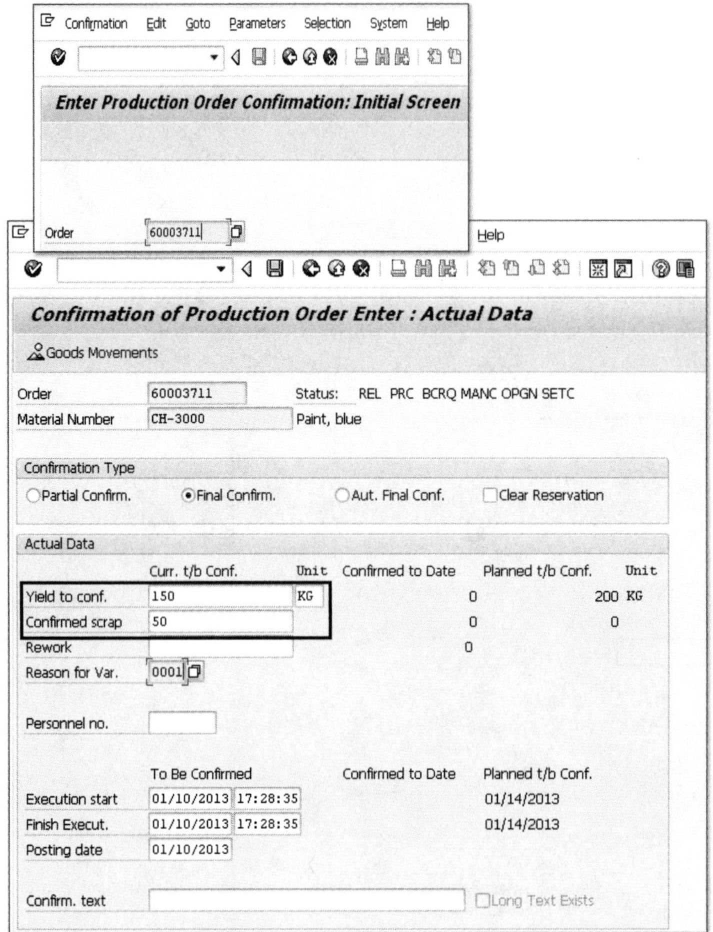

Figure 18.10 Production Order Confirmation with Scrap Quantity

Enter the production ORDER number "60003711", and press [Enter]. In the CONFIRMATION OF PRODUCTION ORDER ENTER: ACTUAL DATA screen, enter the production quantity (YIELD TO CONF.) as "150 KG" and CONFIRMED SCRAP of "50". Save the confirmation by pressing [Ctrl]+[S].

As soon as you save the confirmation, the system issues a pop-up (see Figure 18.11) message to alert the occurred deviation (exception).

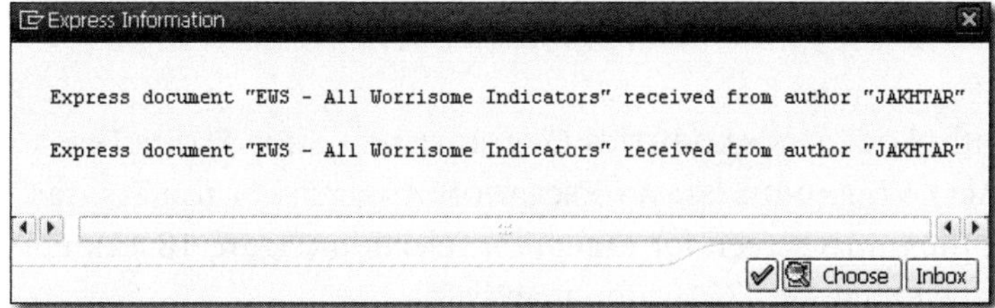

Figure 18.11 Pop-Up Alert Issued by EWS as Soon as an Exception Takes Place

Figure 18.11 shows that two of the three exceptions (deviations) have occurred:

- The planned versus actual production shows a negative trend. That is, that planned production quantity of 200 kg had in fact actual production of 150 kg, hence the negative trend.
- The production scrap recorded in the confirmation was 50 kg, which is greater than the 10 kg or more limit set to issue an EWS alert.

The third exception hasn't occurred, which is if the production quantity during confirmation was less than 80 kg. For this example, the produced quantity (yield) was 150 kg for the production order 60003711.

The system also delivers the same exception message in the SAP Inbox.

Tips & Tricks [+]

While maintaining scheduling to analyze for individual exceptions or exception grouping, make sure to eliminate duplication of information, so that the system doesn't present the same information repeatedly—first in an individual exception and then in an exception grouping. This is achieved by adopting a well-thought-out approach prior to setting up EWS, for example, making a list of all important materials for which you want to monitor parameters for deviations.

18.6 Exception Analysis

Finally, we'll show you how to analyze exceptions with EWS in the shop floor information system. While alert monitoring is certainly a valuable EWS tool to promptly alert you about any deviations and exceptions, you're often required to have detailed information on the exceptions. Fortunately, EWS is supported by

PP, which ensures that comprehensive details related to all exceptions are instantly available.

Follow the menu path, LOGISTICS • LOGISTICS CONTROLLING • SHOP FLOOR INFORMATION SYSTEM • EARLY WARNING SYSTEM • EXCEPTION ANALYSIS, or use Transaction MCYJ to get to the initial screen for exception selection (Figure 18.12). For this example, enter the exception "Z010" for analysis.

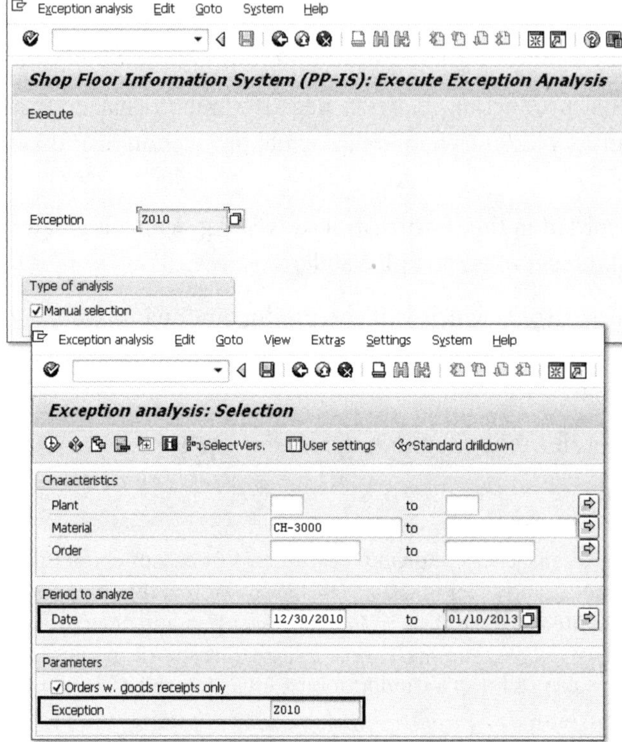

Figure 18.12 Shop Floor Information System: Exception Analysis

Click the EXECUTE icon in Figure 18.12, and the system displays the selection parameter screen. Define MATERIAL "CH-3000", as well as a PERIOD TO ANALYZE from "12/30/2010" to "01/10/2013".

Select the ORDER W. GOODS RECEIPTS ONLY checkbox to denote that the system should only take those production orders into account for which the goods receipts in the warehouse have been undertaken.

When you execute the exception analysis report (by pressing [F8] or clicking the EXECUTE icon) in Figure 18.12, the system displays the results as shown in Figure 18.13.

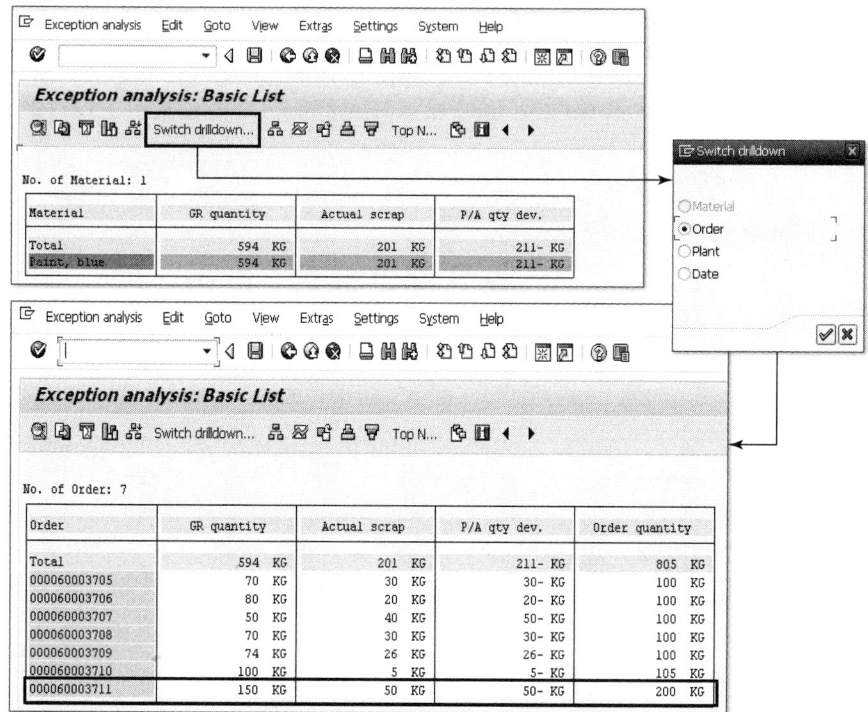

Material	GR quantity	Actual scrap	P/A qty dev.
Total	594 KG	201 KG	211- KG
Paint, blue	594 KG	201 KG	211- KG

Order	GR quantity	Actual scrap	P/A qty dev.	Order quantity
Total	594 KG	201 KG	211- KG	805 KG
000060003705	70 KG	30 KG	30- KG	100 KG
000060003706	80 KG	20 KG	20- KG	100 KG
000060003707	50 KG	40 KG	50- KG	100 KG
000060003708	70 KG	30 KG	30- KG	100 KG
000060003709	74 KG	26 KG	26- KG	100 KG
000060003710	100 KG	5 KG	5- KG	105 KG
000060003711	150 KG	50 KG	50- KG	200 KG

Figure 18.13 Exception Analysis Report of EWS for the Material CH-3000

This exception analysis shows the three defined key figures with data as separate columns. Further, the color is red for the entire row of the MATERIAL column, as we originally defined in Figure 18.5.

As with all standard analyses reports, you can use the SWITCH DRILLDOWN button to view details on a production order basis or on a day (date) basis. The lower half of the figure provides production order details. For this example, the production order 60003711 contains all of the correct details you previously entered in various business transactions, such as production order, confirmation, and goods receipt for production order.

18.7 Summary

This chapter explained how easily a business process owner in any relevant area of working can quickly set up EWS without any extensive functional or technical knowledge. EWS alerts can be modified, deleted, or updated anytime, as deemed necessary to reflect the business need.

The next chapter covers reporting in the SAP ERP system.

Timely, accurate, and comprehensive information helps you make better business decisions. You can immediately start to use a large number of information systems and standard reports available in the SAP ERP system for Production Planning, as well as for logistics.

19 Reporting in SAP

This chapter provides detailed guidance on how to leverage a large number of standard reports that are available in all the logistics components of the SAP ERP system. Not only will the chapter cover the standard reports as information systems available for day-to-day reporting but also a large number of standard analyses available to help business process owners make important and better decisions. After you become familiar with the features and functions for reporting, including tools and icons, it becomes easier to navigate to unexplored areas of reporting.

In this chapter, we specifically cover the information systems of discrete and process manufacturing, including the options to leverage several functions beneficial to users, such as the ability to create various graphs from system data. After we cover the options for reporting with real-time information in the information systems, we'll discuss how to report on historical data with standard analyses available for discrete, process, and repetitive manufacturing (REM), as well as some of the associated features such as time series and ABC analysis. You'll understand how a user can read any SAP ERP system table from the database and be able to manipulate data to an extent. Next, we cover two different ways to enable the user to independently create reports in SAP ERP system: QuickViewer and SAP Query. Finally, we show you how to assign a custom transaction code to a report, created by either of the two query options.

Note [«]

You can access standard analyses not just for SAP ERP Production Planning (PP), but for all of the logistics components of SAP from the central location in the Logistics Controlling node of the SAP menu via LOGISTICS • LOGISTICS CONTROLLING.

You can then navigate to the relevant component's standard analyses option.

19.1 The Basics of Reporting

To fully take advantage of your SAP ERP system, it's essential that you become familiar with the icons and other features frequently available for use in reports. Because these icons and functions are the same in all of the information systems and analyses, you'll begin to navigate faster and be able to quickly get to the desired information or even find tips and tricks to manage the display and availability of data, as needed.

Table 19.1 includes a selective list of icons available when you run the information system or standard analysis. The Details/Application column of this table provides a brief description of when and where an icon is used.

[»] **Note**

The icons in Table 19.1 specifically refer to information systems, analyses (standard and flexible), and reporting tools such as QuickViewer or SAP Query in PP and other Logistics functions. It's important to note that you may find similar icons in master data and transactions such as REM planning table (Transaction MF50) or the stock/requirements list (Transaction MD04), where these icons may have different meanings and functions. For example, in a standard business process, clicking the DETAILS icon brings up the ITEM details of the object. In the reporting function, however, this icon brings up a comprehensive list of ALL KEY FIGURES.

Icon	Icon Detail	Details/Application
	REFRESH	Refresh data
	CHANGE	Change object; for example, process order
	DISPLAY	Display object; for example, process order
Order	CHANGE ORDER	Change production/process order
Order	DISPLAY ORDER	Display production/process order
	MAINTAIN SELECTION OPTIONS	Select operands such as greater than, less than, and so on
	CHOOSE KEY FIGURES	Display key figures in analyses
	GET VARIANT	Choose a previously saved layout
	DESELECT ALL	Unselect all previously selected values

Table 19.1 Selective List of Icons Used in Reporting

Icon	Icon Detail	Details/Application
	Select All	Select all lines or key figures
	Check Entries	Check entered data before execution
	Cancel	Cancel and return to previous screen
Multiple selection..	Multiple selection	Checkboxes to make multiple selections
	Insert line	Insert line between values
	Delete line	Delete a selected line
	Aggregation	Summation
	Sub-total	Applies subtotal, after a total on key figure is first performed
	Do Not Choose All	Deselect all
	Choose All	Select all
	Hide Selected Field	Hide a field or column from display
	Show Selected Field	Display a field or column
	Time Series	Time line of key figures
	Execute (F8)	Run the transaction or report
	Display Key Figures	Display all key figures
	Graph	Graphical representation
Switch drilldown...	Switch drilldown	Switch key figures display
	Sort in Ascending Order	Select the column, and click on the icon to sort in ascending order
	Sort in Descending Order	Select the column, and click on the icon to sort in descending order
	Import from Text File	Import/upload from notepad
	Upload from Clipboard	Upload data from notepad
	Delete Entire Selection Line	Delete the selected entry
	Move	Move field from one location to another
	Mail	Send by email
	Drilldown By	Drilldown by characteristic
	Hierarchical Drilldown	Change drilldown level

Table 19.1 Selective List of Icons Used in Reporting (Cont.)

Icon	Icon Detail	Details/Application
	OTHER INFO STRUCTURE	Switch to another info structure (when already in one info structure)
	DOWNLOAD	Download the data in various formats
	CHANGE LAYOUT	Change or select layout
	FILTER	Set or delete filter
	PRINT	Print or print preview of reports
	MULTIPLE SELECTION	Available on every parameter's selection screen
	FIND/SEARCH	Search for any term
	SEARCH AGAIN	Search the same details/values again
	CALCULATE	Perform calculation

Table 19.1 Selective List of Icons Used in Reporting (Cont.)

Information systems basically present a large amount of data and information into logically structured lists and sections. The following menu paths and transaction codes of both discrete manufacturing and process manufacturing are provided to make the information systems easier to use:

- **Discrete manufacturing**
  Choose LOGISTICS • PRODUCTION • SHOP FLOOR CONTROL • INFORMATION SYSTEM • ORDER INFORMATION SYSTEM, or use Transaction COOIS.
- **Process manufacturing**
  Choose LOGISTICS • PRODUCTION – PROCESS • PROCESS ORDER • REPORTING • ORDER INFORMATION SYSTEM, or use Transaction COOISPI.

19.2 Order Information System

When we talk about the *order information system*, it can be either the production order information system or the process order information system. The former is specific to discrete manufacturing reporting, while the latter is for process manufacturing reporting.

An information system consists of the most up-to-date information of the specific object and is based on the transactions that the user has performed as part of the business process. For example, when you execute the operations list in a production order information system, the system displays the current status of various operations, including those that are completely or partially confirmed. If you again run the same report after six hours, for example, when the user had recorded more confirmations for production orders, at that time, the system shows up-to-date (and different) information. Another example is when the user first uses the Sales Information System (SIS), the system shows the creation of 10 sales order for a specific material. After six hours, when the user again runs the report (SIS), the number of sales orders has increased to 18. This information helps to reflect the accurate and up-to-date status of various activities and business processes.

In the order information system, we've divided the large number of available options for parameters selection into several figures in the following sections for ease of understanding and comprehension.

19.2.1 Selection Screen at the Header Level

In Figure 19.1, which shows the PROCESS ORDER INFORMATION SYSTEM screen (Transaction COOISPI), the user needs to first select the list from the LIST ❶ dropdown menu, so that further parameters selection directly relates to the list.

Based on the list option that you select, you can then define further selection parameters. If you select COMPONENTS from the dropdown list ❷, then you need to give selection parameters in the screen shown in Figure 19.2 (in the next section) at the SELECTION AT COMPONENT LEVEL ❷. If you select OPERATION/PHASES from the dropdown list, then you provide greater details in the SELECT. AT OPERATION LEVEL area ❶.

This all-in-one report caters to the reporting requirements of all major business processes of production. Notice that you can also provide the sales order number with a sales order item number, if you have a make-to-order (MTO) process order (or production order). Similarly, if a process order is with reference to any work breakdown structure (WBS) element, you can enter that information in the selection criteria at the header level.

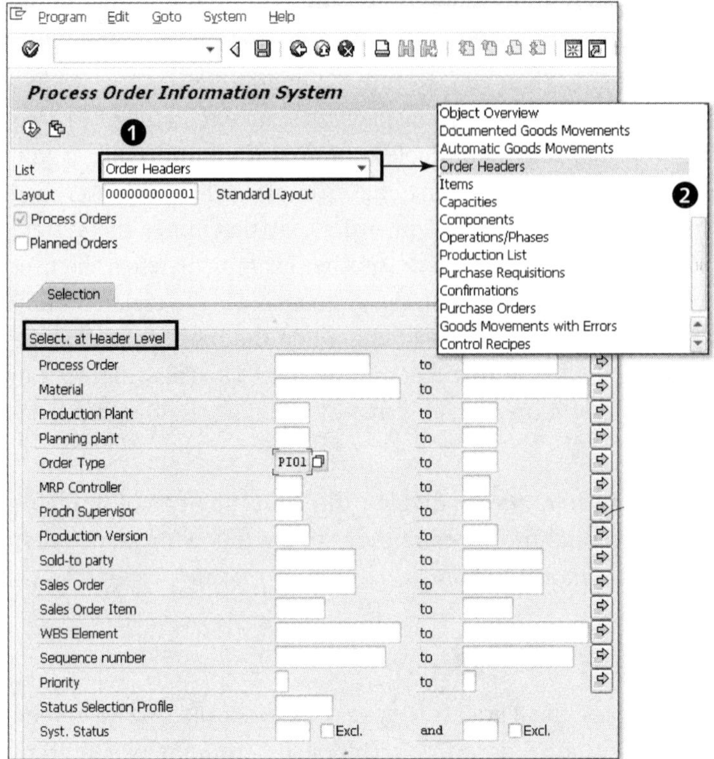

Figure 19.1 Selection Screen for the Process Order Information System

For the example shown previously in Figure 19.1, select ORDER HEADERS ❶ from the dropdown list, and then on the header level, define ORDER TYPE PI01 ❷.

You can select an option from 18 available lists in discrete manufacturing and 14 available lists from process manufacturing. In fact, selection of the list is the first step that you need to ensure before you can run the order information system.

Table 19.2 shows the available lists in the production order information system and in the process order information system.

List	Production Order Information System	Process Order Information System
Object Overview	X	X
Documented Goods Movement	X	X
Trigger Points	X	
Production Resource/Tool	X	
Automatic Goods Movement	X	X
Order Headers	X	X
Items	X	X
Capacities	X	X
Components	X	X
Operations	X	X (phases also)
Production List	X	X
Purchase Requisitions	X	X
Confirmations	X	X
Sequences	X	
Purchase Orders	X	X
Documents Links	X	
Goods Movements with Errors	X	X
Control Instructions	X	X (control recipe)

Table 19.2 Comparison of Available Lists in Information Systems

19.2.2 Selection at the Operations and Components Levels with Options

Scrolling down the screen shown previously in Figure 19.1 leads to several different views where you can enter selection details at the operation and component levels in the SELECT. AT OPERATION LEVEL view as shown in Figure 19.2. You can also select a specific system status (SYST. STATUS) with the EXCL. checkbox selected to exclude information having the specified status. For example, if you don't want to see partially confirmed operations/phases in the report, you can enter "PCNF" as SYST. STATUS and select the EXCL. checkbox.

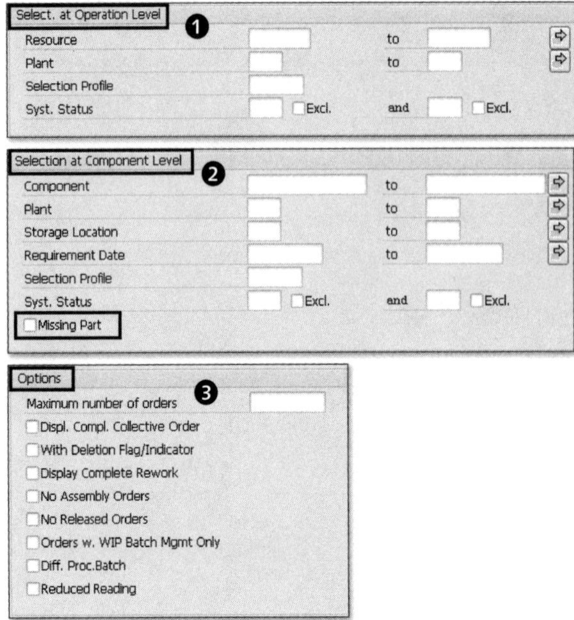

Figure 19.2 Selection Options at Operation and Component Levels

You can select the MISSING PART checkbox so that the system displays the components having the missing parts status in the process orders. The missing parts status is denoted by MSPT.

You can also limit the number of process orders that the system displays so that it's displaying only collective process orders or process orders with deletion flags by selecting the relevant checkboxes.

19.2.3 Selection Screen for Dates

Scrolling further down, you have a large choice of dates. So, if you want your report to show from a certain start date to end date in the process order, you define this in basic start dates, actual start dates, or even according to the release dates of the process orders.

The RELATIVE DATES AT HEADER LEVEL section corresponds to the time interval between today's date and any date, either in the past or future. For example, if you want to see all of the process orders for the past 100 days, as well as for the next 50 days, then you enter "100–" in the RELATIVE BASIC START DATE field and

"50" in the To field. Dates in the past are denoted by the negative sign, after entering the number (e.g., 100–).

19.2.4 Multiple Selection

So far, you may have noticed that on the selection screen of a report, the option is available to enter any parameter in a range or an interval. For example, for the Relative Basic Start Date field in Figure 19.3, you can only enter a selection interval. But what if you want to exclude a certain period within the date range on the selection screen or include two specific dates, apart from the dates given in the dates' interval? All this and several other options are available via the Multiple Selection option available in all information systems and reports of all components.

Click on the Multiple Selection icon in Figure 19.3 ❶ for Process Order. In the resulting screen, four tabs allow you to include or exclude single values or even an entire range. You can enter information in each of the four tabs, and the ensuing output (report) will show the desired results. You'll also find available icons to further facilitate the user, such as copying data from a clipboard, adding or deleting a line within the values, or using the Multiple Selection option. The Multiple Selection icon enables the user to select multiple values by clicking on the requisite checkboxes. Refer to Table 19.1 for some of the functions of each of the icons.

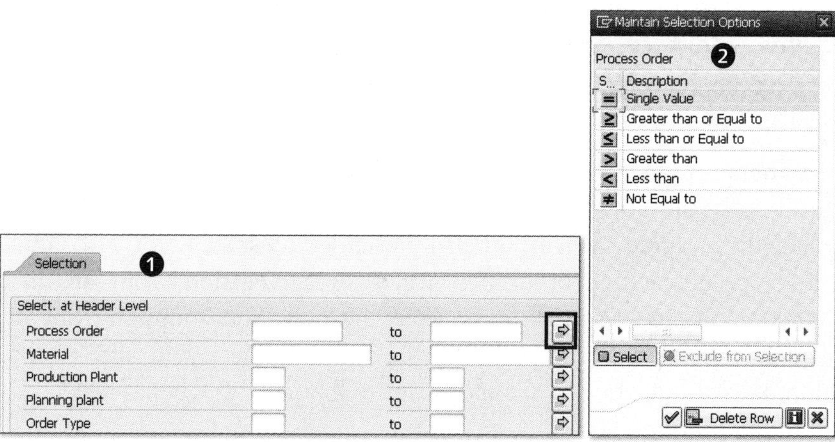

Figure 19.3 Multiple Selection and Maintain Selection Option in the Order Information System

19.2.5 Maintain Selection

A frequently used and available option is the MAINTAIN SELECTION option. When you click the MAINTAIN SELECTION icon, the MAINTAIN SELECTION OPTIONS screen appears as shown earlier in Figure 19.3 ❷. You can use any of the available options in this screen. Alternatively, a shortcut to the MAINTAIN SELECTION OPTIONS screen is to double-click on the relevant field of the selection screen of any report.

19.2.6 Maintain Variant

Maintaining a variant in the SAP ERP system is similar to saving a layout, which is then available for repeated use when needed. You can also set a variant as default by managing your layout options. For example, when you use the order information system, you always use plant 1000, order type PI01, and the dates range of 30 days from today. Instead of entering all of these parameters every time you run the order information system, you can save these details as a variant, which you can then subsequently select.

There are two areas where you can maintain variants. The first one is on the INITIAL PARAMETERS SELECTION screen of any report, information system, or analysis when defining the selection parameters. Simply enter the information once in the parameters selection screen, and then save the variant by choosing GOTO • VARIANTS • SAVE AS VARIANT. On the VARIANT ATTRIBUTES screen that appears, give the variant a name and a short description, and then choose whether to make each field required, hidden, or protected from data entry. With the variant saved, next time you need to choose GOTO • VARIANTS • GET to get the variant. You can also delete incorrect or unwanted variants.

The second option to save a variant is when you've already executed an analysis report in the order information system. The system displays all list fields, and you've made the required changes; for example, any filter set, any sort or column in ascending or descending order performed, or aggregation. You can then choose the SELECT LAYOUT icon and select the SAVE LAYOUT option from the dropdown. In the ensuing pop-up, enter the name of the layout and a short description. Here again, you can change or delete the saved layouts, if needed.

19.2.7 Order Header in the Process

When you execute the report by pressing F8 or choosing EXECUTE, a report such as the one shown in Figure 19.4 appears, which is the ORDER HEADERS report for the process order information system.

List Edit Goto Settings Mass Processing System Help

Order Info System - Order Headers

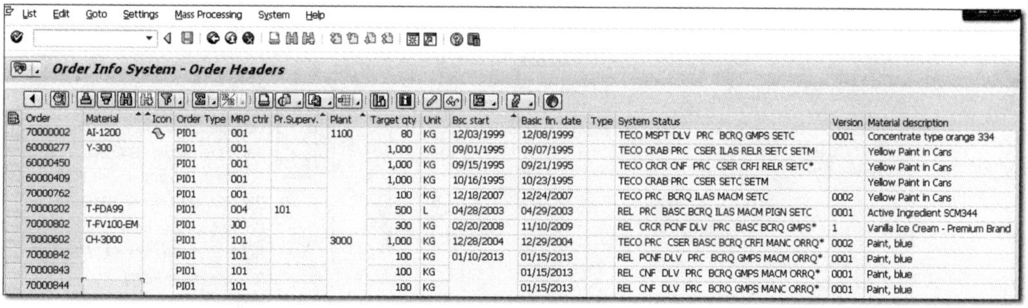

Order	Material	Icon	Order Type	MRP ctrlr	Pr.Superv.	Plant	Target qty	Unit	Bsc start	Basic fin. date	Type	System Status	Version	Material description
70000002	AI-1200		PI01	001		1100	80	KG	12/03/1999	12/08/1999		TECO MSPT DLV PRC BCRQ GMPS SETC	0001	Concentrate type orange 334
60000277	Y-300		PI01	001			1,000	KG	09/01/1995	09/07/1995		TECO CRAB PRC CSER ILAS RELR SETC SETM		Yellow Paint in Cans
60000450			PI01	001			1,000	KG	09/15/1995	09/21/1995		TECO CRCR CNF PRC CSER CRFI RELR SETC*		Yellow Paint in Cans
60000409			PI01	001			1,000	KG	10/16/1995	10/23/1995		TECO CRAB PRC CSER SETC SETM		Yellow Paint in Cans
70000762			PI01	001			100	KG	12/18/2007	12/24/2007		TECO PRC BCRQ ILAS MACM SETC	0002	Yellow Paint in Cans
70000202	T-FDA99		PI01	004	101		500	L	04/28/2003	04/29/2003		REL PRC BASC BCRQ ILAS MACM PIGN SETC	0001	Active Ingredient SCM344
70000802	T-FV100-EM		PI01	J00			300	KG	02/20/2008	11/10/2009		REL CRCR PCNF DLV PRC BASC BCRQ GMPS*	1	Vanilla Ice Cream - Premium Brand
70000602	CH-3000		PI01	101		3000	1,000	KG	12/28/2004	12/29/2004		TECO PRC CSER BASC BCRQ CRFI MANC ORRQ*	0002	Paint, blue
70000842			PI01	101			100	KG	01/10/2013	01/15/2013		REL PCNF DLV PRC BCRQ GMPS MACM ORRQ*	0001	Paint, blue
70000843			PI01	101			100	KG		01/15/2013		REL CNF DLV PRC BCRQ GMPS MACM ORRQ*	0001	Paint, blue
70000844			PI01	101			100	KG		01/15/2013		REL CNF DLV PRC BCRQ GMPS MANC ORRQ*	0001	Paint, blue

Figure 19.4 Order Headers in the Order Information System

In this report, we'll explore a few important options that are available such as aggregation, setting a filter, or viewing the information in a graphical presentation, whereas some of the common and obvious options such as sorting in ascending or descending order of information, and changing or displaying objects aren't covered. Refer to Table 19.1 to gain a better understanding of the icons used.

Fields Selection

Figure 19.4 displays only a limited number of columns and information, whereas you can display and manipulate an enormous amount of information and even save the display layout, if it's routinely used.

When you choose the CHANGE LAYOUT icon, the CHANGE LAYOUT screen appears as shown in Figure 19.5 ❶. The columns or fields already displayed in the report are shown on the left-hand side, whereas the fields or columns that you can display in a report are shown on the right-hand side. To further facilitate the user in selecting fields, you can use several filters to group together relevant fields, so it's easier for a user to choose and display the desired fields in the report. For this example, use the QUANTITIES filter ❶, and then only quantity-relevant fields are available for users to choose from ❷. You can click on the relevant field in Figure 19.5 ❷ and move it with the left-arrow icon, so that the system displays the content in the report.

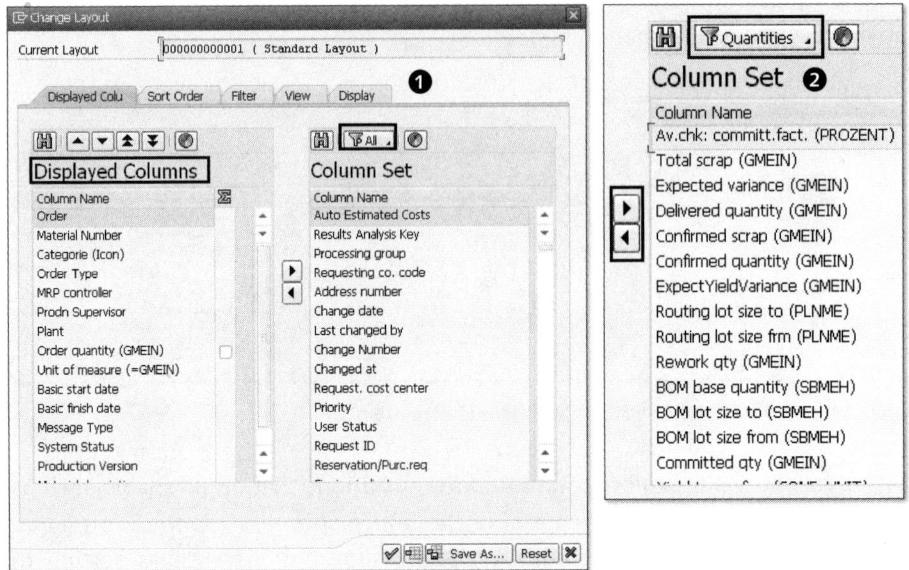

Figure 19.5 Layout and Fields Filter Options in the Order Information System

Quick Menu in Report

Referring back to Figure 19.4, place the cursor on any column and right-click to display all of the available options, not just for that column but for the entire report.

Aggregation

Referring to Figure 19.4 again, you can aggregate or sum up (total) any key figure or quantity. With the Total option in place for a key figure, you can then use the Sub-total option to subtotal any number of characteristics and key figures, as shown in Figure 19.6.

For this example in screen ❶, first use the Total option on the Target quantity column by selecting it and then clicking on the Total icon. Hold down the [Ctrl], key and then select three individual columns—Material, Production supervisor, and Plant—in succession, followed by the Target Quantity column, and click on the SUB-TOTAL icon ❷. The system displays the various summation levels by square dots on the right-hand side. A single dot denotes the summation at the material level, two dots at the production supervisor level, three dots for the plant, and four dots for the target quantity total.

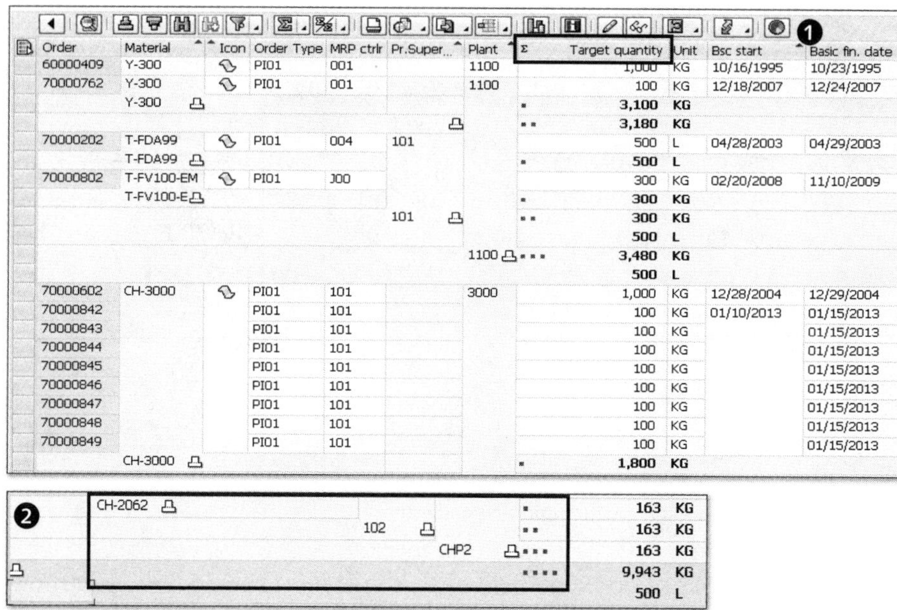

Order	Material	Icon	Order Type	MRP ctrlr	Pr.Super...	Plant	Σ	Target quantity	Unit	Bsc start	Basic fin. date
60000409	Y-300		PI01	001		1100		1,000	KG	10/16/1995	10/23/1995
70000762	Y-300		PI01	001		1100		100	KG	12/18/2007	12/24/2007
	Y-300						▪	**3,100**	**KG**		
							▪▪	**3,180**	**KG**		
70000202	T-FDA99		PI01	004	101			500	L	04/28/2003	04/29/2003
	T-FDA99						▪	**500**	**L**		
70000802	T-FV100-EM		PI01	J00				300	KG	02/20/2008	11/10/2009
	T-FV100-E						▪	**300**	**KG**		
					101		▪▪	**300**	**KG**		
								500	**L**		
						1100	▪▪▪	**3,480**	**KG**		
								500	**L**		
70000602	CH-3000		PI01	101		3000		1,000	KG	12/28/2004	12/29/2004
70000842			PI01	101				100	KG	01/10/2013	01/15/2013
70000843			PI01	101				100	KG		01/15/2013
70000844			PI01	101				100	KG		01/15/2013
70000845			PI01	101				100	KG		01/15/2013
70000846			PI01	101				100	KG		01/15/2013
70000847			PI01	101				100	KG		01/15/2013
70000848			PI01	101				100	KG		01/15/2013
70000849			PI01	101				100	KG		01/15/2013
	CH-3000						▪	**1,800**	**KG**		
	CH-2062						▪	**163**	**KG**		
					102		▪▪	**163**	**KG**		
						CHP2	▪▪▪	**163**	**KG**		
							▪▪▪▪	**9,943**	**KG**		
								500	**L**		

Figure 19.6 Summation (Total and Subtotal) in the Order Information System

You can then also click on the individual summation level by clicking on the SUMMATION icon located at the bottom of the screen.

19.2.8 Filter Settings

While evaluating the information in a report, sometimes it becomes important to filter out some information so that the system displays only the desired information.

Refer to Figure 19.4 once again. In Figure 19.7 ❶, select the TARGET QUANTITY column, click on the FILTER icon, and choose SET FILTER to set the filter. In ❷, you can either double-click on the TARGET QUANTITY field or use the MAINTAIN SELECTION icon, which opens the pop-up window. Select the GREATER THAN OR EQUAL TO option ❸, and choose CONTINUE, which again brings up the screen in ❷.

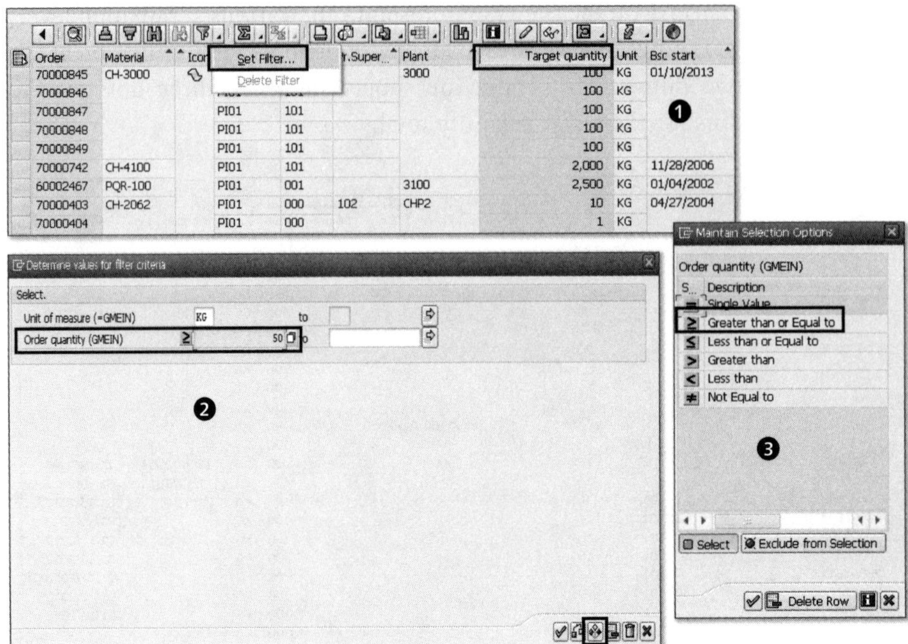

Figure 19.7 Setting Filter in the Order Information System

Enter "50" in the ORDER QUANTITY field so that the report should now only show target quantities equal to or greater than 50. This screen shows that the filter is in place by a little black arrow pointing downward on the TARGET QUANTITY column. The red arrows denote the list is sorted in descending (upward red arrow) or ascending (downward red arrow) order.

[»]

Note

The system will interchangeably show the TARGET QUANTITY or ORDER QUANTITY fields in this report. It's basically the production order quantity (GMEIN field).

19.2.9 Graphs

All information systems as well as standard analyses reports are duly supported by the graphical representation. Furthermore, most of the graph options available in Microsoft Excel, for example, are also available.

In Figure 19.8 ❶, select two columns—MATERIAL and TARGET QTY—and then click on the GRAPH icon, which leads to the initial bar graph ❷. You can change the chart type or other options by right-clicking on the graph and selecting the desired option.

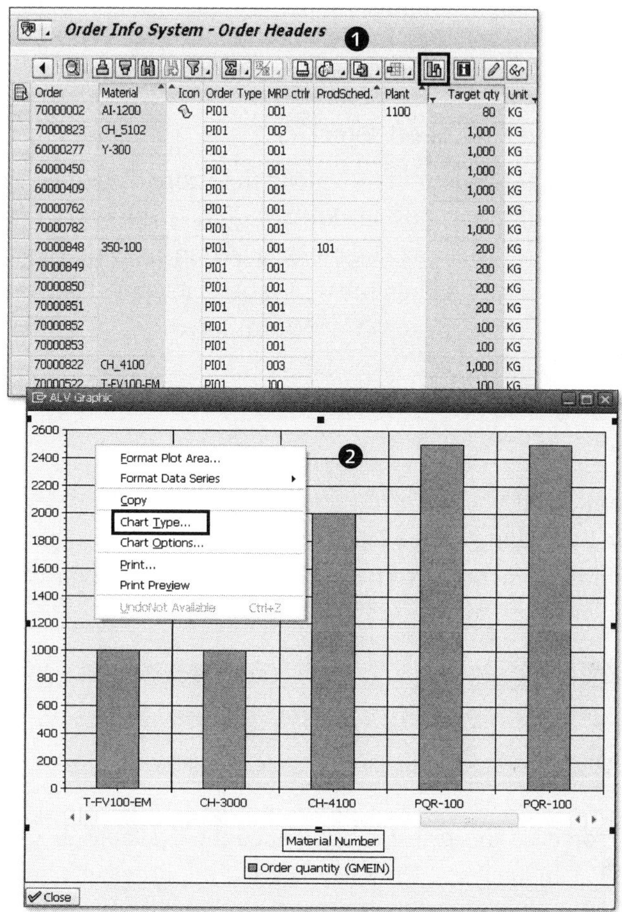

Figure 19.8 Graph Options in the Order Information System

There are a large number of chart types to select from, and you can incorporate the graph title and the titles for the x-axis and y-axis, for example. You can see the print preview of the graph as well as make the actual printout by right-clicking on the graph and selecting the desired function.

19.2.10 Download

When you execute a report or conduct a standard analysis, you can also download the information in one of the several forms available by choosing the DOWNLOAD icon. You can also send the information as an email attachment or even perform an ABC analysis in the information system, although it's also available in the standard analysis reports.

19.2.11 Copy Selective Data to Microsoft Excel

You can copy selected data if you don't want to download the entire data of your SAP ERP system into Excel or a different format. Press Ctrl+Y to select the fields or content within the report. This enables block selection in which you select the block of fields you want to copy. Then press Ctrl+C to copy the data. In Excel or another program (Microsoft Notepad or Word), press Ctrl+P to paste the copied data.

19.2.12 Print

You can see the available options if you want to create a printout of the report. The print option is available by clicking on the PRINT icon within the report. All of the parameters used in the generation of the report in the information system also become part of the printout, including filter, and so on.

[»]

Note

We encourage you to explore other features and functions of each icon as well as the options available on the menu of each report.

We've covered in general some of the features and functions available in the information systems as well as in analyses. Now let's discuss the order information system in more detail. The details covered so far were general in nature and enable you to make effective use of the information systems and analyses in any area of working with your SAP ERP system. We now move on to cover some of the actual lists (reports) of the order information system in the following sections and encourage you to explore the others available. You can access all of the following reports via Transaction COOIS.

19.2.13 Automatic Goods Movement

In automatic goods movement, the system performs one or both of the following functions:

- If the BACKFLUSH functionality is activated, the component's consumption or goods issuance is automatically recorded at the time of confirmation against the process order. This is denoted by movement type 261.
- If the AUTO-GR functionality is activated, the system automatically performs the goods receipt for the process order of the confirmed production quantity when the user performs the confirmation for the process order. This is denoted by movement type 101.

From the LIST dropdown option shown earlier in Figure 19.1 (you'll be referring to this LIST dropdown for each of the report options discussed next), select AUTOMATIC GOODS MOVEMENTS. In Figure 19.9, see the MVMT TYPE (movement type) column for backflush and auto-GR.

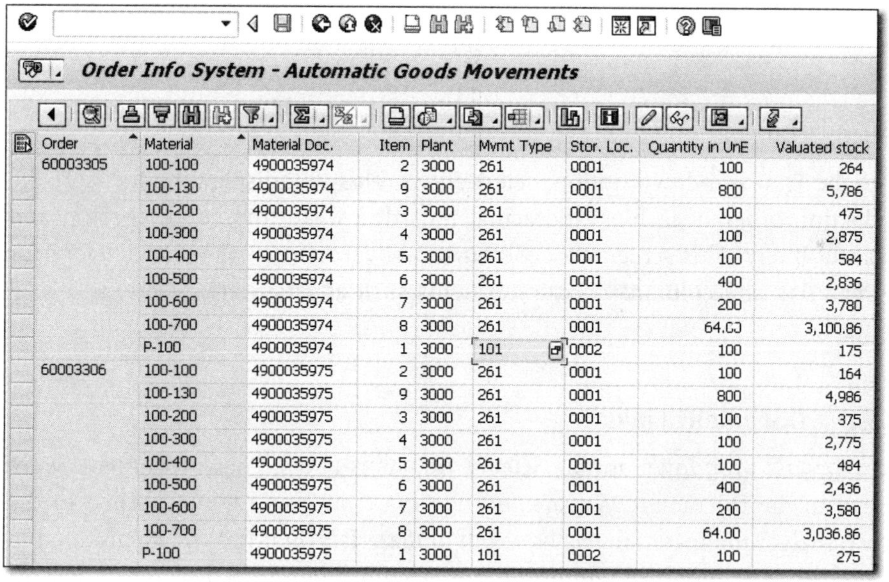

Order Info System - Automatic Goods Movements

Order	Material	Material Doc.	Item	Plant	Mvmt Type	Stor. Loc.	Quantity in UnE	Valuated stock
60003305	100-100	4900035974	2	3000	261	0001	100	264
	100-130	4900035974	9	3000	261	0001	800	5,786
	100-200	4900035974	3	3000	261	0001	100	475
	100-300	4900035974	4	3000	261	0001	100	2,875
	100-400	4900035974	5	3000	261	0001	100	584
	100-500	4900035974	6	3000	261	0001	400	2,836
	100-600	4900035974	7	3000	261	0001	200	3,780
	100-700	4900035974	8	3000	261	0001	64.00	3,100.86
	P-100	4900035974	1	3000	101	0002	100	175
60003306	100-100	4900035975	2	3000	261	0001	100	164
	100-130	4900035975	9	3000	261	0001	800	4,986
	100-200	4900035975	3	3000	261	0001	100	375
	100-300	4900035975	4	3000	261	0001	100	2,775
	100-400	4900035975	5	3000	261	0001	100	484
	100-500	4900035975	6	3000	261	0001	400	2,436
	100-600	4900035975	7	3000	261	0001	200	3,580
	100-700	4900035975	8	3000	261	0001	64.00	3,036.86
	P-100	4900035975	1	3000	101	0002	100	275

Figure 19.9 Automatic Goods Movements in the Order Information System

19.2.14 Capacities

From the LIST dropdown option, select CAPACITIES. Based on the information from the routing of the material, the system automatically calculates the required capacities when the user creates a production order. When the user confirms an operation of the production order, the system correspondingly reduces the capacity requirements of the work center.

This will display the work centers, the operation quantities, and the remaining processing time (in hours) for each of the production orders.

19.2.15 Production Resource/Tool

From the LIST dropdown option, select PRODUCTION RESOURCE/TOOL (PRT). Although not directly involved in the production process, PRT plays a pivotal role in the manufacturing process.

The available PRT categories are material (M), equipment (E), document (D), or miscellaneous (O). The details in the USAGE VALUE denote the duration a PRT is needed during the production process for a production order.

19.2.16 Items

From the LIST dropdown option, select ITEMS. This will display the items of the production order quantities, the actual recorded scrap, the goods receipt, and variance, if any. This screen also contains the overdelivery or underdelivery for the material, and columns containing fields such as QUANTITY, SCRAP, and EXPVARRCPT.

19.2.17 Document Links

From the LIST dropdown option, select DOCUMENT LINKS. These documents were created in the Document Management System (DMS) and then attached to the object, such as the material master or bill of materials (BOM). When a production order for the material containing a DMS document is created or when a BOM containing a DMS document is used in the production order, the system automatically creates the document links. A DMS document consists of a document number, document type, document part, and document version.

19.2.18 Execution Steps

In the production order information system, we get to see the same information that we see in the Control Recipe list from the process order information system. Execution Steps (XSteps) or control recipes are alternatively used in process manufacturing, depending on the option the company chooses as control instructions in Process Management. For discrete manufacturing, the company can only implement the XSteps option in Process Management.

From the List dropdown option, select Control Recipe. This screen shows the status of the control recipe for the specific process order (Created, Sent, Executed, or Terminated), the Batch details, the destination details (Dest. Type, Destinatn, Destination Address columns) of the control recipe, and the in-process (during production) Inspection Lot details.

We've now covered a few reports from the process order information system and the production order information system, and the remaining reports follow the same pattern, while containing information pertaining to a specific area or application.

19.3 Missing Parts Information System

Having access to a standard SAP ERP system report that contains information on missing parts of components in the production or process order is highly valuable to a company's production planner. Production or process orders having missing components are shown with the system status as MSPT. When you execute the report, the system not only shows the requirements quantity but also the committed or available quantity as well as the shortage quantity.

You can access the missing parts information system in all of the different production types, that is, discrete, process, or repetitive. The menu path/transaction for each is listed here:

- **Discrete manufacturing**
  Choose Logistics • Production • Shop Floor Control • Missing Parts Info System or use Transaction CO24.
- **Process manufacturing**
  Choose Logistics • Production – Process • Process Order • Reporting • Missing Parts Info System or use Transaction CO24.

- **Repetitive manufacturing**
  Choose Logistics • Production • Repetitive Manufacturing • Environment • Shop Floor Information System • Environment System • Missing Parts or use Transaction CO24.

19.4 Standard Analysis Reports

The standard analysis reports enable users to draw information from the available standard info structures in the SAP ERP system. The difference between the shop floor information system (reports) and standard analysis is the fact that in information systems, the information is dynamic and changes when the user performs the relevant business functions. In standard analysis, the system draws from historical information to enable the user to make better business decisions. In other words, information from information systems flows into analysis.

In these sections, we cover some of the standard analyses of production, process, and REM. We also show a few features and functionalities available only in analysis and not in information systems, such as time series analysis and ABC analysis.

[»]

Note

Refer to Table 19.3 through Table 19.5 for transaction codes of various available standard analyses in discrete manufacturing, process manufacturing, and REM, respectively. In the following sections, we also provide the relevant SAP ERP system menu path to access the standard analyses of each production type.

19.4.1 Discrete Manufacturing/Production Order

To access the standard analyses of discrete manufacturing, follow the menu path, Logistics • Production • Shop Floor Control • Information System • Shop Floor Information System • Standard Analyses. Table 19.3 lists the transaction codes of various standard analyses available for discrete manufacturing.

Standard Analysis	Transaction Code
Work Center	MCP7
Operation	MCP1

Table 19.3 Standard Analyses for Discrete Manufacturing

Standard Analysis	Transaction Code
Material	MCP5
Production Order	MCP3
Material Consumption	MCRE
Product Costs	MCRI

Table 19.3 Standard Analyses for Discrete Manufacturing (Cont.)

19.4.2 Process Manufacturing/Process Order

To access the standard analyses of process manufacturing, follow the menu path, LOGISTICS • PRODUCTION – PROCESS • PROCESS ORDER • REPORTING • SHOP FLOOR INFORMATION SYSTEM. Table 19.4 lists the transaction codes of various standard analyses available for process manufacturing.

Standard Analysis	Transaction Code
Resource	MCRW
Operation	MCRU
Material	MCP5
Process Order	MCRV
Material Consumption	MCRX
Product Costs	MCRY

Table 19.4 Standard Analyses for Process Manufacturing

19.4.3 Repetitive Manufacturing

To access the standard analyses of REM, follow the menu path, LOGISTICS • PRODUCTION • SHOP FLOOR CONTROL • INFORMATION SYSTEM • SHOP FLOOR INFORMATION SYSTEM • STANDARD ANALYSES • REPETITIVE MANUFACTURING. Table 19.5 lists the transaction codes of various standard analyses available for REM.

Standard Analysis	Transaction Code
GR Statistics	MCP6
Reporting Point Statistics	MCRM

Table 19.5 Standard Analyses for Repetitive Manufacturing

Standard Analysis	Transaction Code
Material Consumption	MCRP
Product Costs	MCRK

Table 19.5 Standard Analyses for Repetitive Manufacturing (Cont.)

19.4.4 Standard Analysis: Work Center

Figure 19.10 ❶ shows a total of 12 plants in the list (Transaction MCP7). No any information has been provided on the selection criteria screen for the standard analysis for work center; we simply executed the report. Choose the SWITCH DRILLDOWN button to open the DRILLDOWN screen. Select the WORK CENTER radio button to change the display of data from plant-level to work center-level. In the DRILLDOWN screen ❷, 96 work centers are listed along with target lead time (TGTLEADTM.) with actual lead time (ACTLEADTM.) details. The smallest/minimum time unit (using shorter intervals can be derived via decimal format) is in days.

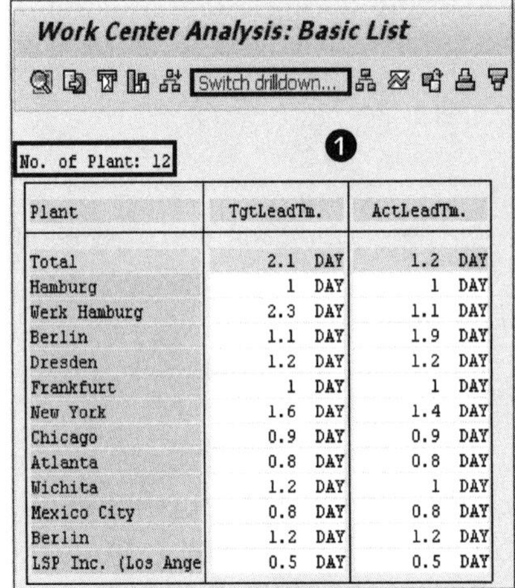

Work Center Analysis: Basic List

Switch drilldown...

No. of Plant: 12 ❶

Plant	TgtLeadTm.	ActLeadTm.
Total	2.1 DAY	1.2 DAY
Hamburg	1 DAY	1 DAY
Werk Hamburg	2.3 DAY	1.1 DAY
Berlin	1.1 DAY	1.9 DAY
Dresden	1.2 DAY	1.2 DAY
Frankfurt	1 DAY	1 DAY
New York	1.6 DAY	1.4 DAY
Chicago	0.9 DAY	0.9 DAY
Atlanta	0.8 DAY	0.8 DAY
Wichita	1.2 DAY	1 DAY
Mexico City	0.8 DAY	0.8 DAY
Berlin	1.2 DAY	1.2 DAY
LSP Inc. (Los Ange	0.5 DAY	0.5 DAY

Work Center Analysis: Drilldown

Switch drilldown...

No. of Work center: 96 ❷

Work center	TgtLeadTm.	ActLeadTm.
Total	2.1 DAY	1.2 DAY
	1 DAY	1 DAY
	0 DAY	0 DAY
Universal Lathe	2.3 DAY	1 DAY
Universal lathe ma	7.4 DAY	1.2 DAY
Drill press I CAP	2.1 DAY	1.1 DAY
Drill press	2.6 DAY	1 DAY
Balancing machine	1.1 DAY	1 DAY
Steel saw	3.5 DAY	1.1 DAY
Annealing furnace	9.7 DAY	1 DAY
Assembly	1.4 DAY	1.4 DAY
Pre-Assembly I	1.1 DAY	1.1 DAY
Pre-Assembly II	3.3 DAY	1.6 DAY
Assembly II	1 DAY	0.5 DAY
Assembly III	1.6 DAY	1.1 DAY
Assembly IV	3.7 DAY	2.2 DAY
Production inspect	2.7 DAY	1 DAY
Production inspect	1.7 DAY	1.1 DAY
Final assembly I	1.6 DAY	1 DAY
Final assembly II	2.2 DAY	1 DAY
Paint shop	1.9 DAY	1 DAY
Press	0.3 DAY	0.5 DAY
Welding	0.3 DAY	0.3 DAY
Inspection work ce	0.3 DAY	0.3 DAY
Fillers	0.3 DAY	0.3 DAY

Figure 19.10 Switch Drilldown in Work Center Analysis

You can display a large number of fields in Figure 19.10 ❷, just as you can in order information systems. To display hidden fields, choose the CHOOSE KEY FIGURES icon (not shown), and the pop-up screen shown in Figure 19.11 appears. The left-hand side of this screen shows the fields already displayed in Figure 19.10, whereas the right-hand side shows a large number of fields for you to choose from. Also notice that there are filters for key figures—ALL, FIGURES (W/O UNIT), and QUANTITIES—which if selected help to quickly identify and select the desired fields in the report.

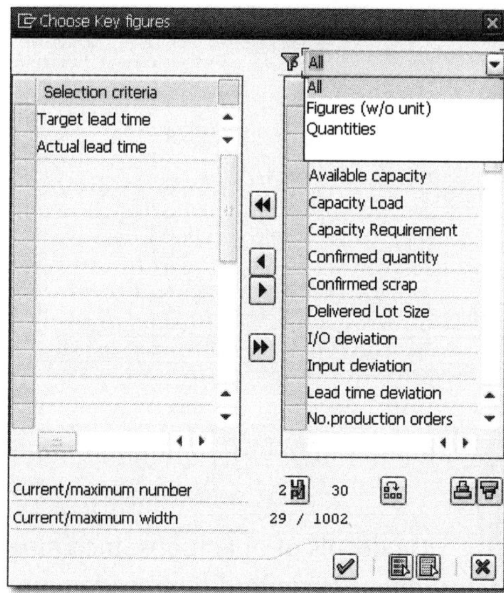

Figure 19.11 Selecting Key Figures in Standard Analysis

For this example, select a few key figures from Figure 19.11 for display. Here you see the input deviation of planned start dates of operation with actual start dates. The values are averaged out. The same concept applies for the output deviation, which is the deviation between the planned end date of operations at the work center with the actual end date entered in the confirmation of the operation. It also shows the quantity deviation of planned versus actual production. Place the cursor on the QUANTITY DEV. (deviation) key figure, and click on the TOP N button.

Figure 19.12 shows the pop-up ❶ to filter out the top 15 work centers having the highest quantity deviations. Enter NUMBER as "15" to then see the TOP 15 QUANTITY DEV. area ❷; the remaining deviations are shown as REST at the bottom of the screen. You can enter any number to view the top numbers of key figures.

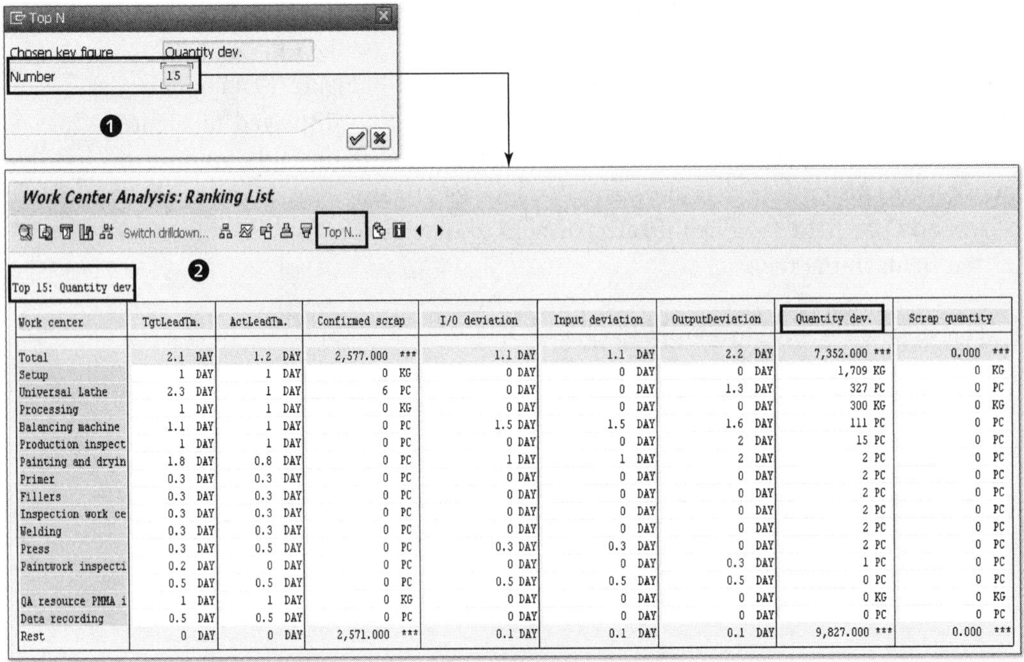

Work center	TgtLeadTm.	ActLeadTm.	Confirmed scrap	I/O deviation	Input deviation	OutputDeviation	Quantity dev.	Scrap quantity
Total	2.1 DAY	1.2 DAY	2,577.000 ***	1.1 DAY	1.1 DAY	2.2 DAY	7,352.000 ***	0.000 ***
Setup	1 DAY	1 DAY	0 KG	0 DAY	0 DAY	0 DAY	1,709 KG	0 KG
Universal Lathe	2.3 DAY	1 DAY	6 PC	0 DAY	0 DAY	1.3 DAY	327 PC	0 PC
Processing	1 DAY	1 DAY	0 KG	0 DAY	0 DAY	0 DAY	300 KG	0 KG
Balancing machine	1.1 DAY	1 DAY	0 PC	1.5 DAY	1.5 DAY	1.6 DAY	111 PC	0 PC
Production inspect	1 DAY	1 DAY	0 PC	0 DAY	0 DAY	2 DAY	15 PC	0 PC
Painting and dryin	1.8 DAY	0.8 DAY	0 PC	1 DAY	1 DAY	2 DAY	2 PC	0 PC
Primer	0.3 DAY	0.3 DAY	0 PC	0 DAY	0 DAY	0 DAY	2 PC	0 PC
Fillers	0.3 DAY	0.3 DAY	0 PC	0 DAY	0 DAY	0 DAY	2 PC	0 PC
Inspection work ce	0.3 DAY	0.3 DAY	0 PC	0 DAY	0 DAY	0 DAY	2 PC	0 PC
Welding	0.3 DAY	0.3 DAY	0 PC	0 DAY	0 DAY	0 DAY	2 PC	0 PC
Press	0.3 DAY	0.5 DAY	0 PC	0.3 DAY	0.3 DAY	0 DAY	2 PC	0 PC
Paintwork inspecti	0.2 DAY	0 DAY	0 PC	0 DAY	0 DAY	0.3 DAY	1 PC	0 PC
	0.5 DAY	0.5 DAY	0 PC	0.5 DAY	0.5 DAY	0.5 DAY	0 PC	0 PC
QA resource PMMA i	1 DAY	1 DAY	0 KG	0 DAY	0 DAY	0 DAY	0 KG	0 KG
Data recording	0.5 DAY	0.5 DAY	0 PC	0 DAY	0 DAY	0 DAY	0 PC	0 PC
Rest	0 DAY	0 DAY	2,571.000 ***	0.1 DAY	0.1 DAY	0.1 DAY	9,827.000 ***	0.000 ***

Figure 19.12 Ranking List in Work Center Analysis

19.4.5 Standard Analysis: Operations

Using Transaction MCP1, you can view day-wise details of operations, including the capacity requirements in hours, the actual queue time, lead time, and execution time. It also shows the target processing time in hours, wait time in hours, and target setup time in hours, among other key figures. Other key fields include CAP. REQMTS (capacity requirements), ACTQUEUETM (actual queue time), and WAIT. TIME.

19.4.6 Standard Analysis: Material

Using Transaction MCP5, you can view the material analysis, which covers the quantity and duration details of various materials. This includes the order quantity, the goods receipt quantity, the scrap, and also the deviations. The duration details consist of delivery time deviation, lead time deviation, and durations such as target queue time and target execution time.

19.4.7 Key Figures

Sometimes, you want to see all of the key figures of one characteristic. Double-click any entry of the characteristic, and the system displays all of the key figures of that characteristic.

19.4.8 Other Info Structures

You can also navigate to the OTHER INFO STRUCTURE screen within the standard analysis as shown in Figure 19.13. To do so, choose the OTHER INFO STRUCTURE icon, and in the ensuing pop-up ❶, enter the name of the new info structure or use the dropdown menu to select one. In this example, INFO STRUCTURE S026 ❷ for material usage is selected.

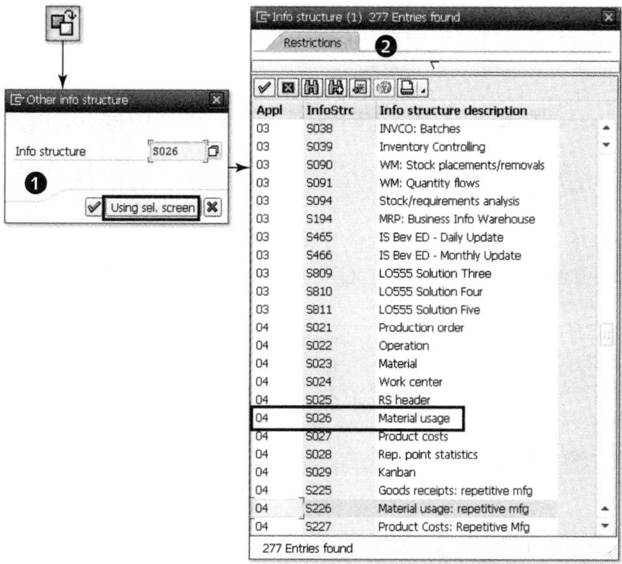

Figure 19.13 Other Info Structures in Production Planning

The standard info structures for PP are shown in Table 19.6.

Information Structure	Information Structure Description
S021	Production order
S022	Operation

Table 19.6 Standard Information Structures Available for Production Planning

Information Structure	Information Structure Description
S023	Material
S024	Work center
S025	Run schedule header
S026	Material usage
S027	Product costs
S028	Reporting point statistics
S029	Kanban
S225	Goods receipts: REM
S226	Material usage: REM
S227	Product costs: REM

Table 19.6 Standard Information Structures Available for Production Planning (Cont.)

19.4.9 Standard Analysis: Goods Receipt in Repetitive Manufacturing

Figure 19.14 ❶ displays the standard analysis of goods receipt in REM (Transaction MCP6). Select the first key figure GR QUANTITY by placing the cursor on it and choosing EDIT • ABC ANALYSIS from the menu bar. In the pop-up that appears, choose the STRATEGY for ABC analysis as TOTAL GR QUANTITY (%) ❷, and press Enter.

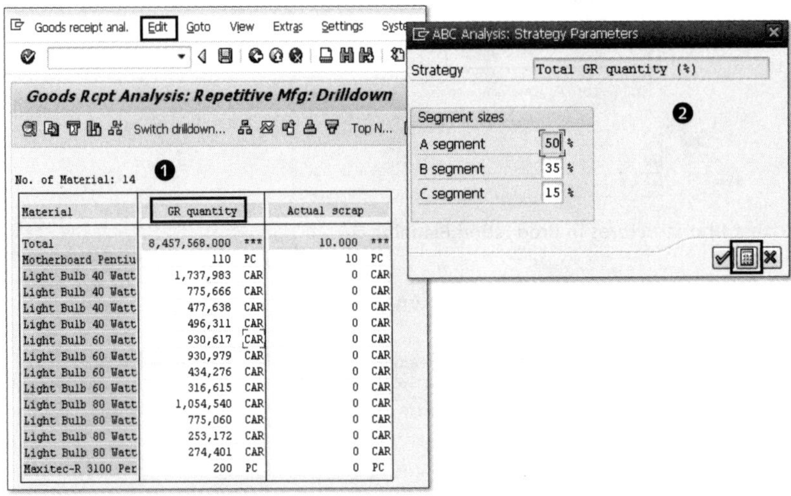

Figure 19.14 Goods Receipt Analysis

Options in ABC analysis are listed here:

- Total GR quantity (%)
  The goods receipt (GR) quantity is divided up by the defined percentage.
- Number Material (%)
  The number of materials is divided up by the defined percentage.
- GR quantity (absolute)
  The actual GR quantity is divided up in either A, B, or C segments.
- Number Material
  The number of materials is divided up in either A, B, or C segments.

ABC analysis enables you to select the right strategy and then divides the key figures into the top three groupings. You also have the flexibility to define what percentage of a key figure to be listed in the A, B, or C segments.

In screen ❷, you have the flexibility to overwrite the system-suggested segment sizes of A, B, or C. For this example, enter A segment as "50", B segment as "35", and C segment as "15". These three entries entail that the system will divide and display the Total GR quantity (%) in such a way that the top 50% will be categorized in segment A, 35% in segment B, and the bottom 15% in segment C in sequential order.

Click on the Calculate icon to open the SAP Business Graphics screen shown in Figure 19.15 ❶. This is the graphical representation of ABC analysis. The blue vertical bar denotes the A, B, or C segment of the GR quantity. Choose the Overview of segments button to open the next screen, which provides the overview of the segments in numbers ❷. Notice that while we've defined the segment to be 50% for A, the system was able to round off the nearest percentage to 55.03% for segment A. The same behavior is observed with segment C, where a user-defined 15% is near-rounded off to 9.99% for segment C.

This happens when there is insufficient data for the system to reflect the results as accurately as the user had defined. In this case, there are only 14 materials for which the system has to perform ABC analysis, so the system rounded off all user-defined values to the nearest percentage points.

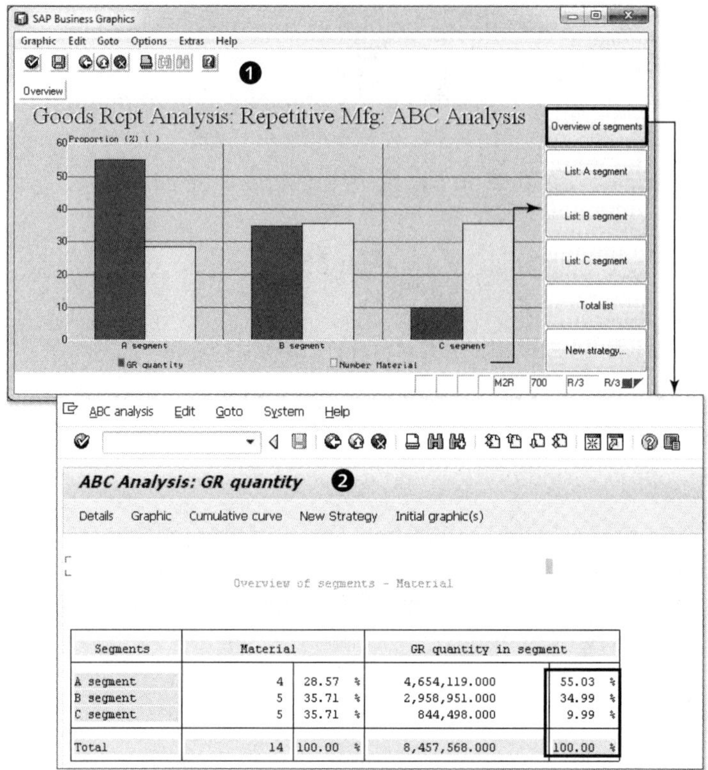

Figure 19.15 ABC Analysis of a Key Figure

19.4.10 Standard Analysis: Product Cost

Using Transaction MRCI, the resulting screen in Figure 19.16 ❶ shows the standard analysis of the product cost in REM. Click on the VALUE COCURR field, and choose the TIME SERIES icon to open the TIME SERIES dialog box ❷.

A *time series* denotes the breakdown of a key figure of a characteristic, which in this case, is MATERIAL. The system shows the characteristic material breakdown of the key figure, VALUE COCURR, for a specific date 10/14/1998. Because the smallest time unit is a day in the product cost report, the system displays a day-wise breakdown of the relevant information.

The next column shows the characteristic material breakdown of the key figure, VALUE COCURR, for a different date, 11/13/1998.

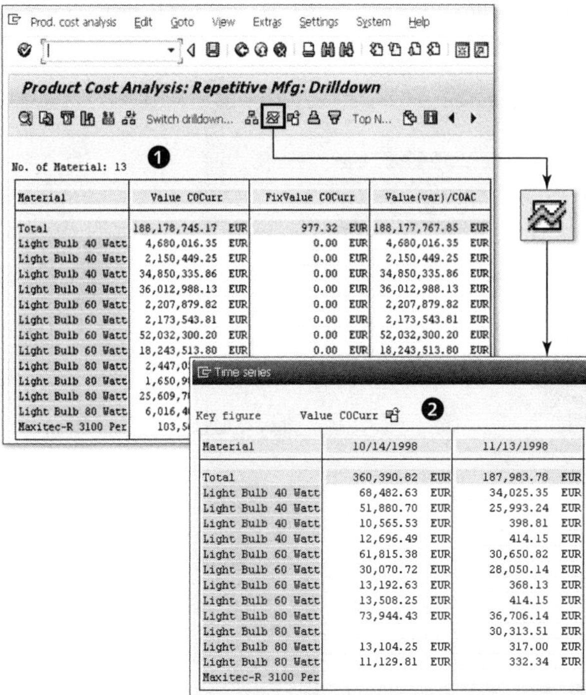

Figure 19.16 Product Cost Analysis and Time Series

19.5 Data Browser

Often the details in the information system of your SAP ERP system or in standard analyses are unable to fulfill the needs of the business process owner. The business process owner needs a direct read from an SAP table. While information systems and standard analyses in the system hold massive information for you to use, often they don't contain all of the fields of a specific table. To attend to this, you can use the Data Browser (use Transaction SE16).

Figure 19.17 ❶ shows the initial screen of the Data Browser where you enter the name of the table that you want to read from an SAP table. For this example, enter the PP TABLE NAME "AFKO", which is for the order header of a production order, and press Enter to bring up the TABLE FIELDS FOR SELECTION pop-up ❷. You can choose the fields for selection, in which you then enter the data and execute for the system to bring up the relevant information only. For this example, select four fields either by their technical name or text: AUFNR, GLTRP, GSTRP, and GAMNG. Choose CONTINUE or press [Enter].

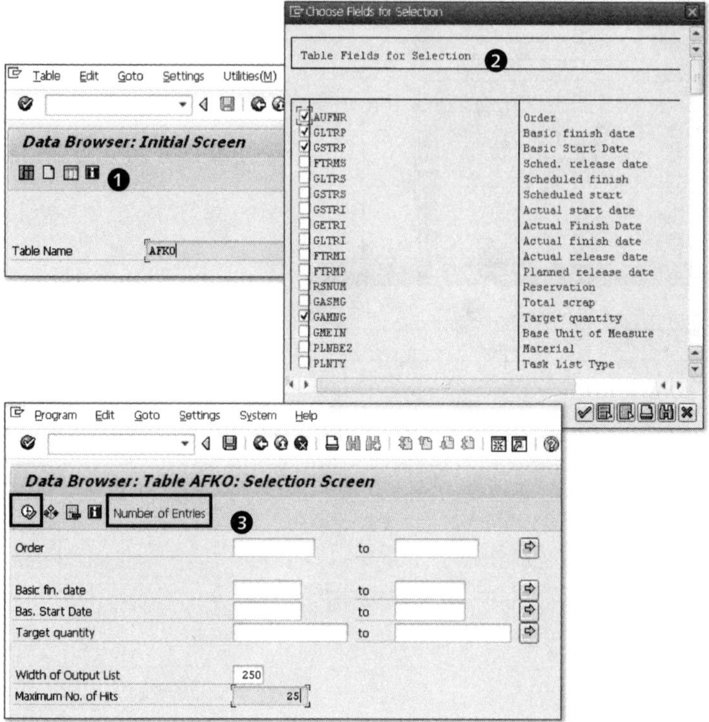

Figure 19.17 Data Browser

In the screen that appears ❸, you can click on the NUMBER OF ENTRIES button to see how many entries will be read by the system if the user runs Data Browser. For this example, enter "25" for the MAXIMUM NO. OF HITS or entries for the system to display on execution. Choose EXECUTE or press [F8] to go to the next screen.

Figure 19.18 shows the selection of only a few fields from the entire table AFKO and also has the option of several standard features to manipulate the data, such as filter, aggregation, and download.

On the initial screen of the Data Browser (Transaction SE16), you can also set user-specific settings by clicking SETTINGS • USER SETTINGS from the menu bar to open the screen.

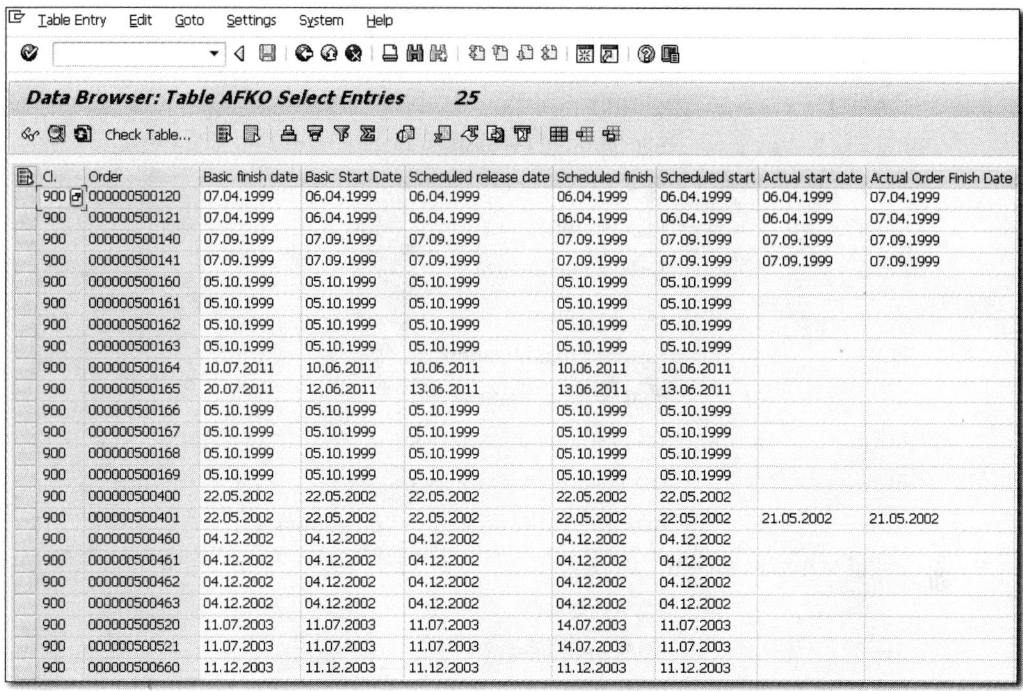

Cl.	Order	Basic finish date	Basic Start Date	Scheduled release date	Scheduled finish	Scheduled start	Actual start date	Actual Order Finish Date
900	000000500120	07.04.1999	06.04.1999	06.04.1999	06.04.1999	06.04.1999	06.04.1999	07.04.1999
900	000000500121	07.04.1999	06.04.1999	06.04.1999	06.04.1999	06.04.1999	06.04.1999	07.04.1999
900	000000500140	07.09.1999	07.09.1999	07.09.1999	07.09.1999	07.09.1999	07.09.1999	07.09.1999
900	000000500141	07.09.1999	07.09.1999	07.09.1999	07.09.1999	07.09.1999	07.09.1999	07.09.1999
900	000000500160	05.10.1999	05.10.1999	05.10.1999	05.10.1999	05.10.1999		
900	000000500161	05.10.1999	05.10.1999	05.10.1999	05.10.1999	05.10.1999		
900	000000500162	05.10.1999	05.10.1999	05.10.1999	05.10.1999	05.10.1999		
900	000000500163	05.10.1999	05.10.1999	05.10.1999	05.10.1999	05.10.1999		
900	000000500164	10.07.2011	10.06.2011	10.06.2011	10.06.2011	10.06.2011		
900	000000500165	20.07.2011	12.06.2011	13.06.2011	13.06.2011	13.06.2011		
900	000000500166	05.10.1999	05.10.1999	05.10.1999	05.10.1999	05.10.1999		
900	000000500167	05.10.1999	05.10.1999	05.10.1999	05.10.1999	05.10.1999		
900	000000500168	05.10.1999	05.10.1999	05.10.1999	05.10.1999	05.10.1999		
900	000000500169	05.10.1999	05.10.1999	05.10.1999	05.10.1999	05.10.1999		
900	000000500400	22.05.2002	22.05.2002	22.05.2002	22.05.2002	22.05.2002		
900	000000500401	22.05.2002	22.05.2002	22.05.2002	22.05.2002	22.05.2002	21.05.2002	21.05.2002
900	000000500460	04.12.2002	04.12.2002	04.12.2002	04.12.2002	04.12.2002		
900	000000500461	04.12.2002	04.12.2002	04.12.2002	04.12.2002	04.12.2002		
900	000000500462	04.12.2002	04.12.2002	04.12.2002	04.12.2002	04.12.2002		
900	000000500463	04.12.2002	04.12.2002	04.12.2002	04.12.2002	04.12.2002		
900	000000500520	11.07.2003	11.07.2003	11.07.2003	14.07.2003	11.07.2003		
900	000000500521	11.07.2003	11.07.2003	11.07.2003	14.07.2003	11.07.2003		
900	000000500660	11.12.2003	11.12.2003	11.12.2003	11.12.2003	11.12.2003		

Figure 19.18 Table AFKO in Data Browser

Table 19.7 contains a selected list of standard PP tables, which you can access or read not only in Data Browser but also in various tools for query development (QuickViewer or SAP Query).

Table	Description
Work Center	
CRHD	Work center header
CRHH	Work center hierarchy
CRHS	Hierarchy structure
CRTX	Text for the work center or PRT
CRCO	Assignment of work center to cost center
KAKO	Capacity header segment
CRCA	Work center capacity allocation

Table 19.7 PP Tables in SAP ERP

Table	Description
TC24	Person responsible for the work center
Routing	
MAPL	Allocation of task lists to materials
PLAS	Task list: selection of operations/activities
PLFH	Task list: PRTs
PLFL	Task list: sequences
PLKO	Task list: header
PLKZ	Task list: main header
PLPH	Phases/suboperations
PLPO	Task list: operation/activity
PLPR	Log collector for task lists
PLMZ	Allocation of BOM items to operations
Bill of Materials (BOM)	
STKO	BOM header
STAS	BOM item selection
STPO	BOM item
STPN	BOM follow up control
STPU	BOM subitem
STZU	Permanent BOM data
PLMZ	Allocation of BOM: items to operations
MAST	Material to BOM link
KDST	Sales order to BOM link
Production Order	
AUFK	Production order headers
AFIH	Maintenance order header
AUFM	Goods movement for production order
AFKO	Order header data PP orders
AFPO	Order item
RESB	Order components
AFVC	Order operations

Table 19.7 PP Tables in SAP ERP (Cont.)

Table	Description
AFVV	Quantities/dates/values in the operation
AFVU	User fields of the operation
AFFL	Work order sequence
AFFH	PRT assignment data for the work order (routing)
JSTO	Status profile
JEST	Object status
AFRU	Order completion confirmations
Planned Order	
PLAF	Planned orders
Reservations	
RESB	Material reservations
RKPF	Header
Capacity Requirements Planning (CRP)	
KBKO	Header record for capacity requirements
KBED	Capacity requirements records
KBEZ	Additional data for table KBED (for individual capacities/splits)
Planned Independent Requirements (PIRs)	
PBIM	Independent requirements for material
PBED	Independent requirement data
PBHI	Independent requirement history
PBIV	Independent requirement index
PBIC	Independent requirement index for customer requirements

Table 19.7 PP Tables in SAP ERP (Cont.)

Tips & Tricks [+]

To quickly find SAP tables that contain the field you want to use in a query, place the cursor on the field, and press F1. When the system brings up the help details, click on the TECHNICAL INFORMATION icon to see extensive details about the field and the SAP table in the ensuing pop-up.

19.6 QuickViewer

You can leverage the knowledge gained so far in reporting to quickly develop reports and other output by creating queries. To do this, you'll use two tools known as QuickViewer and SAP Query.

It's quite easy to create queries in the system, as long as you know the basics of reading SAP tables and can connect various tables or simply use logical databases to generate user-defined reports. This also lessens the need for and the dependency on engaging an ABAP resource to develop relatively simple reports. There are a few features or options that are available in SAP Query but not in QuickViewer, so we'll cover those in the next sections.

To create a query using QuickViewer, follow the menu path, Tools • ABAP Workbench • Utilities • QuickViewer, or use Transaction SQVI. For this example, begin creating a QuickView by joining two relevant PP tables: order header (table AUFK) and reservations (table RESB).

Figure 19.19 ❶ shows the initial screen to create a QuickView, wherein you enter "PP" as the QuickView as an example, and choose the Create button. In the pop-up that appears, provide a Title and add Comments ❷. Then select the Data source as Table join to join two or more SAP tables. If you choose the Table option, then the subsequent information will be limited to one table only. In other words, you have the same information as already covered in the previous section on Data Browser, with the difference that the display information in the query can be intuitively controlled. The Logical database option enables you to select and choose information that is logically connected. For example, all production, procurement, or sales information is logically structured in the database for easy access. Finally, if you choose the SAP Query InfoSet option, the InfoSet must first exist before selection is possible.

[»] **Note**

The InfoSet is covered in the next section on SAP Query.

Select the Basic Mode radio button as well (which is hidden behind the Data source dropdown). The other available option is the Layout Mode (not shown), which enables the user to see a preview of the layout of the report.

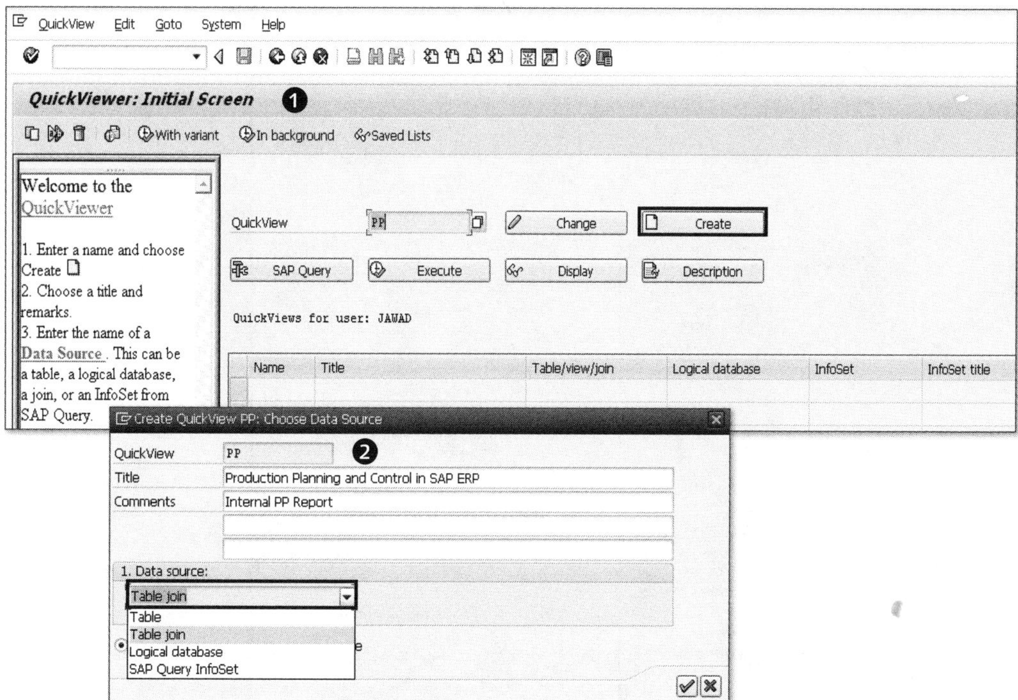

Figure 19.19 Data Source Options Available in the QuickViewer

Choose CONTINUE to open the next screen, and click on the ADD TABLE icon. In the ensuing pop-up, enter the table "AUFK", and choose CONTINUE. The screen then lists all of the fields for order header master data. Click on the ADD TABLE icon, enter the second table "RESB", and press Enter to open the screen shown in Figure 19.20.

The system automatically connects the key fields, which form the logical connection between the two tables. For this example, there is one key field—AUFNR (the ORDER NUMBER). You can also manually connect the key fields by selecting the key field from the first table and dragging it to the second table to form a connection. You must select at least one logically connected field between the two tables to enable the system to completely and correctly read the record of two tables and display the results accordingly.

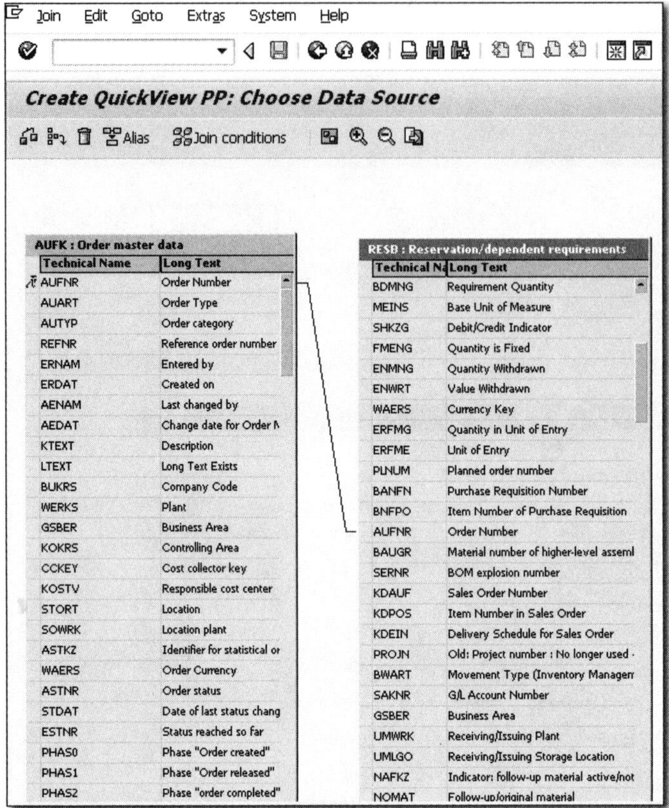

Figure 19.20 Tables AUFK and RESB as Data Sources

Click on the BACK icon, and the system prompts you to generate and save the InfoSet. The InfoSet generation in QuickViewer is automatic, whereas in SAP Query, you have to create the InfoSet. Confirm by choosing CONTINUE. This brings up the screen shown in Figure 19.21, which lists all of the fields of the two tables, AUFK and RESB, that are available for selection. You can make two types of field selections: SELECTION FIELDS or LIST FIELDS. SELECTION FIELDS enable you to provide the necessary selection information on the initial screen of QuickViewer. LIST FIELDS are the fields you want to see displayed with information after executing the report in QuickViewer. You can have fewer, more, or different selection fields than list fields, or vice versa. For example, you may only want plant and order type as selection fields, whereas for list fields, you want the system to display material, plant, order text, target quantity (production quantity), and MRP controller.

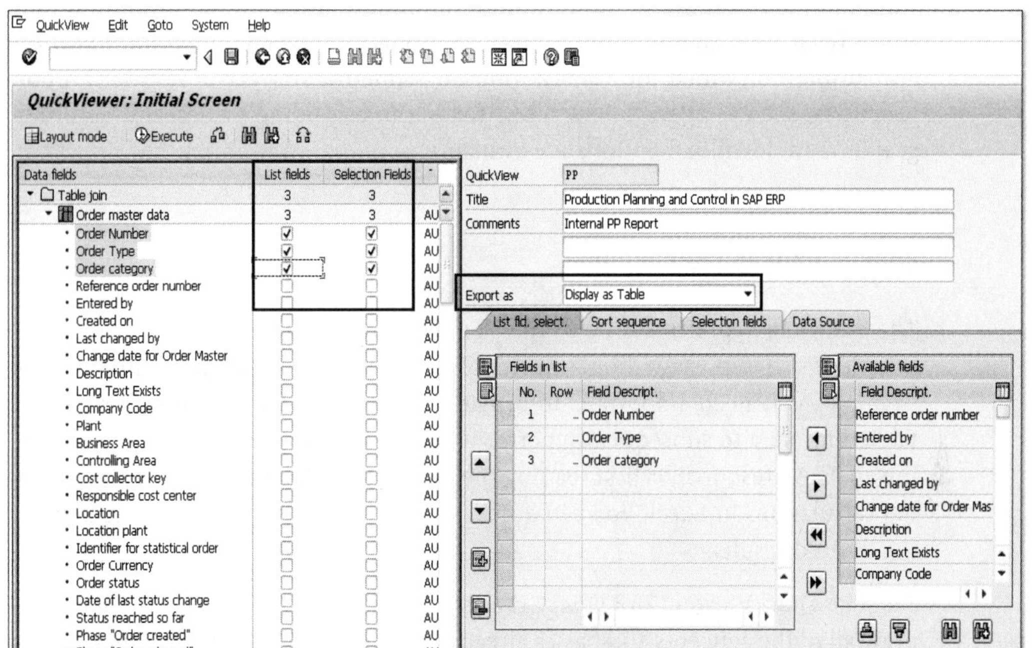

Figure 19.21 Selection of List Fields and Selection Fields in QuickViewer

In the screen shown in Figure 19.21, choose three fields as both LIST FIELDS and SELECTION FIELDS: ORDER NUMBER, ORDER TYPE, and ORDER CATEGORY. Scrolling down the same screen, some more list fields are selected (not shown). Also, on the right-hand side of this screen, you can choose to display the output of the report in various formats. For this example, choose to display the output as DISPLAY AS TABLE in the EXPORT AS field. You can also change this option later on, if needed.

Choose EXECUTE, which leads to the initial selection screen of the query in QuickViewer. The custom title of the query, PRODUCTION PLANNING AND CONTROL IN SAP ERP, as well as the three fields previously selected as SELECTION FIELDS, are available for input. You can also change the output format of the report, if needed.

For this example, set order TYPE as "PI01" and order category (CAT) as "40" so that the resultant report only displays process orders. Choose EXECUTE or press F8, which leads to the output of the query created in QuickViewer.

Apart from the fields selected as SELECTION FIELDS, several additional fields are also displayed, which are the previously defined LIST FIELDS. Standard report functions, such as aggregation, filter, sorting in ascending or descending order, graphs, and download options are available.

19.7 SAP Query

The SAP Query application helps the business process owner quickly create a custom-designed report without the involvement of an ABAP programmer. For example, you need important information that resides in three different SAP tables and then to subsequently perform additional calculation. Instead of engaging an ABAP resource or first reading the tables, downloading, and then performing calculations in Excel, you can quickly create a query in the system to achieve the desired results.

Just like QuickViewer, SAP Query is also a reporting tool. To create a query, you draw all of the concepts and basics already covered in the preceding QuickViewer section. An advantage for opting to use SAP Query is that you can also add local fields, if you want to perform any calculation or simply want to display additional information within the query-developed report. For example, you can add "percentage invoiced" as a local field, which contains a mathematical formula that reflects the difference between the sales order value and the invoice value. The local field option isn't available in Data Browser or in QuickViewer.

Creating a query entails three steps in sequential order, which are explained in further detail in the following sections:

1. Maintain the InfoSet.
2. Create a user group and assigning the InfoSets.
3. Create the query.

19.7.1 Maintain InfoSets

Creating an InfoSet entails defining whether the data source is a single SAP table, a table join, a logical database, or a reference to another InfoSet. You don't have to define an InfoSet in QuickViewer.

To create an InfoSet, use the menu path, TOOLS • ABAP WORKBENCH • UTILITIES • SAP QUERY • INFOSETS, or use Transaction SQ02. Then follow these steps:

1. Select the query area by clicking on ENVIRONMENT • QUERY AREAS • STANDARD AREA (CLIENT-SPECIFIC).
2. On the initial screen of InfoSet creation, name the InfoSet, and choose CREATE.
3. In the ensuing pop-up, provide a title for the query, and select one of the radio buttons to denote whether it's a table join, direct read of table, and so on. Depending on the option you select, you're asked if you want to include all table fields, include key fields, or create empty field groups.
4. The next step for selecting both list fields and selection fields is the same as already covered in Section 19.6. To add fields to the field group, click on the desired field on the left-hand side, and drag it to the right-hand side of the InfoSet screen to make it part of the InfoSet group.
5. Choose the GENERATE icon to generate the InfoSet, which is also confirmed by a system message.

Tips & Tricks [+]

You can also change the standard heading of any field as a list or display field by simply clicking on the field and changing its long text.

19.7.2 Create User Groups

Now you need to define a user group and assign the InfoSet to the group. A user group helps to control the authorization of who is allowed to execute the query. To create a new user group, follow the menu path, TOOLS • ABAP WORKBENCH • UTILITIES • SAP QUERY • USER GROUPS, or use Transaction SQ03. Then follow these steps:

1. Select the query area by clicking on ENVIRONMENT • QUERY AREAS • STANDARD AREA (CLIENT-SPECIFIC).
2. On the initial screen of user group creation, give a name, and choose CREATE.

3. Click on the ASSIGN USERS AND INFOSETS icon, and enter the name of the SAP users for the user group.
4. Click on ASSIGN INFOSETS to assign the previously created InfoSet to the SAP users. Save the InfoSet.

19.7.3 Create Queries

You're now ready to start creating queries from the assigned InfoSets. To create a new query, follow the menu path, TOOLS • ABAP WORKBENCH • UTILITIES • SAP QUERY • QUERIES, or use Transaction SQ01. Follow these steps:

1. Select the query area by clicking on ENVIRONMENT • QUERY AREAS • STANDARD AREA (CLIENT-SPECIFIC).
2. Click on the OTHER USER GROUP icon, and in the pop-up that appears, select the name of the user group created in the previous step.
3. Enter the name of the query you want to create, and choose CREATE. Assign a title to the report, and enter any notes. Other available options are the same as covered in the previous section on QuickViewer.
4. Click on the NEXT SCREEN icon.
5. Select the FIELD GROUPS created in the InfoSet, and again click on the NEXT SCREEN icon. The list of fields appears to choose from.

[»] **Note**

Any field that you want to use for calculation or other purposes in a local field must be given a *short name*. The option to enter a short name is available in front of every field. The short name can be alphanumeric. We suggest that you assign simple short names to fields so that it's easier for you to use them in calculations later. For example, just give a short name to a field as "A" and another as "B" so that it's easier to use them in a formula such as A + B.

6. From the menu bar, select EDIT • LOCAL FIELD • CREATE (Figure 19.22). Enter the SHORT NAME of the field, the FIELD DESCRIPTION, and the HEADING; select the FIELD GROUP; and define the PROPERTIES of the field. Define the field to be a CALCULATION FIELD with NUMBER OF DIGITS as "8" and DECIMAL PLACES of up to "2". Also set the CALCULATION FORMULA as "'RQMT' – 'WDRN'".

Both the fields, RQMT and WDRN, are standard SAP fields, which were given short names. These short names were then used in the local field for calculation purpose. Also, when entering a calculation formula, make sure that all short fields are enclosed in single punctuation marks. You can also give the CONDITION, concerning when a particular formula or local field should work or be displayed.

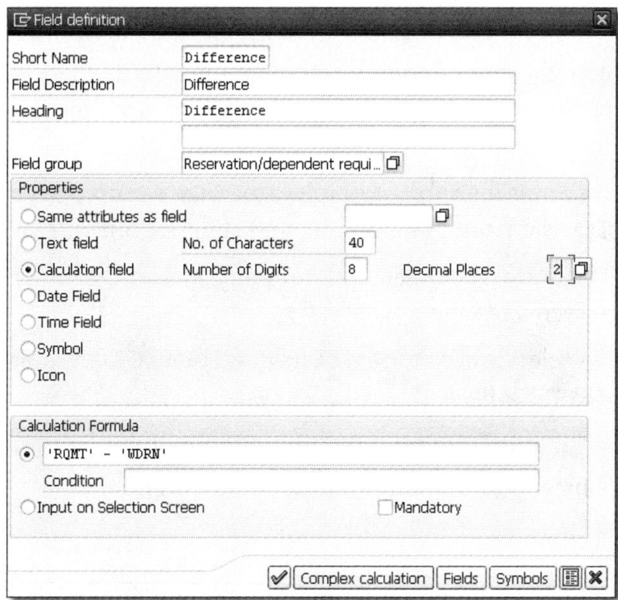

Figure 19.22 Local Field in SAP Query

Apart from calculations, you can also use the local field for entering additional text, dates, times, symbols, or icons.

The remaining steps are all the same as already covered in the QuickViewer section. In SAP Query, you also have the flexibility of creating a basic list, a ranking list, or statistics. In QuickViewer, you can only define a basic list. You can also test the results of the query in both SAP QuickViewer and in SAP Query.

19.8 Assign a Transaction Code to a Query

If you frequently use a report that has been created by either QuickViewer or SAP Query, you can assign a custom transaction and then provide the authorization of that transaction to all relevant business process owners. Doing so will prevent users from accessing the query transactions and mistakenly making any changes.

When you're already running the query-developed report, follow these steps:

1. From the SAP ERP system menu, go to SYSTEM • STATUS. In the pop-up that appears, copy the content of the PROGRAM field on the left-hand side under the SAP DATA tab.
2. Use Transaction SE93, enter the transaction code on the initial screen, and choose CREATE. This transaction code is custom so you can enter any alphanumeric value.
3. In the next pop-up window, enter the short text, select the TRANSACTION WITH PARAMETERS (PARAMETER TRANSACTION) radio button, and choose CONTINUE.
4. Enter the details of the PROGRAM generated by the query. Select the other checkboxes noted in this screen and save.
5. The system prompts you to select where to save. Click on LOCAL OBJECT (not shown), enter the package as "$TMP", and save it.

Next time, simply enter the custom transaction code in the command bar, and it will automatically lead to the parameter selection screen of the query.

19.9 Summary

In this chapter, we covered the large number of features and functions of standard reports as well as the standard analysis available. The missing parts information system is a helpful tool for the production planner. Data Browser offers a quick option to view details of any specific SAP table. QuickViewer is highly intuitive and easy to use as well as quick to set up. SAP Query offers even greater options to manipulate and evaluate the data in the report. You can also create your own custom transaction to the report created in QuickViewer or SAP Query.

The next chapter covers the integration of PP with other components in SAP ERP.

Through integration of different SAP ERP components, people and business processes can work in complete cohesion. The SAP ERP system enables the flow of data and information from one component to another. This chapter covers the basics of how you can integrate SAP ERP Production Planning with other SAP ERP components to create even more possibilities for your business.

20 Integration of Production Planning with Logistics Functions

Hardly any business process owner works in the SAP ERP system without coordinating and collaborating with multiple components. These components reflect different business functions, processes, or needs of the company. In this chapter, we address limited integration aspects of SAP ERP Production Planning (PP) with five other components in SAP ERP—Quality Management (QM), Materials Management (MM), Sales and Distribution (SD), Project System (PS), and Plant Maintenance (PM)—to help you understand the cross-component nature of cooperation and coordination needed. We also briefly include SAP Manufacturing Execution (SAP ME) as well as SAP Manufacturing Integration and Intelligence (SAP MII) integration with the SAP ERP system.

Note [«]

This chapter doesn't cover SAP ERP system integration with SAP Advanced Planning & Optimization (APO), SAP ERP Human Capital Management (SAP ERP HCM), SAP ERP Financials (FI), or SAP ERP Controlling (CO). However, the information, the approach, the "way forward," and the concepts covered are sufficient to form the basis for such integration initiatives.

In this chapter, we'll provide limited and straightforward examples of how the PP component integrates with components such as QM, MM, SD, or PS. Further, as we progressively move forward in this chapter, we start to limit our coverage of the configuration, master data, and other information that you need to perform integration testing. For example, in the first PP-QM integration, we broadly cover the configuration and master data of the QM component and then run the PP-QM integration cycle. In the next example, which is PP-MM integration, we provide broad hints on running material requirements planning (MRP) and how MRP results reflect in MM.

20.1 Integration Prerequisites

Before you get started, you should make sure that the following integration points and prerequisites are covered:

1. Prepare a list of all of the important business processes for which end-to-end testing is required. For example, order-to-cash, procure-to-pay, produce-to-sell, engineer-to-order (ETO), and so on. This list should be prepared on the basis of identifying two or more SAP ERP components (and correspondingly different departments of the company) involved in the business process.
2. Create a process flowchart of each and every step of how the business is practically performing these steps in the company. For example: Does the planning process starts with a sales forecast? Who runs the MRP and how often? How often are changes to the MRP results made? Are quality checks only performed on the finished good or also during the production process, and so on? Is there reduced availability of production capacity due to annual plant maintenance?

   You can also use the business blueprints that your SAP consulting team prepared during the SAP implementation to refer to the business process flowcharts, as-is/to-be process mapping, and gap analyses.

[+] **Tips & Tricks**

To begin with, the business process owners can consider how they are currently performing the tasks. For example: Which document (issue requisition, issue slip, etc.) forms the basis of goods issue for the cost center from the warehouse? What factors do they consider when they prepare the annual maintenance plan? What are the levels of documents approval, and how are those approved (electronically or manual signatures)?

3. List the relevant departments/users involved in each process step.
4. List the SAP ERP components in which the relevant process steps are covered. It should also include the exact process step along with the transaction code.

For example, create a production order with Transaction CO01, create a sales order with Transaction VA01, post goods issue with Transaction MIGO, and so on. It's best to prepare a script with each step listed in sequential order. The sequence of steps should match the process flowchart.

5. Ensure that the relevant master data needed to test the processes is already in place in the SAP ERP system as well as on the test scripts. In other words, agree with the relevant stakeholders on the master data on which the integration team will run the end-to-end integration cycle.
6. Ensure that the entire integration team is aware of the organizational levels, such as company code, plant, sales organization, and distribution channel, that they will be using during the integration testing. It's better to make it part of the integration test scripts for easy and quick reference.
7. Assemble all of the business process owners and SAP ERP consultants at the same location (if possible), and conduct the integration steps in the same sequence as listed in the test script. Record the results (take snapshots of the results/errors), note any deviations, discrepancies, or other errors, and engage the SAP ERP consultants to attend to them.

Tips & Tricks [+]

The business process owners of the company should be taking lead in integration testing. The SAP ERP consulting team should only be engaged for guidance or any immediate troubleshooting that the business process owners can't manage.

8. Separately, list all of the issues and errors, resolve them, and conduct another dry run to ensure completeness and correctness. Broadly, the nature of issues and errors are missing or incomplete configuration, master data, or missing authorization for a business user to run a transaction.
9. Broaden the scope on integration testing in the SAP ERP system to include ABAP objects if desired, such as scripts or user exits that were developed or configured during the SAP ERP implementation project.
10. Attend to any missing or overlooked process step, based on the feedback and input from the business process owners.
11. Finally, improve on these steps and replicate the same in dry-run exercises on smaller and more focused levels.

[»] **Note**

Closely observe how the business process owners understand the SAP ERP system and consider additional trainings, if deemed necessary. Also keep an eye on where further change management initiatives can help with greater system's adoption.

Now, let's get started with the first integration component.

20.2 Integration Aspects of Production Planning with Quality Management

QM is an integral component of logistics and completely integrates with PP. Quality checks and balances are ingrained in every major business process of the company, whether it's procurement, production, transfer between two physically dispersed locations, or plant maintenance. The quality inspection can be on the procured raw material, in-process (during production) inspection, finished goods inspection, inspection before dispatch of goods to a customer, inspection at receiving plant during interplant (between two plants) stock transfer, or inspection of equipment after maintenance or calibration is performed.

In this section, you'll learn how to configure and integrate QM in PP production processes. You can use QM for inspection during production, in-process inspection, and inspection of semifinished and finished good on a goods receipt for a production order (discrete manufacturing), process order (process manufacturing), and run schedule quantity (repetitive manufacturing [REM]).

Standard solutions and configuration delivered by SAP are sufficient to independently configure and implement QM in the production processes for discrete manufacturing, process manufacturing, and REM both during and after production. After setting up the relevant QM master data, a one-time task, you can then record results of all the important inspection parameters and make UDs for the in-process or finished good.

[»] **Note**

You have the option to integrate either of the two production processes with QM or both, as deemed applicable to the business. For example, if only in-process inspection during the production process is of importance, then QM can be implemented standalone. If only the quality monitoring and inspection of the semifinished or finished good is vital for business, then this standalone option is available. If necessary, you could implement both at the same time.

We've kept the integration process of PP with QM as simple as possible in this chapter.

We cover both options to show how comprehensive the QM role is in the production processes. We begin with configuration activities to ensure the material type for which QM needs to be activated is properly configured. Then, the focus shifts to setting up required QM master data and culminates in running the entire end-to-end business process of the production cycle. This cycle includes the creation of a process order, in-process inspection, its good receipt, and all of the related QM activities, such as results recording, UDs (UDs), and stock posting.

Note [«]

For the purposes of this chapter, we assume the material you intend to include in the QM process has already gone through the production process and has other master data, such as the material master, available in the SAP ERP system. We also assume that the QM view in the material master isn't activated yet.

20.2.1 Configuration Steps

Let's begin with the configuration checks needed to activate QM in the production processes. If your company uses the standard SAP ERP material types such as FERT (finished good) or HALB (semifinished good), create a task list type to material type assignment.

Task List Assignment to Material Types

When you integrate the production processes with QM, the system looks for this information (create a task list type to material type assignment) in the very first step of integration. In addition, if there is a company-specific material type, you need to add it as shown in Figure 20.1.

Here you see the task list type assignment to material types. The task list type (TLType) Q is used for QM along with the material type. To access this list, follow the configuration (Transaction SPRO) menu path, Quality Management • QM Planning • Inspection Planning • Assign Task List Type to Material Type.

Choose New Entries to create a new relationship. Save either by pressing [Ctrl]+[S] or choosing Save.

Change View "Assignment of Material Types"

New Entries

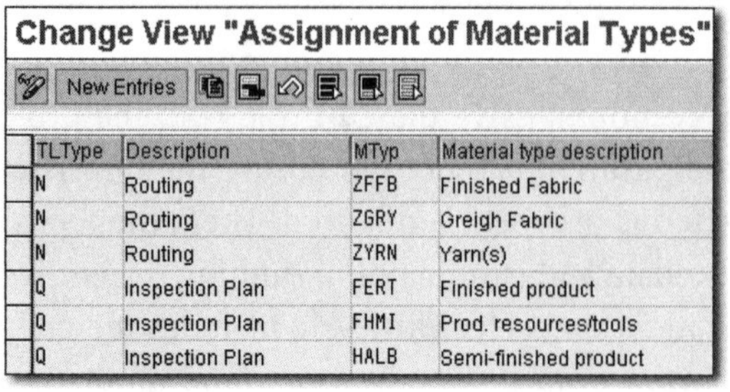

TLType	Description	MTyp	Material type description
N	Routing	ZFFB	Finished Fabric
N	Routing	ZGRY	Greigh Fabric
N	Routing	ZYRN	Yarn(s)
Q	Inspection Plan	FERT	Finished product
Q	Inspection Plan	FHMI	Prod. resources/tools
Q	Inspection Plan	HALB	Semi-finished product

Figure 20.1 Task List Type Assignments with Material Type

Assigning an In-Process Inspection Type to Process Order

The second configuration activity is the assignment of inspection type 03 for in-process inspection to a combination consisting of plant and order type. In this example, assign process order type PI01 and plant 1000 to inspection type 03. An inspection type is a control function that defines how an inspection is performed. Some examples of inspection types include the following:

- 01: Good receipt against purchase order
- 03: In-process inspection
- 04: For good receipt inspection for process and production orders
- 09: For recurring inspection for batch-managed, shelf-life expiry materials
- 89: For manual inspection

To define inspection types, follow configuration (Transaction SPRO) menu path, Production Planning for Process Industries • Process Order • Master Data • Order • Define Order Type Dependent Parameters, or use Transaction COR4 (Figure 20.2 ❶).

Select the line with Plant 1000 and Order Type PI01, and choose the Details icon to bring up the details screen ❷. Assign Inspection Type 03, and save by pressing [Ctrl]+[S] or choosing Save.

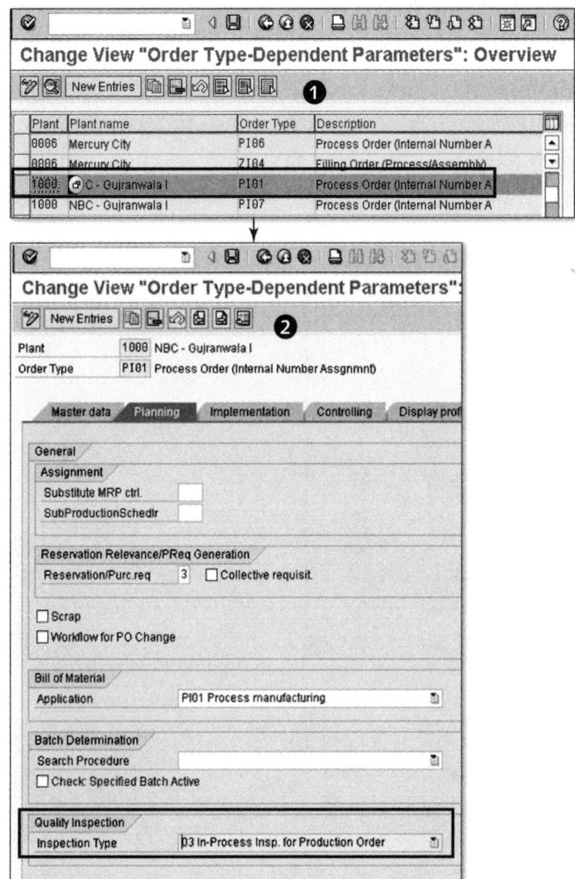

Figure 20.2 Inspection Type 03 Assigned to the Order Type and Plant Combination in the Planning Tab

20.2.2 Quality Management Master Data

Prior to beginning quality checks of production processes, you need to set up the following QM master data in the system:

- QM view in the material master
- Master inspection characteristic (MIC)
- Inspection plan

Quality Management View in the Material Master

While the material is already created, you need to activate its QM view at the organizational level (plant). Follow the menu path, LOGISTICS • MATERIALS MANAGEMENT • MATERIAL MASTER • MATERIAL • CREATE (GENERAL) • IMMEDIATELY, or use Transaction MM01. For this example, enter the MATERIAL number "1980", and choose the SELECT VIEW(S) button (Figure 20.3 ❶).

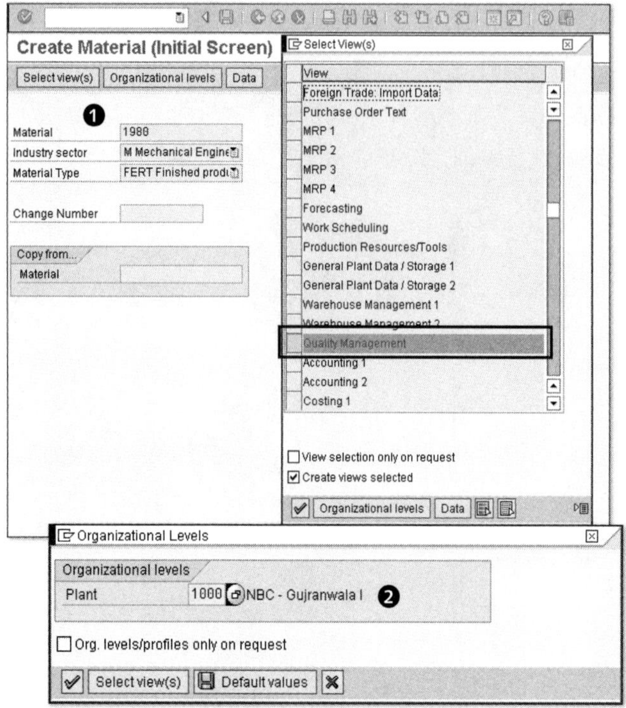

Figure 20.3 Material Master Creation with QM View

Select the QUALITY MANAGEMENT view from the SELECT VIEW(S) list, and then click the ORGANIZATIONAL LEVELS icon to see the organizational level in which the plant is defined in screen ❷. For this example, you can see that material 1980 belongs to plant 1000.

Press [Enter] or choose CONTINUE to get to the QUALITY MANAGEMENT view (Figure 20.4 ❶). From here, select the INSP. SETUP icon to view the inspection setup data in which individual inspection types for the material are defined ❷.

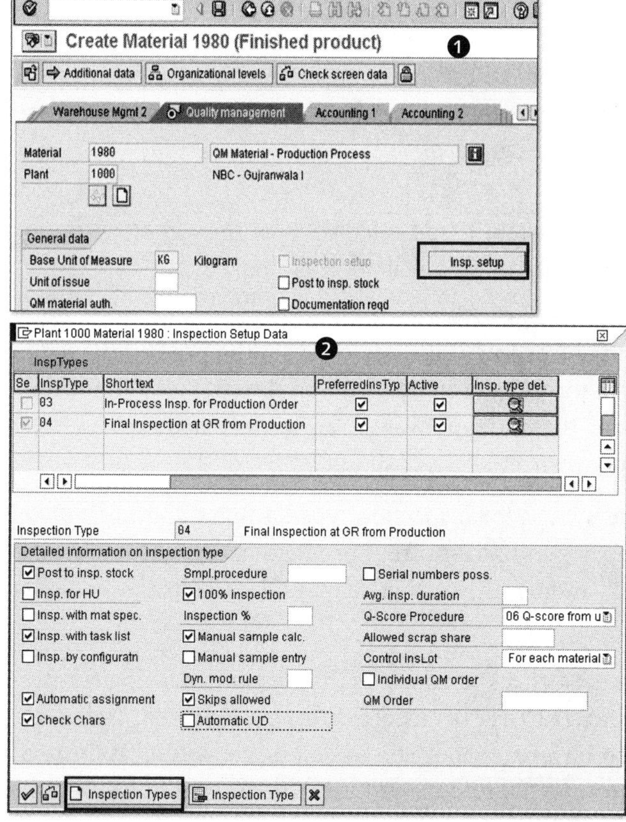

Figure 20.4 QM Inspection Setup in the Material Master

Click the INSPECTION TYPES button. Inspection type 03 is available in standard SAP ERP and is used for in-process inspection. Make sure to select the PREFERREDINSTYP and ACTIVE checkboxes, so that inspection type 03 is active. Click the INSP. TYPE DET. icon to bring up the detailed information on the inspection type section at the bottom of the figure.

Additional important settings in this screen include following:

- INSP. WITH TASK LIST
  Means plant-dependent task lists (inspection plans) must be defined as QM master data. We've covered the creation of an inspection plan in detail.
- CHECK CHARS
  Checks the validity and other details of the MIC before it can be used.

- 100% Inspection
  Ensures that the sample size of all incoming inspections will be 100% (good receipt for process order quantity will be equal to the sample size quantity). In the results recording step, you enter the results found after conducting the physical or chemical tests on the samples.
- Manual sample calc.
  Means that even if there is a system-based calculation of the sample size, this can be overwritten by a manual sample calculation, if needed.
- Skips allowed
  Lets you skip to the results recording if it's deemed that the produced good is of acceptable quality and a direct UD can be made. In the UD step, you record your decision in the system, that is, if you want to accept (or reject) the material and transfer it from quality-inspection stock to unrestricted (free-to-use) stock.

Next, repeat the same process steps for inspection type 04. Inspection type 04 is activated for final inspection at good receipt from production. On successful activation of inspection types 03 and 04, the system displays the results in the screen shown in Figure 20.4 ❷.

Selecting the Post to insp. stock checkbox ensures that all goods receipts for process orders are posted in Quality Inspection stock status until cleared. The Automatic assignment checkbox ensures that as soon as an inspection lot is created, the system checks for relevant quality specifications (in this case, task list or inspection plan) and automatically assigns them to the material's inspection lot. This option is a time-saver because it ensures that only relevant specifications of materials that are to be checked for conformity are assigned.

Choose Continue. The material can now be saved by pressing Ctrl+S or choosing Save.

[»]

Note

Some of the checkboxes shown in Figure 20.4 ❷ are optional depending on whether they are required in a business process. The important aspect to note is that all options selected in checkboxes subsequently act as controlling functions. For example, if the Automatic UD checkbox is deselected, no automatic UDs for inspection lots belonging to this material are possible.

Master Inspection Characteristic

An MIC is the quantitative or qualitative attribute of a material. It has its own set of specifications that needs to be defined as important QM master data. For example, the qualitative MIC can be the texture of a fabric, such as fine, coarse, visible micro-holes, and so on; the quantitative MIC can be the viscosity or density of a raw material.

To create an MIC, follow the menu path, LOGISTICS • QUALITY MANAGEMENT • QUALITY PLANNING • INSPECTION CHARACTERISTIC • CREATE, or use Transaction QS21. For this example, create a quantitative MIC named "QM01" in PLANT 1000 (Figure 20.5 ❶).

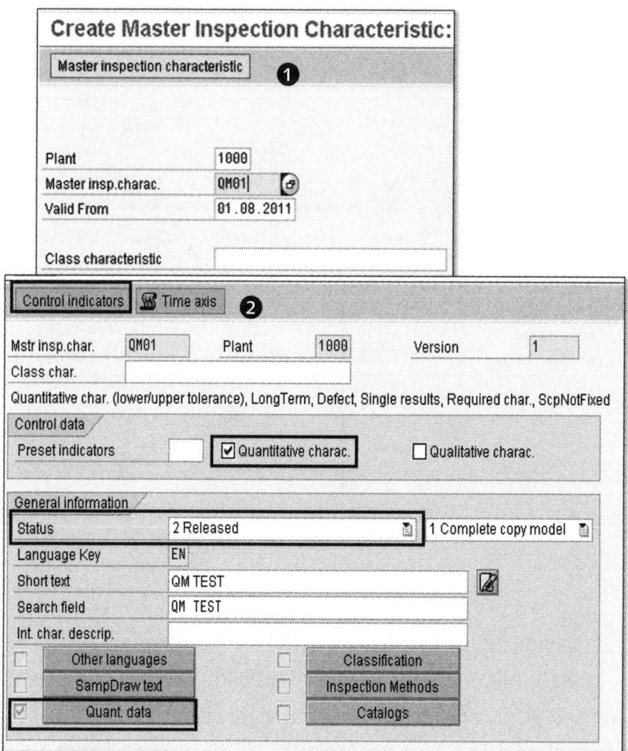

Figure 20.5 Master Inspection Characteristic (MIC) Setup

After defining the MIC, click the MASTER INSPECTION CHARACTERISTIC button at the top of the screen. In the next screen that appears ❷, define the MIC as quantitative by choosing QUANTITATIVE CHARAC., use the SHORT TEXT field to add text, and set the STATUS as 2 RELEASED. Choose the CONTROL INDICATORS button, which leads to the EDIT CHARACTERISTIC CONTROL INDICATORS screen (Figure 20.6 ❶).

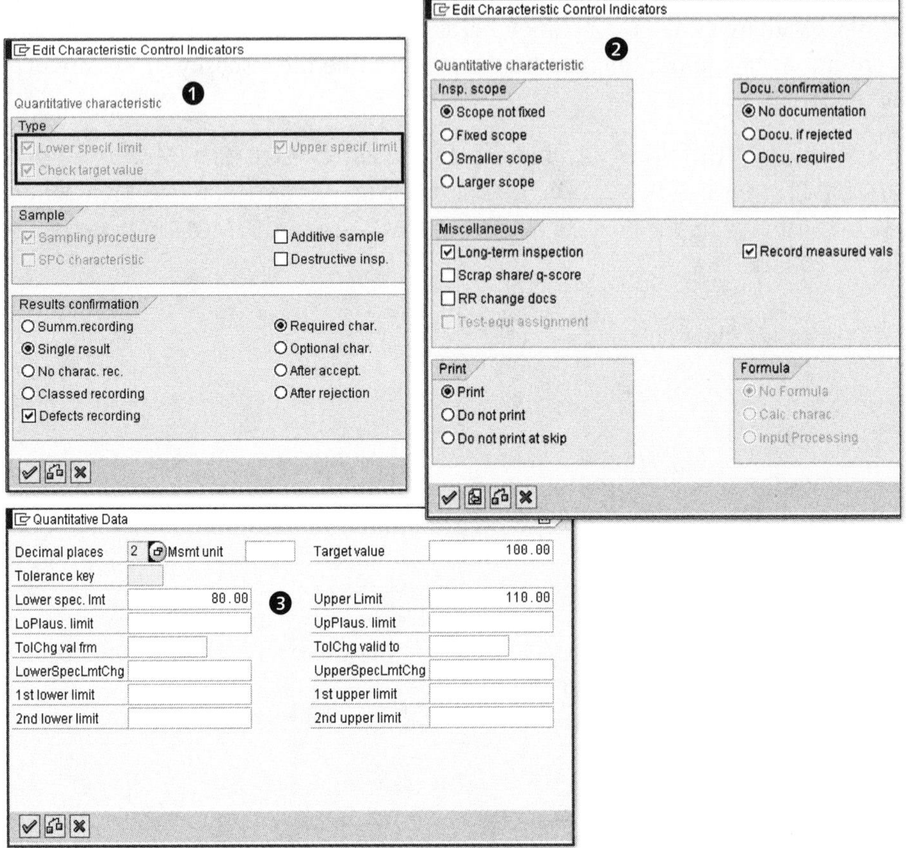

Figure 20.6 Control Indicators of MIC

Here, the control indicators are defined for MIC QM01 with requirements to define lower and upper limits as well as target values. Some of the other control indicators selected are SINGLE RESULT, which mean that each result of a sample needs to be entered individually, and DEFECTS RECORDING, which means that you can record any defects noted in production. Finally, specify that the MIC QM01 is a REQUIRED CHAR. and that the results must be recorded.

Choose CONTINUE to access more control indicators ❷. For this example, define that the MIC QM01 does have SCOPE NOT FIXED, which means you can still change its details later if necessary. Select LONG-TERM INSPECTION, which means that if the results recording of this specific MIC is delayed, the UD can still be undertaken. Finally, choose the option to PRINT the MIC in an inspection report.

Choose CONTINUE to open the QUANTITATIVE DATA screen ❸, in which quantitative data are defined. Choose CONTINUE. The MIC can now be saved either by pressing [Ctrl]+[S] or by choosing SAVE.

Inspection Plan

An inspection plan is the task list referred to whenever an inspection lot for a process order is created. An inspection plan can have different usages such as a separate inspection plan for during-production quality checks in which extensive checking and results recording are performed, or finished good inspection that may have a different set of specifications (MICs) to check. An inspection plan consists of MICs previously created.

To create an inspection plan, follow the menu path, LOGISTICS • QUALITY MANAGEMENT • QUALITY PLANNING • INSPECTION PLANNING • INSPECTION PLAN • CREATE, or use Transaction QP01. For this example, define the MATERIAL as "1980" and PLANT as "1000" for the creation of an inspection plan.

After entering the details for your inspection plan, press [Enter] or choose CONTINUE to enter header details. Set the USAGE as "5", which is for goods receipt, and the STATUS of the inspection plan as "4", which is RELEASED (GENERAL).

Note [«]

A *usage* specifies the purpose for using a specific inspection plan. A material can have different inspection plans for different usages, for example, for goods receipt-specific usage or for universal usage. Universal usage can be for all of the different inspections or for goods issuance usage in a sales order-related quality check before goods are sent to the customer against a sales order.

Next, click the OPERATIONS button to view the OPERATION OVERVIEW screen. Assign the work center "MACH90", on which the inspection is carried out. Normally, a production work center is used to carry out an inspection. A work center is a production or assembly line, and an inspection is carried out by taking a sample from the production line (work center).

Select the line item 0010, and click the INSPECTION CHARACTERISTICS button to open the CHARACTERISTIC OVERVIEW screen (Figure 20.7 ❶). Assign the previously created MICs QM01 and QM03. All other details are automatically copied from the MIC, such as the upper and lower limits and the target value.

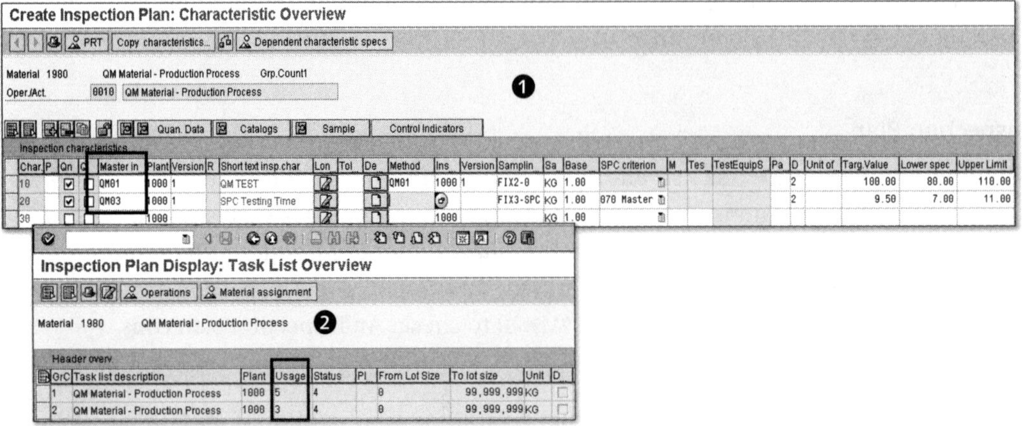

Figure 20.7 Operation View of an Inspection Plan

You can now save the inspection plan by pressing Ctrl+S or choosing SAVE. A message appears saying that the inspection plan for material 1980 for plant 1000 has been saved in group 280. The group number refers to an internal number given to every inspection plan as identification.

To create an inspection plan for in-process inspection (inspection type 03) with the usage type 3 (universal), repeat all of the steps from the screens shown in this section and assign the STATUS as 4 (RELEASED). When these steps are successfully done, two inspection plans with different usages—5 (good receipt) and 3 (universal), both with STATUS 4 (RELEASED)—are created for MATERIAL 1980 and PLANT 1000 ❷.

We're now finished with the QM master data required to implement QM in production processes. However, to explain the actual use of all of the QM master data created so far and how it affects the business, let's discuss the end-to-end production cycle in the next section.

20.2.3 End-to-End Production Process Flow with Quality Management Integration

The end-to-end process flow for two different types of inspections during and after the production process is detailed in Figure 20.8. The process begins with the creation and release of a process order for material 1980 and plant 1000. As soon as the process order is released and saved, an in-process inspection lot is created automatically by the system. You then perform the results recording and UD on the in-process inspection of the during-production run to confirm whether it meets the predefined specifications. When the goods are produced, you perform a goods receipt with reference to the process order, which generates another inspection lot for the finished goods. Similar activities of results recording, followed by UD, are taken. This is then concluded by stock posting of the quality stock. We cover the details of these steps in the following subsections.

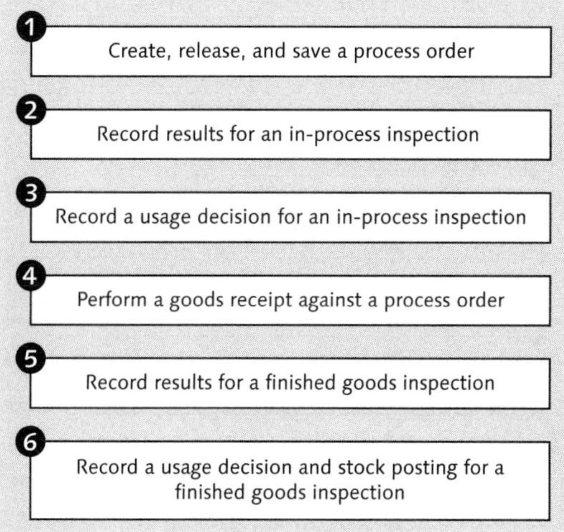

Figure 20.8 End-to-End Process Steps of In-Process and Goods Receipt Inspections

Create, Release, and Save a Process Order

To create a process order, use menu path, LOGISTICS • PRODUCTION – PROCESS • PROCESS ORDER • PROCESS ORDER • CREATE • WITH MATERIAL, or use Transaction COR1. For this example, enter MATERIAL number "1980", production PLANT "1000", and process ORDER TYPE "PI01" in the initial screen. After entering this data, press [Enter] or choose CONTINUE. This leads to the CREATE PROCESS ORDER: HEADER – GENERAL DATA screen. Enter the process order quantity of "100" in the

TOTAL QTY field. Release the process order by clicking the RELEASE icon. As soon as the process order is released, an automatically created inspection lot for in-process inspection is generated, as shown in the ASSIGNMENT tab.

Click the GOODS RECPT tab, which leads to the screen that shows when the good receipt is performed against the process order. The stock automatically moves to QUALITY INSPECTION stock, and an inspection lot is created.

Save the process order by pressing Ctrl+S or by choosing SAVE. For this example, the system generates the process order number 70001062.

Record Results for In-Process Inspection

As soon as the production starts, the results recording for an in-process inspection lot can begin. Follow the menu path, LOGISTICS • QUALITY MANAGEMENT • QUALITY INSPECTION • WORKLIST • INSPECTION • CHANGE DATA, or use Transaction QA32 (Figure 20.9 ❶).

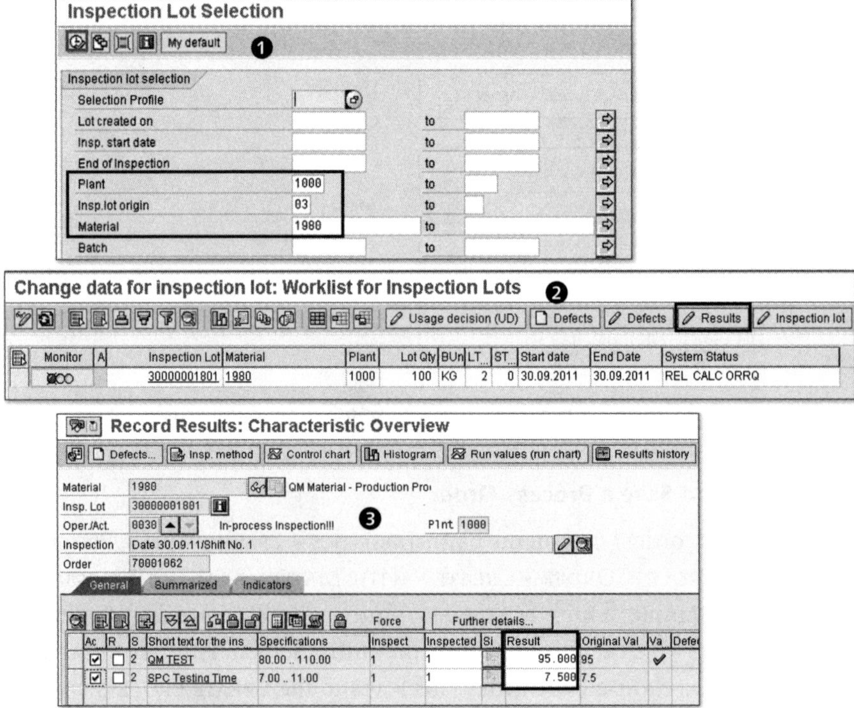

Figure 20.9 Results Recording on In-Process Inspection

Enter the initial parameters on the selection screen, and press [F8] or choose EXECUTE. This leads to the worklist for all the inspection lots for which the results recording or UD hasn't been taken. Select the inspection lot, and click the RESULTS button. This takes you to the RECORD RESULTS screen, in which the individual specifications of a material and its corresponding MICs are available for reference ❷.

Enter the results, and save by pressing [Ctrl]+[S] or choosing SAVE, which takes you back to the first screen.

Record a Usage Decision for an In-Process Inspection

Still in the INSPECTION LOT SELECTION screen shown earlier in Figure 20.9 ❶, select the inspection lot for which the results recording have already been performed (with system status updated to RREC). Click the USAGE DECISION (UD) button to open the screen shown in Figure 20.10 ❶. Here the UD is made by providing the UD CODE "A" for ACCEPT ❷. Save the UD by pressing [Ctrl]+[S] or by choosing SAVE.

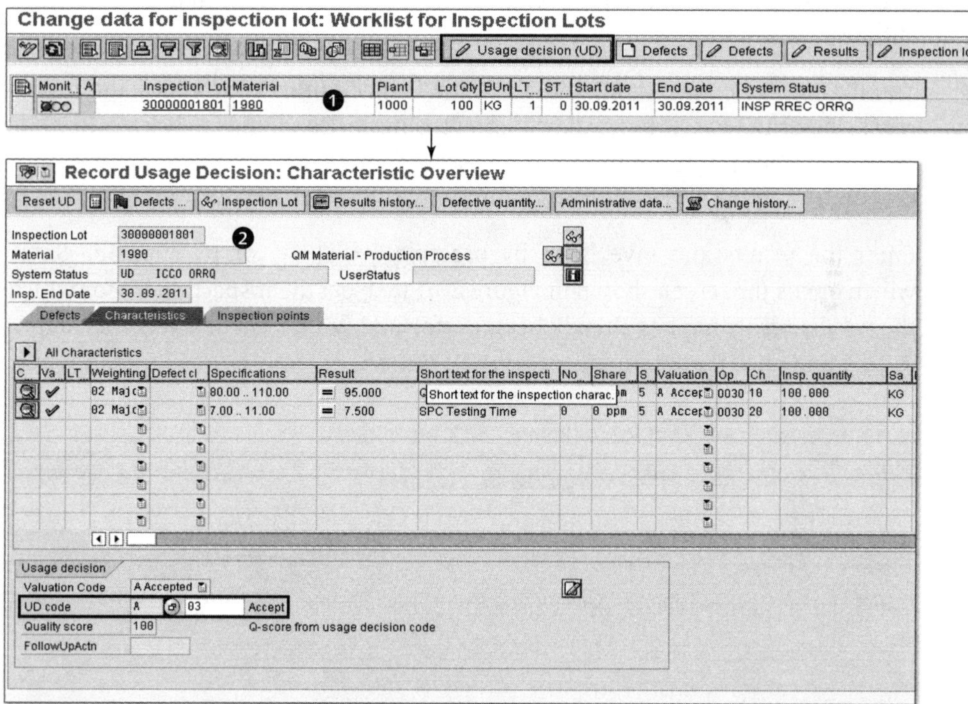

Figure 20.10 Usage Decision for In-Process Inspection

Perform a Goods Receipt against a Process Order

After the good are produced against a process order, you can perform the good receipt. On the goods receipt, an automatic inspection lot is created for which the standard QM processes of results recording, UD, and stock posting can be performed.

Follow the menu path, Logistics • Materials Management • Inventory Management • Goods Movement • Goods Movement, or use Transaction MIGO. Enter the process order number (for this example, enter "70001062"), and press [Enter]. On the goods receipt posting, the stock is posted in Quality Inspection, and a separate and independent inspection lot is created. Save the goods receipt posting by pressing [Ctrl]+[S] or by choosing Save.

Record the Results for Finished Goods Inspection

To record the results, follow the menu path, Logistics • Quality Management • Quality Inspection • Worklist • Inspection • Change Data, or use Transaction QA32. Enter the initial parameters on the selection screen, and press [F8] or choose Execute.

You'll again have the worklist of all of the inspection lots for which the results recording or UD hasn't been taken. Select the inspection lot, and click the Results icon. This leads to the results recording screen, in which the individual specifications of a material and its corresponding MICs are available for reference.

Enter the results and save them by pressing [Ctrl]+[S] or by choosing Save, which opens the screen shown in Figure 20.11. Select the inspection lot for which the results recording has already been performed (with system status updated to RREC), and click the Usage decision (UD) button.

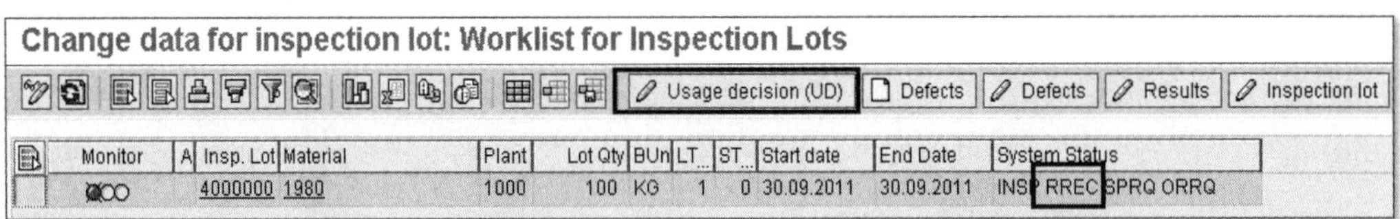

Figure 20.11 Usage Decision and Quality Inspection Stock Posting

Record a Usage Decision and Stock Posting for Finished Goods Inspection

After clicking the USAGE DECISION (UD) button, the RECORD USAGE DECISION: CHARACTERISTIC OVERVIEW screen appears. The UD is made by providing the UD CODE "A" for ACCEPT. If you want to assign a rejection code to a rejected inspection lot, then the relevant options are available in the UD CODE field dropdown list.

Click the INSPECTION LOT STOCK tab. In the inspection lot stock posting screen, the quality inspection stock is transferred or moved to unrestricted-use stock when saved. The unrestricted stock is the quality cleared stock. Save the UD by pressing [Ctrl]+[S] or by choosing SAVE.

20.3 Integration of Production Planning with Materials Management

You can optimize the procurement process by automatically generating vendor delivery schedule lines. For vendors, this option adds value to the procurement process by eliminating the need to create and maintain delivery schedule lines manually.

The procurement process can be enhanced by taking advantage of the MRP functionality, which automatically creates delivery schedule lines for a vendor's scheduling agreement, where applicable. However, to optimize this procurement process, you have to set up the relevant master data appropriately. The MRP runs on the finished product level and reads the bill of materials (BOM) of the finished product. It focuses on materials to be procured externally from vendors whose scheduling agreements are already available in the system.

Note [«]

For in-depth understanding of the concepts and working of MRP, refer to Chapter 11.

In this section, we'll show the step-by-step procedure for setting up the master data in SAP ERP for Materials Management (MM). In addition, we describe how the SAP ERP system uses the BOM of the PP component to calculate the delivery quantities and to generate delivery schedule lines for the scheduling agreement previously created during the MRP run. A schedule line includes the date of delivery and also the quantity of the material that your vendor needs to deliver on that specific date.

20.3.1 Managing Master Data

Setting up master data in MM entails two steps. The first is to activate the source list option for the materials to be procured. A *source list* is a list of various sources for the supply of a given material—that is, the vendors that are allowed to supply a particular material during a specific period. It's made up of outline agreements such as scheduling agreements, quantity or value contracts, or purchasing information records. For those vendors, outline agreements in the form of quantity/value contracts or scheduling agreements exist in the SAP ERP system. An outline agreement is a long-term supply agreement with vendors. We focus on scheduling agreements that are already set up in the SAP ERP system for a vendor. They are the basis for maintaining a source list for procuring materials from various sources. The second step is to maintain the source lists.

[»] **Note**

Instead of a scheduling agreement, you can also use other outline agreements, such as quantity or value contracts, to enable and further integrate the same process of MM with PP. If you do so, then the system will create contract release orders (CROs) with reference to the quantity or the value contract mentioned in the source list.

Activate the Source List Option in the Material Master

Our focus is on the scheduling agreement option as the basis for the source list maintenance. Purchasing information records can be created manually before any purchasing is initiated. Otherwise, when a purchase order for a material is issued to a vendor, the system automatically creates the purchasing information record.

To activate the source list option in the material master, follow the menu path, Logistics • Materials Management • Material Master • Material • Change • Immediately, or use Transaction MM02. In Figure 20.12 ❶, enter the Material number, and click the Select view(s) button. In the pop-up window, select the Purchasing view, and choose Continue.

Next, select the ORGANIZATIONAL LEVELS button at the bottom of the screen. Select the organizational level, that is, PLANT. For this example, enter PLANT "1000" ❷. Press [Enter] or choose CONTINUE to go to the purchasing view of the material master.

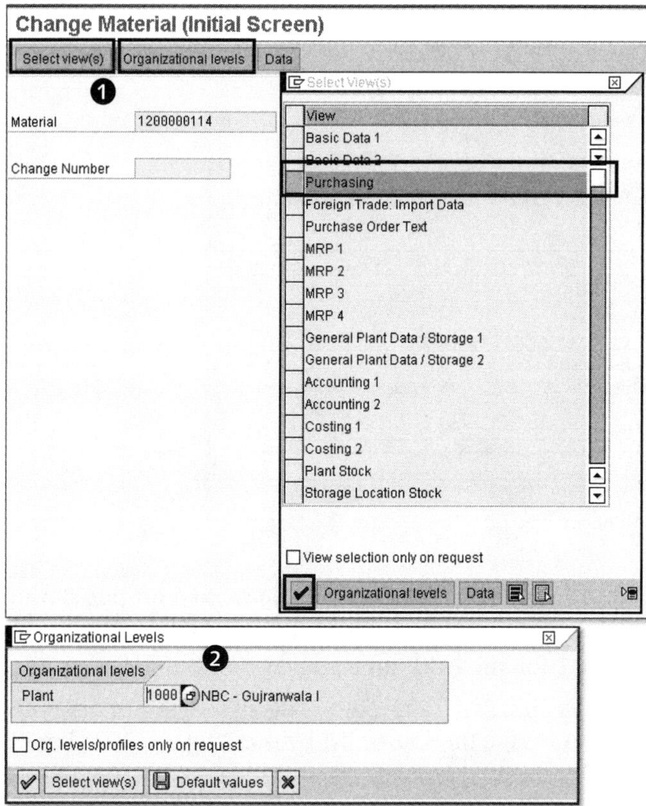

Figure 20.12 Material Master with Purchasing View Selected

Check the SOURCE LIST checkbox at the bottom of the screen to activate it. This ensures that the system checks for the existence of a source list while running MRP and therefore is able to generate procurement proposals.

Maintain the Source List

To maintain the source list, follow the menu path, LOGISTICS • MATERIALS MANAGEMENT • PURCHASING • MASTER DATA • SOURCE LIST • MAINTAIN, or use

Transaction ME01. On the initial screen for maintaining the source list, enter a MATERIAL and PLANT. Press [Enter] or choose CONTINUE. This takes you to the detailed screen for the maintenance of the source list (Figure 20.13).

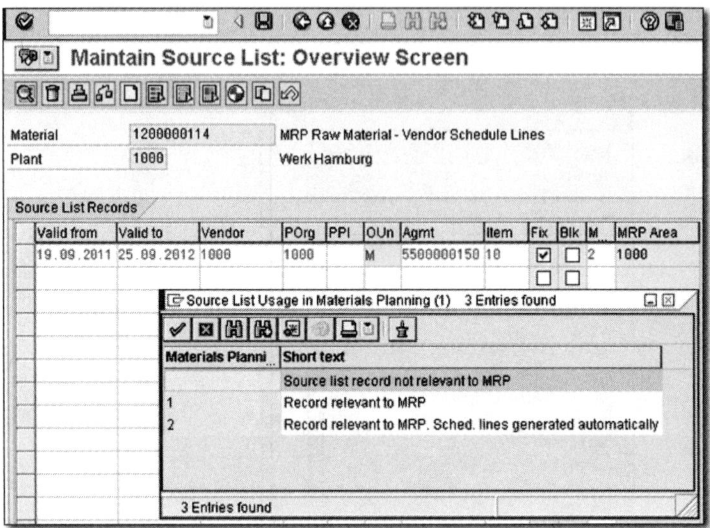

Figure 20.13 Source List Maintenance

Enter all of the required information including validity dates, vendor purchasing organization, scheduling agreement, and line item number of a scheduling agreement. It's important to select 2 for the MRP indicator in the source list, so that vendor delivery schedule lines are generated automatically, as shown by the dropdown menu in Figure 20.13. Save the source list by pressing [Ctrl]+[S] or by choosing SAVE.

Next, click the GENERATE RECORDS icon. This automatically proposes the scheduling agreements that have the material and plant combination as a possible option for the source list. In other words, if the material 1200000114 has three different scheduling agreements with three different vendors, then the system automatically proposes these three scheduling agreements, which you can then accept. You accept them by clicking the ACCEPT icon that is available for selection at that time. This option saves you the effort of individually and manually maintaining the details required in the screen shown in Figure 20.13. Additionally, if any vendor is blocked for any reason, or if the scheduling agreement is invalid, the agreements aren't proposed.

To facilitate the maintenance of source lists of a large number of raw materials, follow the menu path, Logistics • Materials Management • Purchasing • Master Data • Source List • Follow-on Functions • Generate, or use Transaction ME05. Enter the Material and Plant, and then choose the Execute icon or press F8 to generate the source lists. You can perform a test run on generating a source list at this stage to get a preview of which source list records are generated during an actual run.

[«]

Note

The scheduling agreements must be created previously or must exist in SAP ERP prior to creation and maintenance of the source list of a material.

20.3.2 Production Planning Master Data

The PP master data is comprised of a material BOM that serves to fill an important role for the MM component. The BOM of finished good is read during the MRP run, and, at that time, it also performs multi-level explosions to read all assemblies and components of the BOM of the finished good. Proposals on quantities to be procured externally are based on the information given in the BOM. When the MRP is run on a finished good, it reads the information from the BOM of the finished good to calculate the quantities required for external procurement. For example, to produce one quantity of the finished good A, it requires raw material B in a quantity of 3. This information and ratio of 1:3 for A:B (finished good to raw material ratio) is defined in the BOM of the finished good.

Figure 20.14 ❶ shows the display screen for the BOM. To get to the screen, follow the menu path, Logistics • Production • Master Data • Bill of Material • Bill of Material • Material BOM • Display, or use Transaction CS03. Enter the finished good Material code together with the Plant and BOM Usage ("1" for production).

Select the Item button to see next the item overview screen of the BOM display transaction screen ❷. It shows that to produce 1 unit of material 1977, the quantity needed for raw material 1200000114 is 1 meter.

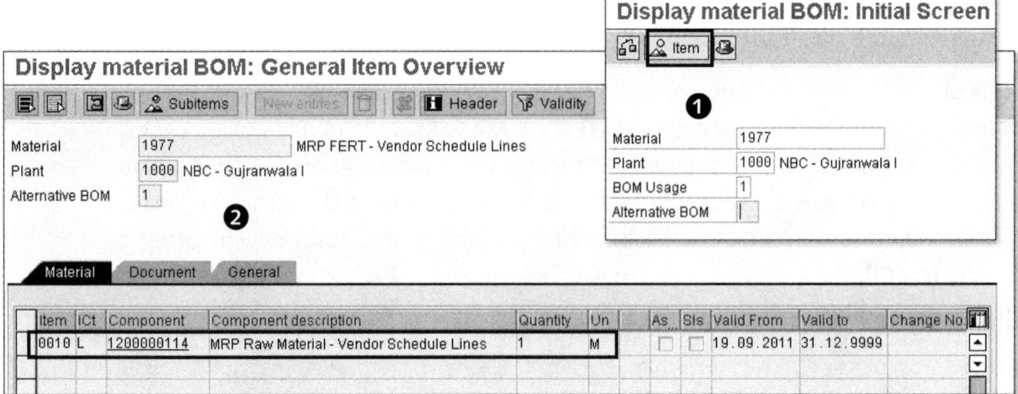

Figure 20.14 Items Overview of BOM

20.3.3 End-to-End Process Flow

We now cover the end-to-end process flow for automatic generation of a vendor's delivery schedule lines (Figure 20.15). The process begins with the creation of the planned independent requirements (PIRs) of the finished good.

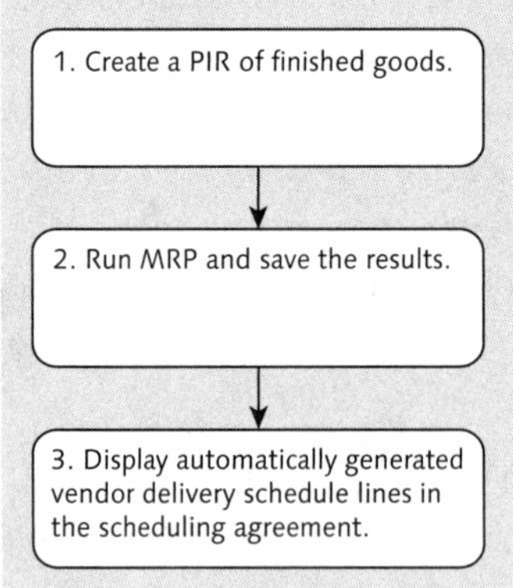

Figure 20.15 An End-to-End Process Flow to Automatically Generate a Vendor's Delivery Schedule Lines

The PIR is the demand planning quantity of the finished good needed to be produced according to the production program. The PIR is set to active version by selecting the ACTIVE checkbox available for selection on the PIR screen. An active version of PIR ensures that it's considered for production and procurement planning during an operational MRP run. You save the PIR by choosing SAVE or by pressing [Ctrl]+[S].

The MRP on the finished good is run next. It calculates the procurement quantity of the raw material from the BOM of the finished good and generates procurement proposals in the form of the vendor's delivery schedule lines. The MRP run also calculates the production quantities of the finished good. However, due to the scope of this chapter, which is focused on raw material procurement, we only cover procurement planning, not production planning.

The scheduling agreement is then checked by the procurement person responsible for managing scheduling agreements with vendors. It's displayed with Transaction ME33L and you can check to see if the automatically generated vendor's delivery schedule lines are incorporated per the results of the MRP.

20.3.4 Display Automatically Generated Vendor Delivery Schedule Lines in the Scheduling Agreement

The outcome and results of the MRP run and automatic generation of the vendor's delivery schedule lines are directly reflected in a scheduling agreement. To display the scheduling agreement, follow the menu path, LOGISTICS • MATERIALS MANAGEMENT • PURCHASING • OUTLINE AGREEMENT • SCHEDULING AGREEMENT • DISPLAY, or use Transaction ME33L. On the initial screen for the display of the scheduling agreement, enter the scheduling agreement number, and click the ITEM OVERVIEW icon to open the ITEM OVERVIEW screen of the scheduling agreement. Select the first line item 10 with MATERIAL 1200000114, and click the DELIVERY SCHEDULE icon.

The detailed view of dates and quantities to be delivered by the vendor for the material 1200000114 appears. MRP causes the CREATION ID column to be shown as B.

The date and quantity information can then be provided to the vendor by the process currently followed by the company. For example, if the company sends a printed version of the delivery schedule to the vendor for its information and action to ensure timely deliveries, then this is the process/procedure that you follow after the automatically generated vendor delivery schedules are created. Finally, when the deliveries against this scheduling agreement are made by the vendor, the system automatically updates the information in the scheduling agreement.

20.4 Integration of Production Planning with Sales and Distribution (Make-to-Order Production)

The sales order of a customer in SAP ERP Sales and Distribution (SD) forms the basis of make-to-order (MTO) production. MTO production stipulates that the production process will only initiate after a customer places an order, which is reflected in the SAP ERP system as a sales order.

The material must be marked as MTO, usually by virtue of its strategy group in the material master. You can then run MRP on the line item of the sales order as a standalone independent activity, or if your company runs MRP on the plant level, then the system will automatically take care of planning the sales order of the customer and the associated production activities. When MRP is run on the sales order item, the system creates the planned order, which is then converted into the production or process order, depending on the production type and industry. You can even control procurement of the raw materials needed for the production process by defining if the procurement should also be specific to sales orders.

We've confined the scope of this section to show how the sales order details are available in the process order, whereas you can then perform all of the subsequent steps of production, QM activities, and goods receipt of the customer stock.

In the following sections, we'll show the master data that you need to maintain, followed by the explaining the associated business processes of MTO.

20.4.1 Managing Master Data

For MTO production, you need to enter the strategy group into the material master (use Transaction MM02). Enter the material number and the plant, and select the MRP 3 view. Then, enter the Strategy group as "20", and save the material (see Figure 20.16).

Planning			
Strategy group	20	Make-to-order production	
Consumption mode		Bwd consumption per.	0
Fwd consumption per.	0	Mixed MRP	
Planning material		Planning plant	
Plng conv. factor		Planning matl BUnit	
Availability check			
Availability check	02	Tot. repl. lead time	0 days
Cross-project			

Figure 20.16 Strategy Group 20 in Make-to-Order Production

20.4.2 Sales Order Creation

For this example, create a sales order of the same material, which was previously declared as MTO production. Figure 20.17 ❶ shows the initial screen for the creation of a sales order (Transaction VA01). Enter the customer number "17" in the SOLD-TO-PARTY field, enter the MATERIAL "124", and the ORDER QUANTITY "10" ❷.

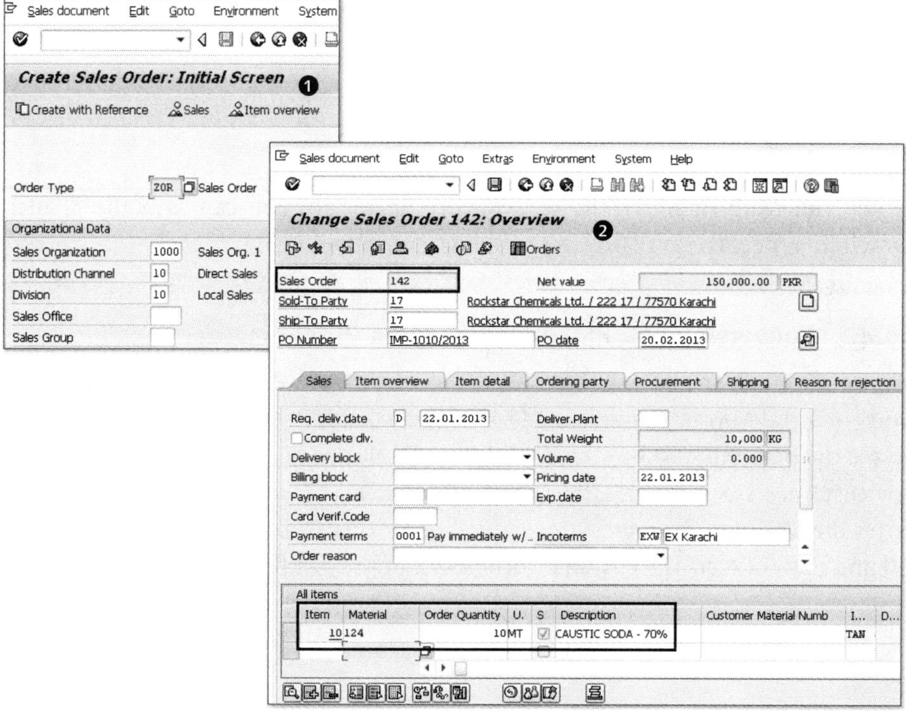

Figure 20.17 Sales Order Creation

[»]

Note

Order type ZOR is a custom-defined order type for an actual company. For your business scenario, you can use the order type that your company uses to create sales orders.

Saving the sales order generates order number 142. The next step is to run the MRP on sales order 142 and line item 0010 (the first item of the order), which will create a planned order for future conversion to a process order.

20.4.3 Material Requirements Planning Run on Sales Order Line Item

To run MRP, use Transaction MD50. Enter the sales order number as "142", enter the SALES ORD. ITEM as "10", and enter any other planning and scheduling parameters that are needed. Press Enter twice. You can run the MRP on a single level or multi-level of the material. For a single level (not shown here), you can select CANCEL PLANNING RUN after the planning of the first material. If you want multi-level planning of the same material, that is, if you want the components of the material to be planned in the MRP run, then you need to select the relevant option available during MRP run, such as PLAN UP UNTIL NEXT STOPPING POINT or PROCEED WITHOUT STOPPING. Alternatively, you can choose to deselect the DISPLAY RESULTS BEFORE THEY ARE SAVED option in the MRP run parameters, and the system will perform top-to-bottom planning without stopping.

If you run MRP interactively (e.g., on one material), you can save the results, which then become available in the stock/requirements list.

20.4.4 Conversion of a Planned Order to a Process Order

Figure 20.18 ❶ shows the STOCK/REQUIREMENTS LIST: INITIAL SCREEN, wherein you enter the MATERIAL number "124" together with PLANT "Z101". Press Enter to see the planning results after MRP is run ❷. Notice that the system first looked for customer stock (CUSTST field) for sales order 142 and didn't find anything. After the MRP run, it creates the planned order (PLDORD field) 903 with quantity 10 and then reflects the customer order 142 in the last line.

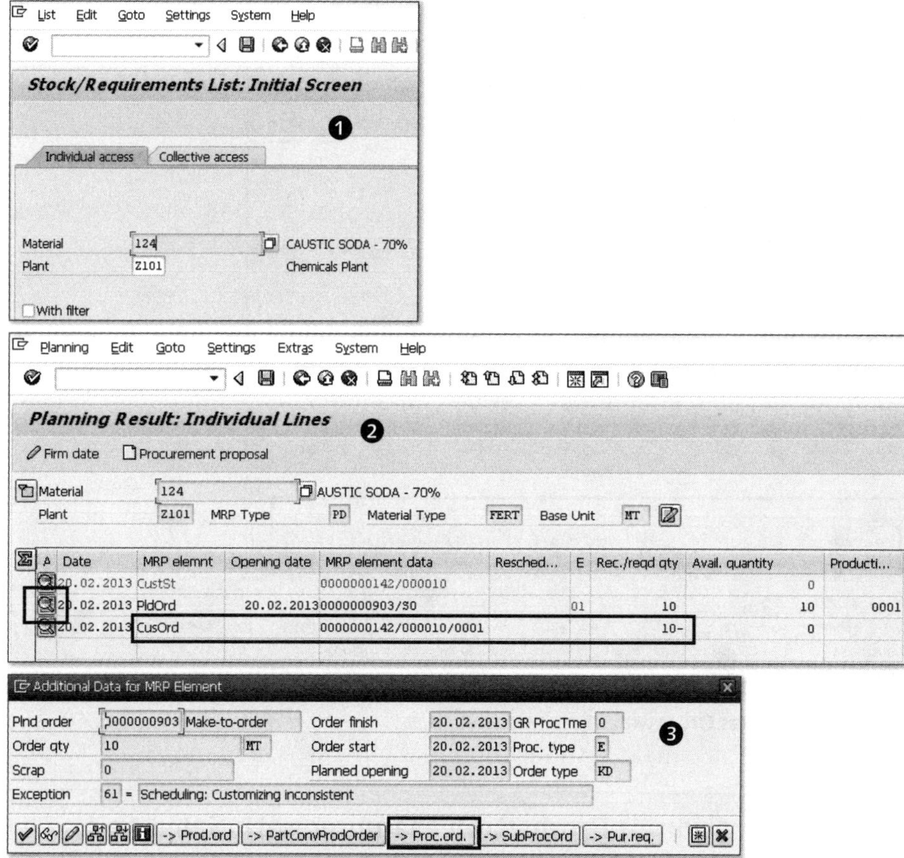

Figure 20.18 Conversion of a Planned Order to a Process Order

Click on the ITEM DETAILS icon shown in Figure 20.18 ❷, to convert planned order 903 into a process order. In the ADDITIONAL DATA FOR MRP ELEMENT screen that appears ❸, click on the ->PROC.ORD. button to convert the planned order into a process order. This takes you to the screen shown in Figure 20.19 ❶.

This process order creation screen also contains details of the sales order number 142 as well as the customer's details. Click on the MATERIALS button ❶ to open the MATERIAL LIST ❷, which contains the details of the components needed to produce the sales-order-specific stock. The column for account assignment (ACCT) shows that even the procured material will be specifically done for the customer.

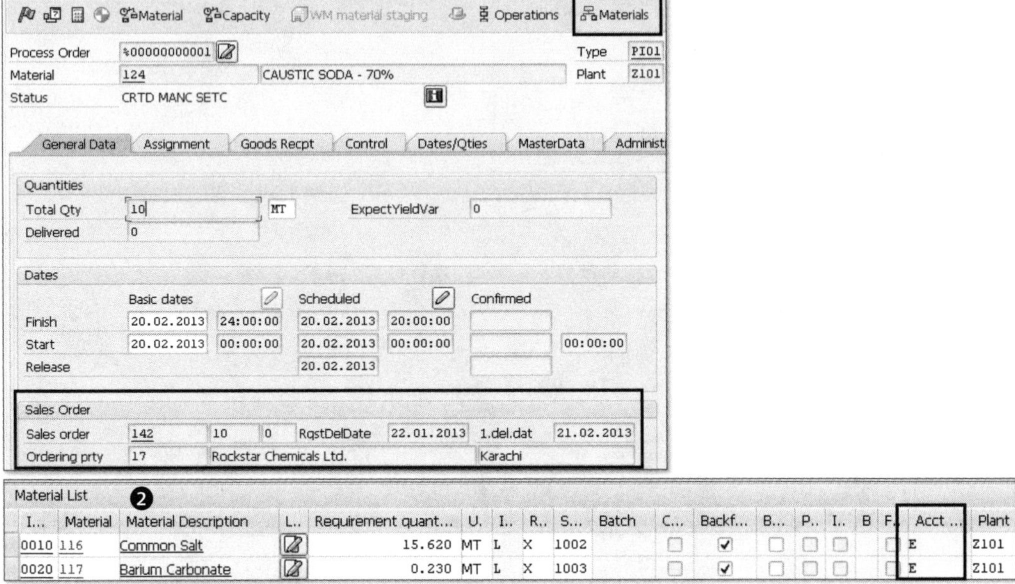

Figure 20.19 Process Order with Sales Order Details

20.5 Integration of Production Planning with Sales and Distribution (Assembly Processing)

The business processes in SD is generally the starting point in the entire chain of business events. The SD business process starts when the customer places an inquiry for the product. On the basis of inquiry, the company issues a quotation to the customer consisting of selling price, discounts or surcharges, and tentative delivery time. If the customer accepts the quotation and place a purchase order, the company proceeds to create a sales order with reference to the quotation. It contains the delivery date as well as payment terms, including advance payment, if any. If the sales order consists of products that are produced using the make-to-stock (MTS) production method, the company performs post goods issue (PGI). In MTS, the stock of the produced goods is already available in the warehouse, in anticipation of impending customer order/demand. The PGI process creates financial entries and the customer becomes liable to pay for the goods delivered, and the company reflects this in its Accounts Receivable (AR) subcomponent of the FI component in the SAP ERP system. This is an integration point between FI and SD components. When the customer makes payment against the delivered goods, it's adjusted from the customer's receivables. From the initial stage of inquiry until the sales process is complete, the system maintain logical links to all the reference and subsequent documents in *document flow*.

As mentioned earlier, in a make-to-order (MTO) production method, the creation of a sales order triggers the production process, and all production and procurement activities can be sales order-specific. This is an integration point between the PP and SD/MM components. A comprehensive trail is maintained for all the business processes involved, and the produced good is stored in warehouse and categorized as special stock (sales order stock).

In assembly processing (or assemble-to-order [ATO]), when you create a sales order, the system automatically creates a production order (or process order). You can also choose for the system to automatically create a planned order, instead of a production order. You need to enter the STRATEGY GROUP as "82" in the MRP 3 view of the material master for the automatic creation of a production order. The strategy group 81 is available to enable the system to create a planned order from a sales order. Similarly, strategy group 86 is available to create a process order from a sales order. To control how the system plans and then creates procurement elements (such as planned orders or production orders), you can use Transaction OVZG.

The requirement class 200 is available for assembly processing for planned orders, and the requirement class 201 is used for assembly processing for production orders. In requirement class 201, you can define the order type that the system uses for production order creation. The default setting for ORDER TYPE is PP04, which you can change to suit your business needs. It helps if you select the AUTOMATIC PLANNG. checkbox to ensure that the system performs automatic planning. Additional checks, such as capacity planning and availability checks, are also available. You also need to ensure that the order type that you select in Transaction OVZG has the SCHEDULING TYPE set as BACKWARD SCHEDULING and the ADJUST BASIC DATES checkbox selected in the scheduling parameters for order type.

[»]

Note

Refer to Chapter 3 to learn how to make configuration settings for a production order, including scheduling parameters.

In assembly processing, the procured raw material is for sales orders only, and the system reflects the procured raw material as sales order stock.

[+]

Tips & Tricks

The system creates a new production order if you make changes to the sales order. For example, when you first created a sales order for a customer, the system maintained a credit block for the customer. Later, when you change the sales order to remove the credit block, the system creates a new production order. To avoid the creation of a new production order each time you make changes to the sales order, coordinate with your SD team to ensure they assign item category "CN" in the sales-related settings of MRP.

20.6 Integration of Production Planning with Project System (Engineer-to-Order Production)

The engineer-to-order (ETO) production type attends to the complexities and challenges that a sales order-based MTO production method is unable to fulfill. In ETO, the PP component integrates with the SAP ERP Project System (PS) component. In the MTO production method, the system is unable to make a distinction between the predecessor–successor relationships in the production process; for example, a material's production can't initiate (successor) until the production of the previous product (predecessor) is ensured. In ETO, the system makes use of work breakdown structure (WBS) and a network for scheduling and coordinating the production processes and also managing Cost Accounting. All produced goods are specific to the project, and the system maintains project-based inventory. In fact, even the procurement of all components needed in the production process can also be project-specific. It's equally possible to maintain a budget for the project, which then forms the basis of monitoring with the actual costs incurred on the project.

In ETO production, you need to assign the in-house produced material either to a network or to a WBS element of the project in the PS component. The advantage of integrating PP with PS in the ETO production is that the entire procurement or production is specific to the project and offers better controls. Such controls can be that all procurement and production, including inventory movements such as goods issue and receipt, are done with reference to the project's network element or WBS element. The system accumulates all costs associated with the project that you can eventually settle.

The network element and WBS element are two of the important elements in PS. The network element controls the sequence of different activities and the predecessor–successor relationship. The WBS element reflects the logical breakdown of a large number of internal and external activities associated with a project in PS.

In the same way, you can also assign the materials that you want to procure for your project to the network or WBS element.

For in-house produced material, it must be marked as MTO/project settlement by virtue of its strategy group in the material master. In the same material master, you also must assign individual requirements of the material. You can then run MRP on the entire project on the WBS element with or without hierarchical explosion. If your company runs MRP on the plant level, then the system automatically takes care of the planning of the in-house produced materials (or externally procured materials). When MRP is run, the system creates a planned order, which is then converted into a production or process order, depending on the production type and industry. You can even control procurement of the raw materials needed for the production process by defining whether the procurement should also be specific to project stock.

We begin this section by identifying the relevant setting in the material master to declare the material as MTO/assembly settlement material. We then create an activity of the same material in the project and run MRP on the material. This creates the planned order, which we convert into a production order. We've confined the scope of this section to showing how the WBS element's details are available in the production order, whereas you can then perform all of the subsequent steps of production, the QM activities involved, and the goods receipt of the project stock.

20.6.1 Managing Master Data

For project-based production, access the change material master screen via Transaction MM02, enter the material number and plant, and then select the MRP 3 view (Figure 20.20). Enter the Strategy Group as "21", and click on the MRP 4 view of the material master.

Planning				
Strategy group	21	Make-to-order prod./ project settlement		
Consumption mode			Bwd consumption per.	0
Fwd consumption per.	0		Mixed MRP	
Planning material			Planning plant	
Plng conv. factor			Planning matl BUnit	

Figure 20.20 Strategy Group for Assembly Order

You also need to ensure that the Individual/Coll. field is set to 1 in the MRP 4 view of material master, so that the system creates individual proposals (planned orders during the MRP run) for the project-based planning.

After these settings are in place, you can save the material by choosing Save.

20.6.2 Assigning a Material to the Project

To assign a material to the project for this example, use Transaction CJ20N, and use an already-created project in the PS component, C-13-KH-0006. Then assign the Material "84" as an activity (in-house material) to the Network 1.

Figure 20.21 ❶ shows the initial screen for the creation or change of project structure. On the left-hand side of the screen, place the cursor on the Network 1 field, and right-click to bring up the Activity element (work) option. On the right-hand side of this screen, enter the Material as "84", the Plant as "Z101", the Requirement Qty as "100", and the item category (Item column) as "L". The system brings up the Purchase Requisition pop-up ❷.

In the pop-up screen, you can select one of the five options to cater to various business scenarios. You can choose to have reservation at the network or the WBS element level. For externally procured material, you can simultaneous generate purchase requisition and create an in-house reservation, so that the procured material is only used in the specific project. For this example, select Reserv. for network, and save the project structure.

Additionally, the system has assigned the identification as 4000055 and item as 0020 as shown in the Identification and view selection box at the top-right area of Figure 20.21. After the MRP run, this identification will be available for reference in the planned order as well as the production order.

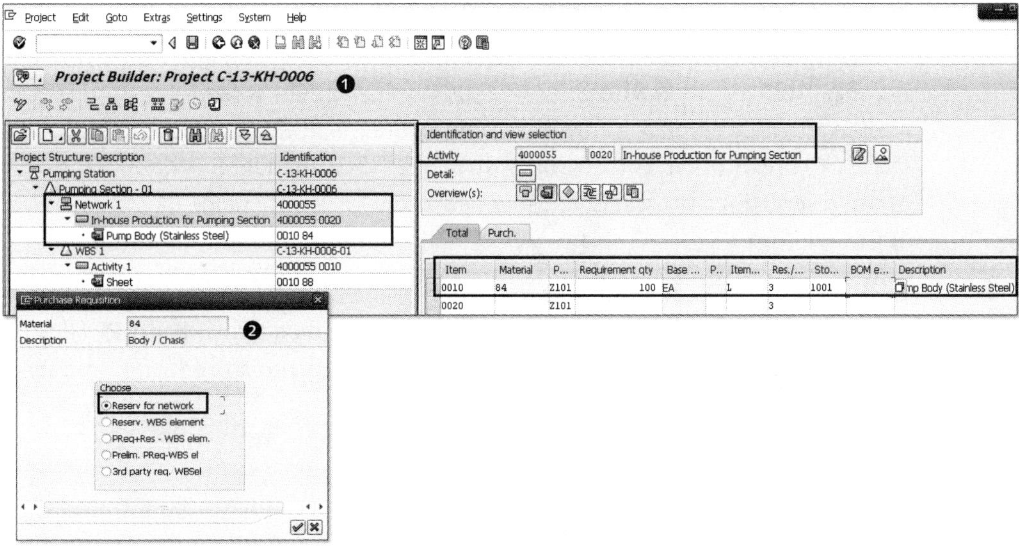

Figure 20.21 Project with Component Assignment in the Network

20.6.3 Material Requirements Planning Run on Material for Project-Based Production

You can run PS-based MRP with Transaction MD51, provide the project structure and WBS element (with or without hierarchical explosion), and assign planning parameters to it, so that the system creates planned orders of in-house produced materials for future conversion to production orders.

Alternatively, you can run MRP on the specific material to generate a planned order for conversion into a production order. Transaction MD02 is used for the MRP run on a single item with multiple levels. Finally, if you run MRP on the plant level, then the system automatically takes care of the planning of the materials (in-house or external) belonging to the project.

For this example, use Transaction MD02 to run MRP on material 84. In the MRP CONTROL PARAMETERS tab, enter "3" for the CREATE PURCHASE REQ. field. Choose to denote the creation of planned orders during the MRP run and save the results.

The next step is to evaluate the results of MRP in the stock/requirements list and convert the created planned order into a production order.

20.6.4 Conversion of a Planned Order to a Production Order

Access the stock/requirements list via Transaction MD04, and enter the MATERIAL number "84" together with PLANT "Z101" on the initial screen. Press Enter to see the planning results after MRP is run (Figure 20.22 ❶). The screen reflects the PROJST C-13-KH-0006 as the project structure, then the ORDRES 4000055 as the order reservation. It creates the planned order (PLDORD field) 912 of quantity 100.

Click on the ITEM DETAILS icon ❶ to convert planned order 912 into a production order. In the ADDITIONAL DATA FOR MRP ELEMENT screen that appears ❷, click on the ->PROD.ORD. button to convert the planned order into a production order.

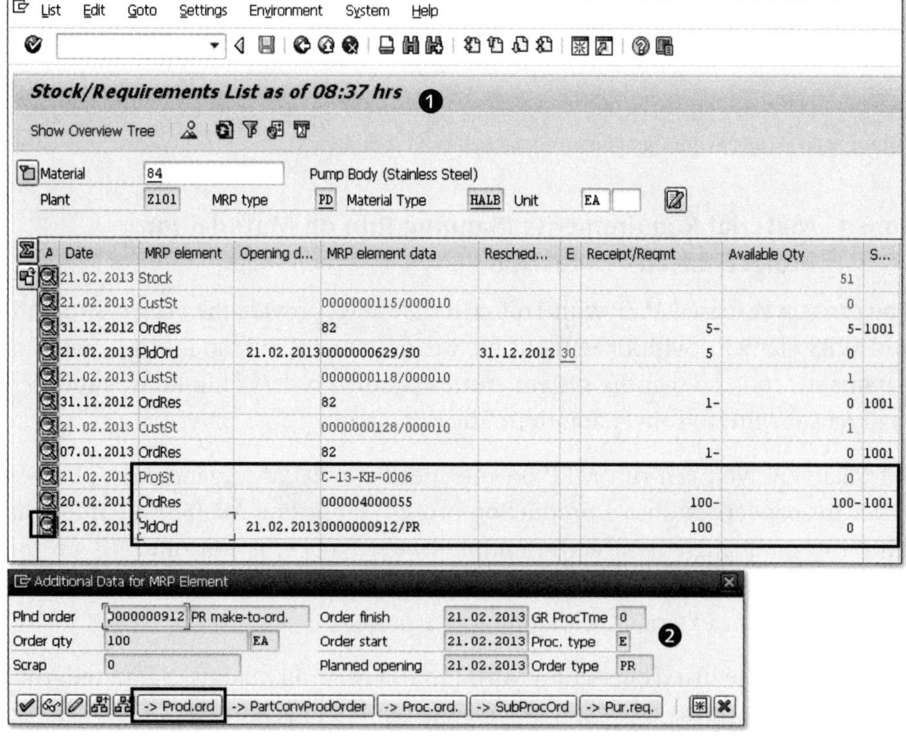

Figure 20.22 MRP Results and Conversion of Planned Order to Production Order

Figure 20.23 ❶ shows the order header screen of the production order and already contains all of the details and other data from the planned order 912.

Click on the ASSIGNMENT tab to see the ASSIGNMENTS settings, in which the same WBS ELEMENT, C-13-KH-0006, is automatically assigned to the production order.

Finally, click on the COMPONENTS OVERVIEW icon ❶ to see the screen that shows the project stock must be procured and then consumed (issued) against the production order ❷.

Save the production order, and it generates an internal number.

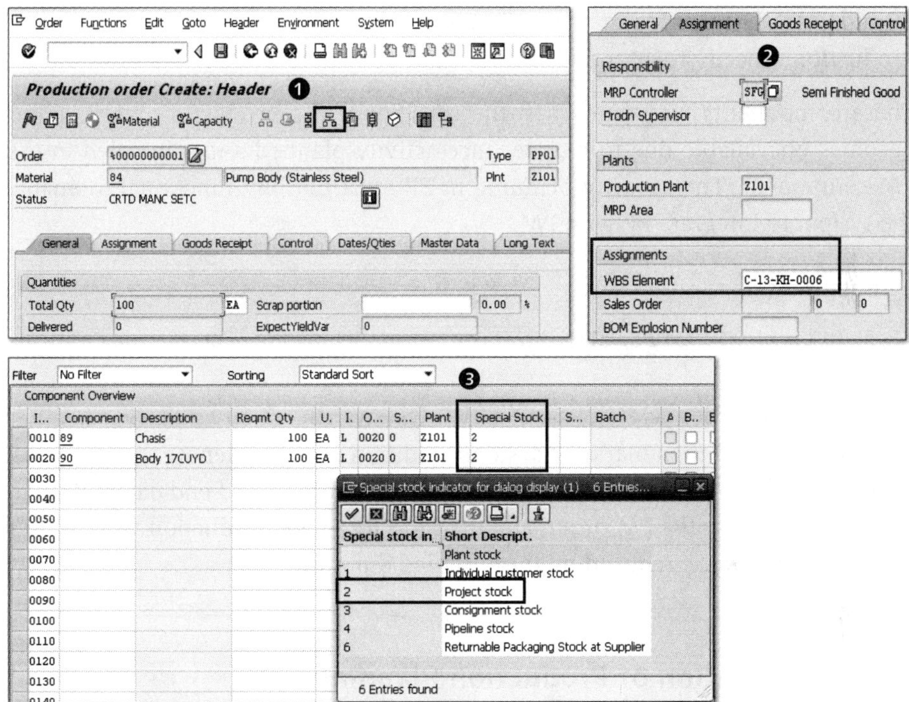

Figure 20.23 Production Order with Project Details

All of the remaining production steps are the same, that is, goods issuance, confirmation, and goods receipt. On goods receipt, the project stock for material 84 will be available in stock.

20.7 Integration of Production Planning with Plant Maintenance

Maintaining optimum plant conditions is one of the important key performance indicators (KPIs) of the maintenance management department of the company. However, a collective endeavor ensures that the maintenance planning, scheduling, and execution also balance with PP to ensure that production doesn't suffer at the cost of maintenance or vice versa. The main integration point between the PP and SAP ERP Plant Maintenance (PM) components is the limited availability of production capacity to attend to maintenance activities.

These components integrate when the capacity at the production work center becomes unavailable due to maintenance activity planned and scheduled in the PM component. The capacity planning in PP must be fully functional to enable the system to integrate PP with PM.

[»] **Note**

Refer to Chapter 14 for a detailed discussion of capacity requirements planning.

Create the PM order (via Transaction IW31), enter the production work center and the plant, and then mark the system condition as "0" to denote that the work center is NOT IN OPERATION. You also need to give the start and end dates/times in the PM order. Save the PM order, and the capacity of the production work center will automatically be reduced by the specified duration.

20.8 Integration of Production Planning with SAP Manufacturing Execution

SAP Manufacturing Execution (SAP ME) is a comprehensive, integrated manufacturing operations solution—a single manufacturing environment to better plan, define control, manage, and execute operations—providing real-time configuration data capture and global visibility across a manufacturing line, plant, or enterprise.

SAP ME integrates with the SAP ERP system to control shop floor manufacturing processes and enable production to do the following:

- Collect a complete record of serialized components in a mixed-mode manufacturing environment.
- Maintain detailed product records, nonconformance information, product yields, and quality data that are available in real time to the entire organization.
- Manage and track complex work in process (WIP) processes—order splits, merges, and configuration changes.

You can perform integration by using SAPMEINT software and SAP MII as a platform (middleware). This helps to bring tremendous improvement in business processes by enabling real-time decision making.

20.9 Integration of Production Planning with SAP Manufacturing Integration and Intelligence

SAP Manufacturing Integration and Intelligence (SAP MII) integrates top-floor systems, for example, enterprise resource planning systems such as SAP ERP, with shop floor systems, such as Manufacturing Execution Systems (MES), Data Historians, and Supervisory Control and Data Acquisition (SCADA) systems. The SAP ERP system downloads the data, such as BOMs, routing, and production orders, and sends it to MES using SAP MII. Production confirmations and goods movement information is sent from MES to SAP MII and then uploaded to SAP ERP to close the loop.

To provide integration to SAP ERP, SAP MII provides connectors such as Java Connector (JCo), Java Resource Adapters (JRA), and web services for SAP Enterprise Services. You use these connectors to send and receive data from SAP ERP (push from SAP ERP and pull from SAP MII). Using the connectors, you can build a complete scenario to download the information from SAP ERP, send it to a shop floor system, read it from the shop floor system, and send it to the SAP ERP system.

SAP MII includes some of the following business scenarios:

- Shop floor production integration
- Shop floor quality integration
- KPIs and alerts
- Content versioning

- Ad hoc reporting
- Transactions for Business Application Programming Interface (BAPI) consumption
- Trend chart visualization
- Plant information catalog
- SAP BusinessObjects integration (SAP BusinessObjects and SAP Crystal Reports)

20.10 Summary

In this chapter, we covered the importance of cross-component integration aspects of PP with some of the other components dealing with logistics in SAP ERP, including MM, QM, SD, PM, and PS. The intracomponent flow of processes, data, and information makes it important for all of the concerned components or departments of the company to work together to map end-to-end business processes.

SAP MII provides the basis for better planning and execution of activities on the shop floor level. Implementing SAP MII is the next logical step for companies to take when their business processes in SAP ERP Production Planning have matured and there's a need for granular and in-depth planning, monitoring, and execution.

As you've seen, the PP component in SAP ERP is quite comprehensive and offers in-depth and practical solutions to a large number of business processes and scenarios for a diverse range of industries. However, the actual return on investment for the SAP ERP implementation can only be ensured when you strive to understand the integration of various SAP ERP components with each other. To achieve flawless integration during a SAP ERP implementation, you need to maintain extensive collaboration and intense coordination among the various stakeholders of the company.

Appendices

A Comparison Table of Production Types

This table provides selective comparison among the three types of production that are highlighted in this book; namely discrete, process, and repetitive manufacturing. Here you can find the typical properties of each production type. In a few cases, there might be exceptions that are very rare or special and aren't worth mentioning here.

Characteristics	Discrete Manufacturing (SFC)	Process Manufacturing (PP-PI)	Repetitive Manufacturing (REM)
General Characteristics			
Industry types	Automobiles, pumps, engines, computers, toys, and electronics such as televisions and computers	Chemicals, paints, fertilizers, processed food, pharmaceutical, beverage, and so on	Consumer goods, mechanical, electronics, and so on
Product stability/ complexity	Complex production process with intermediate storages	Complex production process and generally without or bulk intermediate storage (mostly continuous flow and liquid-based production)	High-volume (mass production), highly stable, and without any production complexities
Production flow	Order-based and complex	Order-based and mostly used in producing materials that flow (or can't disassemble)	Lean (simple) manufacturing works without order types, and only through planned orders
Changeover	Frequent	Frequent	Infrequent
Make-to-stock and Make-to-order	Supported	Supported	Supported
Batch management	Yes	Yes (extensive utilization)	Yes
Active ingredient management	Not available	Available	Not available

Characteristics	Discrete Manufacturing (SFC)	Process Manufacturing (PP-PI)	Repetitive Manufacturing (REM)
Material quantity calculation	Not possible	Possible	Not possible
Components staging	With reference to production order	With reference to process order	Components staged periodically and anonymously (order reference is atypical in REM)
Planning strategies: MTO and MTS	Supported	Supported	Supported
Completion confirmation (backflush)	For individual operations or orders	For individual operations or orders	Period-based confirmation with backflush
Order-related production	Yes (production order)	Yes (process order)	Period-based with planned orders
Production-based	Lot size-based production	Lot sized-based production	Period and quantity-based production
Cost Object Controlling	Order-based costing	Order-based costing	Period-based costing (using product cost collector)
Settlement	Settlement is at order level.	Settlement is at order level.	Settlement is at material level. Production-specific Product Cost Controlling is also possible, for better evaluation in the CO component.
Process management	Process integration (= process control), control instruction (= control recipe), operational method sheet (= process instruction sheet)	Control recipe (sent as process instructions)	Operational method sheet (OMS)

Characteristics	Discrete Manufacturing (SFC)	Process Manufacturing (PP-PI)	Repetitive Manufacturing (REM)
Master Data			
Material master	Yes	Yes	Yes (REM checkbox and REM profile maintenance in material master)
Work scheduling view activated	Yes	Yes	No (maintaining REM details in MRP 4 view of material master is necessary/ sufficient)
Bill of materials (BOM)	Bill of materials (Transaction CS01)	List of materials (Transaction CS01)	Bill of materials (Transaction CS01)
Work center	Work center (Transaction CR01)	Resource (Transaction CRC1)	Production line (Transaction CR01)
Work center hierarchy	Work center hierarchy	Resource network	Work center hierarchy is possible in REM planning to get a synchronous overview of several production lines (MF50). Line hierarchies are possible to monitor and plan the load on several capacities along a production line.
Routing	Routing (Transaction CA01)	Master recipe (Transaction C201)	Rate routing (Transaction CS21). REM also completely supports standard routing (Transaction CA01).
Production version	Optional (Transaction C223)	Mandatory (Transaction C223)	Mandatory (Transaction C223)

Characteristics	Discrete Manufacturing (SFC)	Process Manufacturing (PP-PI)	Repetitive Manufacturing (REM)
Transactional Data			
Manufacturing order	Production order (Transaction CO01)	Process order (Transaction COR1)	Run schedule quantity (e.g., Transaction MF50)
Collective availability check	Production orders (Transaction COMAC)	Transaction COHVPI (on the MASS PROCESSING tab, select MATERIAL AVAILABILITY CHECK from the dropdown)	Planned orders (Transaction MDVP)
Process management	Execution Steps (XSteps)	Control Instructions and XSteps	Operational method sheet (Transaction LDE1)
Confirmation	At operation level (Transaction CO11N)	At phase level (Transaction COR6N)	Confirmation with or without reporting point backflush (Transaction MFBF)
Printing	Shop floor papers (Transaction CO04)	Shop floor papers (Transaction COPI)	Production list (Transaction MF51)
Order confirmation	For production order (Transaction CO15)	For process order (Transaction CORK)	For material or for planned orders of material (Transaction MFBF)
Mass processing	Transaction COHV	Transaction COHVPI or CORM	Transactions: COHV and COHVPI. Deselect PRODUCTION or PROCESS ORDER, respectively, and select PLANNED ORDER, and several mass processing functions become available.
Status management	Yes	Yes	No
Confirmation of activities	Yes	Yes	Yes

Characteristics	Discrete Manufacturing (SFC)	Process Manufacturing (PP-PI)	Repetitive Manufacturing (REM)
Separation of final confirmation and goods receipt possible	Yes	Yes	No
In-process (during production) inspection	Possible at operation level only (Inspection type: 03)	Possible at operational and phase level (Inspection type: 03)	Possible (Inspection type: 13)

B Glossary

Activity type Unit in a controlling area that classifies the activities performed in a cost center. Examples: machine hours or finished units in production cost centers.

Actual costs Costs actually incurred in producing a product.

Alternative BOM Used to identify one bill of material (BOM) within a BOM group. One product can be represented by several (alternative) BOMs if, for example, different production processes are used for different lot size ranges. These alternative BOMs are grouped together in a multiple BOM.

Assembly Group of components of a product that forms a technically coherent whole in the production process. A product defined as an assembly can be used again as a component in another assembly.

Automatic reorder point planning Special procedure in consumption-based planning. If available stock falls below the reorder point, then an entry is made in the material requirements planning (MRP) file. Then during the next planning run a procurement proposal is created. The reorder point and the safety stock level are automatically determined by the forecast program.

Availability check This procedure makes sure that there are enough components available for the planned/production orders in production planning and production control.

Available capacity Output or "ability to execute" a capacity within a specified period. The available capacity is specified by the following values: work start and finish times, duration of breaks, rate of capacity utilization in percent, and the number of individual capacities within one defined capacity.

Backflush Non-manual posting of an issue of components sometime after their actual physical withdrawal for an order. The goods issue posting for backflushed components is carried out automatically at the time of order completion confirmation.

Backlog Capacity requirements that have been scheduled, after scheduling or distribution, to periods that lie in the past.

Backorder processing With the backorder processing function, you can process materials for which there are missing quantities. In this process, you can commit requirements up to the amount of the current ATP quantity, or you can reallocate quantities that have already been committed.

Backward scheduling Type of scheduling in which the operations of an order are scheduled backwards starting from the order due date. The scheduled start and scheduled finish of the order are determined via this scheduling type.

Base unit of measure Unit of measure in which the stock of a material is managed. The system converts all quantities entered in other units into the base unit of measure.

Batch Subset of the total quantity of a material held in stock, which is managed separately from other subsets of the same material. Examples: different production lots (such as paints and pharmaceutical products), delivery lots, quality grades of a material.

Batch determination Functionality that can be used for all movements out of the warehouse in order to locate batches in stock using certain selection criteria. Batch determination can be used in: goods movement (inventory management), production orders/process orders, sale orders/deliveries, and transfer orders.

Batch record Record containing all quality-relevant planned and actual data on the production of a batch. A batch record comprises several documents that contain the relevant data of specific SAP ERP system objects and archive files from external systems. They're stored in an optical archive where they cannot be forged.

Batch where-used list Hierarchically structured record of the use or composition of a batch across all production levels. The batch where-used list facilitates both top-down and bottom-up analysis.

Bill of material (BOM) A complete, formally structured list of the components that make up a product or assembly. The list contains the object number of each component, along with the quantity and unit of measure. The components are known as BOM items. You can create the following BOM categories in the SAP ERP system: material BOM, document structure, equipment BOM, functional location BOM, and sales order BOM.

BOM category Classification of bills of material (BOMs) enabling you to represent different objects (such as materials or documents).

BOM component Part of a bill of material (BOM). If an object (for example, material) consists of more than one part, you can store these parts as components in a BOM. The following objects can be stored as BOM components: material master record, document info record, or class.

BOM explosion Function for determining all the components of a bill of material (BOM) and listing them. You can explode a BOM to show the structure of the product or to show the total quantity of each component.

BOM group Collection of bills of material (BOMs) enabling you to describe a product or a number of similar products. The following bills of material comprise a BOM group: all the variants of a variant BOM, and all the alternatives of a multiple BOM.

BOM item Part of a bill of material (BOM) for which information about the quantity and unit of measure are stored. BOM items are separated into various types, and special item data is managed, according to the object reference (for example, material or document) and other criteria (for example, inventory management).

Bottleneck work center A bottleneck work center is defined in capacity requirements planning for purposes of capacity (finite) scheduling. Proceeding from this work center, forward and backward scheduling are carried out.

Business partner A natural or legal person or a group of natural or legal persons not part of the business organization but with whom a business interest exists.

Campaign The uninterrupted execution of process orders of the same kind based on the same master recipe.

Capacity Capacity refers to the ability of a work center to perform a specific task. Capacities are differentiated according to capacity category. They are arranged hierarchically under a work center.

Capacity category A description that classifies a capacity at a work center. A capacity category can only exist once at each work center. However, capacities at different work centers can have the same capacity category.

Capacity load The capacity load is determined from a comparison of capacity requirements and the capacity available. This value is always specified as a percentage of the available capacity.

Capacity requirements planning A tool for determining available capacity and capacity requirements, and for carrying out capacity leveling. Capacity requirements planning supports long-term rough-cut planning, production rate planning, and short-term detailed planning.

Clean-out recipe Recipe that describes the time, resource, and material requirements, as well as the activities that are necessary to clean out a vessel after the productive run of a campaign.

Client In commercial, organizational, and technical terms, a self-contained unit in an SAP ERP system with separate master records and its own set of tables.

Collective order Linking of planned orders or production orders over several production levels.

Company code The smallest organizational unit for which a complete self-contained set of accounts can be drawn up for purposes of external reporting. This includes recording all relevant transactions and generating all supporting documents required for financial statements.

Completion (business) A function that is executed for a production, process, or maintenance order if no more cost postings are expected. Business completion can be performed only when the maintenance order has been technically completed.

Completion (technical) A function that is performed for a production, process, or a maintenance order if the planned tasks it contains have been executed.

Confirmation Function used to record the drawn materials, working hours used, additional materials required, travel costs incurred, technical findings, and changes made.

Confirmation (technical) Recording of technical confirmation data. This data includes, for example, the following:

- Cause of damage
- Location of damage on object
- Data about machine failure
- Data about plant availability

Controlling area Organizational unit within a company used to represent a closed system for cost accounting purposes. A controlling area may include single or multiple company codes that may use different currencies. The associated company codes must use the same operative chart of accounts.

Constant model Model for constant consumption flow. A constant consumption applies if consumption values vary very little from a stable mean value. Any individual variations from the average value are caused by random influences that appear on an irregular basis.

Consumption-based planning Generic term for the procedure within material requirements planning (MRP) for which stock requirements and past consumption values play a central role.

Container Receptacle in which a material is contained.

Control cycle Controls the replenishment of a specific material via Kanban. It determines the replenishment method to be used between the supply and demand source, and the quantity required. Moreover, it determines the number of Kanbans and the quantity of each individual Kanban, the replenishment strategy used to procure the material, the supply source that is to provide the material, and the production supply area (that is, the demand source that requires the material).

Control recipe A recipe containing all process instructions for the execution of a process order by a process control system or a line operator. Control recipes are created from the process instructions of the process order and sent to the responsible process control instance. One control recipe is generated for each control recipe destination defined in the process order.

Control station The link between production control and the production process itself. It has interfaces for plant and machine data collection and can be used to control lead time parallel to production and to monitor released orders. The control station is easy to use (with the mouse), has a graphical user interface (for planning tables and statistics), and is constantly available.

Cost center Organizational unit within a controlling area that represents a defined location of cost incurrence. The definition can be based on functional requirements, allocation criteria, physical location, or responsibility for costs.

Dependent requirement Planned product requirement that is caused by the production of higher-level assemblies. During the planning of product requirements, dependent requirements are automatically created for the components that are necessary for the production of a planned order.

Detailed planning Used within capacity planning for short-term planning of individual capacities or people. Detailed planning uses exact times and dates and is based on a routing.

Digital signature Equivalent to a handwritten signature for the processing of digital data. A digital signature ensures that the signed transaction can only be carried out by users with a special authorization, that the signatory identification is unique and forgery-proof, and that the signatory name is documented along with the signed transaction and the date and time.

Direct procurement Procurement without stockholding. Components are produced directly for the higher-level assembly and consumed. The aim of direct production is to deal with both scheduling and costing procedures for finished products, assemblies, and components in a bill of material (BOM) structure.

Disaggregation Process in sales and operations planning (S&OP) by which the data of a planning hierarchy level is broken down into the data of its respective members.

Dispatching period The planning period is the period of time in which the system takes all the planned orders and plans their sequence according to the selected planning procedure. This term is used in sequencing.

Engineering change management Component that allows you to change objects with history (with date validity) or depending on certain criteria (with parameter effectiveness). The different processing statuses of the object are saved. All changes are made with reference to a change master record. In the SAP system, you can change, for example, the following object types with reference to a change number: bills of material, task lists, documents, and materials.

Engineer-to-order Manufacturing environment in which complex production activities under essentially one-off conditions are undertaken for a specific customer. The production structures are managed using a work breakdown structure. At this level, the following planning steps for the project are carried out: budget management, revenue planning, finance planning, and cost planning. The production processes are managed using a network.

Ex-post forecast Forecast for past periods. The ex-post forecast is used for the evaluation of forecast quality, for model selection, for parameter optimization, and for forecasting missing periods in retrospect.

External processing Operations that are carried out at a different company. Within work order processing, the system creates purchase requisitions for externally processed operations. The purchasing department converts these purchase requisitions into purchase orders.

External procurement Procurement of raw materials, operating supplies (MRO items), trading goods/merchandise, or services from an external supplier for the organizational units within a firm that need such items.

Factory calendar Calendar in which working days are numbered sequentially. The factory calendar is defined on the basis of a public holiday calendar.

FIFO principal A stock removal method whereby the materials first transferred to a bin are the first to be removed from the bin (first in, first out).

Finite scheduling Scheduling type within capacity planning that takes account of the capacity loads that already exist. Finite scheduling calculates the start and finish dates for operations in the order.

Forecast Estimation of the future values in a time series. In the SAP ERP system, the forecast can be carried out using a number of different procedures, such as first-order and second-order exponential smoothing or simple average models.

Forecast-based planning Special procedure in consumption-based planning that is based on future requirements predictions calculated using the forecast. Forecast values for future requirements are determined by the integrated forecasting program. These values then form the basis of the net requirements calculation in the planning run.

Forecast error The difference between the actual consumption values and the forecast values in the last period.

Forecast model States the structure prevalent in a time series. The following forecast models exist: constant model, trend model, seasonal model, and seasonal trend model.

Forecast parameter Generic term for all forecast-related data. A distinction is made between forecast parameters that are independent of the forecast model and forecast parameters that depend on the forecast model.

Goods issue Term used in inventory management to describe a reduction in warehouse stock due to a withdrawal of stock or the delivery of goods to a customer.

Goods movement Physical or logical movement of materials leading to a change in stock levels or resulting in the direct consumption of the material.

Goods receipt Term from the field of inventory management denoting a physical inward movement of goods or materials. The SAP ERP system differentiates between the following kinds of goods receipt:

- Goods receipt with reference to a purchase order
- Goods receipt with reference to a production order
- Other goods receipts (without reference)

Gross requirements planning Special deterministic procedure within material requirements planning (MRP). In gross requirements planning, no comparison is made between the warehouse stock and the (gross) requirements. This means that these requirements will always be covered by order proposals.

Individual capacity To achieve more detailed planning of resources and commitments, you can subdivide capacities into individual capacities (for example, individual machines) for which you can maintain available capacities. You can allocate or dispatch capacity requirements to these individual capacities in the graphical or tabular planning table.

Individual customer requirement Material requirement in a plant that is created by a sales order and transferred separately to material requirements planning (MRP).

In-house production time The time that is required to produce the material in your own plant. It is calculated by adding together the lead times of all operations plus the float before production and the float after production.

Inspection lot Request to a plant to carry out a quality inspection for a specific quantity of material.

Item category Defines items in a bill of material (BOM) according to certain criteria, such as the object type of the component (for example, material master record or document info record). The item category controls the following: screen sequence, field selection, default values, material entry, inventory management, and sub-items.

Kanban A procedure for controlling production and material flow based on a chain of operations in production and procurement. The replenishment or the production of a material is not triggered until a certain production level actually requires the material. The signal for replenishment is issued by a card (kanban) that is sent by a consumer to the supplier.

LIFO principle Stock removal strategy whereby the goods last transferred to a storage type are the first to be transferred out of the storage type (last in, first out).

Line balancing In line design you can create line balances for a line hierarchy. In line balancing you regularly adapt the line hierarchy to a planned production rate by changing the number of takts (processing stations) and individual capacities (persons), and if necessary you can move operations to other line segments.

Line design Integral part of process manufacturing for flow and process manufacturers. Using line design, operations and sub-operations are defined, and their sequence is determined in the form of a rate routing. Line design is also used to structure the production line, that is, to divide it into segments and takts.

Line segment Work center on a production line. You can create line segments in line design by inserting work centers in a line hierarchy. Then you can define the number of takts and the number of individual capacities for the line segments.

Logistics Information System (LIS) The Logistics Information System is made up of the following information systems: Sales Information System, Purchasing Information System, Inventory Controlling, Shop Floor Information System, Plant Maintenance Information System, and Quality Management Information System. The information systems in LIS can be used to plan, control, and monitor business events at different stages in the decision making process.

Long-term planning Simulation of the future stock and requirements situation.

Lot size Quantity to be produced or to be procured. The lot size is used as a criterion for selecting alternatives within a multiple bill of material (BOM), selecting a routing as a basis for a production order, selecting an operation within alternative operations, and pricing during the sale and/or the purchasing of goods.

Make-to-order production Type of production in which a product is manufactured specially for a particular customer. This includes both sales-order-related production and engineer-to-order.

Manual reorder point planning Special procedure in consumption-based planning. If the available stock level falls below the reorder point, then an entry is made in the planning file.

Manufacture of co-products Production of several materials in one process. Unlike with byproducts, some of the production costs are assigned to the joint product in the order settlement.

Manufacturing order Request asking production to manufacture a specific quantity of a material or perform a specific service on a specific date. A manufacturing order defines which work center or resource is used to manufacture a material and which material components are required.

Master production scheduling In master production scheduling, those parts or products that greatly influence company profits or that take up critical resources are planned with special attention. Master schedule items are marked with the material requirements planning (MRP) procedure for master production scheduling.

Master recipe Description of an enterprise-specific process in the process industry that does not relate to a specific order. The master recipe is used for the production of materials or for rendering of services.

Material availability check Automatic check that is carried out to find out whether there are enough materials to cover a proposed withdrawal from stock.

Material BOM Bill of material that is created with reference to a material master. The BOM can contain items of different item categories (such as stock items, non-stock items, document items, and text items).

Material master record Data record containing all the basic information required to manage a material.

Material overhead costs Costs that are not assigned directly to individual materials, such as procurement costs and storage costs for inventories in which different materials are stored.

Material requirements planning (MRP) Generic term for the activities involved in creating a master production schedule or an external procurement plan for all the materials in a plant or company

Material stock Part of current assets. Material stock is managed at the plant or storage location level.

Material type Groups together materials with the same basic attributes; for example raw materials, semi-finished products, or finished products. Along with the plant, the material type determines the material's inventory management requirement—that is, whether changes in quantity are updated in the material master record and whether changes in value are also updated in the stock accounts in financial accounting.

Material valuation Determination of the value of a stock of materials.

Material variant Product variant of a configurable material. The material master record of a material variant contains assigned characteristic values.

Material where-used list Bill of material reporting function that determines which BOMs a material is used in, along with the quantity. We distinguish between the following lists: direct where-used list and multilevel where-used list.

Midpoint scheduling Scheduling type where an order is rescheduled on the basis of changes to dates. Starting from the start time of the operation, all the previous operations are scheduled backwards. Starting from the finish time of the operation, all the previous operations are scheduled forwards. This type of scheduling is used, for example, in production planning and control when the operation dates are changed during capacity leveling (for example, bottleneck planning).

Minimum lot size Minimum quantity that must be reached during procurement. The minimum lot size can be taken into account in lot size calculation.

Minimum range of coverage Minimum number of days that the dynamic safety stock should cover requirements. The minimum range of coverage is defined in the range of coverage profile.

Minimum stock level Lower limit for the dynamic safety stock. The minimum stock level is calculated using the formula minimum range of coverage × average daily requirement.

MRP area Organizational unit for which you can carry out MRP separately. An MRP area can include one or several storage locations of a plant or a subcontractor. You can define MRP areas within a plant. By defining MRP areas, you can carry out MRP specifically for each area.

MRP group Groups certain materials together from an MRP standpoint to allocate them special control parameters for planning. These control parameters include the strategy group, the planning horizon, and the creation indicator for the planning run.

MRP list Overview of the results of the material requirements planning run.

MRP lot size A key that defines which lot sizing procedure the system uses for calculating the quantity to be procured in the material requirements planning run.

MRP type Key that controls the material requirements planning procedure (MRP or reorder point) to be used for planning a material. It contains additional control parameters, for example, for using the forecast for the materials planning, for firming procurement proposals, and so on.

Non-stock item The item category "non-stock item" is used if you enter a material as a component in a bill of material and the material is not kept in stock.

Object dependencies Knowledge that describes the mutual interdependencies between objects. You can also use object dependencies to ensure that the correct bill of material items and operations are selected when an object is configured.

Operative production rate Control parameter for takt-based scheduling in sequencing. Just like the planned rate, the operative rate specifies the quantity per time unit that you can produce on a production line. With the operative rate you can overwrite the planned rate in order to react to short-term changes in supply or requirements.

Operative takt time Takt times are control parameters for the takt-based scheduling of sequencing and are defined for a production line. Just like the planned takt time, the operative takt time is the time interval in which a material enters the production line and a processed material leaves the line. With the operative takt time, you can overwrite the planned takt time, in order to react to short-term changes in requirements and supply.

Order BOM Single-level bill of material that you generate for a sales order from a material BOM, and that you modify specific to the order so that the material BOM remains unchanged. An order BOM is uniquely identified by the sales order number, the sales order item, and the material number.

Order lead time The lead time of an order is the time between the basic order dates minus the order floats.

Order record Record containing all quality-relevant planned and actual data for a process order.

Order-related production Type of production in which the manufactured products are delivered to inventory without reference to a sales order. The order costs are collected on production orders and settled to inventory.

Order split Function in Shop Floor Control with which an existing production order is divided into two production orders that, from a logistical standpoint, are separate from each other.

Order type Order types categorize orders according to their purpose. The order type contains numerous pieces of information that are necessary for managing the orders. Order types are client-specific. This means that the same order type can be used in all controlling areas in one client.

Organizational structure An organizational plan that sorts the tasks in an enterprise into task areas and determines the jobs and departments that are to process those tasks.

Original plan First version that is saved of a planned independent requirement (demand management).

Overhead cost order Internal order used to monitor overhead costs incurred for a restricted period when executing a job, or for long-term monitoring of portions of overhead costs.

Pegged order Function for determining the requirements quantities and dates of intermediate products and finished products that are the source of a fixed and planned receipt at a specific production level. The pegged order is used to determine which assemblies and planned or customer independent requirements are not covered if delivery or production is delayed or incomplete.

Pegged requirement In the evaluations for MRP, the pegged requirement shows the order from the next higher level of the BOM that is the source of the requirement in a lower level.

Period lot sizing procedure Procedure that groups together requirement quantities from one or several periods to form a lot size. Costs incurred from storage, from setup procedures, or from purchasing operations are not taken into account. The number of periods that are grouped together into a procurement proposal can be defined as desired.

Phase A subdivision of an operation in the process industry.

Planned independent requirement Planned requirement quantity for a finished product over a given period of time. It is not based on sales orders.

Planned lot size Lot size value that the system uses as a default during costing. The non-proportional costs refer to this value.

Planned order Request created in the planning run for a plant to trigger the procurement of a plant material for a certain quantity for a specific date.

Planning file entry Material entry in the planning file. The system creates the entry automatically when a material has been changed in a way that is relevant to materials planning. The material is entered in the planning file as soon as it has been created with a valid MRP type.

Planning hierarchy A user-defined combination of characteristics from an information structure that is used in sales and operations planning (S&OP). It allows both top-down and bottom-up planning, and therefore the integration of centralized and decentralized planning functions.

Planning horizon The planning horizon is the period that is set for the "net change planning in the planning horizon." For this type of net change planning, the only materials planned in the planning run are those that have a change relevant to MRP within the period (in work days).

Planning ID A planning identification (ID) is a key that makes possible a grouping of different materials in terms of time and location for planning and evaluation purposes.

Planning material A planning material is required in the planning with planning material planning strategy. Non-variable parts for similar finished products can be planned using this strategy. The planning material is then assigned to the material master record of the finished products to be planned. Planned independent requirements are created for the planning material. Incoming sales orders consume the planned independent requirements of the planning material.

Planning plant Plant in which, after order execution, the goods receipt takes place for the material produced. The planning plant can be used, for example, if Sales and Distribution is to be organized as an independent plant.

Planning run Execution of materials planning for all materials or assemblies that have the necessary entries in the planning file. The planning run is divided into four main work steps: net requirement calculation, lot size calculation, procurement element/type determination, and scheduling.

Planning run type There are three different types of planning runs: regenerative planning, net change planning, and net change planning in the short-term planning horizon.

Planning segment Line segment that controls the display of the order quantity in the planning table.

Planning table The planning table in repetitive manufacturing assists the planner in planning production quantities by lines. The planning table makes it possible to check production quantities at a glance and to determine the actual capacity load utilization of production lines, as well as check the availability situation of materials.

Planning time fence Defined period in which no automatic changes are carried out to the master plan during material requirements planning (MRP). Within the planning time fence, no new planned orders are created automatically, and already existing planned orders are not changed automatically.

Planning version In sales and operations planning (S&OP), one or more versions of the planning data in an information structure can exist. This allows you to maintain multiple plans of the same information structure in parallel.

Plant maintenance Measures taken to maintain operational systems in working order (for example, machines, production installations).

PPC planning calendar The definition of flexible period lengths at a plant level. In Production Planning (PP), the PPC planning calendars can be used to define procurement dates for determining period sizes. The planning calendar can also be used for creating periodicity and aggregating scheduling agreement releases (MM).

Process data documentation A component of PP-PI used to generate lists of quality-relevant production data and store them in an optical archive. In process data documentation, you can archive batch records and order records.

Process data request Process instruction that specifies that process control is to send a process message with actual process data to the SAP ERP system.

Process instruction Structure used to transfer data or instructions from PP-PI to process control. Process instructions are allocated to the phases of the master recipe and the process order.

Process instruction characteristic A characteristic of a characteristics group released for use in process instructions. Process instruction characteristics are allocated to process instructions either directly or via the instruction category. Along with the corresponding characteristic values, they determine the information transferred or requested in a process instruction (for example, the status of a control recipe) and how the requested data is to be processed.

Process Management A component of PP-PI that represents the interface to process control. Process Management comprises the following functions: receiving released process orders from process planning, creating control recipes from process order data, passing on control recipes to the line operator or process control system involved, receiving, checking, and distributing process messages, and manual entry of process messages.

Process manufacturing The processing of gases, granular materials, or liquids. The manufacturing processes involved can be continuous or discontinuous.

Process material Material type that is designed particularly to represent the manufacture of co-products. When manufacturing co-products, you can use process materials as the header material in the bill of material and master recipe.

Process message A structure used to send actual data on a process from process control to one or several destinations of the following types: other SAP ERP components, user-defined ABAP tables, users of the SAPOffice mail system, and external function modules. Process messages are used to update existing data records, as well as to generate batch and production records.

Process order Manufacturing order used in process industries.

Process planning Detailed planning of process orders. This involves the scheduling of operations, the checking of material availability, and the release of process orders for production.

Process structure The task of the process structure is the design of process flows in an enterprise. Process flows in turn are procedures for completing business tasks that may be processed sequentially or in parallel.

Procurement proposal MRP element that is generated if a material shortage occurs. Procurement proposals are saved in the system in the form of planned orders, purchase requisitions, or delivery schedules.

Procurement type Classification determining whether a material is produced in-house, externally, or both.

Product group Groups together products (materials). The criteria by which this grouping takes place can be defined individually by each user.

Production cost collector Object in cost accounting used in repetitive manufacturing and kanban production control. A separate production cost collector can be created for each production version or material. The collected costs are settled to inventory at the end of the period. The functions performed for production cost collectors during the period-end closing process include: work in process (WIP) calculation, variance calculation, and settlement.

Production lot Particular production quantity of an assembly (finished product or semi-finished part) that is planned and produced along with reference to a number. Using this number, you can determine the costs for the production of a production lot.

Production order Production document used for discrete manufacturing.

Production overhead Costs incurred in production that are not or cannot be assigned to particular cost objects.

Production plan Production plans are created in the planning table of sales and operations planning (S&OP). Production targets can be set for materials, product groups, and/or characteristic values from an information structure and derived from sales targets or using self-defined or standard macros.

Production rate Control parameters for takt-based and rate-based scheduling in sequencing. Rates are defined for a production line.

Production resource and tool (PRT) An operating resource used to shape a material (such as a tool or fixture) or to check its dimensional accuracy, composition, and functional efficiency (such as measurement and test equipment). A PRT can also be a document (such as a drawing) or a program. PRTs are assigned in plans to operations.

Profit center Organizational unit in accounting that divides the company with a management orientation, that is, for internal control.

Production series A group of materials of limited duration that share similar characteristics.

Production type Method of production, such as order-related production or repetitive manufacturing.

Production version Determines the various production techniques that can be used to produce a material. The production version specifies the BOM alternative for a BOM explosion, the task list type, the task list group, and the group counter for the allocation to task lists, lot size restrictions, and area of validity.

Pull list Helps determine which components a production line needs and when and where these components should be made available.

Purchase order Request or instruction from a purchasing organization to a vendor (external supplier) or a plant to deliver a certain quantity of material or to perform certain services at a certain point in time.

Purchase requisition Request or instruction for purchasing to procure a certain quantity of a product or a service so that it is available at a certain point in time.

Purchasing info record Source of information on the procurement of a certain material from a certain vendor.

Purchasing organization Organizational unit within Logistics that subdivides an enterprise according to purchasing requirements. A purchasing organization procures materials and services, negotiates purchase conditions with vendors, and bears responsibility for such transactions.

Quality management Broad term for quality-related activities and objectives. Activities for quality planning, quality inspection, quality control, and QM representation.

Queue time A float that can be used to compensate for delays in the production process. It can be maintained in the work center or in the operation.

Rate routing A routing that can be used in repetitive manufacturing for planning production quantities/volumes. You can define the production quantity and a fixed reference point for each operation in a rate routing and therefore determine the production rate.

Recipe The general instructions for the use of a production process. There are manufacturing recipes describing a production process as well as non-manufacturing recipes, which check to see that all functions of a resource are working perfectly, or which carry out the clearing or changeover of a line. Master recipes and control recipes are manufacturing recipes, while changeover recipes, setup recipes, and clean-out recipes are non-manufacturing recipes.

Recipe group Grouping of recipes that describe alternative production processes.

Recipe material list A list containing all materials required to execute a process order as well as their quantity specifications.

Reorder point If the amount of stock on hand of a material falls below this quantity, an entry is automatically set in the MRP file for the material.

Reference rate routing Task list type that defines a sequence of operations that is repeated regularly. A reference rate routing is used to reduce the effort of entering data in rate routings.

Remaining capacity requirements Capacity requirements in a work order or planned order that have not yet been reduced.

Reorder point planning Special procedure in MRP. If the reorder point is greater than warehouse stock, a procurement proposal is created by MRP. A distinction is made between automatic reorder point planning and manual reorder point planning.

Repetitive manufacturing A component in the SAP ERP system for the planning and control of repetitive manufacturing and flow manufacturing. It enables the period-dependent and quantity-dependent planning of production lines, reduces the work involved in production control, and simplifies backflushing (confirmation and goods receipt posting).

Replenishment lead time Total time for the in-house production or for the external procurement of a product. With in-house production, the replenishment lead time covers all BOM levels.

Reporting point An operation in repetitive manufacturing that is flagged as a milestone in the routing. When carrying out a reporting point backflush, the system backflushes all the materials that have been withdrawn and consumed between two reporting points.

Reporting point backflush Type of backflush in repetitive manufacturing in which several operations in the processing sequence can be backflushed automatically. Using the reporting point backflush procedure, you can backflush components close to actual consumption.

Requirement Quantity of material that is required in a plant at a certain point in time.

Requirements grouping Grouping together of the material requirements of different project stock owners (WBS elements) under one WBS element in order to carry out joint requirements planning.

Requirements planning Method of guaranteeing material availability both internally and externally. Requirements planning involves the procurement of goods on time, the monitoring of stock levels, and the automatic generation of order proposals. Requirements planning can be both consumption-driven or requirements driven.

Requirements type Classification of the various independent requirements into customer requirements, planned independent requirements, or warehouse requirements.

Routing Description of the production process used to manufacture plant materials or provide services in the production industry. Routing type that defines one or more sequences of operations for the production of a material.

Run schedule quantity A quantity that is to be manufactured in a certain period. Technically speaking, run schedule quantities are created as planned orders. Unlike the other planned orders, run schedule quantities do not have to be released and converted into production orders.

Safety stock Quantity of stock held to satisfy unexpectedly high requirements in the stocking-up period. The purpose of the safety stock is to prevent a material shortage from occurring.

Scrap Percentage of a material that does not meet quality requirements.

Stock Materials management term for part of a company's current assets.

Stockkeeping unit Unit of measure in which stocks of a material are managed. The system converts all quantities that have been created with a different unit of measure into the stockkeeping unit. The term *stockkeeping unit* is exclusively an SAP Inventory Management term and is synonymous with the term *base unit of measure* used in other applications.

Sales and operations planning (S&OP) A forecasting and planning tool with which sales, production, and other supply chain targets can be set on the basis of historical, existing, and/or estimated future data. Resource planning can also be carried out to determine the amount of work center capacities and other resources required to meet these targets.

Scheduling In scheduling, the system calculates the start and finish dates of orders or of operations within an order. Scheduling is carried out in MRP. The in-house production times and the delivery times specified in the material master record are taken into account.

Scheduling work center Work center that is used for scheduling and capacity planning in flow manufacturing and repetitive manufacturing. If you have not defined a line hierarchy, the production line is the scheduling work center. If you have maintained a line hierarchy for a production line, you define a work center in the line hierarchy as a scheduling work center.

Scrap Percentage of a material that does not meet quality requirements.

Seasonal trend model Model used for a seasonal trend consumption pattern. A seasonal trend consumption flow is characterized by a continual increase or decrease of the mean value.

Secondary resource Resource that is required in addition to the primary resource and can be assigned to an operation or a phase (such as an operator or a transportation container).

Sequence Order of operations that are sorted according to operation number. By defining various sequences in routings or inspection plans, you can create structures that are similar to networks but less complex.

Sequencing Determines the sequence in which planned orders are produced. The order sequence of the finished products for a line is represented graphically in the sequence schedule. Sequencing is a tool for sequence plan scheduling in flow and repetitive manufacturing.

Setup group category Groups together the setup group keys. For example, you can combine lathes with different setup group keys to form a setup group category called "Turning."

Setup group key Key used in routings to group operations with the same or similar setup conditions. The setup group key can be used in capacity leveling to optimize the setup time. This is done by determining the setup sequence with the shortest setup times.

Shift definition The start, finish, and break times of a shift are determined in the shift definition. Shift definitions reduce the work involved in determining available capacity when working hours change because they are maintained centrally.

Shop floor papers Documents required for carrying out a work order, including operation control tickets, job tickets, pick lists, time tickets, and confirmation slips.

Single-level BOM Represents all components that are used to map one or more assemblies. The single-level BOM contains the immediate components of an assembly. Components that represent assemblies by themselves are not further exploded.

Standard available capacity The standard available capacity refers to the available capacity of a certain capacity category. It is valid if no interval of available capacity has been defined.

Standard BOM Bill of material that is used internally only in the following areas: plant maintenance and standard networks. The components of a standard BOM represent frequently occurring structures that are not object-dependent and can be allocated to the activities in a project-independent standard network.

Standard cost estimate The most important type of cost estimate in material costing. Forms the basis for profit planning or product costing where the emphasis is on determining the variances.

Standard costing Type of product costing. Standard costing allows for considering obvious variations to the planned costs of goods manufactured by using MRP-based changes to the valuation approaches without creating a new standard cost estimate.

Standard trigger point A standard trigger point is a reference object used to create trigger points. By using standard trigger points, you can minimize the effort involved in creating trigger points.

Static lot sizing procedure Procedure in which the lot size is calculated using the entered quantities in the material's material master record. Costs incurred from storage, from setup procedures, or from purchasing operations are not taken into account.

Stock Materials Management term for part of a company's current assets.

Stock determination Cross-application function allowing you to determine the stock from which material is to be withdrawn in the course of stock removal, order picking, and staging operations.

Stock/requirements list Up-to-date overview of a material's stock situation, which is generated using a function that draws together all the current and relevant data (production orders, sales orders, and so on). Therefore, it always shows the most up-to-date availability situation for a material, as opposed to the MRP list, which reflects the stock/requirements situation at the time of the last planning run.

Stock transport order Purchase order used to request or instruct a plant to transport material from one plant to another within the same enterprise. The stock transport order allows delivery costs incurred as a result of the stock transfer to be debited to the material that was transported.

Stock type Means of subdividing storage location stock or special stock. The stock type indicates the usability of a material. The storage location stock and special stocks on a company's own premises are subdivided into three different types: unrestricted-use stock, stock in quality inspection, and blocked stock.

Storage costs Costs incurred from the storage of a material, recorded in the material master record as a percentage of the valuation price and are referred to by optimizing lot sizing procedures during the lot size calculation.

Storage location In storage location MRP, the planning run is carried out at the storage location level. The storage location stock is not contained in the available stock at the plant level. Instead, it is planned separately.

Subcontracting The processing (by an external supplier) of materials provided by a customer. The result of this processing is the manufacture by the supplier (subcontractor) of an ordered material or product, or the performance by the supplier of an ordered service.

Sub-item Subdivision of a BOM item. The difference between one sub-item and others is the installation point. A sub-item has no control functions in the BOM.

Supply area Area in the shop floor where a material is provided that can be directly used for production. The supply area is used in Kanban production control and for material staging using Warehouse Management (WM), for example.

Takt Physical areas of a production line where work takes place. A material passes through one takt of the production line in the minimum takt time, where it is processed. In line design you can define the number of takts for the individual line segments of a line hierarchy. The total number of takts therefore determines the length of a production line.

Task list Describes the non-order-related process for implementing an activity. The main objects of a task list are: task list headers, operations, material assignments, production resources/tools, and inspection characteristics.

Time-phased materials planning Type of materials planning in which the materials are planned according to a specific cycle.

Trend model Model for a consumption flow that represents a trend. You have a trend if consumption values fall or rise constantly over a long period of time, with only occasional deviations.

Trend value Part of the forecast model displaying the past level of the forecast model during forecasting. For trend models and seasonal trend models, the system determines the increase in value with the trend value and updates the increase with it.

Trigger point Trigger points are used to trigger certain functions when the status of an operation changes. A trigger point can be assigned to an operation in a routing or an order. The user specifies conditions in a trigger point via parameters.

Trigger point group A combination of standard trigger points. When you assign a trigger point group to an operation, you automatically assign all the standard trigger points in the group. By using trigger point groups, you can minimize the effort involved in creating trigger points.

Valuated stock Stock of a material belonging to a firm that is part of the firm's current assets. Various procedures can be applied to valuate stock. The valuated stock of a material at a plant is the sum of unrestricted-use stock held at all storage locations, stock in quality inspection at all storage locations, and stock in transfer at the storage location and plant levels.

Variable cost Portion of the total cost that varies with the operating rate and the lot size.

Variant BOM Combination of a number of BOMs that enables you to describe a product or several products that have a large proportion of identical parts. The variant BOM describes each object completely. Each variant BOM contains all the components.

Variant configuration Description of complex products that are manufactured in many variants (for example, cars). All variants are defined as one variant product.

Work center An organizational unit that defines where and when an operation should be carried out. The work center has a particular available capacity. The activities performed at or by the work center are valuated by charge rates, which are determined by cost centers and activity types. Work centers can be machines, people, production lines, or groups of craftsmen.

Work center hierarchy The representation of a structure in which work centers and their relationships to each other are displayed in levels. Work center hierarchies are used within capacity planning to cumulate available capacities or capacity requirements. The hierarchy can also be used to locate work centers.

Work in process (WIP) Unfinished products whose costs are calculated in one of two ways:

- By calculating the difference between the actual costs charged to an order and the actual costs credited to an order
- By valuating the "yield confirmed to" date for each milestone or reporting point, less the relevant scrap.

Work order An order that specifies a task to be carried out within the company. The term "work order" is a generic term for the following order types: production orders, process orders, inspection orders, maintenance orders, and networks.

Work scheduler group Key used to differentiate between the departments responsible for planning (work scheduling, inspection planning, and so on). For example, these groups may be responsible for processing specific materials, orders, or routings.

C The Author

Jawad Akhtar works as the SAP Leader for Business Sales and Delivery at IBM Pakistan. He earned a chemical engineering degree from Missouri University of Science and Technology (USA) in 1996. He has more than 18 years of professional experience and has completed several largescale SAP implementations and rollout lifecycles. He has led large teams in his roles as an SAP integration manager and SAP project manager, and has also been proactively involved in business development and solution architect roles. He is the technical adviser to SAPexperts and is a prolific contributor to SearchSAP, where he shares his views on logistics, supply chain management (SCM), product lifecycle management (PLM), and project management.

Index

A

B

C

D

E

F

G

H

I

Q

R

S

T

U

V

W

X

- Configure MM to reflect your unique logistics requirements
- Maintain critical materials data: materials, customer, and vendor records
- Master functionalities like batch management, inventory management, purchasing, and quotation management

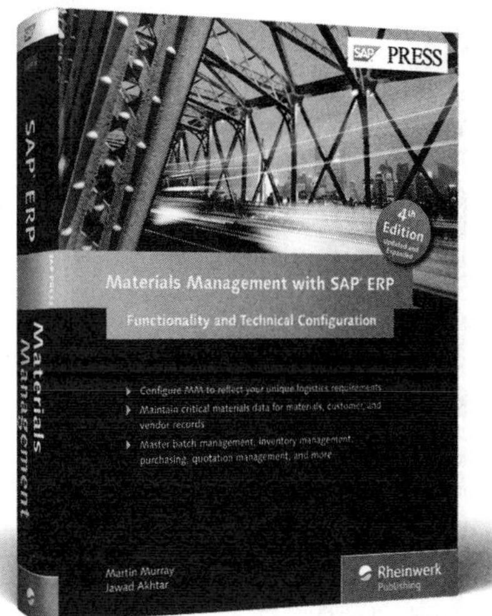

Martin Murray, Jawad Akhtar

Materials Management with SAP ERP: Functionality and Technical Configuration

Get the most out of your Materials Management implementation with this updated, comprehensive guide to configuration and functionality. You'll learn the ins and outs of Materials Management in SAP, from goods receipt and invoice verification to early warning systems and special procurement types. Dive into master data and other configuration tasks to ensure your MM system is optimized for your logistics needs!

739 pages, 4th edition, pub. 02/2016
E-Book: $69.99 | **Print:** $79.95 | **Bundle:** $89.99

www.sap-press.com/4062

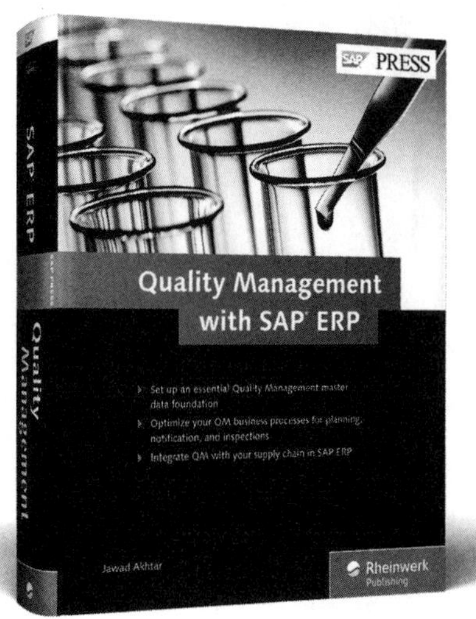

- Set up an essential Quality Management master data foundation
- Optimize your QM business processes for planning, notification, and inspections
- Integrate QM with your supply chain in SAP ERP

Jawad Akhtar

Quality Management with SAP ERP

Get the most out of your Quality Management system in SAP! From QM configuration to business process management to working in the system, this is the resource you need. Get a 360-degree view of the component, learn about QM concepts like samples and certificates, and set up essential master data. Once you've covered the basics, you'll learn how QM works with other components in the supply chain, and learn how to use workflow tools like the Classification System and Engineering Change Management.

883 pages, pub. 03/2015
E-Book: $69.99 | **Print:** $79.95 | **Bundle:** $89.99

www.sap-press.com/3755

www.sap-press.com

- Configure Warehouse Management in SAP ERP for your warehouse requirements
- Use basic and advanced warehouse functionalities for moving stock in, out, and around your warehouse
- Get familiar with ITSmobile and new planning and monitoring tools

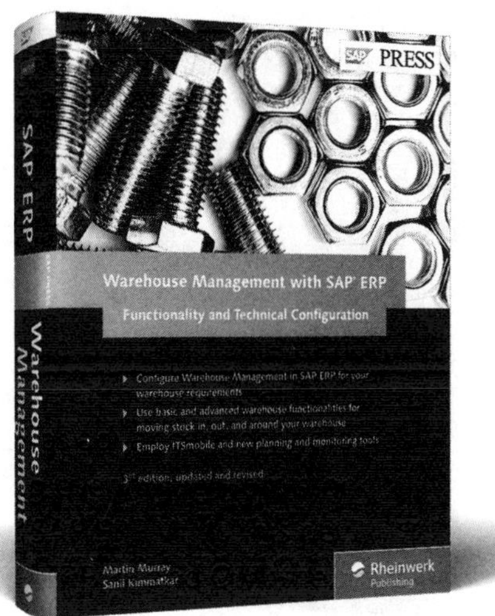

Martin Murray, Sanil Kimmatkar

Warehouse Management with SAP ERP: Functionality and Technical Configuration

Ensure an efficient and orderly Warehouse Management implementation with this comprehensive guide! Learn to customize and use critical functionalities, like goods receipt and goods issue, as well as advanced technologies such as RFID, EDI, and mobile data entry. Covering everything from stock management to picking strategies, you'll master SAP ERP WM. This new edition includes ITSmobile, connections with SAP ERP PP and QM, the warehouse activity monitor, and more.

665 pages, 3rd edition, pub. 05/2016
E-Book: $69.99 | **Print:** $79.95 | **Bundle:** $89.99

www.sap-press.com/4069

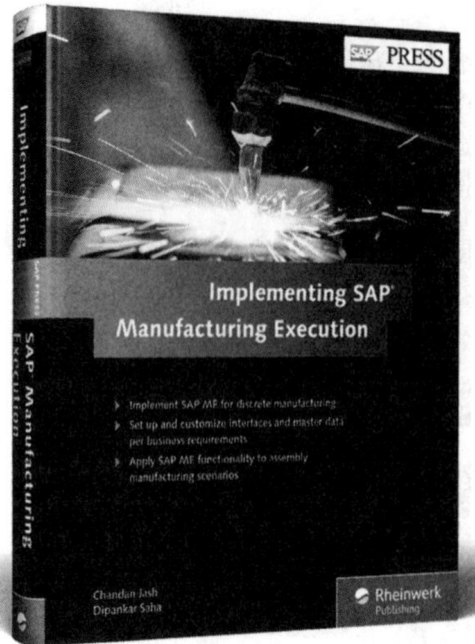

- Implement SAP ME for discrete manufacturing
- Customize interfaces and master data per business requirements
- Apply SAP ME functionality to a car assembly case study

Chandan Jash, Dipankar Saha

Implementing SAP Manufacturing Execution

Get SAP ME up and running! Use detailed instructions to configure SAP ME's routing design, data collection, shop order management, and more. Follow a case study to understand and use the customizations SAP ME offers, including Web Service APIs, advanced reporting, and shop floor systems integration. Ensure your manufacturing execution is in great shape with this guide!

480 pages, pub. 12/2015
E-Book: $69.99 | **Print:** $79.95 | **Bundle:** $89.99

www.sap-press.com/3868

www.sap-press.com

Interested in reading more?

Please visit our website for all new book and e-book releases from SAP PRESS.

www.sap-press.com